ENCYCLOPAEDIA OF
BOTANY

ENCYCLOPAEDIA
OF
BOTANY

(Vol. 1)

Prof. Harmeet Kour

ANMOL PUBLICATIONS PVT. LTD.

NEW DELHI - 110 002 (INDIA)

ANMOL PUBLICATIONS PVT. LTD.
H.O.: 4374/4B, Ansari Road, Darya Ganj,
New Delhi-110 002 (India)
Ph.: 23261597, 23278000
B.O.: No. 1015, Ist Main Road, BSK IIIrd Stage
IIIrd Phase, IIIrd Block,
Bangalore - 560 085 (India)
Visit us at: www.anmolpublications.com

Encyclopaedia of Botany

First Published, 2006

ISBN 81-261-2673-6 (Set)

PRINTED IN INDIA

Published by J.L. Kumar for Anmol Publications Pvt. Ltd., New Delhi - 110 002 and Printed at Mehra Offset Press, Delhi.

Contents

PART-II
Algae-2

Preface

Biology comprises mainly two branches of study, namely, Zoology and Botany. Botany pertains to the study of plants, which have always been of great import since the formation of the planet 'Earth'. Plants play a crucial part in the life of living creatures as well as in balancing the environment. In other words, they are as much necessary for life as air or water is.

Apart from numerous benefits, obtained from plants, they have also been a source of medication since days immemorial. Taking all these dimensions of plants into consideration, they have been classified into different groups and every one of them has been given a unique Botanical name. A plant may have different names in different dialects of different parts of the world, but its botanical name remains one and the same the world over.

This multi-volume aims at furnishing a detailed account of plants from every possible dimension. A ready reference book for scholars and students, this encyclopaedia is an authority on the subject without an iota of doubt.

Editor

PART-I

Algae-1

1

Origin and Evolution

The two chief methods of propagation in plants are asexual and sexual reproduction. Besides, many plants also multiply vegetatively, but it does not involve rejuvenation of protoplast. Asexual reproduction takes place by means of specialized cells, known as spores. It is characterised by the absence of fusion or union of cells at any stage. The spores, which may be motile or non-motile, give rise to new individuals on germination. Sexual reproduction, on the contrary, involves fusion of sex cells, called gametes. The zygote, thus formed, on germination gives rise to a new plant. Individual gametes are, however, not capable of producing new plants.

In blue-greens, the primitive, class of algae, sexual reproduction is absent. These algae propagate mainly by vegetative methods. However, algae belonging to other classes reproduce by asexual as well as sexual methods.

Origin of Sex

The origin of sex is directly related to the origin of gametes; the latter have probably originated from asexual zoospores.

Gametes and zoospores are very much alike structurally, except for the difference in size. This fact can be confirmed by making a comparative study of the morphology of gametes and zoospores in Chlamydomonas, Ulothrix and Cladophora. In this reference it is important to make a mention of Chlamydomonas debaryanum. In this primitive alga, gametes and zoospores are similar in shape, structure and mode of development. The only difference is that zoospores are relatively large than gametes. This difference in size is related to the number of divisions that take place during their formation. If the divisions stop early, the resulting daughter cells are larger and they behave as zoospores. On the contrary, if the divisions continue further, many small daughter cells are formed which do not have the capability to form new individuals (like zoospores). These cells behave as gametes or sex cells.

Another important example of similarity between zoospores and gametes is Ulothrix zonata. It produces the following three types of zoospores.

1. *Quadriflagellate Macrozoospores* : They are formed singly or a few in a sporangium and on germination they form healthy filaments.
2. *Quadriflagellate Microzoospores* : They are produced in large numbers in a cell and on germination they form comparatively weak filaments.
3. *Biflagellate Microzoospores* : They are also formed in large numbers in a cell, but on germination they form very weak filaments.

The gametes of Ulothrix are exactly similar to biflagellate microzoospores in structure. The only difference is the method by which they give rise to new plants; microzoospores develop directly into new plants, whereas gametes first fuse in pairs to form zygote which on germination gives rise to a new plant. In fact, under favourable conditions, zoospores are produced, but under unfavourable conditions, when sufficient food is not available, gametes are formed. These gametes are probably still smaller zoospores which are not self sufficient and by fusing in pairs they regain the strength required to grow into a new plant.

These examples show that there is a fundamental similarity between biflagellate microzoospores and biflagellate gametes. On the basis of this similarity it has been postulated that in algae sexuality has originated as a result of accidental fusion of very small zoospores (which are otherwise incapable of forming new plants). Such fusion usually takes place under unfavourable conditions. It proved useful as it provided new strength and vigour to the individual. This view also gets support from the fact that in the lower groups of plants, sexual reproduction takes place only during unfavourable conditions. Zoospores formed under such conditions are very small due to scarcity of food and are too weak to form new plants. However, zygote formed by the fusion of two such inefficient zoospores (which now behave as gametes) regains the capability to form a new plant. In green algae, the thick walled zygote usually functions as resting spore. Thus sexual reproduction in algae has evolved as an alternate method of reproduction that helps in tiding over unfavourable conditions.

The zygote germinates under favourable conditions to form a new plant which bears asexual spores as usual. Thus formation of zoospores and gametes, is a response to environmental conditions. The gamete as such can be considered as a reduced form of zoospore.

Evolution of Sex

Sexual reproduction in algae originated as a result of fusion of two flagellated gametes. In primitive genera, both gametes of a fusing pair are identical in their morphology and physiology, hence cannot be differentiated into male and female gametes. Such gametes are called isogametes and the mode of their reproduction is known as isogamy. Chlamydomonas debaryanum, Cladophora and Ectocarpus are the common examples of isogamous reproduction. Isogamy is considered to be the most simple and primitive type of sexual reproduction.

In some green algae although the fusing gametes are morphologically identical, they differ in their (physiological) behaviour. For example, in Ulothrix all the gametes are similar in structure, but fusion occurs only between the gametes from different filaments. It shows that morphologically similar gametes sometimes have some inherent physiological disparity. Such forms, though do not have morphologically distinct male and female gametes, carry plus (+) and minus (-) strains. In Spirogyra gametes are morphologically identical and non-motile, but at the time of conjugation one of the two fusing gametes becomes active and motile (male) and the other remains passive and non-motile (female). This difference in physiological behaviour of morphologically identical gametes is a step leading towards anisogamy. This is referred to as physiological anisogamy. In forms like Chlamydomonas braunii, Pandorina, Codium and Ectocarpus, anisogamy is well marked. Here both the gametes of a fusing pair are although motile, they vary in size and are called anisogametes. Micro and macro-anisogametes are formed in micro and macro-gametangium respectively. The macrogamete represents the female and the macrogamete, the male sex.

The next progressive step in the evolution of sexuality can be seen in Chlamydomonas coccifera, C. oogonium and species of Eudorina. In these forms male and female reproductive cells are formed in special structures, known as antheridia and oogonia respectively. The male gametes (antherozoids) are smaller and active and the female gametes (eggs) are relatively large and inactive. Such a mode of reproduction, where distinct male and female reproductive structures are present, is known as oogamous or primitive oogamy.

An evolved type of oogamy can be seen in Oedogonium. Here male gametes (antherozoids) are smaller and flagellated and develop in large numbers within unicellular antheridia. The egg is large and develops singly within oogonium. A large non-motile oogonium has probably evolved to store sufficient food for embryo development.

In green algae like Vaucheria and Chara, oogamy is still evolved. In these forms highly differentiated sex organs are present. Male (antherozoids) and female (eggs) gametes develop in specialized cells which are different from vegetative cells. The oogonium is surrounded by a special covering to protect the egg from high temperature and drought. In Coleochaete, the oogonium has a pseudoparenchymatous covering, whereas in Chara it is surrounded by five tube cells.

Thus isogamy, anisogamy and oogamy represent the successive advanced stages in sexual reproduction.

2

Economic Importance

Algae constitute an important group of plants which mostly occur in aquatic (both freshwater and marine) habitats. They are commonly known as pond scum and sea weeds. Their controlled growth can provide positive help in tackling some present day problems such as disposal of sewage and industrial wastes, conservation of water and soil, and improvement of quality and yield of food products. However, their uncontrolled growth causes pond epidemics, water fouling and fouling of marine vessels. At present many research organizations in India (Indian Agricultural Research Institute, New Delhi; Central Salt and Marine Chemicals Research Institute, Bhavnagar) and abroad are actively engaged in developing marine products. It is all the more important to exploit aquatic resources for the growing food and / or industrial requirements of the world.

Useful or Positive Importance

Algae as Food : In coastal regions, seaweeds are used for human food from ancient times. Population boom in the last decades has forced man to depend upon seaweeds and other algae to supplement his diet. Researches in Japan, Germany and USA have shown that nutritive value and per unit area yield of algae is much more than conventional crops.

Porphyra, Alari, Ulva, Chlorella, Chondrus, Rhodymenia and Codium are some genera which are used as food in various forms.

Porphyra is a red alga which grows in shallow sea water. It is very rich in protein (30-35%) and carbohydrates (40-45%). It is also a good source of vitamin B and C. It is a common food item in Japan (used as a paste called nori), China (locally called 'tsats' ai), Britain (grilled on toast called sloke) and many countries of the pacific region. Porphyra soup is highly prized in Europe and elsewhere. In Japan alone more than 30 million kg of Porphyra is used per year as food.

A gelatinous carbohydrate obtained from Chondrus crispus is used in puddings. It is also used as a stabilizer and clearing agent in the production of beer. A salty confection commonly known as dulse is obtained from Rhodymenia palmata. In Scotland it is chewed as tobacco. Lemanea is considered a delicacy in the Manipur region of India and is locally called nungham.

Food obtained from Laminaria saccharina is known as 'kombu'. It is rich in carbohydrates (57%). Some red algae such as Gelidium, and Gracilaria are used in ice-cream and jelly.

Chlorella, a green alga, is more useful than Porphyra in terms of its nutritive value. Its nutritional value is comparable to that of a mixture of soyabeans and spinach. Several species of Chlorella are high producers of lipids and proteins. They can be cultured easily in the presence of light, carbon dioxide and mineral nutrients. They grow rapidly hence many crops can be obtained in a limited period. An average yield of 13 metric tons per year per acre can be obtained if proper cultural practices are followed. Chlorella contains about 30% carbohydrates, 30% proteins and 15% lipids. However, at present efficient digestibility of the Chlorella cell wall is a problem and researches are underway to convert it in a easily digestible form. Chlorella is useful in space flights and nuclear submarines as food and so , also in recycling and oxygen regenerating system.

In India, at CFTRI, Mysore, research is being conducted on the use of blue green alga, Spirulina as a supplement to diet. The alga is cultured, dried, powdered and then used in the form of one gram tablets. It contains 60% protein, essential vitamins and unsaturated fatty acids. Scenedesmus, besides being rich in proteins, is also a rich source of threonine. It has a high diotic value equal to that of skim milk. The colonies of Nostoc commune are boiled and are used as food in China and Java.

Codium, Monostroma and Ulva are some other green algae used as food. Species of Codium and Ulva are used as salads and Monostroma is used as a common food, aonori, in Japan.

Algae as Fodder

In maritime districts many seaweeds such as Fucus, Laminaria, Ascophyllum and Sargassum are used as fodder. In North Ronaldsay, sheep live on seaweeds for the most part of the year and are pastured only during lambing season. The utility of seaweeds as fodder is due to their vitamin and micronutrient contents. The drying process, storage duration and other similar factors affect their nutrient value. In Scotland and Iceland small industries have been set up for processing seaweeds into feeding meal. Cattle, on diet supplemented by seaweed meal, have increased fat contents in their milk.

Algae also provide food to human beings indirectly through food chain. Many freshwater and marine animals are dependent upon algae for their food. Some of these aquatic animals serve as food for fish and in turn fish is an important source of protein for human beings.

Algae in Industry

Many products of commercial and pharmaceutical importance have been derived from algae. Agar, carrageenan, alginate and diatomite are some important commercial products of algal origin.

1. *Agar-Agar* : It consists of two polysaccharides, agarose and agaropectin. The characteristics of agar are (i) it is insoluble in cold water but readily soluble in hot water, and (ii) 1.5% solution of agar is clear and forms a solid and elastic gel on cooling to 32-39°C; the gel does not dissolve again at a temperature below 85°C.

   Agar is obtained commercially from species of Gelidium, Pterocladia, Gracilaria, Acanthopeltis, Ahnfeltia, Chondrus and Gigartina. Japan and Southeast Asia are the main production centres of agar. Agarophytes (red algae used in production of agar) are collected by diving and then cleaned and bleached in the sun with several washings in freshwater. The material is boiled for several hours and the extract is acidified. This is then frozen and thawed. On thawing, water flows from the agar carrying impurities with it. The agar is then dried and sold in the form of flakes, cakes or powder. In modern methods agar is deodorized using activated charcoal.

   The greatest use of agar is in food, pharmaceutical and cosmetic industry. It is used in the manufacture of processed cheese, puddings, creams and jellies. In preservation and canning of fish and meat it is used as gelling and thickening agent and to reduce the undesirable effects of can. Pharmaceutically, agar is used as a laxative, but more frequently it serves as an inert carrier for the drug which requires a slow release. It is an important constituent of cosmetics' (ointments, lotions, etc.). It is used for almost a century as stiffening agent in culture media.

2. *Carrageenan* : It is a phycocolloid (polysaccharide colloid formed by an alga) similar to agar, but with a higher ash content and requires higher concentration to form gels. The two main constituents of carrageenan are k-carrageenan and λ carrageenan, both negatively charged polymers with high molecular weight.

   Carrageenan is obtained from the cell walls of Chondrus crispus (Irish moss) and Gigartina stellata. In Philippines, Eucheuma is extensively cultured as a source of carrageenan. For extraction, dried plants are washed with fresh water to reduce salt contents and then boiled with excess of water. The soluble carrageenan is separated from insoluble residue by a centrifuge. The soluble fraction on evaporation under vacuum yields carrageenan.

   Carrageenan is extensively used in stabilization of emulsions in paints and cosmetics. In alcohol and sugar industry it is used as a clearing agent. In recent years carrageenan has supplanted agar in food and pharmaceutical industry. It is also utilized in the textile, leather and brewing industries.

3. *Alginate* : These are salts of alginic acid which occur in the cell wall (particularly middle lamella and primary wall) of the brown algae belonging to the order Laminariales. Alginate is extracted from Laminaria, Ascophyllum, Fucus, Macrocystis, Ecklonia, Lessonia and Durvillea. The alginic acid contents of cell walls vary from 14-40%. They are also affected by seasonal factors, yield being the highest in winter and the lowest in summer.

The seaweeds are macerated with acid and then calcium, potassium or magnesium salts are added to get their alginates. Alginates are non-toxic and viscous and readily form gel, hence useful as thickner (in preparation of soup, sauce, cream, printing ink), emulsifier (in ice cream, synthetic cream, polish, emulsion paints) and gelling agent (in confectionery and dental impression powder). Alginic acid stops bleeding effectively. Flame proof fabrics are also prepared from alginates. When mixed with resin, alginates are used for binding panel boards.

4. *Diatomite* : After the death of diatom cells frustules usually dissolve but under certain environmental conditions they may remain intact and accumulate at the bottom of the water body where they occur. If the conditions are exceptionally favourable such accumulation may reach considerable thickness. These deposits of diatomaceous earth are known as diatomite. The most extensive deposits of diatomite are found at Lompoc in California.

   Diatomite is insoluble, chemically inert and shows exceptional physical properties that make it suitable for use in many industries. In oil and chemical industries it is used as a filter. It is heat resistant hence used as insulator in boilers and blast furnaces. Because of its inert nature, diatomaceous earth is used as an absorbant of nitroglycerine. Diatomite is used in the manufacture of car and silver polishing powders, water glass, tooth paste, paints and phonograph records. It is also used as a filler for battery boxes and with bakelite for switch and fuse boxes.

5. *Funori* : A type of glue, known as funori, is obtained from Gloiopeltis furcata. It is used in cosmetic industry for hair curling and dyeing preparations. It is also an important paper and textile sizing agent.

Minerals

Brown algae, particularly Laminariales or kelps are rich in soda, potash, iodine and alginic acid. Ash of dry kelps is a source of soda which is used in the manufacture of soap, glassware and alum.

Kelp ash also contains large amounts of iodine; the average yield is 15 kg per ton. About 7% of the total world production of iodine is obtained from kelps in Japan. Bromine is extracted from some red algae such as Polysiphonia, Rhodymenia, etc. where it may occur in quantities as large as 3-6% of dry weight.

Many important minerals like copper, iron, zinc, cobalt, vanadium, manganese, boron and chromium are present in high proportions in seaweeds. Hence seaweeds are used as a supplement in stockfeed and natural fertilizers. Chlorella vulgaris has the ability of nucleating metallic gold from dilute solutions having gold as gold chloride. Thus, these algae play an important role in the formation of gold deposits in sediments laid down by rivers.

Algae as Bio-fertilizers

Terrestrial algae play an important role in soil biology. Blue-green algae act as nitrogen fixing agent in rice fields. Anabaena cylindrica, Tolypothrix tenuis, Aulosira fertilissima, Oscillatoria princeps, Nostoc commune and many other members of Cyanophyceae have the capability of fixing atmospheric nitrogen. Blue-green algae and Azolla constitute a system which is the main source of algal biofertilizers in south and south-east Asia. By inoculating rice fields with blue-greens, paddy yield can be increased by 30 per cent.

Larger brown and red algae are used as organic manure. They are usually rich in potassium but have relatively low nitrogen and phosphorus proportions. The problem of fertilizers in developing countries can be solved to a great extent if blue-green algae are used in combination with brown and red algae.

Algae also affect soil fertility indirectly. For example, their extracellular products serve as a source of carbon and nitrogen for micro-organisms in the soil. These micro-organisms in turn help in preserving the nitrogen, sulphur and phosphorus elements present in the soil. Besides, extracellular products of some algae are stored as organic or/and inorganic ions which can be easily utilized by crop plants. Some corallina or lime-depositing algae such as Lithothamnion and Lithophyllum arc used for liming crop fields. Blue-green algae also help in reclamation of saline and alkaline soils. The growth of blue-green algae in saline and alkaline water-logged fields results in decrease in pH and increase in phosphorus, nitrogen and organic matter contents of the field and thus converting it into fertile and cultivable land.

Disposal of Sewage

Waterborne domestic and industrial waste is called sewage. It contains organic and inorganic material in soluble and/or suspended form. Although disposal of sewage is mostly an aerobic process, sewage is usually poor in oxygen. It is necessary to treat sewage aerobically before disposal because (i) its anaerobic digestion produces unpleasant odour, and (ii) it is a rich source of many inorganic nutrients like sulphur, nitrogen, phosphorus and potassium and as such properly treated sewage can serve as fertilizer.

Oxidative breakdown of sewage is carried out in small tanks and oxygen required by bacteria for this process (oxidation) is obtained from algal growth. Such tanks support luxuriant growth of many unicellular forms, such as Chlorella, Chlamydomonas, Euglena, Scenedesmus and Pyrobotrys. Oxygen evolved by these algae during photosynthesis is utilized in oxidation by bacteria. The algae in turn utilize ammonia nitrogen as nutrient.

Algae as Antibiotics in Medicines

Many algae such as Chlorella, Cladophora, Lyngbya, Polysiphonia, Laminaria and Halidrys synthesize antibiotic substances. The first such antibiotic, chlorellin, was obtained from Chlorella. Extract of Ascophyllum nodosum is effective against both Gram-positive and Gram-negative bacteria. Water passed through a filter containing Nitzschia palea showed antibiotic property. The filtrate is effective only if the algal filter is in light and photosynthesizing. It suggests that antibiotic substances are formed during photosynthesis.

Members of the order Charales have larvicidal property hence may be useful in controlling mosquitoes.

Extracts from Alsidium, Corallina, Codium, Durvillea, etc. are effective vermifuge.

Tse-ko-Tsoi, an antihelmitic drug, is prepared from a red alga, Digenia simplex, in South China. Some algae are used in the treatment of diseases of kidney, urinary bladder and lungs. Brown algae are used in various goiter medicines because of their high iodine contents. Gelidium is useful in stomach disorders.

Sea weeds are a good source of a number of vitamins. The diatom Nitzschia is fairly rich in vitamin A, riboflavin is present in good amount in Porphyra tenera, Gelidium amansii and Chondrus spp., Furcellania fastigata and Rhodomela subfusca are rich in thiamine, and vitamin C content of Porphryra lacinata is higher than that of oranges.

Algae as a Source of Renewable Energy : In recent times, hydrogen is used as a source of a renewable fuel as it causes no pollution and forms water in contact with oxygen. In view hydrogen (H_2) production is thus an important area of biotechnological research for non-conventional and renewable source of energy. In achieving this objective, the principle used in photosynthetic machinery by green plants is employed. Water is subjected to photolysis, which leads to splitting of water molecules into (i) oxygen, (ii) electrons, and (iii) hydrogen ions (H^+). Hydrogen ions are converted into hydrogen gas (H_2) which can be collected and used as a fuel. Enzymes hydrogenase and nitrogenase convert hydrogen ions (H^+) into hydrogen gas (H_2). Several algae (like Scenedesmus, Chlamydomonas, Dunaliella, Porphyridium and Oscillatoria) possess the enzyme hydrogenase which can be gainfully employed for the production of hydrogen from water.

In blue-green algae such as Anabaena nitrogenase enzyme is the chief hydrogen producing enzyme. Hydrocarbons produced by some algae (e.g., Botryococcus braunii) may also serve as liquid or gaseous fuels.

Algae as Experimental Material

Algae provide valuable experimental materials for research work in plant physiology, genetics and biochemistry. A lot of researches in genetics and cytology have been carried out on Acetabularia. Chlorella has been intensively used for studying the path of carbon during photosynthesis. Valvonia and Halicystis are especially suitable for experiments on membrane permeability.

Harmful or Negative Importance

- ***Algae and Water Supply :*** Luxuriant growth of planktonic algae in ponds and lakes turns drinking water unpotable. Some common problems of potable water associated with algae are as under :

  (1) Decomposition of algae like Anabaena, Dinobryon, Microcystis and Synura in tanks imparts bad taste and odour to water.

(2) Forms like Microcystis, Aphanizomenon and Anabaena produce toxic substances and the water where these algae grow badly affect nervous system, causing depolarization of nerves and brain muscles.

(3) Substances formed by the decomposition of algae in tanks interfere with water treatment process.

(4) Efficiency of water filters at pumping stations is reduced due to the deposition of algae on the filters. This problem is acute during spring and summer seasons when the growth of algae is rapid.

(5) Species of Anacystis and Chaetophora cause corrosion of water pipes and boilers.

At pumping stations and boilers the growth of algae and bacteria is controlled by chemicals like chlorine, copper sulphate and dichlorophen.

Parasitism and Toxicity

Luxuriant growth of planktonic forms in ponds may lead to death of aquatic animals. Toxins secreted by Gymnodinium veneficum, G. brevis, Gonyaulax catenella, G. tamarensis and Pyrodinium phoneus can kill many kinds of fishes. Toxins secreted by Gymnodinium affect the nervous and muscular system of the fish. Species of Gonyaulax secrete endotoxins which accumulate in the digestive system of shellfish feeding on them. When the contaminated shellfish is consumed by human beings death or paralysis may occur.

Algae like Anabaena, Microcystis and Aphanizomenon form 'blooms' in warm water. These algae grow luxuriantly in conditions of oxygen deficiency. In such ponds, during night (when oxygen is not released by photosynthesis), continuous respiration of submerged plants and animals leads to depletion in oxygen level and sometimes it becomes almost zero. Under such conditions mortality rate increases partly due to suffocation and partly due to toxins.

Many toxic planktonic forms enter in human body along with drinking water and may cause diseases. For example, Anabaena and Microcystis cause gastric trouble, Gymnodinium brevis respiratory disease and Lyngbya and Chlorella cause skin diseases. Tetraodontoxin ($C_{16}H_{31}O_{16}$), a toxin produced by Gonyaulax catenella, enters the human body through fish; it causes numbness, nausea, paralysis and ultimately death within 24 hours.

Cephaleuros, a member of Chlorophyceae, occurs as a parasite in many angiospermic plants such as tea, coffee and piper. C. virescens causes red rust of tea.

Fouling of Marine Vessels

Seaweeds growing on the metallic and wooden hull of ships and submarines have a corroding effect. Thick growth of sea weeds increases friction between the hull and water body and results in greater wear and tear.

3

The Ceramiales

The order Ceramiales includes about 160 genera and 900 species, mostly marine. The thallus is multiaxial or polysiphonous. The auxiliary cell is cut off from the supporting cell of the carpogonium just after fertilization. Subsequently, one of the diploid nuclei (formed by the mitotic division of the zygote nucleus) passes into the supporting cell through the auxiliary cell and fuses with the nucleus of the supporting cell. After the fusion of nuclei, the supporting cell gives out a number of gonimoblast filaments. The terminal cells of gonimoblast filaments bear diploid carpospores. The carpospore, on germination, produces a tetrasporophyte, which is similar to the gametophyte and forms tetrahedrally arranged tetraspores. The tetraspores are haploid and on germination give rise to gametophytes. These algae show diplobiontic life-cycle.

The order Ceramiales includes four families, Ceramiaceae, Delesseriaceae, Rhodomelaceae and Dasyaceae. The life-cycle of the genus Polysiphonia (Rhodomelaceae) is described as a representative of the order.

Family Rhodomelaceae

The members of this family are polysiphonous, and the cells of the central axis are uninucleate. The male and female reproductive structures develop on special branches present at the apex of the thallus, called trichoblasts. The trichoblasts, which bear spermatangia, are branched, whereas the female trichoblasts are unbranched.

Polysiphonia

Systematic Position

Class	-	Rhodophyceae
Sub-class	-	Florideae
Order	-	Ceramiales
Family	-	Rhodomelaceae

Occurrence : Polysiphonia is a large genus with some 150 marine species. Most of the species are lithophytic (growing on rocks). Some species are epiphytic (e.g., P. ferulacea, P. urceolata) growing on other algae. P. fastigiata is a semiparasite on the fronds of Ascophyllum nodosum. The genus is represented in India by about 16 species which occur on the Southern and Western coasts. P. ferulacea, P. platycarpa, P. urceolata and P. variegata are common Indian species.

Thallus Structure

The large multiaxial annual or perennial thalli of Polysiphonia form thick tufts or dense bushes of brown or violet-red colour.

Filament Structure

The plant body is heterotrichous, differentiated into a basal prostrate and an erect aerial system. The prostrate system creeps over the substratum. This system is well developed and multiaxial in many species (e.g., P. nigrescens, P. urceolata), and gives out unicellular rhizoids. The distal ends of rhizoids are branched and lobed, which help the prostrate branches to firmly anchor to the substratum. However, in some species (e.g., P. elongata and P. violacea) multiaxial prostrate system is absent and instead the basal cells of erect aerial branches form massive attachment discs.

The prostrate system gives off many multiaxial aerial branches. The main axis and other long aerial branches consist of a central siphon of elongated cylindrical cells, which are surrounded by relatively smaller pericentral cells or pericentral siphon. The cells of the central and pericentral siphons are interconnected through pit connections. In the older portion of main axis pericentral cells become corticated by one or more layers of small cortical cells. The main axis bears two types of lateral branches :

1. *Short Branches or Trichoblasts :* These branches of limited growth arise on the main axis or on long branches in spiral manner. They are uniaxial and lack pericentral siphon. Their cells contain very few chromatophores. In perennial species short branches are shed in the beginning of autumn and develop again in spring season. These branches usually bear male and female reproductive structures, but some short branches are sterile.
2. *Long Lateral Branches :* These branches develop from the basal cells of short branches. Their basal part is polysiphonous but the distal part is usually monosiphonous. They show unlimited growth and do not bear reproductive structures.

Cell Structure

The central and pericentral siphons are made up of elongated cylindrical cells. The cells are uninucleate with many parietal discoid chromatophores which lack pyrenoids. The cell wall is differentiated into an outer pectic and an inner cellulosic layer. The neighbouring cells of the central and pericentral siphons arc interconnected through pit connections. The peripheral cytoplasm contains granules of floridean starch.

Growth : The thallus grows by means of a dome shaped apical cell. It cuts off regular transverse segments towards the axis which elongate and eventually form the central siphon. Some of these cells (usually subterminal) divide vertically and form pericentral cells.

Reproduction

In the life-cycle of Polysiphonia following three types of thalli are found :

1. *Asexual Thallus or Tetrasporophyte :* It develops from diploid carpospore and bears tetrasporangia in which non-motile tetraspores are formed. Tetrasporophytes do not bear sex-organs.
2. *Gametophyte :* It is haploid thallus and bears sex organs. Most species of Polysiphonia are dioecious, i.e., male and female sex-organs are formed on different thalli. The male gametophyte bears spermatangia and the female carpogonia.
3. *Carposporophyte :* It develops by mitotic division of zygote and hence diploid in nature. The carposporophyte is dependent upon female gametophyte and bears diploid carpospores.

Polysiphonia reproduces by asexual and sexual methods.

Asexual Reproduction

Asexual reproduction takes place by means of non-motile haploid tetraspores which are formed in tetrasporangia, present on diploid thalli. Tetrasporangia are small spherical stalked structures and develop from pericentral cells. The fertile pericentral cell, which behaves as tetrasporangial initial, is smaller than other pericentral cells of the same layer. A vertical wall divides this initial into an outer cover cell and an inner sporangial mother cell. The cover cell, by further divisions, forms two or more cover cells.

The sporangial mother cell divides by a transverse wall into a lower stalk cell and an upper sporangial cell. The latter undergoes enormous enlargement and functions as tetrasporangium. The diploid nucleus of the sporangium divides meiotically forming four haploid nuclei, followed by the division of the protoplast into four haploid uninucleate segments. Each segment develops into a haploid spore, and the four spores are arranged tetrahedrally.

Tetraspores are liberated by splitting of the sporangial wall, accompanied by lifting away of the cover cells. Out of the four spores in a sporangium, two produce male gametophytes and two female gametophytes.

Sexual Reproduction

The sexual reproduction in Polysiphonia is oogamous. The plants are dioecious. The male and female reproductive structures develop on the male and female gametophytes, produced by tetraspores.

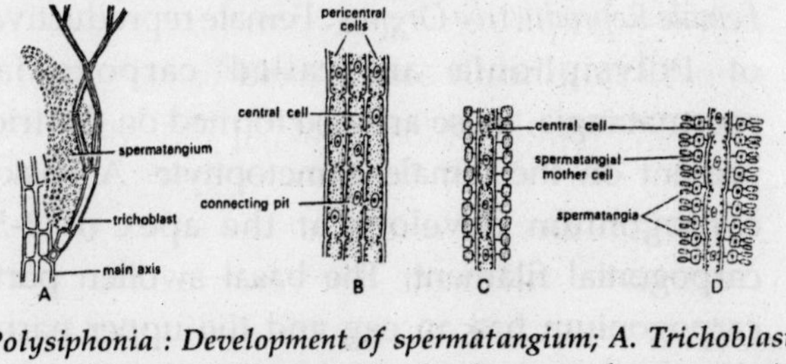

A-D. Polysiphonia : Development of spermatangium; A. Trichoblast with spermatangium, B-D. Stages in the development of spermatangium.

1. *Male Reproductive Organs :* These are called spermatangia or antheridia and develop on fertile trichoblasts present at the tip of the male gametophyte. A fertile trichoblast, when only 2-3-celled, divides dichotomously. In some species both branches of a dichotomy are fertile, whereas in others only one is fertile and the other remains sterile. The latter may divide again forming fertile trichoblasts.

   During the development of spermatangium all the cells of the uniaxial fertile trichoblast, except for few basal ones, divide periclinally forming pericentral cells. The pericentral cells divide again and each forms one or two spermatangial mother cells on their outer faces. These cells arrange themselves in a compact layer on the surface of the trichoblast. The spermatangial mother cells in turn divide to produce varying number of spermatangia. Thus, ultimately a compact cone-shaped cluster of spermatangia is formed.

   The spermatangium is an oblong or globose unicellular structure. Its cell wall is differentiated into an outer thick, a middle gelatinous and an inner refractive layer. The uninucleate protoplast of the spermatangium forms a single spermatium.

   The spermatium is liberated through a narrow slit present at the tip of the spermatangium.

2. *Female Reproductive Organs :* Female reproductive organs of Polysiphonia are called carpogonia. Like spermatangia, these are also formed on the trichoblast present on the female gametophyte. A flask-shaped carpogonium develops at the apex of 2-5-celled carpogonial filament. The basal swollen part of the carpogonium has an egg and the upper narrow and elongated neck forms trichogyne.

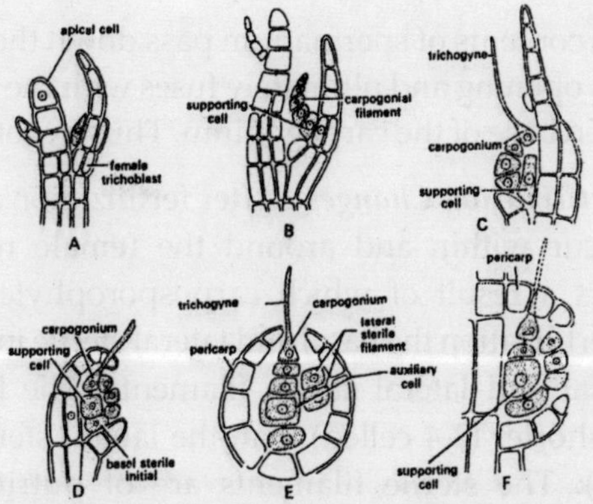

A-F. Polysiphonia : Successive stage in the development of carpogonium; a. Differentiation of supporting cell, B-C. Development of carpogonial filament and formation of carpogonium, D-E. Formation of basal and lateral sterile cells, F. differentiation of auxiliary celll.

A young female trichoblast is uniaxial or unisiphonous, but later becomes multiaxial or polysiphonous by subsequent periclinal divisions. One of the pericentral cells of the adaxial surface of the fertile female trichoblast functions as supporting cell. This cell usually lies 3-5 cells below the apex of the trichoblast. The supporting cell divides transversely to form a 4-5-celled carpogonial filament. The terminal cell of this filament functions as carpogonial mother cell and eventually forms a carpogonium with a relatively long trichogyne.

Meanwhile, the supporting cell cuts off a cell on the basal side which acts as basal sterile initial and a cell on the lateral side which functions as lateral sterile initial. After fertilization, these initials form basal and lateral sterile filaments respectively.

Fertilization

Male gametes (spermatia) of Polysiphonia are non-motile. After their release from spermatangia, they reach the carpogonium with the help of water currents and get attached to the long receptive trichogyne. This is followed by dissolution of the spermatial and trichogynal walls at the point of their contact. The contents of spermatium pass down the trichogyne through this opening and ultimately fuses with the egg nucleus in the swollen base of the carpogonium. Thus zygote is formed.

Post-fertilization Changes : After fertilization a number of changes occur within and around the female reproductive structures as a result of which carposporophyte is formed. Soon after fertilization the basal and lateral sterile initials divide to form basal and lateral sterile filaments. The basal sterile filament is shorter (2-4-celled) than the lateral sterile filament (4-10-celled). The sterile filaments are of nutritive nature. Meanwhile, the supporting cell again divides transversely to form an auxillary cell. It lies at the base of the carpogonium and has a single haploid nucleus. Soon, a tubular connection is established between the auxiliary cell and carpogonium.

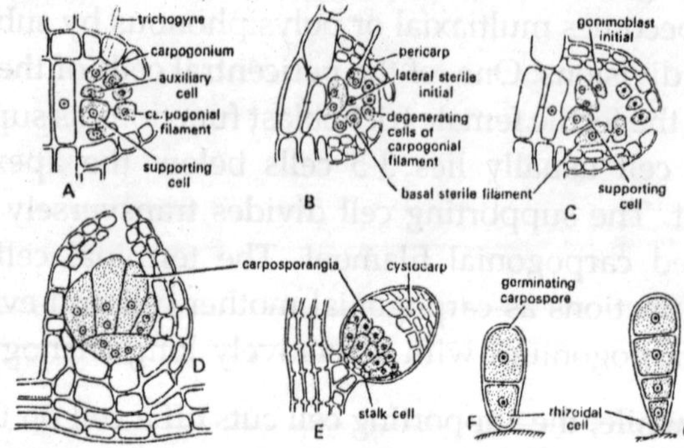

A-G. Polysiphonia : Post-fertilization changes; A. Fertilized carpogonium (note the position of auxiliary cell), B. Division in diploid zygote of carpogonium, C. Development of gonimoblast initial, D. carpospore phyte with young carposporangia, E. Mature carposporophyte (cystocarp), F-G. Germination of carpospore.

The diploid zygote nucleus divides mitotically into two daughter nuclei, one of which remains within the carpogonium and the other passes into the auxiliary cell through the tubular connection. By this time the trichogyne shrivels and the auxiliary cell and the carpogonium fuse with the supporting cell and sterile filaments. This results in the formation of an irregularly-shaped large placental cell. The diploid nucleus of the auxiliary cell divides mitotically forming several diploid nuclei in the placental cell. At this stage a number of tubular outgrowths arise from the placental cell and a diploid nucleus moves into each outgrowth. These outgrowths are cut off from the placental cell and function as gonimoblast initials. The gonimoblast initial divides repeatedly and forms a small gonimoblast filament.

The terminal cell of the gonimoblast filament develops into a carposporangium, which bears a single diploid carpospore. The gonimoblast filaments along with carposporangia get enclosed within a sheath of vegetative filaments and form a large urn-shaped body, called cystocarp or carposporophyte. The sheath surrounding the cystocarp is called pericarp. Carpospores are liberated through a pore (ostiole) at the distal end of the cystocarp. The carpospore, on germination, forms a diploid asexual thallus, called tetrasporophyte. Morphologically, it resembles with the gametophytic thallus.

Alternation of Generation : Polysiphonia is the chief example of triphasic alternation of generation and its life-cycle consists of three distinct stages which are similar in morphological characters.

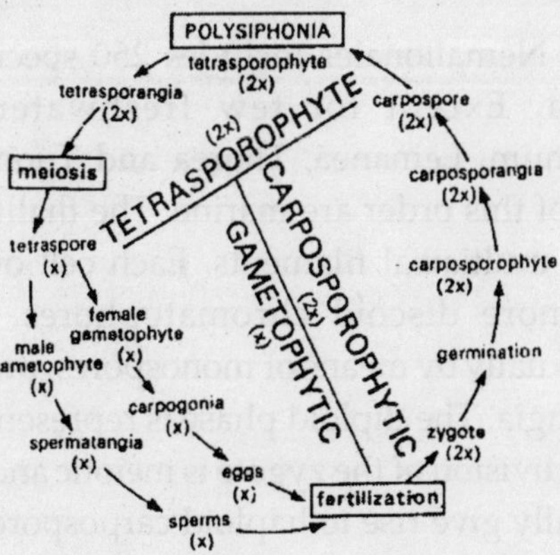

Polysiphonia : Life-cycle (diagrammatic).

(1) Asexual tetrasporophytic stage.

(2) Gametophytic stage.

(3) Carposporophytic stage.

Of these three stages, asexual tetrasporophytic and carposporophytic stages are diploid and gametophytic stage is haploid. In tetrasporophyte, the tetrasporangium forms four haploid tetraspores after reduction division. Two of these haploid spores give rise to male gametophyte and the remaining two to female gametophyte; these bear male (spermatangia) and female (carpogonia) reproductive structures respectively. A diploid zygote formed by the fusion of male and female gametes develops into an urn-shaped fruiting body, known as cystocarp or carposporophyte, through a complex process.

Carpospores formed in the carposporophyte are diploid and give rise to diploid tetrasporophyte on germination. The tetrasporophyte bears only tetraspores. Thus, in the life-cycle of Polysiphonia two diploid phases (tetrasporophyte and carposporophyte) alternate with one haploid (gametophyte) phase. This type of life-cycle is known as triphasic diplobiontic.

Nemalionales

The order Nemalionales includes 250 species, belonging to 35 genera. Except for few freshwater forms like Batrachospermum, Lemanea, Thorea and Tuomeya, most of the members of this order are marine. The thallus is made up of uniaxial or multiaxial filaments. Each cell of the filament has one or more discoid chromatophores. These algae reproduce asexually by means of monospores, produced singly in monosporangia. The diploid phase is represented by zygote only. The first division of the zygote is meiotic and the resultant nuclei eventually give rise to haploid carpospores. Thus these algae show haplobiontic life-cycle.

The order Nemalionales includes eight families: Acrochaetiaceae, Batrachospermaceae, Lemaneaceae, Naceariaceae, Bonnemaisoniaceae, Thoreaceae, Nelmintho-cladiaceae and Chaetangiaceae. The life-history of Batrachospermum is described in detail as a representative of the order.

Family Batrachospermaceae : The family is represented by five freshwater genera which grow in ponds, lakes and slow moving streams. The thallus is uniaxial with many lateral branches, forming whorls. Some members of the family have an alternate dissimilar vegetative phase (juvenile stage) which reproduces only by monospores. This phase gives rise to adult stage. These algae have no sporophytic phase; the diploid phase is represented only by zygote.

Batrachospermum

Systematic Position

Class - Rhodophyceae

Sub-class - Florideae

Order - Nemalionales

Family - Batrachospermaceae

Occurrence : Batrachospennum is a freshwater red alga which grows in slow-moving water of streams, lakes and ponds in the tropical and temperate regions. Some of its species are found attached to stones in swift-flowing acidic water of rivers. Most of the species of Batrachospermum are annual but B. vagum is a perennial species. The genus is represented in India by five species, of these B. moniliforme acid B. vagum are most common.

Thallus Structure

The mature thallus is branched, soft and gelatinous, 15-20 cm long and blue-green, violet or red in colour. The colour of the thallus depends on the intensity of light available. Usually thalli growing in deep and shady ponds and lakes are violet or dark red, whereas those growing in shallow and illuminated ponds are dark green in colour.

The thallus is differentiated into a prostrate and an erect system. The prostrate branches anchor the thallus to the substratum, whereas the branches of the erect system float freely on water.

The thallus is uniaxial and the main or primary axis is made up of a uniseriate row of large cylindrical cells. The axis is differentiated into nodes and internodes and grows by a dome-shaped apical cell. From the nodes of the main axis two types of lateral branches develop : (i) Branches of limited growth - they arise in whorls just below the septa of the axial filament or from the branches of unlimited growth. The whorl of branches of limited growth present at a node is known as glomerule. All branches in a glomerule are of nearly the same length and they are made up of small moniliform cells which give beaded appearance to the thallus. They are further branched or remain unbranched. (ii) Branches of unlimited growth- they develop from nodal cells of the main axis or basal cells of the branches of limited growth. They are also differentiated into nodes and internodes and are enveloped by corticating filaments.

Cortication : The cells of the nodal region in the basal mature part of the axis and the basal cells of the branches of limited growth give rise to some elongated filamentous cells, known as cortical threads. They grow downwards to the next node, forming an envelope around the internode, and this envelope is called corticating layer.

Cell Structure

The cells of the thallus are uninucleate with several discoid chromatophores along the periphery or a diffuse chromatophore. The chromatophore is surrounded by a double layered wall and contains a single pyrenoid. The main pigments are chlorophyll *a*, chlorophyll *d*, *r*-phycoerythrin and *r*-phycocyanin, but due to the excess of red pigment-*r*-phycoerythrin the green colour of chlorophyll is completely or partially masked. Golgi bodies and mitochondria are also present in the cell. The reserve food is in the form of floridean starch.

The cells of the axial filament are interconnected through pit connections.

Growth : The main axis as well as the branches of unlimited growth grow by means of a hemispherical apical cell. The apical cell, by repeated transverse divisions, gives rise to a series of cells towards the posterior side. These derivatives are arranged in a uniseriate row. From the upper end of each derivative (close to the septum), 4-6 protuberances or outgrowths are given out. These outgrowths separate from the parent cell by the formation of septa at their bases and are now called pericentral cells.

The pericentral cells function as initials of the branches of limited growth. After the formation of the branches of limited growth, the pericentral cells act as their basal cells. The branches of unlimited growth may develop from any of these basal cells. The basal cells also give rise to corticating threads. In the older part of the axis, the basal cells separate from each other and give rise to the node. Meanwhile the cells of the central axis elongate and form the internode.

Reproduction

Asexual reproduction takes place by means of monospores. Sexual reproduction is of advanced oogamous type. Species of Batrachospermum are either monoecious or dioecious. The male reproductive structures, are called spermatangia or antheridia and the female carpogonia

Spermatangia

Spermatangia, the male reproductive organs, are unicellular, uninucleate, spherical or globose and colourless structure. They develop singly or in pairs at the distal ends of the branches of limited growth. They are surrounded by a wall which is differentiated into an outer thick, a middle gelatinous and an inner thin layer. Each spermatangium bears a single spermatium.

Development of spermatangium. Spermatangia develop on the branches of limited growth. The terminal or subterminal cell of these branches functions as spermatangium mother cell. Each spermatangium mother cell usually gives rise to one or two spermatangia, but sometimes as many as five spermatangia may develop from a single cell. The spermatangium develops as an outgrowth from the spermatangium mother cell.

The nucleus of the mother cell divides mitotically to form two daughter nuclei, one of these enters into the developing spermatangium and the other remains in the mother cell. At this stage, the spermatangium is separated from the mother cell by a septum. The protoplast of the spermatangium metamorphoses into a single spermatium or male gamete. The spermatangium mother cell may form another spermatangium by repeating the same process.

Carpogonia

Carpogonium, the female reproductive organ, is a flask shaped structure, differentiated into a basal swollen egg cell and a long narrow neck, called trichogyne. The protoplast of the egg cell forms a single uninucleate egg, and the trichogyne receives the male gamete.

Development of carpogonium. Carpogonia develop on special lateral branches, known as carpogonial branches. These branches are only 2-5-celled and develop by repeated transverse divisions of basal cells of the branches of limited growth. The terminal cell of the carpogonial branch forms a flask shaped carpogonium. The subterminal cell, situated just below the carpogonium, is called hypogynous cell or supporting cell.

Fertilization

The spermatia, after their liberation from spermatangia, are carried passively by water currents to the trichogyne of the carpogonium. When a spermatium comes in contact with the trichogyne, the wall between the trichogyne and the spermatium dissolves at the point of their contact. The male nucleus moves into the carpogonium. The male and female nuclei fuse in the basal portion of the carpogonium and form a diploid zygote. After fertilization, the trichogyne shrivels.

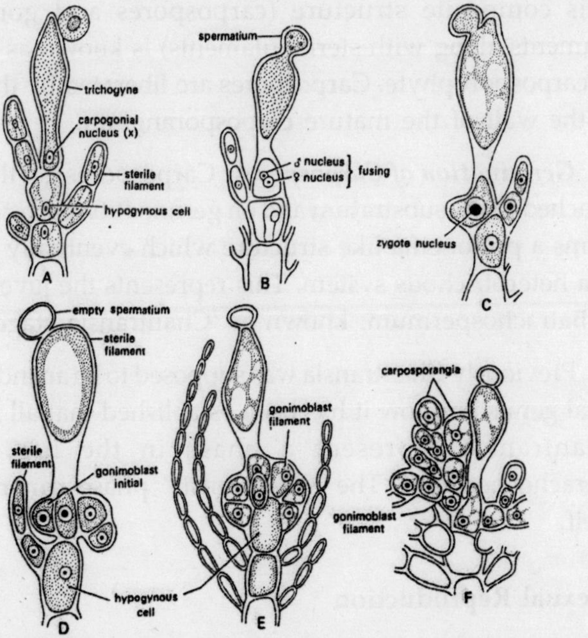

A-F. Batrachospermum : Fertilization and post-fertilization changes; A-B. Fertilisation, C. Zygote, D-F. Post-Fertilization changes.

Post-fertilization Changes : The first division of the zygote nucleus is meiotic, resulting into four haploid nuclei. These nuclei divide repeatedly and form many daughter nuclei. At this stage, many outgrowths arise from the basal part of the carpogonium and a haploid daughter nucleus passes into each outgrowth.

These outgrowths with haploid nuclei are called gonimoblast initials. Repeated transverse divisions of gonimoblast initials give rise to a number of small, unbranched or branched gonimoblast filaments.

The terminal cell of these filaments functions as carposporangium. A single carpospore is formed in each carposporangium. The lower sterile cells of the gonimoblast filaments give rise to many sterile filaments which form an envelope around the carpospores and gonimoblast filaments. This composite structure (carpospores and gonimoblast filaments along with sterile filaments) is known as cystocarp or carposporophyte. Carpospores are liberated by the rupture of the wall of the mature carposporangium.

Germination of Carpospores : Carpospores get themselves attached to the substratum and on germination each carpospore forms a protonema-like structure which eventually gives rise to a heterotrichous system. This represents the juvenile stage of Batrachospermum, known as 'Chantransia stage'.

Previously Chantransia was supposed to be an independent algal genus but now it has been established that all species of Chantransia represent a phase in the life cycle of Batrachospermum. The 'Chantransia' phase can reproduce itself.

Asexual Reproduction

Asexual reproduction takes place by means of uninucleate, haploid and non-motile monospores. These spores are formed in the 'Chantransia' stage, and on germination again give rise to 'Chantransia' stage. The erect branches of this heterotrichous juvenile stage give rise to adult Batrachospermum plants.

Alternation of Generation

The thallus of Batrachospermum is haploid and gametophytic. The male (spermatangia) and female (carpogonia) reproductive structures develop on branches of limited growth.

The haploid egg and spermatium fuse to form a diploid zygote. The zygote undergoes reduction division and the daughter nuclei eventually pass into gonimoblast filaments, arising from the base of the carpogonium. The terminal cell of each of these filaments develops into a carposporangium. The protoplast of the carposporangium forms a single carpospore.

Gonimoblast filaments together with carposporangia are enclosed within an envelope of sterile filaments (arising from the lower sterile cells of gonimoblast filaments). This composite ball like structure is called carposporophyte or cystocarp. It is dependent on the haploid gametophyte.

On germination the carpospore forms a branched heterotrichous filamentous structure, known as 'Chantransia stage'. This stage represents the juvenile phase in the life-cycle of Batrachospermum.

The branched filaments of 'Chantransia' produce monospores which are the means of asexual reproduction. After repeated multiplication by asexual means, the monospore ultimately gives rise to haploid gametophyte of Batrachospermum.

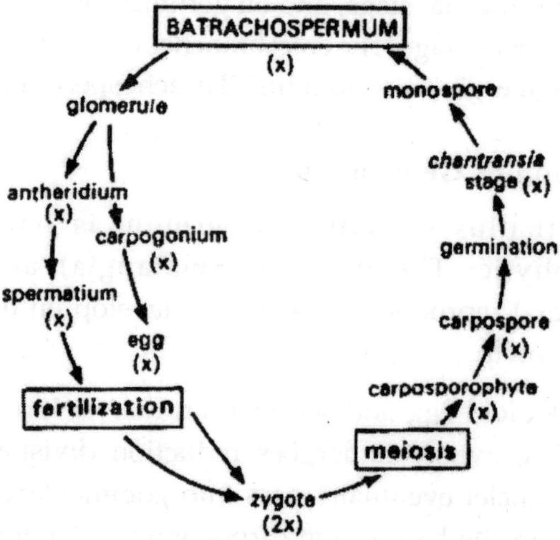

Batrachospermum : Diagrammatic representation of the life-cycle.

Thus, in the life-cycle of Batrachospermum, two haploid phases, gametophyte and carposporophyte, alternate with a diploid phase. Between these two haploid phases, a short-lived diploid phase is represented by zygote. Such a life-cycle with two haploid phases is called triphasic haplobiontic life-cycle.

4

Rhodophyceae

The members of the Class Rhodophyceae are commonly known as red algae, due to the presence of a water soluble red pigment, r- phycoerythrin, which completely masks the green colour of chlorophyll *a*. It is a large group represented by about 831 genera and 5,250 species. Except for 20 fresh water species, all are marine. The fresh water species are found either in fast flowing streams (e.g., species of Batrachospermum, Thorea, Lamanea, Hildenbrandia) or grow in stagnant water with sufficient aeration (e.g., species of Compsopogon, Asterocytis). Amongst marine red algae, larger and fleshy forms occur in the cool temperate areas, whereas species found in the tropical seas are mostly small and filamentous. The members of Rhodophyceae have the ability to live at greater depths in the ocean than do the members of other algal groups.

Porphyridium, an unicellular red alga, is found in terrestrial habitats.

The red algae also exhibit a high degree of epiphytism and parasitism with considerable specificity. The epiphytic species possess normal form and pigmentation, but parasitic species show a great reduction in their form and pigmentation. Polysiphonia lanosa (host A scophyllum nodosum), Ceramium condicola (host-Codium fragile), Ceratocolax hartzii (host Phyllophora membranifolia), Pterocladiophila hemispherica (host-Pterocladia lucida), Gelidiocolax suhriae (host-Gelidium spp.) are some parasitic species found growing on the members of other groups of algae. Species of Ceratocolax, Choreoeolax, Lithothamnium and Rhodochorton are epiphytic on other red algae.

General Characters

(1) Most of the red algae are beautiful, soft and slimy, and the thallus ranges from simple unicellular to complex multiaxial form. In the sub-class Bangioideae the plant body is unicellular (e.g., Porphyridium), filamentous (e.g., Goniotrichum), palmelloid (e.g., Asterocytis) or parenchymatous (e.g., Porphyra, Bangia). But in the sub-class Florideae, the plant body is more advanced and complex and the thallus organization is of the following two types:

(a) *Uniaxial Thallus :* It is characterised by the presence of a single central or axial filament which is usually corticated by many well branched laterals. Examples—Batrachospermum, Acrochaetium, Dumontia.

(b) *Multiaxial Thallus :* Here the thallus is a mass of central or main filaments and each central filament gives out lateral branches. The central filaments together with their laterals form a pseudo-parenchymatous thallus. Examples Helminthocladia, Polysiphonia.

The flagellated motile stages are totally absent.

(2) The members of the sub-class Florideae grow by an

apical cell, whereas those of the sub-class Bangioideae show diffused growth.

(3) The cell wall is differentiated into an outer pectic and an inner cellulosic layer. The mucilagenous material of the outer pectic layer mainly consists of agars and carrageenans (both are made up of repeating units of galactose or its derivatives). These mucilagenous matters constitute up to 70% of the dry weight of the cell wall. In the inner layer microfibrils of cellulose are distributed in a granular matrix. They form the framework of cell wall.

(4) The septa between the adjacent cells have 'pit connections'.

(5) There is much variation in the shape, size and number of nuclei in these algae. All members of the sub-class Bangioideae are uninucleate and the nucleus has a single nucleolus. In the sub-class Florideae, a few species are uninucleate but majority of them show multinucleate condition. In Griffithsia each cell has as many as 4000 nuclei.

(6) The shape of nuclei depends upon their position in the cell. Usually centrally situated nucleus is spherical, whereas the peripheral nuclei are elliptical.

(7) The primitive forms (sub-class Bangioideae) have a single stellate chromatophore with a central pyrenoid, whereas in advanced forms (sub-class Florideae) there are many discoid and parietal chromatophores without any pyrenoid.

(8) The photosynthetic pigments present in the chromatophore include chlorophyll *a* and *d*, α - and β-carotene, xanthophylls (lutein, zeaxanthin, violaxanthin, flavoxanthin, taraxanthin) and biliproteins, such as r-phycoerythrin and r-phycocyanin.

β-phycoerythrin is present in the primitive red alga, Porphyridium.

(9) Chromatophores have no chromatophore ER.

(10) The cells contain organelles like endoplasmic reticulum (ER), mitochondria and dictyosomes.

(11) The accumulated product of photosynthesis is floridean starch and galactoside floridoside; apparently it has the same function as sucrose in higher plants and green algae. The long- term storage product is floridean starch, occurring in the form of grains outside chromatophores in the cytoplasm. In primitive forms starch grains are found clustered as a sheath around the pyrenoid area of the chromatophore, whereas in more advanced forms starch grains are scattered in the cytoplasm.

(12) Many red algae (e.g., Liagora, Galaxaura, and the members of the family Corallinaceae) have extracellular deposition of $CaCO_3$ on their cell walls. Those red algae which have the highest rate of calcification also have the highest rate of photosynthesis. Classification of thallus is two to three times more rapid in light than in dark.

(13) Reproduction is mainly by asexual and sexual methods. Vegetative multiplication of thallus occurs only in unicellular forms.

(14) Asexual reproduction takes place by means of the following type of aplanospores:

(a) *Monospores :* When a single large spore is produced in each sporangium, it is called monospore, and the sporangium bearing it is known as monosporangium. They are found in Compsopogon and Batrachospermum.

(b) *Neutral Spores :* These spores are not borne in sporangia but are formed by direct metamorphosis or division of vegetative cells. They are diploid in nature and on germination give rise to the same generation (sporophytic), that is why they are called neutral spores.

(c) *Carpospores :* These spores develop from the zygote. In some genera (e.g., Batrachospermum) they are formed by reduction division of the zygote nucleus. The nuclei pass into the gonimoblast initials, which develop into gonimoblast filaments. The terminal cell of each filament bears a single haploid carpospore. In other genera (e.g., Polysiphonia) the zygote nucleus directly develops into a single diploid carpospore. Such spores on germination give rise to diploid (sporophytic) plants.

(d) *Tetraspores :* In the life-cycle of some red algae (e.g., Polysiphonia) four haploid spores are formed in each sporangium. These spores are called tetraspores, and the sporangia bearing them as tetrasporangia. These spores are always produced in diploid sporophytes (known as tetrasporophytes) and they give rise to haploid gametophytes on germination.

(15) Non-motile male reproductive cells, called spermatia (singular = spermatium), are borne singly in spermatangia.

(16) The female reproductive organ, called procarp, consists of a dilated basal carpogonium and a usually narrow gelatinous elongated tip, trichogyne. The procarp is sessile in the sub-class Bangioideae, whereas it is stalked in the sub-class Florideae.

(17) In the sub-class Bangioideae, the first division of the zygote nucleus is meiotic, as a result of which four (sometime up to 64) haploid non-motile spores, called carpospores, are formed. These spores on germination

give rise to an alternate dissimilar phase (juvenile stage). In the sub-class Florideae, the fertilization is followed by the production of specialized filaments, called gonimoblasts. These filaments arise either directly from the carpogonium or from one or more auxiliary cells (present at the base of the carpogonium) into which the zygote nucleus or some of its descendants have migrated. The terminal cell of the gonimoblast filament functions as a carposporangium and it contains a single haploid or diploid carpospore. The haploid carpospore on germination gives rise to gametophytic thallus, whereas the diploid carpospore forms sporophytic thallus.

Classification

Fritsch (1935), on the basis of thallus structure and post-fertilization changes, classified Rhodophyceae into two sub-classes :

Sub-class Bangioideae

This sub-class includes primitive forms of Rhodophyceae. They lack pit-connections and apical growth (instead show diffused intercalary growth) and have pigments like allophycocyanin. The reproductive structures are relatively simple. The zygote nucleus directly gives rise to carposporangium after division.

This sub-class includes only a single order—Bangiales. Examples— Porphyra, Porphyriditum, Bangia.

Sub-class Florideae

The members of this sub-class have uni- or multiaxial thallus which shows apical growth. The adjacent cells are connected by pit connections. The reproductive structures are more developed and complex than those of Bangioideae. The zygote on germination forms gonimoblast filaments and the terminal cell of these filaments develops into a carposporangium.

The sub-class Florideae has been divided into six orders :

Order 1. Nemalionales. Examples Batrachospermum, Chantransia, Thorea.

Order 2. Gelidiales. Example - Gelidium.

Order 3. Cryptonemiales. Example - Cryptonemia.

Order 4. Gigartinales. Example - Gigartina.

Order 5. Rhodymeniales. Example - Rhodymenia.

Order 6. Ceramiales. Example - Ceramium, Polysiphonia.

Smith (1955) placed the class Rhodophyceae in the Division Rhodophyta. He divided the Rhodophyceae into two sub-classes, Bangioideae and Florideae, following Fritsch (1935).

Round (1973) also placed the Class Rhodophyceae in the phylum Rhodophyta, Class Rhodophyceae was further divided into two sub-classes-Bangiophycidae and Florideophycidae. An outline of the classification proposed by Round is given below :

Comparison of the sub-class Bangioideae and Florideae

Bangioideae	*Florideae*
1. Freshwater, marine or terrestrial.	Mostly marine.
2. Thallus simple, unicellular or multicellular.	Thallus structure is relatively complex, uniaxial or polyaxial.
3. Cells always uninucleate.	Number of nuclei per cell is much variable and sometimes it is as high as 3000-4000.
4. They show diffused intercalary growth.	Growth takes place with the help of a single apical cell.

Contd...

	Bangioideae	*Florideae*
5.	Each cell has a single, centrally located, stellate chromatophore.	There are many discoid chromatophores in each cell, present along the periphery.
6.	Each chromatophore has a single pyrenoid.	Chromatophores lack pyrenoids.
7.	Pit connections absent.	Pit connections present between the adjacent cells.
8.	Zygote divides to form haploid carpospores and these spores on germination give rise to alternate dissimilar vegetative phase.	Zygote divides to form diploid or haploid gonimoblast filament and the terminal cell of the filament develops into a carposporangium which bears a single carpospore.

Phylum Rhodophyta
Class Rhodophyceae

Sub-class

Bangiophycidae	Florideophycidae
(=Bangioideae)	(=Florideae)
Order 1. Bangiales	Order 1. Nemalionales
	2. Gelidiales
	3. Cryptonemiales
	4. Gigartinales
	5. Rhodymeniales
	6. Ceramiales

Laminariales

The members of order Laminariales, commonly known as kelps are of very large size and complex structure. Some plants reach up to 50-60 m. Most of the members show parenchymatous construction. The plant body may be differentiated into stipe and blade (lamina) and the meristematic region is located in between them. Asexual reproduction takes place generally by zoospores formed in unilocular filamentous and dioecious sporangia. The gametophytec body is microscopic. Sexual reproduction is oogamous type. Plants exhibit heteromorphic alternation of generation. There are about 30 genera and 100 species, mainly inhabitants of the colder oceans of the arctic and antarctic regions.

The members are distinguished from the Fucales by their intercalary growth and heteromorphic alternation of generation. The order Laminariales is usually divided into four families, viz. Chordaceae, Laminariaceae, Lessoniaceae and Alariaceae.

Family Laminariaceae

The sporophytic plant body exhibits a high degree of morphological and anatomical differentiation well divisible into holdfast, stipe and blade. The sporangia are arranged in distinct groups (sori) on both the surfaces of blade. The gametophytes are much reduced, heterotrichous and dioecious. Members possess diplohaplontic heteromorphic alternation of generations.

Life history of Laminaria the representative genus of Laminariaceae is described here in details.

Laminaria

Systematic Position

Class - Phaeophyceae

Order - Laminariales

Family - Laminariaceae

Occurrence : The genus Laminaria with some 40 species is exclusively marine and is widely distributed in the waters of north temperate and arctic zones. L. flavicans, L. saccharina,

L. digitata, L. flexicaulis and L. japonica are some common species of the genus. The plants may be annual (L. ephemera) or perennial (L. cloustoni). In the annual species, a new sporophyte is produced each spring.

Thallus Structure

External Structure : The plant body of Laminaria is diploid and sporophytic attaining a length of two metre or more. It is differentiated into a hold fast, stipe and blade (lamina). The hold fast is an attaching organ variable in form. It may be in the form of a solid disc or profusely branched cluster of cylindrical root-like organs (haptera). The stipe is cylindrical or some what flattened unbranched portion of the thallus that arises from the basal holdfast and bears terminally either a single blade (L. saccharina) or a number of vertically divided segments (L. digitata, L. cloustoni). The lamina is rather tough and leathery in texture. The lamina is renewed at the base each year by the activity of the intercalary meristem. The lamina is flat, long and ribbon-shaped and the margins may be smooth or wavy.

Internal Structure

1. *Stipe* : In transection, the stipe is differentiated into three distinct regions : (i) meristoderm, (ii) cortex, and (iii) medulla.

The meristoderm (named so due to meristematic nature of its cells) is one or two cells in thickness which are small and cuboidal in form. It remains surrounded by a layer of mucilage.

The cortex lies within the meristoderm and surrounds the central axis which constitutes the medulla. The cortex usually remains differentiated into an outer and an inner cortex. The cells of the outer cortex are more or less radially arranged, vertically elongated with pointed ends and thin walls. They show horizontal septation. On the contrary, the cells of the inner cortex are comparatively squarish at the ends and show gelatinisation of the longitudinal walls. The cells possess thread like outgrowths or cross connections and hyphae. The cortical cells are regarded as storage and mechanical elements.

The medulla is composed of loosely arranged cells, the spaces between them are filled with mucilage. Cross connections are present in the cells, formed as a result of fusion of papillose outgrowths of the longitudinal walls. Regular formation of cross connections in the medullary region and the inner part of the inner cortex, give it an appearance of a tangled network of filamentous growth. An another kind of protrusions, known as hyphae also arise from the medullary cells.

The hyphae differ from the cross connections in that they do not fuse with the corresponding outgrowth. Besides, certain cells of the medulla considerably swell at their ends. Such swollen cells are called the trumpet hyphae. The trumpet hyphae in the younger stages become much attenuated and become extremely thick-walled and filled with callus. The medulla constitutes the main conducting system of the plant body.

2. *Blade (Lamina)* : Like stipe, the blade is also differentiated into outer meristoderm, middle cortex and central medulla regions.

The meristoderm is generally single layered, the cells of which are small with many chromatophores. Just beneath the meristoderm lies the cortex. It consists of isodiametric cells. Mucilage canals are present in the cortex. Medulla occupies a rather narrow tangential zone and consists of a network of cross connections and medullary cells. The cross connections are displaced with the enlargement of the blade.

Growth : The plant body grows in length by the activity of an intercalary meristem, situated at the junction of the stipe and the blade.

Reproduction

Laminaria reproduces by vegetative, asexual and sexual methods.

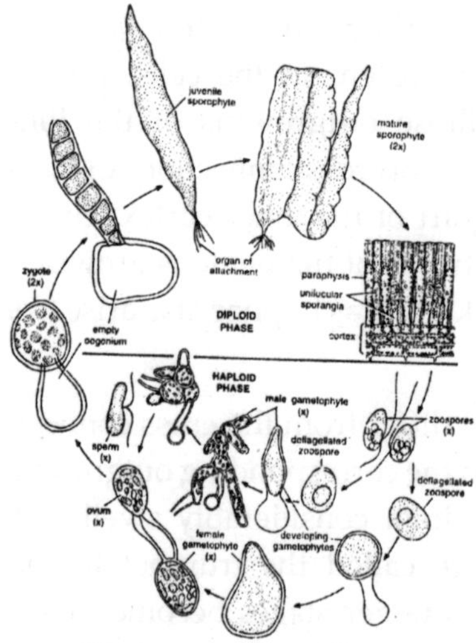

Laminaria : Life-cycle.

Vegetative Reproduction : Some species of Laminaria (e.g., L. longipes, L. rodreguezu, L. sincelarii) reproduce vegetatively by means of stolons. The stolons arise horizontally from the holdfast and then they turn upwards to grow into new plantlets.

Asexual Reproduction

The asexual reproduction takes place by zoospores produced within club-shaped unilocular sporangia on the sporophytic plant body. The sporangia are arranged in sori on both sides of the blade, covering its considerable area. Interspersed with sporangia are present numerous paraphyses.

The sporangia arise from the superficial cells of the blade. Each sporangium is a long, sac-like structure. They are tightly pressed, one each between two adjacent paraphyses. The sporangia are diploid structures. The nucleus of the sporangium first divides by a meiotic and then by successive mitotic divisions to form 32 (sometimes 64) haploid nuclei in each sporangium. The cleavage of the protoplast follows and haploid, uninucleate bits of the protoplast are formed. Each protoplast bit then metamorphoses into a motile zoospore. Each zoospore is biflagellate and possesses a single chromatophore occupying the base. The two laterally inserted flagella are unequal in length.

The sporangia dehisce due to the pressure of the gelatinous paraphyses on the sporangia. The zoospores are thus liberated and swim actively for sometime. Usually half of the zoospores develop into male gametophytes and rest half into female gametophytes.

After swimming for sometime, the zoospore settles down to germinate. It withdraws its flagella, becomes rounded and secretes a wall. It then produces a small protuberance which increases in length and its distal end becomes bulbous. The germling appears dumb-bell shaped at this stage. Ultimately, the bulbous part produces a filamentous gametophyte. The gametophytes consist of elongated cells with numerous discoid chromatophores. Sometimes, the gametophytes are branched and show a tendency towards heterotrichy. The male and female gametophytes develop male and female sex organs respectively.

Sexual Reproduction

The sexual reproduction is oogamous. The male sex organs are called antheridia and the female oogonia. Male and female sex organs are produced on separate gametophytes. The gametophytes usually do not develop sex organs when the temperature of surrounding water is above 15°C.

1. *Male Gametophytes and Antheridia* : The male gametophytes are comparatively short, filamentous and heterotrichous. Antheridia are produced at the tips of one-or two-celled lateral branches either singly or in

groups. In each antheridium, a single biflagellate antherozoid is formed. Each antherozoid is an ellipsoidal or a pyriform structure provided with two laterally inserted flagella. It is liberated near the apex of the antheridium.

2. *Female Gametophytes and Oogonia* : The female gametophyte is also a multicellular (generally reduced to a few cells) filamentous structure. Sometimes, it may consist of only one cell, which may function as an oogonium. Any cell of the female gametophyte may function as an oogonium initial. It enlarges considerably and assumes a pyriform shape. Its contents round off to form a single egg (oosphere). At maturity, each oogonium is surrounded by three layers and the oogonial wall at the distal end becomes thickened.
3. *Fertilisation* : At maturity, a slit develops in the thickened apical portion of the oogonium and the egg emerges out through this slit, and rests on a cushion formed by the localized gelatinisation of the oogonial wall. Fertilisation takes place in this condition. Liberated antherozoids swim to the ovum. Only one antherozoid penetrates the delicate membrane of the ovum. The male and female nuclei fuse to form a diploid zygote.
4. *Germination of Zygote* : The zygote germinates immediately after fertilization while still attached to the tip of oogonium. The zygote undergoes considerable elongation and it divides first mitotically by a transverse division. Further divisions result in the formation of a 4-10 celled filament. Later, both vertical and transverse divisions take place in the upper and middle part of the filament and as a result, a leaf-like expansion is differentiated. The lower cell, however, remains undivided and elongates to give rise a rhizoid-like process. Later, additional rhizoids develop which help

in anchoring the developing sporophyte to the suitable substratum. The female gametophyte disintegrates after the differentiation of three or more rhizoids. Finally, a sporophytic plant body differentiated into holdfast, stipe and blade is developed.

Alternation of Generation : The life-cycle of Laminaria exhibits an advanced type of heteromorphic type of alternation of generation. The main plant body is sporophytic, differentiated into the holdfast, cylindrical stipe and broad blade (lamina). On both sides of the lamina, sporangia are produced in groups. The diploid sporangia undergo reduction division to form haploid zoospores. Usually 32 zoospores are formed per sporangium. Half of them develop into male and the rest into female gametophytes.

The gametophytic body is microscopic, and consists of few celled filamentous body. Thus, there is sharp morphological differentiation between the sporophytic and gametophytic plant body. The reproductive organs-antheridia and oogonia, are produced on male and female gametophytes respectively. The haploid male and female gametes fuse to form diploid zygote. The zygote germinates into a diploid sporophytic thallus. Thus, in the life-cycle of Laminaria, the macroscopic sporophytic plant body alternates with microscopic filamentous male and female gametophytic plant body.

5

The Dictyotales

There are 20 genera and 180 species in the order Dictyotales. The members occur both in temperate and tropical seas but attain their greatest development in the warmer oceans of the tropics and subtropics. Plant body is erect, flat, ribbon-like, well branched and attached to the substratum with the help of a rhizomatous disc. Growth is apical and the branching is usually dichotomous. Asexual reproduction takes place by non-motile tetraspores formed in unilocular sporangia. Sexual reproduction is oogamous type, and the plants exhibit isomorphic alternation of generations.

Family Dictyotaceae

Members are flat, ribbon-like and dichotomously divided thalli which are sometimes fan-shaped. Growth occurs by an apical meristem, or by an apical cell. The sporophytic and gametophytic generations are morphologically similar.

Dictyota is the type genus of the family which is described here in some details.

Dictyota

Systematic Position

Class - Phaeophyceae

Order - Dictyotales

Family - Dictyotaceae

Occurrence : The genus Dictyota is represented by about 35 species, found chiefly in the warmer seas of the tropics. In India, the genus is represented by 12 species which occur along the coasts of Okha, Dwarka, Tuticorin, Mumbai, Porbandar and Mahabalipuram. D. dichotoma is a common species found along the Indian coast. It grows attached to the rocks in the intertidal zone.

Thallus Structure

External Structure : The thallus of Dictyota is macroscopic, about 30 cm in height, erect, ribbon-like and dichotomously branched. The plant body remains attached to the substratum with the help of a discoid or irregularly branched holdfast. All the branches of the plant body lie in the same plane and lack a mid-rib. Both the sporophytic and gametophytic plants are similar morphologically though cytologically quite different.

Internal Structure : Internally, the thallus is simple in organisation and consists of three layers of cells. There is an upper and a lower assimilatory or photosynthetic layer. Between these two layers occurs a median layer consisting of large, colourless thin-walled medullay cells. These cells have a few or no chromatophores. They possess fucosan vesicles and refractive granules. Many uniseriate clusters of colourless hairs develop on both the surfaces of the thallus.

Growth : Growth of the thallus takes place with the help of a single biconvex apical cell situated at the tip of each branch of the plant body. The apical cell cuts off a series of segments at its posterior face. Each segment again divides by a curved wall in a plane parallel to the thallus surface, the larger derivative again divides in the same way. As a result, the thallus becomes three cells thick.

At the time of the initiation of dichotomy, the apical cell divides by a vertical wall into two equal cells. Each daughter cell functions as an apical cell. Besides cells on the margins of dichotomy may also act as apical cells and their division results in the formation of adventitious branches.

Reproduction

Dictyota reproduces by vegetative, asexual and sexual methods.

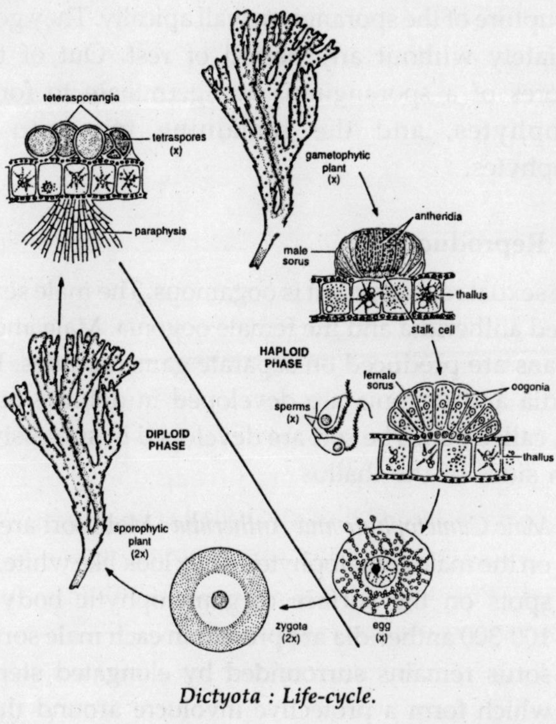

Dictyota : Life-cycle.

Vegetative Reproduction : Vegetative reproduction takes place mainly by fragmentation or by the formation of adventitious branches.

Asexual Reproduction

Asexual reproduction takes place by tetraspores developed in unilocular sporangia (tetrasporangia) on sporophytic plants. The sporangia are borne singly or in groups on both the surfaces of the thallus. The sporangial initial first divides transversely into two cells; the lower cell functions as the stalk cell, whereas the upper one enlarges and develops into a tetrasporangium. The diploid nucleus of the sporangium divides meiotically and thus four haploid tetraspores are formed. The tetrasporcs are non-motile and remain arranged tetrahedrally in the sporangium. The tetrasporcs are liberated by the rupture of the sporangium wall apically. They germinate immediately without any period of rest. Out of the four tetraspores of a sporangium, two germinate to form male gametophytes, and the remaining two into female gametophytes.

Sexual Reproduction

The sexual reproduction is oogamous. The male sex organs are called antheridia and the female oogonia. Male and female sex organs are produced on separate gametophytes. Both the antheridia and oogonia are developed in definite projecting groups, called sori. The sori are developed in successive crops on both sides of the thallus.

1. *Male Gametophyte and Antheridia :* Male sori are present on the male gametophytes. They look like white, shining spots on the surface of gametophytic body. About 100-300 antheridia are present in each male sorus. Each sorus remains surrounded by elongated sterile cells which form a protective involucre around the sorus.

Each antheridium develops from a superficial cell of the thallus which is known as antheridial initial. The antheridial initial divides by a transverse wall to form a small basal stalk cell and an upper large primary antheridial cell. The former does not divide further whereas the latter first divides by a vertical wall followed by another vertical wall at right angles to the first. The cells of the quadrant thus formed divide by repeated transverse and vertical divisions to form the antheridial body.

The mature antheridium is a shortly stalked multicellular body; the protoplast of each cell metamorphoses into a single antherozoid. Each antherozoid is a pyriform structure with a single forwardly directed lateral flagellum which is of tinsel type. However, electron microscopic studies reveal the presence of a second flagellum which is rudimentary. The antherozoids are liberated from the antheridium as a result of the gelatinization of the antheridial walls.

2. *Female Gametophyte and Oogonia* : Female sori are present on the female gametophytes. They are distinguished from the male sori by their deep brown colour, and in the absence of involucre. Each female sorus contains 25-50 or more oogonia.

   Each oogonium develops from a single superficial cell of the thallus, which functions as the oogonial initial. The oogonial initial divides by an asymmetrical transverse division to form a basal stalk cell and an upper primary oogonial cell. The latter enlarges considerably to form the oogonium. Its contents develop into a single egg (ovum). Thus, the mature oogonium is a large, sac-like structure containing an egg. Egg is discharged by the gelatinization of the oogonial wall.

3. *Fertilization. Fertilization is External* : The eggs and the

antherozoids are liberated early in the morning in the surrounding water. A large number of antherozoids surround the liberated eggs. They attach to the eggs with the help of the flagellum. Ultimately, only one antherozoid penetrates the egg. Male and female nuclei fuse together and it results in the formation of a diploid zygote.

4. *Germination of Zygote* : The zygote germinates immediately without undergoing any resting period. It germinates directly into the diploid sporophytic plant of Dictyota without any reduction division.

Alternation of Generation : The life-cycle of Dictyota exhibits isomorphic alternation of generation. Both sporophytic and gametophytic plant body are morphologically similar, but differ in their genetic constitution and functions. The sporophytic plant body is diploid and is concerned with asexual reproduction. The diploid sporangia produced on the sporophytic body undergo reduction division to form haploid zoospores. The latter germinate to form gametophytic body. Half of the zoospores develop into male plant and rest half into the female plant. The reproductive organs- antheridia and oogonia, are produced on male and female gametophytes respectively. The haploid male and female gametes (antherozoids and eggs) fuse to form a diploid zygote. The zygote germinates into a diploid sporophytic thallus.

Fucales

There are 40 genera and 3000 species in the order Fucales. They are cosmopolitan in distribution. Species occurring in the arctic and north temperate seas differ considerably from those of the antarctic and south temperate waters. They are mostly found attached to the substratum by discoid holdfasts. The plant body is complex both morphologically and anatomically.

These algae reproduce by vegetative and sexual methods; asexual spores are absent. Sexual reproduction is oogamous and the sex organs are borne in special cavities, known as conceptacles. The special branches bearing conceptacles are called receptacles. Conceptacles bear microsporangia or macrosporangia which function as antheridia and oogonia respectively. Both micro-and macrosporangia are unilocular.

The members of Fucales do not show distinct alternation of generation. The sporophytic thallus is diploid and the haploid stage is confined only to gametes.

The order Fucales has been divided into seven families : (1) Fucaceae, (2) Himanthaliaceae, (3) Cystoseiraceae, (4) Sargassaceae, (5) Hormosiraceae, (6) Durvilleaceae, (7) Ascoseiraceae.

The life-histories of Fucus and Sargassum are described here in details.

Family Fucaceae

The plant body is usually flattened and dichotomously branched; axes subterate to algae with mid-rib, but not foliar.

Fucus

Systematic Position

Class - Phaeophyceae

Order - Fucales

Family - Fucaceae

Occurrence : The genus Fucus, represented by more than 100 species, is found attached to rocky substrata in the intertidal belt of the littoral zone. The species occur predominantly in cold water and grow luxuriantly in the North Atlantic and Pacific shores. F. vesiculosus, F. serratus, F. furcatus and F. edentatus are some common species of Fucus. However, no species is recorded from India.

Thallus Structure

External Structure : Plant body of Fucus is well branched, ribbon-like, leathery, slimy to touch and dark brown reaching up to 50 cm in length. Externally, the plant body is differentiated into a basal discoid holdfast which anchors it to the substratum, a short stipe and a flattened, dichotomously branched frond (blade).

The stipe is continued in the form of a midrib in the frond of several species. Each branch of the frond is bifurcated at the tip. The ribbon-like thallus may be smooth (F. vesiculosus) or serrate (F. serratus). In some species (e.g., F. vesiculosus), a number of air bladders occur in the upper portion of the frond that give buoyancy to the plant. Numerous small openings are seen throughout the surface of the frond with naked eye. These are sterile conceptacles called cryptoblasts or cryptostomata. There are several hairs or sterile paraphyses in the cryptoblasts which peep out through the openings. Besides, at the terminal parts of the branches occur fertile conceptacles, known as receptacle. Mid-rib is absent in the receptacular region.

Internal Structure : Internally, the thallus is well differentiated into three distinct regions, viz. (i) meristoderm, (ii) cortex, and (iii) medulla.

1. *Meristoderm :* It is the outermost layer and functions as epidermis. It is made up of compactly arranged columnar cells. These cells are densely packed with chromatophores and fucosan, a reserve food material.
2. *Cortex :* It is composed of relatively large thin-walled parenchymatous cells, size of which gradually increases toward the centre. The cortex is sometimes differentiated into outer and inner cortex.
3. *Medulla :* The central part of the frond forms medulla. It is made up of thick walled narrow and elongated cells. Spaces between the cells are filled with mucilage. Along with medullary cells, running longitudinally are

elongated cells, called hyphae which give mechanical support to the thallus.

Growth : The growth of the plant body takes place with the help of a pyramidal apical cell with three cutting faces - two lateral and one basal. The apical cell remains situated in the terminal portion of the lamina. The lateral segments derived from the apical cell initially divide periclinally; of the resulting two cells, the outer one, by repeated divisions in various planes, organizes the epidermis and the cortex whereas the inner one, in association with the basal segment cut off by the apical cell, forms the medulla.

Reproduction : Fucus reproduces by vegetative and sexual methods. Asexual reproduction is absent.

Vegetative Reproduction : It takes place by fragmentation or by formation of adventitious branches.

Sexual Reproduction

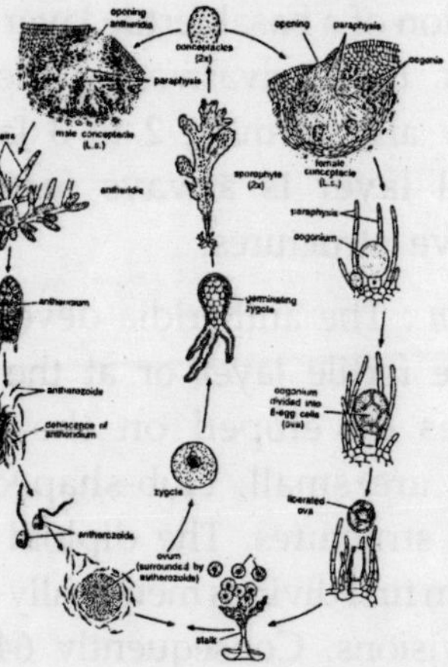

Fucus : Life-cycle.

Sexual Reproduction is Oogamous : The sex organs are produced in special flask-shaped cavities, known as conceptacles. The conceptacles are borne on the receptacles. The receptacles are the tips of the branches which become swollen. The male sex organs are called antheridia and the female oogonia. In homothallic (monoecious) species (e.g., F. spiralis) antheridia and oogonia are present in the same conceptacle or in different conceptacles on the same plant. In dioecious species (e.g., F. vesiculosus), male and female sex organs develop in different conceptacles on separate plants.

1. *Development of Conceptacles* : The conceptacle develops from a superficial cell of the receptacle, which is easily distinguished from the adjoining cells by its larger size and prominent nucleus. It divides at a slower rate than the adjoining cells. Consequently, the initial cell, which was at the surface of the thallus in the beginning, is gradually pushed at the base of the flask shaped cavity. The conceptacle initial then divides by a transverse wall forming an outer tongue cell and an inner basal cell. The tongue cell does not divide further and gradually degenerates. The basal cell, however, undergoes repeated anticlinal divisions resulting into the formation of a basal fertile layer of the conceptacle. Sometimes, the derivatives of the basal cell divide periclinally also forming 2 to 3 layers of which the superficial layer is always fertile giving rise to reproductive structures.
2. *Antheridium* : The antheridia develop either from the cells of the fertile layer or at the base of branched paraphyses developed on the fertile layer. The antheridia are small, club-shaped, shortly-stalked, unicellular structures. The diploid nucleus of young antheridium first divides meiotically and then by several mitotic divisions. Consequently 64 or more haploid nuclei are formed in the developing antheridium. Thereafter, the multinucleate protoplast of the young antheridium divides into many uninucleate segments and each uninucleate segment metamorphoses into a biflagellate antherozoid.

The mature antheridium remains surrounded by a two layered wall. The outer wall layer, the exochite, is firm, whereas the inner wall layer, the endochite, is gelatinous. At maturity, the antheridial wall gets gelatinized when in contact with water and the flagellated antherozoids are liberated.

3. *Oogonium* : Like antheridium, an oogonium initial is also a superficial cell of the fertile layer. Each initial cell divides by a transverse wall. The smaller basal cell forms the stalk cell and the upper larger cell, the oogonial cell. The latter gradually enlarges and becomes almost spherical and densely cytoplasmic. The diploid nucleus of the oogonial cell undergoes a meiotic division and the four nuclei thus formed divide mitotically. Consequently, eight haploid nuclei are formed in the oogonium. These eight nuclei serve as eggs or ova.

   The oogonial wall is differentiated into three layers, the outer exochite, the middle mesochite and the inner endochite. At maturity, the oogonium absorbs water with the result the exochite ruptures and eggs slip out of it, still surrounded by mesochite and endochite and approaches the ostiole. Soon, the endochite also imbibes water and as a result, the mesochite also breaks at its apical portion and ultimately, the eggs are set free in the water.

   Those cells of the basal fertile layer of the conceptacle which do not form reproductive structures, develop into long hair-like paraphysis. Some of the peripheral cells near the ostiole also form unbranched hair-like appendages which émerge from the ostiole in the form of a tuft.

4. *Fertilization* : The fertilization is external, i.e., the antherozoids and the eggs come in contact outside the conceptacle in the water. A large number of antherozoids approach a single egg. After the

attachment of the antherozoid flagella to the egg surface, the beating of the flagella makes the egg rotate for some time. Ultimately, one of the antherozoids finds its way into the egg, leading to the formation of the zygote. About after an hour of fertilization a gelatinous wall is secreted around the zygote.

5. *Germination of Zygote* : The zygote germinates after a short period of rest (about a day) and first, it develops a rhizoidal protuberance on one side of it. After the development of the rhizoidal protuberance, the zygote divides mitotically in a plane transverse to the axis (as defined by the rhizoidal protuberance). The upper cell of the 2-celled embryo divides vertically where as the lower divides transversally, thus forming a 4-celled embryo. The upper two cells ultimately give rise to a tissue differentiated into the medulla and cortex, whereas the lower most cell of the quadrant from which the primary rhizoid originates also forms secondary rhizoids. Ultimately, a multicellular embryo with an apical cell is differentiated.

Alternation of Generation : The life cycle of Fucus does not show any alternation of generation. The plant body is diploid and sporophytic. The reproductive organs, antheridia and oogonia, are produced within special flask-shaped conceptacles. The reduction division takes place at the time of gamete formation.

Thus, the haploid phase, in the life-cycle of Fucus, is represented only by gametes (antherozoids and eggs). The haploid male and female gametes fuse to form diploid zygote. The zygote germinates into a diploid sporophytic thallus.

Family Sargassaceae

The members are very much complicated in their morphology and anatomy and are characterized by the presence of lateral branch system. The family includes five genera. Of these Sargassum is widely known, and the most abundant sea weed in India.

Sargassum

Systematic Position

Class	-	Phaeophyceae
Order	-	Fucales
Family	-	Sargassaceae

Occurrence : The genus Sargassum is represented by about 150 species, found chiefly in the tropical and subtropical seas of the southern hemisphere. The plants form huge floating masses in the Atlantic ocean, off the African continent between 20° and 35° north latitude. Hence this part of the Atlantic ocean is called the Sargasso sea. Species of Sargassum are also found along the coasts of Australia, India and Ceylon. In India, the genus is represented by 16 species which occur along the western and southern coast. Some common Indian species of Sargassum are. S. cinereum, S. carpophyllum, S. duplicatum, S. christifolium, S. tenerrium, S. ilicifolium, S. plagiophyllum, S. wightii and S. myriocystum.

Thallus Structure

External Structure : The thallus of Sargassum is diploid and sporophytic. It is erect and branched and is differentiated into holdfast and the main axis; the latter bears many leaf-like lateral branches and air bladders.

The holdfast is a discoid or warty structure with the help of which the thallus remains attached to the substratum. In some species the holdfast is stolon-like. The main axis or 'stem' is erect, cylindrical or flattened, varying in height from 15-25 cm. But sometimes it attains a height of more than a meter as in S. filipendula. The main axis bears large number of primary laterals in a spiral phyllotaxy of 2/5. The primary laterals have unlimited growth, hence they are also called long shoots. The main axis as well as the primary laterals bear flat leaf-like branches, known as secondary laterals or 'leaves'.

The 'leaves' are flat and simple with distinct mid-ribs. Sometimes the mid-rib is absent as in S. enerve. The margins of the 'leaves' are serrate, dentate or entire. On both surfaces of the 'leaves' small pores, known as ostioles, are present. These pores open in small flask-shaped sterile cavities, called cryptostomata or sterile conceptacles.

The 'leaves' present in the lower part of the primary lateral (long shoots) bear rudimentary axillary branches and those present in the upper portion have axillary branch systems. One or more basal members of the axillary branch are modified into globular or spherical air bladders. The air bladders are filled with air and help the plant in floating by increasing its buoyancy. In some species the air bladder terminates into a 'leaf'. According to some algologists, the air bladders help in respiration. The subsequent members of the axillary branch system develop into long repeatedly branched cylindrical or flattened structures, called receptacles. The receptacles bear reproductive organs in special flask-shaped structures which are known as conceptacles.

Internal Structure

1. *Internal Structure of the Main Axis* : In transverse section the main axis is almost circular in outline and is differentiated into three distinct regions : (i) meristoderm, (ii) cortex, and (iii) medulla.

   The meristoderm (named so due to meristematic nature of its cells) is the outermost layer and functions as epidermis. It is made up of compactly arranged columnar cells. These cells are densely packed with chromatophores and fucosan, a reserve food material.

They form assimilatory region of the thallus. In the mature part of the axis meristoderm is covered with a thin layer of mucilagenous cuticle.

The cortex forms the most part of the axis and it has narrow and elongated cells. These cells are loosely arranged with well developed intercellular spaces. They are filled with reserve food material, and as such the cortex forms the storage region of the axis.

The central part of the main axis forms medulla. It is made up of thick walled narrow and elongated cells. Sometimes these cells have scalariform thickenings. The function of medulla is to transport water and metabolites to different parts of the thallus.

2. *Internal Structure of the Leaf* : Just like the main axis, the leaf is also differentiated into meristoderm, cortex and medulla.

The meristoderm consists of small radially elongated, meristematic cells. These cells have plenty of reserve food and chromatophores. The meristoderm is followed by cortex which consists of thin walled cells, rich in reserve food material. The cortex is the thickest in the midrib region and gradually narrows towards the blades. The central part of the 'leaf' is made up of thick walled cells which make up medulla. In structure and function it is similar to the medulla of the main axis.

On both the surfaces of the 'leaf' are present many sterile conceptacles, known as cryptoblasts or cryptostomata. The cryptoblast is a flask-shaped structure and it opens on the surface of the 'leaf' by a small pore, known as ostiole. The wall of the cryptoblast is made up of sterile cells which have chromatophores. Many unbranched multicellular filaments arise from the wall of the conceptacles and protrude through the ostiole. These filaments are called paraphysis.

3. *Internal Structure of Air-bladder :* The air bladder is differentiated into meristoderm and cortex only. Medulla is absent. The meristoderm is made up of radially elongated narrow cells, enveloped with a mucilagenous layer. It is followed by a 4-8-layered parenchymatous cortex. The central part of the air bladder is occupied by a large air cavity.

Growth : The growth of Sargassum thallus is accomplished by a quadrangular apical cell.

Reproduction : Sargassum reproduces by vegetative and sexual methods. Asexual reproduction is absent.

Vegetative Reproduction : Fragmentation is the chief method of vegetative multiplication. Thallus may break into segments due to death and decay of the older parts or due to injury. These segments grow into new thalli. In some free floating species like S. natans and S hystrix, it is the exclusive method of multiplication.

Sexual Reproduction : Sexual reproduction is oogamous. The sex organs are produced in special flask-shaped cavities, known as conceptacles. Conceptacles are borne on the receptacular branches of the sporophytic thallus. The male sex organs are called antheridia and the female oogonia. Male and female sex organs are produced on separate conceptacles; the conceptacles bearing antheridia are called male conceptacles and those having oogonia female conceptacles. In homothallic (monoecious) species antheridia and oogonia are present in the same conceptacle or in different conceptacles on the same plant. In dioecious species male and female sex organs develop in different conceptacles on separate plants. Most of the species of Sargassum are monoecious.

The development of fertile and sterile conceptacles is similar. The only difference is that the basal cells in sterile conceptacles form paraphysis, whereas in fertile conceptacles they give rise to antheridia and oogonia.

1. *Development of Conceptacles :* The conceptacle develops from a single superficial cell on the receptacular branch. This flask-shaped conceptacle initial can be easily distinguished from the adjoining cells by its larger size and prominent nucleus. It divides at a slower rate than the adjoining cells. Consequently, the initial cell, which was at the surface of the thallus in the beginning, is gradually pushed at the base of the flask shaped cavity. The conceptacle initial divides by a curved septum, forming an outer tongue cell and an inner basal cell. The tongue cell elongates and divides transversely to form a small filament which soon disappears. The basal cell, however, undergoes repeated vertical divisions and the resulting cells form the basal fertile layer of the conceptacle. The reproductive organs develop from the cells of this layer.

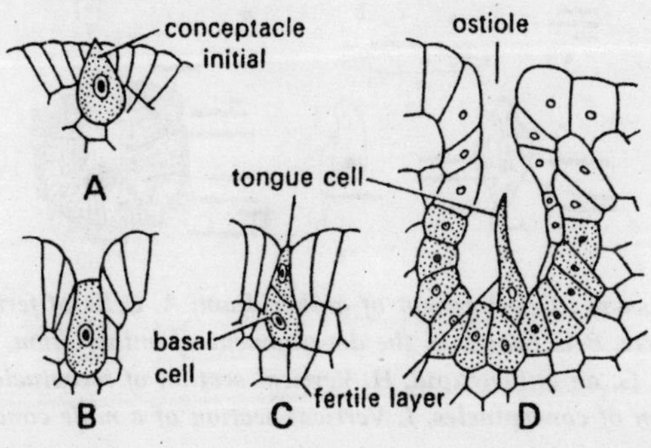

A-D. Sargassum : Stages in the development of conceptacle.

2. *Development of Antheridium :* As mentioned above, antheridia develop from the cells of the fertile layer of the conceptacle. The antheridial initial is densely cytoplasmic and grows into a papilla-like outgrowth. It divides by a transverse wall into a lower stalk cell and an upper antheridial cell. Each of these cells has a diploid nucleus. The antheridial cell develops into an antheridium. It is pushed on one side due to rapid growth of the stalk cell. The growing stalk cell again

divides into a basal stalk cell and upper antheridial cell. This process is repeated several times and consequently a paraphysis with many antheridia is formed. The terminal cell of the paraphysis is usually sterile, it does not develop into an antheridium:

The antheridial initial has only one diploid nucleus which undergoes a meiotic and several mitotic divisions. Consequently, 32-64 haploid nuclei are formed in the developing antheridium. The multinucleate protoplast of the young antheridium divides into many uninucleate segments (the number of these segments is equal to the number of nuclei) and each uninucleate segment metamorphoses into a biflagellate antherozoid.

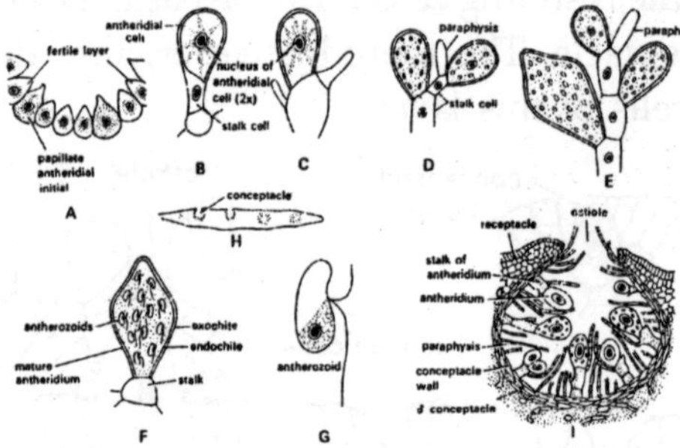

A-I. Sargassum : Development of antheridium; A. Cells of fertile layer in conceptacle, B-E. Stages in the development of antheridium, F. Mature antheridium, G. an antherozoid, H. Vertical section of receptacle showing the position of conceptacles, I. Vertical section of a male conceptacle.

The mature antheridium is an oval structure surrounded by a two layered wall. The outer wall layer, the exochite, is firm, whereas the inner wall layer, the endochite, is gelatinous. The antheridium remains attached to the base of the conceptacle with the help of a stalk cell. At maturity, the antheridium is detached from the stalk and comes out of the conceptacle through the ostiole. The antheridial wall gets gelatinized when in contact with water and the flagellated antherozoids are liberated.

3. *Development of Oogonium* : Oogonia also develop from the basal fertile layer of the conceptacle. The cells of this layer function as oogonial initials. Each initial cell divides by a transverse wall into two unequal cell. The smaller basal cell forms the stalk cell and the upper larger cell, the oogonial cell. The latter gradually enlarges and becomes almost spherical. It is densely cytoplasmic and has oil droplets. The diploid nucleus of the oogonial cell undergoes a meiotic division and the four nuclei thus formed divide mitotically. Consequently, eight haploid nuclei are formed in the oogonium. Out of these, seven degenerate and the remaining nucleus functions as the egg nucleus. This functional nucleus is located in the centre of the oogonium, whereas the degenerating nuclei can be seen along the periphery.

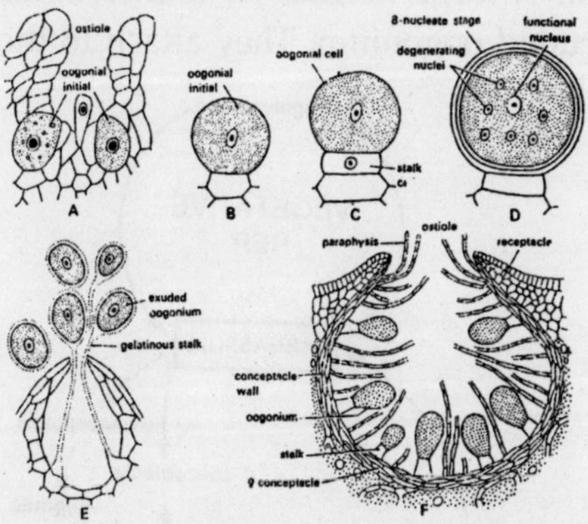

A-F. Sargassum : Development of oogonium; A. Oogonial initials in the fertile layer of conceptacle, B-d. Stages in the development of oogonium, E. Discharge of oogonia from conceptacle, F. Vertical section of a female conceptacle.

The oogonial wall is differentiated into three layers, the outer exochite, the middle mesochite and the inner endochite. The mature oogonia are discharged from the conceptacle through the ostiole but they remain attached with the conceptacle base by long gelatinous stalks formed by the exochite. According to some algologists, the exochite of the oogonium ruptures at the time of its discharge from the conceptacle and the gelatinous stalk is formed by the mesochite. The discharged egg is thus surrounded by the endochite only.

Those cells of the basal fertile layer of the conceptacle which do not form reproductive structures, develop into long hair-like paraphysis. Thus the conceptacle has many paraphysis in addition to reproductive structures.

4. *Fertilization* : At the time of fertilization although the oogonium is outside the conceptacle, it still remains attached with the conceptacle base by a gelatinous stalk. A large number of antherozoids surround the extruded oogonium. They attach to the oogonial wall

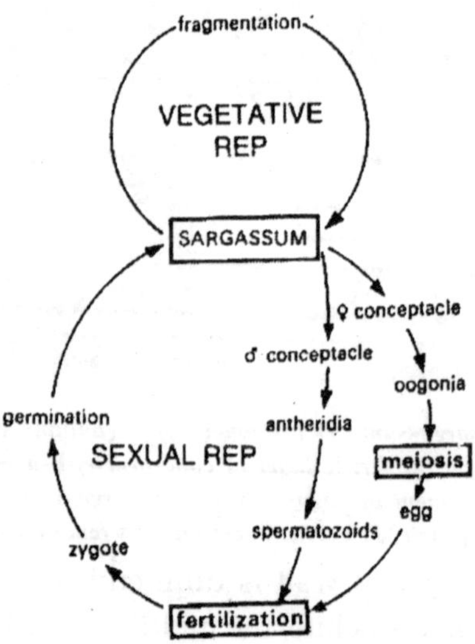

Sargassum : Schematic representation of life cycle.

with the help of the anterior flagellum and swim in water by the posterior flagellum. Only one antherozoid penetrates the oogonial wall. After fertilization the male and female nuclei lie side by side for sometime and then they fuse to form a diploid zygote.

5. *Germination of Zygote* : The zygote germinates immediately after fertilization while the oogonium is still attached to the conceptacle wall by a gelatinous stalk. However, after sometime there is gelatinization of oogonial wall and the zygote is liberated. It now gets attached to some solid substratum in sea water. The zygote divides by a transverse wall into an upper and a lower cell. The lower cell forms rhizoids, whereas the upper cell undergoes many anticlinal and periclinal divisions to form a diploid thallus.

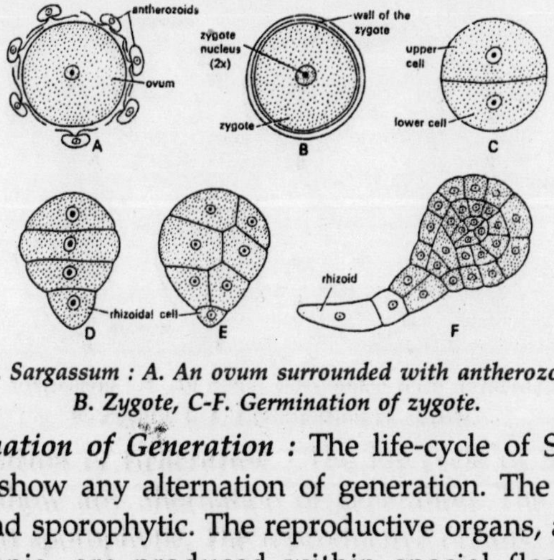

A-F. Sargassum : A. An ovum surrounded with antherozoids, B. Zygote, C-F. Germination of zygote.

Alternation of Generation : The life-cycle of Sargassum does not show any alternation of generation. The thallus is diploid and sporophytic. The reproductive organs, antheridia and oogonia, are produced within special flask-shaped conceptacles. Both these structures are diploid and produce antherozoids and egg after meiosis. Thus the haploid phase, in the life-cycle of Sargassum, is represented only by gametes. The haploid male and female gametes fuse to form diploid zygote. The zygote germinates into a diploid sporophytic thallus.

6

The Ectocarpales

The order Ectocarpales includes 60 genera of brown algae which are found in the cold sea water of the temperate and polar regions. Most of the members of this order are heterotrichous with a prostrate creeping disc, which functions as holdfast and an erect monosiphonous filamentous foliose system. These algae show trichothallic growth with intercalary divisions confined to certain areas in the filament. Asexual reproduction is by zoospores which are produced in uni- or plurilocular sporangia. Sexual reproduction is isogamous but in some species it is oogamous. Reproductive structures are borne laterally or in uniseriate rows.

Fritsch (1935) divided the order into 12 families- Ectocarpaceae, Myrionemataceae, Elachistaceae, Leathesiaceae, Mebogloeaceae, Aenotrichaceae, Spermatochnaceae, Splachnidiaceae, Punctariaccae, Asperococcaceae, Encoeliaceae and Dictyosiphonaceae. But most algologists restrict Ectocarpales to a single family Ectocarpaceae characterized by simple, heterotrichous members exhibiting isogamous sexual reproduction and isomorphic alternation of generations.

Family Ectocarpaceae

The members of the family are characterised by simple heterotrichous thallus organisation, isogamous type of sexual reproduction with isomorphic alternation of generation. The life history of Ectocarpus is described here as a representative of the family.

Ectocarpus

Systematic Position

Class	-	Phaeophyceae
Order	-	Ectocarpales
Family	-	Ectocarpaceae

Occurrence : Species of Ectocarpus are cosmopolitan in distribution, growing in littoral and supralittoral regions. They are more common along the Atlantic coast than the Pacific coast. In India they are found on the Western coast, growing, epiphytically on sea plants or attached to rocks. Epiphytic species, such as E. coniferus and E. breviarticulatus, grow on larger algae like Fucus and Laminaria. Ectocarpus fasciculatus, an epizoic species, grows on fins of certain fishes. E. dermonematus is an endophytic species. The genus is represented in India by about 13 species. Of these, Ectocarpus coniferus, E. indicus, E. arabicus, E. geminifructus and E. dermonematus are common.

Thallus Structure

The thallus of Ectocarpus is heterotrichous, differentiated into a prostrate rhizoidal system and an erect branched system. The prostrate system is creeping and branched, and the thallus remains attached to the substratum with the help of this system. It is more developed in attached forms than in free floating species. In some species rhizoidal part is epiphytic, i.e., it grows within the supporting plant.

The erect part of the thallus is usually irregularly branched and the branches arise just beneath the transverse septa of the cells. The filaments are several centimeters long and monosiphonous, i.e., only one cell in thickness. Sometimes the filaments may become polysiphonous due to vertical divisions of the cells. In some species (e.g., E. granulosus) the basal parts of the main erect filaments become corticated by ascending rhizoidal branches. The distal part of each filament terminates into mucilagenous hairs or erect threads.

Cell Structure

Filaments of Ectocarpus are made up of uninucleate rectangular or cylindrical cells. The cell wall is differentiated into an outer gelatinous layer and an inner firm cellulose layer. The gelatinous layer is mainly composed of algin and fucoiden, characteristic of phaeophycean cell wall. A thin layer of cytoplasm is present inner to the cell wall from which a number of cytoplasmic strands grow toward the centre of the cell. There is a central nucleus and many chromatophores in the cytoplasm. The number and shape of chromatophores varies in different species. They are discoid or band-shaped, and each chromatophore has a projected pyrenoid.

The main cell pigments are chlorophyll-*a*, chlorophyll-*c*, carotene and xanthophyll. Xanthophyll is present in much more amounts than chlorophyll. Consequently, the green colour of chlorophyll is completely masked.

The cytoplasm also contains some vacuoles with polyphenol. These vacuoles are called physodes and they probably function like lysosomes. In the life-cycle of Ectocarpus two distinct vegetative phases, haploid and diploid, are found. Both the phases are filamentous but the diploid filaments have longer cells than the haploid filaments.

Growth : The prostrate part of Ectocarpus thallus shows apical growth, whereas it is intercalary in the erect part. At the base of the erect filament there are few meristematic cells which divide transversely. The newly formed cells are smaller than those present at the distal end of the filament. In some species the meristematic cells are located in the middle of the filament (e.g., E. elachistaeformis) or at the base of the distal mucilagenous hair (e.g., E. geminifructus). Meristematic cells at these locations cut off segments in both upward and downward directions. The derivatives in upward direction increase the length of the terminal hair, whereas those in downward direction form vegetative photosynthetic cells of the filament. This type of growth, characteristic of Ectocarpales, is known as trichothallic.

Reproduction

Ectocarpus reproduces by asexual and sexual methods.

Asexual Reproduction

Asexual reproduction takes place by bilagellate zoospores. The sporophytic (diploid) plant produces two types of sporangia : (i) unilocular zoosporangia, and (ii) plurilocular or neutral zoosporangia. Both of these types of zoosporangia are borne on the same plant or on different plants. The unilocular sporangia form haploid zoospores and the plurilocular, diploid zoospores.

1. *Unilocular Sporangia :* These sporangia develop singly at the tips of small branchlets. The terminal cell of the branchlet enlarges in size and functions as sporangial initial. The first division of diploid nucleus is meiotic and it is followed by mitotic divisions. This results in the formation of 32 or 64 haploid nuclei. The nuclear divisions are not accompanied by wall formation and as such the sporangium remains unilocular. The protoplast of the sporangium divides into as many segments as the number of nuclei. Each segment

metamorphoses into a pyriform, uninucleate zoospore. The zoospore has a chromatophore and an eye spot. The two flagella are laterally inserted and are unequal; the anterior is longer and pantonematic and the posterior is shorter and acronematic.

Zoospores are discharged between 0600 and 1200 hours. They ooze out of the sporangium in a gelatinous matrix through a small terminal perforation. After their discharge they remain in a spherical mass at the apex of the sporangium for a while and then become free and swim off individually. They seldom remain motile for more than 30 minutes. They withdraw their flagella and attach themselves to the substratum by their anterior ends. They germinate within 2-3 hours and produce haploid filaments.

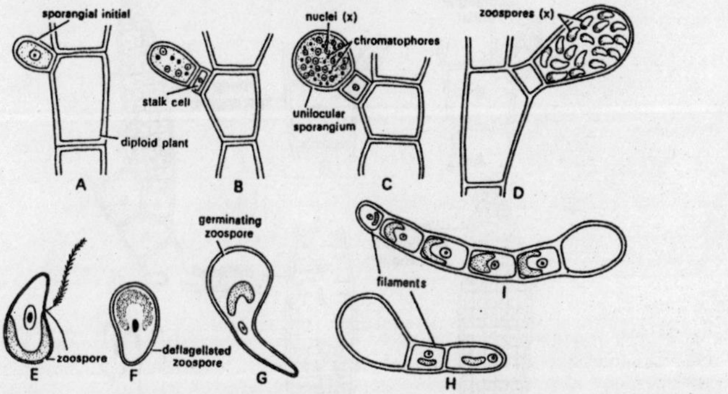

A-I. Ectocarpus : A-D. Successive stages in the development of unilocular sporangium; E. a zoospore; F-I. Stages in the germination of zoospore.

Sometimes after the liberation of zoospores new sporangium initiates within the same sporangial wall.

2. *Plurilocular or Neutral Sporangia :* Like unilocular sporangia, these sporangia also develop from terminal cells of the branchlets of diploic thalli. The terminal cell, which acts as sporangia initial, enlarges and becomes spherical and accumulates numerous chromatophores. The sporangial initial now undergoes

repeated transverse and vertical divisions, forming as many as 660 cubical cells arranged in definite tiers. This multicellular structure is known as plurilocular sporangium. Each diploid uninucleate cell of the sporangium produces a single biflagellate zoospore. The only difference between the zoospores produced by unilocular and plurilocular sporangia is that the former are haploid and the latter are diploid.

Mature zoospores are liberated from the sporangium by a terminal or lateral pore. They remain motile for 3-5 hours and then germinate to form diploid asexual thalli. At maturity these thalli bear both unilocular and plurilocular sporangia. Thus diploid zoospores help only in the multiplication of the sporophyte. They do not play any part in alternation of generation.

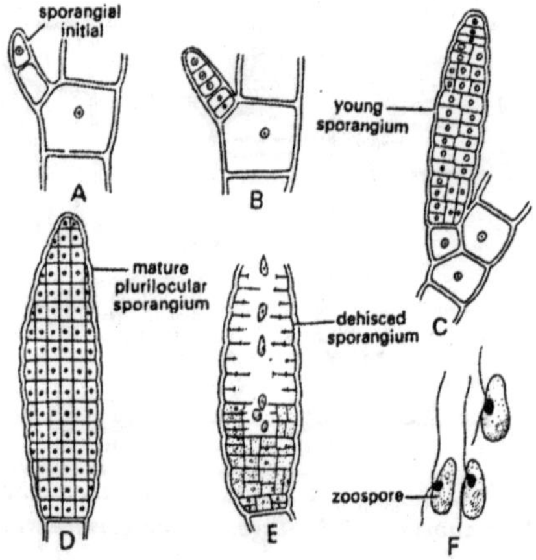

A-F. Ectocarpus : A-C. Development of plurilocular sporangium; d. Mature plurilocular sporangium; e. Liberation of zoospores from sporangium; F. zoospores.

The formation of unilocular and plurilocular sporangia by a species is affected by temperature and salinity of water. For example, E. siticulosus produces unilocular sporangia at 13°C, plurilocular at 19°C and both unilocular and plurilocular at 16°C.

Sexual Reproduction

Sexual reproduction may be isogamous, anisogamous or oogamous, but most of the species of Ectocarpus are anisogamous. The anisogamy may be physiological (e.g., E.siliculosus) or morphological (e.g., E.secundus).

1. *Gametangia and Gametes :* Gametes are formed in plurilocular gametangia which are similar to plurilocular sporangia in their structure and development.

   The gametangia may be sessile or shortly stalked. They develop singly from the terminal cells of the lateral branchlets. The gametangial initial divides transversely to form 6-12 cells, arranged in a uniseriate row. Repeated transverse and vertical divisions of these cells produce hundreds of cubical cells which are arranged in 20-40 transverse layers. The uninucleate protoplast of each cell is transformed into a biflagellate pyriform gamete.

   In E. secundus, which shows morphological anisogamy, two types of gametangia are produced. The megagametangia, which have larger loci, produce larger gametes and the microgametangia, which have smaller loci, produce smaller gametes. After their discharge from the gametangia, unequal gametes fuse in pairs to form zygospores. In E: siticulosus, which shows physiological anisogamy, gametes from the gametangia developing on filaments of different strains fuse to form zygospores. The fusing gametes are morphologically identical but are distinct physiologically. The gametes of one strain are more active and behave as male gametes, whereas those of the other strain are less active and behave as female gametes.

2. *Fertilization :* In isogamous species the fusing gametes are similar both morphologically and physiologically. In such species two gametes from the same

gametangium or from the two different gametangia of the same thallus fuse to form zygospore. But in physiologically anisogameus species, gametes from two filaments of different strains fuse. In such species, the female gamete settles down about 5 minutes after liberation and secretes a volatile sexual attractant, sirenine. The male gametes, which may remain motile for about 8 hours, are attracted towards the female gamete by this substance. A large number of male gametes cluster around a female gamete and attach to it by their long anterior flagella. After sometime one of the male gametes contracts its flagellum and consequently comes in contact of the female gamete. These two gametes fuse to form zygospore. The process of fusion takes about 20 seconds, and after fusion the zygospore loses its attraction for male gametes. As such the remaining male gametes detach themselves from the female gamete and swim freely. The process in which many male gametes cluster around a female gamete and one male gamete fuses with the female gamete to form zygospore is called clump formation.

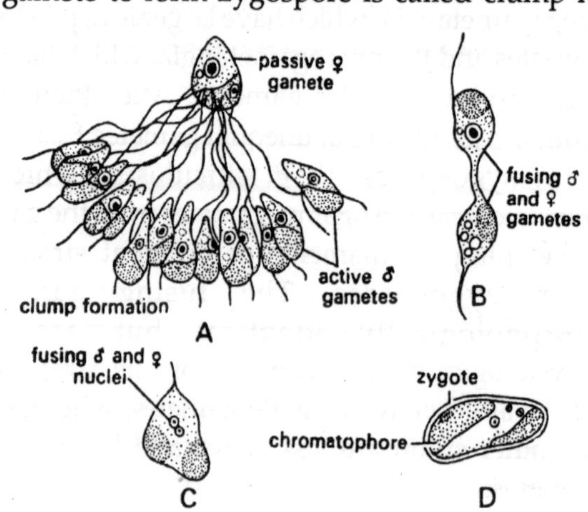

A-D. ectocarpus siliculosus : A. Cluster of male gamete around a female gamete (clump formation); B-C. Fusion of male and female gametes; D. Zygote.

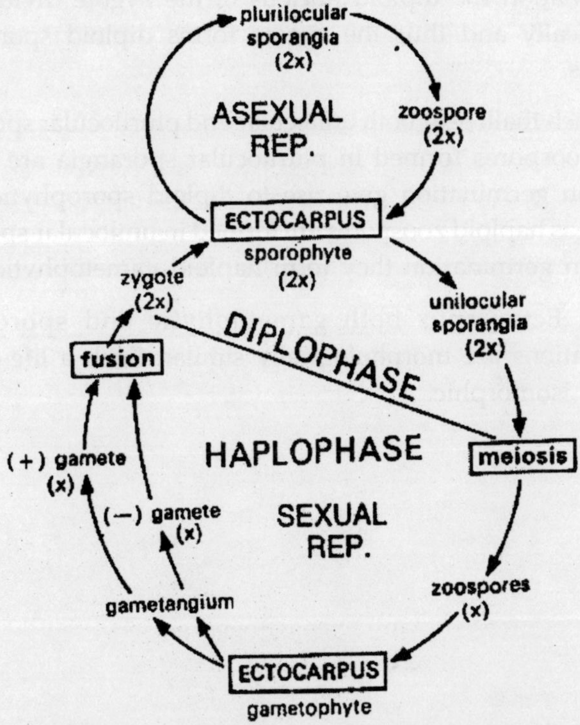

Ecotocorpus* : *Schematic representation of the life-cycle.

3. *Germination of Zygospore* : The zygospore germinates within 2-3 days after fertilization. At the time of germination the diploid zygospore nucleus divides mitotically hence it gives rise to a diploid thallus. This thallus bears both unilocular and plurilocular sporangia.

Some unfused female gametes have the ability to germinate parthenogenetically and they form haploid filaments.

Alternation of Generation

The sexual thalli of Ectocarpus are dioecious and haploid. They produce only plurilocular gametangia. The iso-or anisogametes fuse to form diploid zygote. At the time of germination the diploid nucleus of the zygote divides only mitotically and thus the zygote forms diploid sporophytic thallus.

Such thalli bear both unilocular and plurilocular sporangia. The zoospores formed in plurilocular sporangia are diploid and on germination give rise to diploid sporophytic thalli, whereas haploid zoospores are formed in unilocular sporangia and on germination they form haploid gametophytic thalli.

In Ectocarpus both gametophytic and sporophytic generations are morphologically similar. Such a life-cycle is called isomorphic.

7

The Phaeophyceae

Class Phaeophyceae is represented by about 240 genera and 15000 species. They are characterised by the presence of a brown coloured carotenoid, fucoxanthin in large amounts, which masks the green colour of chlorophyll pigment. This gives a characteristic brown colour to these plants, hence the name brown algae.

Except few fresh water forms like species of Pleurocladia, Heribaudiella, and Bodanella most of the brown algae are marine. They grow in intertidal belt and upper littoral region. They dominate colder waters but members of Ectocarpales, Dictyotales and species of Sargassum grow in warm waters.

In the tropics members of Phaeophyceae are most abundant in the Sargasso sea of the Atlantic. They are usually found attached to rocks with a discoid or branched holdfast. Many of these algae have air bladders which increase their buoyancy.

In India brown algae are commonly found on the western and southern coasts.

The important characters of the class Phaeophyceae are as follows .

(1) The brown algae range in structure from simple to complex parenchymatous forms. They include : (i) Heterotrichous forms, e.g., Ectocarpus; (ii) Multiaxial construction forms, e.g., Myriactula pulvinata; (iii) Uniaxial pseudo-parenchymatous forms, e.g., Hecatonema sargassicola; and (iv) Parenenymatous forms, e.g., Laminaria, Fucus.

Besides, Postelsia palmaeformis assumes the form of a miniature tree and is commonly known as sea palm.

(2) The plant body in the members of Laminariales and Fucales is differentiated into holdfast, stem-like stipe and flattened leaf- like blades. Many species possess air bladders which keep them afloat.

(3) The tissues of the stipe and blade are differentiated into epidermis, cortex and central medulla.

(4) Growth of the thallus is apical (Sphacelariales, Dictyotales, Fucales), intercalary (Laminariales) or trichothallic (Ectocarpales, Sporochnales).

(5) Phaeophycean cell wall is differentiated into two layers. The outer mucilagenous layer has alginic and fucinic acid, whereas the inner layer is mainly cellulosic. Alginic acid is of commercial importance and is obtained from Laminaria, Sargassum and Ascophyllum. It is used in the manufacture of adhesives and artificial silk.

(6) The cytoplasm contains many small vacuoles and refractive bodies. The latter are called fucosan vesicles. They are abundant in the cytoplasm of metabolically active cells.

(7) Chromatophores are ellipsoidal and discoid. They contain chlorophyll *a*, chlorophyll *c*, and *b*-carotenes and xanthophyll (e.g., lutein, flavoxanthin,

violaxanthin). Besides, they also contain large amount of a brown pigment, fucoxanthin ($C_{40}H_{56}O_6$). Fucoxanthin masks green colour of chlorophyll pigments and that is why these algae are brown in colour.

(8) Chromatophores have three or four thylakoids per band and are surrounded by a chromatophore envelope and two membranes of endoplasmic reticulum. The chromatophore has a single, stalked, projected and naked pyrenoid.

(9) The long-term storage product is laminarin. However, the accumulation product of photosynthesis is D-mannitol, a sugar alcohol.

(10) The cells are uninucleate with one or more nucleolus. In some brown algae (e.g., Fucus) the resting nuclei show chromocentres which are not found in other groups of algae.

(11) The flagellated structures have a pair of laterally inserted unequal flagella, of which the larger one is anterior and pantonematic, whereas the smaller one is posterior and acronematic.

(12) They reproduce by vegetative, asexual and sexual methods.

(13) Fragmentation is the most common method of vegetative multiplication. Besides, in the members of Sphacelariales special reproductive branches are found which are called propagules. These branches, when detached from parent plant, grow into new plants.

(14) Except for Tilopteridales, Dictyotales and Fucales, all brown algae reproduce asexually by zoospores.

(15) In Ectocarpales and Sphacelariales zoospores are formed in unilocular or plurilocular sporangia.

(16) The zoospores produced in unilocular sporangia are

haploid and those in plurilocular sporangia are diploid. Both the haploid and diploid spores are similar in structure. They are pyriform, biflagellate and have chromatophores, contractile vacuoles and eye spot.

(17) In some members of Tilopteridales the protoplast of the sporangium is transformed into a large quadrinucleate spore, called monospore.

(18) Phaeophycean algae show much variation in their sexual reproduction. It is isogamous in Ectocarpales and Sphacelariales, anisogamous in Cutleriales and Tilopteridales and oogamous in Fucales, Laminariales and Dictyotales.

(19) Gametes are more or less similar to zoospores and are also biflagellate.

(20) In most brown algae fertilization is external, i.e., gametes fuse outside the gametangium in water. There is no reduction division in zygote; it forms a diploid thallus on germination.

(21) The life cycle of brown algae may be isomorphic, heteromorphic, or diplontic.

Classification

Fritsch (1935), on the basis of vegetative structure and sexual reproduction, divided the class Phaeophyceae into nine orders : (1) Ectocarpales, (2) Tilopteridales, (3) Cutleriales, (4) Sporochnales, (5) Desmarestiales, (6) Laminariales, (7) Sphacelariales, (8) Dictyotales, and (9) Fucales.

Smith (1955) recognized the following three classes and 12 orders in the division Phaeophyta :

Isogeneratae

Members of this class are characterised by alternation of two isomorphic generations. It has five orders differing in vegetative structure, mode of growth and structure of reproductive organs : (1) Ectocarpales, (2) Sphacelariales, (3) Tilopteridales, (4) Cutleriales, (5) Dictyotales.

Heterogeneratae

Those brown algae which show alternation of heteromorphic (i.e., morphologically distinct) generations are included in this class. It has six orders : (1) Chordariales, (2) Sporochnales, (3) Desmarestiales, (4) Punctariales, (5) Dictyo siphonales, (6) Laminariales.

Cyclosporeae

The brown algae included in this class have a life cycle in which, there is no alternation of free-living multicellular generations. This class includes only a single order, Fucales.

Pennales

The members of the order Pennales are characterised by the presence of a bilateral symmetry. A pseudoraphe or a raphe is usually present in the median line of the valve. They show gliding movement. Sexual reproduction, if present, is of amoeboid isogamous type. Auxospores are normally formed by conjugation. They occur mostly in fresh water habitats. The order is classified according to the presence or absence and the number and morphology of the raphe on the valves.

Family Naviculoideae : In the members of this family the raphe is present on both the valves. It occupies the apical axis and is usually without a keel.

Navicula

Systematic Position

Class	-	Bacillariophyceae
Order	-	Pennales
Family	-	Naviculoideae

Occurrence : Navicula, a fresh water diatom, is cosmopolitan and ubiquitous in distribution, found in ponds, lakes, etc. However, a few species occur in terrestrial habitats. The species of Navicula are very important as primary producers in the food web of aquatic ecosystems. The most common species of the genus occurring in India is N. halophila.

Thallus Structure

The plant body is unicellular with an isobilateral symmetry. The cells usually occur in isolated form. The cell wall (frustule) consists of two overlapping halves, the upper half is said to be epitheca and the lower half hypotheca. Each theca is further divided into two parts : the main surface called valve and the incurved margins, known as connecting band. The two connecting bands of the two thecas are fitted together. The connecting band of the epitheca overlaps that of the hypotheca and the two bands remain united in the overlapping region (called girdle) by a cementing organic substance present between them. A cell can be seen from two different aspects, the valve view and the girdle view. In valve view the cell appears boat-shaped and in girdle view more or less rectangular. The cell is several times longer than its breadth.

The cell wall is made up of crystalline silica and an organic component of unknown nature. In N. pelliculosa, the wall consists of a silica layer with an outer coating of organic material called organic skin. There is a longitudinal slit known as raphe. The raphe is interrupted in the middle by a central nodule formed by the internal thickening of the valve. Each pole of the valve contains a polar nodule, formed in the same manner as the central nodule. The valve markings consist of fine pores bridged from the inside by sieve membranes also containing numerous fine pores. The raphe situated in the central rib is called apical rib. The rows of pores situated on the lateral sides of the apical rib and separated one from the other by solid silica ribs are known as trans-apical ribs. The pores within a row are separated by siliceous connections which link the adjacent trans-apical ribs.

The protoplast is surrounded by a thin plasma membrane, which lies close to the inner face of the cell wall. There are two brownish yellow chromatophores, adpressed to the girdle side in the resting cell. The nucleus is situated in the centre of a dense bridge of cytoplasm which connects the two chromatophores. The cytoplasm contains mitochondria, golgi bodies and endoplasmic reticulum. The reserve food mateiial is in the form of fatty oils volutin globules.

Reproduction

Navicula reproduces vegetatively and by the formation of auxospores.

Vegetative Reproduction : There is rapid multiplication by successive divisions, which usually occur in the evening. First, one nucleus divides into two and then cytoplasm by an ingrowing cleavage furrow.

Prior to cell division there is a slight increase in the size of the nucleus. The cell also increases in volume, the valves loosen a little to accommodate the increase in the cell volume. The nucleus, which is present in the centre of the cell, moves toward the girdle side, and in the process it loses its spherical outline and becomes somewhat compressed. Now the nucleus divides mitotically in a plane perpendicular to the valves. It is followed by the division of the protoplast in two by invagination of the plasmalemma. In girdle view cytokinesis is seen commencing as a small cleavage which progresses centripetally towards the centre of the cell. After the completion of cytokinesis each daughter protoplast secretes a new valve on its inner side.

During the nuclear division chromatophores move away from the girdle and are adpressed to valves, as such with the completion of cytoplasmic cleavage each daughter cell receives one chromatophore. The single chromatophore divides transversely to form two chromatophores which take their usual position on the girdle side.

The two valves of the parent cell (each with a daughter protoplast) become epithecas of the daughter cells. But they produce new hypotheca.

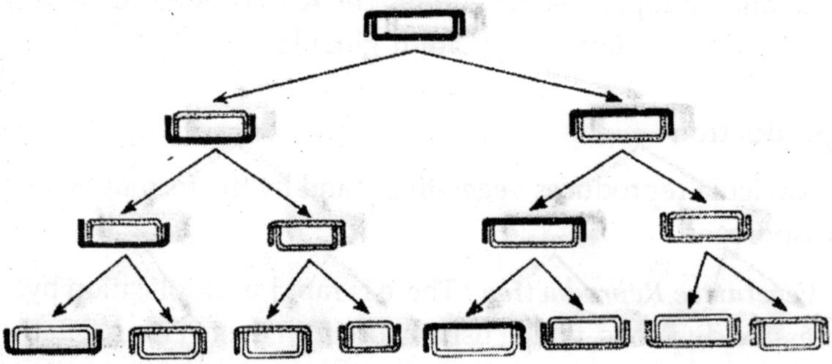

***Navicula* : Diagrammatic representation of the vegetative reproduction (cell division).**

The daughter cell which retains the epitheca of the parent cell, attains nearly the same size as the parent cell. But the one which receives the hypotheca of the parent cell (now epitheca of the daughter cell) is smaller than the parent cell. Thus, there are two lines of descendants in cell division, one line retains the original size, whereas in the other there is diminution in size in successive generations.

Auxospore Formation : The diatom, after reaching a certain minimum size due to repeated cell division, begins to form auxospores. Auxospores are in fact special rejuvenescent cells, formed as a compensatory measure for the vegetatively reproducing diatom in regaining its normal size. The auxospores are usually the result of sexual process in Navicula.

Two cells come very close to each other, adhere by their valve surfaces or occasionally by their girdle faces and then get enveloped within a common mucilagenous mass. The diploid nucleus of each cell undergoes reduction division and forms four haploid nuclei. Of these, two degenerate and two remain functional. The functional nuclei ultimately metamorphose into two amoeboid gametes. The two gametes show some amoeboid movements inside the valves and then round up. The valves of the mother cell are pushed apart at this stage and the two gametes of each pairing cell become free within their respective valves. A single chromatophore is present in each gamete.

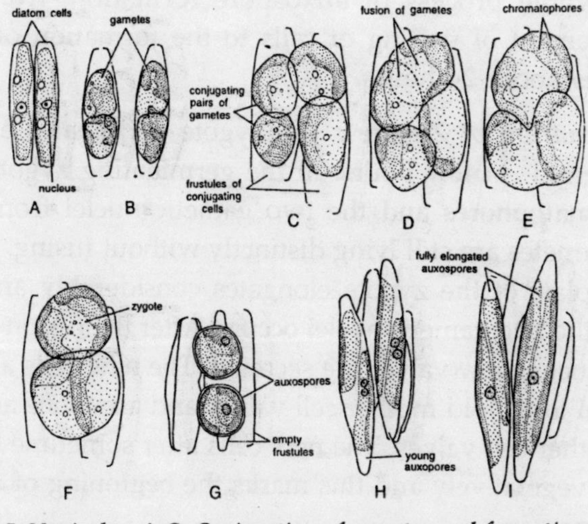

A-I. Navicula : A-G. Conjugation of gametes and formation of the auxospore, H-I. A stage in the enlargement of auxospore, I. fully elongated auxospore (After Subrahmanyam, 1946).

The two gametes formed in one cell move towards the gametes of the other cell, and when they come in contact they fuse in pairs. The two zygotes lie in the cell where pairing of gametes took place. The two gametes which are more active and show movements, may be considered as physiologically male, while the other two, which are more or less stationary, may be regarded as physiologically female. Hence, both the gametes formed in a cell are either male or female. The fusion of the two pairs of gametes is not simultaneous. First one of the male gametes moves towards the female gamete that is directly opposite to it and fuses with it and then the second male gamete follows that and fuses with the second female gamete.

Gamete formation invariably takes place late in the evening and actual fusion occurs soon after night-fall.

The zygote is large and irregular to begin with, but later it contracts and becomes spherical. It secretes a wall round itself and is now known as auxospore.

The whole process of auxospore formation, from the commencement of pairing of cells to the formation of new cells, takes about 24 hours.

Germination of Zygote : The zygote germinates after a resting period of few hours. In the germinating zygote, the two chromatophores and the two gametic nuclei from two pairing gametes are still lying distinctly without fusing. Then, the protoplast of the zygote elongates considerably and the fusion of the two gametic nuclei occurs. After the fusion of the gametic nuclei, new valves are secreted. The new cells always lie parallel to the old mother cell valves and are twice as long as the mother cell valves. The new cells after sometime began to divide vegetatively and this marks the beginning of a new generation.

The four discarded valves of the two mother cells are often seen attached to the newly formed frustules.

8

The Bacillariophyceae

The class Bacillariophyceae is a large group of algae, consisting of over 10,000 species belonging to 200 genera. The members are popularly known as diatoms. Fine geometrical expressions and intricate sculpturing in the cell wall make these organisms one of the most beautiful microscopic objects. They occur in almost every aquatic habitat as free-living photosynthetic autotrophs, colourless heterotrophs or photosynthetic symbionts. Diatoms are also found in terrestrial habitats and form a substantial proportion of the soil flora. In tropical rain forests, they grow in association with blue-green algae on leaves of trees. In India, the group is represented by 569 species distributed over 92 genera. Pinnularia graciloides, Niedium gracile, Cymbella affinis, Eunotia pectinalis, Melosira jurgensii and Coscinodiscus radiatus are some common Indian species of diatoms.

Important distinguishing characters of Bacillariophyceae are as follows :

(1) Diatoms are mostly unicellular but some colonial species form filaments, loose chains or mucilagenous colonies.

(2) The cells are microscopic and are variable in shape. They may be triangular (e.g.,Triceratium), boat shaped (e.g., Gyrosigma, Cymbella), rod shaped (e.g., Bacillaria), oval (e.g., Cocconeis placentula) or spherical (e.g., Coscinodiscus excentricus).

(3) The cell shows radial or bilateral symmetry.

(4) The cell is surrounded by a rigid two-part overlapping cell wall, called frustule. It is made up of pectic substances but is impregnated with high amounts (sometimes up to 95%) of siliceous substances.

(5) The vegetative cell is uninucleate and diploid. It usually has two large parietal or many discoid plastids. It is olive green or golden in colour, and its main pigments are chlorophyll- *a*, chlorophyll-*c* and xanthophyll (such as fucoxanthin, diatoxanthin, diadinoxanthin).

(6) The storage products of photosynthesis are oil and chrysolaminarin.

(7) Motile cells (antherozoids) have one pantonematic flagellum.

(8) The cell shows gliding movements.

(9) Vegetative reproduction takes place by cell division.

(10) Sexual reproduction is isogamous or oogamous.

(11) The zygote formed as a result of gametic union develops into a special type of spore, called auxospore. Continued reproduction by cell division results in the reduction of cell size and auxospores help in regaining the original cell size. Thus, auxospore formation is a restorative or rejuvènation process.

Classification

Diatoms have been classified into two orders on the basis of their symmetry:

Order 1. Pennales : The members of this order are characterized by a bilateral symmetry. They show gliding movement. Sexual reproduction, if present, is amoeboid isogamy. Examples-Fragilaria, Synedra, Eunotia, Navicula, Pinnularia, Epithemia.

Order 2. Centrales : This order includes radially symmetrical forms. They lack gliding movements. Sexual reproduction is anisogamous or oogamous and the gametes are motile. Examples-Cyclotella, Melosira, Rutilaria, Biddulphia, etc.

A general account of diatoms is provided in the following pages.

Occurrence

Diatoms are found in almost every aquatic environment. Genera like Synedra, Melosira, Navicula and Asterionella are strictly freshwater, and Sceletonema, Eucampa, Biddulphia, Fragilaria, Hemidiscus, Corethron, etc., are found in marine habitat. Some aquatic forms occur as plankton of ponds and lakes. The algal flora of roadside ponds and pools always includes some diatoms. Species of Pinnularia, Navicula, Frustulia, Amphora and Caloneis are found in terrestrial habitat. Such forms are able to withstand desiccations for a very long period. Some diatoms are epiphytic on other algae. For example, species of Cymbella, Gomphonema, Eunotia and Cocconeis grow on the filaments of Cladophora, Rhizoclonium and Vaucheria. Licmophora is an endozoic diatom.

Thallus Structure

Diatoms are unicellular and cells are usually found in isolated form. However, isolated cells sometimes get imbedded in a common gelatinous matrix to form simple colonies. Such colonies do not have complex plant bodies and thus can not be regarded as multicellular organisms. Colonial diatoms occur either as uniseriate filaments (e.g., Melosira,) or as branched filaments, having many radiating arms at the free end (Licmophora flabellata,). In some diatoms the colony is formed due to interlocking of bristles, present on the cell wall (Chaetoceras diadema,).

Cell Structure

Cell wall and protoplast are the two main constituents of the cell.

1. *Cell Wall :* The characteristic feature of Bacillariophyceae is the presence of a cell wall laden abundantly with silica. The empty wall, called frustule, is made up of two almost equal overlapping halves or theca, the smaller fitting into the larger like a pair of Petri dish. The outer or upper half is epitheca and the inner or lower hypotheca. Each half-wall (theca) has two main parts :

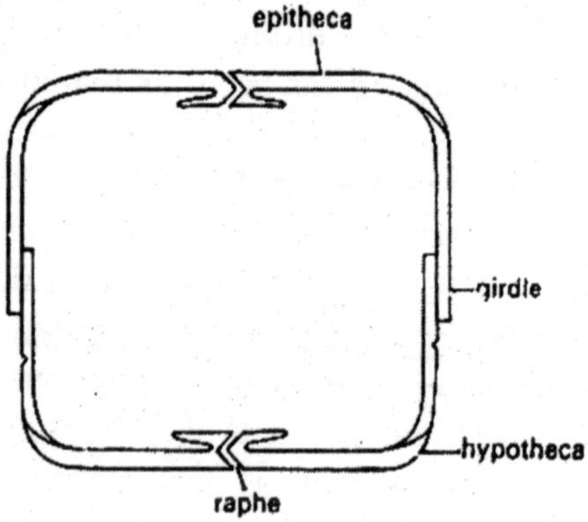

Pinnularia viridis : Transverse section.

(i) valve, a more or less flattened plate like structure that forms the main surface of the theca, and (ii) connecting band (or cingulum), the incurved margins of the valve. The connecting bands of epitheca and hypotheca are joined together by an organic connecting material and form a girdle. Occasionally, there are one or more additional bands between the valve and the girdle and such bands are known as intercalary bands. The girdle bands often have minute teeth which hold the two valves together. When a frustule lies in such a way that valve side is uppermost, it is known as valve view or top view (Fig. A) and when connecting bands are uppermost it is called girdle view or side view.

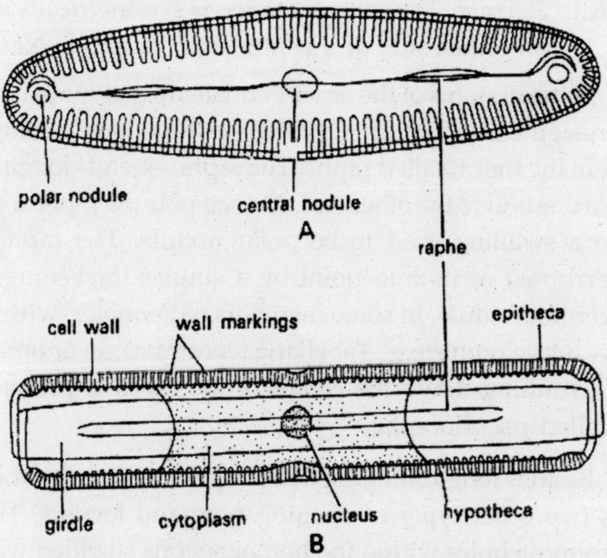

A-B. Pinnularia: A. Valve view, B- Girdle view.

Frustules may be perfectly symmetrical, asymmetrical or with intergrades between the two.

On the basis of this character, diatoms are divided into two orders-Centrales (radially symmetrical diatoms) and Pennales (bilaterally symmetrical diatoms).

The frustule is made up of a pectin material. A characteristic feature of diatoms is that the valve portion of the frustule is silicified. The siliceous material is laid down in certain regular patterns that make the wall ornamented. The ornamentation is of the following four basic types:

(a) *Centric and Radial* : The ornamentation is based around a central point. Example Coscinodiscus.

(b) *Trellisoid* : The ornamentation is uniform without reference to a point or central line (Fig.B). Example-Eunotia.

(c) *Gonoid* : The ornamentation is dominated by angles. Example-Triceratium.

(d) *Pennate* : The ornamentation is symmetrically arranged on either side of a central line. Example-Navicula.

Some diatoms of the order Pennales (e.g., Pinnularia viridis, Gyrosigma attenuatum, Cymbella cistula) have a longitudinal slot in the theca, called raphe. The raphe extends longitudinally from one end to the other end. At each pole the raphe terminates into a swelling, said to be polar nodule. The raphe is also interrupted at its mid point by a similar thickening, known as central nodule. In some members of Pennales, which do not have a true raphe (e.g., Tabellaria fenestrata), an unornamented area running down the center of the theca is present and it is called pseudoraphe.

Besides longitudinal slits (raphe), the cell wall of the diatom has two other types of openingspore and loculus. The pores are simple holes within the homogeneous silicified wall. They have ribs or costae which provide them mechanical strength. The locules, on the other hand, are more complex openings. They are in fact polygonal chambers in the cell wall. At the one end of the chamber there is a sieve membrane (also called pore membrane or velum) with many small pores and a large hole, called foramen. The sieve membrane may be on the outerside (e.g., Coscinodiscus wailesii) or on the innerside (e.g., Coscinodiscus lineatus).

2. *Protoplast* : It is surrounded by a thin plasma membrane, lying close to the inner face of the cell wall. Within plasma membrane a thick layer of cytoplasm surrounds

a large central vacuole. The cell is uninucleate and the oval or spherical nucleus is usually embedded in the parietal cytoplasm. Sometimes the nucleus is suspended in the centre of the cell by cytoplasmic strands.

The cytoplasm has many mitochondria, Golgi bodies and chromatophores. In the members of the order Centrales the cells contain many discoid, stellate or irregular chromatophores, while in Pennales there are one or two large parietal plate-like chloroplasts in each cell. Besides, plastids of specific shapes occur in certain members. For example, plastids are lobed discoid in Melosira sp., stellate in Grammatophora marina, semistellate in Striatella and 'H' shaped in Gomphonema and Mastogloia.

The chromatophore has a double layered binding membrane, surrounded by two additional layers of chromatophore ER. The thylakoids within the chromatophores are in stacks of 4-6 lamellae. Each chromatophore, has a single, almost centrally situated pyrenoid. The main photosynthetic pigments of diatoms are chlorophyll *a*, C_1 and C_2 β-carotene and xanthophyll. The golden-brown colour of diatoms is due to xanthophylls, such as fucoxanthin, diatoxanthin and diadinoxanthin.

As in other algae, diatoms do not store their photosynthetic products in the form of starch. Their storage product instead is chrysolaminarin, which is located in vesicles.

Movement : The motile diatoms are characterised by the presence of raphe. They show gliding movement. The diatom adheres itself to the substratum in the area of raphe. Certain crystalloid bodies, containing mucilagenous material present in the cytoplasm, release mucilage through polar or central nodule on an appropriate stimulus. Now the mucilage streams along through the raphe in the direction opposite to that in which movement of diatom is to set in. The movement of diatoms may be straight, curved or zig-zag. The rate of gliding varies from 2-14 $\mu m\ sec^{-1}$ at room temperature. The motility is affected by factors like light, temperature and nature of the substratum.

Reproduction

Diatoms reproduce by vegetative, asexual and sexual methods.

Vegetative Reproduction : Cell division is the chief method of vegetative propagation. The cells usually divide at midnight or early morning. Prior to cell division the protoplast of the mother cell increases in size, pushing the epitheca away from the hypotheca (Fig. A). Now the nucleus divides mitotically in a plane perpendicular to the valves. Nuclear division is followed by the division of the protoplast in two by invagination of the plasmalemma (Fig. B, C).

Each of the two uninucleate daughter protoplasts with a single nucleus is retained within one 'valve, i.e., one daughter protoplast is covered on one side by the epitheca and on the other by the hypotheca; the other side of both the protoplast being naked. The direction of the division of protoplast is dependent on the number of plastids in the cell. If the cell has only a single plastid, the division is usually longitudinal and if the cell contains two or more plastids, the division may be longitudinal or transverse.

The two valves of the parent cell (each with a daughter protoplast) become epithecas of the daughter cells. However, each daughter cell produces a new hypotheca (Fig. D). The daughter cell which retains the epitheca of the parent cell, attains nearly the same size as that of the parent cell. But the other daughter cell which receives the hypotheca of the parent cell (now epitheca of the daughter cell) is smaller than the parent cell. Thus there are two lines of descendants in cell division; one line retains the same size, whereas in other there is diminution in size in successive generations.

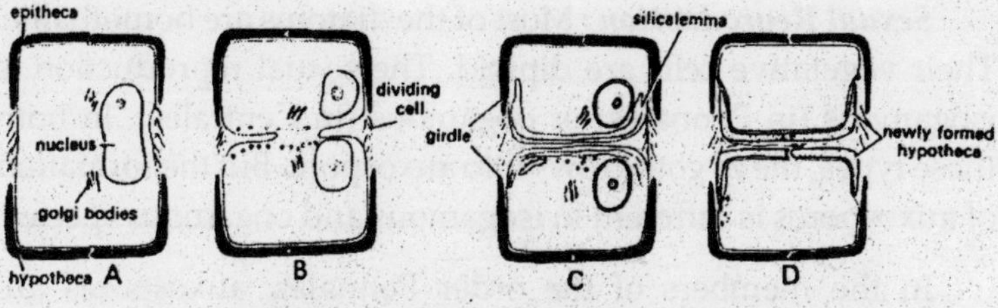

A-D. Schematic representation of the stages of frustule formation in Gomphonema parvulum.

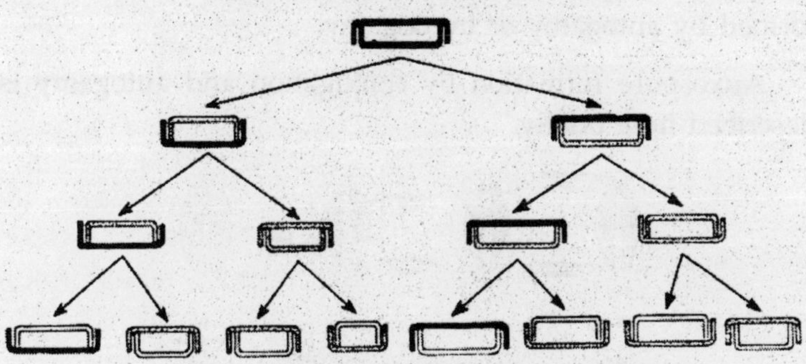

Diagrammatic representation of cell division in diatoms (note reduction in size in successive generations).

Asexual Reproduction : It takes place by means of auxospores. Continued multiplication by cell division results in gradual diminution in cell size in successive generations.

The reduction in cell size can continue only up to a certain limit and beyond this limit either the cell is to die or convert into an auxospore. In fact, auxospore formation is a restorative process through which the smaller cell restores the original size. Prior to auxospore formation, the cell secretes a large amount of mucilage which pushes the two valves apart.

The protoplast is thus set free. It grows considerably and attains the normal size. Then it secretes around itself a silicified membrane, the perizonium. Eventually new valves are secreted inside the perizonium and the auxospore becomes an independent cell.

Sexual Reproduction : Most of the diatoms are homothallic. Their vegetative cells are diploid. The sexual reproduction is isogamous (in Pennales) or oogamous (in Centrales). In both these types, the zygote grows into auxospore. But the formation of auxospores is different in isogamous and oogamous species.

In the members of the order Pennales, auxospores are formed either by the fusion of conjugating cells, by autogamy or by parthenogenesis, whereas in Centrales auxospores are formed by autogamy or by oogamy.

Auxospore formation by conjugation and autogamy is described here briefly.

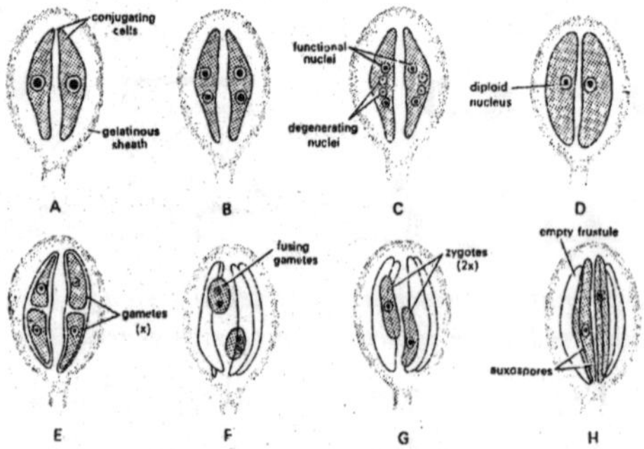

A-H. Successive stages in the development of auxospores by conjugating cell.

1. *Development of Auxospores by Conjugating Cells :* Auxospores are formed by this method in diatoms like Cymbella, Pinnularia and Navicula. This process is initiated when the size of the cells is reduced to a certain critical point after successive cell divisions. Two such cells come close to each other and get enveloped within a common mucilagenous mass (Fig.A) The diploid nuclei of these cells undergo meiosis and as such the diploid nucleus of each cell produces four haploid nuclei. Of the four nuclei in a cell, two degenerate and the other two remain functional (Fig. B,C). Meanwhile the protoplast of the cell divides

into two segments, each carrying a single haploid nucleus. These segments now function as gametes (Fig. D, E). Two gametes from different cells fuse to form a zygote (Fig. F,G). The two zygotes thus formed grow in size to form auxospores. The auxospore secretes a perizonium, and later siliceous valves are formed (Fig.H). The four discarded valves of the two mother cells are often seen attached to the newly formed frustules.

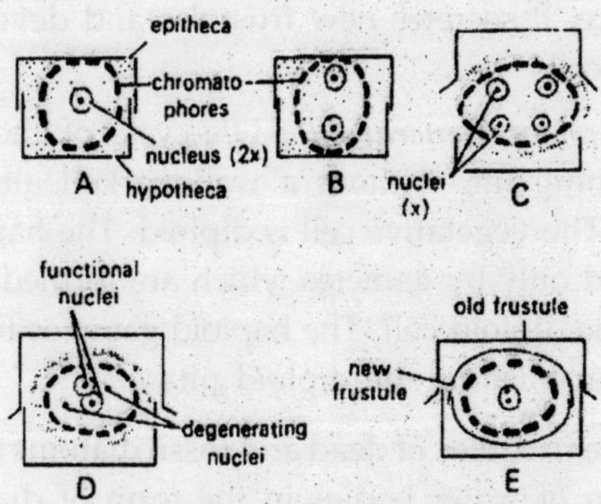

A-E. Successive stages of auxospore formation by autogamy.

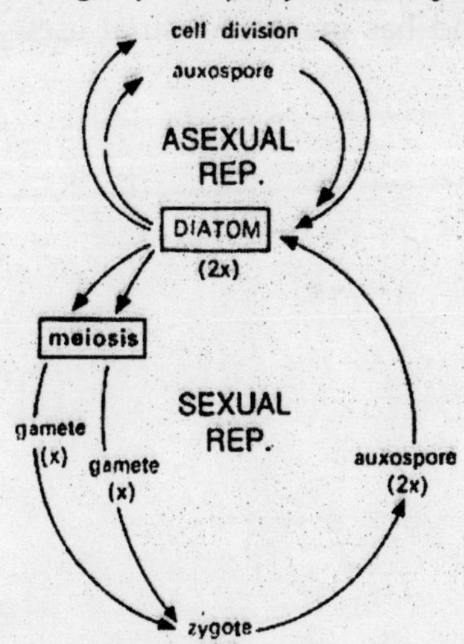

Life-cycle of a diatom.

2. *Development of Auxospores by Autogamy* : In Cyclotella meneghiniana and Melosira nummuloides, auxospores are formed by autogamy. In this process, the cell secretes some mucilage and as such pushes the epitheca and hypotheca slightly apart. Now the diploid nucleus of the cell divides meiotically forming four haploid nuclei. Two of these nuclei degenerate and the remaining two fuse to form a diploid nucleus. The protoplast with diploid nucleus functions as an auxospore and grows in size. It secretes new frustules and develops into a new cell.

Alternation of Generation : Life-cycle of the diatom is diploid-predominant, without a well-marked alternation of generations. The vegetative cell is diploid. The haploid phase is represented only by gametes which are formed by meiotic division of the diploid cell. The haploid gametes fuse to form zygotes, re-establishing the diploid phase.

Note : The frustules of dead and fossil diatoms accumulate at the bottom of water bodies in the form of diatomaceous earth or kieselguhr. Such deposits are found in various parts of the world and has many industrial uses.

9

The Charales

The systematic position of the order Charales is debated and it has been treated differently by various workers. Fritsch (1935) included it in the class Chlorophyceae, Smith (1938) gave it the rank of a class, Charophyceae, and Bold (1956) and McLean and Cook (1956) included it in a separate phylum Charophyta, intermediate between algae and bryophyta. The treatment of Fritsch is followed here.

The order includes 6 genera and 300 species, mostly found in shallow freshwater lakes with sandy and muddy bottoms. Some species also grow in brackish water. The thallus is well organized and is differentiated into rhizoids and an erect branched axis. The axis is divided into nodes and internodes. At the nodes of the main axis whorls of branches of unlimited growth are present, whereas the nodes of lateral branches bear branches of limited growth. Such branched axis gives it the appearance of an angiospermic plant.

Sexual reproduction is advanced oogamous. Both male and female sex organs, known as globules and nucules respectively, are macroscopic. These structures are contained within a sterile envelope, similar to that of bryophytes. Asexual spores are, however, absent.

The order includes only one family, Characeae, represented in India by five genera (Chara, Nitella, Lychnothamnus, Nitellopsis and Tolypella) and 65 species. The life-history of Chara is described here as the representative form.

Chara

Systematic Position

Class - Chlorophyceae

Order - Charales

Family - Characeae

Occurrence : Chara has about 180 species, most of the species are freshwater, found submerged in shallow water ponds, tanks or lakes. A few species grow in alkaline water (e.g., C. baltica) and hot springs (e.g., C fragilis). Species of Chara growing in heavy water become encrusted with calcium and magnesium carbonate and hence the common name stoneworts is given to it. These deposits make their body rough and brittle.

In India, the genus is represented by about 30 species, of these, C. braunii, C. gracilis, C gymnoptis and C. zeylanica are common.

Structure

Thallus : The plant body of Chara is multicellular and macroscopic. It is 20-30 cm in height and is differentiated into rhizoids and main axis.

The rhizoids are white, threadlike, multicellular, uniseriate and branched structures. They arise from the base of the plant or from the peripheral cells of the lower nodes of the main axis. They are characterised by the presence of oblique septa. The ends of adjacent cells of rhizoids are protruded in opposite directions at the septum so as to form a kind of joint. At this joint, protrusion from the upper cell cuts off a segment which divides to form a quadrant. It is known as rhizoidal plate, and it gives rise to a tuft of branched rhizoids. The rhizoids, besides serving as organ of attachment, also help in the absorption of minerals.

The main axis is epigeal, long, branched and is differentiated into nodes and internodes. The internode is made up of two types of cells. Its central part consists of a single elongated cell, called axial cell or internodal cell. The axial cell is ensheathed or corticated by a layer of vertically elongated cells of much smaller diameter. These cells, originating at the, nodes, are called cortical cells (Fig. D). Half of the cortical cells originating at a node grow upward (ascending filaments) and half downward (descending filaments). Thus, the lower half of an internode is covered by ascending filaments and the upper half by descending filaments (Fig. A). But, in some species (e.g., C. corallina), the cortical cells do not ensheathe the internodes.

The node consists of a central pair of cells, surrounded by 6-20 peripheral cells (Fig. E). The following three types of appendages arise from a node :

1. *Branches of Limited Growth :* They arise in whorls of 6-16 at the nodes of the main axis and branches of unlimited growth. They cease to grow after forming 5-15 nodes and internodes, hence called branches of limited growth. They are also called primary laterals, 'leaves' or branchlets. The internodes of these branches are shorter than those of the main axis. At the nodes of these branches are present unicellular filamentous secondary laterals or stipulodes. Reproductive structures are also borne at the nodes of these branches.

2. *Branches of Unlimited Growth* : These branches arise from the axils of the branches of limited growth, and are also called axillary branches or long laterals. They are also differentiated into nodes and internodes and bear a whorl of primary laterals at the node. They grow indefinitely like the main axis.

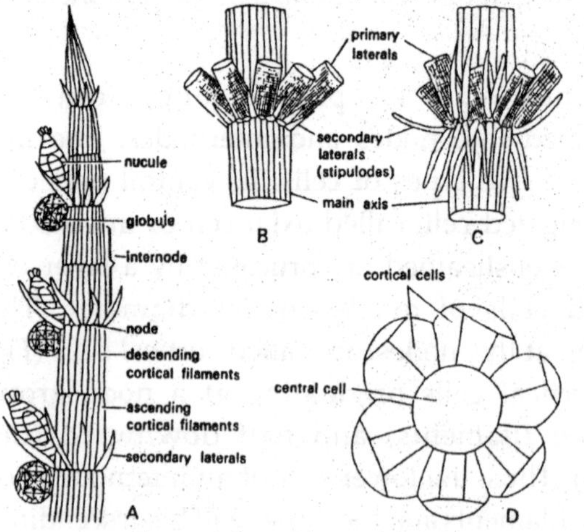

A-D. Chara : Aerial axis; a. Branch of limited growth, B-C.Appendages at nodes, D. Transection of internode.

3. *Stipulodes* : The basal nodes of the branches of limited growth develop some unicellular outgrowths, called stipulodes. Some species of Chara (e.g., C. braunii, C. nuda), are unistipulate, i.e., the number of stipulades at each node is equal to the number of primary laterals present at that node.

   But most of the species are bistipulate, i.e., the number of stipulodes at each node is twice the number of primary laterals. If stipulodes are present in one whorl at each node, the condition is known as haplostephanous. (as in C. braunii) and if in two whorls it is called diplostephanous (as in C. tomentosa, C. baltica). In C. pashani, the stipulodes are, however, absent.

Cell Structure

The nodal cells are uninucleate and densely cytoplasmic with few small ellipsoidal chloroplasts. The internodal cells are, however, much elongated, each with a large central vacuole, many nuclei and innumerable discoid chloroplasts. Cytoplasmic streaming is particularly evident in the large internodal cells. The cytoplasm in these cells is divided into an inner endoplasm, which lies next to the vacuole, and an outer ectoplasm containing chloroplasts. Only endoplasm is capable of streaming. It continually streams up and down along the longitudinal plane. The streaming of the cytoplasm is due to alternating contraction and expansion of the protein fibrils which remain fixed to the cell wall.

Growth : In Chara growth of the main axis and its branches takes place by a dome-shaped apical cell.

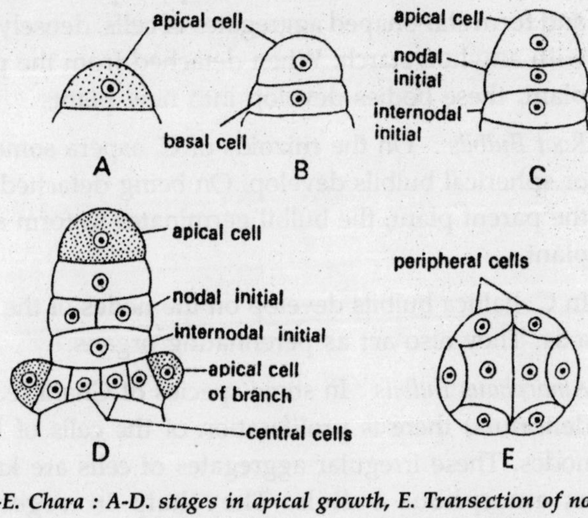

A-E. Chara : A-D. stages in apical growth, E. Transection of node.

Successive transverse divisions of the apical cell produce a row of disc shaped cells (Fig. A, C). Each of these cells divides transversely into two daughter cells. The upper biconcave daughter cell functions as nodal initial and the lower biconvex cell as internodal initial. As a result of many vertical divisions at right angles to each other in nodal initial, a pair of central cells and many peripheral cells are formed. The peripheral cells act as mother cell for the primary laterals and behave in the same manner as the apical cell. The peripheral cells of the basal node form rhizoids. The internodal initial, however, does not divide further but elongates considerably.

Reproduction

Chara reproduces by vegetative and sexual methods. Asexual spores are not found.

Vegetative Reproduction : The following structures formed in Chara help in vegetative propagation.

1. *Amylum Stars* : In some species of Chara (e.g., C stelligera) basal nodal cells develop in a peculiar fashion and form star-shaped aggregates of cells, densely filled with amylum starch. When detached from the parent plant, these bodies develop into new plants.
2. *Root Bulbils :* On the rhizoids of C. aspera some oval or spherical bulbils develop. On being detached from the parent plant, the bulbil germinates to form a new plant.

   In C. baltica bulbils develop on the nodes of the main axis. They also act as perennating organs.
3. *Amorphous Bulbils :* In some species of Chara (e.g., C. delicatula) there is proliferation of the cells of lower nodes. These irregular aggregates of cells are known as amorphous bulbils. They help in vegetative propagation.
4. *Secondary Protonema :* These are tubular structures which develop from primary protonema or the basal cell of the rhizoid. Like primary protonema, they also form new plants.

Sexual Reproduction

Sexual reproduction in Chara is of highly advanced oogamous type. Structurally, sex organs of Chara are most complex amongst algae. The male sex organs are called antheridium or globule and the female oogonium or nucule. Both, these organs are macroscopic. Most of the species of Chara are homothallic or monoecious and the male and female sex organs are borne on the same branch at juxtaposed position. Some species of Chara (e.g., C. wallichii) are, however, heterothallic or dioecious, i.e., they bear male and female organs on different plants.

The sex organs are borne on the branches of limited growth (primary laterals) and the nucule is located above the globule. Development of globule and nucule is almost simultaneous, but in some species globule matures before nucule.

1. *Antheridium or Globule :*

(a) *Internal Structure of Mature Antheridium* : Mature antheridium is a large, macroscopic, spherical and conspicuously red or yellow structure. Its wall is made up of eight cells which are called shield cells. The outer wall of the shield cells has many infoldings which form various types of ornamentations on the surface. A rod shaped cylindrical cell arises from the centre of each shield cell which is known as manubrium. At the distal end of each manubrium there is one or more small spherical or globose cells which are called primary capitulum cells. These cells are in direct contact of manubrium. There is also a second row of capitulum cells, known as secondary capitula. Each secondary capitulum cell bears 2-4 long unbranched antheridial filaments or spermatogenous filaments at its tip. Each filament has 25-250 cells and each cell forms an antherozoid. Thus, each antheridium produces 20,000-50,000 antherozoids.

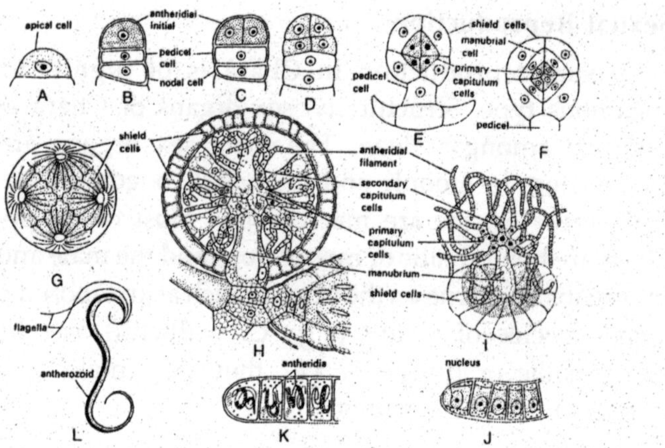

A-L. Chara : Development of globule and antherozoids; a-F. Successive stages in the development of globule, G. Mature globule, H. Globule in longitudinal section, I. A shield cell with manubrium, primary and secondary capitulum cells and spermatogenous filaments, J-K. Stages in spermatogenesis, L. An antherozoid.

(b) *Development of Antheridium (Globule):* As mentioned earlier, globules develop at the nodes of the branches of limited growth. A peripheral adaxial cell of the node, which functions as antheridial initial, divides transversely into a basal pedicel cell and a terminal antheridial mother cell (Fig.A-B). The pedicel cell does not divide further and forms a small stalk (pedicel) in the mature antheridium. The antheridial mother cell enlarges in size and then divides by two vertical divisions to form a quadrant (Fig.C, D)- Each of these four cells divides by a transverse wall and thus eight cells are formed (Fig. E). This stage is called octant stage. Each cell of the octant divides periclinally as a result of which two layers of eight cells each are formed. The cells of the inner or outer layer again divide periclinally. In this way three radial layers of eight cells each are formed. The cells of the outermost layer enlarge and transform into eight shield cells (Fig. G). The cells of the middle layer elongate toward the centre of the antheridium and eventually form eight rod shaped manubrial cells. The cells of the inner layer present at the distal end of the manubrial cells form primary capitulum cells. Each primary capitulum cell divides to form six secondary capitulum cells (Fig. H-I). Sometimes secondary capitulum cells may divide again to form tertiary capitulum cells. Each ultimate capitulum cell divides repeatedly to form 2-4 long, multicellular, thread-like filaments, called antheridial filaments or spermatogenous filaments. An antheridial filament has up to 250 uninucleate cells.

The cells of the antheridial filament function as sperm mother cells. The protoplast of the sperm mother cell gets metamorphosed into a single, spirally coiled biflagellate antherozoid (Fig. J, L). The process of antherozoid development is known as spermatogenesis.

(c) *Liberation of Antherozoids :* At maturity, shield cells of the antheridium separate from each other, exposing the antheridial filaments. The sperm mother cells gelatinize in water and the antherozoids are liberated.

2. *Oogonium or Nucule :*

(a) *Structure of the Mature Oogonium :* The nucule of Chara is macroscopic oval structure with a short stalk. It is borne at the node of the primary lateral. In homothallic species, the nucule lies just above the globule. Five spirally twisted tube cells cover the nucule except at the tip, where there are five corona cells. In the centre of the nucule, there is a single uninucleate egg. The food material is stored in the egg in the form of oil drops and starch grains. At the tip of the egg there is a hyaline area, the receptive spot.

Presence of an envelope of sterile cells around the nucule is an advanced character amongst algae. Such a sterile jacket can be compared with the jacket of the archegonium of bryophytes.

(b) *Development of Oogonium (Nucule)* : The oogonium develops from the upper peripheral cell derived by the division of basal nodal cell, situated at the base of the antheridium.

The oogonial initial undergoes two transverse divisions to form three cells in a row. The terminal cell functions as the oogonial mother cell, the middle as the sheath initial and the basal cell elongates to form the pedicel cell, supporting the developing nucule (Fig. A, C).

The oogonial mother cell elongates and then divides transversely into a small basal stalk cell and a large distal oogonium cell (Fig. E). The latter enlarges considerably, accumulates food and differentiates into an uninucleate egg.

The sheath initial divides vertically to form five sheath cells. The latter divide transversely to form two tiers of five cells each. The cells of the upper tier do not elongate much and form the corona at the tip of the oogonium. The cells of the lower tier, known as tube cells, elongate considerably to form a protective covering around the oogonium (Fig. F).

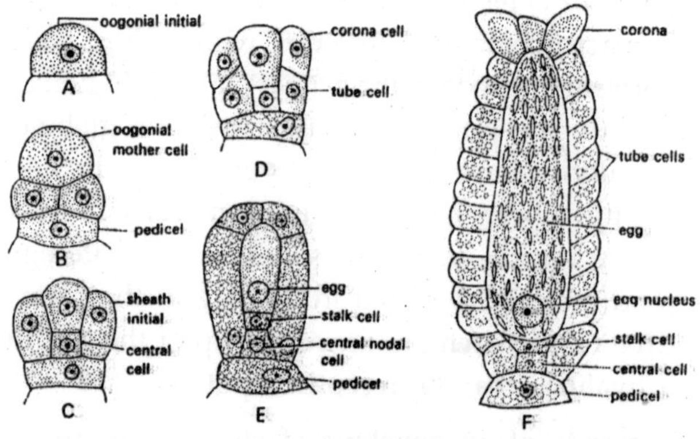

A-F. Chara : Successive stages in the development of nucule (oogonium).

Fertilization

At the maturity of the oogonium, the five tube cells separate from one another and narrow slits are formed in between them. Antherozoids swim through these slits and penetrate gelatinized wall of the oogonium. Many antherozoids may enter into an oogonium but only one of them fuses with the egg to form the oospore.

Oospore

The Oospore Secretes a Thick Wall : The inner walls of the tube cells also become thickened but the other portions of the sheath wall decay. The oospore thus looks like a thick walled nut. It is yellow, red or brown in colour depending on the species. The oospore, with surrounding remains of the sheath, falls to the bottom of the pool. It germinates after a resting period. The duration of resting period varies with environmental conditions.

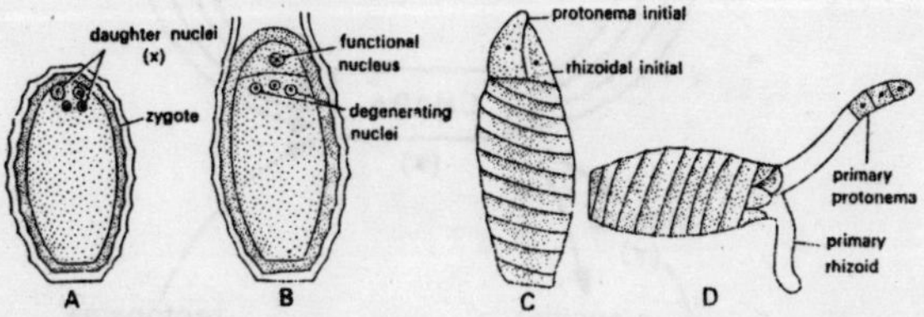

Germination of oospore. Oospores of Chara do not germinate in dark; they require white or red light for germination. At the end of the resting period, the diploid nucleus moves to the anterior end of the oospore and divides meiotically forming four haploid nuclei (Fig.A). A transverse wall is now laid down near the anterior end forming two cells. The upper cell is small, lenticular and uninucleate and the lower cell is large and trinucleate. The upper cell divides vertically to form a rhizoidal initial and a protonemal initial. Both these cells grow in opposite directions. The rhizoidal initial shows positive geotropism and forms primary rhizoid, whereas the protonemal initial shows negative geotropism and forms primary protonema. The primary protonema differentiates into nodes and internodes. Rhizoids and secondary protonema arise from the peripheral cells of the basal node. The peripheral cells of the upper nodes give rise to lateral branches.

Alternation of Generation : The plant body of Chara is haploid. The diploid condition is found only in oospore. At the time of germination, the diploid nucleus of the oospore divides meiotically and thus haploid state is regained. The life-cycle of Chara is thus of haploid predominant type.

Taxonomic Position of Chara

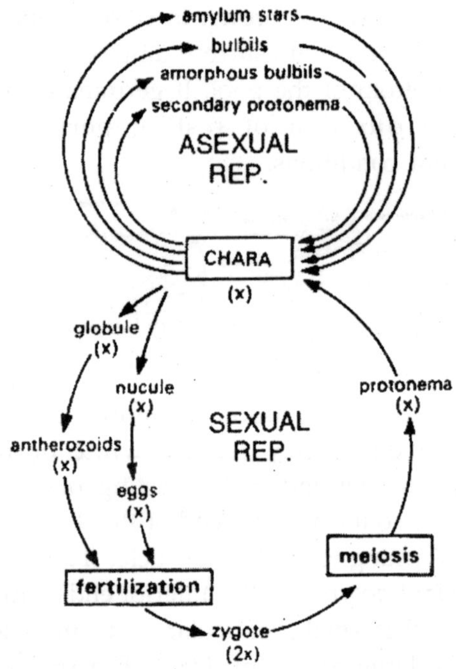

Chara : Diagrammatic representation of the life - cycle.

The Systematic Position of Charales has been Controversial : Fritsch (1935) included Chara in the order Charales under the class Chlorophyceae. Smith (1938, 55) placed the order Charales in a separate class- Charophyceae of the division Chlorophyta, whereas Round (1973) placed the order Charales in the phylum Charophyta.

The photosynthetic pigments, the nature of cell wall and food reserves, the structure of flagella and the life cycle of Chara is typically chlorophycean. However, there is a marked advance in vegetative organisation (differentiation of the axis into nodes and internodes, apical growth) and a great elaboration of the sexual reproductive organs (presence of jacket cells around sex organs, spiral nature of antherozoids, and mode of germination of zygote) of Chara over other members of Chlorophyceae.

Some algologists are of the view that the order Charales has evolved from Chaetophorales, especially from forms like Draparnaldiopsis. Chara resembles with Draparnaldiopsis in the presence of nodes and internodes, corticating filaments, formation of lateral branches of limited growth and participation of these laterals in reproduction. Thus, Chara shows basic characters typical of Chlorophyceae with some advanced features.

10

The Siphonales

The order Siphonales is represented by about 50 genera and 400 species, mostly marine, but there are some fresh water representatives in the families Protosiphonaceae, Phyllosiphonaceae and Vaucheriaceae. The members of the order are characterised by a coenocytic, siphonaceous and branched thallus.

In some marine members like Biopsies, Caulerpa and Acetabularia the thallus is differentiated into prostrate and erect branches. Transverse walls or septa are usually absent in vegetative thallus; they are laid down only at the time of the formation of reproductive structures or at the site of injury. There is a large central vacuole which is filled with cell sap. Protoplasm occurs in the form of a thick layer in between the wall layer and the vacuole.

Many discoid chromatophores and nuclei are present in the protoplast. The chromatophores always lie in the peripheral and the nuclei in the central region. In addition to chlorophyll, the members of this order contain two other pigments, siphonein and siphonoxanthin.

The members of Siphonales reproduce by vegetative, asexual and sexual methods. Fragmentation and akinetes are the two main methods of vegetative multiplication. Asexual reproduction is by zoospores or aplanospores. Sexual reproduction is usually oogamous, but in some species it is isogamous or anisogamous.

Classification

Fritsch (1935) included the following nine families in the order Siphonales :

(1) Protosiphonaceae, example - Protosiphon.

(2) Caulerpaceae, examples - Caulerpa, Biopsies.

(3) Derbesiaceae, example - Derbesia.

(4) Dasycladaceae, examples - Dasycladus, Acetabularia.

(5) Codiaceae, example - Codium.

(6) Valoniaceae, examples - , Valonia, Siphonocladus.

(7) Chaetosiphonaceae, example - Chaetosiphon.

(8) Phyllosiphonaceae, example - Phyllosiphon.

(9) Vaucheriaceae, example - Vaucheria.

Smith (1955) recognised the following five families in the order : (1) Bryopsidaceae, (2) Caulerpaceae, (3) Halicystidaceae, (4) Codiaceae, (5) ichotomosiphonaceae.

Family Codiaceae

The members of Codiaceae are freely branched filamentous algae, abundantly growing in tropical waters. The thallus consists of tubular branches of spongy nature, made up of densely interwoven filaments. The filaments are characterised by the presence of constrictions with stratified annular thickenings, especially near the points of branching. They show typical siphonaceous coenocytic structure. The plastids lie at the peripheral position, and in most of the genera they lack pyrenoids. Vegetative reproduction occurs by means of detachable propagules. Sexual reproduction is anisogamous; the gametes are biflagellate and are produced in ovoid cup-shaped gametangia.

The family includes more than 120 species representing some 16 genera. Rhipidodesmis, Avrainvillea, Rhipilea, Codium and Udotea are some common genera of the family. Here the life-history of Codium is described as the representative of the family.

Codium

Systematic Position

Class - Chlorophycene

Order - Siphonales

Family - Codiaceae

Occurrence : Codium is a widely distributed non-calcareous genus with more than 45 marine species. The species of Codium occur from low-tide mark up to the depth of 70 meters in tropical as well as temperate waters. Codium tomentosum, C. bursa and C. fragile are some common tropical species of the genus.

Thallus Structure : The thallus is differentiated into prostrate and erect systems. The prostrate system, with the help of which the thallus is firmly anchored to the substratum is either disc-like (e.g., C. tomentosum) or branched, with hair-like rhizoids (e.g., C. adhaerens). The erect system varies considerably in form in different species. It is repeatedly forked bunch of long tubular branches (C. tomentosum), a flat cushion or crust (C. adhaerens) or a large round hollow ball, 10-20 cm in diameter (C. bursa). In the latter, within the hollow ball, are found many blue-green algae and small sea animals. This association with blue-green algae enables these species of Codium to fix atmospheric nitrogen.

The erect branches have a central medulla, composed of interwoven colourless forked filaments, and a peripheral cortex of compactly arranged inflated club-shaped branches, known as utricles or cortical vesicles. Each utricle represents one of the forks of longitudinal medullary filament which has become swollen. Species restricted to shady places (e.g., C. megalophysum, C. papenfusii) and deep waters (e.g., C. cranwelliae) usually have large utricles. There appears to be some relationship between utricle size and light intensity. On the outer surface of each utricle small deciduous hairs are present. These hairs are produced periodically and provide a tomentose character to the plant. At the base of the utricle and also just below the point of forking of the medullary filaments characteristic strong annular thickenings develop, leaving a fine pore for intercommunication.

The utricle has a thick peripheral layer of cytoplasm, surrounding the large central vacuole. Many small discoid chloroplasts (without pyrenoids) are crowded in the outer portion of the cytoplasm and several small diploid nuclei are present in the interior.

Growth : Codium shows both apical and intercalary growth. The apical growth is brought about by the elongation of medullary filaments, which later form cortical portion by the forking. Intercalary growth takes place by continuous addition of new utricles between the pre-existing ones. In cushion-like forms (e.g., C. adhaerens) growth is mainly intercalary

Reproduction

Codium reproduces by vegetative and sexual methods.

Vegetative Reproduction : It takes place by means of small club-shaped propagules which develop on utricles (Fig. A, C).

When detached from the parent plant, the propagule grows into a new thallus.

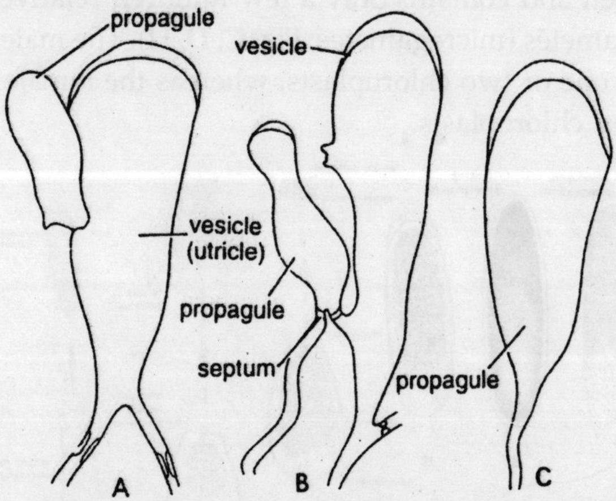

A-C. Codium : Development and structure of propagule.

Sexual Reproduction : Sexual reproduction is anisogamous and takes place with the help of distinct male and female gametes. Some species are strictly dioecious (e.g., C. fragile), while others are monoecious.

Gametes are produced in ovoid male and female gametangia which develop laterally on utricles. A gametangium initiates as a tubular outgrowth on the lateral walls of the utricle. The outgrowth has compact cytoplasm with evenly distributed diploid nuclei and the plastids are crowded at the distal end (Fig.A).

The young gametangium elongates and is cut off from the utricle by a strong annular thickening that develops at its base. Some nuclei of the multinucleate gametangium degenerate and the remaining functional nuclei enlarge and divide meiotically to form haploid nuclei. Now the protoplast of the gametangium divides into many uninucleate segments. Each uninucleate segment develops into a gamete.

The mature male gametangium is golden yellow and

contains a few thousand small pyriform male gametes microgametes; Fig. G), whereas the female gametangium is dark green and contains only a few hundred relatively large female gametes (microgametes; Fig. C, D, H). The male gamete contains one or two chloroplasts, whereas the female gamete has many chloroplasts.

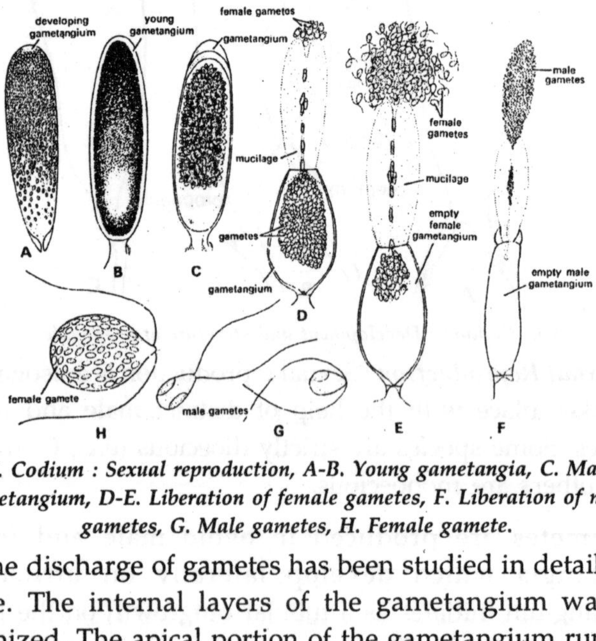

A-H. Codium : Sexual reproduction, A-B. Young gametangia, C. Mature gametangium, D-E. Liberation of female gametes, F. Liberation of male gametes, G. Male gametes, H. Female gamete.

The discharge of gametes has been studied in detail in C. fragile. The internal layers of the gametangium wall are gelatinized. The apical portion of the gametangium ruptures and a large mass of mucilage is extruded through it.

The mucilage mass is almost of the same shape and size as the gametangium, and it has a central canal through which gametes move out (Fig. D, G). Initially, gametes lack flagella and are carried out passively due to the hydrostatic pressure within the gametangium.

Soon after the discharge, each gamete develops two flagella on its anterior side and swims freely with their help. Ejection of gametes occurs very rapidly and all gametes are released within a minute or so.

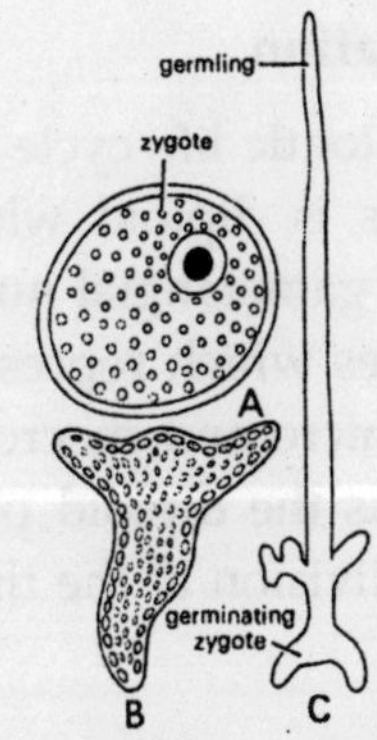

A-C. Codium : Germination of zygote; A. Zygote, B. Lobed germling, C. Young germling.

Fertilization takes place when motile male and female gametes come close to each other. The male gamete gets attached to the side of the female gamete and loses its flagella. The gametic union results in the formation of a diploid zygote, which propels for sometime with the help of the flagella of the female gamete. As the zygote settles on some substratum, it withdraws its flagella, assumes a spherical shape and secretes a wall (Fig. A). The zygote germinates immediately and forms a lobed germling. The germling soon gets differentiated into an erect aerial and a horizontal disc like rhizoidal system. The entire aerial system (medulla and cortex) is formed by the ramification of the single initial filament.

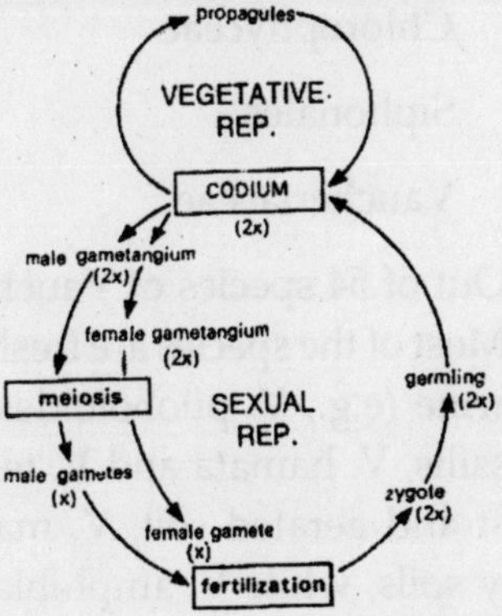

Codium : Diagrammatic representation of the life cycle.

Alternation of Generation

Codium shows diplontic life-cycle with no alternation of generation. The thallus is diploid which bears micro- and macrogametangia. The gametangial nuclei divide meiotically to form haploid gametes which represent the haploid phase in the life-cycle. The micro-and macrogametes fuse to form diploid zygote and thus the diploid phase is re-established. There is no reduction division at the time of the germination of zygote.

Family Vaucheriaceae

The members of this family are found in freshwater habitats in temperate regions. The aseptate, siphonaceous and coenocytic thalli are found attached to the substratum by a branched hapteron. They reproduce asexually by means of zoospores or aplanospores. Sexual reproduction is of oogamous type.

Life-history of Vaucheria is described here as the representative genus of the family.

Vaucheria

Systematic Position

Class - Chlorophyceae

Order - Siphonales

Family - Vaucheriaceae

Occurrence : Out of 54 species of Vaucheria known, about 19 occur in India. Most of the species are freshwater or terrestrial but a few are marine (e.g., V. piloboloides). Many terrestrial species like V. sessilis, V. hamata and V. terrestris form green mats on the moist and aerated soil. V. mayyandensis grows on exposed clayey soils, while V. amphibia is an amphibious species. V. jonesii is found on the snow covered surfaces.

Some common Indian species of Vaucheria are V. amphibia, V. polysperma, V. uncinata, V. geminata and V.sessilis.

Thallus Structure : Vaucheria has a sparingly branched, coenocytic and siphonaceous thallus. The filaments show apical growth and they attain a length of several centimeters. They arè aseptate but septa or cross walls are formed at the time of the formation of reproductive structures or at the site of injury. However, in V. pseudohamata the filaments are septate. Terrestrial species remain attached to the soil surface with the help of branched hapterons or rhizoids. In V. mayyanadensis, the thallus is differentiated into a well-developed underground rhizoidal system and erect aerial branches.

The wall of the filament is relatively thin and lacks elasticity. It is differentiated into an outer pectic and an inner cellulosic layer. There is a well-developed siphon-like central vacuole within the filament that runs throughout the length of the thallus except immediately behind the growing tips. The vacuole is filled with a cell sap that contains lipids, degenerated chloroplasts and crystals. The protoplast is present in the form of a continuous layer between the cell wall and central vacuole. The nuclei lie toward the centre and chromatophores towards the periphery in the protoplast. The cytoplasm shows streaming movements which involve nuclei, mitochondria and dictyosomes.

Although Vaucheria has been placed in Chlorophyceae by Fritsch (1935), it has pigments like chlorophyll *a*, chlorophyll *e*, carotenoid and xanthophyll which are characteristic of Xanthophyceae. Chlorophyll *b*, a typical pigment of Chlorophyceae, is absent in this genus. The two dominant pigments of the order Siphonales, siphonein and siphonoxanthin do not occur in Vaucheria.

The reserve food material in Vaucheria is oil instead of starch. It occurs as small colourless droplets in the cytoplasm.

But filaments growing in continuous light accumulate food in the form of starch.

Reproduction : Vaucheria reproduces by vegetative, asexual and sexual methods.

Vegetative Reproduction : Vegetative multiplication takes place by fragmentation of the vegetative filaments. Transverse septa are formed that block the injured portions of coenocyte.

Asexual Reproduction : Vaucheria reproduces by a variety of asexual spores. Formation of these spores depends upon environmental factors. Following asexual spores are found in Vaucheria:

1. *Zoospores* : Asexual reproduction in aquatic species of Vaucheria usually takes place by means of multiflagellate and multinucleate zoospores. They are produced singly in club-shaped sporangia which are formed at the tip of the filament and the central vacuole gradually decreases in size (Fig. A). Consequently, the tip becomes club-shaped and dark green in colour. It is cut off from rest of the filament by a transverse septum. A central vacuole develops and there is reversion in the position of nuclei and chloroplasts. The nuclei which were toward the centre in the vegetative filament come to occupy the peripheral position and the chromatophores move from the periphery to the centre. The sporangial protoplast now shrinks slightly and a pair of flagella, which are of unequal length, develops opposite each nucleus. This multinucleate and multiflagellate sporangial mass behaves as a zoospore (Fig. C). It has no eye spot or pyrenoids. At the apex of the zoosporangium the wall gelatinises, forming a narrow aperture. The zoospore escapes through this pore, usually in early morning hours.

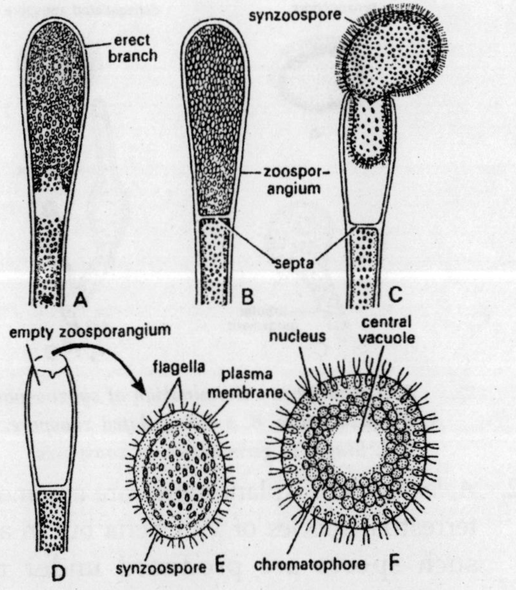

A-F. Vaucheria: Development and liberation of zoospore; A-B. Development of zoospore, C-D. Liberation of zoospores. E. synzoospore, F. Transverse section of a synzoospore.

Zoospores of Vaucheria are different from those of other green algae due to their multinucleate and multiflagellate nature (Fig. E, F). Some algologists consider that they are compound structures in which protoplast fails to break into uninucleate segments. They prefer to call such multinucleate and multiflagellate zoospores as synzoospores.

Germination of zoospores. The zoospores swim for about 15 minutes after their release from the zoosporangium and then come to rest. They withdraw their flagella and secrete a thin wall.

Thus, before germination the zoospore becomes deflagellated, non-motile and sluggish (Fig. A, B). Soon one or two tubular siphonaceous outgrowths come out (Fig. C). In terrestrial species one of the outgrowths attaches itself to the substratum and functions as holdfast, whereas the other forms erect branched system.

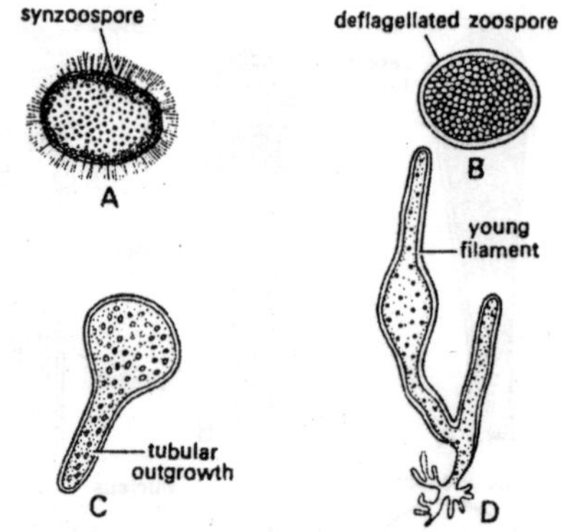

A-D. Vaucheria : Germination of synzoospore; A. A synzoospore, B. a deflagellated zoospore, C-D. Stages in germination of zoospores.

2. *Aplanospores :* Aplanospores are commonly formed in terrestrial species of Vaucheria but in aquatic species such spores are produced under unfavourable conditions. Like zoospores, aplanospores are also formed singly within aplanosporangium; the latter develops as a club-shaped structure at the tip of the young vegetative branch. A transverse septum cuts off the terminal sporangium from the rest of the filament (Fig. A, B).

   The protoplast of the sporangium metamorphoses into a single multinucleate aplanospore. Normally, the conversion of sporangium protoplast into zoospore or aplanospore depends upon external factors. For example, aquatic species, which commonly produce zoospores, can be induced to produce aplanospores by keeping them in dim light or quiet water.

   Similarly, terrestrial species can be induced to produce zoospores by keeping the vegetative filaments moist for few days.

   The aplanospore is a thin walled club-shaped or globose structure. It is liberated by rupturing the apical wall of the sporangium (Fig. B, C). Sometimes, the protoplast of the aplanosporangium divides into a number of spores. These small thin walled aplanospores are called micro-aplanospores. The aplanospore or microaplanospore germinates to form a new thallus (Fig. D).

3. *Akinetes* : Some terrestrial and aquatic species, when subjected to unfavourable conditions, contents of their filaments divide into small segments by thick gelatinous walls. These thick walled multinucleate segments are called akinetes, cysts or hypnospores. Their protoplast is often laden with oil droplets.

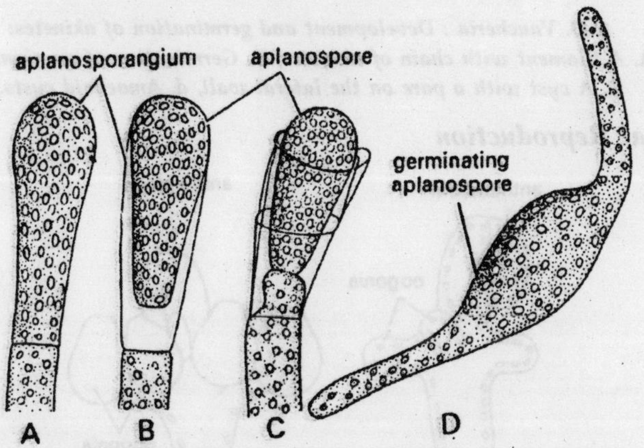

A-D. Vaucheria : Development and germination of aplanospore.

Cysts germinate under favourable conditions, each forming a new thallus. Sometimes the protoplast of the cyst may divide into many uninucleate amoeboid segments (Fig. D). These segments come out through a pore in the lateral wall of the cyst.

At this stage, Vaucheria thallus resembles to that of another genus, Gongrosira, and therefore, this stage is called as gongrosira stage. Each amoeboid mass secretes a thin wall and eventually germinates to form a new thallus.

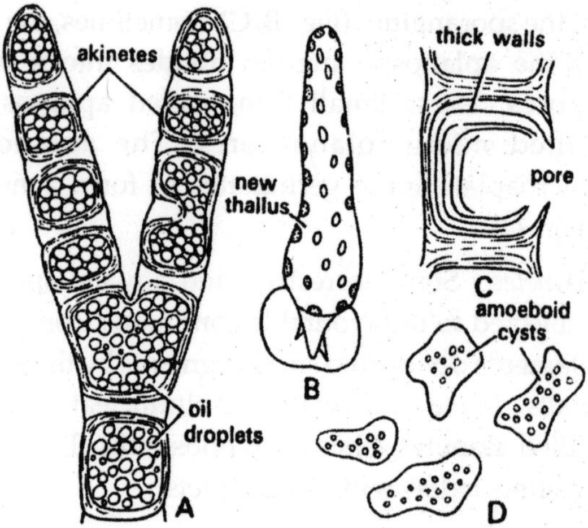

A-D. Vaucheria : Development and germination of akinetes; A. A filament with chain of akinetes, B. Germination of an akinete, C. A cyst with a pore on the lateral wall, d. Amoeboid cysts.

Sexual Reproduction

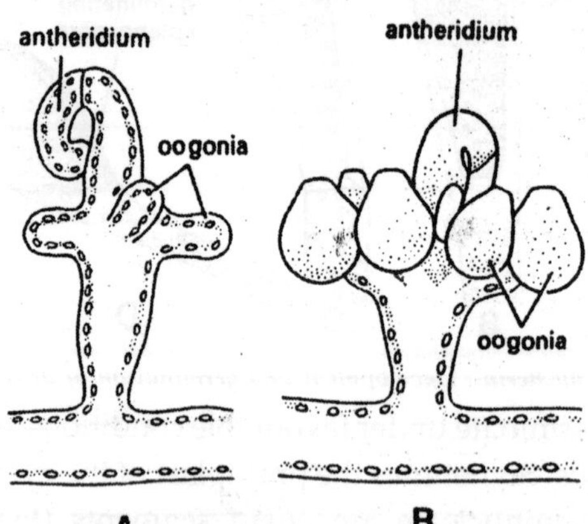

A-B. Vaucheria geminata : Central stalked antheridium surrounded by many developing (A) and mature (B) oogonia.

Sexual reproduction in Vaucheria is oogamous. It takes place with the help of specialized sex organs, called antheridium and oogonium. Most of the species are homothallic or monoecious (Fig. A, B), but some like V. dichotoma, V. litorea and V. mayyanadensis are heterothallic or dioecious.

Sex organs usually develop on filaments growing in damp soil or quiet water; they are infrequent on filaments growing in flowing water. In homothallic species antheridia and oogonia are borne adjacent to one another on a common lateral branch or on adjacent lateral branches. The sex organs are stalked or sessile (V. sessilis). Some species show variation in the relative position of antheridia and oogonia, for example, in V. geminata a central stalked antheridium is surrounded by 3-4 oogonia (Fig. A, B) and in V. gardneri each antheridium has many oogonia at its base.

1. *Antheridium :* The antheridia develop on main filament or its lateral branches as strongly curved cylindrical structures. Large number of chromatophores and nuclei accumulate in this region and this curved part is then cut off from the rest of the filament by a septum (Fig. A, D). The nuclei present in this region divide mitotically and then the protoplast divides to form many uninucleate segments. Each segment metamorphoses into a spindle shaped biflagellate antherozoid (Fig. E, F). The two flagella are of unequal length, dissimilar and laterally inserted. The antherozoids do not have chloroplast or eye spot.

   The antherozoids are liberated through an aperture at the anterior end of the antheridium (Fig. G). In V. debaryana there are usually several pores on the lateral wall of the antheridium, through which the antherozoids liberate.

2. *Oogonium :* The oogonium initiates as a small protuberance at the base of the antheridium, which is formed due to the accumulation of cytoplasm and large number of nuclei. This dense multinucleate mass of cytoplasm is called wanderplasm (Fig. A). It gradually increases in size and forms an ovoid or rounded oogonium which is separated from the subtending filament by a septum (Fig. B, F).

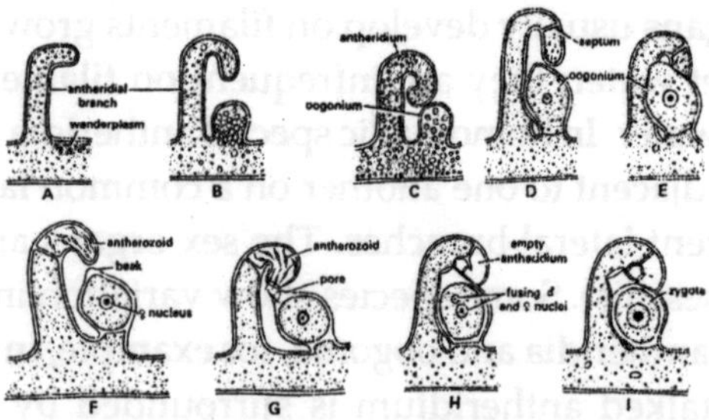

A-I. Vaucheria : A-F. Development of antheridium and oogonium, G-H. Liberation of antherozoids and fertilization, I. Zygote.

The mature oogonium is spherical or sub-spherical with an apical beak. It has only a single nucleus, the other nuclei present in the young state degenerate after the septum is laid down. However, some workers believe that all nuclei except one pass out of the oogonium before the formation of the septum. The uninucleate protoplast of the mature oogonium forms an egg.

At the anterior end of the egg there is a hyaline area which acts as a receptive spot.

3. *Fertilization :* Antheridia and oogonia mature almost simultaneously. The tip of the apical beak of oogonium gelatinizes to form an aperture (Fig. G). Simultaneously, large number of antherozoids are liberated from the adjacent antheridium. A drop of colourless cytoplasm comes out of the apical beak of the egg, and a number of antherozoids get entangled in this drop. Only one of the many antherozoids enters the oogonium and finds its way into the egg through receptive spot. The nucleus of antherozoid is smaller than that of the egg (Fig. H). After entering the egg, the male nucleus gradually increases in size and becomes nearly equal to the egg nucleus. Then the two nuclei fuse to form a diploid zygote (Fig. I). The zygote secretes a 3-7 layered thick wall. The protoplast of the zygote is laden

with oil droplets. The zygote is green in colour initially but it gradually turns red due to degeneration of chlorophyll. It remains in the oogonium until it is liberated by decay of the oogonial wall. The zygote remains dormant for a few months before germinating. The dormant state helps in tiding over unfavourable conditions like desiccation and high temperature.

Germination of zygote. After the expiry of dormant period, the zygote germinates to form a new thallus. Before germination, its diploid nucleus divides to form four haploid nuclei. Repeated divisions of these nuclei result in the formation of a coenocytic thallus.

Taxonomic Position of Vaucheria

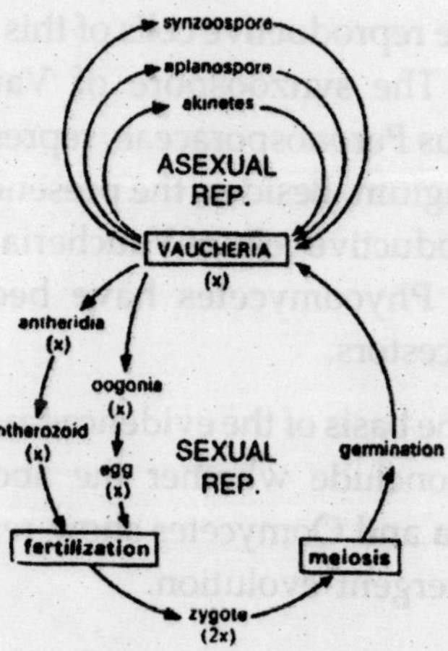

Vaucheria : life-cycle.

The taxonomic position of Vaucheria has been debated in the past. Fritsch (1935, 48) included it in the order Siphonales of the class Chlorophyceae, whereas Smith (1938, 55) placed it in a separate class Xanthophyceae. It was placed in the class Chlorophyceae on the basis of : (i) filamentous siphonaceous thallus organisation, (ii) asexual reproduction by synzoospores, and (iii) specialised oogamous type of sexual reproduction.

However, it differs from other members of Chlorophyceae in (i) the absence of macerised cellulosic cell wall, which is typical of Chlorophyceae, (ii) the absence of chlorophyll *b*, (iii) predominance of oil as reserve food material, (iv) the absence of pyrenoid, and (v) unequal flagella of synzoospores. On the basis of these differences Smith removed Vaucheria from Chlorophyceae and placed it in the class Xanthophyceae, with which it shares the following features : (i) presence of chlorophyll *e*, a pigment characteristic of Xanthophyceae, (ii) absence of two xanthophyll pigments, siphonein and siphonoxanthin in zoospores of Vaucheria, and (iii) unequal flagella in antherozoids.

Davis (1904) and Williams (1926) traced phylogenetic relationships between Vaucheria and Oomycetes. The development of the reproductive cells of this alga is analogous with Oomycetes. The synzoospore of Vaucheria, like the gonidium of various Peronosporaceae, represents the contents of an entire sporangium. Besides, the presence of two unequal flagella in the reproductive cells of Vaucheria also lend support to the view that Phycomycetes have been derived from Vaucheria-like ancestors.

However, on the basis of the evidences available at present, it is difficult to conclude whether the above resemblances between Vaucheria and Oomycetes show real affinities or are the result of convergent evolution.

11

The Conjugales

Algae belonging to the order Conjugales are Commonly known as 'pond silks' or 'water silks'. They are confined to fresh waters only. They differ from other Chlorophyceae in the absence of flagellated reproductive cells and in the structure of chloroplast. These algae vary in their structure from unicellular to unbranched filamentous forms. The unicellular forms are commonly known as desmids. The cell wall is composed of an outer pectic and an inner cellulosic layer. The cells are uninucleate and the nucleus is centrally located. The chloroplasts are of three types (i) one or more peripheral spirally twisted ribbon-like bands as in Spirogyra; (ii) an axial plate extending the length of the cell as in Mesotaenium and Mougeotia; and (iii) two stellate axial chloroplasts as in Zygnema and Cylindrocystis.

Vegetative multiplication in unicellular forms takes place by simple cell division and in filamentous forms by fragmentation. Motile asexual spores are not found in these algae. They reproduce asexually with the help of aplanospores. Sexual reproduction takes place by the fusion of non-flagellate amoeboid gametes through conjugation. The zygospore formed by the fusion of two gametes secretes a thick wall with characteristic ornamentation.

Fritsch (1935) divided the order into five families Mesotaeniaceae, Zygnemaceae, Mougeotiaceae, Gonatozygaceae and Desmidiaceae.

Family Zygnemaceae

Members of Zygnemaceae are among the most common unbranched filamentous algae, favouring small stagnant bodies of water. They are especially abundant in spring months. The cells of the filament are cylindrical and more or less permanently united. The cell wall lacks pores. Union of two aplanogametes usually takes place by the establishment of a conjugation tube.

The family is represented by 3 genera, Spirogyra, Debarya and Zygnema. The life-history of Zygnema is described here as the representative of the family.

Zygnema

Systematic Position

Class	-	Chlorophyceae
Order	-	Conjugales
Family	-	Zygnemaceae

Occurrence : Zygnema is a freshwater alga forming bright green free-floating masses in stagnant water. Species growing in swiftly running streams develop holdfasts, particularly when coming in contact of any rough surface. Z. terrestre is a terrestrial species. The genus has about 100 species, of these 47 are recorded from India. Some common Indian species are Z. himalayensis, Z. czurdae, Z. giganticum and Z. globosum.

Thallus Structure : It is a free floating unbranched

filamentous alga and the plant body consists of a uniseriate row of cylindrical cells. All the cells of the filament are alike, but in attached forms the basal cell develops into an attaching organ. The cells are two to four times longer than breadth.

The cell wall is composed of two layers, the outer layer is of pectose and the inner of cellulose. The pectose layer is relatively thin than in other members of Zygnemaceae. This layer gives the filaments a slippery touch, hence the name 'pond silks' or 'water-silks' has been given to these algae.

Each cell contains two stellate (star-shaped) chloroplasts which lie axial to each other. A single pyrenoid is present in each chloroplast. The cell is uninucleate; the nucleus lies between the two chloroplasts in the center of the cell.

Growth : Growth of the filament takes place by repeated transverse divisions of the cells. All the cells of the filament (except the holdfast in attached forms) are capable of divisions.

Reproduction

Zygnema reproduces by vegetative, asexual and sexual methods.

Vegetative Reproduction : Vegetative multiplication is brought about by the following methods :

1. *Fragmentation :* The filament may break into small segments or individual cells due to the dissolution of middle lamella. Such fragmentation is usually brought about by certain changes in the habitat. Besides, water currents and aquatic animals feeding upon plants may also break the filaments into small segments. Some of the dissociated cells or fragments may die, while others form new filaments.

2. *Akinetes :* In some aquatic (e.g., Z, giganteum, Z. spontaneum, Z. sterile) as well as terrestrial species (Z. terrestre) certain cells of the filament develop into

thick walled akinetes. Their formation is induced by unfavourable environmental conditions. During the formation of akinetes, the vegetative cell divides into two similar halves (each with a single chloroplast) which accumulate food reserves. Thus akinetes are half the length of the vegetative cell. They germinate under favourable conditions and form new filaments by repeated transverse divisions.

Asexual Reproduction : Like other members of the order Conjugales, Zygnema reproduces asexually by means of non-motile spores. The following two types of non-motile spores are found in Zygnema:

1. *Aplanospores* : They are formed singly within the vegetative cell in many species of Zygnema (e.g., Z. himalayensis, Z. kumaoense, Z. terrestre). The protoplast of the cell retracts from the cell wall and forms a non-motile thick walled spore, called aplanospore. The aplanospore has a wall distinct from the parent cell wall.
2. *Parthenospores* : Sometimes gametes fail to fuse and secrete a thick wall. These are called parthenospores, azygospores or azygotes.

Aplanospores and parthenospores germinate under favourable conditions and give rise to new filaments.

Sexual Reproduction : Sexual reproduction is isogamous or rarely anisogamous. In most of the species, the fusing gametes are morphologically as well as physiologically similar. However, some species show physiological anisogamy. Any cell of the filament may function as gametangium. It produces a single non-flagellate, non-motile amoeboid gamete. Gametic union takes place by the formation of a tubular conjugation tube between the two cells— a characteristic feature of the order Conjugales.

Conjugation is of two types : (i) Lateral conjugation, and (ii) Scalariform conjugation.

1. *Lateral Conjugation* : When conjugation takes place between two adjacent cells of the same filament, it is called lateral conjugation. Thus, lateral conjugation involves only a single filament (Fig. A, D).

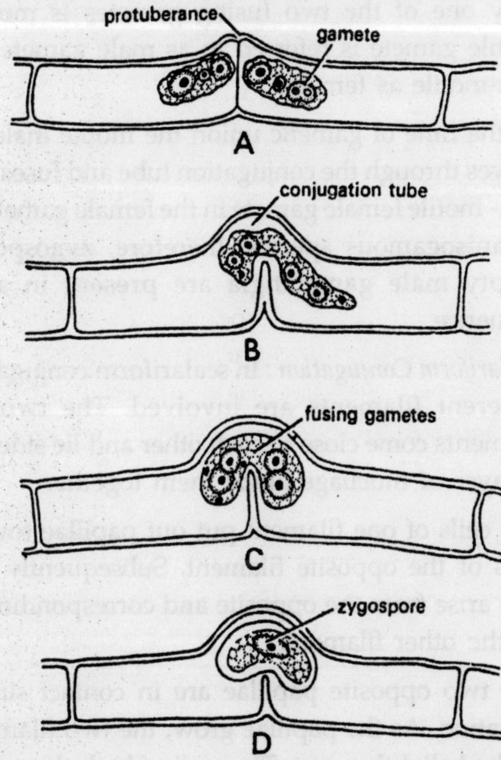

A-D. Zygnema : Sexual reproduction; isogamous lateral conjugation.

At the time of the formation of conjugation tube between two adjacent cells (gametangia), longitudinal walls of these cells protrude on either side of the septum. This results in the formation of two outgrowths, and eventually a part of the septum close to the outgrowths dissolves forming a passage which serves as conjugation tube.

Most of the species of Zygnema (e.g., Z. czurdae, Z. himalayense, Z. terrestre) are isogamous and in these species both the gametes are equally active and move into the conjugation tube.

Fusion of gametes takes place in the region of dissolved septum. But in anisogamous species (e.g., Z. stellinum) only one of the two fusing gametes is motile. The motile gamete is referred to as male gamete and the non-motile as female.

At the time of gametic union the motile male gamete moves through the conjugation tube and fuses with the non- motile female gamete in the female gametangium. In anisogamous species, therefore, zygospores and empty male gametangia are present in alternate sequence.

2. *Scalariform Conjugation :* In scalariform conjugation two different filaments are involved. The two mating filaments come close to each other and lie side by side. A layer of mucilage binds them together.

The cells of one filament put out papillae toward the cells of the opposite filament. Subsequently papillae also arise from the opposite and corresponding points on the other filament.

The two opposite papillae are in contact since their initiation. As the papillae grow, the two filaments are pushed slightly apart. The wall of both the papillae at the point of their contact dissolves to form a tubular passage, known as conjugation tube.

In the mean time the protoplasts of the conjugating cells contract to form gametes. In isogamous species gametes are alike and both the gametes (one from each filament) move into the conjugation tube and fuse to form a zygote.

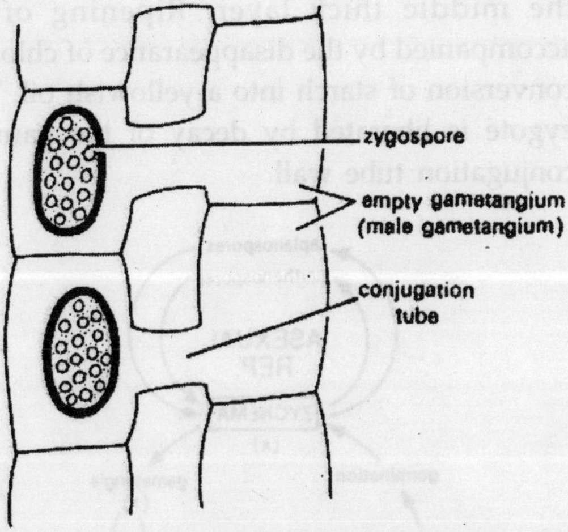

Zygnema : Zygospores (note the empty gametangia).

However, in anisogamous species gametes can be distinguished into male and 'female on the basis of their motility. The motile male gamete moves into the female gametangium through the conjugation tube and fuses with the female gamete to form a zygote.

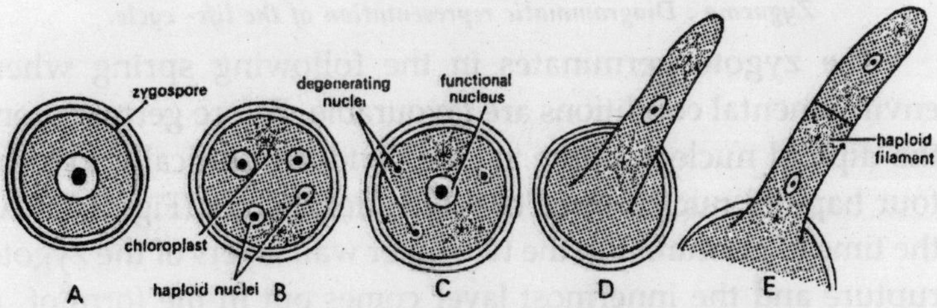

A-C. *Zygnema : Germination of zygospore; A. Zygote, B-C. Reduction division in the zygote.*

3. *Germination of Zygote* : The zygote, formed by the fusion of two haploid gametes, represents the diploid phase in the life-cycle of Zygnema (Fig. A). The zygote secretes a three layered wall; the outermost thin layer (exospore) is made up of cellulose or pectose, and the middle thick layer (mesospore) and the inner thin layer (endospore) are cellulosic. Ornamentation and coloration occur in

the middle thick layer. Ripening of zygote is accompanied by the disappearance of chlorophyll and conversion of starch into a yellowish oil. The mature zygote is liberated by decay of the gametangial or conjugation tube wall.

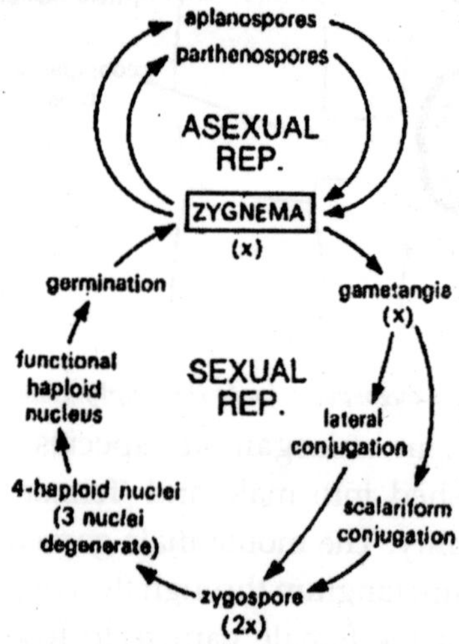

Zygnema : Diagrammatic representation of the life- cycle.

The zygote germinates in the following spring when environmental conditions are favourable. Before germination, the diploid nucleus of the zygote divides meiotically to form four haploid nuclei, three of which degenerate (Fig. B,C). At the time of germination the two outer wall layers of the zygote rupture and the innermost layer comes out in the form of a long germ tube. This germ tube undergoes repeated transverse divisions to form a new (haploid) filament.

The life-cycle of Zygnema is thus primarily haploid, zygote being the only diploid stage.

Family Desmidiaceae

The members of this family are placoderm desmids, characterised by the presence of cells with two distinct halves (semicells) separated by a median constriction (sinus). The cell wall is highly ornamented and has pores. Most of the desmids are unicellular, but a few are joined end to end in filamentous colonies or united in amorphous colonies. The cell has a central nucleus and two chloroplasts, one in each semicell.

Desmids multiply vegetatively by cell division and sexual reproduction takes place by conjugation. The family is represented by 23 genera. Desmidium, Staurastrum, Closterium, Xanthidium, Euastrum and Cosmarium are some common genera. The life history of Cosmarium is described here.

Cosmarium

Systematic Position

Class - Chlorophyceae

Order - Conjugales

Family - Desmidiaceae

Occurrence : Cosmarium is a free floating placoderm desmid. It occurs in fresh water ponds, pools and ditches, either singly or in amorphous colonies. Late spring and early summer seasons are most favourable for its growth. The genus includes over 800 species, of these Cosmarium bengalensis, C. connatum, C. distichum, C. granatum and C. tenne are common in India.

Thallus Structure : The plant body consists of a single haploid flat cell with two distinct symmetrical halves (or semicells). The two halves are separated by a deep median constriction, called sinus, and a connecting band, known as isthmus, joins them. The cell appears differently in different views; it is semicircular, ellipsoidal or reniform in front view, elliptical in vertical view and oval in side view.

The cell wall is either smooth or heavily ornamented with

granular projections. The wall of a mature cell is differentiated into two concentric layers. The inner layer is smooth and mostly consists of cellulose. The outer layer is more thick and firm than the inner layer and is made up of cellulose, impregnated with pectic and iron compounds. Both layers are perforated by minute and simple or large specialized pores, usually grouped symmetrically. The pore is in the form of a simple canal in the inner cell wall layer, but in the outer layer the canal is surrounded by a specially differentiated cylindrical noncellulosic zone. The entire pore structure is called pore-organ. Mucilage secreted through these pores forms a thin coat on the outer surface of the cell.

The cell of Cosmarium is uninucleate and the nucleus is embedded in a small median mass of cytoplasm. In most species there are two axile chloroplasts, one in each semicell, but in C. reniforme and C. ovale each semicell has two parietal chloroplasts in the form of bands.

Reproduction

Reproduction in Cosmarium takes place by vegetative and sexual methods.

Vegetative Reproduction : Vegetative reproduction takes place by cell division. The cell may multiply vegetatively for long periods without conjugating at all. The process of cell division is peculiar due to bipartite structure of the cell wall. It initiates with the elongation of the connection zone (isthmus) of the two semicells. The two halves thus move slightly apart (Fig. A, B). At this stage the nucleus of the cell divides mitotically into two daughter nuclei. Then a transverse wall is laid down in isthmus region which divides the parent cell into two daughter cells. Each daughter cell possesses a portion of the isthmus which grows into a semicell. Thus in the new cell, formed by the division of the parent cell, one semicell is obtained from the parent cell and the other semicell is formed by the portion of the isthmus region. In the meantime, chloroplast of the parent semicell divides and one of the daughter chloroplasts passes into the new semicell.

The membrane of the young semicell is quite smooth initially but subsequently the outer membrane develops ornamentation which is typical of a species. The young semicells usually remain in contact by their apices until they are fully grown (Fig. D).

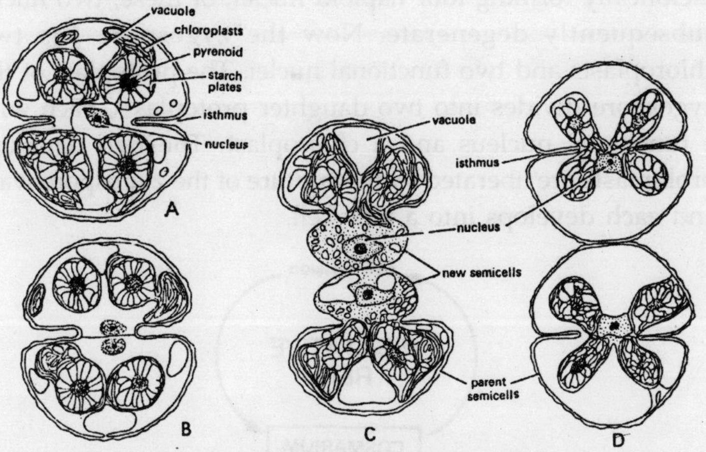

A-D. Cosmarium : Stages in cell divisions.

Sexual Reproduction : Sexual reproduction takes place by conjugation of isogametes. Conjugation usually takes place between two fully developed cells of opposite strains. An immobile cell of one strain pairs with the mobile cell of opposite strain (Fig.A). A sexually active substance is presumably secreted by the immobile cell which is responsible for chemotactic movement of the mobile cell. The paired cells are enveloped in a common mucilagenous sheath. The semicells of both the conjugating cells split open at isthmus and their protoplasts are set free (Fig. B, C). The entire protoplast of each cell acts as a gamete. The amoeboid gametes fuse within minutes and form an irregularly-shaped zygospore. The zygospore soon rounds off and secretes a smooth wall (Fig.D).

Ornamentation in the wall begins to appear within an hour after the fusion of gametes (Fig. E).

Germination of Zygospore : The zygospore settles at the bottom of the pond. At this stage it has four chloroplasts and a diploid nucleus (formed by the fusion of two haploid nuclei of the gametes). Immediately after fusion, two of the four chloroplasts degenerate. Germination of the zygospore starts after a period of rest. Diploid nucleus of the zygospore divides meiotically forming four haploid nuclei; of these, two nuclei subsequently degenerate. Now the zygospore has two chloroplasts and two functional nuclei. The protoplast of the zygospore divides into two daughter protoplasts, each with a functional nucleus and a chloroplast. The two daughter protoplasts are liberated by the rupture of the zygospore wall, and each develops into a new cell.

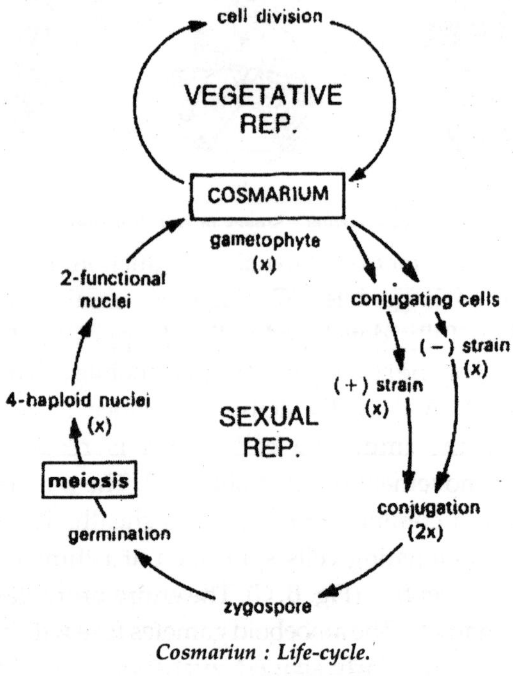

Cosmariun : Life-cycle.

12

The Chaetophorales

The members of Chaetophorales are characterised by the presence of a heterotrichous thallus, that is, the thallus possesses a prostrate and an erect system. In some genera, one of the two systems is suppressed; for example, in Drapenaldia, Drapernaldiopsis and Microthamnion only erect system is present whereas in Coleochaete and Ulvella there is only prostrate system.

Certain members of the order have setae (e.g., Coleochaete) or hairs (e.g., Stigeoclonium) and hence the name Chaetophorales (Chaetos = hair, phorales = plant; plants with hair or setae).

Most of the members of Chaetophorales occur in fresh water, but a few have a marked terrestrial tendency (e.g., Fritschiella tuberosa). The asexual reproduction takes place by quadri-or biflagellate zoospores, aplanospores or akinetes. The sexual reproduction is primarily isogamous (e.g., Fritschiella, Stigeoclonium), occasionally anisogamous (e.g., Aphanochaete) or oogamous (e.g., Coleochaete).

Fritsch (1935) divided the order Chaetophorales into the following five families: (1) Chaetophoraceae, (2) Trentepohliaceae, (3) Coleochaetaceae, (4) Chaetosphaeridiaceae, and (5) Pleurococcaceae.

Family Chaetophoraceae

The members of the family have branched filaments with uninucleate cells that contain a single parietal chloroplast. The filaments show basal-distal differentiation. The branches terminate in sharp points or in setae which are several cells in length. Most members exhibit heterotrichy. Asexual reproduction is by zoospores and sexual reproduction usually of isogamous type.

Fritschiella

Systematic Position

Class	-	Chlorophyceae
Order	-	Chaetophorales
Family	-	Chaetophoraceae

Occurrence : This genus, established by Prof. M.O.P. Iyengar, was named after the great algologist Prof. F.E.Fritsch. The type species, Fritschiella tuberosa, was first reported from South India. It was subsequently collected from Nepal, Burma, Sudan and Japan. It is a terrestrial alga which grows on moist soil undergoing drying after monsoon rains. Another species of the genus, F. simple, *x* was recorded from East Pakistan (now Bangladesh) in 1962.

Thallus Structure : It is a heterotrichous alga and the thallus is differentiated into four distinct systems.

1. *The Rhizoidal System :* It is made up of one or more downwardly directed rhizoid-like elongated colourless cells arising from the prostrate system. This system is, however, absent in F. simplex.

2. *The Prostrate System :* It consists of a number of rounded or irregular swollen clusters of relatively small cells. In mature thallus this system forms short congested branches which are differentiated into nodes and internodes. The cells are thin walled and rich in reserve food (starch, protein, etc.) and, therefore, probably their function is storage.

   The prostrate system varies considerably in the extent of its development. In young plants, the prostrate system is not well formed, but it gradually increases in size as the plants grow. Mature plants invariably show a very well developed prostrate system.
3. *The Primary Projecting System :* It is the upright system which emerges from the prostrate system. It consists of uniseriate or biseriate filaments which may be simple or branched. The cells of the filament are small and rounded like those of the prostrate system. The filaments are green and besides photosynthesis, they also help in conduction.
4. *The Secondary Projecting System :* This system consists of freely branched uniseriate filaments with elongated cells. In the natural habitat the secondary projecting system grows as a tuft above the surface of the substratum. This system is poorly developed in the thalli growing in exposed areas, because of high rate of transpiration. It is absent in F. simplex.

Of the above four systems, the rhizoidal and prostrate systems are completely underground. The primary projecting system is partly embedded in the soil and partly above the ground. The secondary projecting system emerges completely above the soil surface.

Cell Structure : The cells are thin walled and uninucleate. A curved plate-like chloroplast with 2-8 pyrenoids is present in the cells of the secondary projecting system. In primary projecting and prostrate system the cell contents are dense and the chloroplast is poorly developed and has only 2-4 pyrenoids. In rhizoidal system, however, chloroplasts are totally absent.

Perennation

During dry conditions the primary and the secondary projecting systems and the rhizoidal system disappear and the prostrate system acts as perennating structure. The cells of the prostrate system are capable of tiding over unfavourable conditions without any structural modification. On return of favourable conditions the cells germinate directly and give rise to new plants. Sometimes the cells of the nodal region are detached from the thallus due to death of the internodal cells. These cells, besides serving as perennating bodies, also help in vegetative propagation.

Another means of perennation is the formation of small (about 1 mm in length) funnel-shaped and tuber-like bodies on the prostrate system. These dark coloured bodies have a thick layer of cuticle. They germinate under favourable conditions and form new thalli.

Reproduction

Fritschiella reproduces by asexual and sexual methods. The methods of perennation described above also help in vegetative propagation of the thallus.

Asexual Reproduction : The asexual reproduction takes place by biflagellate and quadriflagellate zoospores. The zoospores are formed in the diploid filaments ($2n = 8$) on the prostrate system (F. tuberosa) or the primary projecting system (F. simplex).

The filaments which form zoospores develop club-shaped endings that are covered with a thick pellicle of dark-brown colour. The cells of these club-shaped structures are similar to the other cells of the filament in their shape and size, but their contents are more homogeneous and brick-red in colour. When conditions are favourable and sufficient moisture is available, the nucleus of each of these cells undergoes meiosis and the contents form four haploid (n=4) zoospores. These zoospores are at first colourless or dull brick-red in colour but after liberation they assume usual green colour.

There are 2-4 biflagellate zoospores per cell, or only one quadriflagellate zoospore is formed in a cell. Both types of zoospores are produced simultaneously on the sporophyte plant.

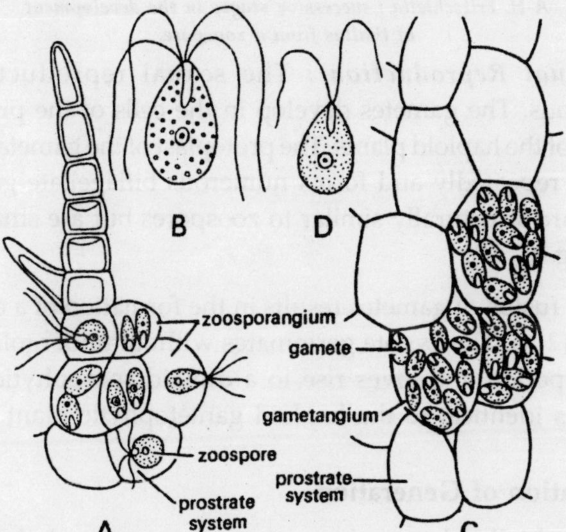

A-D. Fritschiella : Reproduction; A. Part of the thallus showing quadriflagellate zoospores, B. A quadriflagellate zoospore, C. Part of the prostrate system showing gametangia, d. A biflagellate gamete

The zoospores are liberated by the rupture of the sporangial wall and on germination they give rise to new plants. The plants produced by quadriflagellate zoospores are diploid sporophytic plants, while those produced by biflagellate zoospores are haploid gametophytic plants. Both haploid and diploid plants are however, morphologically similar (Fig. A, F).

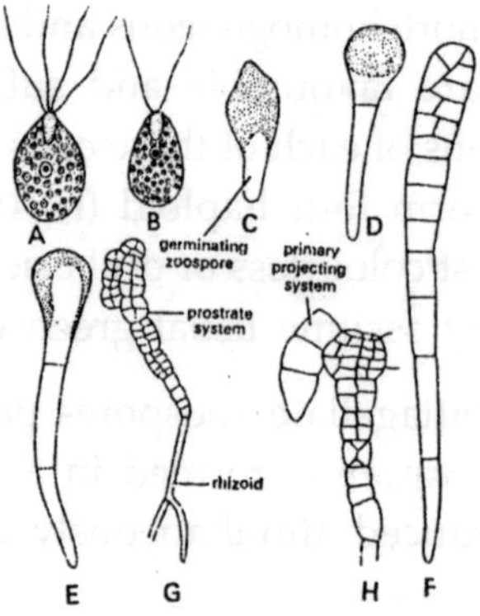

A-H. Fritschiella : successive stages in the development of thallus from a zoospore.

Sexual Reproduction : The sexual reproduction is isogamous. The gametes develop in the cells of the prostrate system of the haploid plants. The protoplast of the gametangium divides repeatedly and forms numerous biflagellate gametes which are structurally similar to zoospores but are smaller in size (Fig. C, D).

The fusion of gametes results in the formation of a diploid zygote (2 X). The zygote germinates without undergoing any resting period and gives rise to a diploid sporophytic plant which is identical to the haploid gametophytic plant.

Alternation of Generation

Fritschiella shows alternation of two morphologically similar generations (isomorphic alternation of generation). The haploid and diploid filaments are alike morphologically. The diploid filaments produce haploid zoospores which give rise to haploid gametophytic filaments on germination. The cells of the prostrate system produce gametes. Similar gametes fuse to form zygote (2 x) and the latter gives rise to diploid filaments on germination.

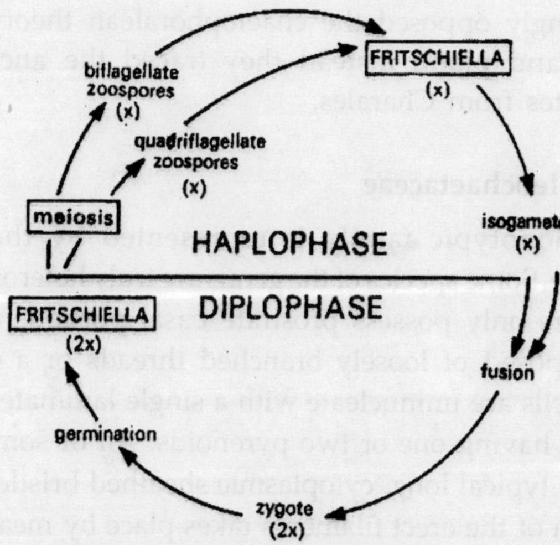

Fritschiella : Life-cycle.

Fritschiella and Origin of Land Plants : There are certain points of interest in the life history of Fritschiella which throws light on the supposedly algal ancestry of higher plants. The tendency of terrestrial or amphibious habit, marked division of labour as shown by photosynthetic subaerial portion, the prostrate system serving as an organ for reproduction and perennation, and the underground rhizoidal system performing the function of anchorage are some remarkable features of Fritschiella.

Fritsch (1935) speculated that the members of Chaetophorales represent the surviving descendants of forms which migrated to the land in the remote past. He was of the opinion that the prostrate gametophyte and the erect aerial sporophyte of land plants evolved from some heterotrichous terrestrial green alga, very similar to the present day Fritschiella. He further remarked that the high degree of thallus differentiation exhibited by Fritschiella and other allied taxa certainly indicate the existence of a strong potentiality for further elaboration in the direction of evolution of an early algal transmigrant form. On the contrary, Stewart and Mattox

(1975) strongly opposed the chaetophoralean theory of the origin of land plant, instead they traced the ancestry of archegoniates from Charales.

Family Coleochaetaceae

This monotypic family is represented by the genus Coleochaete. Some species of the genus are truly heterotrichous, while others only possess prostrate basal portion, which is either composed of loosely branched threads or a compact disc. The cells are uninucleate with a single laminate parietal chloroplast having one or two pyrenoids. All or some of the cells bear a typical long, cytoplasmic sheathed bristle or seta. The growth of the erect filaments takes place by means of an apical cell, and the basal cushion possesses a marginal meristem. Asexual reproduction takes place by means of biflagellate zoospores formed singly and sexual reproduction is of oogamous type.

Coleochaete

Systematic Position

Class - Chlorophyceae

Order - Chaetophorales

Family - Coleochaetaceae

Occurrence : Coleochaete is a freshwater alga which grows epiphytically on submerged parts of several aquatic angiosperms like Hydrilla, Nymphaea, Typha, Sagittaria and Trapa. Thalli of Coleochaete are sometimes epiphytic on Oedogonium and Vaucheria. Some species are endophytic in the members of the order Charales, e.g., C. nitellarum grows endophytically within the cells of Nitella and Chara.

The genus is represented in India by three species, C. orbicularis, C. scutata and C. nitellarum.

Thallus Structure : The thallus is multicellular, heterotrichous and branched. Heterotrichous habit is best represented by C. pulvinata which has well-developed prostrate and erect systems. In the other species one of the two systems (prostrate or erect) is either absent or reduced, e.g., in C. nitellarum and C. divergens there is a well-developed erect system but the prostrate system is filamentous or reduced and in C. scutata and C. orbicularis erect system is absent and the lateral branches of the prostrate system form discoid thalli which may be irregular (e.g., C. scutata) or regular (C. orbicularis) in outline.

Cell Structure : The cells of the thallus may be quadrangular, polygonal or cylindrical. They are uninucleate and each cell has a large laminate parietal chloroplast with a single pyrenoid. Some of the cells of the thallus have a long unbranched cytoplasmic thread, called seta. The base of the seta is covered with a sheath of gelatinous material. This sheath remains attached with the thallus even after the seta breaks off.

Growth : The branches of erect system show apical growth. The terminal cell of the erect filament divides transversely and adds to its height. The growth of the prostrate discoid system takes place by means of a marginal meristem. The erect branches originate as lateral outgrowths of the prostrate system.

Reproduction

Coleochaete reproduces both asexually and sexually.

Asexual Reproduction : The asexual reproduction takes place by means of biflagellate zoospores. They are formed usually in spring and beginning of summer season. Any vegetative cell of the thallus may function as zoosporangium. A single large zoospore is formed in each zoosporangium.

The zoospore is ovoid, biflagellate and uninucleate. It has a large parietal chloroplast but there is no eye spot. The zoospore escapes through a pore at the apex of the zoosporangium. It loses its flagella soon after liberation and secretes a wall.

At the time of germination the spore divides by a transverse or vertical wall. In C. scutata (which is a discoid form) the first division is transverse, the upper cell forms seta and the lower divides repeatedly and forms the discoid thallus (Fig. D- E). In heterotrichous species (e.g., C. pulvinata) the first division is vertical and both the daughter cells divide repeatedly to form prostrate system. The erect or projecting system develops from the prostrate system.

During unfavourable conditions the zoosporangium, instead of zoospore, forms fairly thick walled aplanospores. The aplanospores germinate under favourable conditions and form new thalli.

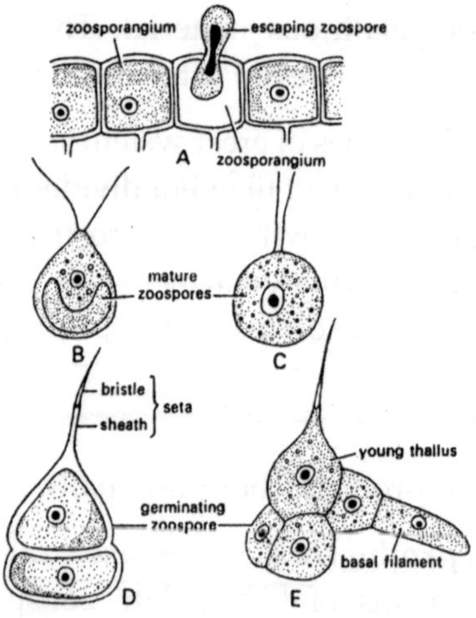

A-E. Coleochaete : Stages in asexual reproduction.

Sexual Reproduction : The sexual reproduction is of advanced oogamous type. The species of Coleochaete are homothallic (e.g., C. pulvinata) or heterothallic (e.g., C. scutata). The male and female sex organs are called antheridia and oogonia (or carpogonia) respectively.

1. *Antheridium* : The antheridia are borne at the tips of the erect branches, usually in groups. In discoid forms (e.g., C. scutata) the antheridia develop midway between the periphery and the center of the thallus. The antheridium develops as an outgrowth of the terminal cell of the lateral branch. This outgrowth is separated from rest of the filament by a septum and develops into an antheridium. Each antheridium forms a single,. uninucleate and biflagellate antherozoid which may be green or colourless. The antherozoid is liberated by the rupture of the antheridial wall.

2. *Oogonium or Carpogonium* : The oogonia also develop at the tips of the erect branches. But the cell lying just below the oogonium continues the growth of the branch and as such the oogonium appears lateral in position. The oogonium is a flask-shaped structure with a long colourless neck, known as trichogyne. However, in discoid species trichogyne is absent and the neck is absent and the neck is represented by a short papilla. The basal swollen part of the oogonium contains a single nucleus and one or more chloroplasts. The protoplast of the basal part rounds off to form a single egg.

3. *Fertilization* : Just before fertilization the tip of the trichogyne breaks down and some colourless protoplasm is extruded from the broken tip. Although many antherozoids are attracted towards the oogonium, only one enters the oogonium. The male and female nuclei do not fuse immediately. The male nucleus is smaller than the female nucleus. The former gradually increases in size and as soon as it attains the size of the female nucleus, both the nuclei fuse to form a diploid zygote.

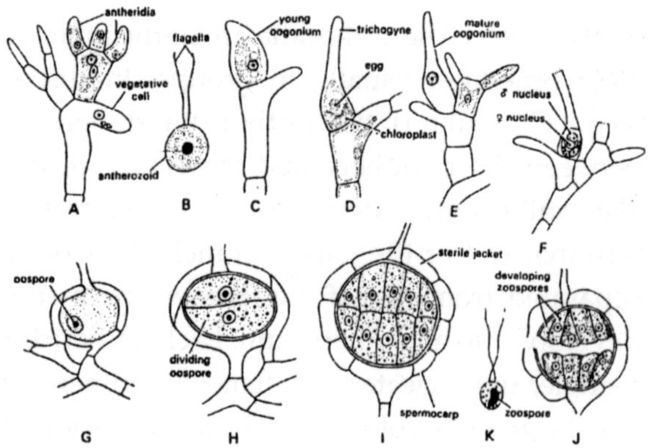

A-K. Coleochaete : Sexual reproduction; A. A branch with antheridium, B. Antherozoid, C-E. Successive stages in the development of oogonium, F. Fertilization, G. Oospore, H-J. Stages in the germination of oospore. K. Zoospore.

Post-fertilization Changes

After fertilization, the basal swollen part of the oogonium is separated from the trichogyne by the formation of a septum. The zygote greatly increases in size and secretes a thick wall. Now the cells below and around the oogonium are stimulated and form a pseudoparenchymatous layer around the oogonium. The oogonium with its sheath layer is called spermocarp. The spermocarp is reddish-brown in colour. It is shed in water by degeneration of the basal cell and remains dormant throughout winter.

Germination of spermocarp. On the return of favourable conditions the contents of the spermocarp again turn green. The zygote nucleus divides meiotically. The first division of the zygote is by a transverse wall and the subsequent divisions are at right angles to each other. Thus 8-32 daughter cells are formed. Each cell metamorphoses into a single biflagellate zoospore. The zoospores are liberated by breaking of the spermocarp and the zygote walls into two halves. They swim freely for a short time, then settle down and eventually develop into new thalli.

Alternation of Generation

A distinct alternation of generation is not found in Coleochaete. The haploid thallus bears male and female gametes on erect branches which fuse to form diploid zygote. The zygote nucleus divides meiotically, forming 8-32 haploid zoospores. The zoospores, on germination, again give rise to haploid thalli. Thus diploid phase is represented only by zygote. Such a life cycle where diploid phase is restricted only to zygote is known as haplontic.

Affinities of Coleochaete

The photosynthetic pigments-chlorophyll *a* and *b*, reserve food starch, non-jacketed unicellular sex organs and haplontic life cycle are typically chlorophycean features of Coleochaete. It shows resemblance with the members of the order Chaetophorales in its cell structure and heterotrichous thallus organisation. However, elaborate oogamous reproduction, retention of the zygote on the haploid plant, presence of a protective layer of vegetative cells around the mature zygotes, elaborate post-fertilization events (development of spermocarp) are some of the marked advanced features exhibited by Coleochaete.

The occurrence of a discoid thallus in some species of Coleochaete (e.g., C. scutata) reminds one of the characteristics of flat thalloid gametophytes of liverworts, such as Riccia. Bower (1908) compared the spermocarp of Coleochaete to a simple or primitive sporophyte of early land plants. The more recent studies of Marchant and Pickett- Heaps (1973), Graham and McBride (1979), Frederick et al. (1973), Graham (1982), Graham and Wilcox (1983) have pointed out that Coleochaete possesses certain features like the presence of phragmoplast during cell division, presence of glycolate oxidase and peroxisomes enzymes, and parenchyma in some species, which indicates its affinities with early land plants.

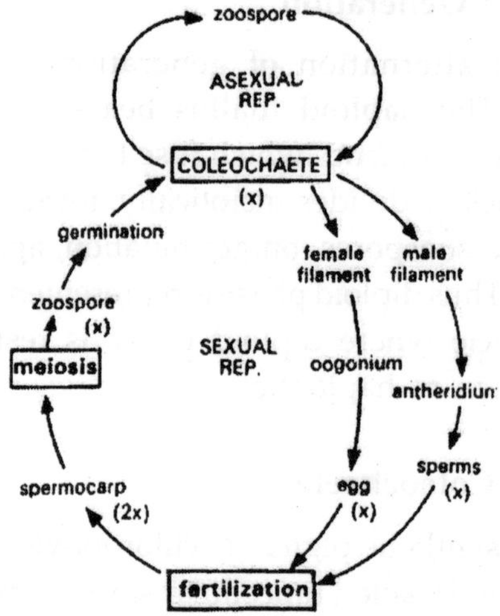

Coleochaete : Life-cycle.

Hence, Stewart and Mattox (1975) suggested that land plants might have originated from Coleochaete like ancestors.

There are remarkable analogies between the sexual reproduction of Coleochaete and the Nemalionales (e.g., Batrachospermum) among the red algae. The structure of the oogonium of Coleochaete markedly resembles the carpogonium of Rhodophyceae. Furthermore, in Coleochaete the oospore is retained within the oogonium and development of sheath around it forms a sort of a fructification (spermocarp) which is analogous to the cystocarp of Batrachospermum. Pringsheim (1860) and Lambert (1910) reported that the swarmers (meiozoospores) of Coleochaete produce several generations of dwarf asexual plants before sexual plants are produced. A more or less similar type of life cycle is found in Batrachospermum. These similarities, however, do not throw any light on the phylogenetic relationship of Coleochaete and Rhodophyceae but simply indicate homoplasy (parallel development).

13

The Oedogoniales

The order Oedogoniales is represented by a single family Oedogoniaceae which includes three genera, Oedogonium, Oedocladium and Bulbochaete. Oedogonium and Bulbochaete are exclusively freshwater forms, usually present in permanent water bodies, while species of Oedocladium also occur in terrestrial habitats. These algae may also grow on other algae and aquatic plants as epiphytes. They are filamentous forms; the filaments are branched in Bulbochaete and Oedocladium and unbranched in Oedogonium. Bulbochaete has long colourless setae with swollen bases. These algae exhibit apical-basal polarity which is maintained even if the filament breaks up and becomes free floating.

The cell possesses a reticulate chloroplast which occupies the whole cell. Adjacent cells maintain cytoplasmic connections through plasmodesmata. Asexual reproduction takes place by zoospores and sexual reproduction is of oogamous type. The plants growing in running water do not form reproductive structures.

Oedogonium

Systematic Position

Class	-	Chlorophyceae
Order	-	Oedogoniales
Family	-	Oedogoniaceae

Occurrence : Oedogonium is a freshwater filamentous alga with more than 400 species, usually present in permanent water bodies of water, such as ponds, lakes and shallow tanks. The filaments are attached to rocks, logs, etc. or are epiphytic on aquatic plants and other algae. Some species (e.g., O. terrestris, O. randhawe) are terrestrial, growing on moist soils. The genus is represented in India by more than 200 species, the common ones being O. areolatum, O. armigerum, O. cardiacum, O. aster and O. elegans.

Thallus Structure : The thallus of Oedogonium is multicellular and filamentous. The filaments are unbranched and uniseriate. All cells of the filament, except the basal and apical ones, are cylindrical and alike. The basal cell, which acts as holdfast, is devoid of chloroplast. Its proximal end is extended into finger like or disc like projections which help in attachment to the substratum. The terminal or apical cell is rounded, elongated or acuminate. Thus, the filaments show clear apical-basal polarity.

Cell Structure : The cell has a fairly thick rigid cell wall, which is differentiated into an outer chitin, a middle pectin and an inner cellulose layer. The centre of the cell is occupied by a large vacuole. The protoplast forms a thin layer in between the central vacuole and the cellulose layer. The chloroplast is reticulate, extending in the form of a sheet throughout the protoplast. Most of the strands of the chloroplast are parallel to the long axis of the cell. Many pyrenoids are present at the intersections of the chloroplast reticulum. The cells are uninucleate. The nucleus lies in the middle region of the cell in between the chloroplast reticulum.

A characteristic feature of Oedogonium is the presence of distinctive transverse bands at the distal end of some cells. The band, formed at the time of cell division, is called apical cap, and the cell with apical cap is known as cap cell.

Growth and Cell Division : Growth in Oedogonium takes place as a result of the divisions in the intercalary cells. All cells of the filament, except holdfast and apical cell, are capable to divide, but divisions are confined to only few cells.

The cell division in Oedogonium is a complex process. The important events during the cell division are as follows :

(1) The cell division is initiated by the formation of a ring of wall material in the upper part of the cell, which gradually increases in thickness (Fig. A-B).

(2) The nucleus migrates to the centre of the cell and divides mitotically.

(3) During telophase an unattached floating septum is laid down between the two daughter nuclei. The septum is formed due to continuous accumulation of vesicular material (microtubules, etc.) from the cytoplasm.

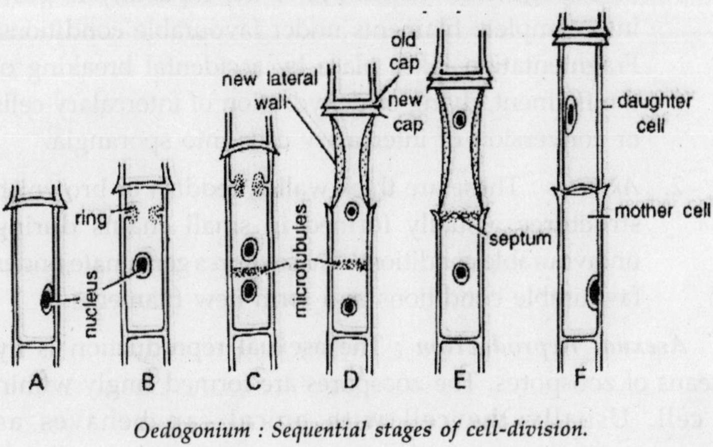

Oedogonium : Sequential stages of cell-division.

(4) The ring of the wall material now elongates, causing a split in the parent wall near the apical ring. In this process a part of the ruptured parent wall remains attached to the anterior end of the new daughter cell in the form of a cap (Fig. D). It is known as the apical cap, and the cell having a cap is called the cap cell. The other portion of the ruptured parent wall can be seen at the proximal end of the new cell.

(5) The floating septum gradually migrates upward to the base of the daughter cell and develops into a mature cross wall (Fig. E, F).

Thus, each division of the cell results in the formation of an apical cap. The number of apical caps in a cell indicates the number of times it has divided.

Reproduction

Oedogonium reproduces by vegetative, asexual and sexual methods.

Vegetative Reproduction : The vegetative multiplication takes place by fragmentation and akinetes.

1. *Fragmentation :* Like many other algae, small fragments of Oedogonium filament have the capability to grow into complete filaments under favourable conditions. Fragmentation takes place by accidental breaking of the filament, dying or dehydration of intercalary cells or conversion of intercalary cells into sporangia.
2. *Akinetes :* These are thick walled reddish or brownish structures, usually formed in small chains during unfavourable conditions. These spores germinate under favourable conditions and form new filaments.

Asexual Reproduction : The asexual reproduction is by means of zoospores. The zoospores are formed singly within a cell. Usually the cell with apical cap behaves as zoosporangium. During zoospore formation cell contents contract slightly from the cell wall. A small lens-shaped hyaline area develops on one side of the protoplast which eventually becomes the anterior end of the zoospore. At the base of this hyaline area a ring of basal granules appears, and from each granule a single flagellum arises. The basal granules remain connected with each other by a fibrous strand.

The mature zoospore is ovoid spherical or pyriform. It is uninucleate and contains a chloroplast. Sometimes the zoospore also possesses an eye-spot. At maturity of the zoospore, the wall of the zoosporangium splits near the cap region, and the adjacent cell moves apart to make passage for the liberation of the zoospore. It is believed that a mucilagenous substance is secreted at the base of the zoospore in zoosporangium which probably aids in the extrusion of the zoospore after absorbing water. The zoospore comes out of the zoosporangium in a delicate mucilagenous vesicle which soon gets dissolved and the zoospore is liberated (Fig. D, E).

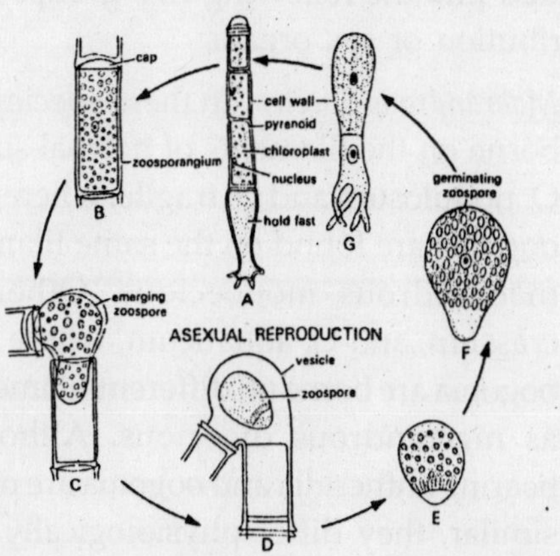

A-G. Oedogonum : Asexual reproduction; A. A filament, B. Zoosporangium, C-d. Liberation of zoospore, E. A zoospore, F-G. Germination of zoospore.

Germination of Zoospores : After liberation, the zoospore swims for about an hour and then settles on any solid substratum with its anterior end downwards. It retracts its flagella and elongates considerably. A transverse septum separates the basal hyaline holdfast from the apical green cell (Fig. E-G). The apical cell divides repeatedly to form a new filament.

Sexual Reproduction : The sexual reproduction is of advanced oogamous type. It is more common in the forms growing in stagnant water than those in flowing streams.

It takes place with the help of male and female gametes. These gametes, known as antherozoids and eggs, are produced in antheridium and oogonium respectively. The male and female gametes differ morphologically as well as physiologically. Certain factors, such as alkaline nature of water and deficiency of nitrogenous substances in the medium promote sexual reproduction. In some species sexual reproduction can also be induced by placing the mature filaments in an atmosphere saturated with carbon dioxide.

1. *Distribution of Sex Organs :* Species of Oedogonium are divided into the following two groups on the basis of distribution of sex organs:

   (a) *Macrandrous Species :* In these species antheridia are borne on the filaments of normal size. Species like O. nodulosum and O. fragile, where antheridia and oogonia are found on the same filament, are called macrandrous monoecious. Others such as O. crassum, and O. aquaticum, where antheridia and oogonia are borne on different filaments, are known as macrandrous dioecious. Although filaments bearing antheridia and oogonia are morphologically similar, they differ physiologically.

   (b) *Nannandrous Species :* In nannandrous species filaments bearing antheridia and oogonia show morphological distinction. The male filaments, which are much smaller than the female filaments, are called dwarf male or nannandrium.

Nannandrous species are always dioecious, i.e., antheridia and oogonia are borne on different filaments.

2. *Antheridium* : Macrandrous species form terminal or intercalary antheridia by the division of antheridial mother cell (Fig. A). Any cell of the filament with apical cap can function as antheridial mother cell. This cell divides transversely into an upper smaller antheridial cell and a lower larger sister cell (Fig. B). The latter undergoes several transverse divisions to form a row of 2-40 flat uninucleate antheridia. The nucleus of the antheridium divides mitotically into two nuclei, each gets surrounded by some cytoplasm and metamorphoses into an antherozoid (Fig. D). Thus, each antheridium produces two antherozoids. But in O. cardiacum four antherozoids are formed in an antheridium.

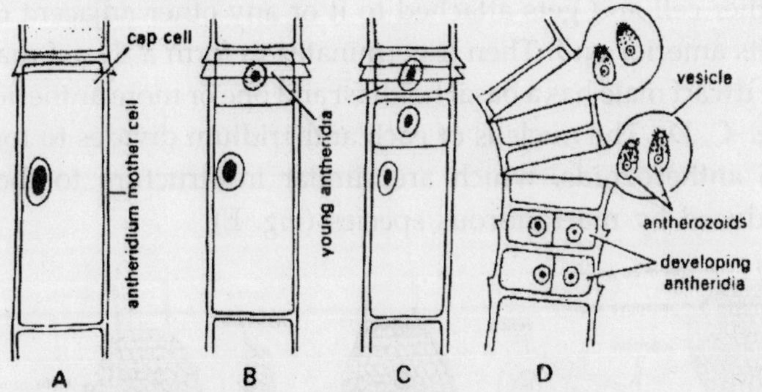

A-D : Oedogonium : Development of antheridium.

The antherozoid is a unicellular, uninucleate and multiflagellate structure. It is similar to the zoospore but for its smaller size and fewer flagella. It is yellowish in colour due to its reduced plastids. Like zoospores, the antherozoids are also liberated by transverse splitting of the wall of the antheridium. The antherozoids are also enclosed in a hyaline vesicle at the time of liberation.

As mentioned earlier, in nannandrous species antheridia develop in specialised, small, 2-4-celled filament, known as dwarf male. The dwarf males develop from androspores, formed in androsporangia (Fig. A). Some nannandrous species (e.g., O. concatenatum), where androsporangia and oogonia develop on the same filament, are called gynandrosporous and those (e.g., O. conferatum), where the two are borne on different filaments, are designated as idioandrosporous. The structure of androsporangia is similar to those of antheridia in macrandrous species, except that only one androspore is produced per androsporangium.

The nucleus of androsporangium does not divide and the protoplast metamorphoses into a single uninucleate multiflagellate androspore. The androspore, when liberated from androsporangium, is also enclosed in a thin vesicle. The vesicle soon dissolves and the androspore swims freely for sometime. It is attracted chemotactically towards the oogonial mother cell and gets attached to it or any other adjacent cell by its anterior end. Then it germinates to form a dwarf male. The dwarf male has a basal holdfast and one or more antheridia (Fig. C, D). The nucleus of each antheridium divides to form two antherozoids, which are similar in structure to those produced by macrandrous species (Fig. E).

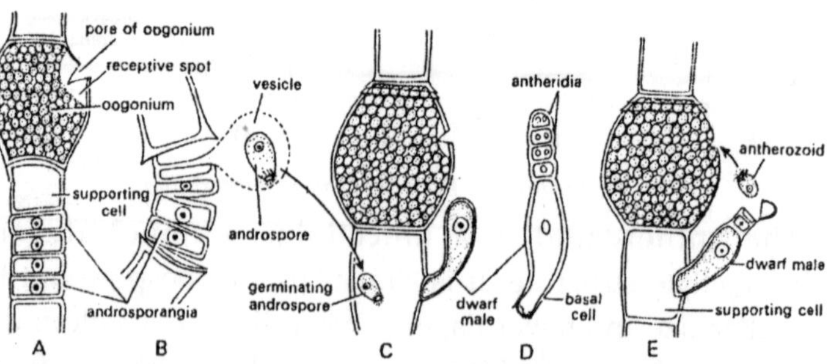

A-E. Oedogonium : Development of androspore and dwarf male : A-B. Development of androspore, C. Germination of androspore and the formation of dwarf male, d. a dwarf male with a chain of antheridia, e. Liberation of antherozoids.

3. *Oogonium* : In monoecious species antheridia and oogonia are present on the same filament, but in dioecious species they are borne on different filaments. Development of oogonium is, however, similar in both macrandrous and nannandrous species. Any vegetative cell with apical cap may function as oogonial mother cell. It divides transversely into an upper oogonium and a lower suffulatory or supporting cell (Fig.A, B). The supporting cell may again function like oogonial mother cell and forms another oogonium and supporting cell. In such cases two or more oogonia are found in a row. Initially, oogonium is slightly wider than the vegetative cells, but at maturity it swells and becomes globose and its protoplast forms a single uninucleate non-motile egg.

   The young egg is green in colour due to the presence of chloroplast. At maturity, a small pore or transverse slit develops on the oogonial wall. Just opposite to this slit a hyaline area is present in the egg which is known as receptive spot. The egg nucleus moves under the receptive spot before fertilization. In some species, prior to fertilization, a gelatinous material is extruded through the pore to facilitate the entry of the antherozoid into the oogonium.

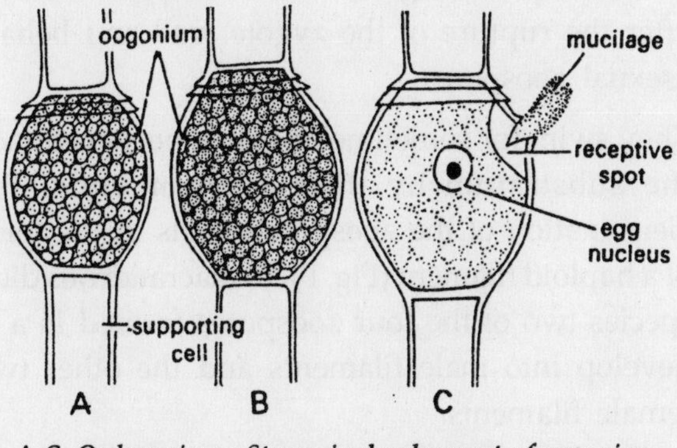

A-C. Oedogonium : Stages in development of oogonium.

4. *Fertilization* : The process of fertilization is similar in both macrandrous and nannandrous species. Antherozoids are attracted chemotactically towards the mature oogonium. Usually a single antherozoid enters into the oogonium through the opening present on the oogonial wall. If more than one antherozoids enter the oogonium, only one of them fuses with the egg to form a diploid zygote (Fig. A).

5. *Zygote* : Soon after fertilization the zygote separates from the oogonial wall and secretes a thick three-layered wall of its own. A newly formed zygote contains chloroplast and hence green in colour, but as it matures there is accumulation of reddish oil in the protoplasm and it becomes reddish brown in colour.

6. *Germination of Zygote* : The zygote is liberated by decay of the oogonial wall and settles down at the bottom of the pond. It undergoes a period of rest. This dormant state helps in tiding over unfavourable period. The zygote germinates under favourable conditions. The diploid nucleus of the zygote divides by meiosis to form four haploid nuclei, followed by the division. of the protoplast into four segments. Each uninucleate segment metamorphoses into a multiflagellate zoospore (zoomeiospores; Fig. D). These zoospores are liberated after the rupture of the zygote wall and behave like asexual zoospores.

   They swim for sometime and then become attached to the substratum by their anterior ends (Fig. E). Germination of the zoospore results in the formation of a haploid filament (Fig. F). In macrandrous dioecious species two of the four zoospores formed in a zygote develop into male filaments and the other two into female filaments.

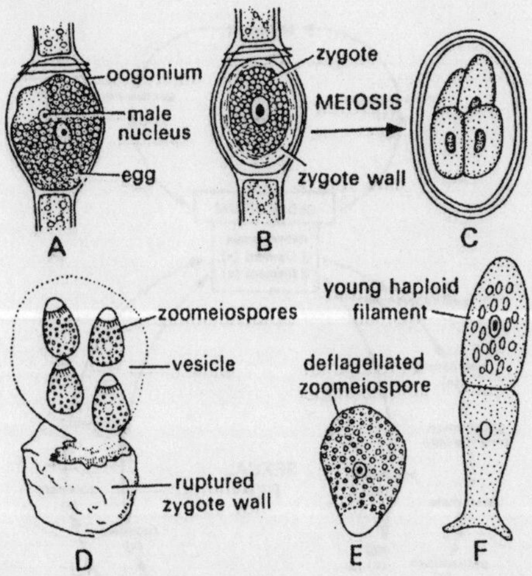

A-F. Oedogonium : Fertilization and germination of zygote; A. Fusion of antherozoid and egg nucleus, B. Zygote , C-D. Reduction division of diploid zygote nucleus and formation of zoomeiospores. E-F. Germination of zoomeiospore.

Alternation of Generation

In Oedogonium diploid stage is short lived. It is found only in the zygote formed by the fusion of haploid egg and antherozoid. The diploid nucleus of the zygote forms four haploid zoospores by meiotic divisions, which grow into haploid filaments. Therefore, the life-cycle of Oedogonium is haploid predominant type.

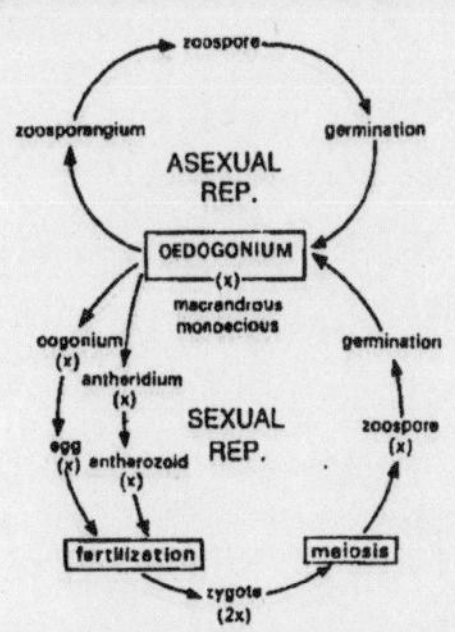

Life-cycle of macrandrous monoecious species.

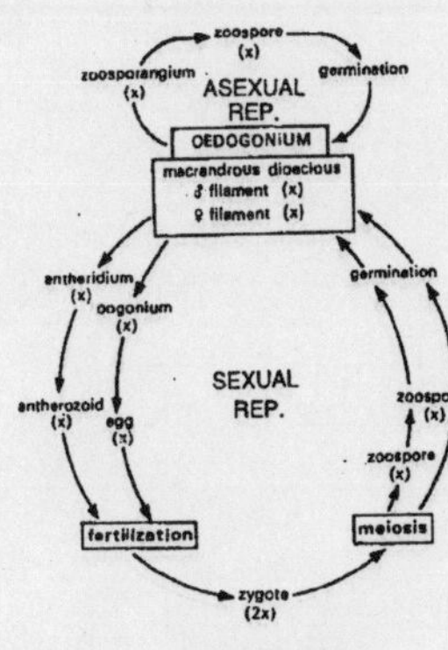

e-cycle of macrandrous dioecious species.

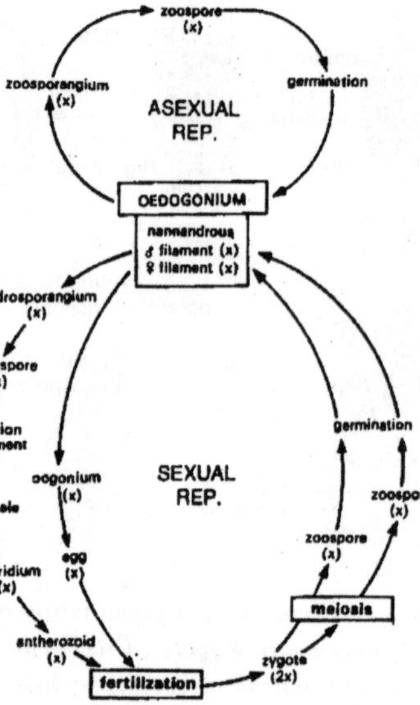

Life-cycle of nannandrous species.

14

The Cladophorales

The order Cladophorales includes about 12 genera and 350 species, occurring in both marine and freshwater habitats. The thallus is branched (sometimes unbranched) and filamentous, and remains attached to the substratum by a rhizoidal cell. The cells of the filament are usually elongated and coenocytic with reticulate chloroplasts.

The asexual reproduction is by means of zoospores and aplanospores and the sexual reproduction is iso- or anisogamous. The gametes are motile and biflagellate. Most of the members of this order show isomorphic alternation of generation.

The order comprises of a single family, Cladophoraceae (Fritsch, 1935). Cladophora, Chaetomorpha, Pithophora, Rhizoclonium and Urospora are some important genera of the family. The life-history of Cladophora is described here as the representative of the order.

Cladophora

Systematic Position

Class	-	Chlorophyceae
Order	-	Cladophorales
Family	-	Cladophoraceae

Occurrence : The genus Cladophora is represented by about 160 species which occur in both fresh and saline waters. The optimum temperature for the growth of freshwater species is 15-25°C. Usually high concentrations of nitrates and phosphates in the medium promote their growth. Most of the freshwater species are found attached to stones or rocks by branched septate rhizoids. C. glomerata is common in streams and C. crispata is found on the walls of water tanks and also on the shells of molluscs (epizoic). Some species are epiphytic, growing on higher aquatic plants. The filaments of some free floating species (e.g., C. holosatica, C. profunda) form ball-like growth. Such species are called aegagropilous forms and they have been placed in a separate section Aegagropila.

The genus is represented in India approximately by 17 species, of these C. crispata, C. glomerata, C. fracta and C. bengalensis are common in fresh waters.

Thallus Structure

The thallus of Cladophora is made up of profusely branched multicellular filaments. It remains attached to the substratum by septate rhizoidal branches which develop from the basal cells of the filament. In some species (e.g., C. kosteri) attachment rhizoids form a prostrate thallus, while in others (e.g., C. basiramosa, C. pygmaea) rhizoids are replaced by a single disc-like holdfast. In many species prostrate rhizoidal system, with their cells filled with reserve food material, helps in perennation.

Branching : The branches of Cladophora arise in a characteristic manner. They always originate just below the septa between the cells in the form of small lateral protuberances. Some cell contents pass into the protuberance which is cut off from rest of the filament by a septum along the long axis of the filament, close to the point of its origin. At this stage the septum is at right angle to the end wall of the cell. As the protuberance grows upward and outward (due to active growth of the parent cell membrane just beneath the protuberance), the angle between the septum and the end wall becomes more than 90°. The growth of the new branch pushes aside the parent axis and as such the branching appears to be dichotomous (which in fact originate as lateral outgrowth). This process of pushing the parent axis by a new lateral branch is called evection (Fig. A-C). Sometimes two lateral branches are formed in a single cell and then the branching appears to be trichotomous.

Cell Structure : The cells of the filament are cylindrical, multinucleate and 5-20 times longer than their breadth. They have thick stratified cell wall that is differentiated into an outer chitinous, a middle pectic and an inner cellulosic layer. The outer chitinous layer gives a coarse and rough look to the filament.

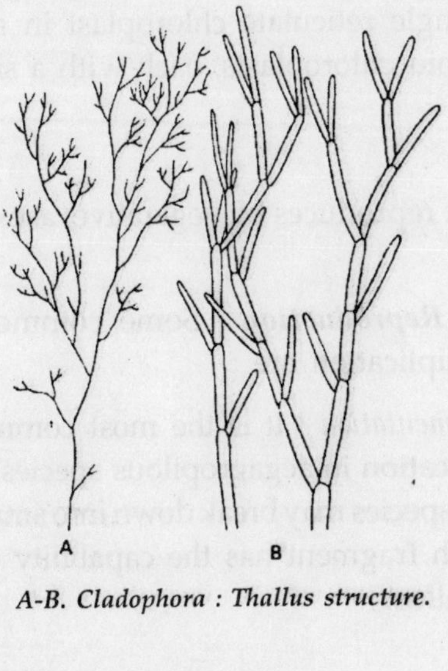

A-B. Cladophora : Thallus structure.

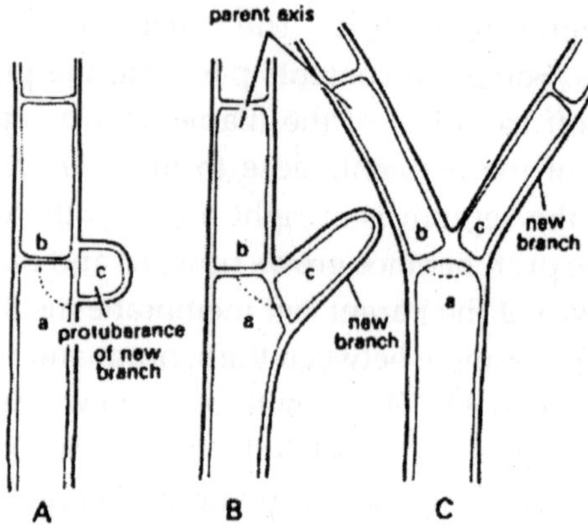

A-C. Cladophora : Stages in the development of a new branch.

The central part of the cell is occupied by a large vacuole. The space between the cell wall and the vacuole is filled with a fairly thick layer of cytoplasm. The cytoplasm contains a reticulate chloroplast with numerous pyrenoids irregularly distributed in the reticulum. The cells are multinucleate and the number of nuclei is probably related to the volume of the cell. Sometimes uninucleate or binucleate cells are also found. Occasionally, there are cells with several small vacuoles instead of a single large central vacuole. Some algologists believe that instead of a single reticulate chloroplast in a cell there are numerous discoid chloroplasts, each with a single pyrenoid.

Reproduction

Cladophora reproduces by vegetative, asexual and sexual methods.

Vegetative Reproduction : Some common methods of vegetative multiplication are :

1. *By Fragmentation* : It is the most common method of multiplication in aegagropilous species. The filaments of these species may break down into smaller fragments, and each fragment has the capability to grow into a new thallus.

2. *By Storage Cells* : Storage cells are found in the perennial species of Cladophora, e.g., C glomerata and C. ophiophila. These cells develop in filaments formed by the division of rhizoidal cells, and they store food material. During unfavourable conditions all the cells of the thallus, except the storage cells, die. The storage cells help in tiding over unfavourable conditions. They germinate to form new thalli on the return of favourable conditions.

3. *By Stolons* : The aegagropilous species of Cladophora have stolon-like rhizoidal branches. During favourable period these branches give rise to some erect branches which form new thalli.

Sometimes the cells of the rhizoidal system disarticulate and these individual cells grow into new thalli.

Asexual Reproduction : Asexual reproduction takes place by following methods.

1. *By Akinetes* : In some fresh water species of Cladophora, certain cells of erect filaments become laden with food material and secrete thick walls. They become pear shaped and function as akinetes. But unlike akinetes of other algae these thick walled cells divide to form new cells without being detached from the filament.

2. *By Zoospores* : Reproduction by zoospores is common in species like C. glomerata, C. crispata and C. paniaudii. The zoospores may be biflagellate (C. crispata) or quadriflagellate (C. parriaudii). Usually the terminal or subterminal cell of vigorously growing diploid filament functions as zoosporangium. Prior to zoospore formation, the nuclei of the zoosporangium divide repeatedly. These divisions are usually meiotic, but in C. glomerata the nuclear divisions are mitotic (hence the zoospores of this species are diploid). Following

nuclear divisions, the protoplast of the zoosporangium also divides and many uninucleate segments are formed. Each segment eventually forms a bi- or quadriflagellate zoospore. The zoospores escape through a pore at the distal end of the sporangium. They swim freely for sometime, then lose their flagella and become attached to the substratum by their anterior ends. They form diploid or haploid filaments (depending upon the nature of the zoospores).

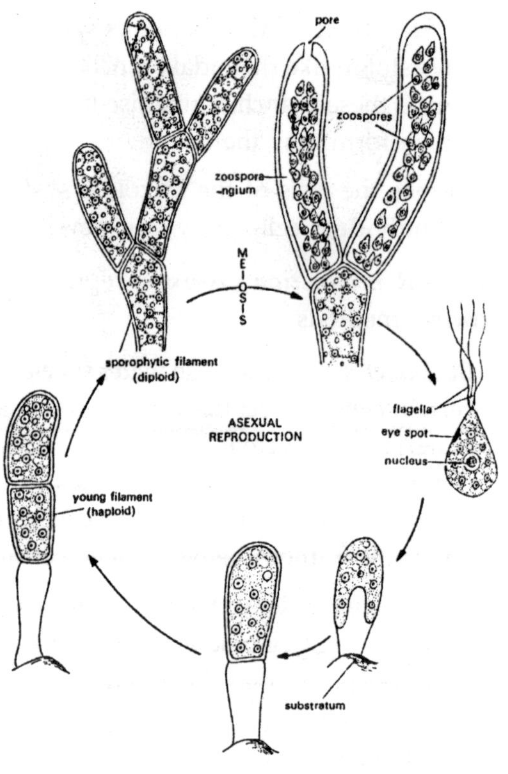

Cladophora : Asexual reproduction.

Sexual Reproduction : The sexual reproduction is isogamous. Most of the species of Cladophora are dioecious or heterothallic. Gametes are formed in the haploid filament almost in the same manner as zoospores in the diploid filament.

The nuclei of those cells of haploid filament which function as gametangia divide repeatedly. This is followed by the division of the protoplast, and as such many uninucleate segments are formed. Each segment forms a single gamete.

In C. glomerata, gametes are formed in the diploid filaments. Hence, the reduction division takes place prior to gamete formation.

The gametes are uninucleate, biflagellate and pyriform. It has an eye spot and reticulate chloroplast with many pyrenoids. The gametes are liberated through a pore at the tip of the gametangium. In some species gametes have a distinct beak at their tips.

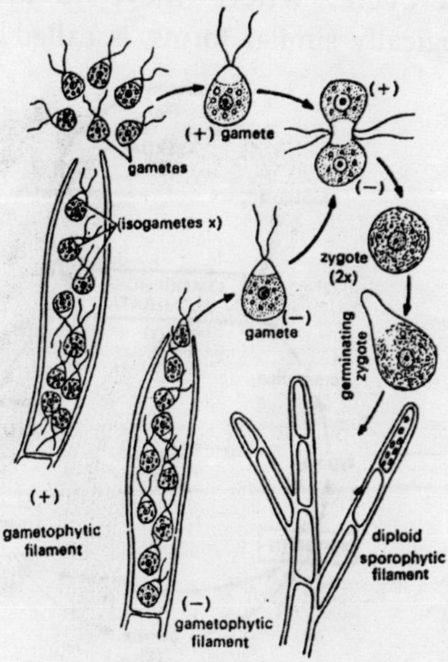

Cladophora* : *Sexual reproduction.

Although gametes are identical in their appearance (such gametes are called isogametes), fusion takes place when two gametes of different strains (+ and -) meet. Zygote so formed is diploid in nature. It germinates immediately and gives rise to a diploid sporophytic thallus.

Alternation of Generation

In the life-cycle of most of the species of Cladophora haploid gametophytic and diploid sporophytic filaments are formed. The diploid sporophytic filaments, after meiosis, produce bi- or quadriflagellate haploid zoospores which give rise to haploid thalli on germination. The cells of the haploid filament function as gametangia and produce haploid isogametes. The gametes from two different filaments (of + and - strain) fuse to form a diploid zygote. The zygote gives rise to a diploid sporophytic thallus on germination. The haploid gametophytic and the diploid sporophytic filaments are morphologically similar. Such life-cycle, where there is alternation of two morphologically similar forms, is called isomorphic.

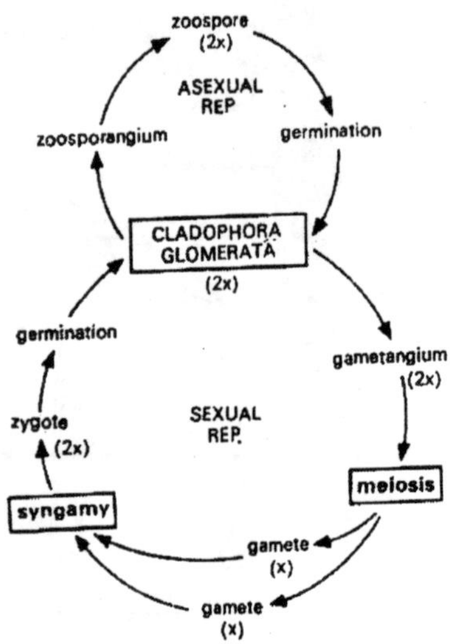

Cladophora : Isomorphic life-cycle.

Life-cycle of Cladophora Glomerata : The life - cycle of C. glomerata (and a few other species) is different from the one described above. Here diploid zoospores are formed in zoosporangia and these zoospores give rise to diploid thalli again on germination. During sexual reproduction there is a reduction division and as such haploid gametes are formed.

The gametes of two different strains fuse to form a diploid zygote. Thus in C. glomerata, the haploid phase is represented by gametes only, and haploid gametophytic thalli, like those in other species, are absent.

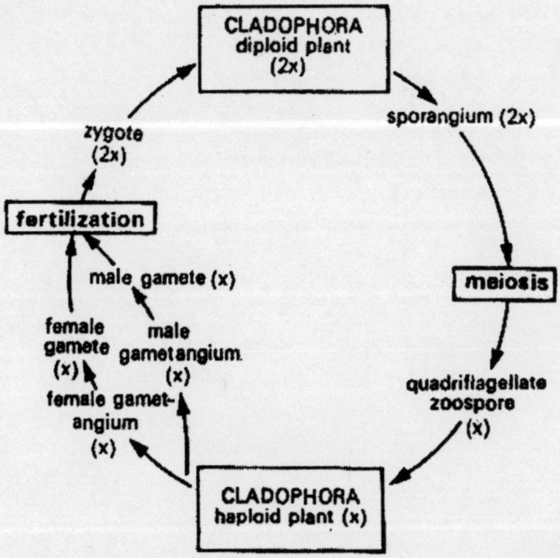

Cladophora glomerata : Diagrammatic representation of life-cycle.

PART-II

Algae-2

15

The Ulotrichales

Members of the order Ulotrichales occur mostly in fresh water or terrestrial habitats but a few forms are marine. They include unbranched filamentous and parenchymatous forms showing little structural or functional specialization among vegetative cells. The lowest cell of the thallus is, however, 'modified to serve as the organ of anchorage. The cells are usually uninucleate and each cell contains a single parietal chloroplast with one or more pyrenoids. Asexual reproduction takes place by the formation of bi- or quadri-flagellate zoospores, aplanospores or hypnospores. Sexual reproduction is isogamous, rarely anisogamous or oogamous.

Fritsch (1935) divided the order into six families : Ulotrichaceae, Microsporaceae, Cylindrocapsaceae, Ulvaceae, Prasiolaceae and Sphaeropleaceae.

Family Ulvaceae

In the members of the family Ulvaceae, the plant body is an expanded sheet, one or two cells in thickness, or a hollow cylinder with a wall one cell in thickness, or a ribbon two or more cells broad. The cells are uninucleate and with a single laminate chloroplast. Asexual reproduction takes place by means of quadriflagellate zoospores. Sexual reproduction is isogamous or anisogamous. Common examples of the family are Enteromorpha, Monostroma and Ulva.

Ulva

Systematic Position

Class	-	Chlorophyceae
Order	-	Ulotrichales
Family	-	Ulvaceae

Occurrence : Ulva, commonly known as 'sea-lettuce' or 'green laver', has a cosmopolitan distribution. It is exclusively a marine alga, usually found attached to rocks in intertidal zones of the sea. It has preference for waters enriched by nitrogenous wastes. Some species occur in polluted estuaries and brackish water. The genus is represented by some 30 species, the common ones are : U. fasciata, U. lactuca, U. lobata, U. reticulata and U. stenophylla.

Species of Ulva are eaten as salad and are also used in soups, especially in Scotland. The chemical composition of dried U. lactuca is 15% protein, 5% sugar and starch, less than 1% fat, and 11% water. It forms a good roughage for the human digestive system.

Thallus Structure

The large expanded thallus is an expanded sheet of parenchymatous tissue two layers of cells in thickness. The thallus is narrowed into a short basal stalk and is attached to the substratum with the help of a holdfast, composed of rhizoidal outgrowths from the lower cells. These colourless rhizoidal outgrowths extend downwards through the spaces present in between the two layers of the thallus and intertwine freely with one another. They emerge from the thallus near the point of its attachment to the substratum and become closely appressed to one another to form a pseudo-parenchymatous holdfast. The emergent portions of rhizoids are transversely septate, multinucleate and contain chloroplasts. The holdfast of the thallus is perennial and proliferates into new blades each spring.

The cells of the thallus are isodiametric or vertically elongated. Their walls are more or less confluent with one another and form a tough gelatinous matrix.

Cell Structure : The cells are uninucleate and the nucleus variously located in the interior half of the cell. A cup-shaped or laminate chloroplast lies towards the outer face of the cell. The chloroplast contains a single pyrenoid.

Growth : Ulva shows diffused growth. All cells of the thallus have the capacity to divide but the plane of the division is always at right angles to the thallus surface.

Reproduction

Ulva reproduces by vegetative, asexual and sexual methods.

Vegetative Reproduction : Vegetative multiplication usually takes place by fragmentation. Any part of the thallus, if detached accidentally, may regenerate into new thallus. Fragmentation is quite common in plants growing in quiet waters of estuaries. But plants growing in open hardly show any vegetative multiplication.

Asexual Reproduction : The asexual reproduction takes place by quadriflagellate zoospores, produced in diploid plants. Any vegetative cell of the thallus may act as zoosporangium. But zoospores are first formed in the cells near the margin and later in almost all other cells of the thallus. At the time of zoospore formation, the protoplast of the vegetative cell divides and redivides to form 4-8 daughter protoplasts. The first division is a reduction division, and as such the daughter protoplasts are haploid structures. Each haploid daughter protoplast metamorphoses into a uninucleate, quadriflagellate zoospore. Zoospores are usually released in the morning when the thallus is wet by the water of incoming tides. They swarm for an hour or so, then come to rest on some suitable substratum, withdraw their flagella and germinate to produce haploid gametophytic thalli.

Sexual Reproduction : The sexual reproduction is usually isogamous, and takes place through gametes which develop in the cells of haploid plants. The plants are heterothallic and only gametes from two different plants fuse. Reproductive areas are formed near the margins of the thallus and the fertile portions change from green to olive-green to brownish-green. Any cell of the fertile region can act as gametangium. Its protoplast divides repeatedly to form 32-64 biflagellate gametes. The first division of the protoplast is parallel to the surface of the thallus, followed by a second division at right angle to the first one. Prior to cleavage of the protoplast, each cell develops a beak-like outgrowth at its outer face which extends to the thallus surface. The tip of this outgrowth eventually develops a pore through which gametes are liberated. The gametes are positively phototactic and biflagellate, and each has a parietal chloroplast with a pyrenoid and an eye spot.

Some species of Ulva like U. lobata are, however, anisogamous. In these species, it is possible to distinguish between male and female gametophytes by differences in the colour of the fertile portion. Further more, the male gametes are narrower and smaller with a yellowish green chloroplast and indistinct pyrenoid, whereas the female gametes are larger with a bright green chloroplast and distinct pyrenoid.

Gametes are usually liberated in the morning, at the beginning of a series of spring tides. Two gametes from two different plants fuse to form a diploid quadriflagellate zygote. The zygote swarms for a short while, and then comes to rest, looses its flagella, and secretes a wall. Within a few days, it divides by a mitotic division. Out of the two daughter cells thus formed, one develops into a rhizoid and the other into a blade. Thus the sporophytic plant body is formed.

In some species, parthenogenetic development of the gamete gives rise to a new plant. There is a periodic fruiting pattern in Ulva, controlled by the lunar cycle. Gametes are released a few days before zoospores and fruiting occurs during a series of neap tides (e.g., U. pertusa) or during a series of spring tides (e.g., U. lobata).

Alternation of Generation

There is an alternation of two morphologically similar generations (isomorphic alternation of generation). The haploid and diploid thalli are morphologically similar. The diploid thalli produce haploid quadriflagellate zoospores which give rise to haploid gametophytic plant body on germination. The haploid thalli produce gametes, which fuse to form a diploid zygote. The zygote gives rise to sporophytic plant body on germination.

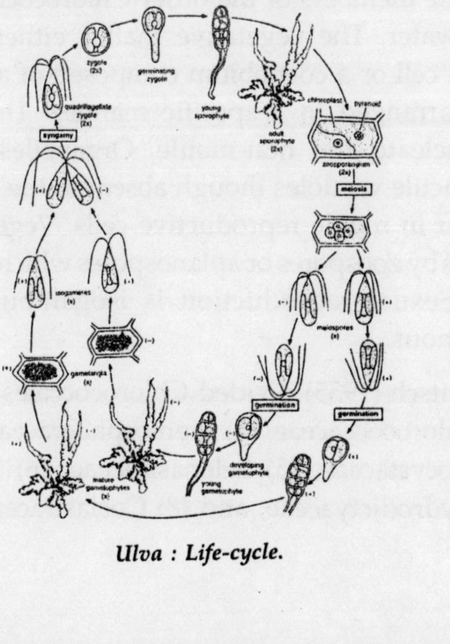

Ulva : Life-cycle.

Taxonomic Position of Ulva

The taxonomic position of Ulva has been debated. The family Ulvaceae to which the genus Ulva was assigned was included by Fritsch (1935, 48) to the order Ulotrichales of the class Chlorophyceae. The inclusion of the family Ulvaceae in the order Ulotrichales is supported by the fact that young stages that of Ulva are uniseriate filaments closely resembling the simple filamentous plant body of Ulothrix. This treatment also receives support in the similarity of the cell structure, gametes and zoospores in Ulva and Ulothrix. Chapman and Chapman (1962, 73) and Lee (1980) even go as far as to state that Ulva is merely an elaborate form of Ulothrix.

On the other hand, Smith (1938, 55), Round (1965), Morris (1968) and Prescott (1969) do not subscribe to this view. They place the family Ulvaceae in a separate order Ulvales on the basis of elaboration of the thallus structure and the existence of a distinct alteration of generation in the life-cycle. This treatment is upheld by most of the recent algologists.

Chlorococcales

The members of the order Chlorococcales are exclusively freshwater. The vegetative thallus either consists of only a single cell or a coenobium composed of a definite number of cells arranged in a specific manner. The vegetative cell is uninucleate and non-motile. Organelles like eye spot and contractile vacuoles though absent in the vegetative cell, may appear in motile reproductive cells. Vegetative reproduction occurs by zoospores or aplanospores which resemble the parent cell. Sexual reproduction is isogamous, anisogamous or oogamous.

Fritsch (1935) divided Chlorococcales into eight families : (1) Chlorococcaceae, (2) Eremosphaeraceae, (3) Chlorellaceae, (4) Oocystaceae, (5) Selenastraceae, (6) Dictyosphaeriaceae, (7) Hydrodictyaceae, and (8) Coelastraceae. He included the genus Scenedesmus in the family Coelastraceae along with Coelastrum, Hofmania and Tetradesmus. Smith (1955) also recognised eight families in Chlorococcales but placed Scenedesmus in a separate family, Scenedesmaceae.

Family Chlorellaceae

The family Chlorellaceae includes non-motile, solitary (e.g., Chlorella) or colonial (e.g., Radiococcus) forms. The colonial forms show indefinite orientation of cells. The uninucleate cell has a single chloroplast, usually without pyrenoid. Reproduction takes place by autospores.

Chlorella

Systematic Position

Class - Chlorophyceae

Order - Chlorococcales

Family - Chlorellaceae

Occurrence : Species of Chlorella grow in a wide variety of habitats. They occur in fresh water, brackish water and also in terrestrial habitats. Most of the species are free living but some grow within the cells and tissues of invertebrates, forming symbiotic or parasitic association. Chlorella parasitica occurs as a parasite in Spongilla and Ophryodium; C. conductrix forms symbiotic association with invertebrates like Hydra, Stentor and Paramecium and C. lichina is an algal symbiotic partner of a lichen-Calicium chlorina.

The genus has about eight species. Chlorella ellipsoidea, C. parasitica and C. gonglomerata are common Indian species.

Structure : Chlorella is a unicellular non-motile alga. The cells are small, spherical to broadly ellipsoidal in shape and 2-12 μ m in diameter. The cell wall is made up of cellulose. The cell has a single cup-shaped parietal chloroplast, usually without pyrenoid. The chloroplast occupies major part of the cell. In the chloroplast, photosynthetic lamellae or thylakoids usually do not show grana-like organization, but in C. pyrenoidosa thylakoids are arranged in granal stacks. The cup-shaped cavity of the chloroplast is filled with colourless cytoplasm. The cytoplasm has a single nucleus and other cell organelles like endoplasmic reticulum, mitochondria, Golgi bodies and vacuoles. Changes in environmental conditions, like duration, intensity and quality of light, temperature and chemical composition of the growth medium greatly influence the micromorphology of the chloroplast, pyrenoid and cell wall.

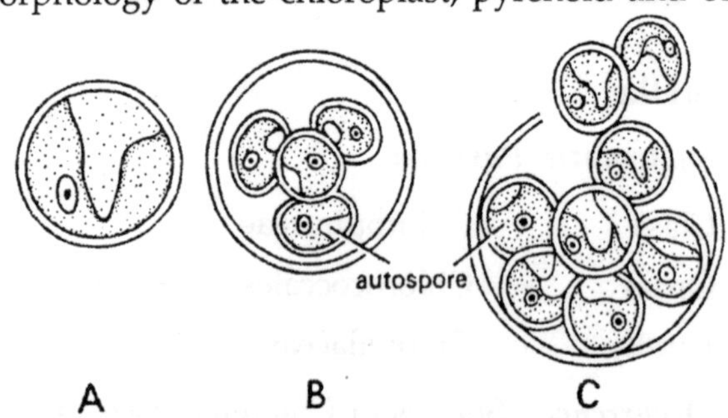

Chlorella : Formation of autospores; A. A vegetative cell, B-C. Cell with autospores.

Reproduction : Chlorella reproduces only asexually by means of autospores (also called aplanospores). Motile spores or gametes are not formed.

During the formation of autospores, the cell grows in size and its contents divide into 2-16 daughter protoplasts. These divisions may occur both in light as well as in dark. Each daughter protoplast secretes a wall and is now exactly similar to the parent cell in its shape. The autospores are liberated by the rupture of the parent cell wall.

The formation of autospores and the factors governing the process have been studied in detail in C. ellipsoidea, and the following four phases have been identified in the life cycle of Chlorella.

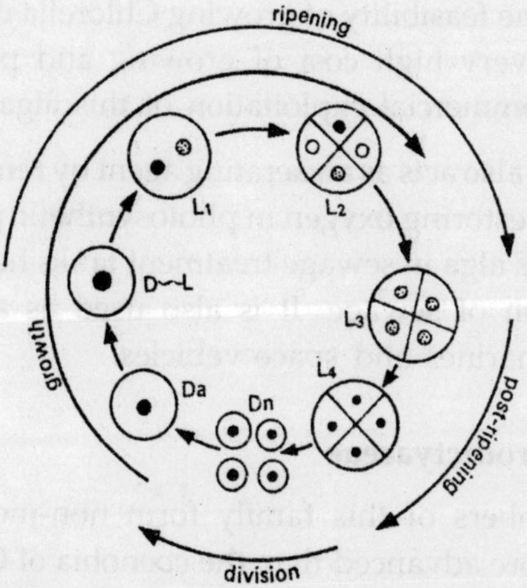

Chlorella ellipsoidea; Four phases in the formation of autospores.

1. *Growth Phase* : During this phase autospores grow in size by utilizing photosynthetic products.
2. *Early Ripening Phase* : In this phase growing cells prepare themselves for divisions.
3. *Post-ripening Phase* : In this phase the cell actually divides twice in dark or in light. Autospores formed in dark are known as dark nascent cells (Dn). These in turn give rise to photosynthetically active uninucleate cells, designated as dark active cells (Da). During transition from dark to light phase, cells grow in size.
4. *Division Phase* : During this phase the parent cell wall ruptures and autospores are liberated.

Economic Importance of Chlorella : Chlorella grows and multiplies fast in a variety of conditions and, therefore, it is a favourite material for the study of biochemical processes, particularly photosynthesis. It has high nutritive value. It contains proteins (50%), fats (20%) and carbohydrates (20%), besides minerals, vitamins and amino acids. It also contains an antibiotic, chlorellin, effective against many bacteria. Thus several developed countries like Japan, Germany and USA are working on the feasibility of growing Chlorella on commercial scale. But a very high cost of growing and processing has prohibited commercial exploitation of this alga so far.

Chlorella also acts as an aerating agent by removing carbon dioxide and restoring oxygen in photosynthetic process. Thus, growth of this alga in sewage treatment tanks helps in aerobic decomposition of sewage. It is also used as air purifier in nuclear submarines and space vehicles.

Family Hydrodictyaceae

The members of this family form non-motile coenobia which are more advanced than the coenobia of Coelastraceae. They reproduce asexually by zoospores. The zoospores become apposed to each other within the parent cell wall and as such a miniature colony is formed within the parent wall. The sexual reproduction is isogamous and the gametes are motile.

Hydrodictyon

Systematic Position

Class - Chlorophyceae

Order - Chlorococcales

Family - Hydrodictyaceae

Occurrence : Hydrodictyon or 'water net' is a free floating fresh water alga. It grows luxuriantly during late summer and spring season and sometimes almost completely covers the pond. The genus is represented by five species. Hydrodictyon indicum and H. reticulatum are common Indian species which occur in temporary or permanent water reservoirs.

Thallus Structure : Hydrodictyon is a macroscopic coenobial alga forming free floating 'water nets' in fresh water reservoirs. The mature coenobium is sausage-shaped and hollow in the centre, usually measuring 20-80 cm in length. The coenobium consists of few hundred to several thousand cells. The long cylindrical and pencil-shaped or ovoid cells are joined at the ends forming a design of hexagons and pentagons.

All the cells of a colony have the same general structure. Each cell is surrounded by a firm cell wall of cellulose. Inner to the cell wall, there is a thin layer of cytoplasm which encloses a large central vacuole. The young cell is uninucleate with a parietal chloroplast, but at maturity it becomes multinucleate and has a reticulate chloroplast. There are several pyrenoids in a mature cell.

The number of cells in a coenobium is fixed when it is young. Further growth of the coenobium is entirely due to the increase in cell size and not the number of cells.

Reproduction **:** Hydrodictyon reproduces mainly by asexual and sexual methods, but vegetative multiplication may also take place by fragmentation of the coenobium into small segments.

Asexual Reproduction **:** Asexual reproduction takes place by means of small, uninucleate and biflagellate zoospores (or swarmers). All the cells of a coenobium are capable of forming zoospores which eventually develop into new coenobia. The multiplication is very rapid. Before the formation of zoospores, pyrenoids disappear in the cell and starch grains are accumulated in the chloroplast. The multinucleate protoplast divides into 7,000-20,000 uninucleate segments (Fig. A, B). Each segment gets surrounded by a small amount of cytoplasm and a limiting membrane. It eventually develops into a biflagellate zoospore. Due to sudden contraction zoospores form a dense column in the centre of the mother cell.

The real swarming starts once the zoospores tear loose from the central column; then they swim freely and haphazardly inside the mother cell (Fig. C). The wall of the mother cell, however, does not rupture and the zoospores remain inside the mother cell. They buzz about rapidly within the mother cell. After a period of rapid movement, the zoospores loose their flagella and settle down on the inner side of the cell wall. They arrange themselves into characteristic hexagonal or pentagonal fashion to form a new net (Fig. D).

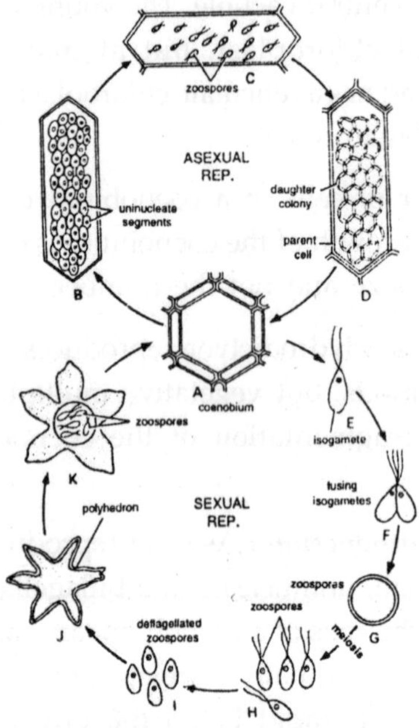

A-K. Hydrodictyon : A-D. Stages in asexual reproduction, E-K. Stages in sexual reproduction.

Although swarmers move quite haphazardly without any definite direction, the position of cells at the end of their movement decides the configuration of the future net.

The young coenobium is liberated by softening of the wall of the mother cell. In a young coenobium the cells grow mainly along their long axis.

Sexual Reproduction : Sexual reproduction is isogamous with the exception of H. africanum which shows anisogamy.

Any vegetative cell of the mature coenobium is capable to form gametes. The protoplast of the cell undergoes repeated cleavage and eventually 30,000-100,000 segments are formed. Each segment metamorphoses into a motile biflagellate gamete. The gametes are smaller than the zoospores (Fig. E). They are liberated from the parent cell through a pore in its wall.

Species of Hydrodictyon are monoecious (homothallic). Two isogametes from the same coenobium fuse laterally to form a zygospore. The gametes loose their motility after fusion and secrete a thick wall (Fig. F, G). The zygospore has the capability to tide over low winter temperature. At the onset of spring season, the diploid nucleus of zygospore divides meiotically into four (or rarely eight) nuclei. This is followed by cleavage of protoplast into as many segments as the number of nuclei. Each uninucleate segment develops into a biflagellate zoospore (Fig. H). The zoospore is slow moving and has many discoid chloroplasts. After a short period of motility, the zoospore looses its flagella and comes to rest (Fig. I). It secretes a thick spiny wall and develops into a polyhedral cell, called polyhedron (Fig. J). The haploid nucleus of a polyhedron divides to form 200-300 zoospores (Fig. K). These zoospores eventually arrange themselves into a new daughter net within the polyhedron. The young net is liberated by the rupture of polyhedron wall.

Since the net of Hydrodictyon is formed by joining together of a large number of zoospores (each of which acts as an individual cell of the net), some investigators consider it as a collection of a number of individual plants.

Family Coelastraceae

These algae form two, four, or eight celled coenobia. The cells are uninucleate, variable in shape and specifically arranged with respect to one another. The members of the family reproduce only by means of autospores.

Scenedesmus

Systematic Position

Class	-	Chlorophyceae
Order	-	Chlorococcales
Family	-	Coelastraceae

Occurrence : Scenedesmus, a common freshwater alga has more than 100 species, abundant in ponds and lakes. It occurs as an almost pure culture in aquaria and in jars where water is left standing for sometime. Scenedesmus obliquus, S. dimorphus, S. quadricauda and S. denticvlatus are some common species of the genus.

Thallus Structure : The thallus of Scenedesmus is 4-,8- or sometime 16- celled coenobium. The cells are attached to each other by mucilage pads. The morphology of the coenobium can be modified considerably by varying the composition of the culture medium. In a medium with low phosphorus or low salt concentration, Scenedesmus can be induced to grow as unicells. The shape of the cells varies from fusiform (e.g., S.obliquus and S. dimorphus) to ellipsoidal (e.g., S. quadricauda and S. denticulatus, Fig. C, D).

The cell wall is smooth (e.g., S. bijuga,) or have processes that are elaborations of the cell envelope and are of taxonomic importance. For example, in S. quadricauda and S. armatus, the terminal cell bears long spines; in S. denticulatus and S. hystrix small spiny projections are present on the entire cell surface and in S. acuminatus, the cell has a tuft of bristles at each end. The presence of these processes is probably related to the planktonic mode of life and they help in floatation of the colony.

Each cell of the coenobium is uninucleate and possesses a single laminate chloroplast, containing one pyrenoid. The chloroplast occupies the entire cell cavity in a mature cell.

Reproduction : Reproduction in Scenedesmus is accomplished by cell division. Each cell of the coenobium is capable of giving rise to a new daughter colony, but all the cells of a colony rarely divide simultaneously. The protoplast undergoes successive transverse and vertical divisions forming four daughter protoplasts. Further divisions may also occur to give rise to eight or rarely sixteen daughter protoplasts. The protoplasts of the last generation give rise to autospores. Within the parent cell wall, the autospores undergo necessary adjustments according to the arrangement of cells in the coenobium. Fully developed coenobium is liberated by longitudinal rupture of the parent cell wall (Fig. A, F).

Sexual reproduction is known to occur in S. obliquus, where it takes place by the formation of biflagellate gametes.

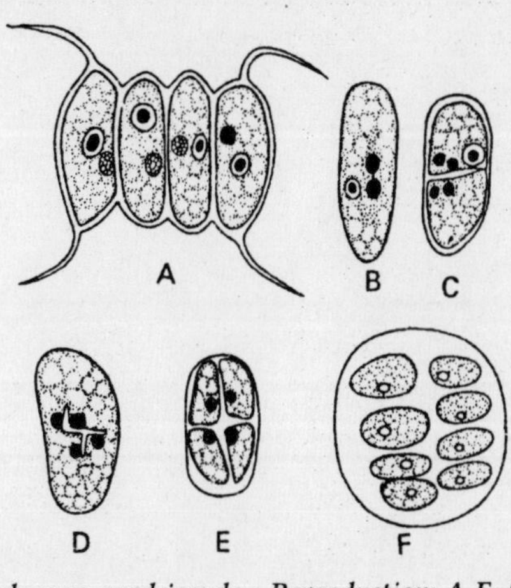

A-F. Scenedesmus quadricauda : Reproduction; A-E. Successive stages of cell division, F. A full developed coenobium.

16

The Volvocales

The order Volvocales includes about 50 genera and 350 species. They mostly grow in freshwater and are abundant in waters rich in nitrogenous compounds. They are characterised by the presence of flagellated motile vegetative cells. There are usually two or sometimes four, flagella of equal length. The members are unicellular (e.g., Chlamydomonas, Dunaliella) or multicellular ; in multicellular colonial forms the number of vegetative cells is in the multiple of two (e.g., 4-32 biflagellate cells in Pandorina, 4 or 16 in Gonium, 16-64 in Eudorina, 500-60,000 in Volvox). All the cells of a colony are surrounded by a common gelatinous sheath, but they are functionally independent. The cells are uninucleate and the nucleus divides only during reproduction. Each cell has a massive cup-shaped chloroplast with usually a single pyrenoid.

They reproduce both asexually and sexually. The asexual reproduction in the unicellular forms takes place by the division of protoplast into two or more segments, each forming a zoospore. In the colonial forms, cells undergo repeated divisions to form daughter colonies. The sexual reproduction ranges from isogamous to oogamous. In the isogamous and anisogamous forms both the fusing gametes swim freely in water, but in the oogamous forms the egg is located in the oogonium.

Fritsch (1935) divided this order into three sub-orders.

Order—Volvocales

Sub- orders

Chlamydomonadineae	Tetrasporineae	Chlorodendrineae
Families		
1. Chlamydomonadaceae	1. Tetrasporaceae	1. Chlorodendraceae
2. Sphaerellaceae	2. Palmellaceae	
3. Polyblepharidaceae		
4. Phacotaceae		

Family Chlamydomonadaceae : This family includes unicellular forms of the Volvocales. The cells are uninucleate with a definite cell wall. Each cell has two or four flagella of equal length and a cup-shaped chloroplast. The asexual reproduction is by zoospores and the sexual reproduction is from isogamous to oogamous.

Chlamydomonas, Chloromonas, Cymbomonas, Platymonas, and Carteria are common examples of this family. Chlamydomonas, a typical representative of the family, is described in details.

Chlamydomonas

Systematic Position

Class - Chlorophyceae

Order - Volvocales

Family - Chlamydomonadaceae

Occurrence : Chlamydomonas is one of the simplest unicellular fresh-water alga. It has about 500 species, occurring in fresh water ponds, lakes and tanks, which are rich in nitrogenous matter. It provides green turbidity to the ponds and lakes.

There are also some terrestrial species which are found along the banks of lakes and pools on the moist soil. Some species of Chlamydomonas occur in unusual habitats, e.g., C. yellowstonensis grows on the snow covered mountains of Yellow Stone National Park of the U.S.A., imparting them yellowish green colour, and C. nivalis, which is found on snow in alpine and arctic regions, imparts red colour to the snow due to the presence of a red coloured pigment (haematochrome). C. halophila and C. ehrenbergii occur in saline water.

In India, the genus is represented by about 18 species, of these C. moewusii, C. monadina, C. subcaudata, C grandistigma and C. iyengarii are common.

Thallus Structure : Chlamydomonas is a unicellular motile alga. The cell is biflagellate, spherical, ellipsoidal or pear-shaped; about 30 μ m in length and 20 μ m in diameter. The anterior end of the cell is usually papillate and the posterior relatively broader but in C. caudata the posterior end is pointed.

The cell is bounded by a thin but firm cellulose wall. Glycoproteins rich in hydroxyproline are the main structural component of the cell wall. In some species, the cellulose wall is surrounded by a distinct gelatinous sheath. The protoplast is surrounded by a thin semipermeable plasma membrane. Most of the species of Chlamydomonas have an massive cup-shaped parietal chloroplast.

It usually occupies almost entire posterior broader part of the cell. However, the chloroplast is 'H' -shaped in C. biciliata, discoid in C. alpina, reticulate in C. reticulata, ridged in C. stenji and axile in C. eradians. The chloroplast has a single pyrenoid, lying in its posterior part along the median line. But in C. kleinii, C. debaryanum and C. sphagnicola there are more than two pyrenoids in each chloroplast.

In C. reticulata the chloroplast is, however, without pyrenoid. The pyrenoid has a central protein core surrounded by starch plates, and it functions as the centre for starch synthesis.

The cup-shaped structure, formed by the chloroplast in the anterior part of the cell is filled with cytoplasm. The nucleus lies in the centre of the cytoplasm. In those species where chloroplast is other than cup-shaped, the nucleus is axial or eccentric. At the anterior pointed end of the cell there are two contractile vacuoles which are concerned with the osmoregulation of the cell. Each cell has an eye spot or stigma, which lies near the anterior end. The shape and position of the eye spot varies in different species. In C. pluristigma there are 2-3 eye spots in each cell. The eye spot has a colourless biconvex photosensitive lens and a curved pigmented plate. It is a photoreceptive organ and functions as a primitive eye.

The anterior papillate part of the cell bears two flagella which are equal in length. The flagella are of whiplash type and are associated with neuromotor apparatus of the cell. The neuromotor apparatus consists of three main parts : (i) blepharoplast, a small granular structure present at the base of each flagella; (ii) paradesmos, a transverse filament connecting the two blepharoplasts, and (iii) rhizoplast, a thin and soft descending thread connecting one of the blepharoplasts with the centrosome of the nucleus. The centrosome remains connected with the nucleolus by many fine fibrils.

Fine Structure of Chlamydomonas Cell : Ultrastructure of Chlamydomonas has been studied by Roberts et al. (1972) and Hill et al. (1973). Under electron microscope the cell wall appears multilayered, made up of cellulose fibrils. The plasma membrane, which lies inner to the cell wall, is differentiated into an outer and an inner proteinaceous layer and a middle lipid layer. There are many ribosomes, mitochondria and endoplasmic reticulum in the cytoplasm, present between the plasma membrane and chloroplast. The cytoplasm, which fills the cup-shaped cavity of the chloroplast, contains a nucleus, volutin granules and golgi apparatus, besides other organelles. The two layered nuclear membrane has many pores. Although a distinct nucleolus is present in the nucleus, chromosomes are not visible.

The cavity of the chloroplast, bounded by a double layered limiting membrane, is called stroma. It is filled with a homogeneous granular matrix. The stroma has many stacks of tubular photosynthetic lamellae. The lamellae are lipoproteinaceous in nature and are called thylakoids. Each stack has 2-6 thylakoids. The thylakoids are always oriented in the direction of the long axis of the cell. The pyrenoid is also a lamellate structure, and starch plates are synthesized on its periphery. The granular matrix of the chloroplast also contains many ribosomes, but they are smaller than those present in the cytoplasm. Ribosomes are absent in the pyrenoid.

Electron microscopic studies reveal that the eye spot consists of two or three parallel rows of fat globules (Fig. B). They are laden with carotenoids and lack limiting membrane.

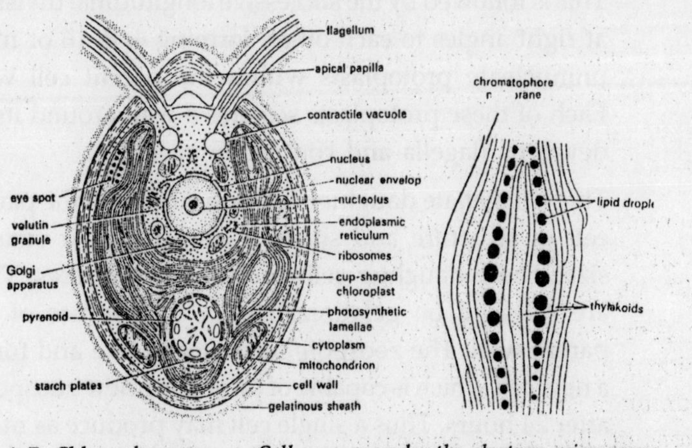

A-B. Chlamydomonas : a. Cell as seen under the electron microscope, B. Eye spot (note the presence of thylakoids).

Nutrition : Most species of Chlamydomonas are phototrophs, i.e., they synthesize their own food in the presence of chloroplast, light, water and carbon dioxide. A part of the sugar synthesized is utilized in the metabolic processes of the cell and the rest is stored around the pyrenoid in the form of starch grains. However, C. dysosmos is a facultative heterotroph and can grow in dark in the presence of acetate as a carbon source.

Reproduction : Chlamydomonas reproduces by asexual and sexual methods.

Asexual Reproduction : Asexual reproduction takes place mainly by zoospores. But some other types of asexual spores, such as aplanospores, hypnospores and palmella stage are also found.

1. *By Zoospores :* Zoospores are formed under favourable conditions, usually during the night. At the time of zoospore formation, the cell withdraws its flagella and becomes non-motile. The contractile vacuoles disappear, and the protoplast withdraws from the cell wall and divides longitudinally into two daughter protoplasts. This is followed by the successive longitudinal divisions at right angles to each other, forming 4, 8, 16 or more uninucleate protoplasts within the parent cell wall. Each of these protoplasts secretes a wall around itself, develops flagella and contractile vacuoles.

   These flagellate daughter cells are similar to the parent cell in structure and shape, except for their smaller size. These daughter cells are called zoospores. They are released on gelatinization or by rupture of the parent wall. The zoospore increases in size and forms a new cell which is capable of producing new zoospores after 24 hours. Thus a single cell may produce as many as 200,000 daughter cells within a week.

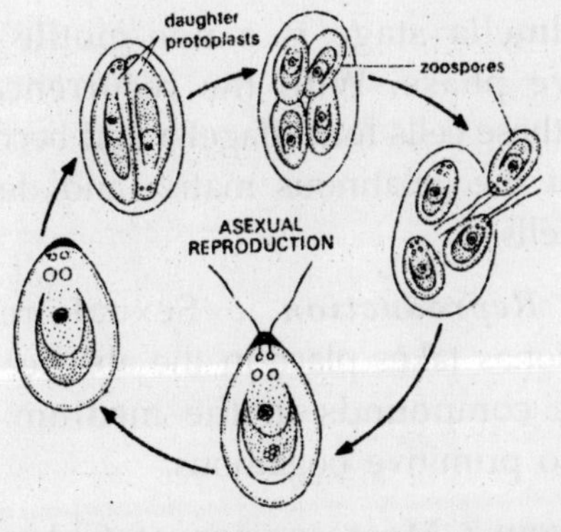

Chlamydomonas* : *Asexual reproduction; Formation of zoospores.

Sometimes zoospores fail to escape from the gelatinized parent wall and this stage is known as palmella stage.

2. *By Aplanospores :* These spores are formed during unfavourable conditions. Under conditions of drought, the cells withdraw flagella and their protoplasts divide into 2-16 daughter protoplasts. Each protoplast secretes a thin wall but does not develop flagella. These non-motile cells are called aplanospores. In some species (e.g., C. nivalis), aplanospores secrete a thick wall under conditions of severe drought. These thick walled non-motile spores are called hypnospores. They germinate under favourable conditions and form new cells directly or their protoplasts divide to form zoospores.

3. *Palmella Stage :* Under adverse conditions (but not very dry) the protoplast of the parent cell divides to form 4-8 daughter cells. These cells, however, do not develop flagella and are non-motile and remain within the matrix formed by the gelatinization of the parent wall. This temporary colony of hundreds or thousands of cells, in a common gelatinous matrix, is called palmella stage. In C. nivalis, all the cells of the palmella stage secrete a thick wall and form hypnospores. The wall of the hypnospore is provided with haematochrome, a red. pigment, imparting red colour to the hypnospore.

The palmella stage is a non-motile and temporary reproductive phase. With the recurrence of favourable conditions, these cells form flagella and become motile. They come out of the gelatinous matrix and develop into large vegetative cells.

Sexual Reproduction : Sexual reproduction in Chlamydomonas takes place in the absence or deficiency of nitrogenous compounds in the medium. It varies from isogamous to primitive oogamous.

1. *Isogamy* : Most species of Chlamydomonas are isogamous. The sexual process starts with the division of the cell protoplast into 8, 16, 32 or 64 biflagellate gametes. The gametes are usually' naked, without any cell wall. In many isogamous species the vegetative cells directly function as gametes without undergoing any division (hologamy). The isogamous species are characterised by structural similarities between the fusing gametes. In homothallic species (e.g., C. debaryanum, C. longistigma) the gametes produced in the same parent cell fuse to form the zygote (Fig.), whereas in heterothallic species (e.g., C. moewusii, C. reinhardii) gametes from the cells of two different strains fuse.

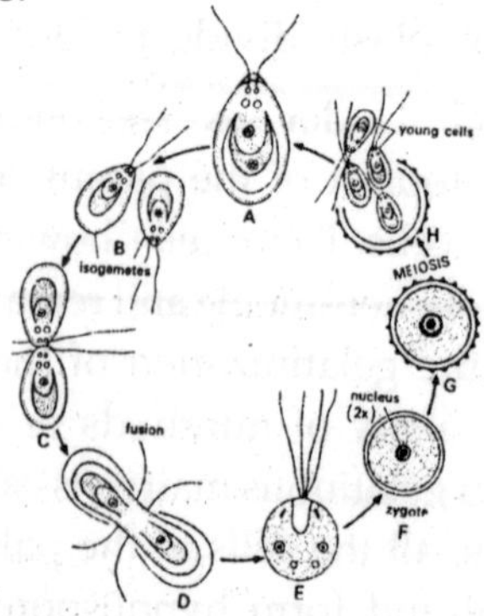

A-H. Chlamydomonas longistigma : Isogamy (homothallic); A. Vegetative cell, B. Isogametes, C-E. Fusion of isogametes, F-G. Zygote, H. Germination of zygote.

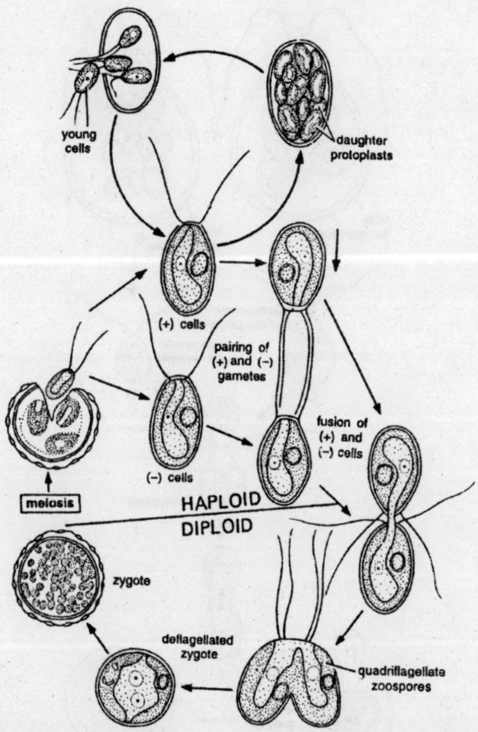

***Chlamydomonas moewusii* : Isogamy (heterothallic); life-cycle.**

The flagella of gametes are covered by agglutins, the chemical substances involved in the recognition of gametes of the opposite strains. These substances are not present in the flagella of the vegetative cells. In heterothallic species when gametes of 'plus' strain come in contact with those of 'minus' strain, the flagella of two gametes of opposite strains adhere because of agglutins. Initially, the gametes clump in groups of up to 50 with their flagella towards the center.

In each clump the number of plus and minus gametes is variable. Eventually, gametes of opposite strains fuse in pairs at their anterior ends. The flagella become free and the paired gametes swim away from the clump.

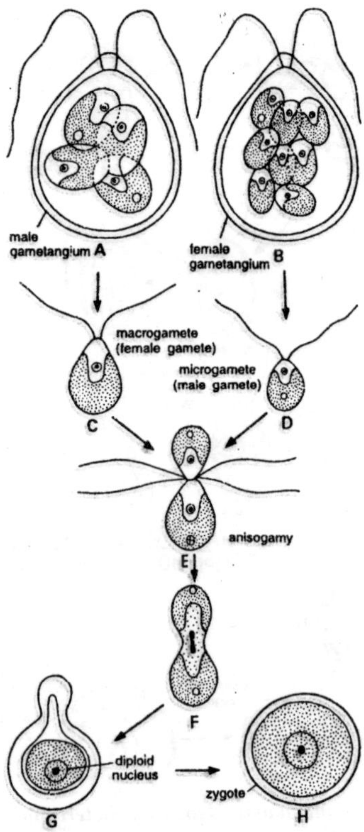

A-H. Chlamydomonas braunii : Anisogamy; A. Female gametangium, B. Male gametangium, C. female gamete, D. Male gamete, E-F. fusion of gametes, G-H. Zygote.

2. *Anisogamy* : C. braunii and C. suboogama are the typical examples of anisogamy. Here the fusing gametes (male and female) are of different sizes. In the female gametangium only two or four macrogametes (female gametes) are formed which are similar to the mother cell in shape and size (Fig. A, C). In the male gametangium 8-16 microgametes (male gametes) are formed which are nearly half the size of the mother cell and are much narrower and elongated (Fig. B, D). The microgametes are more active than the macrogametes, and the latter come to rest sooner than microgametes.

The active male gamete comes close to the female gamete and both fuse at their anterior ends to form a zygote (Fig.E-H).

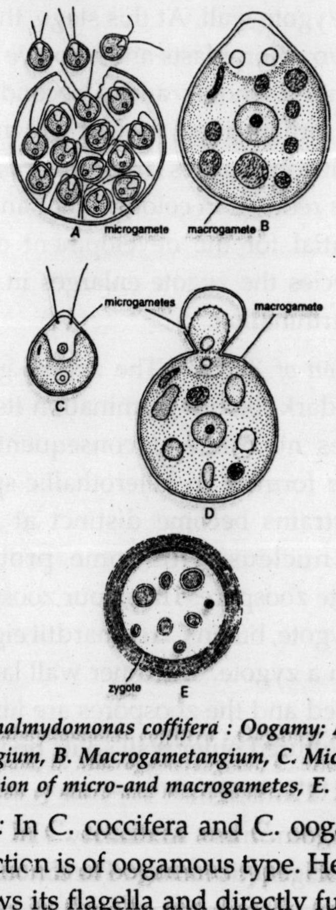

A-E. Chalmydomonas coffifera : Oogamy; A. Micro gametangium, B. Macrogametangium, C. Microgamete, D. Fusion of micro-and macrogametes, E. Zygote.

3. *Oogamy* : In C. coccifera and C. ooganum, the sexual reproduction is of oogamous type. Here the female cell withdraws its flagella and directly functions as a non-motile macrogamete (it is equivalent to egg or ovum). The protoplast of the male cell divides repeatedly to form 16 segments (Fig. A). Each segment develops into a biflagellate micro or male gamete. The flagellated male gamete is active and comes close to non-motile female gamete. The two fuse at their anterior ends to form a zygote (Fig. C, H).

4. *Zygote :* The wall at the point of fusion between the two gametes soon dissolves. The quadriflagellate zygote swims for a while before it settles down. It secretes a primary zygote wall. At this stage, the zygote has two nuclei, two chloroplasts and two eye spots. The nuclei fuse within 24 hours and a secondary zygote wall, which is ornamented, is laid down. As the zygote matures, it accumulates large amounts of oils and starch and turns reddish in colour. Light and carbon dioxide are essential for the development of the zygote. In some species the zygote enlarges in size enormously before germination.

5. *Germination of Zygote :* The zygote germinates inside water in dark. Before germination its diploid nucleus undergoes meiosis and consequently four haploid nuclei are formed. In heterothallic species 'plus' and 'minus' strains become distinct at this stage. Each haploid nucleus with some protoplast forms a biflagellate zoospore. Thus, four zoospores are formed in each zygote, but in C. reinhardtii eight zoospores are formed in a zygote. The inner wall layer of the zygote is dissolved and the zoospores are liberated when the outer sculptured outer wall splits open. Each zoospore develops into a new cell.

In Chlamydomonas sexuality is controlled by environmental and genetic factors. The amount of ammonium nitrogen in the growing medium is the most important. Its deficiency in the medium inhibits sexuality in Chlamydomonas and the gametes behave simply as vegetative cells.

Besides this, the intensity of light and temperature, and the presence of calcium in the medium also affects mating of gametes. In high light intensity larger number of zygotes are formed than in low light intensity.

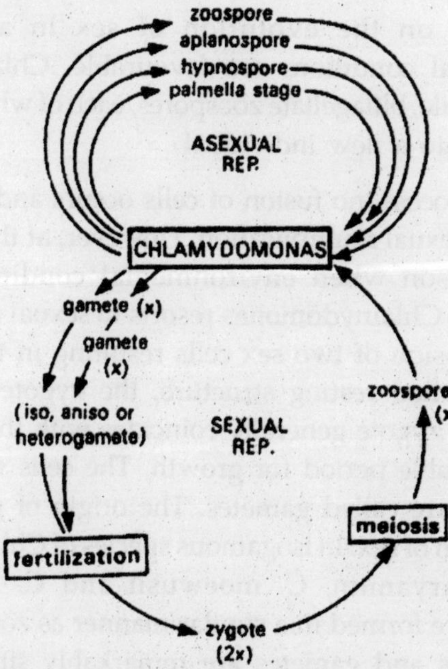

Chlamydomonas : Diagrammatic representation of life-cycle.

The fusion occurs between the gametes of opposite strains (+ and -) in heterothallic species. It has been observed that non-flagellate mutant forms of Chlamydomonas are completely asexual. This suggests that flagella are important in mating processes. The presence of agglutin coating on the surface of flagella probably is essential for sexual processes.

Alternation of Generation

The haploid phase is dominant in the life-cycle of Chlamydomonas. The vegetative cell is haploid and it produces haploid gametes. The diploid phase is attained in the zygote by gametic fusion. The diploid nucleus of the zygote divides meiotically before it germinates. Thus the diploid phase occurs for a short duration.

Origin and Differentiation of Sex : The various reproductive processes exhibited by different species of Chlamydomonas throw light on the evolution of sex in algae. When environmental conditions are favourable, Chlamydomonas produces motile, biflagellate zoospores, each of which is capable to develop into a new individual.

In this process, no fusion of cells occurs and hence it is a method of asexual multiplication. However, at the close of the growing season when environmental conditions become unfavourable Chlamydomonas resorts to sexual reproduction. It involves fusion of two sex cells resulting in the formation of a thick-walled resting structure, the zygote. The resting period of the zygote generally coincides with the duration of the unfavourable period for growth. The cells which fuse to form zygote are called gametes. The origin of gametes is as such tne origin of sex. In isogamous species of Chlamydomonas like C. debaryanum, C. moewusii and C. longistigma, isogametes are formed in a similar manner as zoospores. Both the zoospores and gametes are remarkably similar in their form, structure, development and mode of liberation. They differ only in their size and subsequent behaviour. The gametes are invariably smaller in size and are incapable of growing into a new individual alone.

The marked resemblances between the gametes and zoospores suggest that isogametes have originated by the subdivision of swarmers (zoospores) into still smaller swarmers in response to unfavourable environmental conditions. These smaller swarmers (reduced zoospores) have become too small and weak to grow into new plants by themselves due to insufficiency of stored food and perhaps some other factors. Most of these perish in their wanderings, but a few meet by chance and fuse in pairs. This pairing increases their vitality and vigour to give rise to a new individual. This clearly illustrates that gametes are just reduced zoospores which are incapable to develop into new individuals. These reduced zoospores transformed into gametes and their union brings about sexual reproduction: The sex in Chlamydomonas is thus originated by accidental fusion of undersized zoospores in response to unfavourable environmental conditions. Since the process proved to be advantageous, it was maintained.

During further advancement gametes became of dissimilar size as in C. braunii. This was a step towards anisogamy. The next step in the evolution of sex is demonstrated by C. coccifera. Here, a vegetative cell instead of dividing enlarges and loses its flagella and motility. This cell acts as an ovum (egg cell), and fertilized by an active male gamete (antherozoid) formed in a manner as zoospores. Of these two gametes, one (the ovum) becomes sluggish and stores food material to support the new plant until it is established on its own, and the other (the male gamete or antherozoid) attains motility to reach the immotile ovum for fertilization. Thus, C. coccifera provides an example of the most advanced level of sexuality that can be traced in Chlamydomonas.

Family Volvocaceae

This family includes those flagellated forms of Volvocales which form colonies. The number of cells in a colony is in multiple of two, and each genus has a definite number of cells. Such colonies with definite number of cells are called coenobia (singular = coenobium). The shape of the coenobium and the arrangement of cells are specific for a genus. The members of Volvocaceae are fresh water forms. They reproduce asexually as well as sexually. The sexual reproduction is isogamous, anisogamous or oogamous.

Volvox

Systematic Position

Class - Chlorophyceae

Order - Volvocales

Family - Volvocaceae

Occurrence : Volvox is the most highly evolved coenobial alga, both in its morphology and reproduction. It has about 20 species which occur in fresh waters of temporary or permanent pools as minute spherical or ovoid balls of pin-head size. The colonies multiply so rapidly during rainy season that the entire pond is covered by them and looks greenish. The genus is represented in India by five species; of these V. globator, V. aureus, V. prolificus and V africanus are fairly common.

Thallus Structure : Volvox is a colonial alga, and the colonies are oval to spherical in shape, each with 500-60,000 cells. The cells are biflagellate and are arranged in a single layer within the periphery of the gelatinous colonial envelope (Fig. A). Each cell of the colony has its own gelatinous sheath and the sheaths of the individual cells may be distinct or confluent. In most of the species adjacent cells are connected by cytoplasmic strands which are formed during cell division. In some species (e.g., V. mononae, V. tertius) cytoplasmic strands are absent. The central hollow cavity of the colony is filled with a watery (e.g., V. globator) or gelatinous (e.g., V. aureus) substance.

The cells in a colony are usually pyriform in shape with broad posterior and narrow anterior end. They are eukaryotic and mostly show a typical chlamydomonad structure. But in V. globator and V. rousseletii it is of Sphaerella type. The proloplast is enclosed within a semipermeable plasma membrane, which forms the outer boundary of the cytoplasm. There is a single nucleus, a cup-shaped chloroplast, two contractile vacuoles, two flagella and an eye spot in each cell.

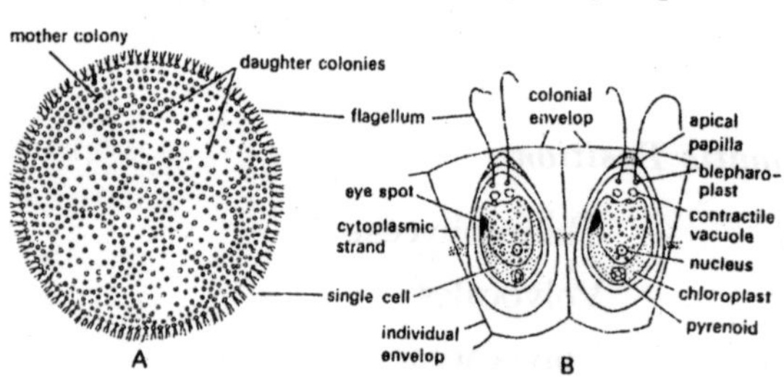

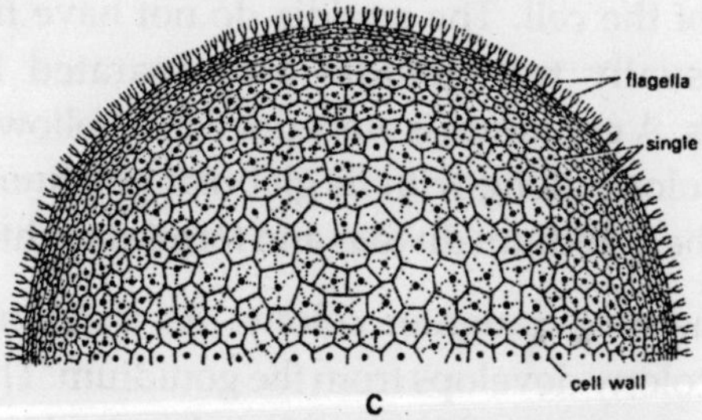

A-C. Volvax : A. A coenobium, B. A part of coenobium, C.A part of coenobium

The nucleus is connected with the flagella through the neuromotor apparatus of the cell. The eye spot is towards the external face of the cell. It has a colourless biconvex photosensitive lens and a curved pigmented plate. The massive cup-shaped parietal chloroplast occupies the posterior broader part of the cell. It has a single pyrenoid made up of a central protein core and peripheral starch plates.

Most cells of a colony are vegetative and only a few are reproductive. The reproductive cells do not have flagella and are relatively large than the vegetative cells. There are only a few or no reproductive cells at the anterior end of the colony, whereas the posterior end has many reproductive cells. The cells at the anterior end of the colony have relatively large eye spots than those at the posterior end. Thus, the colony of Volvox shows polarity.

Reproduction : Volvox reproduces by asexual and sexual methods. Asexual reproduction usually takes place during spring and summer months, whereas sexual reproduction starts towards the end of the summer season.

Asexual Reproduction : Asexual reproduction takes place by special type of cells situated at the posterior end of the colony. These cells, called gonidia or parthenogonidia, are about ten times larger than the vegetative cells. They are spherical and highly vacuolated. They develop numerous pyrenoids within their chloroplast. The nucleus is suspended in the center of the cell. The gonidia do not have flagella and eye spots. Usually two gonidia are separated by several vegetative cells. A gonidium is embedded in a hollow gelatinous cavity of the colony along with its globular gelatinous sac. The protoplast of the gonidium divides and forms daughter colonies.

Development of daughter colony. As mentioned above, the daughter colony develops from the gonidium. The enlarged and spherical gonidium undergoes three longitudinal divisions (with respect to the colony) forming eight cells. These cells arrange themselves in a curved plate and this stage is called plakea stage.

Each of these eight cells again divides longitudinally forming 16 cells. At this stage the cells form an almost complete sphere except for a small opening, the phialopore, directed towards the periphery of the colony. The phialopore eventually becomes the posterior pole of the daughter colony. These 16 cells divide repeatedly and may form as many as 60,000 cells.

However, such large number of divisions usually do not occur and divisions stop after a fixed number of cells have been formed. At this stage the cells are naked and are in close contact with each other. The anterior papillate part of each cell is directed towards the center of the hollow sphere. The normal orientation of the cells of a colony is achieved by inversion of cells of this sphere.

The inversion starts with the formation of a constriction at the posterior end of the sphere (i.e., opposite the phialopore) which pushes the cells of the posterior end towards the phialopore.

In the mean time the phialopore becomes broader and the cells of the posterior part are pushed through it. This process is known as inversion. At the end of inversion, the anterior papillate ends of the cells face towards the periphery of the colony (Fig.A, I).

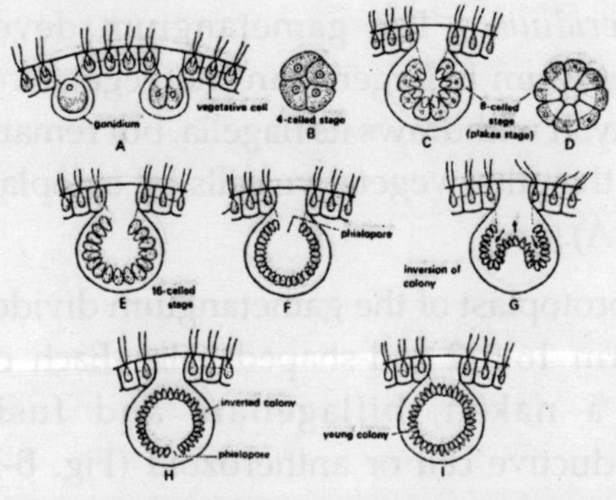

A-I. Volvox: Asexual reproduction (development of daughter colony); A-D. Formation of plakea stage, E. 16-celled stage, F-H. Stages of inversion, I. Daughter colony.

After inversion of the colony is completed the cells secrete gelatinous cell wall and develop flagella. In this way a daughter colony is formed inside each gonidium. Many daughter colonies may develop within a parent colony at the same time and they swim freely in the gelatinous matrix of the parent colony. The daughter colonies are released after the rupture or disintegration of the parent colony. The parent colony usually degenerates after the release of daughter colonies, but in V. africanus and V. carteri two to four generations of daughter colonies can be found within the parent colony.

Sexual Reproduction : Sexual reproduction in Volvox usually takes place towards the end of the growing season. It is of oogamous type. The species of Volvox are either monoecious (e.g., V. globator) or dioecious (e.g., V. aureus). Sometimes colonies of V. aureus have both antheridia and oogonia. The monoecious species are protandrous, i.e., antheridia mature before oogonia. Usually a colony involved in sexual reproduction does not have asexual daughter colonies. The cells developing into reproductive structures are located in the posterior part of the colony, and these cells are called gametangia. Usually the gametangia are in limited number. The male and female gametangia are called antheridia and oogonia respectively.

1. *Antheridium* : The gametangium developing into antheridium is larger than the vegetative cells of the colony. It withdraws its flagella, but remains connected with the other vegetative cells by cytoplasmic strands (Fig. A).

   The protoplast of the gametangium divides repeatedly to form 16-512 rod-shaped cells. Each cell develops into a naked, biflagellate and fusiform male reproductive cell or antherozoid (Fig. B-E).

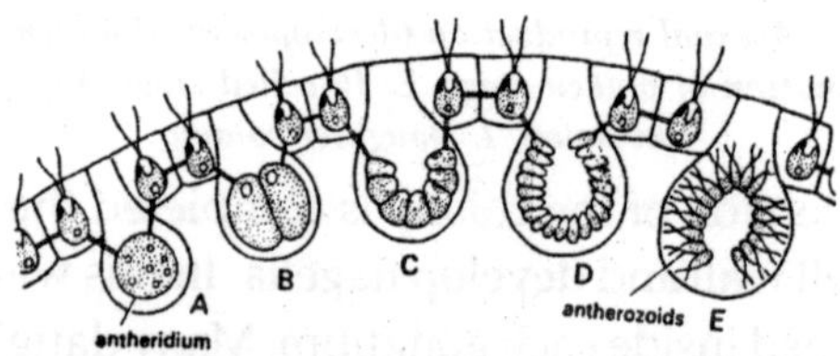

A-E. Volvox :Development of antherozoids.

   In some species, all antherozoids in an antheridium form a bundle and are also released in bundle form. In other species (usually those in which the protoplast of antheridium divides to form more than 128 antherozoids), the antherozoids are arranged in a hollow sphere which shows inversion like asexual daughter colonies. Here the antherozoids are released individually.

2. *Oogonium* : The gametangium developing into oogonium is also larger than the other cells of the colony. It increases in size and eventually becomes flask-shaped. The entire protoplast of the oogonium finally forms a single uninucleate, non-flagellate and spherical egg.

   The egg has a parietal chloroplast and as such is green in colour. It is laden with reserve food material. The beak of the flask-shaped oogonium opens at the surface of the colony (Fig. C, D).

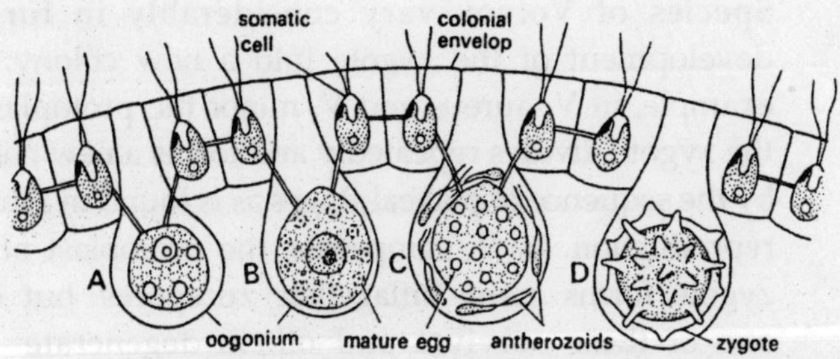

A-D. Volvox : A-B. Development of oogonium, C. Fertilization, D. Zygote.

3. *Fertilization :* The antherozoids, after liberation from the colony, swim in the medium. If they come in contact with a vegetative colony, they simply swim over, but when they come in contact with a female colony, they get attached to its surface and break down the colonial matrix of the colony in the area of their attachment. This task is probably accomplished by a proteolytic enzyme secreted by the antherozoid bundle. When antherozoids are in a bundle, the bundle breaks down and the antherozoids are liberated in the gelatinous matrix of the hollow female colony. One antherozoid fuses with each egg to form the zygote (Fig.C, D).

   The zygote secretes a three layered thick wall which may be smooth (e.g, V. globator) or spiny. The wall contains haematochrome which imparts red colour to the zygote. The zygote may remain dormant for quite some time and is then liberated by disintegration of parent colony.

4. *Germination of Zygote :* The zygote usually does not germinate until the following spring. Before germination, the diploid nucleus of the zygote undergoes a meiotic division forming four haploid cells (Fig. A, C). These cells undergo further divisions. The outer layer of the zygote splits and the two inner layers protrude in the form of a vesicle. The cells migrate into this vesicle.

Species of Volvox vary considerably in further development of the zygote into a new colony. For example, in V. aureus and V. minor the protoplast of the zygote divides repeatedly and forms a new colony by the sequence of plakeal stages as is found in asexual reproduction. In V. campensis, the protoplast of the zygote forms many biflagellate zoospores, but only one of these survives and others degenerate. The surviving zoospore comes out of the vesicle and swims freely in the medium. It divides to form about 500 cells which arrange themselves into a hollow sphere. The hollow sphere undergoes inversion as in asexual daughter colony and develops into a new colony (Fig. E, F). In V. rousseletii, the entire protoplast of the zygote forms a single biflagellate zoospore which divides repeatedly and forms a new colony after inversion. The colonies developed by this method have relatively less number of cells.

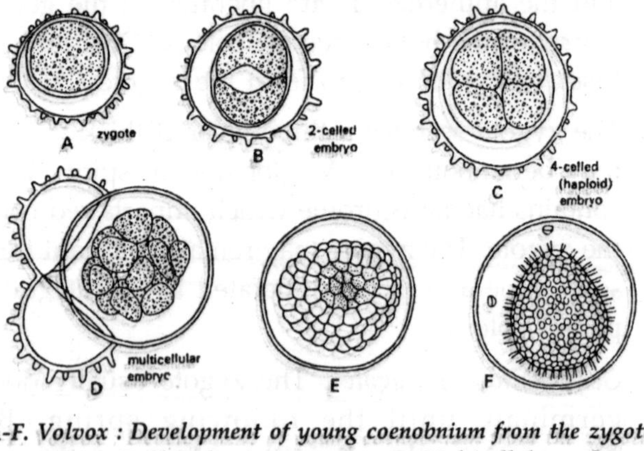

A-F. Volvox : Development of young coenobnium from the zygote; A. Zugpte. B-C/ Reduction in zygote, D. Multicellular embryo, E. Inversion, F. Young coenobium.

Alternation of Generation

The life cycle of Volvox is of haplontic type; the zygote being the only diploid stage in the life-cycle. Its nucleus divides meiotically before the formation of zoospores which intimately develop into Volvox colony with haploid cells.

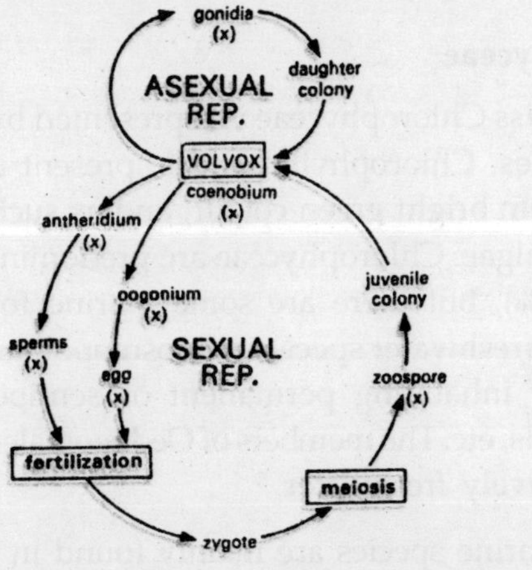

Volvox : Diagrammatic representation of life-cycle.

Thus, there is an alternation of a haploid phase with diploid zygote.

Functional Specialization

(Division of Labour)

Volvox shows several features of advancement over Chlamydomonas. The primitive unicellular motile Chlamydomonas cell performs all vital functions, both vegetative and reproductive. Volvox is an assemblage of similar and independent cells of chlamydomonad type. Each cell of the colony is independent and carries out its own functions. At the vegetative phase, all cells of a colony are alike. But at the reproductive stage, certain cells of the posterior end of the colony act as asexual (gonidia) or sexual (gametangia) reproductive structures. These cells become distinct by their larger size, conspicuous nucleus and denser contents. The adult colony thus exhibits a degree of polarity and differentiation of cells is accompanied by division of labour. Furthermore, Volvox shows an elaborate type of oogamous sexual reproduction. Thus, it represents the most elaborate type of coenobial form, with considerable advancement.

Chlorophyceae

The class Chlorophyceae is represented by 425 genera and 6500 species. Chlorophyll *a* and *b*, present in their plastids, impart them bright green colour, and as such they are called the green algae. Chlorophyceae are predominantly freshwater (about 90%), but there are some marine forms also (about 10%). The freshwater species are cosmopolitan in distribution, commonly inhabiting permanent or semipermanent pools, lakes, ponds, etc. The members of Oedogoniales and Conjugales are exclusively freshwater.

The marine species are mainly found in the tropical sea; members of Siphonales and Ulvaceae are predominantly marine. Certain forms of Volvocales, Cladophorales and Chaetophorales are found both in the fresh and marine waters. Species of Ulothrix, Vaucheria and Zygogonium are sub-aerial, and are usually found on damp soil in the form of sheets. Species of Protococcus and Trenteopohlia are epiphytic, found on tree trunks, damp woodwork and shady leaves. Many species of Cladophora and Characium are epizoic (growing on animal bodies), whereas a few unicellular forms (e.g., Chlorella) are endophytic. Cephaleuros and Rhodochytrium are parasitic, and they lack green colour. Cephaleuros causes red-rust disease of tea leaves.

Some species occur in unusual habitats; for example, Chlorella has some thermophilic species, found in hot springs, Chlamydomonas yellowstonensis and Haematococcus nivalis occur on the snow covered mountain tops and Chlamydomonas ehrenbergii inhabits in saline water. Some green algae in, association with certain fungi constitute lichens.

They following are important distinguishing features of the green algae .

(1) They exhibit a wide range in their thallus structure and morphology.

(i) Unicellular motile form e.g., Chlamydomonas, Sphaerella.

(ii) Unicellular non-motile form e.g., Chlorella, Chlorococcum.

(iii) Coenobium : e.g., Pandorina, Eudorina, Volvox, Pediastrum, Hydrodictyon.

(iv) Palmelloid : e.g., Tetraspora.

(v) Dendroid : e.g., Ecballocystis.

(vi) Filamentous unbranched e.g., Ulothrix, Spirogyra.

(vii) Filamentous branched e.g., Cladophora.

(viii) Heterotrichous e.g., Fritschiella, Coleochaete, Stigeoclonium, Draparnaldiopsis.

(ix) Siphonaceous e.g., Protosiphon, Vaucheria, Codium.

(x) Parenchymatous e.g., Ulva, Enteromorpha.

(2) The cell structure is eukaryotic.

(3) The cell wall generally has cellulose as the main structural polysaccharide except for siphonaceous forms, which have mannan or xylan. In Chara, the cell wall is impregnated with calcium carbonate.

(4) The protoplast is bounded by a thin and semipermeable plasma membrane. The cytoplasm possesses many small vacuoles or there is a large central contractile vacuole. The diastolic and systolic movements of the contractile vacuole possibly have osmoregulatory function. The large central vacuole in the mature cell pushes the protoplasm along the plasma membrane in the form of a thin layer, called primordial utricle.

(5) Like all other eukaryotes, the cells of the green algae

have mitochondria, golgi bodies, ribosomes, endoplasmic reticulum, membrane bound plastids and other cell organelles.

(6) Their main pigments are chlorophyll *a* and *b*, but α — and β-carotenes and xanthophylls are also present. The pigments are found in the lamellae (thylakoids) of well organized chloroplast.

(7) The shape of the chloroplast shows much variation; it is cup shaped (e.g., Chlamydomonas), girdle shaped (e.g., Ulothrix), reticulate (e.g., Cladophora, Oedogonium), parietal (e.g., Draparnaldiopsis), stellate (e.g., Zygnema), spiral (e.g., Spirogyra) or discoid (e.g., Vaucheria, Chara).

(8) The chloroplast normally contains pyrenoids surrounded by a sheath of starch. In Derbesia, pyrenoids disappear when the alga is grown in diffuse light.

(9) They have well organized nucleus. Usually there is a single nucleus in each cell, but the members of Siphonales and Cladophorales are coenocytic. The position, size and shape of nuclei are quite variable. The nucleoli and chromosomes remain suspended in the granular matrix of the nucleus. In general, there is a single nucleolus per nucleus but in Conjugales, there are several nucleoli in a nucleus.

(10) Most of the flagellate cells have a photosensitive red eye spot or stigma in the anterior portion, near the flagellar base.

(11) Their photosynthetic pathways are similar to those of higher plants. The reserve food is starch, composed of amylose and amylopectin.

(12) The motile vegetative or reproductive cells (zoospores or gametes) have two (rarely four) equal, smooth, whip-like flagella. However, in Oedogoniales the motile cells

have a crown of flagella. In a few instances (e.g., Haematococcus), there are two pleuronematic flagella of equal length. Structurally, when viewed under electron microscope, flagella show a typical 9+2 fibrillar organization.

(13) They reproduce by vegetative, asexual and sexual methods. The vegetative propagation takes place by cell division and fragmentation. Akinetes, zoospores and aplanospores are the main asexual spores. Zoospores are often formed during night and are then liberated in morning.

(14) Sexual reproduction occurs in all members, except some Chlorococcales. It may be isogamous, anisogamous or oogamous.

(15) The zygote or oospore secretes a thick wall and undergoes a resting period. It germinates under favourable conditions and forms a new thallus.

Classification of Chlorophyceae : Fritsch (1935), on the basis of morphology and reproductive characters, recognised the following nine orders in Chlorophyceae :

(1) Volvocales (2) Chlorococcales

(3) Ulotrichales (4) Cladophorales

(5) Chae:ophorales (6) Oedogoniales

(7) Conjugales (8) Siphonales

(9) Charales.

Smith (1955) classified the green algae into two classes-Chlorophyceae and Charophyceae. The class Chlorophyceae was divided into 12 orders, and only a single order was included in Charophyceae.

Class 1. Chlorophyceae

Order (1) Volvocales

(2) Tetrasporales

(3) Ulotrichales
(4) Ulvales
(5) Schizogoniales
(6) Cladophorales
(7) Oedogoniales
(8) Zygnematales
(9) Chlorococcales
(10) Siphonales
(11) Siphonocladales
(12) Dasycladales

Class 2. Charophyceae
Order (1) Charales

Mclean and Cook (1956) divided Chlorophyceae into the following eight orders :

(1) Volvocales
(2) Chlorococcales
(3) Chaetophorales
(4) Ulotrichales
(5) Oedogoniales
(6) Conjugales
(7) Siphonales
(8) Siphonocladales

The members of the order Charales were placed by them in a separate division Charophyta.

17

The Nostocales

This is the largest order of the class Cyanophyceae. The members of this order are non-heterotrichous filamentous forms without branching, but some members show false branching. Heterocysts are found in most of the members. Some genera show polarity with heterocysts and/or akinetes at the base and a hyaline hair at the apex of the trichome. Cyanophyceae reproduce by hormogonia, akinetes, exospores, endospores or hormospores.

Desikachary (1959) divided this order into five families : Oscillatoriaceae, Microchaetaceae, Nostocaceae, Scytonemataceae and Rivulariaceae.

Family Oscillatoriaceae

This is the largest family of the class Cyanophyceae. The members of this family are filamentous; the filaments are unbranched and uniseriate. All the cells of a filament are alike. In most of the members the trichomes have an envelope of homogeneous mucilagenous substance. All cells of a trichome are capable of division. They reproduce by hormogonia.

Oscillatory movements of trichomes and the absence of heterocysts are the two main characteristic features of the family. The common examples of this family are : Oscillatoria, Lyngbya, Spirulina, Arthrospira, Phormidium and Trichodesmium.

Oscillatoria

Systematic Position

Class - Cyanophyceae

Order - Nostocales

Family - Oscillatoriaceae

Occurrence : Oscillatoria is found in a wide variety of freshwater (such as rain water pools, streams, ponds and lakes, and even in polluted water) and sub-aerial habitats. The alga forms extensive sheets along the muddy banks of rivers and ponds, rocky cliffs and on damp soil. It also occurs in the form of a thin layer at the bottom of shallow ponds. O. prolifica is a planktonic species and O. terebriformis is found in hot water springs. O. selina and O. limosa occur in saline waters. O. rubescens forms dense deep-water blooms in temperate lakes and ponds and its migration and colonization in deep water lakes is often associated with a high level of eutrophication. Some species of Oscillatoria have specific colours; for example, O. prolifica is red, O. formosa is blue, and O. angusta is colourless. In India this genus is represented by about 105 species, of these O. formosa, O. angusta, O. sancta, O. princeps, O. chlorina and O. decolorata are common.

Thallus Structure : Oscillatoria is an unbranched filamentous alga. The filaments (or trichomes, as they are called) occur singly or are matted together to form thin or thick sheets. Each trichome is made up of many cells, arranged in uniseriate fashion. There are no constrictions between adjacent cells. All cells of a filament are similar in shape and structure. Their breadth is usually greater than their height, but species with narrow trichomes have cylindrical cells (e.g., O. splendida). The trichome is slightly different at the anterior end and shows a definite polarity. The free end is mostly rounded or convex (e.g., O. annae, O. limosa;) but it is acuminate in O. acuminata and O. brevis and discoid in O. princeps. In some species (e.g., O. proboscidea;) the apical cell of the filament carries a thickened membrane, known as calyptra. The above mentioned differences are due to variations in turgor pressure of the terminal cell.

The trichomes are naked or are enveloped in a thin gelatinous sheath.

Cell Structure : All cells of a trichome have similar structure. The cell is surrounded by a thin and firm cell wall which lies external to plasmalemma. The wall is differentiated into four layers, designated as LI, LII, L III and LIV; L I being next to plasmalemma. Each layer is about 10 nm in thickness. In structure and composition the cell wall is similar to that of Gram-positive bacteria. It is composed of mucopeptide together with carbohydrates, amino acids and fatty acids. The L I and L III layers are electron transparent, whereas the L II and L IV layers are electron opaque. Contractions in the fibrils of the cell wall are responsible for gliding movements of Oscillatoria.

The cell exhibits a typical prokaryotic structure. The protoplast is differentiated into the peripheral pigmented chromoplasm and the central centroplasm (Figs.). The chromoplasm is principally composed of thylakoids and their associated structures, phycobilins. The thylakoids are lamellate structures with two membranes joined at the ends, and enclosing an intrathylakoid space of variable width. Photosynthetic pigments are found on the surface of thylakoids. Thus membrane bound plastids, like those of eukaryotic cells, are absent. The amount of c-phycoerythrin and c-phycocyanin is higher than any other pigment.

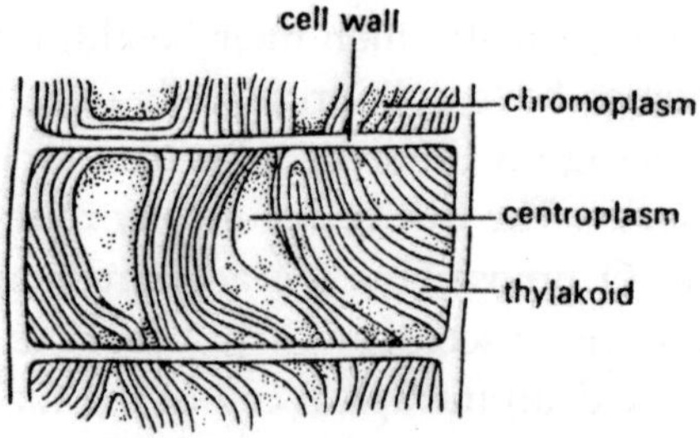

Oscillatoria : Cell structure as seen under light microscope.

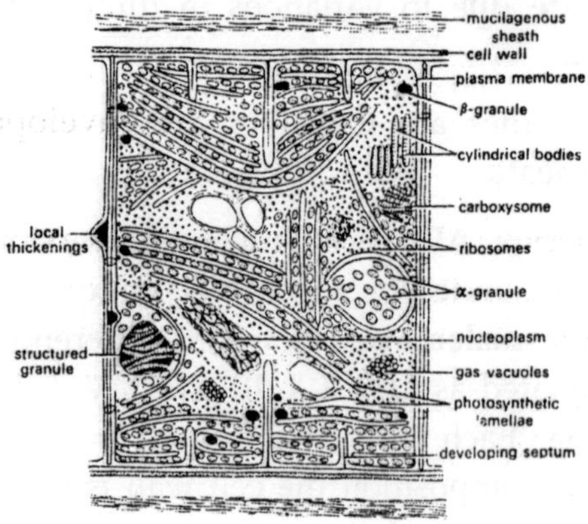

Oscillatoria : Cell structure as seen under electron microscope.

In floating species, rounded or hollow cylindrical and tubular gas vesicles are found in the cytoplasm. The wall of the vesicle is exclusively composed of protein ribs or spirals. Since species with gas vacuoles are present near the surface of water, the vacuoles probably help in light shielding, and/or provide buoyancy. The gas vacuoles are usually formed in low light intensity when filaments are near the bottom of the pond. Increase in buoyancy brings trichomes to the surface of water where they burst due to high light intensity, and again go deep into water.

Some angular polyhedral bodies are found near the centroplasm. They store carboxysomes which contain CO_2 fixing enzyme ribulose 1,5-bisphosphate carboxylase. In addition to these, some small spherical polyphosphate bodies containing stored phosphate are also found. They are absent in the young cells but are prominent in the older cells. In actively photosynthesizing cells polyglucon granules are found in between the thylakoids. These are carbohydrate granules, composed of 14-16 glucose units.

The central region of the cell which consists of the nuclear material is termed as centroplasm or nucleoplasm. Although this region stains like the nucleus of eukaryotes, nuclear membrane and nucleoli are absent. It is not sharply differentiated, and commonly interfingers with the peripheral chromoplasm.

Gaidukov Phenomenon

Gaidukov (1902) observed that species of Oscillatoria develop different shades when exposed to lights of different wave lengths. For instance, a species of Oscillatoria can assume a green colour in red light, a reddish - blue in green light, and a bluish-green in yellow light. Such a capacity to change colour complementary to that of the light is known as chromatic adaptation or Gaidukov phenomenon. This adaptation is of definite advantage since it enables the alga to maximally absorb the available light for photosynthesis.

Growth : Oscillatoria exhibits intercalary growth. All the cells of a trichome are capable of division and they divide at right angles to the long axis of the trichome. Cell division occurs by median constriction of the cell. The two inner wall layers, L I and L II, grow inward in a ring like fashion, pushing the thylakoids inwards, and gradually cutting across them as the new septum is formed. When growth is rapid, most of the cells of the filament divide simultaneously. Usually the septum is not complete, and it leaves a fine interconnection between the two adjacent cells, called plasmodesmata.

This type of cell division is considered by some as primitive mitotic, while others prefer to call it amitotic.

Movement : Oscillatoria exhibits spontaneous movements. The rate of movement differs with the species. It is also affected by temperature, pH, viscosity of the medium and light intensity. The following two types of movements are recognised :

Gliding or Creeping Movement : When the trichome exhibits rhythmic movements on any solid surface without the help of any visible organ it is called gliding movement. This type of movement is characterised by the absence of a distinct change in the shape of the organism. It takes place within the mucilage sheath, the sheath sticking to the substrate and then being left behind by the advancing trichome. Gliding is always along the longitudinal axis of the trichome. In some species gliding movement is accompanied by rotation of the trichome.

Oscillatory Movement : The trichomes of Oscillatoria move to the left and right side of the axis just like the pendulum of a wall clock. Because of the presence of such oscillatory movements, the genus is named Oscillatoria.

Several views have been put forward to explain the movement in Oscillatoria. Some important ones are :

(1) The movement is due to invisible cilia present at the apex of the trichomes.

(2) The movement is accomplished by the secretion of mucilage (through the small pores present on the surface of the trichome) against substrate. The physical changes in the extruded mucilage also help in the movement.

(3) The undulating surface of the trichome is responsible for the movement.

(4) The movement results from rhythmic contraction waves in helical fibrils on the cell surface.

Nutrition

Main photosynthetic pigment is chlorophyll a, and the accessory pigments are c-phycocyanin and c-phycoerythrin.

These pigments trap the solar energy and synthesize starch and glycogen, which are later converted into cyanophycin starch and glycoprotein.

Reproduction : Oscillatoria reproduces vegetatively by the following methods :

By Hormogonia : During vegetative growth of the trichome, some cells are filled with mucilage and become dead. These cells are called necredia (singular = necredium). They loose their turgidity, and due to the pressure of transverse walls of the adjacent cells become biconvex.

Such dead cells function as separation discs and they provide weak links in the trichome. The cells of the trichome between the two necredia form a hormogonium. Soon after separation into hormogonia, the terminal cell of the broken end becomes spherical, conical or tapered, depending on the species. The hormogonia are capable of movement, and by repeated cell divisions they form new trichomes.

By Fragmentation : Sometimes the trichome divides into small pieces due to insect bite or mechanical injury. The fragments, so formed, have the capability of growing into long filaments by cell division, just like hormogonia.

Spirulina

Systematic Position

Class	-	Cyanophyceae
Order	-	Nostocales
Family	-	Oscillatoriaceae

Occurrence : Spirulina occurs in a wide variety of fresh water habitats, often found densely entangled with the Oscillatoria filaments. Some species are found in brackish water, and can tolerate very high salt concentrations. For example, S.

subsalsa can grow and multiply in solutions that are more concentrated than 3M sodium chloride. Some common species of Spirulina are S. albida, S. major, S. princeps, S. subsalsa and S. versicolor.

Species of Spirulina are rich in proteins, essential vitamins and unsaturated fatty acids. They are receiving considerable attention for their use as a supplement to diet.

Structure : It is an unicellular alga, the trichome consists of a single, cylindrical cell, wounded into a loose or close helix.

The cell exhibits a typical prokaryotic structure. The protoplast is differentiated into a peripheral pigmented chromoplasm and a central centroplasm. Cyanophycin granules are, however, absent.

Movement : Spirulina shows spontaneous movements. The motion is usually creeping or gliding in the direction of the longitudinal axis and may take place both backward and forward. In some species, gliding movement is accompanied by the rotation of the trichome, taking place in the direction of coiling.

Reproduction : Spirulina reproduces vegetatively by the process of fission.

Family Nostocaceae

The filamentous forms included in this family occur singly or in a common mucilagenous matrix. The trichomes are unbranched and uniseriate, and all the cells of the trichome are similar. They do not exhibit polarity, i.e., they are not differentiated into basal or apical ends.

Heterocysts are found in all members of the family; they may be terminal or intercalary in position. The reproduction takes place by akinetes and hormogonia.

Nostoc

Systematic Position

Class - Cyanophyceae

Order - Nostocales

Family - Nostocaceae

Occurrence : Species of Nostoc occur in freshwater as well as in terrestrial habitats. The fresh water species form macroscopic colonies on the surface of water which vary from few millimeters to 8 cm or more in diameter (Fig. A). Each colony is bounded externally by a pellicle like membrane. N. commune, a terrestrial species of alpine regions, forms leathery or rubbery sheet on the damp soil. Some species of Nostoc are endophytic and they occur in symbiotic association within other plants. For example, N. punctiforme is found in the coralloid roots of Cycas and Anthoceros thallus, and N. sphaericum and N. collema are the components of lichens and occur in symbiotic association with fungi.

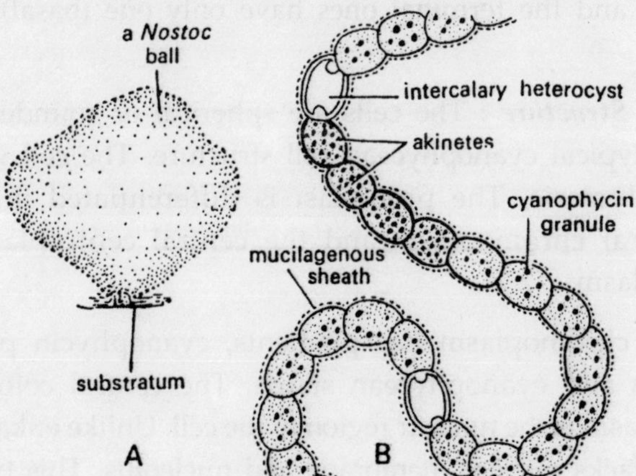

A-B. Nostoc : A. Gelationous colony, B. Trichome.

Some species of Nostoc are grown in rice fields. They have the capability to fix atmospheric nitrogen, thus helping in increasing fertility of the soil.

In India, Nostoc is represented by about 26 species, of which N. calcicola, N. endophyllum, N. ellipsoporum, N. muscorum and N. punctiforme are common.

Thallus Structure

The trichomes of Nostoc are uniseriate, usually contorted and twisted in various ways. An important feature is the presence of prominent constrictions between the adjacent cells which give a characteristic moniliform appearance to the trichome (Fig.B). Each trichome is enveloped in a gelatinous sheath which is transparent, hyaline or coloured. Usually many trichomes aggregate together, and their gelatinous envelopes dissolve to form colonies of various shapes and sizes.

The trichomes of Nostoc are characterised by the presence of heterocysts which are intercalary or occasionally terminal, usually solitary and of much the same size and shape as the vegetative cell. The intercalary heterocysts have two polar nodules and the terminal ones have only one (basal) polar nodule.

Cell Structure : The cells are spherical or rounded and exhibit typical cyanophycean cell structure. The cell wall is mucopolymeric. The protoplast is differentiated into the peripheral chromoplasm and the central centroplasm or nucleoplasm.

The chromoplasm has pigments, cyanophycin protein granules and cyanophycean starch. The central colourless centroplasm is the nuclear region of the cell. Unlike eukaryotic cells, it lacks nuclear membrane and nucleolus. This type of nucleus is called incipient nucleus.

Reproduction : Nostoc reproduces only vegetatively. Sexual reproduction is completely absent. The vegetative propagation takes place by the following methods.

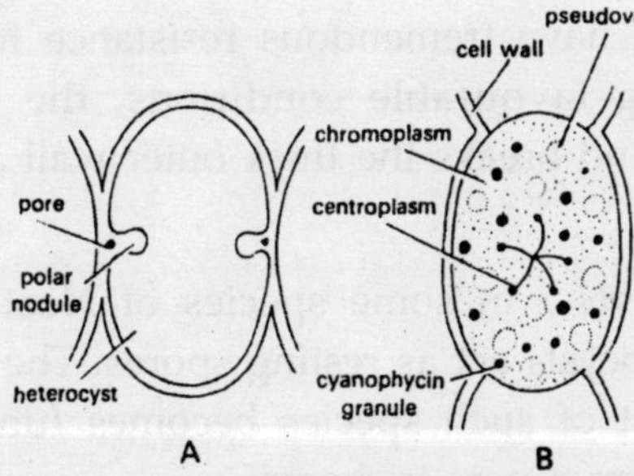

A-B. Nostoc : A. Heterocyst, B. Cell structure as seen under light microscope.

By Fragmentation : The colony may break into small fragments due to mechanical, physiological or other factors. Each fragment has the capability to develop into a new colony.

By Hormogonia : The trichome breaks into small segments due to the degeneration of intercalary vegetative cells or because of the presence of intercalary heterocysts. Multicellular fragments so formed are called hormogonia. They come out of the gelatinous sheath of the colony, grow rapidly and form new colonies. Very often, the hormogones fail to come out of the parent colony and divide inside the gelatinous sheath of the parent colony. This results in a large number of trichomes inside the parent colony.

By Akinetes : During unfavourable period, some cells of the trichome are transformed into resting spores or akinetes. They have an additional three layered coat outside the normal cell wall. The outer layer is variously sculptured. The protoplasm is full of food reserves.

Development of akinetes normally takes place in a mature colony. It is always the vegetative cell adjacent to the heterocyst which is the first to metamorphose into akinete. Subsequently, many vegetative cells are transformed into akinetes. It is not unusual to find all the cells between the two successive heterocyst transformed into akinetes. Metamorphosis of the vegetative cell into akinete is probably stimulated by certain chemical substances secreted by heterocysts.

The akinetes have tremendous resistance for cold and drought. During favourable conditions, the protoplasm becomes active and breaks the thick outer wall and forms a new trichome.

By Heterocysts : In some species of Nostoc (e.g., N. commune) heterocysts act as resting spores. The protoplasm of the heterocyst of such species becomes functional and germinates to form a new trichome.

At the time of germination the protoplast divides by a transverse wall to form two cells. Both these cells divide again to form a 4-celled germling (Fig. A, C). The thick wall of the heterocyst ruptures at this stage and the germling comes out and develops into a new trichome.

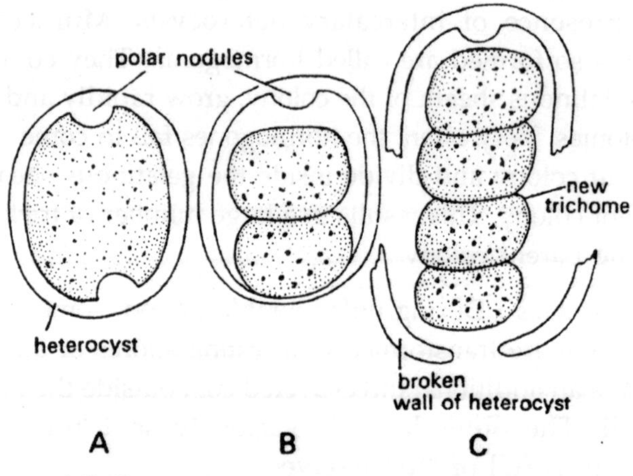

A.C. Nostoc : Stages of reproduction by heterocyst.

By Endospore : In some species of Nostoc (e.g., N. commune, N. microscopicum) the protoplasm of the heterocyst divides successively to form endospores.

Unlike akinetes, the endospores are thin walled and hence can not be regarded as resting spores. Disintegration of the capsule wall results in their liberation, and the rounded spores later germinate to form new trichomes.

Fusion of Filaments

Although true sexual reproduction and sex organs are absent in Nostoc as in other blue-greens; yet fusion and anastomosis of trichomes have been observed in N. muscorum. This phenomenon can be compared with somatogamy.

Family Scytonemataceae : The members of this family are filamentous forms with false branches. Except Plectonema, all members of the family possess spherical or oval heterocysts which are usually intercalary. They reproduce mostly by hormogonia. In India, the family is represented by members like Scytonema, Plectonema, Tolypothrix and Scytonematopsis.

Scytonema

Systematic Position

Class - Cyanophyceae

Order - Nostocales

Family - Scytonemataceae

Occurrence : This filamentous blue-green alga includes about 40 freshwater or terrestrial species. They are found in brackish water, stagnant freshwater ponds, as well as on the moist soil, tree-barks and walls in the form of bluish-grey scum. S. caldarium has been reported from hot springs and S. ocellatum is the most common terrestrial species. The terrestrial species have a tremendous capacity of absorbing moisture from the atmosphere. About 33 species of Scytonema are recorded from India. S. fritschii, S. amplum, S. iyengari and S. pascheri are some common Indian species.

Thallus Structure : The plant body is filamentous, and the filaments intertwine to form a scum of bluish-green, blackish-green or yellowish-green colour. Each filament, made up of spherical, oval or squarish cells, is surrounded by an individual mucilagenous sheath. The sheath may be hyaline or coloured, smooth, homogeneous or lamellated. The terrestrial species, in particular, possess pigmented sheaths due to the presence of a pigment, scytonemin.

Cell Structure : The cell shows a typical cyanophycean structure. The cell wall is typically polymeric in nature, composed of mucopolypeptides and muramic acids. The protoplast is differentiated into the peripheral pigmented chromoplasm and the central centroplasm. The chromoplasm, in addition to chlorophyll- *a*, β-carotene and xanthophylls, also contains two important phycobilin pigments, c-phycoerythrin and c-phycocyanin. The centroplasm is relatively dense and it represents the (incipient) nucleus of the cell. It does not have the limiting nuclear membrane and nucleolus.

Branching : A characteristic feature of Scytonema is the presence of false branches, which are formed by the breaking of the filament. When a filament breaks, one or both the broken ends protrude out of the mucilagenous sheath and grow like branches. If both the broken ends come out of the sheath and grow like branches, it is called geminate branching, and if only one, then it is said to be false branching.

The false branches are formed by the following methods.

(a) *By Loop Formation :* Certain parts of the filament increase in length due to rapid cell division. Such localized growth results in loop formation that protrudes through the sheath. Eventually the loop breaks in the middle, cell division continues and two false branches are formed (Fig. A).

(b) *By Formation of a Separation Disc :* Sometimes one of the intercalary cells becomes dark coloured due to the diffusion of chromatophores in the protoplast, and its cell wall becomes thin. This cell forms a biconvex separation disc due to the pressure of the adjacent cells. The disc ultimately degenerates, resulting in the breaking of the trichome. One of the broken ends protrudes through the sheath as false branch (Fig. D).

(c) *By Rupture of Filament near the Intercalary Heterocysts :* In trichomes with intercalary heterocysts a zone of weak adhesion lies where the vegetative cell and the heterocyst abut each other. If the trichome breaks on one side of the heterocyst, a single branch is formed and if it breaks on both the sides of the heterocyst, the branching is geminate.

(d) *By Degeneration of one or more Intercalary Cells :* This is a very common method of the formation of false branches. One or more intercalary cells degenerate and leave gaps in the trichome. One or both the broken ends of the trichome may protrude through the sheath and develop into branches.(Fig. B).

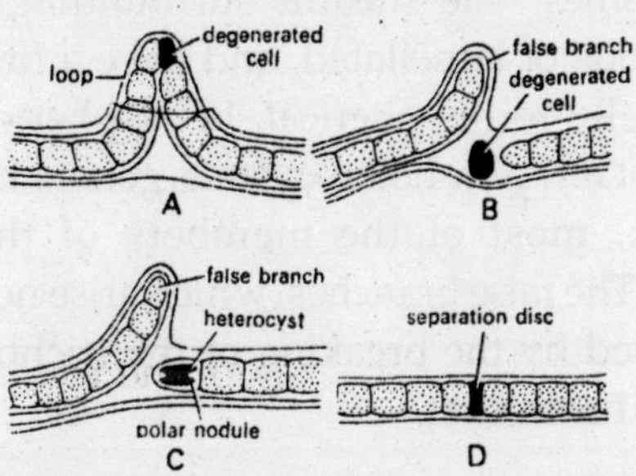

A-D. Scytonema : Modes of branching: A. Loop formation, B. Degeneration of intercalary cell, C. Breaking of filament near the intercalary heterocyst, D. Formation of separation disc.

Growth : In Scytonema growth occurs by repeated transverse divisions in the apical or sub-apical cell. In terrestrial species, the apical cell remains active for sometime and then degenerates. Thereafter the sub-apical cell functions as apical cell. Besides the apical and sub-apical cells, some intercalary cells also divide transversely and add to the growth of the trichome.

Reproduction : Multiplication of trichomes in Scytonema takes place by the formation of hormogonia or akinetes.

1. *By Hormogonia :* Many factors, such as the presence of separation discs and heterocysts, and disintegration of intercalary cells help in the fragmentation of the trichome. These fragments are called hormogonia, and each hormogonium is capable of growing into a new trichome.
2. *By Akinetes :* Some species of Scytonema (e.g., S. rivulare) form thick walled perennating cells, called akinetes. These resting cells, besides multiplication, also help in tiding over the unfavourable conditions. On return of favourable conditions, the akinetes germinate into a new trichome.

Family Rivulariaceae

This family includes freshwater forms with uniseriate tapering trichomes. The sheaths surrounding the trichomes are homogeneous or lamellated, and have a firm texture. The trichomes usually form spherical, hemispherical or irregular colonies. Except few non-heterocystous genera like Amphithrix and Calothrix, most of the members of this family are heterocystous. The false branches, which arise near the tapering ends, are formed by the breaking of the trichome just below an intercalary heterocyst.

Gloeotrichia

Systematic Position

Class - Cyanophyceae

Order - Noslocales

Family - Rivulariaceae

Occurrence : It is a fresh water alga, found floating in pools and ponds in the form of spherical mucilagenous colonies. A few species are attached to submerged plants. In India, the genus is represented by about 10 species, of these G. indica and G. ghosei are common.

Thallus Structure : The plant body of Gloeotrichia consists of whip-like uniseriate trichomes which are broader at the base and narrow towards the apex. The trichome terminates into an erect or curved multicellular hair. At the base of each trichome there is a spherical or cylindrical heterocyst with thick wall and homogeneous contents.

The heterocyst has only one polar nodule. Just above the heterocyst, there are one or more elongated resting spores, known as akinetes. When there are more than one akinetes, they usually occur in a chain. Sometimes two akinetes are separated by a vegetative cell.

The heterocyst, akinetes and some basal cells of the trichome remain covered with a gelatinous sheath. The gelatinous sheaths of several neighbouring trichomes fuse together to form small spherical colonies. In a colony the heterocysts of all the trichomes are pointed towards the centre and the narrow distal ends towards the periphery.

The cell shows typical cyanophycean structure. The growth of the trichome takes place by active transverse divisions in the intercalary cells, located just below the distal hair. These intercalary cells cut segments both on the upper and the lower side. This type of growth is called trichothallic.

Reproduction : The reproduction in Gloeotrichia is brought about by hormogonia and akinetes.

1. *By Hormogonia :* The hormogonia are small segments of the trichome formed by fragmentation. Each segment is capable of growing into a new trichome.
2. *By Akinetes :* The akinetes occur singly or in chains, and are formed under unfavourable conditions. They secrete a thick wall which help them in tiding over the unfavourable conditions. Under the favourable conditions the akinetes germinate, each forming a new trichome.

Rivularia

Systematic Position

Class	-	Cyanophyceae
Order	-	Nostocales
Family	-	Rivulariaceae

Occurrence : Rivularia (L. rivulus = a small brook) is found in a wide variety of fresh water and terrestrial habitats. It grows on damp soil near river beds, moist rocks, or even epiphytically on some submerged plants. R. atra is found in salt marshes, and species like R. bullata, and R. polyotis occur in supra-littoral and upper intertidal zones of sea coasts.

Thallus Structure : The unbranched trichomes of Rivularia are arranged in a radial fashion to form a spherical, hemispherical or irregular colony (Fig. A). The radial arrangement of trichomes in a colony is due to repeated false branching in their basal portions. The evidence for false branching is found only in the young colony and is lost in a mature colony due to large displacement of trichomes. The trichomes in a colony remain partially or completely surrounded by a confluent sheath. The sheath is of very firm consistency, tough and very frequently heavily encrusted with lime.

The trichome is broad at the base and gradually tapers into a hair towards the apex. At the base of the trichome, a single heterocyst with one polar nodule is present (Fig.C).

Cell Structure : The cell exhibits a typical cyanophycean structure. The mucilagenous components of the cell wall contain arabinose and glucose. The protoplast is differentiated into a peripheral pigmented chromoplasm and a central centroplasm. The chromoplasm contains two important phycobilin pigments, c-phycoerythrin and c-phycocyanin: The centroplasm is relatively dense and represents the incipient nucleus of the cell.

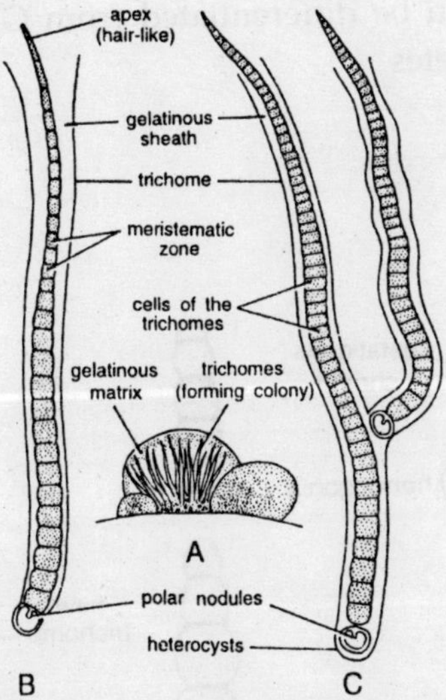

A-C. Rivularia : A. Colonies growing on a substratum, B. A colony showing trichome arrangement, C. Trichomes.

Growth : The growth of the trichome in Rivularia is trichothallic. It takes place by the activity of meristematic cells in the intercalary meristematic zone located at the base of the terminal hair. The meristematic cells are comparatively small in size and have denser granular contents.

Reproduction : Rivularia reproduces vegetatively by the formation of hormogonia and rarely by the heterocyst which gives rise to new trichome on germination. Sexual reproduction is not known.

At the time of the formation of hormogonia, the trichome usually sheds its hair. The delimitation of hormogonia starts in the meristematic zone and gradually proceeds towards the basal portion of the trichome. Hormogonia are formed either singly or in series. In the latter case, they are separated by biconcave separation discs. At the time of germination, one of the two end cells of a hormogonium is modified into a basal heterocyst, and the other cell divides repeatedly in a single plane to form a long trichome (Fig. A, C).

Rivularia can be differentiated from Gloeotrichia in the absence of akinetes.

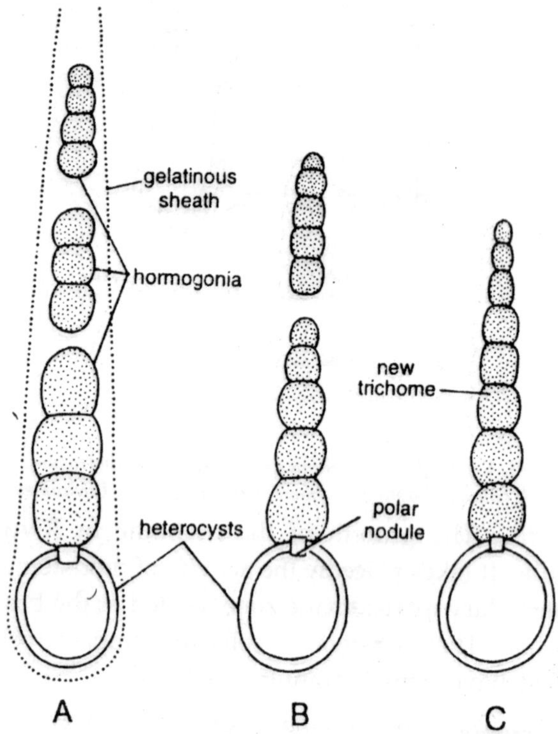

A-C. Rivularia : Hormogonia formation; A. Portion of a trichome with hormogonia in a series, B-C. Two stages in the germination of a hormogonium.

Chroococcales—Gloeocapsa : The order Chroococcales includes the most primitive members of the class Cyanophyceae. Almost all the members of the order are fresh water forms which are unicellular and are often organised into regular colonies; but they never show a trichome organisation. They reproduce chiefly by fission or by endospore formation. Heterocysts and hormogonia are absent.

The order comprises two families, Chroococcaceae and Entophysalidaceae.

Family Chroococcaceae

Members of the family are unicellular and free-living forms. They occur either singly or in colonies. The individual cells are usually spherical, ovoid, ellipsoidal or cylindrical. Multiplication takes place by fission or sometimes by nannocytes. The common examples of the family are Anacystis, Aphanocapsa, . Chroococcus, Gloeocapsa, Microcystis and Synechococcus.

Gloeocapsa

Systematic Position

Class	-	Cyanophyceae
Order	-	Chroococcales
Family	-	Chroococcaceae

Occurrence : Gloeocapsa (Gr. gloia, glue; capsa, a box) is a terrestrial or sub-aerial alga, generally found on damp soils, wet rocks, banks, etc., as well as in the water reservoirs and lakes. Some species are marine, they occur as a black encrusting film on rocks at the upper limit of the high-.tide mark in the littoral zone. Some species of the genus are endophytic. G. alpina constitutes algal component of lichens and occurs in symbiotic association with fungi, whereas G. gelatinosa is found on the bark of Hevea brasiliensis.

Some species of Gloeocapsa have the capability to fix atmospheric nitrogen and thus help in increasing fertility of the soil.

Thallus Structure : It is a colonial alga, and the colonies are spherical, irregular or expanded mass of mucilage. The number of cells in a colony ranges from 2-8. The cells are characteristically spherical, and each cell is surrounded by a conspicuous, variously coloured, homogeneous or stratified, concentrically lamellate gelatinous sheath. The sheath also surrounds the entire colony.

Cell structure. The cell exhibits typical cyanophycean structure. The protoplast is well differentiated into an outer pigmented chromoplasm and an inner centroplasm. The latter represents incipient nucleus; true nucleus is lacking. The chromoplasm possesses a special pigment said to be gloeocapsin.

In most species of Gloeocapsa photosynthetic thylakoids are arranged in rows at right angles to the plasma membrane, the outer thylakoids are directly connected with the plasma membrane.

Reproduction : Gloeocapsa reproduces only vegetatively by means of fission or by the formation of nannocytes.

Fission : The fission takes place in three planes at right angles to one another resulting in more or less spherical daughter cells. At the time of fission, individual cells are released from the colony by the disorganisation of the parent gelatinous sheath. Each cell then grows into a new colony by simple cell divisions. The chromatin material in the centroplasm divides by a constriction into two tiny masses, followed by the division of the cytoplasm, resulting in two cells. The process is further repeated and as such a colony is formed.

Nannocytes : In some species such as G. granosa, G. crepidinum, nannocytes are formed. They are modified endospores, which are represented by small naked protoplasts formed by repeated divisions of the cell contents. On liberation from the parent cell nannocytes germinate immediately without undergoing any resting stage.

Microcystis

Systematic Position

Class - Cyanophyceae

Order - Chroococcales

Family - Chroococcaceae

Occurrence : Microcystis (Gr. micro = small, kystis= bladder) is a fresh water colonial alga. Most species are euplanktonic and are aided by pseudovacuoles. In tropical countries, species of Microcystis multiply so prolifically that they impart a bluish-green colour to the pond water which appears like 'pea soup'. Some species of Microcystis, like M. aeruginosa, cause water blooms in lakes.

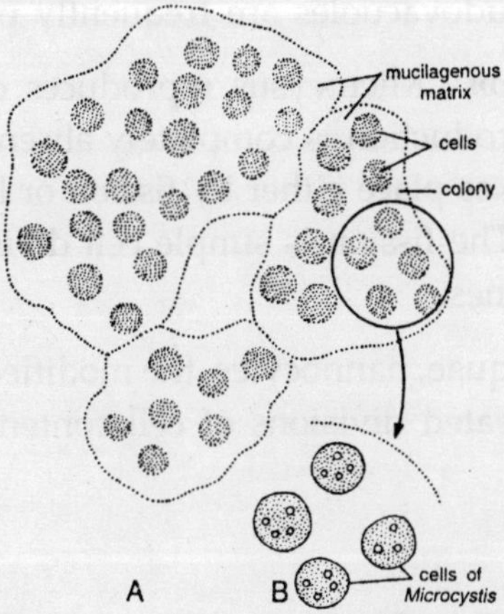

A-B Microcystic : A. A Colony, B. A part of the colony (magnified).

The prolific growth of some species of Microcystis (e.g., M. aeruginosa) in a water body often results in severe depletion of oxygen. This leads to mass mortality of fish and other aquatic animals due to suffocation. Certain species of Microcystis (e.g., M. aeruginosa) produce a toxin, called microcystin, which is highly toxic to fish and aquatic birds. The toxin is also fatal to domesticated animals that drink contaminated water. M. toxica is highly poisonous and is responsible for the death of thousands of sheep and cattles. In humans, some species of Microcystis cause gastric trouble, if they are ingested along with drinking water or during swimming. There are several species of Microcystis which inhibit the growth of bacteria (such as Staphylococcus, Clostridium), zooplanktons and several other algae.

Thallus Structure : The colonies of Microcystis may be round and definite in shape or irregular or sometimes clathrate, depending upon the species. There are hundreds or thousands of small, marble-like cells in each colony. The spherical or elongated cells in a colony float freely in a colourless, homogeneous mucilaginous matrix. They lack individual sheath, and exhibit typical cyanophycean cell structure. Numerous pseudovacuoles are frequently present in a cell.

Reproduction : Microcystis reproduces only vegetatively and sexual reproduction is completely absent. The vegetative propagation takes place either by fission or by fragmentation of the colony. The fission is simple cell division which takes place in all planes.

In M. flos-aquae, nannocytes, the modified endospores are formed by repeated divisions of cell contents.

18

The Cyanophyceae

There are about 160 genera and 1500 species in Cyanophyceae. In India, the class is represented by 98 genera and 833 species. The members of Cyanophyceae (Myxophyceae) are commonly known as blue-green algae as their principal pigment is a bluish-green c-phycocyanin. They also contain chlorophyll *a*, β-carotene and c-phycoerythrin. They differ from other algae : (i) in the absence of well organised cell organelles, (ii) the pigments are not restricted to definite chromatophores, but are distributed throughout the peripheral cytoplasm (chromoplasm), and (iii) the nucleus is of primitive type which lacks nuclear membrane and nucleolus. These characters show typical prokaryotic nature of the members of this group. These algae do not have flagella and their movement is brought about by gliding action.

Studies during the last three decades have clearly shown that the cell structure of blue-green algae is entirely different from the members of other algal groups. On the basis of their prokaryotic structure, many microbiologists consider blue-green algae as bacteria. Christensen (1962), on the basis of prokaryotic cell structure of blue-green algae and bacteria, placed both of them in a common phylum Prokaryota. Blue-green algae have also been named as Cyanobacteria. Although there are many similarities between bacteria ana blue-green algae, yet they show differences in their evolution.

Phycologists regard any organism with chlorophyll *a* and the thallus not differentiated into roots, stem and leaves to be an alga. Therefore, most of the algologists include Cyanophyceae in algae, rather than in bacteria.

The blue-green algae are found in a wide variety of habitats. Most of the species are fresh water, a few like Dermocarpa and Trichodesmium are marine. Species of Nostoc and Oscillatoria are found in terrestrial habitats. Some species of Nostoc and Anabaena grow as endophytes in roots of Cycas, leaves of Azolla and thalli of Anthoceros. Species of Chroococcus, Gloeocapsa, Nostoc, Scytonema and Stigonema are the main algal component (phycobiont) of lichens. The blue-green algae, growing in organically rich permanent waters, form planktons. The polluted water of lakes and ponds, exposed to intense sunlight and high temperature, provides a suitable habitat for the growth of planktonic blue-greens. They are also found on the bottom of ponds (benthos) and increase fertility of the underlying soil.

General Characters

Some important distinguishing features of blue-green algae are as follows :

(1) They are unicellular (e.g., Chroococcus), colonial (e.g., Gloeocapsa, Gloeotricha) or form chains of cells - trichomes (e.g., Oscillatoria, Scytonema). In colonial forms, all cells of a colony occur in a common gelatinous matrix. The cells of the colony may divide to form definite (e.g., Merismopedia) or indefinite and irregular (e.g., Aphanotheca) assemblage of cells. In filamentous

forms (a trichome surrounded by a sheath is called a filament), each trichome has its own gelatinous sheath. The filaments are usually unbranched and uniseriate (e.g., Anabaena, Nostoc, Lyngbya) or may sometimes show pseudo-branching (e.g., Scytonema, Tolypothrix).

(2) Their cell structure is prokaryotic. They lack membrane bound organelles and the nucleus is of incipient type.

(3) The cell wall is composed of microfibrils held together in a mucilage sheath. The orientation of fibres may vary in different species, and may be reticulate (e.g., Nostoc) or long intertwining parallel (e.g., Lyngbya). It helps in sloughing off parasites and in retaining water, thus avoiding desiccation (particularly in those species which do not produce resting spores).

(4) The cell wall (external to plasmalemma) is differentiated into four layers. It is composed of mucopeptide, together with carbohydrates, amino acids and fatty acids.

(5) The principal pigments are chlorophyll *a*, c-phycocyanin and c-phycoerythrin, and they impart characteristic blue-green colour to these algae.

(6) They lack membrane bound plastids. Instead, pigments are found embedded within lamellae composed of two membranes joined at the ends. These structures are known as thylakoids.

(7) They show gliding or jerky movements; flagella being altogether absent.

(8) The reserve food material is a special type of starch (cyanophycean starch) and protein (cyanophycin granules).

(9) In some blue-green algae (e.g., Nostoc, Scytonema) specialised structures, known as heterocysts, are present. They may be terminal (e.g., Gloeotricha) or intercalary (e.g., Nostoc) in position. Definite function

of heterocysts is not known, but they are supposed to be the centre of N_2-fixation.

(10) Reproduction takes place by vegetative and asexual methods. Sex organs, gametes and flagellated zoospores are altogether absent. However, in some (e.g., Anabaena, Nostoc, Synechococcus, Cylindrospermum) genetic recombination has been reported.

(11) Vegetative reproduction takes place by cell division, fragmentation and hormogonia formation.

(12) Asexual reproduction takes place by akinetes, endospores, exospores, heterocysts and nannocytes.

Cell Structure

The blue-green algae have prokaryotic cell organisation. The internal membranes which separate DNA from the cytoplasm and isolate enzymatic machinery of photosynthesis and respiration in specific organelles are absent. The cell as such is devoid of organised nuclei, chromatophores and mitochondria. The cell wall contains a specific mucopeptide as its strengthening component.

Sheath and Cell Wall : The presence of the mucilagenous sheath is a constant feature of blue-green algae. The sheath consists of three layers of microfibrils reticulately arranged within an amorphous matrix. This gives a homogeneous appearance to the sheath. The mucilagenous sheath retains absorbed water which is useful during the periods of desiccation and for sloughing off parasites. A double-layered cell wall is present inside the sheath.

The inner layer of this wall contains mucopeptide and muramic acid. When viewed under electron microscope, the wall shows four layers, designated as LI, LII, LIII and LIV. Each layer is about 10 mm in thickness, and the LI lies next to plasmalemma. In structure and composition the cell wall of blue-green algae is similar to that of Gram-positive bacteria; the cell wall in both is composed of mucopeptide together with carbohydrates, amino acids and fatty acids. The LII layer has mainly peptidoglycan. The LI and LIII layers are electron transparent, whereas the LII and LIV layers are electron opaque.

A cytoplasmic membrane (plasmalemma) is present inner to the cell wall and this membrane consists of two electron opaque layers separated by a translucent layer. The cytoplasmic membrane invaginates inside the cell, and these invaginations are considered sites of biochemical functions normally associated with mitochondria, endoplasmic reticulum and Golgi bodies in eukaryotic cells.

Photosynthetic Lamellae : Electron microscopic studies have revealed the presence of a complex lamellar system (thylakoid) which is functionally analogous to the plastids of eukaryotic cells. The lamellae in cyanophycean cell are not separated from the remainder of the cytoplasm in a membrane-bounded organelle. They appear as elongated, flattened sacs consisting of two unit membranes, each about 75 Å thick. Adjacent thylakoids are separated from each other by a space of 50 nm, occupied by contiguous rows of discoidal phycobilisomes. These phycobilisomes contain photosynthetic pigments, c-phycocyanin and c-phycoerythrin.

Intracytoplasmic Inclusions : Various kinds of subcellular organelles have been observed in cyanophycean cells and these include cyanophycin granules, polyhedral bodies, ribosomes, polyglucoside bodies, polyphosphate bodies, α-granules, β-granules and gas vacuoles.

The significance and functions of these organelles are not fully understood. Ribosomes are involved in protein synthesis, cyanophycin granules are regarded as reserve food, and polyhedral bodies (now called carboxysomes) contain an important enzyme, ribulose diphosphate carboxylase. Several planktonic species of blue-green algae contain numerous small gas vacuoles (pseudovacuoles). These vacuoles are not tonoplast bound and contain metabolic gases. They are as such a provision for buoyancy, regulation. Under electron microscope, the gas vacuoles appear to consist of packed arrays of cylindrical vesicles. The gas vacuoles are produced in low light intensity and disappear in high light intensity. They refract light and as such cells appear purplish, brown or even black under microscope.

Nucleoplasm : The central region of a blue-green algal cell appears somewhat transparent and contains most of the genetic material. An organised nucleus with a nuclear membrane and nucleolus is absent. Since DNA is not associated with protein material (histones or protamines), organized chromosomes, characteristic of the eukaryotic cell, are not found in the cyanophycean cell. RNA is present in addition to DNA.

Heterocysts

These are specialized structures found in the members of Nostocales (except Oscillatoriaceae) and Stigonematales. They arise from somewhat enlarged vegetative cells, and are characterised by the presence of conspicuous pore at one or both poles. In terminal heterocysts, the pore is usually present only at one end, whereas intercalary heterocysts have pores at both ends. The position of heterocysts is a specific character; they are terminal in Gloeotricha, intercalary in Nostoc and Scytonema and rarely lateral as in Nostochopsis. They may be solitary or in pairs (e.g., Anabaenopsis). Intercalary heterocysts are usually spaced regularly along the filament.

Structure : Heterocysts are thick walled, pale yellow and barrel-shaped structures. The wall is differentiated into an outer fibrous, a middle homogeneous and an inner thin lamellar layer. It is characterised by the presence of pores either at one or both the poles. The wall layers become thicker in the polar region, and the pores are plugged with a refractive material, called polar granules (nodules).

Heterocysts contain dense and homogeneous cytoplasm. They contain mostly carotenoids, hence yellowish in colour. In the absence of phycobilins (which are the principal light absorbing pigments in oxygen-evolving photosynthetic reactions) and chlorophyll *a*, there is no CO_2 fixation or O_2 production in heterocysts.

The photosynthetic lamellae (thylakoids) are tightly packed, with reduced interlamellar spaces. They are concentrated along the periphery in a complex reticulate manner. These lamellae contain two lipids-glycolipids and acyl lipids, which are not found in vegetative cells. Three types of granules (structural granules, crystalline granules, small granules) are associated with the peripheral lamellae, but they decrease in number and eventually disappear as the heterocyst matures. The mature heterocyst may remain attached to the filament for varying periods.

Development : Heterocysts develop from vegetative cells. One or both the daughter cells of a recently divided vegetative cell may behave as proheterocyst. A proheterocyst undergoes a number of changes to form heterocyst. It enlarges in size and forms a multilayered envelop outside the cell wall. The wall becomes differentiated into an outer fibrous, a middle homogeneous and an inner lamellated layer. The outer fibrous layer is uniformly thick, whereas the middle and inner layers are thicker at the polar ends.

The pores appear at one or both poles, and through these pores cytoplasmic continuity is maintained with adjacent vegetative cells. There is gradual dissolution of storage granules and breakdown of photosynthetic thylakoids. Sometimes the lamellae may form swollen vesicles and they become concentrated along the periphery and at the pores. In a mature heterocyst the polar pores become plugged with a refractive material (also known as polar nodules), due to which their cytoplasmic connections with adjacent cells are broken.

The frequency of heterocyst production is affected by the following external factors :

(1) The blue and green light inhibit heterocyst development, whereas the white and red light support it.

(2) The concentration of phosphate salts in the medium stimulates heterocyst development, whereas the absence of Mg and Fe ions in the medium inhibits heterocyst development.

(3) The presence of combined nitrogen in the medium inhibits heterocyst development. The formation of heterocyst is inversely related to the amount of nitrogen in the medium.

(4) The cellular C : N ratio plays an important role in inducing and controlling the differentiation of vegetative cells into heterocysts.

Functions : Functions of heterocysts are not definitely known, and several views have been put forward from time to time. Early workers considered heterocysts as simply dead cells, thus having no functions. But this view is no more valid as heterocysts have been shown to respire and germinate just like living cells. Brand (1903) considered them as sporangial or spore like structures, and Steinecke (1932) regarded them as secondary reproductive structures. There are evidences to suggest that heterocysts control sporulation.

If heterocysts are removed from the trichome, sporulation is inhibited. Besides this, the vegetative cells adjacent to heterocysts sporulate earlier than those away from it. Canabaeus (1929) found a sort of connection between the formation of gas vacuoles and heterocysts. Fogg (1944) considered them as the sites of salt accumulation. Fritsch (1951) suggested that they secrete certain substances which stimulate growth and cell division.

Fay et al. (1968) considered heterocyst as N_2-fixing organ in blue-green algae. This view gets support from the fact that most of the nitrogen fixing blue-green algae possess heterocysts. The factors inhibiting nitrogenase enzyme system also inhibit the formation of heterocyst, and experiments show strong reducing conditions within heterocysts.

Now it has been conclusively shown that heterocysts are the main sites of N_2-fixation in blue-greens, particularly under aerobic conditions. Under anaerobic conditions, the normal vegetative cells have the capability to fix nitrogen, but under aerobic conditions the nitrogenase enzyme becomes inactive due to the presence of oxygen. Under such conditions, the thick-walled heterocysts provide anaerobic environment, necessary for nitrogenase activity. Stewart (1970) suggested that the main physiological roles of heterocysts are :

(i) control of sporulation,

(ii) involvement in growth and nitrogen metabolism, and

(iii) archaic reproductive units.

The various functions attributed to heterocysts clearly suggest that probably they have varied functions under diverse conditions.

Nitrogen Fixation by Blue-green Algae : An interesting aspect of blue-green algae is the capacity of several forms to fix atmospheric nitrogen. This ability of blue-greens makes them truly autotrophic. Beside blue-greens, the ability to fix atmospheric N_2 is found in few other prokaryotes like heterotrophic and photosynthetic bacteria.

The ability to fix nitrogen is restricted to about 22 filamentous species of Cyanophyceae. All these forms are characteristically heterocystous, e.g., species of Anabaena, Aulosira, Nostoc, Tolypothrix, Cylindrospermum, Mastigocladus, Anabaenopsis, Calothrix and Scytonema. Under anaerobic condition, certain non-heterocystous forms (e.g., Plectonema boryanum) also fix atmospheric nitrogen.

The blue-green algae, fixing nitrogen from the air, must possess an efficient mechanism to check oxygen inhibition of the enzyme nitrogenase. Thick-walled heterocysts provide a good locus for an efficient anaerobic functioning of nitrogenase, even in aerobic conditions. It has been shown that in heterocystous blue-greens, under aerobic conditions, nitrogen fixation occurs only in heterocysts. But under anaerobic conditions, both heterocysts and vegetative cells possess active nitrogenase enzyme.

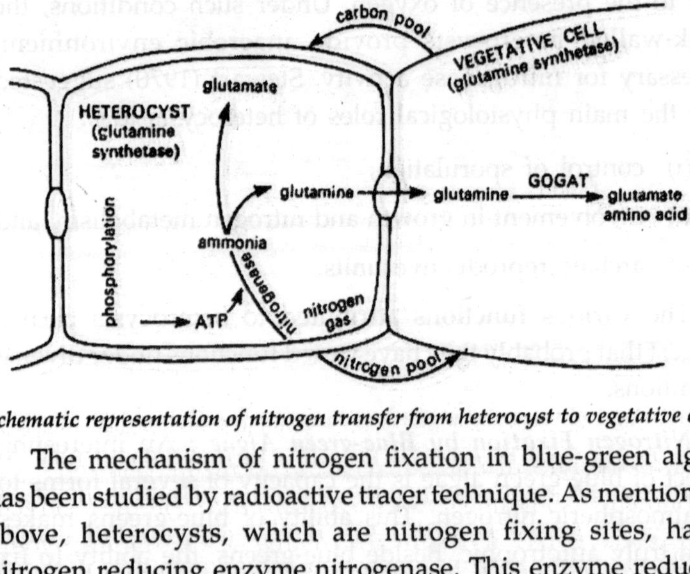

Schematic representation of nitrogen transfer from heterocyst to vegetative cell.

The mechanism of nitrogen fixation in blue-green algae has been studied by radioactive tracer technique. As mentioned above, heterocysts, which are nitrogen fixing sites, have nitrogen reducing enzyme nitrogenase. This enzyme reduces nitrogen gas into ammonia, which in turn is metabolized to glutamine by the enzyme glutamine synthetase (GS). Glutamine is then transported from heterocyst to the adjacent vegetative cell, where it is converted into glutamate by the enzyme GOGAT. Both glutamine and glutamate, by various transamination reactions, form other amino acids. The sequence of reactions in heterocysts and vegetative cells is shown in figure. The fixed nitrogen may be utilized in a number of ways: (i) it is assimilated by the blue-greens themselves, (ii) soluble nitrogenous compounds liberate from healthy cells into the culture medium, and (iii) after death and decay the nitrogenous compounds break into ammonia, which is eventually converted into nitrate by nitrifying bacteria, present in the soil.

In certain blue-green algae (e.g., Anabaena cylindrica) the nitrogenase enzyme is also present in vegetative cells, in addition to heterocysts. Besides this, in Gloeocapsa, although heterocysts are absent, it has the ability of fixing atmospheric nitrogen. Therefore, it shall be unfair to conclude that only hetcrocysts are the sites of nitrogen fixation.

In tropical countries, blue green algae have an important role in increasing the fertility of rice fields. According to Professor R.N. Singh, the fertility of rice fields in India is increased tremendously by species of Anabaena, Tolypothrix and Aulosira. Tolypothrix tenuis could fix nitrogen at the rate of 406 kg per year per acre. This can increase rice production by about 20%.

Phylogeny and Affinities of Cyanophyceae : There are many evidences which indicate that Cyanophyceae are the most primitive of algal organisms. Their prokaryotic cell structure, absence of division of labour and flagellate forms, and lack of sexuality are some of the characters which support their primitive nature. Fossil evidences indicate their origin in early Precambrian period. Their prokaryotic cell structure and absence of flagella lead to believe that they originated from unicellular forms devoid of flagella. Since most of the members of Cyanophyceae are terrestrial, many investigators believe that they originated from terrestrial ancestors rather than aquatic.

Similarity of blue-green algae with red algae in certain characters, such as the presence of similar phycobilins and complete absence of flagellated reproductive cells suggest some affinity in these two groups. However, some other important characters of blue greens, such as prokaryotic cell structure, and the presence of heterocysts, hormogonia and akinetes are not found in red algae. This indicates that there does not appear any close relationship between these two groups.

Since the recognition of blue-greens as a biological group in the early 19th century, they have been treated as a class or division of algae. But around 1960, when fundamental differences in cellular organisation between prokaryotes and eukaryotes were fully elucidated, it became evident that they are more closely related to the prokaryotic bacteria than to the eukaryotic algae. However, a detailed study of the nature of their photosynthetic pigments and capacity to perform oxygenic photosynthesis show that the only unifying character in Cyanophyceae and bacteria is prokaryotic structure of the cell.

The affinities of Cyanophyceae with Rhodophyceae and bacteria are summarised below:

Similarities with Red Algae

(1) Flagellated motile cells are absent in both groups.

(2) The blue (c-phycocyanin) and red (c-phycoerythrin) pigments present in Cyanophyceae are chemically similar to those found in Rhodophyceae (r-phycocyanin and r-phycoerythrin).

(3) Pit connections are present in the family Stigonemataceae of the class Cyanophyceae, and similar structures are also found in Rhodophyceae.

(4) The mechanism of fatty acid synthesis is similar in both groups.

Dissimilarities with Red Algae

(1) The blue-green algae show prokaryotic cell structure, whereas it is typical eukaryotic in the red algae.

(2) Cells of the blue-greens are covered with a mucilagenous sheath, which is absent in the red algae.

(3) Presence of hormogonia and heterocysts is a characteristic of the most blue-greens, but these structures are not found in the red algae.

(4) Sexual reproduction is absent in the blue-green algae, whereas the red algae show an advanced type of sexual reproduction.

Affinities with Bacteria

(1) Both, bacteria and blue-greens show typical prokaryotic structure.

(2) Like bacteria, the blue-green algae have an incipient nucleus (nuclear material without nuclear membrane), and lack membrane bound plastids.

(3) The mucilagenous sheath surrounding blue-green cells and the capsule present in many bacteria have a similar structure, as both are made up of extremely fine fibrils.

(4) Some bacteria (e.g., Thiothrix) are structurally similar to hormogonia, found in the blue-green algae.

(5) Beggiatoa, a sulphur bacterium, resembles with Oscillatoria in shape and movement.

(6) Like bacteria, the blue - green algae are also sensitive to antibiotics.

(7) Many metabolic processes (e.g., sulphur and nitrogen metabolism) are similar in both groups.

(8) Recombination of genetic material, similar to bacteria, has been reported in some blue-greens (e.g., Anacystis nidulans).

Classification of Cyanophyceae : Cyanophyceae has been divided into five orders (Fritsch, 1945). Papenfuss (1955) also recognised five orders of the blue green algae, but included them in the class Schizophyceae along with bacteria (class Schizomycetae) in the division Schizophyta. These five orders have been distinguished on the basis of morphological features, cell form, type of cell division and type of resulting colony, and some features of cell structure, such as presence of inclusions,, polar granules, gas vacuoles and specialized cells (akinetes and heterocysts).

1. *Order: Chroococcales :* Fresh water, marine or terrestrial, unicellular or colonial algae; colonies without polarity; reproduction by fission or endospores; heterocysts and hormogonia absent. Common examples : Aphanocapsa, Chroococcus, Gloeocapsa, Microcystis.
2. *Order : Chamaesiphonales :* Mostly marine, a few fresh water, unicellular or colonial algae; colonies with distinct polarity; reproduction by endospores or exospores; heterocysts and hormogonia absent. Example : Dermocarpa.
3. *Order : Pleurocapsales :* Fresh water or marine, filamentous algae; reproduction by endospores; heterocysts not formed. Example : Pleurocapsa.
4. *Order : Nostocales :* Mostly fresh water, a few marine, filamentous algae; filaments not differentiated into prostrate and upright systems; multiplication by hormogonia; may or may not have heterocysts; spores and akinetes present. Common examples: Oscillatoria, Spirulina, Nostoc, Anabaena, Rivularia, Lyngbya, Cylindrospermum, Gloeotricha, Scytonema, Tolypothrix.
5. *Order : Stigonematales :* Freshwater, filamentous algae with distinct prostrate and upright filaments, showing true branching; reproduce by hormogonia or rarely by akinetes; heterocysts present. Example : Stigonema.

19

The Classification

Chlorophyll is present in all groups of algae, but in some groups pigments other than chlorophyll are dominant and mask the green colour of chlorophyll. On the basis of their colour following four groups were recognized.

(1) Myxophyceae or Cyanophyceae (blue - green algae) : dominant pigment - c -phycocyanin.

(2) Chlorophyceae (green algae) : dominant pigment - chlorophyll *a* and *b*.

(3) Phaeophyceae (brown algae) : dominant pigment - fucoxanthin.

(4) Rhodophyceae (red algae) . dominant pigment -r - phycoerythrin.

In the last few decades a lot of information has become available on the structure, reproduction and physiological processes of algae which has been used to develop more natural systems of classification. The basic characters employed in the classification of algae are the physiological characteristics of vegetative cells (e.g., chemical nature of the cell wall, reserve food materials and types of pigments) and the morphology of motile reproductive cells (e.g., kind, number and position or insertion of flagella).

International Code of Botanical Nomenclature has recommended the following suffixes for the different categories of algae :

Division	–	phyta
Sub-division	–	phytina
Class	–	phyceae
Sub-class	–	phycidae
Order	–	ales
Sub-order	–	inales
Family	–	aceae
Sub-family	–	oideae

Some Important Classifications

Classification Proposed by Fritsch : The first most comprehensive and authorative classification of algae was given by F.E. Fritsch (1935,48) in his book 'The Structure and Reproduction of the Algae'. His classification was based on such criteria as pigmentation, types of flagella, assimilatory products, thallus structure and methods of reproduction. He divided algae into the following 11 classes :

1. *Class Chlorophyceae (Isokontae, Green Algae) :* The main pigments are essentially identical to those found in higher plants. They are chlorophyll *a* and *b*, carotenoid and xanthophylls. The chloroplast usually has pyrenoids. The food reserve is starch, which frequently aggregates around the pyrenoids. The cell wall is composed of cellulose, and the cells are typically eukaryotic. Flagella, if present, are of equal length and are usually inserted at the anterior end. Reproduction

takes place by vegetative, asexual and sexual (iso-, aniso-, and oogamous) methods.

The class Chlorophyceae has been divided into nine orders : (i) Volvocales (e.g., Volvox), (ii) Chlorococcales (e.g., Chlorella), (iii) Ulotrichales (e.g., Ulothrix), (iv) Cladophorales (e.g., Cladophora), (v) Chaetophorales (e.g., Fritschiella), (vi) Oedogoniales (e.g., Oedogonium), (vii) Conjugales (e.g., Zygnema), (viii) Siphonales (e.g., Vaucheria), and (ix) Charales (e.g., Chara).

2. *Class Xanthophyceae (Heterokontae, Yellow Green Algae) :* The main pigments are chlorophyll *a* and *e*, β-carotene and xanthophylls. The plastids are without pyrenoids and the reserve food material is oil. The cell wall is mainly composed of pectic substances with little cellulose. The cell is typical eukaryotic with two anteriorly inserted unequal flagella. The shorter flagellum is of whiplash type and the longer one of tinsel type. Reproduction takes place by vegetative, asexual and sexual (isogamous) methods.

   This class has been divided into four orders : (i) Heterochloridales (e.g., Heterochloris,. Chloramoeba), (ii) Heterococcales (e.g., Myxochloris, Halosphaera), (iii) Heterotrichales (e.g., Tribonema, Microspora), and (iv) Heterosiphonales (e.g., Botrydium).

3. *Class Chrysophyceae :* Phycochrysin is the dominant pigment which imparts brown or orange colour to these algae. The chromatophores have naked pyrenoid-like bodies. The food reserve is chrysolaminarin and leucosin. The cell wall is silicified or calcified and non-cellulosic. The motile cells have two anteriorly inserted equal or unequal flagella. Sexual reproduction is very rare; when present then it is isogamous.

The class includes three orders (i) Chrysomonadales (e.g., Chrysococcus, Chrysodendron, Chromulina), (ii) Chrysosphaerales (e.g., Chrysosphaera, Echinochrysis), (iii) Chrysotrichales (e.g., Nematochrysis, Chrysoclonium).

4. *Class Bacillariophyceae (Diatoms, Yellow or Golden-brown Algae)* : The members of this class are characterised by the dominance of golden-brown pigments, viz. fucoxanthin, diatoxanthin and diadinoxanthin. The chromatophores have pyrenoids and the photosynthetic products are fat and volutin. The cell wall is pectic and silicified and variously ornamented. Usually it consists of two halves which are radially or bilaterally symmetrical. The motile cells usually have a single flagellum. The sexual reproduction takes place by fusion and by the formation of gametes or auxospores.

   The class includes two orders : (i) Centrales (e.g., Cyclotella, Chaetoceras), and (ii) Pennales (e.g., Grammatophora, Navicula, Pinnularia, Denticula, Epithemia).

5. *Class Cryptophyceae* : The main pigment is xanthophyll which imparts brown or red colour to these algae. Pyrenoid-like bodies are present but they are often independent of chromatophores. The photosynthetic products are starch and/ or oil. The motile cells are dorsiventral with two anteriorly inserted unequal flagella. A complex vacuolar system is present in the cell. Sexual reproduction is rare and when present, it is of isogamous type.

   The class includes two orders : (i) Cryptomonadales (e.g., Cryptomonas, Rhodomonas, Cyanomonas), and (ii) Cryptococcales (e.g., Tetragonidium).

6. *Class Dinophyceae (Peridinieae)* : These algae contain many discoid chromatophores and the main pigment

is xanthophyll which imparts brown or red colour. The food reserves are starch and fats. The cell wall is cellulosic. Most of the members of this class are motile unicells with two flagella. Many species are colourless saprophytes or show holozoic mode of nutrition. Sexual reproduction is rare and is of isogamous type, if present. They also multiply by cell division.

In this class six orders are included : (i) Desmomonadales (e.g., Desmocapsa, Pleromonas, Desmomastix), (ii) Thecatales (e.g., Exuviaella, Prorocentrum), (iii) Dinophysiales (e.g., Dinophysis, Ornithocercus, Phalacroma), (iv) Dinoflagellata (e.g., Amphidinium, Blastodinium, Ceratium, Heterocapsa), (v) Dinococcales (e.g., Dinastridium, Cystodinium, Dissodinium), and (vi) Dinotrichales (e.g., Dinothrix, Dinoclonium).

7. *Class Chloromonadineae :* The members of this class have numerous discoid chromatophores; they provide bright green tint due to the presence of excess of xanthophyll. Pyrenoids are lacking and the reserve food is fat and oil. The motile forms are with two almost equal flagella. Sexual reproduction is absent; multiplication takes place by longitudinal division of cells.

   This is a small class with only one order, Chloromonadales, comprising of highly specialized unicellular flagellates (e.g., Trentonia, Vacuolaria).

8. *Class Euglenineae :* These are highly specialized unicellular green flagellates. The cell has usually many chromatophores. Pyrenoid-like bodies are present in some forms. The main pigment is chlorophyll, and the product of photosynthesis is a polysaccharide, paramylon. Most of the members have one or two flagella which arise from the base of a canal like invagination at the anterior end. These algae possess

a complex vacuolar system. The multiplication is by cell division. Sexual reproduction is known only in few forms and is of isogamous type.

This class includes three families : (i) Euglenaceae (e.g., Amphitropis, Ascoglena, Euglena), (ii) Astasiaceae (e.g., Astasia, Distigma, Menoidium), and (iii) Peranemaceae (e.g., Anisonema, Entosiphon, Petalomonas).

9. *Class Phaeophyceae (Brown Algae)* : The chromatophores, beside other pigments, contain fucoxanthin. Some lower forms have naked pyrenoid-like bodies. The food reserves are in the form of alcohol (mannitol), polysaccharides (laminarin) and fats. The cell wall is mainly composed of cellulose with alginic and fucinic acids. The motile reproductive cells have two lateral or sub-apical flagella, one directed forward and the other backward. The sexual reproduction ranges from isogamous to oogamous type.

   The class comprises of nine orders (i) Ectocarpales (e.g., Ectocarpus, Halothrix, Punctaria), (ii) Tilopteridales (e.g., Tilopteris), (iii) Cutleriales (e.g., Cutleria), (iv) Sporochnales (e.g., Sporochnus, Carpomitra), (v) Desmarestiales (e.g., Desmarestia, Arthrocladia), (vi) Laminariales (e.g., Laminaria, Lessonia, Macrocystis), (vii) Sphacelariales (e.g., Dictyota, Padina), and (viii) Fucales (e.g., Fucus, Sargassum, Durvillea).

10. *Class Rhodophyceae (Red Algae)* : Most of these algae are marine with uniaxial or multiaxial thalli. The chromatophores, besides other pigments, contain r-phycoerythrin and r-phycocyanin which impart red colour to these algae. Pyrenoid-like bodies occur in the lower forms of this class. The food reserve is a solid polysaccharide, the foridean starch. The outer cell wall is pectic and the inner cellulosic. The reproductive cells are non-motile; the sexual reproduction is of advanced oogamous type.

The class is divided into seven orders : (i) Bangiales (e.g., Bangia, Porphyra, Porphyridium), (ii) Nemalionales (e.g., Batrachospermum, Lemanea, Nemalion, Chaetangium, Naccaria), (iii) Gelidiales (e.g., Gelidium), (iv) Cryptonemiales (e.g., Dumontia, Corallina, Gloeopeltis), (v) Gigartinales (e.g., Ahnfeltia, Chondrus, Gigartina, Gracilaria), (vi) Rhodymeniales (e.g., Champia, Rhodymenia, Gastroclonium), and (vii) Ceramiales (e.g., Ceramium, Polysiphonia, Lophosiphonia).

11. *Class Myxophyceae (Cyanophyceae, Blue-green Algae) :* These algae are characterised by the presence of a very rudimentary nucleus and they do not have well organized chromatophores (i.e., cell organization is prokaryotic). The chief pigments are chlorophyll *a*, β-carotene and c-phycocyanin. The cell wall is made up of mucopolymers. The food reserve is cyanophycean starch. Sexual reproduction is absent (recently genetic recombination has been reported); asexual reproduction is by hormogonia or akinetes.

    This class comprises of five orders (i) Chroococcales (e.g., Chroococcus, Gloeocapsa, Microcystis), (ii) Chamaesiphonales (e.g., Chamaesiphon, Dermocarpa), (iii) Pleurocapsales (e.g., Pleurocapsa), (iv) Nostocales (e.g., Gloeotrichia, Nostoc, Oscillatoria, Rivularia, Scytonema, Spirulina), and (v) Stigonematales (e.g., Stigonema, Mastigocladus, Nostochopsis).

Classification Proposed by Smith : The classification of algae proposed by Smith (1933, 51, 55) is based on the physiological characteristics of vegetative cells and the morphology of motile reproductive cells. He divided algae into seven divisions and then related classes were included in each division. According to him, the number of divisions necessary for complete classification of algae is less than the number of classes as those classes which show close affinity should be placed in the same division. For example, Xanthophyceae, Chrysophyceae and Bacillariophyceae show certain resemblances in the structure and composition of the cell wall, flagellation and nature of food reserves and despite differences in their pigments, there is enough ground for placing them together in the same division, Chrysophyta.

The seven divisions of algae recognised by Smith are as follows :

Division 1. Chlorophyta : This division includes about 6,750 species, of these 90% are fresh water and the remaining 10% are marine. Chlorophyll *a* and *b* are the dominant pigments; the reserve food is starch. Motile reproductive cells are usually bi- or quadriflagellate; flagella are equal, of whiplash type and anteriorly inserted.

The following two classes were included in this division.

Class 1. Chlorophyceae (grass-green algae; e.g., Volvox, Ulothrix, Oedogonium).

Class 2. Charophyceae (stoneworts; e.g., Chara).

Division 2. Euglenophyta : There are 450 fresh-water or terrestrial species in this division. The dominant pigments of these algae are chlorophyll a, b and β-carotene, and their reserve food is paramylum and fats. Motile cells may be uni-, bi-, or triflagellate; flagella anterior, inserted into a narrow gullet. Multiplication takes place usually by cell division.

This division has only one class.

Class Euglenophyceae (euglenoids; e.g., Euglena, Colacium).

Division 3. Pyrrophyta : There are about 1,030 species, mainly unicellular and rarely colonial forms. The pigments are chlorophyll a, chlorophyll c, β-carotene and xanthophylls; the food reserve is starch and/ or oil. The cell wail is cellulosic.

Motile cells are usually with two anteriorly inserted unlike flagella. Sexual reproduction is rarely present.

This division has been divided into the following two classes.

Class 1. Desmophyceae (dinophysids; e.g., Exuviaella).

Class 2. Dinophyceae (dinoflagelloids ; e.g., Dinophysis, Dinastridium).

Division 4. Chrysophyta : This division is represented by more than 6,000 species, about three-fourth of which are fresh-water and one-fourth marine. These algae are characterised by the predominance of carotenes and xanthophylls. The food reserve is leucosin and oil. The cell wall is usually composed of two overlapping silicified halves. Sexual reproduction is iso-, aniso-, or oogamous.

This division includes the three classes:

Class 1. Chrysophyceae (golden brown algae; e.g., Chromulina).

Class 2. Xanthophyceae (yellow-green algae; e.g., Botrydium).

Class 3. Bacillariophyceae (diatoms; e.g., Pinnularia).

Division 5. Phaeophyta (Brown Algae) : There are about 1,000 species of brown algae, mostly marine. The dominant pigments are phycophein and fucoxanthin; the assimilatory products are laminarin (polysaccharide) and mannitol (alcohol). The cell wall is cellulosic with fucinic and alginic acids. Motile reproductive cells are pyriform with two laterally inserted flagella, one of which is of tinsel type. Sexual reproduction is iso-, aniso-, or oogamous type.

This division has been divided into the following three classes.

Class 1. Isogeneratae (e.g., Ectocarpus, Sphacelaria, Dictyota).

Class 2. Heterogeneratae (e.g., Myrionema, Laminaria).

Class 3. Cyclosporeae (e.g., Sargassum, Fucus).

Division 6. Cyanophyta (Blue-green Algae) : This division is represented by 1,500 species, mostly fresh water. Some species are free-living, while others grow on larger algae or within the tissue of other plants. The cell is prokaryotic; the nucleus lacks nuclear membrane and nucleolus. In addition to other pigments, they contain a blue (c-phycocyanin) and a red (c- phycoerythrin) pigment. The food is stored in the form of cyanophycean starch. Motile stages are absent but vegetative filaments of some forms show gliding movements. Sexual reproduction is absent; asexual reproduction takes place by hormogonia, fragmentation, akinetes, etc.

The members of this division are placed in a single class, Myxophyceae or Cyanophyceae (e.g., Nostoc, Anabaena, Oscillatoria).

Division 7. Rhodophyta (Red Algae) : This division includes about 2,500 species, mostly marine. They have a predominance of r-phycoerythrin which masks the other pigments to give them a distinctive red colour. They store food in the form of floridean starch. The thallus is non-motile and complex. Sexual reproduction is oogamous; motile reproductive cells are not found.

This division contains only one class, Rhodophyceae (e.g., Batrachospermum, Polysiphonia, Chondrus, Gracilaria).

Classification Proposed by Papenfuss : G.F. Papenfuss (1955), on the basis of phylogeny, recognised 8 phyla and 12 classes in algae as listed hereunder.

Phyla	*Classes*
1. Chlorophycophyta	Chlorophyceae
2. Charophycophyta	Charophyceae
3. Euglenophycophyta	Euglenophyceae

Contd...

Phyla	*Classes*
4. Chrysophycophyta	Chrysophyceae
	Xanthophyceae
	Bacillariophyceae
5. Pyrrophycophyta	Cryptophyceae
	Chloromonadophyceae
	Dinophyceae
6. Phaeophycophyta	Phaeophyceae
7. Rhodophycophyta	Rhodophyceae
8. Schizophycophyta	Schizophyceae

Classification Proposed by Chapman : V.J. Chapman (1962) divided algae into the following four phyla on the basis of morphology, pigmentation and biochemical characters.

Phylum 1. Euphycophyta

Class 1. Charophyceae

Class 2. Chlorophyceae

Class 3. Phaeophyceae

Class 4. Rhodophyceae

Phylum 2. Myxomycophyta

Class 1. Myxophyceae (Cyanophyceae)

Phylum 3. Chrysophycophyta

Class 1. Chrysophyceae

Class 2. Xanthophyceae

Class 3. Bacillariophyceae

Phylum 4. Pyrrhophycophyta

Class 1. Cryptophyceae

Class 2. Dinophyceae

Classification Proposed by Prescott : G.W. Prescott presented his system of classification of algae in 1969 in his book, 'The Algae : A Review.' He was of the opinion that pigment similarities are fundamental and imply kinship. He has taken into consideration the following criteria for the classification of algae.

(1) Photosynthetic pigments - their kinds, chemical composition and relative amounts.

(2) Photosynthetic products - their chemical nature.

(3) Flagella - their types, number and position of insertion.

(4) Cell wall - structure and chemical composition.

(5) Presence or absence of true nucleus.

(6) Life cycle and reproduction - morphology of the reproductive structures and the methods of reproduction.

He classified algae into nine phyla:

Phylum	*Classes*
1. Chlorophyta (green algae; chlorophyll predominant pigment, food reserve starch).	i. Chlorophyceae ii. Charophyceae
2. Euglenophyta (euglenoid algae; chlorophyll predominant pigment, food reserve paramylum).	
3. Chrysophyta (yellow-green algae; carotenoids and xanthophylls predominant pigments, food reserve leucosin and oil).	i. Chrysophyceae ii. Bacillariophyceae iii. Heterokontae (Xanthophyceae)
4. Pyrrhophyta (dinoflagellates; carotenes, xanthophylls and peridinin	i. Desmokontae (Desmophyceae) ii. Dinokontae

Contd...

Phylum	*Classes*
predominant pigments, food reserve starch and oil).	(Dinophyceae)
5. Phaeophyta (brown algae or brown seaweeds; fucoxanthin predominant pigment, food reserve soluble carbohydrate).	i. Isogeneratae ii. Heterogeneratae iii. Cyclosporeae
6. Rhodophyta (red algae; phycoerythrin dominant pigment; reserve food floridean starch).	i. Sub-phylum Bangioideae ii. Sub-phylum Florideae
7. Cyanophyta (blue-green algae; phycocyanin predominant pigment).	i. Sub-phylum Coccogoneae ii. Sub-phylum Hormogoneae
8. Cryptophyta (blue and green flagellates; compressed slipper shaped flagellated cells; xanthophyll or chlorophyll predominant pigment; food reserve starch-like carbohydrate).	
9. Chloromonadophyta (chloromonads; chlorophyll predominant pigment, food reserve oil)	

Classification Proposed by Round : F.E. Round (1973) divided algae into two major groups—Procaryota and Eucaryota. Procaryota are characterized by the absence of delimiting nuclear membrane, nucleolus and other membrane bound cell organelles, whereas Eucaryota have well organized nucleus and other cell organelles. The following is the outline of Round's classification :

Some Salient Features of Major Classes of Algae

Characters	*Cyanophyceae*	*Chlorophyceae*	*Bacillariophyceae*	*Phaeophyceae*	*Rhodophyceae*
1. Habitat	fresh water, marine or terrestrial.	fresh water, marine or terrestrial.	fresh water, marine or terrestrial.	only four genera are fresh water, rest marine.	mostly marine; a few are terrestrial.
2. Thallus form	unicellular, colonial or filamentous.	unicellular, colonial, coenobial, filamentous, thalloid or siphonaceous.	unicellular or colonial.	filamentous, branched, usually heterotrichous.	unicellular, filamentous, pseudoparenchymatous.
3. Cell structure	prokaryotic, membrane bound nucleus and other cell organelles like plastids, mitochondria, etc., are absent.	eukaryotic, well organised nucleus and other cell organelles present.	eukaryotic.	eukaryotic.	eukaryotic.
4. Cell wall	mucopeptides and muramic acid are two main constituents of the cell wall.	cell wall is made up of cellulose and pectic material.	cell wall is made up of silica and pectic material.	outer mucilaginous layer of cell wall consists of alginic and fucinic acids and the inner layer is made up of cellulose.	outer layer is pectic in nature whereas inner layer is cellulosic.
5. Flagella	non flagellate motile cells are absent.	motile cells with 2, 4 or more flagella; flagella anteriorly inserted, equal in length and whiplash type.	usually 1, or sometimes 2, anteriorly inserted.	2, laterally inserted, of unequal length; one of them whiplash and another of tinsel type	non flagellate, motile cells are absent.
6. Dominant pigments	chlorophyll *a*, β-carotene, xanthophyll and c-phycocyanin.	chlorophyll *a*, and *b*, α - and β-carotene, xanthophyll.	chlorophyll *a*, and *c*, β-carotene, xanthophyll. (fucoxanthin).	chlorophyll *a* and *c* β-carotene, xanthophyll (violaxanthin, fucoxanthin).	chlorophyll *a*, and *d*, β-carotene (lutein), xanthophylls, r-phycoerythrin.

Contd...

Characters	*Cyanophyceae*	*Chlorophyceae*	*Bacillariophyceae*	*Phaeophyceae*	*Rhodophyceae*
7. Nucleus	nuclear membrane absent, DNA not associated with histones, therefore well-organised chromosomes are not found.	well organised nucleus with chromosomes and nucleolus is present.	as in Chlorophyceae.	as in Chlorophyceae.	as in Chlorophyceae.
8. Eye spot	absent	attached with chloroplast but it has no relation with flagella	absent	in male gametes, eye spot is present, attached to chromatophores, it is related with flagella	absent
9. Pyrenoid	absent	single or compound; present either on the surface or embedded in chloroplast.	simple or compound; one or more.	single, stalked or projected	simple or compound; embedded, sometimes stalked.
10. Reserve food material	special type of starch (cyanophycean starch) and protein (cyanophycin protein).	starch and fats.	oil, volutin and chrysose.	laminarin, mannitol, sucrose.	floridean starch.
11. Reproduction	multiplication only through asexual methods; sexuality completely absent.	reproduction by vegetative, asexual and sexual means; sexual reproduction isogamous, anisogamous or oogamous.	reproduction by vegetative (cell division) and sexual means; sexual reproduction iso-, aniso-, or oogamous.	reproduction by vegetative (fragmentation, adventitious branches), asexual (motile and aplanospores) sexual reproduction. iso-, aniso or oogamous.	reproduction by vegetative (cell division, fragmentation), (monospores. neutral spores, tetraspores, carpospores) and sexual means; sexual reproduction complex and advanced oogamous.

Major groups	*Phylum*	*Pigments*
1. Procaryota	1. Cyanophyta	blue-green pigments
2. Eucaryota	2. Euglenophyta	
	3. Chlorophyta	
	4. Charophyta	green pigments
	5. Parsinophyta	
	6. Xanthophyta	
	7. Haptophyta	
	8. Dinophyta	
	9. Bacillariophyta	brown pigments
	10. Chrysophyta	
	11. Phaeophyta	
	12. Rhodophyta	red pigments
	13. Cryptophyta	green-red pigments

Classification Proposed by Chapman and Chapman : V.J. Chapman and D.J. Chapman (1973) have also employed the Procaryota-Eucaryota concept in the classification of algae. In Procaryota they placed Cyanophyta, and the remaining 10 divisions were included within Eucaryota. Their divisions were based on the structure of organelles, such as flagella, and biochemical characteristics. A brief outline of their classification is given below.

	Division	*Class*
Procaryota :	Cyanophyta (blue-green algae)	Charophyceae
Eucaryota :	1. Rhodophyta (red algae)	Rhodophyceae
	2. Chlorophyta (green algae)	Chlorophyceae Prasinophyceae Charophyceae

Contd...

Division	Class
3. Euglenophyta (euglenoids)	Euglenophyceae
4. Chloromonadophyta (chloromonads)	Chloromonadophyceae
5. Xanthophyta (yellow-green algae)	Xanthophyceae
6. Bacillariophyta (diatoms)	Bacillariophyceae
7. Chrysophyta (golden-brown algae)	Chrysophyceae Haptophyceae
8. Phaeophyta (brown algae)	Phaeophyceae
9. Pyrrhophyta (dinoflagellates)	Dinophyceae Desmophyceae
10. Cryptophyta (cryptomonads)	Cryptophyceae

Classification Proposed by Lee : Robert Edward Lee (1980) divided algae into six divisions :

Division	*Class*
1. Cyanophyta (prokaryotic cells)	1. Cyanophyceae (blue-green algae)
2. Glaucophyta (eukaryotic cells with no chloroplasts with endosymbiotic blue-greens (cyanelles)	
3. Chromophyta (algae with chloroplast,	1. Euglenophyceae (euglenoids) 2. Dinophyceae (dinoflagellates)

Contd...

Division	*Class*
E.R.,storage product out-side chloroplast envelope)	3. Cryptophyceae (cryptophytes) 4. Chrysophyceae (golden-brown algae) 5. Prymnesiophyceae (haptophytes) 6. Bacillariophyceae (diatoms) 7. Rhaphidophyceae (chloromonads) 8. Xanthophyceae (yellow-green algae) 9. Eustigmatophyceae 10. Phaeophyceae (brown algae)
4. Rhodophyta (chlorophyll *a* and *d*, phycobiliproteins, no flagellated cells, storage product floridean starch)	1. Rhodophyceae (red algae)
5. Chlorophyta (chlorophyll *a* and *b*, storage product starch found inside the chloroplast)	1. Chlorophyceae (green algae)
6. Charophyta (Chlorophyll *a* and *b*, sex organs surrounded by an envelope of sterile cells)	1. Charophyceae (stone worts)

20

General Characters

Algae comprise a group of chlorophyll containing thalloid plants of the simplest type, having no true roots, stems, leaves or leaf like organs. They are placed in the Division Thallophyta along with fungi. They differ from fungi in the presence of photosynthetic pigment - chlorophyll and in their mode of nutrition. Although most of the algae are autotrophic, i.e., they synthesize their food by themselves, yet heterotrophic and holozoic forms are not uncommon. The important distinguishing features of algae are as follows :

(1) They are mostly chlorophyll containing autotrophic thalloid plants.

(2) They occur in a variety of habitats, but majority of them are aquatic.

(3) The plant body does not show differentiation into various tissue systems.

(4) They have mostly unicellular sex organs without a jacket of sterile cells around them. Jacket cells, if present, have different initials. There is a progressive complexity in reproduction.

(5) Embryo is not formed after gametic union.

(6) They show distinct alternation of generation.

Occurrence

Algae are of universal occurrence and they are found in a variety of habitats, such as freshwater, sea water, on snow, on rocks and on/or within the plant and animal bodies. Of these, aquatic forms are most common. On the basis of habitat, they may be classified into the following three groups:

(1) Aquatic algae

(2) Terrestrial algae

(3) Algae of unusual habitats

Aquatic Algae : Aquatic forms are found in fresh water or in saline water of the sea.

Fresh Water Forms : These forms occur in fresh water (low salinity water) of ponds, pools, lakes, rivers, etc. Some fresh water forms like Cladophora, Oedogonium, Ulothrix and Chara are found in slow running water, whereas others like Chlamydomonas, Volvox Hydrodictyon and Spirogyra occur in stagnant water.

Marine Forms : These forms occur in saline water of the sea and are represented by the members of Phaeophyceae (e.g., Ectocarpus, Sargassum, Fucus) and Rhodophyceae (e.g., Polysiphonia).

The aquatic algae are either free-floating (e.g., Chlamydomonas, Volvox, Spirogyra) or are attached to a substratum with the help of an attachment disc, known as holdfast (e.g., Oedogonium, Ulothrix). Many free-floating algae, together with other similar organisms, form colonies on the surface of water which are called water blooms or phytoplanktons.

The algae present in phytoplankton are dependent upon the chemicals present in water. Usually marine planktons have a large variety of species than fresh water planktons. Planktonic species often have a relatively slow rate of sinking, as they contain gas vacuoles or oil globules which increase their buoyancy. Motility of planktonic colonies is mainly by means of flagella.

Algae like Dictyosphaerium, Fragilaria, Cosmarium, Volvox, Asterionella and Golenkinia are commonly found in fresh water planktons, whereas Actinocyclus, Chaetoceros and Coscinodiscus are common in marine planktons.

During night planktons utilize oxygen dissolved in water and as such submerged aquatic organisms do not get sufficient oxygen for respiration. Besides, phytoplanktons also prevent atmospheric oxygen from dissolving in water, thus making the habitat unsuitable for submerged aquatic life.

Many algae are found attached to rocks along the edges of lakes and seas and these forms are called phytobenthos.

Terrestrial Algae : Some algae are found in terrestrial habitats like soils, rocks, logs, etc. Forms like Vaucheria, Botrydium, Fritschiella and Euglena, which are found on the surface of the soil, are known as saphophytes, whereas many blue-green algae (e.g., Nostoc, Anabaena), which occur under the surface of the soil, are called cryptophytes.

Some algae are also found on tree trunks and moist walls, and they absorb carbon dioxide and water from the atmosphere (e.g., Protococcus, Scytonema).

Algae of Unusual Habitats : In addition to aquatic and terrestrial habitats, some algae also occur in uncommon habitats.

Halophytic Algae : There are many algae which occur in saline water of seas, but some others can withstand high concentration of salts and occur in salt lakes. They are called halophytes, and include Chlamydomonas ehrenbergii, Dunaliella and Stephanoptera.

Epiphytic Algae **:** They grow on larger algae or on bryophytes and angiosperms. For example, species of Bulbochaete, Oedogonium and Microspora are found attached to the larger species of Cladophora, Rhizoclonium and Vaucheria. Bryophytic and angiospermic flora of rivers and lakes harbour many algal members on their surface. Algae with mucilagenous thalli like Chaetophora, Oedogonium and Zygnema are epiphytic on Achnanthes, Eunotia, Synedra, etc.

Epizoic Algae **:** Many algae grow on animals like snails, fishes and tortoise, and they are known as epizoic. For example, Cladophora crispata grows on snails, and species of Stigeoclonium are found in the gills of fishes. Similarly, species of Characiopsis and Characium are found on the legs of Branchipus.

Endozoic Algae **:** Algae which occur in the tissues of animals are known as endozoic. For example, species of Zoochlorella are found in Hydra viridis. Besides, there are some 15 species of the family Oscillatoriaceae (blue-green algae) which occur in the respiratory and digestive tracts of vertebrate animals.

Symbiotic Algae **:** Several members of Chlorophyceae and Cyanophyceae form symbiotic association with fungi, bryophytes, gymnosperms and angiosperms. Lichens are important example of symbiotic life and their algal components belong to Cyanophyceae (e.g., Nostoc, Gloeocapsa, Microcystis) or Chlorophyceae (e.g., Coccomyxa, Chlorella, Protococcus). Colonies of Nostoc and Anabaena are found in symbiotic association in the thallus of Anthoceros and the coralloid roots of Cycas.

Cryophytic Algae **:** Algae growing on ice or snow are called cryophytic. These algae provide attractive colours to snow covered mountains. The alpine and arctic mountains become red due to the growth of Haemotococcus nivalis. Green snow in Europe is due to Chlamydomonas yellowstonensis and species of Mesotaenium. Sometimes snow becomes black due to Scotiella nivalis and Raphidonema brevirostri and brownish purple due to Ancyclonema nordenskioldii.

Lithophytic Algae : Algae growing on moist rocks and stones are known as lithophytic. Many blue - green algae like Nostoc, Rivularia and Gloeocapsa commonly grow on moist and shady rocks. Scytonema is common on moist walls during rainy season.

Parasitic Algae : Algae also grow as parasites on many plants and animals. Cephaleuros virescens (Chlorophyceae) causes red rust in tea and coffee plantations in Assam and adjoining areas. Phyllosiphon (Chlorophyceae) is a parasite on the leaves of Arisarum vulgare. Polysiphonia fastigiata (Rhodophyceae) is a semiparasite on Ascophyllum nodosum.

Thermophytes (Thermal Algae) : Many blue-green algae (e.g., Oscillatoria brevis, Synechococcus elongatus, Heterohormogonium sp., Haplosiphon lignosum) are found in hot water springs (50-70°C), where normal life is not possible. They are able to survive such at high temperatures due to the absence of well organised nucleus. Since most hot springs are alkaline, some algologists believe that tolerance to high temperature is also associated with this factor.

Range of Thallus Structure : The vegetative structure of algae shows a wide variety and it ranges in form from unicellular to complex multicellular thalli. Their size ranges from one micron to several meters. On the basis of thallus organisation, algae are divided into the following five groups :

(1) Unicellular forms

(2) Colonial forms

(3) Filamentous forms

(4) Siphonaceous forms

(5) Parenchymatous forms

Multicellular forms have been derived by repeated divisions of unicellular forms. Colonial forms are developed by the aggregation of the products of cell division within a mucilage mass. Filamentous forms are formed by repeated transverse divisions of cells without separation of daughter cells. Repeated nuclear divisions, without cross wall formation, give rise to siphonaceous forms. Parenchymatous thalli are formed by the division of cells of a filament in two or more planes.

Unicellular Forms : Simple unicellular forms are found in all groups of algae except Charophyceae and Phaeophyceae. These forms are sometimes referred to as acellular since they function as complete living unit without any cellular differentiation. Unicellular forms can be classified into the following four sub-groups.

Rhizopodial Unicells : These forms lack a rigid cell wall. They possess cytoplasmic projections which help them in amoeboid movement. In the ahsence of rigid cell wall, these forms are periplastic. Examples - Rhizochrysis (Chrysophyceae), Rhizochloris (Xanthophyceae), Chrysamoeba (Chrysophyceae).,

Flagellated Unicells : Flagellated vegetative cells are found in all groups of algae except Cyanophyceae, Phaeophyceae and Rhodophyceae. Such cells show similarity with motile gametes and zoospores. They vary in the number and type of flagella. For example, in Chlorophyceae, vegetative cells, zoospores and gametes have two equal flagella. Their number may sometimes vary, but is always in multiple of two. In Dinophyceae and Xanthophyceae there are two unequal flagella; in the former they emerge in different planes and in the latter in the same plane. In Euglenophyceae there is only one flagella, inserted at the anterior end.

Flagellated unicells may be periplastic without cell wall (e.g., Euglena) or with a distinct cell wall (e.g., Chlamydomonas). In Phacotus (Chlorophyceae), there is a thick calcareous covering (capsule) around the cell wall.

Spiral Filamentous Unicells : Some unicellular algae form spiral or coiled structures. e.g., Spirulina (Cyanophyceae).

Non-motile Unicells : These are non-motile coccoidal algae which do not possess flagella, eye spot, etc., meant for locomotion. The simplest non-motile forms are found in Cyanophyceae where they do not possess well organised nucleus and plastids (prokaryotic), e.g., Chroococcus. The non-motile unicells of Chlorophyceae possess nucleus and plastids (e.g., Chlorella;). The non-motile unicells of Bacillariophyceae, such as diatoms, are made up of two halves or theca, joined by a girdle band.

Colonial Forms : The colonial habit is achieved by aggregation of the products of cell divisions within a mucilagenous mass, by aggregation of motile cells or juxtaposition of cells subsequent to cell divisions. Therefore, all members of a colony have similar structure. These associations are usually loose, and as such a colony may break into smaller pieces. In some colonial forms, all members of a colony are connected with each other by cytoplasmic connections, hence they cannot break into segments (e.g., Volvox). On the basis of morphology, colonial organisations may be divided into the following four types :

Coenobial : A colony with a definite shape, size and arrangement of cells is known as coenobium. The number of cells in a coenobium is determined at the juvenile stage and subsequently there is only increase in size. Coenobia may be motile or non motile; in the former cells are flagellated (e.g.,

Pandorina, Eudorina, Volvox;) and in the latter without flagella (e.g., Hydrodictyon;).

Palmelloid : Unlike coenobial forms, in a palmelloid colony the number of cells, their shape and size is not definite. The cells remain irregularly aggregated within a common mucilagenous matrix, but they are independent and function as individuals. In some palmelloid forms it is a temporary phase (e.g., Chlamydomonas), whereas in others it is a permanent feature as in Tetraspora (Chlorophyceae), Aphanotheca (Cyanophyceae;).

Dendroid : Here the colony looks like a microscopic tree. As in palmelloid colony, the number, shape and size of the cells is also indefinite in dendroid colonies. A mucilagenous thread is present at the base of each cell, and the threads of different cells are united to form a branched structure. It gives a tree like appearance to the whole colony (e.g., Chrysodendron, Chrysophyceae;).

Rhizopodial : In a rhizopodial colony the cells are united through rhizopodia (e.g., Chrysidiastrum, Chrysophyceae;).

Filamentous Forms : Filamentous forms are developed by repeated transverse divisions of cells. The daughter cells do not separate and they remain attached one upon the other in a definite sequence to form a filament. The filaments may be branched or unbranched, and the cells in a filament may be arranged in a single row (uniaxial) or more than one rows (multiaxial).

Unbranched Filaments : Simple unbranched filaments are found only in a few groups of algae. Such filaments may be free floating (e.g., Spirogyra;), attached to some substratum (e.g., Zygnema, Ulothrix Oedogonium;) or form colony (e.g., Nostoc, Oscillatoria;).

Branched Filaments : Branched filaments are formed by repeated transverse divisions of lateral outgrowths of cells.

The branching of filaments may be false or true. A false branch does not arise as lateral outgrowth, but the trichome may break either due to death or decay of intercalary cells or at the point of heterocyst. The broken end emerges out of the mucilagenous sheath in the form of a branch, ,e.g., Scytonema (Cyanophyceae;). False branches in Scytonema arise almost always in pairs at some distance from the heterocyst.

True branches, which arise as lateral outgrowths, may result in the following three types of filaments :

(a) *Simple Filaments* : Simple branched filaments remain attached to the substratum by a basal cell. In such filaments branches may arise from any cell except the basal cell. In Cladophora, branches arise just below the septa between two adjacent cells.

(b) *Heterotrichous :* In this type, the thallus is very much evolved and differentiated into prostrate and erect systems (e.g., Fritschiella, Ectocarpus, Draparnaldiopsis, Coleochaete, Stigeoclonium). Both the prostrate and the erect systems may be well developed (e.g., Fritschiella, Ectocarpus;) or there is progressive elimination of prostrate (e.g., Draparanaldiopsis) or erect (e.g., Coleochaete) system.

(c) *Pseudoparenchymatous* : In many filamentous forms one or more central or axial filaments, together with their branches, form a 'parenchymatous' structure. If a pseudoparenchymatous thallus is formed by the branches of only one filament, it is called uniaxial (e.g., Batrachospermum), and if branches of more than one filaments are involved, it is said to be multiaxial (e.g., Polysiphonia;).

Siphonaceous Forms : Here the thallus is made up of branched, aseptate, coenocytic, tubular filaments as the nuclear divisions are not accompanied by wall formation, e.g., Vaucheria and Botrydium.

Parenchymatous Forms : In this type, the flat foliose or tubular thalli are formed by the divisions of cells in two or more planes. The daughter cells do not separate from the parent and give rise to parenchymatous thalli of various shapes, like flat (Ulva;), tubular (Scytosiphon, Phaeophyceae) or complex (Sargassum;). The growth of such thalli is apical (e.g., Fucus, Dictyota), intercalary (e.g., Laminaria) or trichothallic (e.g., Porphyra, Rhodophyceae).

Cell Structure in Algae : On the basis of their organization, algal cells may be differentiated into prokaryotic, mesokaryotic and eukaryotic types. The prokaryotic cell organization is found in Cyanophyceae which is characterized by : (i) the presence of incipient nucleus, (ii) the absence of membrane bound organelles like plastids, endoplasmic reticulum, Golgi bodies and mitochondria, (iii) the absence of basic proteins-the histones in DNA, (iv) the presence of mucopeptide in the cell wall, and (v) the absence of mitosis.

A eukaryotic cell, on the other hand, is characterized by the presence of a well organized nucleus and membrane bound organelles like plastids, mitochondria and Golgi bodies. Majority of algae show this type of cell organization. An intermediate type of cell organization, i.e., mesokaryotic is found in the members of Dinophyceae, where although the nucleus has a nuclear membrane and chromosomes (eukaryotic characters), basic proteins are absent (prokaryotic character).

Cell Wall : Most of the non-motile unicellular and multicellular algae possess a typical cellulosic cell wall of non-living matter outside the cytoplasmic membrane. It is composed of pure or mixed carbohydrates (cellulose, mucilage, pectin, etc.), and substances like alginic acid, fucoidin, fucin and hemicelluloses are present in brown algae (Phaeophyceae), pectin in red algae (Rhodophyceae), calcium carbonate in coralline algae (red algae) and silica in diatoms (Bacillariophyceae).

Electron microscopic studies have revealed that cellulosic cell wall is composed of microfibrils of cellulose which vary in thickness from 30 to 200 μm. The microfibrils remain embedded in a smooth or granular matrix and are variously oriented. They may be arranged in two layers at right angles to each other, in three layers or in a random manner. The cell wall of diatoms is more complex, having pores with perforated membranes.

In blue-green algae (Cyanophyceae), the cell wall is mucopolymeric, mainly consisting of glucosamine, amino acids, muramic acid and diaminopimelic acid. A true cell wall is, however, absent in some algae like Euglena, Gymnodinium and Pyramimonas. They simply have a bounding membrane of cytoplasm, known as pellicle.

Flagella : Motile vegetative or reproductive cells are found in all groups of algae except Cyanophyceae and Rhodophyceae. Their motility is due to small filiform (thread like) protoplasmic appendages, called flagella. The number of flagella varies from one to four to many (Oedogonium, Vaucheria). They are mainly of the following two types :

1. *Whiplash or Acronematic Flagella* : Such flagella have a smooth surface.
2. *Tinsel or Pleuronematic Flagella* : The surface of these flagella is covered with fine filamentous appendages, known as mastigonemes or flimmers. They are further divided into three categories on the basis of arrangement of mastigonemes.

   (a) *Pantonematic* : In this type mastigonemes are arranged in two opposite rows or show radial arrangement.

   (b) *Pantocronematic* : A pantonematic flagellum with a terminal fibril is known as pantocronematic.

   (c) *Stichonematic* : Here mastigonemes develop only on one side of the flagellum.

A motile cell may have either one or both the above mentioned types of flagella. It is a specific character. If all flagella of a cell are similar, it is known as isoknot and when dissimilar, it is called heteroknot.

The motile stages of Chlorophyceae possess two or four anteriorly inserted whiplash flagella of equal length, whereas the members of Phaeophyceae and Xanthophyceae have one whiplash and one tinsel flagellum of unequal length.

A transection of flagellum reveals that it consists of nine peripheral doublet and two central singlet fibrils. All fibrils are :

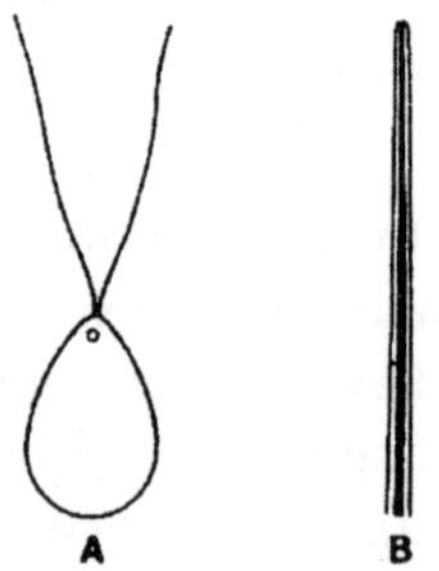

A-B. Acronematic flagella A. An algal cell with acronematic flagella, B. A single acronematic flagellum.

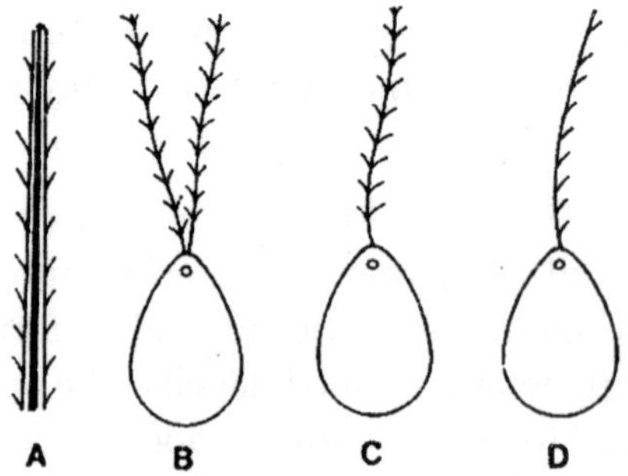

A-D. Pleuronematic flagella: A. A pleuronematic flagellum showing mastigonemes, B. A cell with two pantonematic flagella, C. A cell with pantocronematic flagellum, D. A cell with stichonematic flagellum.

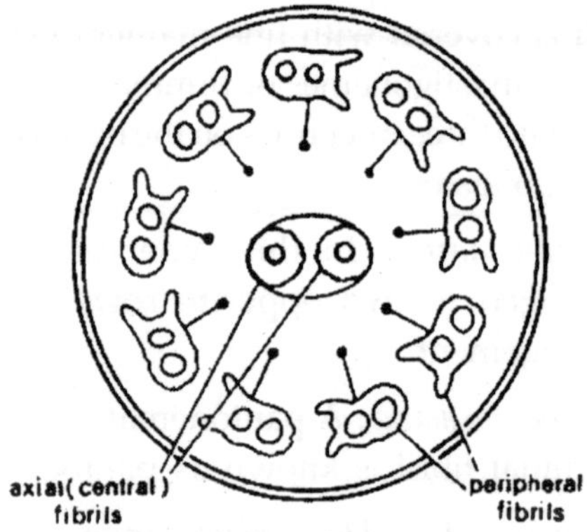

Transection of a flagellum.

enclosed within a common covering, formed by the extension of plasma membrane, but the two central fibrils have an additional covering of their own. The peripheral as well as central fibrils extend almost throughout the length of the flagellum. The nine peripheral fibrils at the proximal end are attached to a hollow basal body (which is separated from the flagellum by a diaphragm). The two central fibrils terminate just short to the diaphragm.

Nucleus : Almost all algae, with the exception of Cyanophyceae, have a well organized eukaryotic nucleus. In Cyanophyceae, the genetic material is not found within the membrane bound nucleus and the DNA strands do not combine with histones to form chromosomes. Such a nucleus is called prokaryotic or incipient. The nucleus in Dinophyceae is also not truely eukaryotic, although it is membrane bound, but well organised chromosomes are not present.

In uninucleate algae the nucleus generally lies alongside the cell wall but sometimes it is suspended in the centre of the cell by fine cytoplasmic threads (e.g., Spirogyra, Zygnema). In multinucleate members, the nuclei usually occur between the vacuoles and the plastids in the cytoplasm. The nucleus varies in size from less than 1 μm to 80 μm.

The nuclear membrane, which bounds the nucleus, is two layered, and in some members the two unit membranes are separated by a perinuclear space. The outer membrane is continuous with the membranes of endoplasmic reticulum. Each nucleus contains one or more dark stained nucleoli or endosomes. The four types of nucleolar structures found in algae are : (i) a single nucleolus per nucleus, (ii) two or more distinct nucleoli per nucleus, (iii) a complex nucleolar mass, and (iv) a linear association of large number of small nucleoli.

There is a great variation in the morphology of chromosomes. They are small and spherical in desmids, and elongated and thread like in Oedogonium. Chromosomes may have a localized (median or sub-terminal) or diffused centromere. The size of chromosomes varies form 0.25 μm or less (e.g., Ulotrichales) to more than 12 μm (e.g., Oedogoniales and Charales). The number of chromosomes also shows much variation, the lowest number, n=2, is found in Porphyra linearis (Rhodophyceae) and the highest, n = 592, in Netrium digitalis (Chlorophyceae).

Golgi Bodies or Dictyosomes : Golgi bodies are present in all algal cells except blue-green algae, and can be seen under electron microscope. They may be found in the region of the nucleus (e.g., Chlamydomonas), near plastids (e.g., diatoms and Bulbochaete), associated with the flagellar bases (e.g., Chlamydomonas) or without any apparent association with any particular organelle (e.g., Euglena, Fucus). Golgi bodies are composed of 2-20 flat vesicles which are arranged in stacks. Each stack is called a dictyosome (all dictyosomes collectively form the Golgi apparatus).

The exact role of Golgi bodies in algae is not known. They are associated with the synthesis of cell metabolites, and have also been shown to contribute to the plasma membrane as in higher plants.

Mitochondria : Well organised mitochondria are present in all algal cells with the exception of blue - green algae. A mitochondrion is bounded by a double membrane. The inner membrane is projected into lumen and forms finger like projections, called cristae. There are usually few cristae in Chlorophyceae and many in Chrysophyceae, Phaeophyceae and Xanthophyceae. The lumen, surrounded by bounding membranes, is filled with a granular matrix which contains nucleic acids (RNA and DNA). Hence, these structures can be called independent cell organelles. Their shape and number per cell varies in different groups of algae. There are usually more than one mitochondria per cell, but in Micromonas (Chlorophyceae) each cell contains a single mitochondrion.

The respiratory enzymes are located in mitochondria, and hence mitochondria act as respiratory centres of the cell. Mitochondria are also the sites of enzyme action in protein synthesis and amino acid interconversions.

Endoplasmic Reticulum (ER) : The cytoplasm of the algal cell is traversed by a system of interconnecting tubules, known as endoplasmic reticulum. It is connected with the outer membrane of the nuclear envelope, but does not penetrate other cell organelles like chloroplast or pyrenoid. The surface of ER is studded with ribosomes, the sites of protein synthesis in the cell.

Eye-spot or Stigma : The motile vegetative and reproductive cells of algae have a pigmented spot, known as eye-spot or stigma. It may be located in the anterior, middle or posterior part of the cell. On the basis of their position and structure, five types of eye-spots are recognised.

Type A : Eye-spot is located in the chloroplast and it has no association with flagella, e.g., Chlorophyceae and Cryptophyceae.

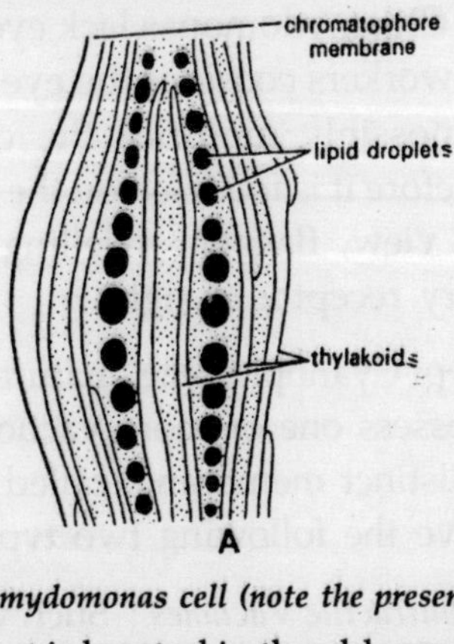

Eye-spot : of Chamydomonas cell (note the presence of thylakoids).

Type B : Eye-spot is located in the chloroplast and associated with a swollen flagellum, e.g., members of Chrysophyceae, Phaeophyceae and Xanthophyceae.

Type C : Eye-spots are independent clusters of osmophilic granules and are situated at the anterior side of the cell, near the flagellar swelling, e.g., Euglenophyceae.

Type D : Eye-spots having osmophilic granular structure with membranous lamellae and are situated near flagellar bases, e.g., Dinophyceae.

Type E : Eye-spot is made up of a lens, retinoid and pigmented cup. These are the largest and most complex type of eye-spots and are found in some Dinophyceae.

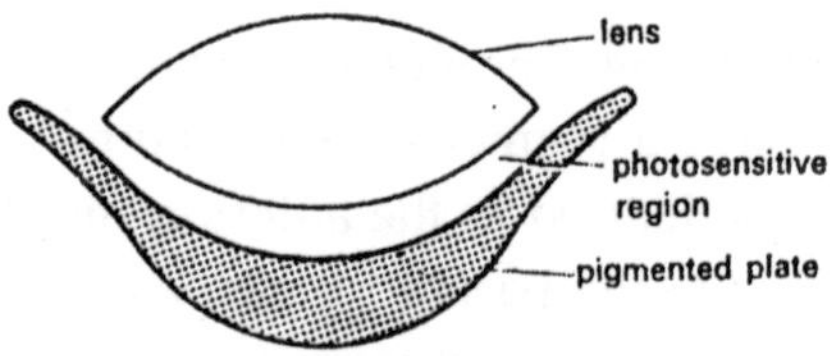

B

B : Eye-spot of Volvax.

Although the eye spot is considered to be a light sensitive organelle which directs the movements of swimming cells, certain mutants of Chlamydomonas lack eye spot and are still phototactic. Some workers consider the eye spot as a shading organelle which possibly modifies the optical quality of transmitted light before it is focussed on the flagellar swelling. According to this view, flagellar swelling, rather than eye spot, is the primary receptor of light.

Vacuoles : Except Cyanophyceae, mature cells of almost all classes of algae possess one or more vacuoles. Each vacuole is bounded by a distinct membrane, called tonoplast. Motile algal cells may have the following two types of vacuoles :

1. *Simple or Contractile Vacuoles :* Such vacuoles, found in the members of Chlorophyceae, show periodic contractions and throw out the waste products out of the cell.

2. *Complex Vacuoles :* A complex vacuole consists of : (i) a tube like cytopharynx, (ii) a large reservoir, and (iii) a group of vacuoles of varying sizes. All these parts work in close coordination. These vacuoles are characteristic of Dinophyceae and Euglenophyceae.

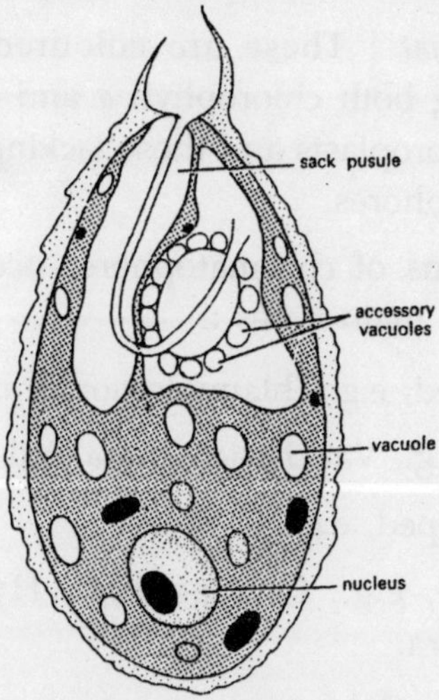

Vacuolar apparatus of Phalacroma cell.

Vacuoles act as the main osmoregulatory organ in the cell and help in regulating the absorption of water and solutes. In holozoic forms, the cytopharynx of complex vacuoles serves as gullet. Sometimes, vacuoles also store reserve food materials such as laminarin and chrysolaminarin.

Gas Vacuoles : These gas containing cavities are characteristic of the mature healthy cells of Cyanophyceae. They occur as stacks of small transparent cylinders of uniform diameter. Their walls are freely permeable to gases. They render buoyancy to planktonic forms. They also act as a screen against intense light.

Pigments : Algal cells have a characteristic colour due to the presence of a combination of pigments, specific to each class. In all classes, except Cyanophyceae, these pigments are present within membrane bound organelles, known as plastids. In blue-greens, the pigments are concentrated in the peripheral cytoplasm, known as chromoplasm. Plastids are of the following two types :

1. *Leucoplast* : These-are colourless plastids.
2. *Chromoplast* : These are coloured plastids; those containing both chlorophyll *a* and chlorophyll *b* are called chloroplasts and those lacking chlorophyll *b* as chromatophores.

Different forms of chromatophores occur in algae. The following are the main types :

(1) cup-shaped, e.g.,Chlamydomonas and Volvox.

(2) discoid, e.g., Vaucheria, Chara and centric diatoms.

(3) girdle-shaped, e.g., Ulothrix.

(4) reticulate, e.g., Oedogonium, Hydrodictyon and Cladophora.

(5) spiral, e.g., Spirogyra.

(6) stellate, e.g., Zygnema.

The various types of pigments found in the algal cell are given below.

1. *Chlorophyll* : There are five types of chlorophylls, viz. chl *a, b, c, d* and *e*. Of these, chlorophyll *a* is present in all groups of algae, chlorophyll *b* only in Chlorophyceae and Euglenophyceae, chlorophyll *c* largely in algae of marine habitats (Phaeophyceae, Cryptophyceae, Bacillariophyceae and Crysophyceae), chlorophyll d in some red algae only as a trace constituent and chlorophyll e in certain Xanthophyceae such as Vaucheria hamata and Tribonema bombycinum.
2. *Xanthophyll* : More than 20 types of xanthophylls are known. They are formed by the incorporation of

molecular oxygen into carotene molecule. Many xanthophylls, common in higher plants (lutein, violaxanthin, and neoxanthin), are found in the members of Chlorophyceae and Phaeophyceae. Fucoxanthin is the main xanthophyll pigment of Phaeophyceae and diatoms whereas myxoxanthophyll, myxoxanthin and oscilloxanthin are found only in Cyanophyceae.

3. *Carotenes* : These are oxygen free alicyclic compounds, composed of isoprene units. The five types of carotenes occur in algae are : α-carotene in Chlorophyceae; Cryptophyceae and Rhodophyceae; β-carotene in all algal groups, except Cryptophyceae, c-carotene in Chlorophyceae; E-carotene in Bacillariophyceae, Cryptophyceae, Phaeophyceae and Cyanophyceae and flavacene in members of Cyanophyceae.

4. *Phycobilins* : These are water soluble complexes of protein and bile pigments, present in the photosynthetic tissue of plants. Phycobilins are red (phycoerythrin) and blue (phycocyanin) pigments which are confined to Rhodophyceae and Cyanophyceae respectively. They act as light harvesting pigments in photosynthesis and the light absorbed by them is transferred to chlorophyll *a*. Thus, like carotenoids, phycobilins are also accessory pigments.

Pyrenoids : Pyrenoids are proteinaceous bodies present in chromatophores. These organelles are considered to be associated with the synthesis and storage of starch. In members of Chlorophyceae, pyrenoids are surrounded by starch plates. But in some algae such as diatoms, they accumulate lipids instead of starch.

A chromatophore may have one (e.g., Chlamydomonas) or more than one (e.g., Oedogonium) pyrenoids. These pyrenoids are embedded in the chromatophore (e.g.,

Chlamydomonas) or attached on its surface with the help of a stalk (e.g., Ectocarpus). The embedded pyrenoids are usually traversed by large number of chromatophore lamellae.

In some algae, pyrenoids are transient structures,. found only at certain stages. Their presence, in such cases, is associated with the photosynthetic activity of the cell and availability of stored food material.

Reproduction

The propagation in algae takes place by vegetative, asexual and sexual methods.

***Vegetative Reproduction* :** The vegetative reproduction includes all those processes of propagation in which portions of the plant body become separated off to give rise to new individuals without any obvious changes in the protoplast or in other words, it does not involve rejuvenation of the protoplasts. It is the most common method of reproduction in algae and takes place by the following means:

1. *Cell Division or Fission* : It is the simplest method of propagation and is commonly found in unicellular algae (e.g., Euglena, Chlamydomonas), desmids and diatoms. In this process, the unicellular algal cell divides mitotically to form two daughter cells, each eventually grows into an independent organism.
2. *Fragmentation* : In filamentous forms like Ulothrix, Oedogonium, Spirogyra and Zygnema, the thallus often breaks into small fragments. Each fragment has the capability to grow independently and forms a new thallus. The fragmentation of filaments may be due to mechanical pressure, dissolution of transverse walls or difference in turgor pressure between adjoining cells. In colonial blue-green algae, vegetative propagation takes place by fragmentation of larger colonies.

3. *Hormogonia* : It is a characteristic method of reproduction in blue-green algae. Under unfavourable conditions, the trichomc breaks into motile segments of varying length, called hormogonia. The fragmentation of parent filament into hormogonia may be due to the formation of intercalary heterocysts, specialized separation discs or necridia or due to death and decay of intercalary cells of the trichome. Hormogonia are commonly found in Nostoc, Oscillatoria and Cylindrospermum.
4. *Formation of Adventitious Branches* : Adventitious branches are formed in some large thalloid forms of algae. These branches, when get detached from the parent thallus, develop into new plants (e.g., Dictyota, Fucus). Adventitious branches like protonema, formed on the internodes of Chara or stolon of Cladophora glomerata, are found.
5. *Tubers* : On the rhizoids and the lower nodes of Chara some tuber like structures are formed due to storage of food material. When detached from the parent plant, the tuber produces an independent plant. Similar structures are also found in Cladophora.
6. *Budding* : In some algae (e.g., Protosiphon) vegetative propagation takes place by budding. Bud like structures are formed due to proliferation of vesicles. They eventually get separated from the parent plant by the formation of the septum, and have the capacity to form plants.

Asexual Reproduction : Asexual reproduction involves rejuvenation of the protoplasts and the reproductive bodies are either naked or provided with a newly synthesized wall of their own. Further, it is accomplished by using specially differentiated cells that are capable of multiplying directly the alga without fusing with other cells. In some aspects, it is very similar to vegetative reproduction as both are accessory means of propagation and are not concerned with bringing together in one cell lineage the genes from different cell lineages. In general, it is a process whereby the protoplast is released from the algal cell and germinates to form a new plant. Asexual reproduction takes place by a variety of motile or non-motile spores (cells that germinate without fusing and form new individuals). These spores may be differentiated into the following broad categories on the basis of their structure.

1. *Zoospores* : These are motile, naked structures with two (e.g., Chlamydomonas, Ectocarpus), four (e.g., macrozoospores of Ulothrix) or many (e.g., Oedogonium, Vaucheria) flagella. The flagella are usually inserted anteriorly, but are lateral in some brown algae. The zoospore has a bit of chloroplast and an eye spot. The cell forming zoospore is known as zoosporangium. The protoplast of zoosporangium may form a single zoospore (e.g., Oedogonium, Vaucheria) or undergoes repeated divisions and each segment forms a zoospore (e.g., Cladophora). Zoospores are haploid or diploid. They are liberated by dissolution of parent cell wall or through a terminal pore. They are usually formed during the night and are liberated early in the morning. After liberation, they rest for sometime and then germinate to form new individuals.
2. *Aplanospores* : These are non-motile spores, commonly found in terrestrial algae, but some aquatic algae (e.g., Ulothrix, Microspora) also form them during drought conditions. They differ from zoospores in having a distinct wall and in the absence of flagella. Each cell may form a single aplanospore or its protoplast may divide to form many aplanospores.
3. *Hypnospores* : Aplanospores of some algae like Pediastrum and Sphaerella secrete thick walls to overcome prolonged period of desiccation. Such thick

walled aplanospores are called hypnospores. Under favourable conditions, hypnospores germinate and grow into new individuals or their protoplast may form zoospores. The hypnospores of Chlamydomonas nivalis are red in colour due to the deposition of a pigment, haematochrome, in their walls.

4. *Tetraspores* : Diploid plants of some algae (e.g., Polysiphonia) form aplanospores which are called tetraspores. They arc formed within tetrasporangium, usually by meiosis, and are haploid. They germinate to form haploid plants.

5. *Autospores* : The aplanospores with a structure similar to the parent cell are called autospores. In Scenedesmus and many members of the order Chlorococcales (e.g., Chlorella) the aplanospores acquire all distinctive features of the parent cell before their liberation from the sporangium., The so formed autospores are in fact replica of the parent cell, the only difference is that they are smaller in size.

6. *Akinetes* : In some algae vegetative cells develop into thick walled spore like structure with abundant food reserves. These are called akinetes. Unlike aplanospores, akinetes always have additional wall layers around the protoplast which are fused with the parent wall. They are resistant to unfavourable environmental conditions. The formation of akinetes, besides other factors, is affected by the availability of carbohydrates and light. They are found in many blue-green and green algae.

7. *Exospores and Endospores* : In many blue-green algae the protoplast divides to form special type of aplanospores, known as exospores and endospores. The cell that produces endospores becomes somewhat enlarged and its contents divide successively in three planes, forming four to many endospores (e.g., Dermocarpa). The

exospores are formed externally; the protoplast of the cell comes out through a terminal pore and successively cuts spherical spores (e.g., Chamaesiphon).

Sexual Reproduction : All groups of algae, except Cyanophyceae, reproduce sexually when gametes fuse to form zygote. In contrast to vegetative and asexual reproduction, it leads to the creation of new combinations of genes by pooling together in one line of descent, the genes derived from the different parents, thus resulting in a reshuffling of the gene material. On the basis of the structure and physiological behaviour of sex organs and their complexity, the following six types of sexual reproduction are recognised in different groups of algae.

1. *Autogamy* : When two gametes of the same mother cell fuse to form a diploid nucleus, it is called autogamy. In this process there is only karyogamy (fusion of two gametic nuclei). The autogamy lacks incorporation of external genes. Hence, plants developing as a result of autogamy do not show new characters. Diatoms are the common example of autogamy.
2. *Hologamy* : In some unicellular forms (e.g., Chlamydomonas, Dunaliella) the vegetative cells of different strains (+) and (-) (female and male) behave as gametes and fuse to form zygote. From evolutionary point of view hologamy is more advanced than autogamy, as it involves fusion of two cells having different genetic constitution. But hologamy is an inefficient process from the point of view of propagation, as two vegetative cells fuse to form only one zygote.
3. *Isogamy* : In isogamy the two gametes which fuse to form zygote, are morphologically and physiologically similar. Such gametes are called isogametes, as they are indistinguishable into plus and minus strains. They

are usually motile and flagellate. Isogametes are found in Ulothrix, Chlamydomonas eugametos, etc.

4. *Physiological Anisogamy* : In some algae like Spirogyra, Zygnema and Ectocarpus, though gametes are similar in morphological characters, they show physiological variation with plus (+) and minus (-) strains.
5. *Anisogamy* : In anisogamy, fusion takes place between morphologically and physiologically distinct gametes (anisogametes). The male or microgametes are smaller and more active, whereas the female or macrogametes are larger and sluggish. Chlamydomonas braunii and Pandorina are common examples of anisogamy.
6. *Oogamy* : This is the most advanced type of sexual reproduction. In this process a large non-motile egg or ovum fuses with a small motile sperm or antherozoid (in Rhodophyceae, sperms are non- motile). Egg is formed within the oogonium and sperms within the antheridium. Volvox, Oedogonium, Chara, Vaucheria, Sargassum, Batrachospermum and Polysiphonia are common examples of oogamy.

Zygote and its Germination : Post-fertilization Events : Soon after sexual fusion, the zygote forms a thin wall, which later becomes as thick as 3-layered. The zygote in this condition can undergo complete desiccation without harm. Usually the zygote undergoes a period of rest of a shorter or a longer duration before germination. In marine forms, the zygote generally germinates immediately and does not develop a thick wall.

The two gametic nuclei generally fuse immediately after fusion but in many cases (e.g., Zygnema) they do not fuse until just before the germination of the zygote.

Meiosis generally takes place during the germination of zygote and thus a haploid plant body is formed (e.g.,

Chlamydomonas, Volvox, Ulothrix, Spirogyra, Zygnema, Oedogonium, Vaucheria, etc.). However, in a number of forms (e.g., Sargassum), the zygote nucleus does not undergo reduction division but divides equationally and gives rise to a diploid plant body, reduction division does not take place until just before reproduction.

Perennation

Perennation means the survival from year to year by vegetative means. It includes the methods employed by alga to pass through periods unfavourable for their vegetative growth. Under such conditions, the protoplast of the cell becomes dormant by reducing its activity to its minimum. Under stressing (unfavourable) circumstances when life processes can not run smoothly, the alga develops some stage in life cycle which is protected by a thick resistant wall.

Most fresh water and sub-aerial algae arc exposed to the dangers of desiccation and extremes of temperature. They, therefore, develop certain devices which enable them to carry through the period of severe conditions in the next growing season. These protective devices constitute the means of perennation. Hypnospores, akinetes, tubers, mucilagenous sheath, zygote and cysts are some common perennating structures found in algae.

Types of Life-cycle

The growth and development of an alga passes through a number of distinct morphological and cytological stages. The sequence of these orderly changes is said to be life cycle or life history. In other words, the sequence of all the different phases or events, through which an organism passes from the zygote of one generation up to the zygote of next generation, is known as its life-cycle. The union of male and female gametes results in the formation of zygote. As there is doubling of chromosomal complement in zygote, it represents diploid state.

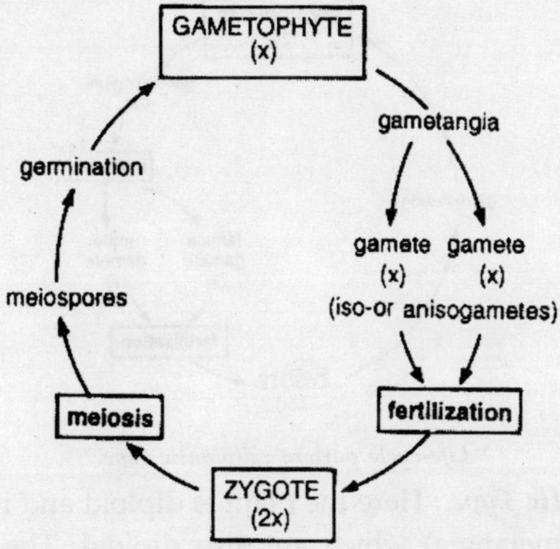

Life-cycle pattern : haplontic type.

In some groups, the original chromosome number is achieved by meiotic division of the zygote nucleus. Thus on germination, the zygote produces haploid plants. In others, the zygotic nucleus divides mitotically and the resulting cells form a diploid phase, which later produces cells which are capable of meiotic division.

The various types of life-cycle met among algae can be broadly classified into the following five categories .

Haplontic Type : In this type, plant is haploid and bears haploid gametes in the gametangium. The gametic fusion results in the formation of a diploid zygote which is the only diploid phase in the life cycle. The zygote nucleus divides meiotically to produce four meiospores, each of these develops into a new individual.

Thus there is an alternation of a haploid plant with a diploid zygote. This type of life-cycle occurs in the motile unicellular forms (e.g., Chlamydomonas) and most multicellular green algae like Ulothrix, Oedogonium, Spirogyra, Zygnema and Chara.

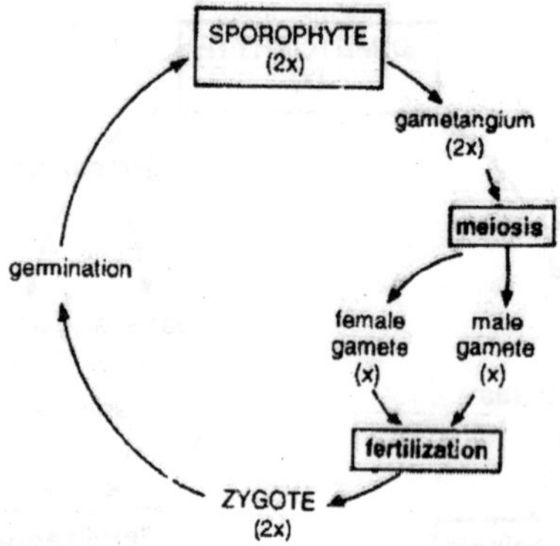

Life-cycle pattern : diplontic type.

Diplontic Type : Here the plant is diploid and it bears sex organs (gametangia) which are also diploid. The reduction division takes place at the time of the formation of gametes and as such the gametes are haploid. After gametic fusion diploid stage is re-established in the form of zygote. The zygote does not undergo any reduction division and it gives rise to a diploid plant body. Thus, there is an alternation of a diploid plant with haploid gametes. The gametes representing the only haploid stage in the life cycle of the plant. This type of life-cycle is found in some members of Chlorophyceae (e.g., Cladophora glomerata) and Phaeophyceae (e.g., Fucus, Sargassum) and nearly all Bacillariophyceae (diatoms).

Isomorphic Type

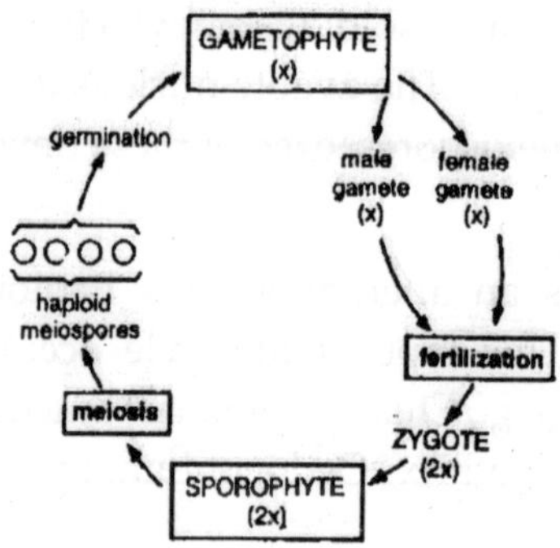

Life-cycle pattern : isomorphic type.

In this type of life-cycle, there is an alternation of two generations which are externally similar, but one is haploid (gametophyte) producing gametes and the other diploid (sporophyte) producing zoospores. The zygote germinates directly into a diploid plant without undergoing reduction division. Sporangia develop on the diploid plant body and reduction in the number of chromosomes takes place prior to the formation of zoospores. The haploid zoospores thus formed grow into haploid plants. Sex organs develop on the haploid plants and these give rise to haploid gametes. The haploid gametes fuse to form diploid zygote. Isomorphic life cycle has been reported in a number of green algae (e.g., Ulva, Enteromorpha, Cladophora, Draparnaldiopsis, Fritschiella) and brown algae (e.g., Ectocarpus).

Heteromorphic Type : In this type of life-cycle, both diploid (sporophytic) and haploid (gametophytic) plant bodies are morphologically distinct (heteromorphic) and they alternate with each other. Mostly the sporophyte is large and the gametophyte small. Heteromorphic life cycle is seen mainly amongst brown algae (e.g., Dictyota, Laminaria, etc.). In Dictyota, there is a difference in cell size between the gametophyte and sporophyte, whereas in Laminaria, the haploid plant body consists of minute filaments which produce gametes, whilst the diploid plant body is a large macroscopic plant (attaining a length of several metres) bearing zoosporangia which produce zoospores after reduction division.

The haploid zoospores germinate and produce haploid gametophytic plants. Haploid gametes are produced in sex organs formed on gametophytic plants. The gametes fuse to form a diploid zygote, which directly germinates into a diploid sporophytic plant. Besides heteromorphism of gametophyte and sporophyte generations, there is also slight heteromorphism of the male and female gametophytes, expressed as a difference in cell size and degree of branching.

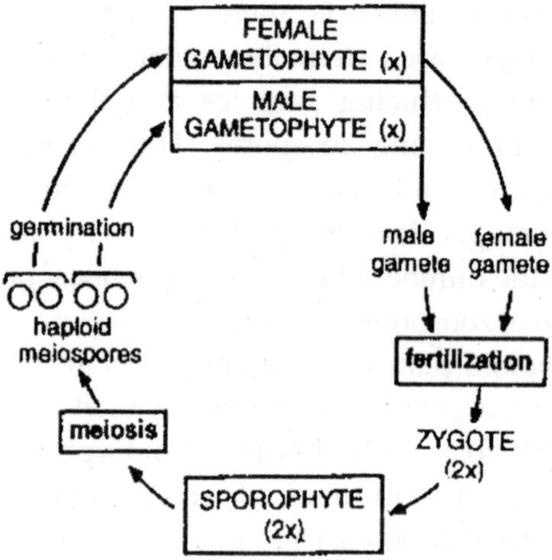

Life-cycle pattern : heteromorphic type.

In genera like Cutleria and Urospora, however, haploid gametophytic plants are larger whereas diploid sporophytic plants are smaller in size.

Triphasic Type : In triphasic life-cycle, there is a succession of three generations. It may be of the following two types .

1. *Haplobiontic :* In some Rhodophyceae like Batrachospenmum and Nemalion two well developed haploid phases are present in the life-cycle. This type of life cycle is hence called haplobiontic triphasic. The diploid phase is represented only by the zygote. The plant body of Batrachospermum is gametophytic which bears sex organs (i.e., spermatangia and carpogonia). Haploid gametes (spermatium and egg) are formed in these sex organs. A diploid zygote is formed by the fusion of two gametes. Gonimoblast filaments are formed from the basal part of the carpogonium; the uppermost cell of these filaments functions as carposporangium. The first division in the zygote nucleus is meiotic and as such the gonimoblast filaments are haploid. The haploid carposporangium bears a

haploid carpospore. The gonimoblast filaments, carposporangia and carpospores are enveloped by numerous sterile filaments, and together represent carposporophyte generation which is haploid. On liberation, carpospores germinate into heterotrichous chantransia stage. The projecting filaments of the latter develop into new gametophytic plant body.

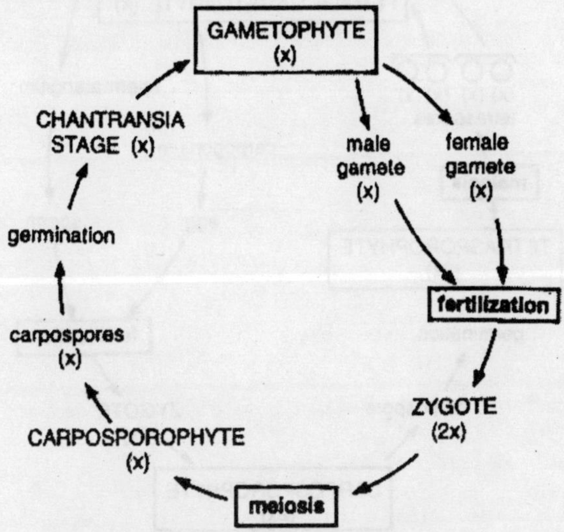

Life-cycle pattern : triphasic haplobiontic type.

2. *Diplobiontic* : In diplobiontic triphasic life-cycle there are two diploid phases (carposporophyte and tetrasporophyte) alternating with haploid gametophytic phase. This type of life-cycle is well exemplified by Polysiphonia (Rhodophyceae). Here, the haploid phase is represented by male and female gametophytic plants, sex organs and gametes. The male and female gametes fuse to form a diploid zygote. The zygote produces gonimoblasts which represent an additional diploid phase (carposporophyte). Diploid carpospores are formed in the carposporophyte. On liberation, the carpospore germinates to produce a diploid

tetrasporophytic plant. The tetrasporophytic plant produces diploid tetrasporangia. Four haploid tetraspores are formed in a tetrasporangium by reduction division. Tetraspores give rise to haploid gametophytic plants. The life cycle is thus completed.

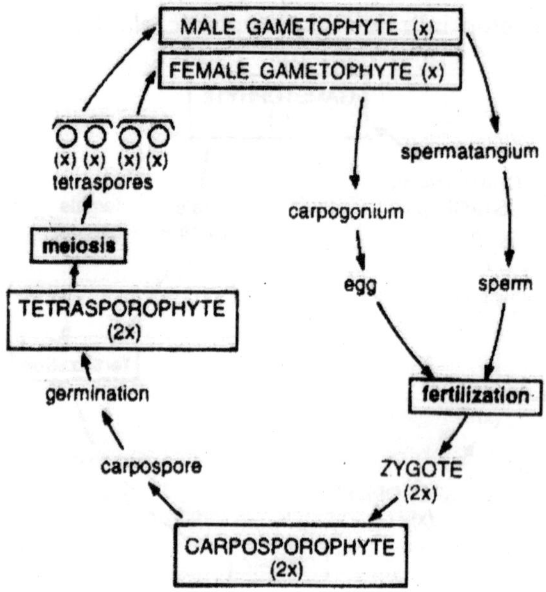

Life-cycle pattem triphasic diplontic type.

Bibliography

Adil, A. : *Manual of Practical Algae,* Mohit Pub., New Delhi, 2003.

Aggarwal, S.K. : *Vistas in Mycology and Plant Pathology,* APH, New Delhi, 2000.

Ahuja, B. : *Plant Anatomy,* Rawat Pub., New Delhi, 2000.

Antony, T. : *Quarry : Closing in on the Missing Link,* Free Press, New York, 1993.

Arora, J.K.: *Biopesticides and Pest Management,* Ankit Pub., New Delhi, 2003.

Aslam, M.S. : *Genetics-Behavioural and Molecular,* Campus Books, New Delhi, 2003.

Bacon, Mark : *Water use Efficiency in Plant Biology,* Blackwell, London, 2004.

Bahadur, Rai K. Ranga : *A Manual of Botany for India,* Atlantic Publishers, New Delhi, 1994.

Bayleybolfour, I. :*Plant Geography Upon A Physiological Basis,* ABS, London, 2004.

Behnke, H.D. : *Sieve Elements : Comparative Structure, Introduction and Development,* Springer, London, 1990.

Bentley, Robert : *Study of Structural Morphological and Physiological Botany,* Atlantic Publishers, New Delhi, 1994.

Bilgrami, C : *Fundamentals of Botany,* Vikas Publishing House, New Delhi, 2003.

Bilgrami, K.S. : *Physiology of Fungi,* Vikas Publishing House, New Delhi, 2003.

Boodle, L.A. and Sinnott, E.W. : *Student Handbook of Botany and Plant Science,* Atlantic Publishers, New Delhi, 1995.

Bose, J.C. : *Irritabiligy of Plants,* Discovery Publishing House, New Delhi, 1999.

Brooks, F.T., and Smith, M. : *Dictionary of Fungus Diseases of Plants,* Campus Books, New Delhi, 2003.

Busch, L. : *Plants, Power and Profit : Social, Economic, Lawrence Busch, and Ethical Consequences of the New Biotechnologies,* Blackwell, London, 1992.

Chrispeels, M.J. : *Plants, Genes and Agriculture,* Jones & Bart., London, 1994.

Coulter, John Gaylord : *Fundamentals of Plant Science,* Atlantic Publishers, New Delhi, 1994.

Dayal, R. : *Predaceous Fungi,* Campus Books, New Delhi, 2000.

———— : *Predaceous Fungi,* Commonwealth Publishers, New Delhi, 2000.

Dhaliwal, G.S. and Singh, Balwinder : *Pesticides and Environment,* Campus Books, New Delhi, 2000,

Dube, H.C. : *Introduction to Fungi,* Vikas Publishing House, New Delhi, 2003.

Dyson, Freeman : *Origin of Life,* Oxford, New York, 1999.

Frank, S. Smythe : *Flowering Plants of Himalaya,* Commonwealth Publishers, New Delhi, 1994.

Fuller, Harry J. : *University Botany,* Atlantic Publishers, New Delhi, 1994.

Gaston, Kevin J. and spice, John I. : *Biodiversity : An Introduction,* Blackwell, 2004.

Grewal, R.C. : *Medicinal Plants*, Campus Books, New Delhi, 2003.

———— : *Dictionary of Botany*, Arihant Pub., New Delhi, 2003.

Hall, Marion and Halliday, Tim : *Behaviour and Evolution*, Springer, London, 1998.

Hanan, James S. : *The Algorithmic Beauty of Plants*, Springer, London, 1990.

Hardie, D.G. : *Protein Phosphorylation : A Practical Approach*, Oxford, New York, 1999.

Johri, Lata and Tyagi : *A Textbook of Bryophyta*, Dominant Publishers, New Delhi, 2004.

———— : *A Textbook of Fungi*, Dominant Publishers, New Delhi, 2004.

———— : *A Textbook of Gymnosperm*, Dominant Publishers, New Delhi, 2005.

Kains, M.G. and Mcquesten, L.M. : *Fundamentals of Plant Science*, Campus Books, New Delhi, 2002.

———— : *An Introduction to Plant Science*, Atlantic Publishers, New Delhi, 1995.

Kumar, H.D. : *General Ecology*, Vikas Publishing House, New Delhi, 2003.

Kumar, Suresh : *A Textbook of Plant Taxonomy*, Campus Books, New Delhi, 2004.

———— : *Economic Botany*, Campus Books, New Delhi, 2002.

Lang, W.H. : *A Textbook of Botany*, Atlantic Publishers, New Delhi, 1994.

Leung, D.Y.M. : *Superantigens : Molecular Biology, Immunology, and Relevance to Human Disease*, Dekker, London, 1997.

Lozzo, Renato V. : *Proteoglycans : Structure, Biology, and Molecular Interactions*, Dekker, London, 2000.

Madigan, M.T. : *Brock : Biology of Microorganisms,* Prentice Hall, New York, 2000.

Magner, Lois N. : *A History of the Life Sciences,* Dekker, London, 1994.

Magurran, Anne E. : *Measuring Biological Diversity,* Blackwell, London, 2003.

Mishra, S.R. : *Bacterial Plant Diseases,* Discovery Publishing House, New Delhi, 2003.

———— : *Morphology of Plants,* Discovery Publishing House, New Delhi, 2004.

Nairne, Alexander : *Scientific Classification of Flowering Plants,* Atlantic Publishers, New Delhi, 1994.

Narasaiah, M.L. : *Biodiversity and Environment,* Discovery Publishing House, New Delhi, 2004.

Nover, L. : *Plant Promoters and Transcription Factors Results and Problems in Cell Differentiation,* Springer, London, 1994.

Oliver, Daniel : *Study of Indian Botany,* Atlantic Publishers, New Delhi, 1994.

Panday, S.N. : *Economic Botany,* Vikas Publishing House, New Delhi, 2003.

Panday, S.N. and Sinha, B.K. : *Plant Physiology,* Vikas Publishing House, New Delhi, 2003.

Panday, S.N. and Trivedi, P.S. : *A Textbook of Botany,* Vikas Publishing House, New Delhi, 2003.

Pfleiderer, I. : *The Life of Indian Plants,* Atlantic Publishers, New Delhi, 1992.

Prasad, S.N. : *Environmental Biology,* Campus Books, New Delhi, 2003.

———— : *Cell Biology,* Campus Books, New Delhi, 2003.

Pugnaire, Ed. F.I. : *Handbook of functional Plant Ecology,* Dekker, London, 1999.

Rai, Mahendra and Mares, Donatella : *Plant-Derived Antimycotics : Current Trends and Future Prospects,* Blackwell, London, 2003.

Rashid, A. : *An Introduction to Pteridophyta,* Vikas Publishing House, New Delhi, 2003.

Rashid, A., Kumar, Santosh and Tembhre, Manju : *Anatomy and Physiology of Fishes,* Vikas Publishing House, New Delhi, 2003.

Roy, E. Biles, Seymour, E.L.D. : *Encyclopaedia of Horticulture,* Commonwealth Publishers, New Delhi, 1994.

Sandhu, G.S. : *Textbook of Cytology,* Campus Books, New Delhi, 2002.

Sandhu, G.S. and Kumar, Suresh : *Cyto-Genetics,* Campus Books, New Delhi, 2003.

Sarkar, Dilip De : *The Silkworm: Biology, Genetics and Breeding,* Vikas Publishing House, New Delhi, 2003.

Sharma Rajni : *An Introduction to Plant Morphology,* Campus Books, New Delhi, 2004.

———— : *Plant Pathology,* Campus Books, New Delhi, 2004.

———— : *Plant Tissue Culture,* Campus Books, New Delhi, 2003.

Shukla, S.K. and Srivastava, P.R. : *Encyclopaedia of Environmental Pollution,* Commonwealth Publishers, New Delhi, 1994.

Singh, Mahinder : *A Textbook of Biochemistry,* Dominant Publishers, New Delhi, 2005.

Singh, S.K. : *Plant Physiology,* Campus Books, New Delhi, 2000.

Sinha, Asha *Microbes and Plants,* Campus Books, New Delhi, 2001.

Sinha, R.K. :*Biodiversity–Global Concerns,* Commonwealth Publishers, New Delhi, 1996.

Sinnott, E.W. : *Botany and Plant Science*, Commonwealth Publishers, New Delhi, 2000.

Srivastava, M.B.: *Introduction to Forestry*, Vikas Publishing House, New Delhi, 2003.

Stiles, Walter :*Principles of Plant Physiology*, Atlantic Publishers, New Delhi, 1995.

Sundara, Rajan, S. : *Introduction to Cell Biology*, Vikas Publishing House, New Delhi, 2003.

Talbot, Nicholas J. : *Plant-Pathogen Interactions*, Blackwell, London, 2001.

Tassy, Pascal : *The Message of Fossils*, McGraw-Hill, London, 1993.

Transeau, E.N. : *General Botany*, Atlantic Publishers, New Delhi, 1995.

Tripathi, G. and Tripathi, Y.C. : *Biological and Biotechnological Resource*, Campus Books, New Delhi, 2002.

———— : *Bioresource and Environment*, Campus Books, New Delhi, 2002.

Waisel, Yoav : *Plant Roots : The Hidden Half*, Dekker, London, 1996.

Warren, L. : *Bound Carbohydrates in Nature*, Oxford, New York, 1994.

Wrigglesworth, J. : *Energy and Life : Modules in Life Sciences*, T & F, New York, 1997.

Zaidi, N. : *An Introduction to Algae*, Ankit Pub., New Delhi, 2002.

ENCYCLOPAEDIA OF BOTANY

ENCYCLOPAEDIA
OF
BOTANY

(Vol. 2)

Prof. Harmeet Kour

ANMOL PUBLICATIONS PVT. LTD.
NEW DELHI - 110 002 (INDIA)

ANMOL PUBLICATIONS PVT. LTD.
H.O.: 4374/4B, Ansari Road, Darya Ganj,
New Delhi-110 002 (India)
Ph.: 23261597, 23278000
B.O.: No. 1015, Ist Main Road, BSK IIIrd Stage
IIIrd Phase, IIIrd Block,
Bangalore - 560 085 (India)
Visit us at: www.anmolpublications.com

Encyclopaedia of Botany

First Published, 2006

ISBN 81-261-2673-6 (Set)

PRINTED IN INDIA

Published by J.L. Kumar for Anmol Publications Pvt. Ltd., New Delhi - 110 002 and Printed at Mehra Offset Press, Delhi.

Contents

Preface

Biology comprises mainly two branches of study, namely, Zoology and Botany. Botany pertains to the study of plants, which have always been of great import since the formation of the planet 'Earth'. Plants play a crucial part in the life of living creatures as well as in balancing the environment. In other words, they are as much necessary for life as air or water is.

Apart from numerous benefits, obtained from plants, they have also been a source of medication since days immemorial. Taking all these dimensions of plants into consideration, they have been classified into different groups and every one of them has been given a unique Botanical name. A plant may have different names in different dialects of different parts of the world, but its botanical name remains one and the same the world over.

This multi-volume aims at furnishing a detailed account of plants from every possible dimension. A ready reference book for scholars and students, this encyclopaedia is an authority on the subject without an iota of doubt.

Editor

PART-I

Fungi

1

Economic Importance

Fungi play a significant role in our daily life. They are our greatest friends as well as foes. Some fungi are prized for their usefulness to mankind, while others are highly injurious as they are responsible for many diseases in plants, animals and human beings. Some important useful and harmful activities of fungi are discussed below.

Useful Activities

Fungi are beneficial to us, directly or indirectly, in many ways. They keep our environment clean by removing organic debris; they provide food, medicines and several other useful products, and they play an important role in many industries.

Fungi as Food : Fungi have been used as food since times immemorial and esteemed as delicacies. The fructifications of several mushrooms and morels (*e.g.*, Agaricus bisporus, Amanita vaginata, Boletus edulis, Clavatia gigantea, Lentinus edodes, Morchella, Pleurotus, Volvaria, Volvariella volvacea) are used as food. Agaricus campestris is cultivated in many areas of north and south India and Morchella esculenta is grown in Punjab and Kashmir. Mushrooms are preferred for food because, of their fairly large protein contents (21-30%). They are also rich in vitamins, carbohydrates, minerals and amino acids.

Yeast is an important source of vitamin B and D. Saccharomyces, Endomyces, Rhodotorula and Torulopsis are particularly rich in proteins and hence 'they are mixed in incomplete livestock. Similarly, a popular food, 'sufu' is produced from species of Mucor and Antimucor.

Some foods like soybeans and cassava although rich in nutrients, can not be easily digested by man. These are made easily palatable by fermenting fungi. Soybeans are fermented by species of Rhizopus (*e.g.*, R. oligosporus, R. oryzae and R. arrhizus) to prepare tempeh, a food which is more digestible and tasty. Similarly, 'Incaparina' foods developed by the Institute of Nutrition of Central America and Panama, consist of corn (26%), cotton seed flour (38%), sorghum (26%), food yeast (3%) and minerals and vitamins. Northern Utilisation Research and Development Division of USDA has developed protein cake by combining wheat, barley, oats, rice, and soybean flour cooked and fermented with the help of Rhizopus oligosporus. This cake is rich in niacin and riboflavin. Single-cell protein (SCP), obtained from yeast, species of Aspergillus, Penicillium, Fusarium, Neurospora and Candida is a complete substitute for conventional protein foods.

Fungi as Medicine **:** Several fungi are used in the production of medicines.

1. *Ergot* : Ergot is prepared from the sclerotia of Claviceps purpurea. It contains some alkaloids which are used to induce uterine contraction for abortion, in the treatment of menstrual disorders and to check haemorrhages. However, excess of ergot leads to serious

convulsions or a gangrenous condition that may result in loss of limbs.

2. *Ephedrine :* It is synthesized from benzaldehyde by the action of yeast and is used in the treatment of asthama and nasal troubles.

3. *Steroids :* Steroids are complex organic compounds, effective against rheumatoid arthritis, allergic, dermatologic and other diseases. They are also used as anaesthetic and antifertility agents. Steroids are adrenal cortical and gonadal hormones and their derivatives. Their extraction from biological system is highly expensive. A wide variety of fungi have the capacity of synthesizing many steroids; for instance, cortisone is prepared by fermentation of plant glycosides by Rhizopus nigricans and Aspergillus niger.

4. *Vitamins :* Vitamins are accessory micronutrients, which are required by living organisms for their proper growth. Several yeasts, including Saccharomyces cerevisiae, are good source of vitamin β–complex. Ergosterol, a precursor of vitamin D, is synthesized from some molds and yeasts. Eremothemium ashbyii is a rich source of vitamin B_{12}, whereas vitamin A is extracted from Rhodotorula gracilis.

5. *Antibiotics :* Antibiotics are metabolic substances produced by some living organisms which are injurious to other living beings. Alexander Fleming, in 1944, for the first time extracted the wonder drug penicillin from Penicillium notatum. Since then several other antibiotics have been extracted from fungi. Some important antibiotics obtained from fungi and their range of action are listed in Table.

Several fungi are active against human pathogens; for instance, spread of Candida albicans is prevented by the extract of Dricholoma saponaceum. Mushrooms like Lentinus edodes,

Flammulina velutipes, Agaricus bisporus, Volvariella volvaceae, and Boletus edulis possess anti-tumour properties. Many edible mushrooms (*e.g.*, Lentinus edodes, Agaricus bisporus) have the ability to lower blood cholesterol.

Fungi in Production of Organic Acids **:** Biochemical activities of several fungi are utilized in the commercial production of organic acids.

1. *Citric Acid :* It is produced by fermenting sucrose and molasses by Aspergillus niger and A. wentii. Citric acid is used in soft drinks and other foods and medicinal preparations. It is a superior sequestering agent and is used in the manufacture of ink, dyeing, electroplating and leather tanning.
2. *Itaconic Acid :* It is produced by fermentation of sugars by Aspergillus itaconicum and A. terreus. It is widely used in the manufacture of alkyl resins, synthetic fibres and plasticizers.
3. *Gluconic Acid :* It is produced by fermenting sugars by Aspergillus niger and Penicillium purpurogenum. It is

Some Antibiotics Obtained from Fungi and their Range of Action

Antibiotics	*Source*	*Range of action*
1. Penicillin	Penicillium notatum, P. chrysogenum	Bacteria
2. Proliferin	Aspergillus proliferans	Tuberculosis
3. Isopenicillin N	Penicillium chrysogenum	Bacteria
4. L-sarein	Aspergillusgiganteus	Tumour
5. Ramycin	Mucor remannianus	Bacteria
6. Campestrin	Psalliota campestris	Bacteria
7. Ustilagic acid	Ustilago maydis	Fungi
8. Chaetomin	Chaetomium cochloides	
9. Ergotine	Claviceps purpurea	
10. Citrinin	Penicillium chrinum	
11. Jawaharene	Aspergillus niger	Bacteria, tumours

Contd...

Antibiotics	*Source*	*Range of action*
12. Baccatin A	Gibberella baccata	Bacteria, fungi
13. Cephalosporin	Cephalosporium spp.	Gram '+ve' bacteria
14. Fumagillin	Aspergillus fumigatus	Amoeba, staphylococcal bacteriophage
15. Griseofulvin	Penicillium nigricans, P. urticae,	Fungi P. griseofulvum
16. Trichothecin	Trichothecium roseum	Virus, fungi
17. Gliotoxin	Trichoderma spp.	
18. Puberulic acid	Penicillium puberulum	

used in pharmaceutical, food, textile, leather and photographic industries.

4. *Kojic Acid :* It is produced by fermentation of sugars by Aspergillus oryzae. It possesses insecticidal and antibiotic properties and is used in the preparation of pyridines, ethers, esters, azodyes, etc.
5. *Gallic Acid :* It is synthesized by the hydrolysis of gallotannin present in gall nuts by Penicillium glaucum and Aspergillus gallomyces. It is used in the manufacture of inks and dyes.
6. *Fumaric Acid :* It is obtained by fermentation of sugars by Rhizopus stolonifer. It is used in the manufacture of alkyl resins and wetting agents.

Besides, several species of Mucor and Rhizopus are used in the production of lactic, oxalic and succinic acids.

Fungi in Industry

1. *In Brewery :* Alcoholic fermentation with the help of fungi is the basis of brewing industries. Wine is produced by fermenting rice with Aspergillus oryzae. This method is popular in Japan and many European countries.

Saccharomyces cerevisiae is used in the production of beer.

2. *In Baking Industry* : Fermentation of carbohydrates by Saccharomyces cerevisiae produces ethyl alcohol and CO_2.

$$C_6H_{12}O_6 \xrightarrow{\text{yeast}} 2C_2H_5OH + 2CO_2$$
(alcohol)

Carbon dioxide liberated in this process is used in the preparation of several bakery products such as breads and cakes.

3. *In Cheese Industry* : Some species of Penicillium (*e.g.*, P. camembertile, P. rocaueforti, P. candidum) are used for maturation of cheese. They impart specific flavours to cheese.

Fungi in Enzyme Production

Many intra- and extracellular enzymes are found in fungi and some are extracted on commercial scale.

1. *Invertase* : The yeast Saccharomyces cerevisiae is used for extraction of the enzyme invertase. This enzyme hydrolyses sugars into glucose and fructose and is used in confectionery and paper industry.

2. *Zymase* : The enzyme zymase obtained from Saccharomyces cerevisiae in used in the preparation of ethyl alcohol by fermentation of carbohydrates.

$$C_6H_{16}O_6 \xrightarrow{\text{zymase}} 2C_2H_5OH + 2CO_2$$
(alcohol)

3. *Amylase* : Aspergillus niger and A. oryzae are used in the production of the enzyme amylase. This enzyme is used in alcohol industry, in the manufacture of dextrinised starch and in medicines.

4. *Cellulase :* Trichodenna reesh is used in the production of the enzyme cellulase, which is used in the preparation of cheese and in saccharification of cellulosic and lignocellulosic wastes together with recycling of hydrolysate for the production of single cell protein, microbial biomass, vitamins (like β–carotene), etc.

Fungi are also used in the production of several other important enzymes like lipase, pectinases, glucose oxidase, proteases and lactase.

Fungi in Agriculture : Fungi play an important role in agriculture in various ways.

As Scavengers : The ratio of carbon dioxide in the atmosphere is maintained by decomposition of plants and animals debris by fungi and bacteria. In the absence of these scavengers, the surface of the earth would have covered with the accumulating remains of dead animals and plants.

In acidic soils the activity of bacteria becomes very slow and under such conditions decomposition is mainly carried by saprophytic fungi. The enzymes secreted by these fungi decompose complex organic substances into their inorganic components and thus increase soil fertility.

In Biological Control : Fungi play an important role in biological control of diseases; for instance, Trichoderma lignorum suppresses the growth of root rot fungus, Pythium and the growth of Rhizoctonia solani can be checked by Penicillium vermiculatum and Rhizoctonia oryzae. Gliocladium roseum is used to control Sclerotinia diseases. Trichoderma harzianum is used to control blight of tomato and peanuts caused by Sclerotium rolfsii. Fungal pathogens play an important role in nature in the reduction of weeds.

The strategy of using fungal plant pathogens in biological control involves a classic tactic and a bio-herbicide tactic. The first approach involves the introduction of a foreign plant pathogen, while the latter envisages the development of endemic pathogens and using them as microbial weed killers (bio-herbicides). Pathogens are multiplied in vitro and applied as herbicides so that each individual weed plant is heavily inoculated with infectious propagules. For instance, Septogloeum gillis, Wallrothiella arecuthdoii and Colletotrichum gloeosporioides have the ability to attack mistletoes (Arecuthobium spp.) at all stages of development.

Similarly, the growth of the weed, Eupatorium adenophorum can be controlled by Cercospora eupatorii and Colletotrichum destructive can be used as a bio-control agent for dodder. Several fungi are also utilized for controlling soil borne organisms like amoeba and nematodes. For instance, Nematophthora gyrophila is capable to control effectively Heterodera avenae, a cereal cyst nematode.

Importance as Mycorrhiza : A symbiotic relationship between fungal hyphae and roots of higher plants is known as mycorrhiza. Several fungi, like species of Rhizoctonia, Phoma, Tricholoma, Boletus, Phallus, Scleroderma and Amanita, form mycorrhizal relationships with higher plants. The fungal partner of the mycorrhiza obtains food from the roots of higher plant and in return it supplies mineral elements to the other partner. Mycorrhiza are important organs of water absorption for the plants growing in dry conditions and plays a dominant role, particularly in the uptake of less-mobile plant nutrients.

In Soil Aggregation and Soil Fertility : Some fungi, such as species of Absidia, Aspergillus, Cladosporium, Chaetomium, Mucor, Penicillium, and Rhizopus, have soil - binding properties. The mucilagenous substances secreted by them are helpful in soil aggregation. Yeasts (such as species of Rhodotorula and Saccharomyces) and several phylloplane fungi have nitrogen fixing capabilities, thus increasing soil fertility. In a forest ecosystem, the natural mushroom flora greatly helps in biodegradation of woody wastes. The ultimate end product in the form of humus is quite useful for the growth of other plants.

As Growth Hormones : Gibberellin, produced by Gibberella fujikuroi, is an important plant hormone. It is used to accelerate growth of many crops. The hormone trisporic acid is obtained from Mucor mucedo and Choanephora trisphora.

As Insecticides : Many insect pests can be controlled by the use of fungi Aschersonia aleyroidis, Beauveria bassiana, Cordyceps melonthae, Empusa sepulchralis, Fusarium oxysporum and Metarhizium anisopliae.

Some soil fungi (*e.g.*, Trichoderma viride, Fusarium roseum f. sp. cerealis, Rhizopus nigricans) act as 'soil nibblers' since they attack and kill some roots but do not cause excessive damage. Their overall effect is to promote more fibrous root development and to increase total area of root absorption.

Fungi as Test Organisms : Several fungi are used as important research tools for the study of various fundamental biological processes. They grow very fast and require a short period to complete their life cycle. Furthermore, experimentation with fungi requires less space and inexpensive equipments. For instance, Neurospora has become these days an ideal material for genetical studies and Physarum polycephalum is a very good material for the study of DNA synthesis, morphogenesis and mitotic cycle.

Neurospora crassa is used as a test organism to detect the presence and quantity of vitamin B in a given sample. Aspergillus niger is used for the detection of trace elements like zinc, nickel and copper, even when they are present in very minute quantities.

Several fungi such as species of Aspergillus, Absidia, Mortierella, Penicillium, Torulopsis, Endomyces, Rhodotorula are capable of synthesizing fats and fatty substances. Many alkaloids are also produced by fungi; for instance, fumigaclarin is produced by Aspergillus fumigatus and Claviceps purpurea produces a number of important alkaloids like ergotinine, ergotinic acid, ergobasine and ergotetrine. Several fungi which possess coloured spores and mycelium are utilized as dyes. A red pigment obtained from Monascus purpureus is used for colouring rice and Cercosporina kikuchii produces a reddish-violet novel pigment, neocercosporin.

A similar maroon pigment is obtained from Aspergillus fumigatus. A brown pigment atromentin is obtained from Paxillus atromentosus and a yellow pigment citrinin from Penicillium citrinum. Some Basidiomycetes like Armillaria mellea show bioluminescence, thus their fruiting bodies are used as ornamentals and path markers during night.

Harmful Activities

There are many fungal activities which are harmful to mankind. Fungi incite many diseases in plants, animals and human beings. They cause considerable loss to food, vegetable and fruit crops. They destroy valuable timbers and cause spoilage of food and many other useful articles.

Fungi as Pathogen

Plant Diseases : Most of the parasitic fungi cause diseases in plants, especially to flowering plants. The common fungal diseases of plants are smuts, rusts, mildews, blights, rots and wilts. They are responsible for destroying crops worth billions of rupees. They may result even into a catastrophe if allowed to run their course unchecked. Disastrous Irish potato famine of 1845-49 caused by Phytophthora infestans was responsible for taking lives of approximately one million people. Similarly, tobacco yield was reduced by over 60% in North Africa and Middle East in 1962 because of the infection of Peronospora tabacina. In England, over five million elm trees were destroyed during 1967-77 because of the infection of Ceratocystis ulmi.

Some important plant diseases caused by fungi are listed in Table.

Animal Diseases : Several fungal species live parasitically on/in animals and they cause various diseases.

Some Common Plant Diseases Caused by Fungi

Fungi	*Plant disease*
1. Synchytrium endobioticum.	Wart disease of potato
2. Pythium aphenodermatum	Stem rot of papaya
3. Phytophthora infestans	Late blight of potato
4. Erysiphe polygoni	Powdery mildew of peas
5. Erysiphe graminis	Powdery mildew of wheat
6. Peronospora pisi	Downy mildew of peas
7. Albugo candida	White rust of crucifers
8. Sclerospora graminicola	Green ear of bajra
9. Puccinia graminis tritici	Black rust of wheat
10. Urocystis tritici	Flag smut of wheat
11. Ustilago tritici	Loose smut of wheat
12. Ustilago hordei	Covered smut of barley
13. Ustilago nuda	Loose smut of barley
14. Ustilago kolleri	Covered smut of oat
15. Ustilago scitamineae	Whip smut of sugarcane
16. Sphacelotheca sorghii	Grain smut of jowar
17. Tolyposporium penicillari	Smut of bajra
18. Altemaria solani	Early blight of potato
19. Cercospora personata	Tikka disease of groundnut
20. Colletotrichum fulcatum	Red rot of sugarcane
21. Pyricularia oryzae	Blast disease of rice
22. Fusarium udum	Wilt of arhar
23. Plasmodiophora brassicae	Club root disease of crucifers
24. Protomyces macrosporus	Stem gall of coriander
25. Haemelia vastatrix	Leaf rust of coffee

Some Diseases of Animals Caused by Fungi

Fungus	Animal disease
1. Tinea rubum	Athlete foot
2. Trichophyton, Microsporum,	Ringworm Epidermophyton
3. Mucor, Rhizopus	Mucomycosis
4. Aspergillus	Aspergillosis
5. Penicillium	Penicıllosis diseases.

Some important animal diseases caused by fungi are given in Table.

Actinomycosis and black leg disease in animals are also caused by fungi.

Members of Entomophthorales (*e.g.*, species of Entomophthora) are parasites on insects and other animals. Similarly, members of Saprolegniales (*e.g.*, species of Achlya and Saprolegnia) are parasitic on the fish. The fish of domestic aquaria are commonly infected by Saprolegnia ferax and S. parasitica. Ichthyophorus haferi is commonly known as a fish destroying fungus.

Some Common Diseases of Human beings caused by Fungi

Fungus	Human disease
1. Aspergillus fiavus, A. fumigatus,	Aspergillosis A. niger
2. Blastomyces denmatitidis	Blastomyeosis
3. Candida albicans	Candidiasis
4. Cladosporium corriond	Chromomycosis
5. Coccidiodes immitis	Coccidiomycosis
6. Lipomyces neoformans	Cryptococcosis
7. Geotrichum candidum	Geotrichosis
8. Histoplasma capsulatum	Histoplasmosis
9. Mucor pusillus	Neuritis
10. Trichophyton purpureum	Onychomycosis
11. Sporotrichum schenkii	Sporotrichosis

Diseases of Human Beings : Several fungi infect human beings, causing diseases of skin, respiratory tract, lungs, viscera, nasal sinuses, eye, liver and kidney. Some common diseases caused by fungi in human beings are listed in Table.

Wind-borne spores of several fungi (*e.g.*, Alternaria, Aspetgillus, Chaetomium, Helminthosporium, Monilia, Phoma, Trichoderma) are allergic and cause distressing symptoms.

Spoilage of Food Stuffs : A large number of food articles, if not properly stored, are spoiled by fungi, like Mucor, Rhizopus, Aspergillus, Penicillium and yeast. Species of Mucor and Rhizopus are commonly seen on the bread and pickle. Dairy products are spoiled by the species of Mucor, Oidium, Torula, Penicillium and Cladosporium. Mucor mucedo and Rhizopus stolonifer spoil frozen meat by causing black spot disease. Some fungi can infect food stuffs even at a very low temperature; for instance, Cladosporium herbaceum can grow on meat stored at 6°C.

Several species of Alternaria, Aspergillus and Rhizopus cause post-harvest diseases in fruits and vegetables thus shortening their storage life. Oranges are destroyed by Penicillium digitatum and dried prunes and figs by Zygosaccharomyces, Debromyces and Zigopichia. Some common diseases of fruits in storage are listed in Table.

Some Common Diseases of Fruits in Storage caused by Fungi

Fungus	*Storage disease*
1. Alternaria spp.	Alternaria rot of apple
2. Aspergillus fumigatus	Aspergillus rot of apple
3. Alternaria tenuis	Alternaria rot of banana
4. Rhizopus stolonifer	Soft rot of tomato
5. Rhizopus arrhizus	Soft rot of potato

Deterioration of Articles : Many important paper documents, valuable leather articles, textiles, plastic, objects, photographic films, electronic goods, rubber, painted surfaces, etc. are spoiled by fungi. Chaetomium globosum, Aspetgillus niger, Stachybotrys atra and species of Alternaria and Cephalosporium are some common fungi which spoil paper. Penicillium purpurogenum and species of Fusarium and Stachybotrys cause spot print on leather goods. Moulds are responsible for the damage of cellulose fibers. They weaken and discolour textiles and alter their affinity for dyes. Myrothecium verrucaria is notorious for its ability to degrade cellulose.

The proteins of wool and silk are also degraded by moulds. Paints and painted surfaces, like most organic substrates, are subject to attack by various fungi. The fungi mainly responsible for disfigurement or deterioration of painted surfaces are Pullularia pullulans, Phoma glomerata and species of Alternaria and Cladosporium. Several species of moulds are responsible for damaging glass components of optical equipments, electrical cables, telephone poles, etc. Amorphotheca resinae may plug the pipes and valves of the fuel tanks of air crafts and might prove dangerous, especially during flight. This fungus also grows luxuriantly on aluminium alloy and may corrode the metal.

Jute articles are destroyed by Chaetomium globosum, Memnoniella echinata and Stachybotrys atra. Rubber products are spoiled by the species of Aspergillus and Penicillium.

Destruction of Timbers : Several fungal species like Polyporus spp., Serpula lacrymans, Penicillium divarianum, Fusarium negundi, Coniophora cerebella and Lentinus lapidens are responsible for the destruction of valuable timbers. These fungi break down the components of wood and reduce mechanical strength of the wood. Armillaria mellea causes wood rot diseases in several trees. Wood degradation by fungi is usually of two types- white rots and brown rots. White rots cause degradation of brownish lignin, leaving a white spongy cellulosic mass in the wood. On the other hand, in brown rots, degradation of cellulose takes place and a brown pinkish mass with a predominance of lignin is left behind.

Some fungi like Ceratostomella, Chlorosplenium, Fusarium, Lasiosphaeria, Penicillium and Trullula grow in sap wood and stain it. Chlorosplenium aeruginosum and Penicillium divaricatum impart characteristic green and yellow stain to hard wood and Lasiosphaeria pezizula imparts greyish olive stain to heart and sapwood.

Hallucinogenic Fungi : Some fungi, such as Amanita phalloides, A. verna and Boletus santanus, are highly poisonous and if ingested they may prove fatal.

Several mushrooms, such as Amanita muscaria, Psilocybe mexicana and Panaeolus spp. secrete hallucinogenic substances like psilocybin and psilocin. These substances may destroy brain cells and perception power of human beings.

Mycotoxins : Many fungi produce mycotoxins which are responsible for food, poisoning and other distressing symptoms. These mycotoxins may be broadly classified into the following three categories.

Food Toxins : These toxins are mainly responsible for food-poisoning. The toxin production can occur in most plant products, but cereal and oil seed crops are mostly contaminated. The following four groups of food toxins are mainly associated with human diseases:

(a) *Aflatoxins :* Aflatoxins, the most potent carcinogens, are produced by Aspetgillus flavus, A. fumigatus, A. parasiticus and Penicillium islandicum, when they, infest dried food and groundnut meal. They are highly oxygenated low molecular weight, heat stable, heterocyclic compounds. Eight forms of aflatoxins, B_1,

B_2, G_1, G_2, M_1, M_2, B_{2a} and G_{2a} have been recognised. Of these, aflatoxins B_1, B_2, G_1 and G_2 are the most important. Aflatoxins are derivatives of furanocoumarin. They bind with DNA and prevent its transcription and as such protein synthesis is inhibited. They cause liver cancer in animals and human beings. The clinical symptoms include jaundice, rapidly developing ascites, portal hypertension, etc.

(b) *Ochratoxins* : Ochratoxins, closely related derivatives of isocoumarin linked to 1-β–phenylalanine, are mainly produced by Aspergillus ochraceus and Penicillium viridicatum, when they infest stored maize, pea nuts, beans and mixed animal feeds. Of the nine ochratoxins, ochratoxin A is the most important. It causes tubular necrosis of the kidney, mild degeneration of the liver and enteritis of the small intestine in ducklings.

(c) *Zearalenone* : Zearalenone, a phenolic resorcyclic acid lactone, is produced by several species of Fusarium, such as F. graminearum and F. moniliforme, growing on maize. It causes esterogenic symptoms in swine.

(d) *Trichothecenes* : Trichothecenes are produced by several species of Cephalosporium, Fusarium, Myrothecium, Stachybotrys and Trichoderma. They possess a tetracyclic 12,13-epoxytrichothecoene skeleton. Of the nearly 30 trichothecenes known, T-2 toxin, nivalenol and deoxynivalenol are most important. They are responsible for severe local irritation, inflammation, sub-epidermal haemorrhage and general necrosis.

Ergot Toxins : The sclerotia of Claviceps purpurea contain some poisonous alkaloids, like ergotamine, ergometrimine, ergocrystinine, ergocristine and ergonovin. Ergot poisoning in human beings causes diarrhoea, abdominal pain and vomiting. It also effects nerves and results in psychiatric disturbances.

Mushroom Toxins : Several mushrooms produce mycotoxins which cause diarrhoea and vomiting in earlier stages, but in severe cases, liver damage, kidney failure, complete unconsciousness, and even death may take place. Amanita phalloides produces about ten toxins; of these, phalloidin affects the plasma membrane of liver cells, and α-amanitin causes lesions in stomach and intestine cells. Gyromitrin, a highly fatal toxin is produced by the species of Helvella, such as H. esculenta, H. gigas, H. infula and H. underwoodii. Some species of Inocybe and Clitocybe produce the toxin muscarine. Species of Coprinus produce the toxin coprine, that affects the autonomic nervous system.

2

Edible Fungi

Mushrooms and their Cultivation

Mushrooms are the fruiting bodies of higher fungi - Ascomycetes and Basidiomycetes. They have been used as food by human beings since tune immemorial. Their use, as recorded in the ancient Indian literature, dates back to 3000 B.C. Mushrooms were also known to the ancient Greek and Roman civilizations. Egyptians considered them as God's food and used them only on certain festive occasions. In the ancient times, mushrooms were utilised only for taste and flavour, and were collected from their natural habitats. The techniques for the cultivation of mushrooms were developed at the beginning of the 18th century in France.

There are about 7,000 species of mushrooms, but a little over 100 species are suitable for human consumption. Only a few species of mushrooms are commercially cultivated. The common edible mushrooms belong to Agaricus, Lentinus, Flammulina, Volvariella, Pleurotus, Pholiota, Tremella, Auricularia, Stropharia and Tuber.

The total annual yield of some important edible mushrooms is given in Table.

Aman ta vaginata, A. rubescens, Agaricus campestris, A. arvensis, Lepiota procera, Boletus edulis, B. pinicola, B. luteus, Lactarius deliciosus, Collybia velutipes, Russula cyanoxantha, Marasmius oreades and Cantharellus cibarius are some important edible wild species.

Total Annual Yield of some Edible Mushrooms

Species	*Common name*	*Yield (in tonnes)*
Agaricus bisporus	White button mushroom, European or tropical mushroom	870,000
Lentinus edodes	Shiitake	170,000
Flammulina velutipes	Enokitake	60,000
Volvariella volvacea	Paddy straw mushroom, Tropical mushroom	49,000
Pleurotus ostreatus	Oyster mushroom	32,000
P. sajor-caju	Dhingri	32,000
Pholiota nameko	Nameko	17,000
Tremella fusiformis	Tremella	10,000
Auricularia polytricha	Jew's ear	10,000
Stropharia rugoannulata	Strophaires	2,000
Tuber melanosporum	Truffles	2,000

The cultivable mushrooms can be divided into the following five groups:

(1) Mushrooms that grow on fresh plant residues, *e.g.*, Lentinus, Flammulina, Tremella, Coprinus, Pleurotus, Auricularia and Pholiota.

(2) Mushrooms that grow on only partly composed plant residues, *e.g.*, Volvariella, Stropharia and Coprinus.

(3) Mushrooms that grow on well composed plant residues, *e.g.*, Agaricus.

(4) Mushrooms that grow on soil and humus, *e.g.*, Lepiota, Lepista, Morchella and Gyromitra.

(5) Myeorrhizal mushrooms, *e.g.*, Boletus, Amanita, Morchella, Lactarius and Tuber.

Nutritive Value of Mushrooms : In developed countries, the average consumption of animal protein is about 31 kg per capita per year, whereas it is only 4 kg in India. In India, the plant proteins are more widely used than the animal proteins. Mushrooms are the richest source of vegetable proteins. They contain 21-30% proteins on dry weight basis.

Thus the percentage of proteins in mushrooms is much higher than in cereals, pulses, fruits or vegetables. The proteins of mushrooms contain all essential amino acids and their quantity is higher than in the egg. There is also a good amount of lysine amino acid in mushrooms (about 550 mg/g).

Besides, mushrooms contain sufficient quantities of mineral elements, such as Ca, Na, P and K. They also contain folic acid and are as such very good source of iron. They contain vitamins B. C, D and K, which are not distorted during ripening, drying, freezing or canning process. On the other hand, they contain very little fats (0.35-0.65% dry wt) and starches (0.02% dry wt). Thus mushrooms make an excellent food for diabetic and heart patients.

The chemical composition of three important edible mushrooms is given in Table.

The Chemical Composition of Three Edible Mushrooms

Mushroom	Moisture	Ash	Protein	Fat	Carbohydrate	Energy value (cal)
	(% on fresh weight basis)					
Agaricus bisporus	89.50	1.26	3.94	0.16	6.28	34.4
Pleurotus flabellatus	90.95	0.97	2.78	8.65	5.33	24.4
Volvariella diplasia	90.40	1.10	3.90	0.25	5.51	29.2

Cultivation of Mushrooms

Mushrooms are heterotrophic as they lack chlorophyll. Hence light is not necessary for their growth. The three major requirements for their cultivation are:

(1) Suitable temperature

(2) Good compost

(3) Good quality spawn.

Suitable Temperature : The most suitable temperature for the growth of mushrooms is between 30° and 37°C. Temperature below 15°C and above 45°C is detrimental for the growth of the mycelium and fruiting bodies. In many parts of India, such as U.P., M.P., Punjab, Haryana, Orissa and Maharashtra, the optimum season for their cultivation is from April to September. In the plains of West Bengal, they are usually grown between March and September. In favourable conditions, the first crop of mushrooms is ready within 30-45 days and thus several crops can be taken in a growing season. Although light has no effect on the growth of mushrooms, the shelf life of the mushrooms grown in dark is comparatively more.

Good Compost : Two types of compost — natural and synthetic — are used for cultivation of mushrooms.

Natural Compost : The natural compost is prepared by mixing wheat or barley straw in horse dung. Normally 33 kg straw is mixed with 100 kg of dung. It should be kept in mind that dung of any other animal should not be mixed with horse dung. Stored or rain wet dung is not suitable for compost. A meter high heaps of the fresh and wet dung, obtained from the stable are made in open air. After 3-4 days, when the dung starts producing ammonia due to fermentation, it is turned and stacked again. This process is repeated 4-5 times at the intervals of 5-6 days. During this course, gypsum (25 kg/tonne dung) is added to the dung. Finally, 40 ml nemagon is sprayed on the mixture.

Synthetic Compost : The ingredients of the synthetic compost are:

(i)	Wheat straw — 300 kg	(iv)	Urea — 4 kg
(ii)	Wheat bran — 30 kg	(v)	Potash — 1.5 kg
(iii)	Ammonium sulphate or Calcium ammonium nitrate — 6 kg	(vi)	Gypsum — 30 kg

Wheat straw (cut to 8-12 cm long pieces) is spread on a clean cement concrete ground and wetted properly with water. Then it is mixed with half of the quantities of the other ingredients, except gypsum. This mixture is stacked into about a meter high heaps and the heaps are covered. After an interval of five days, these stacks are scrapped and the remaining half of the ingredients is thoroughly mixed with it and the whole mixture is piled again. This process is repeated six times.

In the third and fourth turning, gypsum is added to the mixture. After the sixth turning, if there is a smell of ammonia, two or more turnings should be given until the smell of ammonia disappears. In the last turning, 10 ml malathion (dissolved in 5 litres water) is added to the mixture. In addition, 1 litre/kg of lintaff is also added to prevent insect infestation. The final mixture has sufficient quantity of moisture and can fill 20 trays of 100 × 50 × 15 cm.

Good Quality Spawn : Good quality spawn is essential for better harvest. Different substrates in which mushrooms were already grown are used as inoculum. Mostly the compost left after the harvest of mushroom crop is used as inoculum. Sometimes small pieces of mushroom stipe are used as inoculum. Inoculum of Agaricus is usually prepared by adding grains of wheat, rye or millets to the substrate.

Mushrooms are grown in beds of various dimensions

(1 × 1 × l m, 3.5 × 3.5 × l m, or 0.6 × 0.6 × l m). Wooden trays of 100 × 50 × 15 cm dimensions are utilised for the cultivation of Agaricus. The wooden trays are filled with the compost and then pressed so that a space of about 3 cm deep is left on the top of the tray.

The inoculum is spread over the compost and then it is covered by a thin layer of compost. The trays are finally covered with old newspapers. They are arranged in vertical stacks in such a way that there is sufficient space for aeration between the two trays. There should be a space of about one meter between the roof of the room and the top of the stacks of trays. The trays are sprayed with water at suitable intervals to keep the substrate moist.

In India, usually three types of mushrooms, *viz.*, Agaricus bisporus (button mushroom), Pleurotus spp. (dhingri) and Volvariella spp. (paddy straw mushroom) are cultivated.

The main operations in the cultivation of Pleurotus sajor-caju (dhingri) are as follows:

Substratum : The substrate is prepared from chopped paddy straw, crushed maize cobs, wheat straw, rye straw, dried and pulvarized grasses, compost, wooden logs, etc.

Method of Cultivation : The substrate is soaked in water in a tank for 8-12 hours. After removal from the tank, it is washed with fresh water and the excess of water is allowed to drain. The substrate is filled in wooden trays of 1 × 0.50 × 0.25 m dimensions. The filling of trays is done in two ways : (i) The trays are first filled up to 9 cm and spawn is sprinkled on the entire surface evenly. Then the substrate is again filled until the final depth of the substrate is 16-18 cm. (ii) The trays are filled with the substrate up to 16-18 cm and then the surface is sprinkled by spawn. After spawning, trays are covered by polythene sheets. They are sprayed with water once or twice a day so that sufficient moisture is maintained. After 10-15 days of spawning, white cottony growth appears on the entire surface and at this stage casing of trays is done.

The polythene sheets are removed and casing is done by covering the surface with 2-2.5 cm thick uniform layer of the soil containing 1 part sand, 1 part garden soil and 1 part resin free saw dust. The soil should be sterilized before casing at 20 pound pressure for two hours or by heating it in a drum at about 90°C or by treating it with 5% formaline solution. The cased trays are continued to be irrigated by spraying water. The first flush of mushroom appears after 10- 15 days of casing. If casing is not possible, the trays should be left as such and polythene covers be removed after the appearance of the first flush. New flushes of mushrooms continue to appear for 30-45 days after the first flush.

For the healthy growth of mushrooms, maintenance of a temperature of 25±5°C and 85-90% relative humidity are essential. Besides, aeration and spraying of water is also necessary. To control insect pests and diseases spray of 0.1% malathion and 0.2% diethane Z-78 is required.

Harvesting : Mushrooms are harvested when the pileus is about 8-10 cm in diameter. The harvesting is done by twisting the stalk so that the broken pieces are not left in the trays as they may become the source for spread of diseases. The space formed by the removal of stalks should be filled by sterilized soil.

After harvesting, the lower portion of the stalk with adhering debris is cut with the help of a clean and sharp knife. Mushrooms can be stored in refrigerator for about a week. The fresh mushrooms can be dried either in the sun or in an oven at 55-60°C for 8 hours. The dry mushrooms are packed and sealed before marketing.

3

The Lichens

Lichens are a small group of curious plants. They are made up of algal and fungal components, living together in an intimate symbiotic relationship. The algal component is known as phycobiont (phykos = alga, bios = life), and the fungal component as mycobiont (mykes = fungus, bios = life). The plant body of lichens neither resembles algae nor fungi. Thus, lichen is an association of a fungus and an algal photosynthetic symbiont, resulting in a stable thallus of specific structure. Some important characteristics of lichens are as follows:

(1) Lichens have a composite thalloid structure, formed by the association of algae and fungi.

(2) The algal component of lichens forms carbohydrates by photosynthesis which are utilised both by the algal and fungal components. The fungal component is responsible for the absorption and retention of water.

(3) On the basis of the structure of thallus lichens have been classified into three broad types : (i) crustose, (ii) foliose, and (iii) fruticose.

(4) They reproduce asexually by the formation of various types of asexual spores, such as oidia and pycnidia.

(5) Only fungal component of lichens are involved in sexual reproduction.

(6) The female sex organ, known as carpogonium, is differentiated into a basal coiled ascogonium and an elongated multicellular trichogyne.

(7) The male sex organs are flask-shaped spermagonia, found on the upper surface of the thallus. They produce spermatia.

(8) The fruiting bodies of lichens are disc-shaped apothecia or flask-shaped perithecia.

(9) The ascus produces eight ascospores. The hypha produced by the germination of ascospore, when comes in contact with a suitable alga, forms a new lichen thallus.

(10) The growth of lichens is extremely slow. They can survive high temperature and dry conditions for a long time. Hence, they can grow easily on dry substratum.

(11) Lichens produce a specific acid, known as lichen acid.

Distribution : There are about 400 genera and 1600 species of lichens, widely distributed in most parts of the world. They usually grow on the barks of trees, dry logs of wood, bare rocks and other similar situations. Some species of lichens (*e.g.*, Cladonia rangifera) even grow in the Arctic and Antarctic regions and others occur on the seashore. In India, lichens are abundantly found in the Eastern Himalayan regions. Some common species are: Cladonia aggregata, Graphis duplicata, G. scripta, Gyrophora cylindrica, Haematomma puniceum, Physcia aspera, Usnea aspera and U. dichotoma.

Nature of Association of Phycobiont and Mycobiont in Lichens : The algal components of lichens belong to

Chlorophyceae or Myxophyceae, whereas the fungal components to Basidiomycetes or Ascomycetes. The algal components of about 75% lichens are the members of Chlorophyceae and the fungal components mostly belong to Ascomycetes, except for four genera where they are from Basidiomycetes.

The inter-relationships between the algal and fungal components of lichens are usually considered to be symbiotic in which both algae and fungi are equally benefited . It is believed that the alga synthesizes organic food' materials sufficient for both the algal and fungal components. In exchange of food, water and minerals are made available to alga by the fungal component. Besides, the fungal component also protects alga from high temperature and desiccation.

However, some workers believe that the fungal component has a partially or wholly parasitic relationship with the algal component. This view gets support from the facts that (i) haustoria of fungi are seen in algal cells of some lichens, and (ii) on separation, the algal component of lichen can live independently, whereas the fungal component can not survive.

According to yet another view, although the association of both the components of lichens is symbiotic, the fungal component shows predominance over the algal component, and the latter is a subordinate partner. Such an association is known as helotism.

Classification of Lichens

There is no natural system of classification of lichens. According to the rules of the International Code of Botanical Nomenclature no significance is attached to the algal components for the classification of lichens. On the basis of the structure of fruiting bodies of fungal components, Zahlbruckner (1907) classified lichens into the following two groups.

Ascolichens : The fungal component of these lichens is a member of the class Ascomycetes. These lichens are divided into two series on the basis of the structure of fruiting body:

1. *Gymnocarpeae :* The fruiting body is a disc-like apothecium. These lichens are also known as discolichens (*e.g.*, Parmelia).
2. *Pyrenocarpeae :* The fruiting body is a flask shaped perithecium. These lichens are also known as pyrenolichens (*e.g.*, Dermatocarpon).

Basidiolichens : The fungal component of these lichens is a member of the class Basidiomycetes. Genera like Corella and Dictyonema belong to this group.

However, Alexopoulos and Mims (1979) classified lichens into the following three groups:

1. *Basidiolichens :* The fungal component is a member of the class Basidiomycetes.
2. *Deuterolichens* : The fungal component is a member of the class Deuteromycetes.
3. *Ascolichens :* The fungal component is a member of the class Ascomycetes.

Structure : The plant body of lichens is thalloid and irregular in shape. It is usually grey or greyish green in colour but some species are yellow, red, orange or brown in colour.

External Morphology : The following three types of lichens have been recognised on the basis of their external morphology.

Crustose Lichen : These are encrusting lichens with an inconspicuous, thin and flat thallus, firm in texture. The thallus is very closely adhered to the substratum and provides a crust-like appearance. It is wholly or partially embedded in the substratum and in the latter case only fruiting bodies are visible above the surface of the substratum. Examples : Graphis,

Haematomma, Lecanora, Lecidia, Rhizocarpon, Strigula and Verrucaria.

Foliose Lichen : These lichens are flat with leaf-like and lobed thallus. They are attached to the substratum with the help of rhizoid-like rhizines. Examples: Parmelia, Physcia, Peltigera, Xanthoria, Gyrophora and Collema.

Fruticose Lichen : These are shrubby lichens with a well developed, shrub-like, cylindrical and branched thallus. They grow erect or hang from the substratum. The plant body is attached to the substratum with the help of a basal mucilagenous disc. Example : Alectonia, Cladonia, Letharia and Usnea.

Internal Structure : The internal structure of lichens is very complex. As mentioned earlier, the thallus is composed of algal and fungal components. Such a type of thallus is known as consortium. In an advanced foliose lichen the following four distinct regions can be recognised in a vertical section.

Upper Cortex : It is the outermost thick and protective zone of the thallus, composed of compactly interwoven fungal hyphae. The hyphae are arranged at right angle to the surface of the thallus. There are usually no intercellular spaces between the hyphae and if at all there are any, they are filled with gelatinous substances. In some lichens (*e.g.*, Parmelia) there are many irregularly arranged breathing pores on the outer surface. These pores help in gaseous exchange.

Algal Layer : The algal layer occurs just below the upper cortex. This layer is also known as gonidial layer (gonidium cells are reproductive cells of alga). The algal cells remain embedded in between the tangled network of loosely interwoven hyphae. In some species (*e.g.*, Collema, Leptogium), algal cells and fungal hyphae are distributed more or less uniformly throughout the thallus. Such species are known as homoisomerous. On the other hand, in members like Parmelia and Physcia, the algal cells form a distinct layer within the thallus. These species are known as heteromerous. Some common green and blue-green algae found in the lichens are the species of Protococcus, Pleurococcus, Cladophora, Trebauxia and Trentepohilia (Chlorophyceae), and Nostoc, Gloeocapsa and Rivularia (Myxophyceae).

Medulla : It is the central part of the thallus, comprised of loosely interwoven fungal hyphae with large spaces between them.

Lower Cortex : This zone is composed of compactly arranged hyphae which run parallel or perpendicular to the surface of the thallus. Some of these hyphae become specialized and extend downward from the lower surface of the cortex and help in the attachment of thallus to the substratum. These specialized hyphae are known as rhizines.

Internal structure of crustose lichens is also more or less similar to foliose lichens. The lower cortex does not occur in fruticose lichens due to their cylindrical structure and medulla forms the central part of the axis.

Reproduction : Lichens reproduce both by asexual and sexual means.

Asexual Reproduction : Asexual reproduction is of common occurrence in lichens. Some common methods of asexual reproduction are as follows.

Fragmentation : Small fragments of thallus are formed by accidental breaking or by the death or decay of the older parts. Each fragment develops into a new thallus, provided it contains both algal and fungal components.

Soredium : Some small bud-like outgrowths, known as soredia, develop on the surface of the thallus. A soredium contains one or few algal cells closely enveloped by a weft of fungal hyphae. The soredia form a granular layer of greyish-white colour on the surface of the thallus. They are detached from the thallus by the impact of wind or rain. Sometimes, as in Parmelia and Physcia, soredia develop in an organised manner in special pustule-like areas. Then they are known as soralia. The soredia germinate on suitable substratum and form new thalli.

Cephalodium : These are small wart-like structures formed on the surface of the thallus. One of the characteristic feature of the cephalodium is that its algal and fungal components differ from that of the thallus.

It is due to the fact that cephalodia develop on the younger parts of the thallus from soredia of some other species. Hence, the cephalodium may be regarded as sterile thallus of some other lichen.

Isidium : Isidia are small, stalked, greyish-black coral-like outgrowths which develop on the upper surface of the thallus. The isidium has an outer cortical layer enclosing the algal and fungal components. It is usually constricted at the base and is easily detachable from the parent thallus. It germinates under favourable conditions and forms new thallus.

In addition to propagation, isidia also help in increasing the photosynthetic surface of the thallus. They vary in shape and may be rod like (*e.g.*, Parmelia), coral-like (*e.g.*, Pelfigera), scale-like (*e.g.*, Collema) or cigar like (*e.g.*, Usnea).

Some lichens (*e.g.*, Physcia, Buellia) develop flask-shaped pycnidia which form pycnidiospores. The latter form fungal hyphae on germination. These hyphae, when come in contact of a suitable alga, develop into a new lichen body. Sometimes fungal hyphae break to form oidiospores.

Sexual Reproduction : In lichens the process of sexual reproduction is performed only by the fungal component. The fungal component of most of the lichens belongs to the class Ascomycetes. Hence, the sexual reproduction is as in Ascomycetes.

The female sex organs are known as carpogonia. A carpogonium is differentiated into a basal coiled ascogonium and an elongated multicellular trichogyne. The ascogonium remains embedded within the algal layer of the thallus, whereas the trichogyne projects over the surface of the thallus.

The male sex organs are flask-shaped spermagonia. They form spermatia which function as male gametes. The spermagonium usually develops close to carpogonium. This enables spermatia to adhere to the projected part of sticky trichogyne. On dissolution of the walls between the spermatium and trichogyne, the nucleus of spermatium migrates into the carpogonium through trichogyne. The male nucleus fuses with the female nucleus.

Several branched ascogenous hyphae develop from the base of the fertilised ascogonium. The terminal or penultimate binucleate cell of the ascogenous hypha develops into an ascus. The two nuclei within the ascus fuse to form a diploid nucleus. The diploid nucleus first divides meiotically and then mitotically to form eight haploid daughter nuclei. Each haploid nucleus metamorphoses into an ascospore.

The asci remain enveloped by paraphysis. The somatic tissue surrounding the asci and paraphysis forms the fruiting body, which may be apothecium (*e.g.*, Pannelia, Anaptychia) or perithecium (*e.g.*, Dermatocarpon, Verrucaria).

The ascospores vary considerably in their shape, size, structure and septation. They are hyaline and greenish or brown in colour. They are released gradually from the ascus. The ascospore produces a hypha on germination and this hypha, when comes in contact with a suitable alga, forms a new lichen thallus.

Economic Importance of Lichens

Lichens are wonderful gift of nature. They are the source of food, medicines, flavours and stains. Several animals and insects also feed on lichens. Besides, lichens are of considerable ecological importance as pioneers in colonisation of rocky habitat. They are the first plants to survive on dry bare rocks. Lichens are also very good indicators of air pollution. Some important useful activities of lichens are discussed below.

As Food and Fodder : Lichens are used as food since ancient times. They are important constituents of food in north Polar Tundra and eastern Siberian regions. Species of Lecanora, Parmelia, Umbillicaria, and Cetraria icelandica (Iceland moss) are some of the lichens which are used as food in many parts of the world. Umbillicaria esculentus is a delicacy in Japan, while the species of Parmelia are used as curry powder in India.

Lichens contain a polysaccharide, lichenin, but lack true starch and cellulose. Evemia prunastri is used by Egyptians for making breads. In France, some lichens are used for making delicious chocolates and pastries. In Japan, Endocarpon miniatum is used as a vegetable.

Several lichens (*e.g.*, Aspicilia calcarea, Lecanora saxicola) are used as food by mites, snails, caterpillars, slugs, termites, etc.

Lichens, such as Lobaria pulmonaria, Evemia prunastri, Ramalina fraxinea and R. fastigiata are used as fodder for animals. They possess great nutritive value due to the presence of lichenin. Cladonia rangifera (reindeer moss) serves as a common food in Tundra regions for animals, specially reindeer and musk ox. Dried lichens are fed to horses and swans.

As Medicine : The medicinal value of lichens is due to the presence of lichenin and some astringent substances. Some lichens are used in the treatment of bile, diarrhoea, fever, nervous disorders, hydrophobia and skin diseases. Parmelia perlata is specially useful in dyspepsia and in the treatment of snake and scorpion bites. Species of Cladonia, Cetraria and

Pertusaria are used in intermittent fever, while Cladonia pyxidata is useful in whooping cough. Species of Usnea are used to stop bleeding. Usnic acid, obtained from the species of Usnea and Cladonia, is a broad spectrum antibiotic used in the treatment of infections. Erythrin, obtained from Roccella montagnei, is used in angina. Lobaria pulmonaria and Cetraria islandica are used in tuberculosis and other lung diseases. Lichens are important constituents of several ayurvedic medicines. Some lichens are used with tobacco because of their hallucinogenic effects.

As Dyes : Dyes obtained from lichens have been used since ancient times for colouring fabrics, etc. Red and purple dyes are obtained from Ochrolechia androgyna and O. tartaria. Orchil, a blue dye obtained from some lichens (*e.g.*, Cetraria icelandica) is used for dyeing woolens. Orcein, a purified and active principle of orchil dye, is used as a stain in histological studies and also for dyeing coir in South India.

Parmelia omphalodes is the source of a brown dye which is used for dyeing wool and silk fibres. Litmus, an important acid-base indicator dye in chemical laboratories, is obtained from Roccella montagnei and Lasallia pustulata.

In Tanning Industry : Some lichens (*e.g.*, Cetraria icelandica, Lobaria pulmonaria) are used as tanning agent in leather industries. Two colouring substances, lecanoric acid and erythrin, found in the thalli of lichens, give orcein and carbonic acid on treatment with ammonia.

In Cosmetics and Perfumery : Several species of Evemia and Ramalina are the source of essential oils, used in the manufacture of cosmetic soaps. Ramalina calicaris is used for whitening hair in wigs. Pseudevemia furfuracea and Evemia prunastri are widely used in the manufacture of perfumes.

Lichens are also used in the manufacture of hawan samagris, dhup and other perfumeries because of their pleasant smell.

They are important ingredients of many antiseptic creams as they have tumour-inhibiting and spasmolytic properties.

Fermentation and Distillation : In Russia, Sweden and Siberia, liquors are manufactured by fermentation and distillation of some lichens, such as Cladonia rangiferina and Ramalina fraxinea.

As Culture Media : Lichens are important constituents of some culture media used for the culture of fungi and bacteria in laboratories.

Ecological Significance : Lichens are pioneer plants which help in colonisation of bare rocky habitats. They excrete some organic acids which disintegrate rocks and form soil and substratum for subsequent establishment of other vegetation types.

Several lichens are harmful to us. They cause a considerable loss due to etching of glass surfaces and marble stones. Some lichens, such as Letharia vulpina (wolf moss), are poisonous.

4

Sex Hormones

Such diffusable substances which play specific role in sexual reproduction in living beings are known as sex hormones. The presence of sex hormones in fungi was first shown by de Bary (1881) in heterothallic species of Achlya, A. bisexualis and A. ambisexualis. Later, Kauffman (1908) observed that in Saprolegnia hypogyna development of antheridial hyphae is stimulated by certain hormones. Synthesis of these hormones is controlled by some inorganic salts. There are several fungi where hormonal regulation of sexual reproduction occurs. Machlis (1972) has classified sex hormones in three groups on the basis of their functions:

1. *Erotaction* : These hormones attract motile gametes.
2. *Erotropin* : They stimulate chemotropic growth in sexual structures.
3. *Erogen* : These hormones control initiation and differentiation of sexual structures.

The above classification of sex hormones is not much useful as one hormone can perform more than one function. For example, in Achlya the hormones which control development of sexual structures also determine the direction of their growth.

Some Important Sex Hormones of Fungi

Sirenin : Sirenin, a sperm attracting hormone, can be obtained from the female hyphae of Allomyces macrogynus

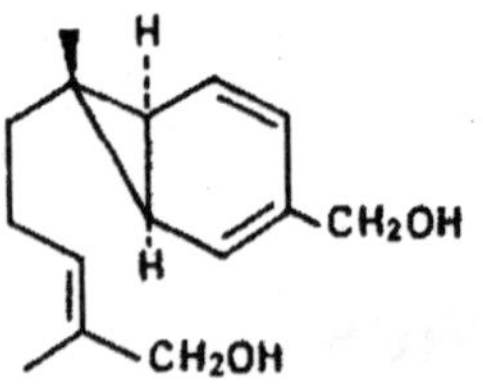

Sirenin.

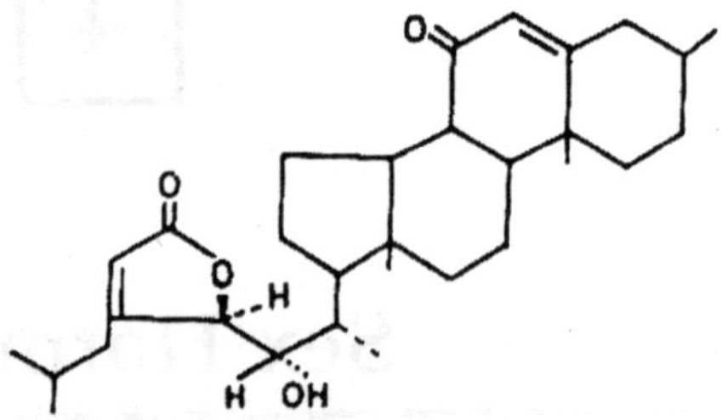

Antheridiol.

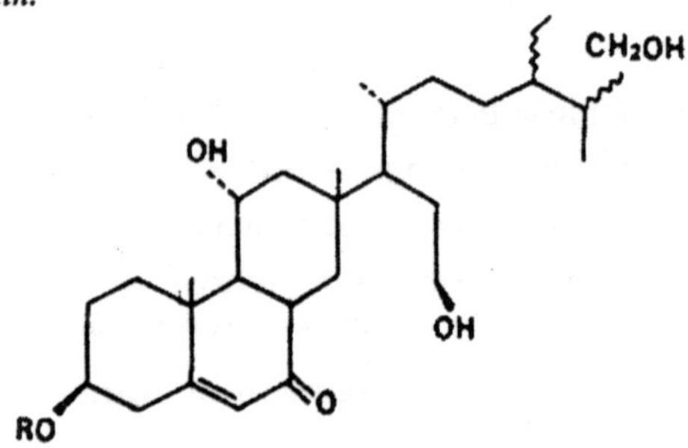

Oogonial.

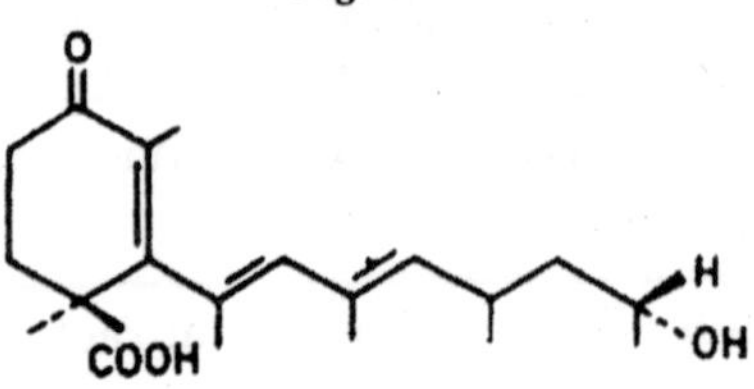

Trisporic acid.

and A. arbuscula. The chemical structure of this hormone was studied by Machlis et al. (1968). It is an oxygenated sesquiterpene which occurs in d-and 1-forms. However, only 1-sirenin is biologically effective. The empirical formula of sirenin is $C_{15} H_{24}O_2$ and its molecular weight is 236 (Fig.).

Antheridiol and Oogonial : Raper (1939-59) studied antheridiol and oogonial hormones in two species of Achlya, A. bisexualis and A. ambisexualis. He envisaged four hormones, A, B, C and D. But now it has been conclusively proved that there are only two hormones - antheridol and oogonial; A and C are antheridiol, and B and D, oogonial.

Antheridiol stimulates four types of reactions: (i) initiation of antheridial hyphae on male plant, (ii) chemotropic stimulation of antheridial hyphae, (iii) stimulation of male hyphae for production of oogonial hormone, and (iv) delimitation of antheridia.

Antheridiol is a crystalline steroid. Its empirical formula is $C_{29} H_{42} O_5$ and molecular weight 470 (Fig.). It is effective even in very dilute form (10^{-10}M). It was isolated by McMorris and Barksdali (1967).

Oogonial is synthesized by male hyphae of Achlya ambisexualis in the presence of antheridiol. It was also isolated from some bisexual strains of Achlya. This hormone stimulates development of oogonium on female hyphae. It is also a crystalline steroid (Fig.).

Hormone-D : There is no direct evidence for the existence of this hormone. Raper (1952), however, suggested that hormone-D delimits oogonium and oosphere. Usually 2-4 hours contact between antheridial and oogonial initials is necessary for the delimitation of oogonium. Hence, it is believed during this period some substances are synthesized in antheridium which delimit oogonium and oosphere. These substances are considered as hormone-D.

Trisporic Acid : Trisporic acid is an unsaturated and oxygenated form of tri-methyl cyclo-hexane. It was isolated from Blakeslea trispora, Phycomyces blakesleeanus and Mucor mucedo. It stimulates progametangia. Three types (A, B and C) of trisporic acid are found. Trisporic acid-C shows 80%

functional activity, trisporic acid-B, 15% and trisporic acid A, 1-2%. The empirical formula of trisporic acid is $C_{18}H_{26}O_4$ and its molecular weight is 306 (Fig.).

Yeast α -Factor : Levi (1950) has suggested that in Saccharomyces cerevisiae sexual reproduction is controlled by some hormones. Two forms of haploid mating types (a and α) occur in Saccharomyces cerevisiae and a diploid cell is formed by their conjugation.

It is believed that some diffusable chemical substances are synthesized in a-type which stimulate the formation of copulatory processes. Duntze et al. (1970-73) isolated α-factor in pure form. It is a peptide with copper ion and its molecular weight is 1400.

Significance of Sex Hormones in Fungi

The significance of sex-hormones, in some representative fungi is described below.

Allomyces : Sirenin hormone is found in two species of Allomyces, A. macrogynus and A. arbuscula. The haploid gametophytic and diploid sporophytic mycelia of these species are similar in form. In gametophytic mycelium, male and female gametangia are developed in pairs, one above the other. The male gametangia are orange coloured and they form male gametes which are motile and comparatively small, whereas the female gametangia are colourless and they from female gametes which are sluggish and about two times larger than the male gametes.

Before its release from the female gametangium, the female gamete secretes sirenin which diffuses outside in the water. The male gametes surround the female gametangium under the influence of sirenin. The number of male gametes surrounding a female gametangium depends upon the concentration of sirenin.

As the female gamete is released from the gametangium, it fuses with the male gamete. Thus a diploid motile zygote is formed by their fusion. The zygote and the female gamete, however, remain unaffected by the hormone.

Achlya : The sex hormones isolated from a dioecious species of Achlya, A. ambisexualis, were studied by Raper (1939-59). They are divided into four groups: (i) A complex (includes A, A_1, A_2 and A_3 hormones), (ii) B, (iii) C, and (iv) D. The hormones A and C are now referred to as antheridiol and hormone B as oogonial. It was observed that if male and female mycelia of A. ambisexualis grow far apart, they do not develop sex organs, but when they grow in close contact, their sexual hormonal system becomes active and sex organs are initiated. The sexual process in A. ambisexualis is divided into the following four steps.

Step 1 : Development of antheridial hyphae on male mycelium. This step is regulated by four hormones (A, A_1, A_2, A_3), collectively known as A-complex. The hormones A and A_2 are secreted by the vegetative hyphae of the female mycelium and A_1 and A_3 by the vegetative hyphae of the male mycelium. The hormones A and A_2, secreted by the female hyphae, can individually or collectively initiate the development of antheridial hyphae on the male mycelium. The hormone A_1, secreted by the male mycelium, quantitatively increases the activity of hormones A and A_2, secreted by the female hyphae, whereas the hormone A_3, also secreted by the male hyphae; reduces the activity of remaining three hormones, A, A_1 and A_2 of A-complex.

Step 2 : Development of oogonial hyphae on female mycelium. As antheridial hyphae develop on the male mycelium, they secrete hormone B (oogonial) which induces the development of oogonia on the vegetative hyphae of the female mycelium.

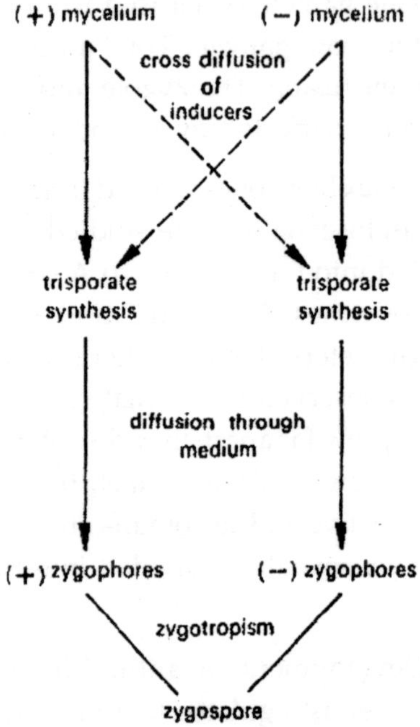

Schematic representation of the effect of sex hormones in the reproduction of Mucor mucedo.

Step 3 : Chemotropic growth of antheridial hyphae towards oogonial hyphae. The hormone C, secreted by the oogonial hyphae, directs the growth of the antheridial hyphae towards the oogonial hyphae, and delimits antheridia after the antheridial hyphae made close contact with the oogonial hyphae.

Step 4 : Delimitation of oogonia and oospheres. This step is regulated by hormone D, which is secreted by the antheridium. This hormone incites the development of a septum at the base of the oogonium which delimits the oogonium. It also stimulates the cleavage of the oogonial protoplast into many small segments, the oospheres.

Mucor : The effect of the sex hormones in Mucor mucedo was studied by Plempel et al. (1957-61). Trisporic acid, which occurs in this species, is synthesized from β–carotene. Trisporic B and C stimulate the development of zygophores in the heterothallic mycelia. The hormone is synthesized only when the mycelia of (+) and (-) strains grow in a normal continuous diffusable medium. The hormone synthesized in (+) strain incites the development of progametangium in (-) strain, whereas the hormone synthesized in (-) strain stimulates the development of progametangium in (+) strain.

The attraction of zygophores of (+) and (-) strains towards each other is a chemotropic response, which is regulated by a volatile substance. This volatile substance also promotes the synthesis of trisporic acid in two opposite mating types.

Parasexuality

Earlier, it was believed that genetic recombination can only be achieved by sexual reproduction, but studies on bacteria have shown that it is also possible through alternative mechanisms. Such an alternative mechanism was observed for the first time by Pontecorve and Raper (1952) in Aspergillus nidulans and they called it as parasexual cycle (from Greek para = beside of sex). In this process, plasmogamy, karyogamy and meiosis do take place, but not at specified time or specified points in the life cycle of the fungus. The phenomenon of parasexuality is of common occurrence in the members of the class Deuteromycetes. Sexual reproduction is absent in these fungi and hence the parasexual cycle is of much significance. Now several fungi are known where both sexual and parasexual cycles occur. The following are some important fungi where the phenomenon of parasexuality has been studied Aspergillus nidulans, A. niger, A. fumigatus, A. oryzae, Penicillium chrysogenum, P. expansum, P. italicum, Cephalosporium mycophyllum, Cochilobolus sativus, Fusarium mysporum, Ustilago maydis, U. hordei, Piricularia oryzae, Coprinus cinereum, Coprinus radiata, Schizophyllum commune and Verticillium alboatrum.

Mechanism of Parasexual Cycle

The parasexual cycle is completed in a number of steps as shown by Pontecorvo (1956, 58) in Aspergillus nidulans. The sequence of these steps is as follows (Fig.).

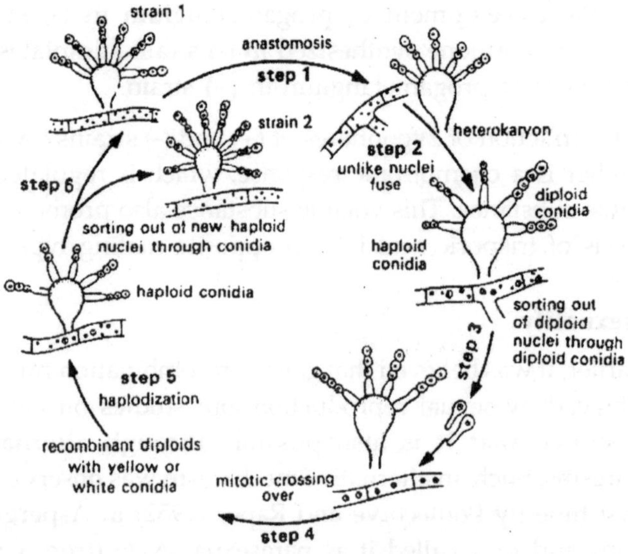

Steps in parasexual cycle as suggested by Pontecorvo (1958).

Formation of Heterokaryotic Mycelium : Formation of heterokaryotic mycelium is necessary for genetic recombination. The anastomosis of somatic hyphae of different genetic constitution is the most common method for the formation of heterokaryotic mycelium. But heterokaryotic mycelium may also be formed by the fusion of some nuclei and their subsequent division or by the mutation of one or more nuclei of the homokaryotic mycelium as in some Ascomycetes.

Nuclear Fusion or Formation of Heterozygous Diploid Nucleus : The fusion of haploid nuclei of similar or dissimilar genotypes results in the formation of homozygous or heterozygous diploid nucleus respectively. If the genotype of unlike nuclei present in the heterokaryotic mycelium is A and

B, then five types of nuclei can be formed by their fusion - two types of haploid nuclei (A and B), two types of homozygous diploid nuclei (AA and BB) and one type of heterozygous diploid nucleus (AB).

***Multiplication of Diploid Nuclei* :** The above mentioned five types of nuclei present in the heterokaryotic mycelium presumably multiply at about the same rate, but the diploid nuclei are present in much smaller number than the haploid nuclei. It has been estimated that the haploid and the heterozygous diploid nuclei are present in the ratio of 1 : 1,000.

***Occasional Mitotic Crossing Over* :** This is the most important step in the parasexual cycle. Occasional crossing over takes place during multiplication of diploid nuclei and this results in the formation of new gene combinations. These recombinations are dependent on the existence of heterokaryosis and they give the fungus some advantages of sexuality within the parasexual cycle. In Aspergillus nidulans, mitotic crossing-over occurs only rarely, with a frequency of 10^{-2} per nuclear division. However, in some fungi (*e.g.*, Penicillium chrysogenum and Aspergillus niger), it is as frequent as during meiosis.

***Sorting of Diploid Nuclei* :** In those fungi which produce uninucleate conidia, sorting of diploid nuclei takes place by their incorporation into conidia. The conidium, which contains diploid nucleus, gives rise to a diploid mycelium on germination. Diploid strains of several imperfect fungi are now known. The conidia of the diploid strains are smaller than those of the haploid strains.

***Occasional Haploidization of Diploid Nuclei* :** Occasionally some hyphae of diploid mycelium form haploid conidia which form haploid mycelia on germination. The formation of haploid conidia by diploid mycelium indicates that haploidization occurs in some diploid nuclei.

Sorting of New Haploid Strains : New haploid strains are sorted out by the incorporation of haploid nuclei in uninucleate conidia. Some of these haploid strains are genotypically different from their parents because of mitotic recombinations.

After the operation of parasexual cycle in the mycelium for sometime, the following types of nuclei may occur : (i) haploid nuclei in both the mycelia, *i.e.*, the mycelia of the two different genetic constitution which form heterokaryosic mycelium, (ii) some haploid nuclei of new genetic constitution by genetic recombinations, (iii) several types of diploid homozygous nuclei, and (iv) several types of diploid heterozygous nuclei.

A comparison of sexual and parasexual cycle is given in Table.

Comparison of Sexual and Parasexual Cycle

Sexual cycle	*Parasexual cycle*
1. Fusion of nuclei of two different genetic constitution takes place, *i.e.*, heterokaryosis is a characteristic feature.	Fusing nuclei are heterokaryotic.
2. Homozygous or heterozygous zygotes are formed by nuclear fusion.	It is not possible to identify whether homozygous zygotes are formed but heterozygous zygotes can be identified on the basis of their colour and nutrition.
3. Zygotes remain present only for one nuclear generation.	Zygotes can remain present for several mitotic divisions.
4. Genetic recombination takes place by meiosis.	Genetic recombination takes place by mitotic crossing over.
5. Crossing over takes place in all chromosomes simultaneously.	Crossing over takes place only in one chromatid of a chromosome at a time.
6. Meiotic products can be easily identified.	Recombinant nuclei occur in the somatic cells, and they can be identified only by suitable genetic sign.

Applications of Parasexuality

The phenomenon of parasexuality is applied in several industrial processes. Most of the fungi which are significant in industries belong to the imperfect group of fungi and they reproduce by parasexual cycle. The better strains of these fungi are obtained by induced mutation or by making suitable modification in the culture medium. The strains of desirable characters can be developed by mitotic recombination. Parasexuality is also applied in the analysis of genetic and physiological processes of perfect and imperfect fungi. There has been great success in obtaining genetic control of pathogenicity and host range in several species of Fusarium by parasexual cycle.

5

The Heterothallism

While studying sexual process in Mucorales, A. F. Blakeslee (1904) observed that some species develop zygospores independently, whereas others form zygospores only when two mycelia of different strains come in contact with each other. On this basis, he divided the species of Mucorales into two groups, *viz.*, homothallic and heterothallic (equivalent to monoecious and dioecious conditions respectively in higher plants).

As dioecious plants and animals require two complementary mating types for sexual reproduction, in the same way the heterothallic fungi also require two complementary mating types for sexual reproduction. However, in fungi the mating types do not show morphological differences as in higher plants and animals. These morphologically similar mating types of heterothallic fungi were said to be as (+) and (-) strains by Blakeslee.

Alexopoulos and Mims (1979) defined heterothallic species as those species consisting of self-sterile (self-incompatible) individuals requiring, therefore, the union of two compatible thalli (mating types) for sexual reproduction, regardless of the possible presence of both male and female sex organs on the same individual. Thus, heterothallism can be defined as condition where two complementary mating types are needed for sexual reproduction.

Blakeslee regarded (+) and (-) strains of heterothallic species equivalent to two different sexes (male and female sexes). Subsequently, several other fungi were discovered where two complementary mating types were necessary for sexual reproduction as in the members of the Mucorales. The two mating sex organs were different in morphological features. Whitehouse (1949) recognised the following two types of heterothallism.

(1) Morphological heterothallism

(2) Physiological heterothallism

Morphological Heterothallism : When morphologically different male and female sex organs (or gametes) are present in two closely associated mycelia, it is known as morphological heterothallism. The two types of sex organs (gametes) are distinct in structure and one of them can clearly be said as male and the other female. This type of heterothallism is found in lower fungi, such as Achyla ambisexualis, A. bisexualis, Blastocladiella variabilis, Dictyuchus monosporus, Phytophthora pelmivora and Peronospora parasitica. In Blastocladiella variabilis, the male and female gametangia are morphologically distinct, the male gamete is comparatively small than the female.

Whitehouse also used the term haplodioecious for morphologically different heterothallic species.

Physiological Heterothallism : In physiological heterothallism, sexual reproduction takes place by two morphologically similar but physiologically different hyphae. The gametangia as well as gametes do not show morphological distinctions. The mating hyphae are different in incompatibility factors. Physiological heterothallism is also known as haploid incompatibility. It is of the following two types.

2-allele Heterothallism : In this type of heterothallism, the nuclei of both mating types are different in genetic characters. The sexual compatibility is controlled by a pair of genetic factors, represented by 'A' and 'a'. Both these factors are present on two different chromatids of a chromosome. Due to the dominance of 'A' over 'a', 'A' and 'a' are also represented by the sign (+) and (-). At the time of meiosis, chromatids with (+) and (-) allele are separated and half of the haploid spores thus formed have (+) and the other half (-) allele. Spores with (+) allele and (-) allele respectively give rise to (+) and (-) mycelia. The mycelia of strains (+) and (+), and (-) and (-) are incompatible. Thus, two complementary mating types (+ and -) are must for sexual reproduction.

The 2-allele heterothallism has been observed in several fungi, such as Ascobolus magnificus, Puccinia graminis, Ustilago kolleri, Mucor mucedo and Neurospora sitophila.

Multiple-allele Heterothallism : In this type of heterothallism, sexual compatibility is determined by more than two alleles. The main advantage of multiple-allele heterothallism is that the chances of mating of compatible strains are increased due to larger number of alleles involved.

The multiple-allele heterothallism may be bipolar or tetrapolar.

(a) *Bipolar Multiple-allele Heterothallism :* Here the compatibility is determined only by one locus and instead of one pair, multiple-allele are present in this locus. For example, if multiple allele L_1, L_2, L_3, ... Ln are present on one locus L, several mating types would be formed by meiotic division (Fig.). The number of mating types will depend on the number of allele present or a locus. The mating type with L_1 allele is compatible to any other mating types except the same mating type. In this type of heterothallism, besides L_1, several other alleles L_2, L_3, ...Ln are present on the locus L, and hence chances of getting complementary mating type

	L_1	L_2	L_3	L_4
L_1	–	+	+	+
L_2	+	–	+	+
L_3	+	+	–	+
L_4	+	+	+	–

Bipolar multiple- allele heterothallism
(–, non-compatible; +, compatible).

are much increased. The bipolar multiple-allele heterothallism is a characteristic feature of the members of Basidiomycetes, except rusts and smuts.

(b) *Tetrapolar Multiple-allele Heterothallism* : This type of heterothallism is very similar to bipolar multiple-allele heterothallism. But here the compatibility is determined by two loci. The compatibility factor, *i.e.*, multiple allele is present on two loci L_1 and L_2 of two different chromatids of a chromosome. At the time of meiotic division, both the loci are separated with the chromatids. There are at least 100 allele on each locus. Any two mating types, which differ in allele present on L_1 and L_2 loci, are compatible. For example, a mating type with A_1 B_1 allele is compatible with any other mating type except the one with the same allele. But the mating type with A_1 B_1 allele is not fully compatible with A_1 B_2 or A_2 B_1 allele. This type of heterothallism occurs in Coprinus firmetarius and Ustilago maydis.

	A_1B_1	A_1B_2	A_2B_1	A_2B_2
A_1B_1	NC	NFC	NFC	C
A_1B_2	NFC	NC	C	NFC
A_2B_1	NFC	C	NC	NFC
A_2B_2	C	NFC	NFC	NC

Tetrapoiar multiple-allele heterothallism (C = compatible; NC=noncompatible; NFC=not fully compatible).

Homothallic Fungi

In these fungi, every thallus is sexually self-fertile and self-compatible. All the thalli in these fungi are of only one strain. Thus, two hyphae of a mycelium developed from the same spore are compatible to each other. Homothallic fungi can be divided into two groups.

Primary Homothallic : In such fungi spores are uninucleate and the nucleus of each spore has only one genotype. Allomyces javanicus, Pyronema omphalodes and Puccinia malvacearum are some common examples of primary homothallic species.

Secondary Homothallic : Basically, these fungi are heterothallic, but nuclei of compatible mating types are present in each spore. Thus, it can be said that each spore of these fungi is heterothallic. Under such circumstances, sexual reproduction can occur in a culture derived from a single (heterokaryotic) spore. The examples of the secondary homothallic fungi are Neurospora tetrasperma and Coprinus ephimerus.

Melanoconiales

Form — order Melanoconiales : The members of this form-order are parasites on plants, causing anthracnose disease. The mycelium is profusely branched, septate and both inter- and intracellular. The asexual fruiting body is sub-epidermal acervulus. Conidiophores are usually unbranched, borne freely, and are closely packed in acervulus. Conidia are borne singly or in chains at the tips of conidiophores.

The form-order Melanoconiales includes only a single form-family Melanoconiaceae, which includes about 100 genera. Some important form-genera of the family are Colletotrichum, Pestalotia, Pestalotiopsis, Monochaetia and Gloeosporium. The life-cycle of Colletotrichum is described here as the representative of the family Melanoconiaceae.

Form-genus Colletotrichum

Systematic Position

Kingdom	—	Mycota
Division	—	Eumycota
Sub-division	—	Deuteromycotina
Form-class	—	Coelomycetes
Form-order	—	Melanoconiales
Form-family	—	Melanoconiaceae

Colletotrichum is a parasitic fungus causing anthracnose disease in plants. The infected tissues develop necrosis and hypoplasia. The genus is represented by some 21 species. Some important species of this fungus are listed below, alongwith the diseases caused by them.

(1)	Colletotrichum falcatum	—	red rot of sugarcane.
(2)	C. corchorum	—	anthracnose of jute.
(3)	C. lagenarium	—	anthracnose of cucurbits.
(4)	C. graminicola	—	red leaf spot of sorghum.
(5)	C. lindemuthianum	—	anthracnose of beans.

(6) C. capsici — die back of chillies.

(7) C. gloeosporioides — anthracnose of citrus and banana.

The following description pertains to Colletotrichum falcatum.

Colletotrichum falcatum is responsible for red rot of sugarcane. The disease is prevalent in India wherever sugarcane is grown. It causes enormous loss to sugarcane crop. It appears on all aerial parts of the plant. The early symptoms of the disease are yellowing and drooping of leaves. The stems show little indication of the disease in early stages but as severity of the disease increases the cane splits lengthwise and red blotches appear throughout the length of the cane. These blotches emit a peculiar smell of alcohol fermentation. The reddening appears mainly in the vascular region. Ultimately the cane becomes dull in appearance, get rotten and shrinks at the internodes.

Vegetative Structure : The mycelium of Colletotrichum is inter- or intracellular, profusely branched and septate. The hyphae are initially hyaline but at maturity they become dark in colour and develop thick walls. The cytoplasm is dense and contains oil droplets. In later stages, the hyphae closely intertwine with one another and form small stromata under the host epidermis.

Reproduction : Sexual stages are lacking and Colletotrichum reproduces asexually by conidia formation. The fruiting body is a plate-like acervulus that develops just below the leaf epidermis. It consists of a stromatic layer, which produces simple hyaline conidiophores from its surface. The conidiophores are arranged compactly like palisade tissue. They are usually unseptate and unbranched and measure about 20μm in length and 8 μm in diameter. Usually a single conidium develops at the tip of each conidiophore but occasionally conidia are produced in acrogenous chains. At maturity conidia are released by the rupture of the overlying host epidermis.

The hyaline crescent or sickle-shaped conidia measure 20-30µm in length and 5-7µm in breadth. An oil drop is present in the centre of each conidium. Alongwith conidiophores many setae are also present in the acervulus. The setae are long, stiff, pointed, septate, unbranched and bristle-like structures. They are 4-5 times longer than the conidiophores and form a protective envelope around the conidiophores.

Germination of Conidia : The conidia are dispersed by rain or by air. Humid atmosphere is necessary for their germination. They remain viable only for a very short period and germinate at once on a suitable substratum. Each conidium produces one or more germ tubes that grow on the surface of the host. The tip of the germ tube swells and acts as appressorium. One or more penetration tubes develop from each appressorium and they penetrate the host epidermis as infection hyphae.

During unfavourable conditions, the germ tube becomes septate and the terminal or intercalary cells transform into chlamydospores. The chlamydospores are thick-walled and green-black in colour. They remain viable for a long time and germinate to produce new mycelia when conditions are favourable.

Perfect Stage of Colletotrichum : The perfect stage of Colletotrichum falcatum is Glomerella tucumanensis (syn. Physalospora tucumanensis), a member of the order Sphaeriales of the class Ascomycetes. It has been observed in the artificial medium as well as on the sugarcane leaves. It has club-shaped asci which measure 70-150 µm in length and 50-70µm in breadth. Ascus has eight ascospores in a biseriate order.

6

Deuteromycotina

The sub-division Deuteromycotina (Fungi Imperfecti) includes those fungi in which only asexual or imperfect stages are known and sexual or perfect stages are not yet recorded. These fungi, as such, can not be placed in well characterized groups like Mastigomycotina, Zygomycotina, Ascomycotina and Basidiomycotina.

This sub-division includes about 600 genera and over 20,000 species. They show resemblances with the members of the Ascomycotina and Basidiomycotina in some features. Several members, which were previously included in Deuteromycotina due to the absence of sexual (perfect) stages, later transferred to an appropriate group when their perfect stages were recorded in nature or in artificial culture media. The following are a few examples :

Imperfect stage	*Perfect stage*
Altemaria solani	Pleospora infectoria
Sclerotium oryzae	Leptosphaeria salvinii
Colletotrichum falcatum	Physalospora tacumanensis

Contd...

Imperfect stage	*Perfect stage*
Sclerotium rolfsii	Corticium rolfsii
Cercospora personata	Mycosphaerella berkeleyii
Helminthosporium oryzae	Cochhobolus miyabeanus
Helminthosporium gramineum	Pyrenophora graminea

Thus, the sub-division Deuteromycotina is a purely artificial and temporary assemblage of fungal species, waiting to be included in an appropriate group after the discovery of their perfect stages. They are defined as the conidial stages of Ascomycotina or more rarely of Basidiomycotina, whose sexual stages have either not been discovered or do not exist at all (Alexopoulos and Mims, 1979).

General Characters

(1) The members are saprophytes or parasites. The parasitic species cause early blight and tikka diseases in plants and athelete's foot, several skin diseases and pulmonary infection in animals.

(2) The mycelium is well developed and profusely branched with perforated septa.

(3) The hyphae may be inter- or intracellular and the cells are multinucleate.

(4) The reproduction takes place by asexual means only and sexual stages are not known.

(5) The asexual reproduction takes place by oidia or conidia formation. The conidia are non-motile structures which develop exogenously on conidiophores. The conidiophores are either free or are formed in some special types of fruiting bodies, such as synnemata, acervuli, sporodochia or pycnidia.

(6) Some members show parasexuality.

The classification of Deuteromycotina is artificial and hence the various taxa are known as form-order, form-family, form-genus, etc.

On the basis of the structure of the asexual fruiting body and the type of conidia, the group has been divided into the following four orders:

1. *Moniliales* : The members of this order reproduce by conidia formation. The conidiophores are usually free but sometimes they are in groups or form sporodochium.
2. *Melanoconiales* : They reproduce by conidia, developed in acervulus.
3. *Sphaeropsidales* : Conidia are formed in pycnidia or modified type of pycnidia.
4. *Mycelia Sterilia* : The mycelium is sterile and conidia are not formed.

However, Ainsworth (1973) divided this group into three classes- Blastomycetes, Hyphomycetes and Coelomycetes.

Class Hyphomycetes

Members are either saprophytes or parasites. Mycelium is well developed, profusely branched and septate. Majority of the members reproduce by conidia. Pycnidia or acervuli are never produced.

Alexopoulos and Mims (1979) recognized two orders Moniliales and Agromycetales under the sub-class Hyphomycetidae.

Form-order Moniliales

This order is represented by about 10,000 form-species. The members are saprophytes or parasites, causing a number of diseases of plants and animals. The mycelium is profusely branched, septate and colourless or black in colour. They reproduce by conidia which are borne on undifferentiated mycelium or conidiophores.

Form-family Dematiaceae

(1) Majority of the members of this family are saprophytes but few are parasites, causing diseases in plants and human-beings.

(2) The mycelium as well as conidia are of dark colour.

The Altemaria, Curvularia, Cercospora, Helminthosporium and Drechslera are some important genera of this form-family. Of these, Alternaria, Cercospora and Helminthosporium are described here.

Form-genus Alternaria

Systematic Position

Kingdom	—	Mycota
Division	—	Eumycota
Sub-division	—	Deuteromycotina
Form-class	—	Hyphomycetes
Form-order	—	Moniliales
Form-family	—	Dematiaceae

The form-species of Alternaria are saprophytes or parasites, and are distributed throughout the world. The saprophytic species grow on the dead organic material and help in its disintegration, whereas the parasitic species cause several diseases in plants. Some important pathogenic species of Alternaria are as follows :

(1) A. solani causes early blight of potato and of other solanaceous plants.

(2) A. cucumarina infects the members of the family Cucurbitaceae like water melon, cucumber and gourd.

(3) A. tenuis causes black point disease in wheat.

(4) A. brassicae infects the members of the family Brassicaceae like mustard and cauliflower.

Symptoms : The disease symptoms appear when the plants are 3-4 weeks old. The early symptoms are in the form of small, yellowish brown spots on the leaves, which enlarge to form concentric rings. These are known as target board symptoms. The spots formed on the leaf reduce the photosynthetic area and eventually the leaf dries and falls off. In severe infection, entire lamina, petiole and stem is badly damaged. Altemaria solani even infects the tubers, and the potato pulp turns brown and becomes corky in texture.

Vegetative Structure : The profusely branched and septate mycelium is light brown in colour. The cells are multinucleate. In the early stage of infection, the hyphae of the parasitic species are intercellular but later on they become intracellular.

Reproduction : Sexual stages are not known in Alternaria and the asexual reproduction takes place by exogenously formed conidia. The conidia are produced singly or in chains at the tips of the branched or unbranched conidiophores. The conidiophores are undifferentiated and emerge through the stomata of the infected leaf.

The dark coloured conidia are multicellular and obclavate, obovoid or uniform in shape. They are both transversely and longitudinally septate. There are about 5-10 septa in each conidium, and the number of septa is probably dependent on environmental conditions. The conidium measures 30-130μm in length and 12-30μm in breadth. It is surrounded by a two layered wall, where the outer layer is pigmented.

Germination of Conidia : The conidia are readily disseminated by wind. On being detached from the conidiophore, they germinate on a suitable substratum. Humid conditions favour their germination. The conidium produces

5-10 germ tubes, usually one from each segment. The germ tubes enter through the stomata of the host leaf and soon develop into new mycelia within the intercellular spaces.

Alternaric acid has been isolated from the culture filtrate of this fungus, and it is regarded that the pathogenecity of the fungus is due to the presence of this acid.

Perfect Stage of Alternaria

According to Webster (1980), the perfect stage of Altemaria is a loeuloascomycetous fungus, Pleospora infectoria.

Form-genus Cercospora

Systematic Position

Kingdom	—	Mycota
Division	—	Eumycota
Sub-division	—	Deuteromycotina
Form-class	—	Hyphomycetes
Form-order	—	Moniliales
Form-family	—	Dematiaceae

Cercospora is represented by over 2,000 saprophytic or parasitic form-species. Most of the species are plant pathogens and the disease caused by them is known as leaf spot or tikka disease. The fungus infects usually the host leaves, but when the infection is severe the stem is also affected. The infection usually starts as pale green spots on the upper surface of the leaf. These spots gradually enlarge and turn brown. Subsequently the cells of the lower epidermis separate from the mesophyll and ultimately the entire leaf dries up and crumbles down. Conidia are formed on the lower surface of the leaf and conidiophores are developed in concentric rings.

Some important pathogenic species of Cercospora are as follows :

(1) C. personata — causes tikka disease of ground nut.

(2) C. beticola — infects sugar beet.

(3) C. gossypina — infects cotton plants.

(4) C. oryzae — infects paddy crops.

(5) C. indica — causes leaf spot disease of Cajanus cajan.

(6) C. calotropidis — infects Calotropis procera.

(7) C. hibiscus — causes leaf spot disease of lady's finger.

Vegetative Structure : The mycelium is profusely branched, hyaline and septate. It is both external and internal. The hyphae grow in between the intercellular spaces of the host tissues and derive food materials with the help of branched haustoria. In C. arachidicola, the mycelium is intercellular to the earlier stages but becomes intracellular after the death of the host cells.

Reproduction : Cercospora reproduces only asexually by means of conidia which develop on long slender conidiophores. At the time of the formation of conidia, hyphae accumulate just below the epidermis of the host and form a compact brown or black pseudoparenchymatous mass, known as stroma. A tuft of short septate, thin walled and unbranched conidiophores emerge out from the stroma.

Each conidiophore produces a conidium at its apex. But the growth of the conidiophore is renewed after the formation of the first conidium. The first formed conidium is thus pushed to one side and another conidium is formed at the apex. This process may repeat several times and as such several conidia are formed on a conidiophore. The conidium is straight or curved, hyaline or pigmented, slender, narrow, tapering and transversely septate. On detachment from the conidiophore, the conidium leaves a small scar at the place of its attachment. The conidia are dispersed by wind or rain-splash. They germinate under favourable conditions by giving rise to one or more germ tubes. Each germ tube ultimately develops into a new mycelium. The optimum temperature for the growth of the fungus is 24-28°C.

Perfect Stage of Cercospora

Perfect or sexual stages of some form-species of Cercospora are known and they are included in the genus Aycosphaerella. For example, M. cerasella is the perfect stage of C. cerasella and M. berkeleyd of C. personata.

Form-genus Helminthosporium

Systematic Position

Kingdom	—	Mycota
Division	—	Eumycota
Sub-division	—	Deuteromycotina
Form-class	—	Hyphomycetes
Form-order	—	Moniliales
Form-family	—	Dematiaceae

Species of Helminthosporium are widely distributed imperfect fungi. Many species are parasitic on cereal crops like wheat, rice and barley. The diseases caused by some pathogenic species of Helminthosporium are as follows:

(1) H. gramineum	—	leaf stripe of barley
(2) H. sativum	—	foot rot of barley
(3) H. teres	—	net blotch of barley
(4) H. sigmoideum	—	stem rot of paddy
(5) H. oryzae	—	leaf spot of rice

Symptoms : The leaf spot symptoms occur on coleoptile, leaves, leaf sheaths and also on glumes. On coleoptile, the spots are small, brown and circular to oval in outline. The infected leaves have conspicuous brown spots. Sometimes the spots may be surrounded by a yellowish halo and may coalesce to form irregular shapes. In severe infections the entire leaf may become brown and eventually dry out. Ears may become distorted and sterile.

Vegetative Structure : The profusely branched and septate mycelium is greyish-brown to dark brown in colour. It is both intra- and intercellular, but there are no haustoria.

Reproduction : Helminthosporium reproduces asexually and sexual reproduction is absent. The asexual reproduction takes place by the formation of exogenously produced spores, known as conidia.

The conidia are borne on stout erect conidiophores, which emerge in groups through the stomata. The conidiophores arise as lateral branches from prostrate hyphae. They are unbranched structures but they sometimes divide at the base. They are characteristically bent and have knee-joints, the points where conidia are attached.

The conidia are produced successively on the growing tips of the sporangiophores. They are 3-10 septate, ellipsoid or cylindrical, straight or slightly curved, smooth-walled and dark-coloured structures. The number of septa in a conidium varies with environmental conditions and physiologic races.

Germination of Conidia : The conidium germinates characteristically from both the ends.

The germ tube is surrounded by a thick mucilagenous substance which enables it to adhere to a hard surface. The tip of the germ tube swells to form an appressorium.

A peg-like infection hypha develops from the appressorium that penetrates the epidermis. The infection hypha then grows into the intercellular spaces of the host tissue and forms a profusely branched mycelium.

Perfect Stages of Helminthosporium : The perfect (sexual) stages of some form-species of Helminthosporium are known. They are included in various genera of the group Ascomycotina. Some perfect stages of Helminthosporium are as follows :

Imperfect stage	*Perfect (sexual) stage*
1. Helminthosporium gramineum	Pyrenophora graminea
2. H. teres	Pyrenophora teres
3. H. sativum	Ophiobolus sativus
4. H. oryzae	Cochliobolus miyabeanus
5. H. sigmoideum	Leptosphaeria salvinii

Form-family - Tuberculariaceae

(1) All the members produce sporodochia.

(2) Conidiophores usually rather short arising more or less radially from sporodochia.

Form-genus Fusarium

Systematic Position

Kingdom	—	Mycota
Division	—	Eumycota
Sub-division	—	Deuteromycotina
Form-class	—	Hyphomycetes
Form-order	—	Montliales
Form-family	—	Tuberculariaceae

Most of the species of Fusarium are facultative parasites, whereas some are saprophytes on various substrates. Various species are responsible for many sérious diseases, such as damping off of seedlings, root rot and wilt disease of several plants, and rots of fruits and vegetables.

Some important pathogenic species of Fusarium are as follows :

(1) F. mysporum f. udum causes wilt of pigeon pea (Cajanus cajan).

(2) F. lini causes wilt of flax (Linum usitatissimum).

(3) F. oxysporum f. vasinfectum causes wilt of cotton (Gossypium spp).

(4) F. solani causes wilt of potato (Solanum tuberosum).

(5) F. oxysporum f. cubense causes panama disease of banana.

Symptoms : The characteristic symptom is wilting of seedlings and mature plants and the host appear as it has suffered from water shortage. The wilting is characterized by the gradual or rather sudden withering and drying up of leaves, followed by drying of the entire plant.

In all the wilts, the fungus blocks the vessels, thus interfering with the ascent of water causing the plants to wilt and ultimately results in death.

Vegetative Structure : The mycelium is profusely branched, septate, hyaline, both inter- and intracellular and remains confined to the vascular tissue and plug their lumen completely.

Reproduction : Sexual stages are lacking and Fusarium reproduces only asexually by conidia formation. The conidia are of two types, *viz.* (i) microconidia, and (ii) macroconidia. The microconidia are borne terminally on short hyphae (conidiophores) within the large vessels.

They are small, elliptical or curved, 1- or 2-celled structures. On the other hand, the macroconidia are formed on sporodochia on the surface of the host near ground level. They are long, curved, pointed at the top and 3-4 septate. The conidia germinate to form new mycelium.

In addition to the above, chlamydospores are also formed over older mycelium. The chlamydospores are globose, thick-walled, 1-2 celled structures produced terminally or intercalary on the older hyphae.

The chlamydospores detach from the parent hyphae and function as resting spores. Under favourable conditions, the chlamydospores germinate by means of germ tubes to form a new mycelium.

7

The Lycoperdon

Class Gasteromycetes

The group Gasteromycetes is an unnatural assemblage of basidiomycotina and the members are commonly known as puff-balls, stinkhorns, bird's nest fungi and earthstars. The members are mostly saprophytes growing on soil, decaying wood and other vegetation, and on dung. Basidiocarps are typically angiocarpous, *i.e.*, the basidia remain arranged in hymenium, enclosed within the basidiocarp and the basidiav are aseptate. The basidiospores are positioned symmetrically on the sterigmata, and they are not discharged forcibly from the basidium. Dring (1973) divided Gasteromycetes into nine orders - Podaxales, Phallales, Hymenogastrales, Lycoperdales, Gautteriales, Tulostomatales, Nidulariales, Melanogastrales and Sclerodermatales.

Order Lycoperdales

The order Lycoperdales includes the common puffballs and some of the earthstars. The young fruiting bodies are first hypoterranean, but they mature above ground. At maturity, the powdery gleba, typically consisting of light-coloured spores and a well-developed capillitium, is surrounded by a two-layered peridium.

Dring (1973) divided the order into four families-Arachinaceae, Mesophelliaceae, Geastraceae and Lycoperdaceae. Lycoperdon, belonging to family Lycoperdaceae is described in details in the following pages.

Family Lycoperdaceae

(1) Members are commonly called puffballs, usually found growing on tree' stumps, on decaying logs, or on the ground in the woods.

(2) Peridium two-layered; the exoperidium cracks at maturity.

(3) Gleba with branched capillitium.

Lycoperdon

Systematic Position

Kingdom	—	Mycota
Division	—	Eumycota
Sub-division	—	Basidiomycotina
Class	—	Gasteromycetes
Order	—	Lycoperdales
Family	—	Lycoperdaceae

Lycoperdon (commonly known as 'puffball'), with some 25 species occurs saprophytically on decaying vegetables, wood, humus soils, etc. Most of the puffballs are edible.

Structure : The structure of Lycoperdon can be studied under the following two heads:

(1) Vegetative mycelium

(2) Fruiting body (Basidiocarp)

Vegetative Mycelium : The vegetative mycelium remains underground and is of two types. The primary mycelium, formed by the germination of basidiospore, is hyaline, septate and monokaryotic. The primary mycelium is short-lived and soon gets transformed into dikaryotic secondary mycelium. The cells of the secondary mycelium is binucleate but no clamp connections are formed. The hyphae of the secondary mycelium twist together to form white hyphal cords, called rhizomorphs. The rhizomorphs are made up of an outer cortex of loose hyphae, subcortex of compact hyphae and a central core of parallel hyphae. From the underground rhizomorphs, globose or bell-shaped fruiting bodies (basidiocarps) come out above ground.

Fruiting Body (Basidiocarp) : The fruiting bodies are globose or bell-shaped about 2-5 cm in diameter. It is developed from a mycelial knot at the tip of the rhizomorphs which remains hidden in the substratum. Each Basidiocarp is made up of a basal sterile portion (stipe) and an apical fertile portion and remains surrounded by a double-layered peridium. The outer layer, known as exoperidium is protective and remains covered with cone-like spines which are sloughed off on maturity, whereas the inner layer, called endoperidium is dry and leathery and is composed of both thin-walled and thick-walled hyphae. The peridium opens by one or more pores at the apex.

The tissue within the peridium is called gleba. The mature gleba consists of numerous irregular chambers separated by tramal tissue (capillitium threads). The chambers are lined with basidia. The basal region of the gleba is called sub-gleba (stipe), which is the sterile region of the fruiting body. It extends into the fertile sporal region in the form of columella.

Basidia and Basidiospores : In the gleba, basidia are arranged in a regular fertile layer constituting the hymenium which lines each cavity. Each basidium is a dikaryotic body.

Both the nuclei of the basidium fuse to form a diploid zygotic nucleus. The fusion is immediately followed by reduction division, resulting into the formation of four nuclei. Then, each basidium develops four basidiospores at its tip 'which are arranged symmetrically on the sterigmata. The sterigmata are of various lengths.

The mature basidiospores are globose with thick ornamented walls and are uninucleate structures. At maturity, the basidiospores fall off the basidia and completely fill the cavities. Later, they are discharged in the form of small spore clouds from the fruiting bodies by large drops of rain or wind borne debris, hence the name is puffball. On a suitable substratum, each basidiospore germinates to form primary mycelium.

Aphyllophorales

The members of the order Aphyllophorales, commonly known as pore fungi, bracket fungi, tooth fungi or coral fungi, are parasitic or saprophytic in nature. The saprophytic species commonly occur on the rotting wood. The mycelium is septate and well branched. Asexual reproduction takes place by means of conidia or oidia. Holobasidia are club-shaped and develop in well defined hymenia. The hymenium is not covered by a velum (gymnocarpous) and thus exposed even when spores are immature.

The development of fruiting body is centrifugal with one-sided hymenophore, or clavarioid with amphigenous hymenium. The basidiocarp has porous or lamellate hymenial layer and is papery, woody or leathery. The hymenophore may be smooth, toothed or tubulate. The nuclear spindle in the basidium is longitudinal thus the basidia are stichobasidia. Basidiospores are forcibly discharged.

The members of the order Aphyllophorales differ from

those of Agaricales in having hard, thick and trimitic basidiocarps and stichobasidia, and in the absence of velum.

Since the members show great heterogeneity in their characters, the order has been variously classified. Alexopoulos and Mims (1979) recognised six families, *viz.*, Hydnaceae, Thelophoraceae, Meruliaceae, Clavariaceae, Polyporaceae and Canthrellaceae within the order, whereas Bessey (1968) divided it into eight families, Exobasidaceae, Thalophoraceae, Canthrellaceae, Hydnaceae, Clavariaceae, Polyporaceae, Meruliaceae and Fistulinaceae. However, Donk (1964, 1971) and Talbot (1973) recognised over 20 families in this order.

Family Polyporaceae

(1) The members of the family Polyporaceae, commonly known as bracket fungi or pore fungi, are wood-inhabiting fungi which attack woods of a number of commercially important trees.

(2) The fruiting body is large and complex, hard, tough and leathery. It may be effused or reflexed but never clavarioid.

(3) The hymenium develops as a lining at the periphery of the pores (or cavities) of the tubes formed on the lower surface of the basidiocarp.

(4) Basidia are aseptate, clavate or sub-uniform; each producing normally four hyaline or cream coloured basidiospores.

Polyporus

Systematic Position

Kingdom	—	Mycota
Division	—	Eumycota
Sub-division	—	Basidiomycotina

Class	—	Hymenomycetes
Sub-class	—	Holobasidiomycetidae
Order	—	Aphyllophorales
Family	—	Polyporaceae

Polyporus, commonly known as bracket fungi, occurs as a saprophyte on dead wood logs, or as a parasite causing serious damage to forest trees. There are about 159 species of the genus known from India. P. sulphureus, commonly known as sulphur mushroom, causes wood rot of oak, whereas P. squamosus causes heart rot of Ulmus and P. betulinus (the birch polypora) causes heart rot of birch. The young basidiocarps of P. frondosus and P. sulphureus are edible. The basidiocarps of P. lucidus are attractive reddish-brown in colour and hence are used as decorative pieces. P. versicolor, a widespread wood totter, causes decay of deciduous woody plants.

Several species of Polyporus (*e.g.*, P. umbellatus, P. frondosus, P. leucomelas) are edible.

Vegetative Structure : The subterranean mycelium is composed of slender, branched and septate hyphae. The primary monokaryotic hyphae, which develop from uninucleate basidiospores, are short lived. They soon get diploidised to form a perennating secondary dikaryotic mycelium which forms fruiting bodies. The mycelium grows within and below the bark and invests the central woody cylinder in case of severe infection. The lignified walls of the host cells are digested by the enzymes secreted by mycelial hyphae. The hyphae bear clamp connections. The secondary mycelium gives rise to a fresh crop of basidiocarps every year.

Reproduction : Polyporus reproduces both by asexual and sexual means.

Asexual Reproduction : Asexual reproduction is very rare in Polyporus. It takes place by conidia which are produced on dikaryotic mycelium or on curious sterile fructifications. They produce dikaryotic mycelium on germination.

Sexual Reproduction : Sex organs are absent in Polyporus. The species of Polyporus are heterothallic. The fusion between the two somatic hyphae of opposite strains (somatogamy) brings about diploidization. Thus secondary dikaryotic mycelium is formed. The dikaryotic mycelium is perennial and may survive for several years. Under suitable conditions, the dikaryotic mycelium forms more or less flat fruiting bodies, known as basidiocarps.

Development of Basidiocarp : Some white spherical hyphal knots develop from the perennial secondary mycelium just below the bark or in the wood. The knots gradually emerge out as small sporophores (basidiocarps). As development continues, the young knot is differentiated into a short stalk (stipe) and a rounded cap-like pileus. In some cases, stipe does not develop. The basidiocarp is soft and pliable at this stage, and is covered by a downy layer of mycelium. It becomes hard leathery or corky later on. Now minute holes appear in the basidiocarp which denote the beginning of the formation of pores of the hymenial tubes. The pores gradually increase in size and cover the entire lower surface of the basidiocarp.

Structure of Mature Basidiocarp : The mature basidiocarp is hard and generally leathery or corky structure. In most of the species it appears as a fan-shaped bracket, shelf or knob which is sessile or stalked (stipitate). In sessile fruiting bodies (*e.g.*, P. adustus, P. borealis) the point of attachment to the substratum is typically lateral. But in P. anceps the basidiocarp is effused reflexed, *i.e.*, the fruiting body lies flat on the substratum for some distance and then the margin becomes reflexed to form a horizontal pileus. In P. squamosus the basidiocarp is sub-sessile, narrowing towards the base. The stipitate basidiocarp (*e.g.*, P. brumalis) is well differentiated into a cap-like structure, known as pileus, and a long, massive stalk, called stipe. The stipe is brown in colour and about 5-15' cm in length. It is attached to the pileus in the central, excentric, lateral or dorsal position. The surface of the basidiocarp has zones of various colours, depending on the species. The upper surface of the pileus is flat and slightly undulating or smooth. The lower surface is smooth and flat. It has no radiating gills, instead there are minute pores which may be round, angular or irregular in shape. The pores are openings of small tubes and they give a porous appearance to the undersurface of the pileus. The pore leads inwards into a hollow tube lined by a fertile layer, called hymenium.

The basidiocarp is made up of three different types of hyphae:

(a) *Generative Hyphae :* These hyphae are thin-walled with dense cytoplasmic contents; they may or may not have clamps.

(b) *Binding Hyphae :* These hyphae are much branched, narrow and thick-walled.

(c) *Skeletal Hyphae :* These hyphae are unbranched, thick-walled and with a narrow lumen; they arise as lateral branches of generative hyphae.

On the basis of the type of hyphae present, three kinds of basidiocarps are recognised in Polyporus- monomitic, dimitic and trimitic. The monomitic basidiocarps consist of only generative hyphae (*e.g.*, P. adustus). In dimitic basidiocarps, generative hyphae are associated with binding or skeletal hyphae (*e.g.*, P. sulphureus). In trimitic basidiocarps all the three types of hyphae ale present (*e.g.*, P. versicolor).

Internal Structure of Basidiocarp **:** A section through the basidiocarp reveals the following regions:

(a) *Context :* It is the outermost region of the basidiocarp, composed of outer thick-walled hyphae with large irregular spaces between them. This zone makes up about four-fifth thickness of the fruiting body. Sometimes the context is differentiated into an upper soft and a lower hard and firm layer and then it is called duplex.

(b) *Trama :* This region is composed of loosely arranged, much branched and septate hyphae. The pores or tubes extend from below the context to the lower surface of the trama. They are vertically placed and run parallel to each other within the fruiting body. The tissue in between the tubes, formed of generative and skeletal hyphae, is called dissepiment.

(c) *Hymenium :* The inner fertile layer lining of pores is known as hymenium. It is composed of club-shaped basidia and sterile paraphyses, cystidia and setae. Basidia and sterile structures grow at right angles to the length of the tube. The basidium is a club-shaped structure and it projects slightly into the cavity of the pore. The paraphyses are usually of smaller diameter than basidia. The cystidia are colourless, usually

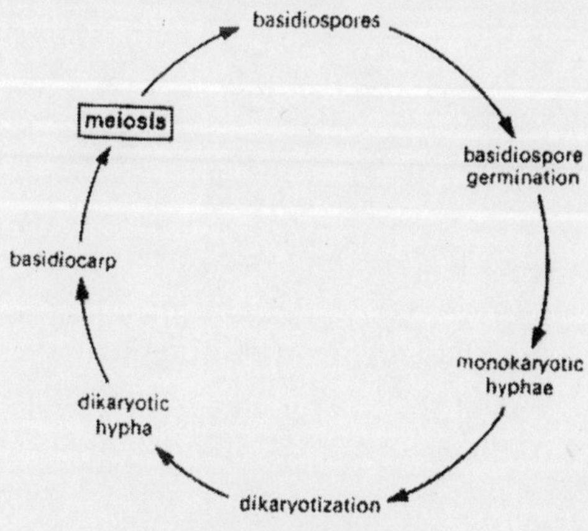

Polmorus : Life-cycle.

conspicuous and larger than basidia. The setae are brown, conspicuous conical structures.

The young basidium is binucleate and the two nuclei fuse as the basidium matures. The diploid nucleus undergoes meiosis and forms four haploid nuclei. The basidium develops four short sterigmata at its free end and from each sterigma a small and oval basidiospore is abstracted. Each of the four basidiospores thus formed receives one haploid nucleus from the basidium.

Dispersal and Germination of Basidiospores : Basidiospores are discharged explosively in the pore sacs. They are formed in very large numbers (in billions) and their discharge continues for months.

When fall over a suitable substratum, the basidiospore germinates into a primary mycelium of plus or minus strain.

8

The Agaricus

Order Agaricales

The order Agaricales includes 197 genera and 4,000 species. Most of the species are saprophytic, growing on dead and decaying wood and leaves and on organic rich soil. Species like Armillariella mellea grow as parasite on woody trees. Some agarics form mycorrhiza with forest trees. Many fleshy agarics are edible, whereas some (*e.g.*, Amanita phalloides, A. muscaria) are highly poisonous. Species of Psilocybe (sacred mushroom) produce hallucination.

The primary mycelium which develops from homokaryotic basidiospore is short lived. The dikaryotic secondary mycelium forms the major part of the life-cycle. It transforms into complex tissues of tertiary mycelium in the form of fruiting bodies. In some species, hyphae become aggregated into rhizomorphs. Asexual reproduction is very rare, but a few species produce oidia, chlamydospores or sclerotia. Sexual reproduction is usually by somatogamy. The members produce well developed and distinct fruiting bodies (basidiocarps). The basidiocarps are fleshy with monomitic type of construction. The nonseptate basidia develop in the form of hymenium on gills or lamellae. The gills develop on the under surface of the fruiting body like spokes of a cycle wheel. Hence, they are also known as gill fungi.

The classification of Agaricales poses a major problem. In the older systems, members of the order were included in either one (Agaricaceae) or two (Agaricaceae and Boletaceae) families. With the increase of our knowledge, the order is divided into several more natural families. On the basis of the presence or absence of annulus, spore colour and ornamentation, mode of attachment of gills to the stipe and composition of trama, Smith (1973) recognised 16 families (Boletaceae, Hygrophoraceae, Tricholomataceae, Entolomataceae, Amanitaceae, Pluteaceae, Lepiotaceae, Agaricaceae, Bolbifiaceae, Strophariaceae, Coprinaceae, Cortinariaceae, Paxillaceae, Gompridiaceae, Russulaceae, Cantharellaceae) within the order. However, Singer (1975, 83) recognised 18 families within the order Agaricales. Agaricus (Agaricaceae) is described in details in the following pages.

Family Agaricaceae

(1) The fruiting body is well differentiated into a stipe and a pileus.

(2) The annulus is typically present on the stipe.

(3) The lamellae are free.

Agaricus (Mushroom)

Systematic Position

Kingdom	—	Mycota
Division	—	Eumycota
Sub-division	—	Basidiomycotina

Class	—	Hymenomycetes
Sub-class	—	Holobasidiomycetidae
Order	—	Agaricales
Family	—	Agaricaceae

Agaricus (= Psalliota) is a saprophytic fungus which occurs on dead decaying leaves, wood logs, manure piles and other similar surroundings in rainy season. These fungi are commonly known as mushrooms. There are about 17 species of Agaricus in India which are locally known as kukarmutta, khumb or dhingri. A. campestris (the field mushroom) and A. bisporus (=A. brunnescens; the cultivated mushroom) are common edible mushrooms, commercially cultivated in various parts of India. Species like Asilvaticus and A. xanthodermus are highly poisonous.

The following description is largely based on Agaricus campestris, a typical species of the genus.

Structure : The structure of Agaricus can be studied under the following two heads.

1. Vegetative mycelium
2. Fruiting body (Basidiocarp)

Vegetative Mycelium : The vegetative mycelium remains underground and is of two types. The primary mycelium, formed by the germination of basidiospore, is hyaline, septate and monokaryotic. Depending on the strain of the basidiospore, it may of (+) or (-) strain. The primary mycelium is short-lived and soon gets transformed into dikaryotic secondary mycelium.

The secondary mycelium, formed by the fusion of two primary mycelia of opposite strains, is perennial, branched and septate. It contains dolipore septa and its cells are binucleate. The hyphae of the secondary mycelium twist together to form white hyphal cords, called rhizomorphs.

The dikaryotic mycelium shows centrifugal growth, *i.e.*, hyphae grow outward from a centre in a circle. A circular colony of hyphae is thus formed in the soil. At maturity, hyphae develop fruiting bodies at their tips in a more or less circular ring. As the outer edge of the colony grows, older parts in the centre degenerate. When fruiting bodies are formed next time, the ring is larger in diameter. Thus successive crops of fruiting bodies are produced in larger and larger rings. These rings are called fairy rings, based on the ancient belief that they marked the path of dancing fairies.

Fruiting Body (Basidiocarp) : The basidiocarp is the aerial part of Agaricus and it develops from rhizomorph. This part is commonly known as mushroom. It is differentiated into a stipe and a pileus. The stipe is about 6-9 cm in height and the pileus is 5-10 cm in diameter. There are about 300-600 gills on the undersurface of the pileus, arranged in radial fashion.

Reproduction : Agaricus reproduces by vegetative, asexual and sexual means.

Vegetative Reproduction :The edible mushrooms are propagated by vegetative means. Small pieces of dikaryotic mycelium are used as inoculum. The pieces are grown in soil rich in organic manure to obtain basidiocarps.

Asexual Reproduction : It is not a common method of propagation in Agaricus. It takes place by chlamydospore formation. The chlamydospores develop in terminal or intercalary positions on the secondary mycelium. They germinate to produce dikaryotic mycelium. In some species oidia are also formed, but these are involved mainly in dikaryotisation rather than developing directly into new mycelia.

Sexual Reproduction : Most species of Agaricus are heterothallic (A. bisporus is homothallic), but they do not have sex organs in traditional sense. Primary mycelia formed by the germination of basidiospores of two different strains act as male and female sex organs. There is somatogamy between the somatic hyphae of opposite strains and this results in diploidization and formation of the secondary mycelium. The secondary mycelium later develops fruiting bodies, known as basidiocarps.

Somatogamy between two primary hypha of opposite strains takes place through the following steps:

(a) *Plasmogamy* : In this step, two primary monokaryotic hyphae of opposite strains come in contact with each other. At the point of their contact, the cell walls are dissolved and a dikaryon is formed. A dikaryotic mycelium develops by successive divisions of the dikaryotic cell. At the time of the division of dikaryotic cell, both nuclei of the dikaryon divide synchronously and form four haploid daughter nuclei, two of (-) and two of (+) strain. Out of these, two nuclei (one of (+) and one of (-) strain) are transferred to the daughter cell by the formation of clamp connection.

The dikaryotic mycelium is perennial and subterranean. It produces basidiocarps under favourable conditions.

(b) *Karyogamy* : In this step, both nuclei of a dikaryon fuse to form a diploid nucleus. Karyogamy takes place in young basidium.

(c) *Meiosis* : Soon after karyogamy, meiosis takes place in the basidium. The basidiospores, thus formed in the basidium, are haploid.

Development of Basidiocarp : The subterranean secondary mycelium takes nutrients from the soil and then forms the fruiting body or basidiocarp. Basidiocarps develop as small white knot like structures at the tips of the underground hyphae. These hyphal knots enlarge gradually and give rise to button stage. At this stage, the developing basidiocarp is differentiated into a basal bulbous part and an apical hemispherical region. The basal part forms the stipe, and the apical hemispherical part, the pileus. Some hyphae at the junction of the stipe and pileus are drawn apart and form a ring-like chamber, called pre-lamellar chamber.

The inner surface of the roof of the pre-lamellar chamber becomes deeply concave and it is lined with alternating radial bands of slow and rapidly dividing cells. The latter form gill primordia, which develop into gill lamellae that hang downward into the pre-lamellar chamber. As pileus expands, there is an increase in radial interspaces between the gills. A membrane, called velum or inner veil, connects the margin of the pileus with the stipe. Due to the elongation of stalk, the buttons are raised on the soil surface. The upper hemispherical region of the button grows more rapidly than the stalk. This causes rupture of the velum and the upper hemispherical region finally expands out as an open umbrella-like structure with numerous gills attached to its lower surface. The gills are exposed by the rupture of the velum. At this stage, remnants of the velum are still attached to the stipe in the form of a ring, called annulus.

In dry season, when soil is hard, buttons do not grow and remain underground. In rainy season, when soil is moist, they grow rapidly and come out of the soil surface. Hence, many basidiocarps can be seen in rainy season on the soil surface.

Structure of Mature Basidiocarp : The mature basidiocarp is an umbrella-shaped structure with a long massive stipe and a broad pileus. The stipe is a thick, fleshy and cylindrical structure, light pink or white in colour. An umbrella-like pileus, 5-12 cm in diameter, is present at the distal end of the stipe. The upper convex surface of the pileus is white, light brown or yellow in colour. About 300-600 radially arranged gills hang down from the inner surface of the pileus. All gills are not of the same length; they may be of full, half or quarter length. The surface of the gill is enveloped by a fertile layer, the hymenium. The gills are of light pink colour when young, but they turn brown or purplish black at maturity.

Internal Structure of Basidiocarp : The basidiocarp is made up of pseudoparenchymatous mass of hyphae.

(a) *Internal Structure of Stipe :* The stipe is composed of numerous longitudinally intertwined hyphae. The hyphae are compactly arranged and form a pseudoparenchymatous tissue in the peripheral region, whereas they are loosely arranged, with large intercellular spaces, in the central region. The peripheral region of the stipe is known as cortex and the central region as medulla.

(b) *Internal Structure of Pileus :* The internal structure of the pileus is similar to that of the stipe. The pileus is also differentiated into an outer cortex and a central medulla. The hyphae of the stipe region extend in the distal end to form pileus.

(c) *Internal Structure of Gills :* The internal structure of the gill is, however, complex and the following three regions are discernible.

   (i) *Trama:* This is the central sterile region of the gill. It consists of many loosely arranged interwoven hyphae.

   (ii) *Sub-hymenium (Hypothecium) :* This region is situated on both sides of the trama. It is formed by the lateral branches of hyphae of trama region. The cells of these hyphal branches are isodiametric and 2-3 nucleate. This is also a sterile zone like trama.

   (iii) *Hymenium :* This is the outermost layer of the gill. It is fertile and composed of hyphae of the sub-hymenium region. In this region the cells are arranged in palisade-like layer. These aseptate fertile cells are known as basidia. Club-shaped sterile paraphysis occur in between the fertile cells.

The young basidium is a dikaryon and as the basidium matures the two nuclei fuse to form a diploid nucleus. This diplophase is ephemeral and after karyogamy the diploid nucleus divides meiotically to form four haploid nuclei. Of these four nuclei, two are of (+) strain and two of (-) strain. At the distal end of the basidium, four peg-like outgrowths are formed. These outgrowths are known as sterigmata. The sterigmata swell at their tips and each forms a single basidiospore. Thus, four monokaryotic haploid basidiospores are formed in a basidium. The young basidiospore is unpigmented but it develops brown or black pigments at maturity. At the junction of the basidiospore and the sterigma, there is a small projection, known as hilar appendices.

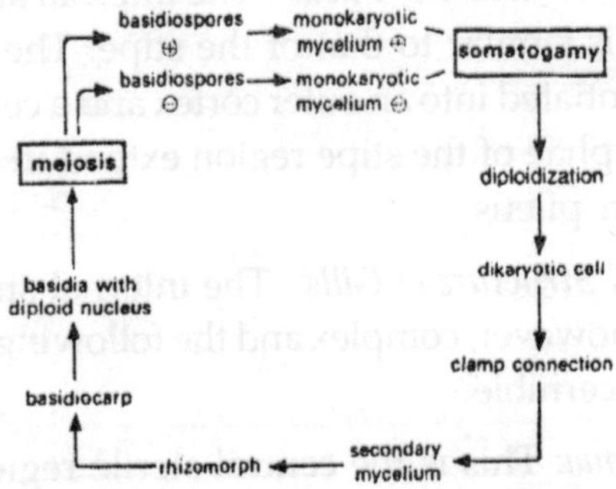

Agaricus : Schematic representation of the life-cycle.

Dispersal and Germination of Basidiospores : As the basidiospore matures, a drop of liquid appears at the hilar appendices which remains surrounded by a limiting membrane. The basidiospore thus lies just above this drop. The drop increases in size gradually and attains a size of about one-fifth of the spore. Then the basidiospore is suddenly shot away from the sterigma. The four basidiospores of a basidium are dispersed in rapid succession.

When the basidiospore falls on a suitable substratum, it germinates by producing a germ tube that grows into a primary monokaryotic mycelium. Depending on the strain of the basidiospore, the mycelium may be of (+) or (-) strain. The primary monokaryotic mycelia form dikaryotic mycelium by somatogamy.

9

The Puccinia

Order Uredinales

The order Uredinales includes about 128 genera and over 4,600 species. The members of this order are popularly known as rust fungi, because of characteristic reddish-brown colour of their spores. They are internal obligate parasites, which infect angiosperms, gymnosperms and some ferns. In recent years, however, it has been discovered that some rust fungi are not true obligate parasites, and species like, Gymnosporangium juniperi-virginianae, Puccinia malvacearum, P. graminis, P. striiformis, P. triticina and Uromyces ari-triphylli could be grown on artificial culture medium.

The mycelium is branched, septate and intercellular, forming distinct haustoria. The septa are simple and develop centripetally, *i.e.,* they taper in thickness towards 'the central 'pore. The members are autoecious (when complete their life cycle on a single host) or heteroecious (when complete their life cycle on two hosts). Most rust fungi are polymorphic, *i.e.,* they produce more than one type of spores to complete their life cycle. Rusts may be microcyclic or macrocyclic. In microcyclic rusts (*e.g.*, Puccinia malvacearum) only one type of dikaryotic spores are formed, whereas in macrocyclic rusts (*e.g.*, Puccinia graminis) more than one type of dikaryotic spores are formed. On germination, the teleutospore produces promycelium, which gives rise to basidiospores. On the basis of the structure of teleutospores, Arthur (1934) recognised two families in the order Uredinales, *viz.*, Melampsoraceae and Pucciniaceae. But Alexopoulos (1962) and Alexopoulos and Mims (1979) recognised three (Pucciniaceae, Melampsoraceae and Coleosporaceae) and Savile (1978) five (Pucciniastraceae, Melampsoraceae, Phragmidiaceae, Raveneliaceae and Pucciniaceae) families within the order.

A detailed description of Puccinia (Pucciniaceae), the most widely known genus of the order, is provided in the following pages.

Family Pucciniaceae

(1) Teleutospores are free and stalked.

(2) Promycelium, formed on germination of teleutospore, is septate.

Puccinia

Systematic Position

Kingdom	—	Mycota
Division	—	Eumycota
Sub-division	—	Basidiomycotina
Class	—	Teliomycetes
Order	—	Uredinales
Family	—	Pucciniaceae

Puccinia is a very large genus with about 1,800 species, practically distributed in all parts of the world. In India, the genus is represented by more than 147 species, which cause serious rust diseases in cereal crops like wheat, barley, oats and maize.

Distinguishing Features of P. Graminis, P. Striiformis and P. Recondita.

P. graminis	*P. striiformis*	*P. recondita*
1. Causes black rust disease in wheat.	Causes yellow rust disease in wheat.	Causes brown rust disease in wheat.
2. Appears in March in northern India and in the last week of November in South India.	Appears in January-February in northern and eastern India.	Appears in January in northern and eastern India.
3. Uredosori appear mainly on stems but in later stages leaves and ears are also infected.	Uredosori mainly appear on leaf-sheaths and ears.	Uredosori are confined to leaves.
4. Uredopustules are large, elongated and red or brown in colour.	Uredopustules are small, oval and are arranged in long strips.	Uredopustules are small, oval or round, scattered and bright orange.
5. Uredospores are oval and brown with 4-round equatorial germ pores.	Uredospores are nearly round and yellow with 6-10 scattered germ pores.	Uredospores are round or oval and bright orange with 3-4 germ pores.
6. Teleutosori are elongated and black in colour; they are confined to the stem.	Teleutosori form black elongated pustules; they are confined to the lower surface of the leaf.	Teleutosori rarely develop; if present, they are black and are confined to the lower surface of the leaf.
7. Telia burst on maturity.	Telia do not burst.	Telia do not burst.
8. Teleutospores are bi-celled, chestnut brown to black; round or conical at the apex.	Teleutospores are bi- celled, cuniform and constricted at the septum; conical at the apex.	Teleutospores are bi-celled and dull black; rounded and thickened at the apex.
9. The alternate host is Berberis vulgaris.	The alternate host is Muehlenbergia hugeli.	The alternate host is Thalictrum falvum.

The species of Puccinia are internal obligate parasites; only spores are seen on the host surface. Obligate parasites can not be grown on artificial culture medium, but Williams (1966), Helfer (1989) and others have been successful in growing some species of Puccinia (*e.g.*, P. coronata, P. hordei, P. recondita, P. striiformis, P. graminis tritici) on culture medium.

Species of Puccinia may be autoecious, *i.e.*, they complete their life cycle on a single host (*e.g.*, P. asparagi, P. butleri, P. menthae, P. pulverulenta), or heteroecious, *i.e.*, they complete their life cycle on two different hosts (*e.g.*, P. graminis, P. coronata).

In India, wheat, crop is infected by the following three species of Puccinia which cause considerable damage to the crop:

(1) P. graminis - black rust or stem rust.

(2) P. striiformis (= P. glumarum) - yellow rust or stripe rust.

(3) P. recondita (= P tritici) - brown rust or leaf rust.

Distinguishing features of these three species are listed in Table.

Puccinia Graminis

Puccinia graminis is the causal organism for black rust disease of wheat and other cereal crops. This disease is also known as stem rust. The pathogen responsible for the disease was earlier considered to be an insect, but Persoon (1797) for the first time reported that the disease is caused by a fungus.

P. graminis is an obligate parasite. It is a heteroecious fungus. There are two phases - dikaryophase and haplophase - in its life cycle. The dikaryophase occurs in its primary host, wheat (Triticum aestivum), whereas the haplophase in its alternate host, Berberis vulgaris. Although P. graminis can survive in the absence of alternate host, its life-cycle is completed only when both hosts are available.

Physiological Specialization : Puccinia graminis has several physiological races, which show physiological specificity towards their hosts. For instance, P. graminis causes rust disease in several cereal crops like wheat, barley and oats, but the strain which infects wheat plants does not infect barley or oats. This is due to the fact that each strain has some metabolic specificity towards the host. The phenomenon where a specific host is infected by a specific pathogen is called biological specialization.

P. graminis is thus a composite species. On the basis of its pathogenecity, size of uredospores and cultural variability it has been divided into the following six strains.

Strain	*Host*
(1) P. graminis tritici	Wheat
(2) P. graminis secalis	Rye
(3) P. graminis avenae	Oat
(4) P. graminis phleipratensis	Festuca
(5) P. graminis agrostis	Agrostis
(6) P. graminis poae	Poa

Each strain has several physiological races. For example, about 300 physiological races of P. graminis tritici have already been isolated.

In India, wheat crops in northern and southern regions are largely affected by P. graminis. The rust usually appears late in the season. It is often not seen in northern India until March, the earring time of the crop. However, in south India the disease appears in November and as such the vegetative growth of the crop is most affected. The infection decreases the rate of photosynthesis, causing a great loss to yield.

Vegetative Structure : There are two types of mycelia - dikaryotic and monokaryotic - in the life-cycle of P. graminis. Both these mycelia are intercellular, septate and branched. There is a simple pore in each septum which maintains protoplasmic connections between the adjacent cells. The cell wall is composed of chitin and glucan.

The dikaryotic mycelium occurs in the primary host, *i.e.*, wheat plants; whereas the monokaryotic mycelium occurs in the alternate host, *i.e.*, barberry bushes. Each cell of the dikaryotic mycelium has a dikaryon and that of monokaryotic mycelium a single nucleus. The granular cytoplasm of the cell also contains vacuoles, glycogen bodies and oil globules. The mycelium takes nourishment from the host cells with the help of spherical haustoria. The haustoria are often closely appressed to the nucleus of the host cell.

Life-cycle

Puccinia graminis is a macrocyclic heteroecious rust. There are five types of spores, *viz.*, uredospores, teleutospores, basidiospores, pycnidiospores and aeciospores in its life cycle. These spores develop in two different hosts in a definite sequence. The sequence of various stages occurring in the primary and alternate hosts are as follows.

Uredospore stage] on primary host
Teleutospore stage] (Triticum aestivum)

Basidiospore stage

Pycnidiospore stage] on alternate host
Aeciospore stage] (Berberis vulgaris)

Stages of Puccinia graminis on Wheat Plant

Uredospore Stage : The dikaryotic mycelium is produced by the germination of aeciospores on wheat plants. It enters through stomata and develops in the intercellular spaces in the tissues of the leaf, stem and glumes of the host. The dikaryotic mycelium, present in the sub-epidermal region, develops many erect hyphae which grow at right angle to the epidermis. A binucleate uredospore develops at the tip of each erect hypha. These spores develop in groups and these groups are known as uredosori.

With the formation of uredospores, disease symptoms appear in the form of reddish-brown pustules or streaks on the stem, leaf and leaf base. These symptoms usually appear in late spring. The host epidermis bursts due to the pressure of developing uredosori and thus uredospores are liberated.

The mature uredospore is a stalked, unicellular, oval and binucleate structure. The spore wall is thick and is differentiated into three layers; the outer layer is relatively thick and spiny, whereas the inner layer has four equatorial thin areas, called germ pores.

The binucleate uredospore functions as a conidium and it has the capacity to germinate immediately after its formation. Uredospores can re-infect wheat plants and hence they are effective in the spread of the disease. However, they are incapable to infect the alternate host, barberry bushes.

Germination of Uredospores : Under favourable conditions, uredospores germinate as soon as they come in contact with fresh wheat leaves. Each produces one or more germ tubes through germ pores. The germ tube grows over the surface of the host epidermis and on reaching a stoma the tip of the germ tube develops into a vesicle, called appressorium. Hyphal branches develop from the appressorium in the intercellular spaces. This dikaryotic mycelium forms a new generation of uredospores which infect healthy plants. This results into a heavy build up of infection on wheat crop. Once formed, the uredospores spread the disease rapidly under favourable conditions.

Teleutospore Stage : At the close of the wheat season, the uredosori also start producing teleutospores in addition to uredospores. The uredosori are ultimately converted into teleutosori and produce teleutospores exclusively. Teleutosori are also produced independently from the mycelium formed by the late infection of uredospores. The teleutosori appear as black raised streaks along leaf sheaths and stems of infected plants. The groups of binucleate cells which give rise to teleutospores are known as telia.

The teleutospore is stalked, bi-celled, spindle shaped structure, constricted slightly at the septum. The wall of the teleutospore is thick and smooth, and the tip is usually pointed or round. Each cell of the teleutospore is binucleate and is provided with a germ pore. The germ pore is at the apex in the upper cell, whereas it is just below the septum in the lower cell. As the teleutospore matures, the two nuclei in a cell fuse to form a diploid nucleus.

Like uredospores, the teleutospores also exert pressure on the overlying epidermis. The epidermis is thus ruptured and the spores are exposed. The teleutospores act as resting spores and may survive most unfavourable conditions. They are incapable of re-infecting wheat plants. They do not germinate until the next spring. Under favourable conditions of high atmospheric humidity and low temperature, they germinate in soil as no host is required for their germination.

Basidiospore Stage : On return of favourable conditions in spring, the teleutospore germinates. It produces one germ tube from each cell. The germ tube has limited growth and is known as promycelium or epibasidium. The diploid nucleus moves into the promycelium and divides meiotically to form four haploid nuclei, two of (+) and two of (-) strain. The promycelium now divides into four cells by the formation of transverse septa. Each cell produces a single basidiospore which is borne asymmetrically on a fine sterigma. The four basidiospores are thus formed on each promycelium, two of (+) and two of (-) strain.

Basidiospores are small, unicellular and thin walled structures, each with a haploid nucleus. They are discharged by an explosive mechanism and are disseminated by wind. They can germinate only on the leaves of alternate host, barberry bushes. Basidiospores can survive only for few days and they perish in the absence of alternate host.

Stages of Puccinia Graminis on Berberis Plant

The haplophase of P. graminis occurs on Berberis plant. This phase begins with the formation of basidiospores.

Pycnidiospore or Spermagonium Stage : Under favourable conditions basidiospores germinate on the leaves of Berberis. Each produces a germ tube which penetrates through the leaf epidermis and grows into the intercellular spaces of the host tissue. The monokaryotic mycelium thus formed may be of (+) or (-) strain, depending on the strain of the basidiospore. Usually several basidiospores of different strains infect the same leaf and produce mycelia of both (+) and (-) strains. After a few days of infection the mycelia become aggregated into pseudoparenchymatous masses beneath the leaf epidermis and form spermagonia or pycnidia.

The spermagonia are yellowish flask-shaped structures, developed on the upper surface of the leaf. They may be of (+) or (-) strain, depending on the strain of the mycelium. The spermagonium opens on the upper surface of the leaf by a minute pore, called ostiole. The ostiole is guarded by a tuft of unbranched, tapering, orange-coloured sterile hairs, known as periphyses. Amongst the periphyses thin-walled branched flexuous hyphae (receptive hyphae) are also present and these hyphae project much beyond the periphyses.

The wall of the spermagonium is lined internally with a palisade-like layer of numerous; uninucleate tapering cells, known as spermatiophores (pycniophores). Each spermatiophore abstracts many small, uninucleate spermatia (pycniospores). Spermatia ooze out through the ostiole and are held by the periphyses in sticky drop of liquid. The spermatia are small, oval to spherical, hyaline and smooth spores. They may be of (+) or (-) strain.

There are numerous spermagonia of both (+) and (-) strains on the leaves of Berberis. Spermatia can neither infect the primary host nor the alternate host. When spores are released, the spermagonium secretes nectar drops which attract insects. The spores of one strain are transferred to the flexuous hyphae of the opposite strain by insects. This process is known as spermatization. At the point of contact of spermatium with the flexuous hypha the walls are dissolved and the nucleus of the spermatium is transferred to the flexuous hypha. A dikaryotic mycelium is formed as the result of spermatization.

During the course of spermogonial formation, some hyphae of each mating type form protoaecidia on reaching the lower surface of the leaf. The protoaecidia appear as globose mass of hyphae. They do not grow further if there is no spermatization, but if there is spermatization, the protoaecidium develops into an aecidium.

Aeciospore Stage : Aecidia are cup-shaped structures formed on the lower surface of the barberry leaf. They develop from the same mycelium which form pycnidia on the upper surface. The mycelium becomes dikaryotic due to spermatization and this dikaryotic mycelium forms the roof of the protoaecidium. The cells of the protoaecidium are known as aecidiophores. Each aecidiophore cuts off numerous binucleate cells which are arranged in a chain. These chains are made up of long and short cells arranged alternately. The long cells mature into aeciospores, whereas short cells, known as disjunctor, remain sterile and soon disintegrate. Simultaneously with the formation of aeciospores, the peripheral cells of aecidium divide to form a thick protective covering, known as peridium.

The development of aeciospore chains inside the aecidium exerts a pressure on the peridium. As the result it bursts towards the epidermis. At about the same time, disjunctor cells present in between aeciospores disintegrate and thus aeciospores are released. Aeciospores are unicellular, binucleate, thin-walled and orange coloured structures. In young stages, aeciospores are polyhedral, but at maturity they become globose by absorbing water. The aeciospores are incapable of infecting barberry plants, but they can infect wheat plants. They are dispersed by wind. They germinate on the surface of the primary host by producing germ tubes. The germ tube penetrates stoma and grows in the intercellular spaces of the host tissue and forms dikaryotic mycelium. The dikaryotic mycelium starts producing uredospores within 10-12 days.

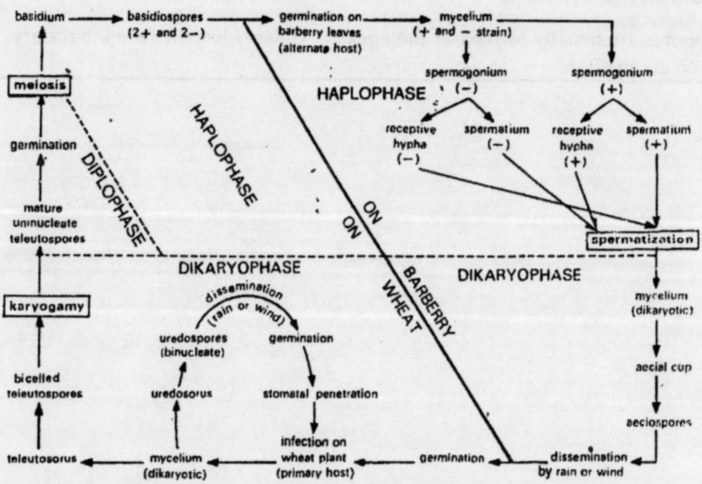

Puccinia graminis : Schematic representation of the life-cycle.

Differences between Rusts and Smuts

Both rusts and smuts are the members of the class Teliomycetes of the sub-division Basidiomycotina. Some important differences between rusts and smuts are enumerated in Table.

Differences between Rusts and Smuts

Rusts	*Smuts*
1. Rusts are generally obligate parasites, can be grown successfully on culture medium.	Smuts are usually facultative saprophytes
2. Rusts are autoecious or heteroecious.	Smuts are always autoecious.
3. Disease symptoms appear in the form of characteristic coloured pustules.	Disease symptoms appear chiefly in the form of black powdery mass.
4. Three types of spores (uredospores, teleutospores, aeciospores) are formed from the secondary dikaryotic mycelium.	Only one type of spore (chlamydospores) is formed from the dikaryotic mycelium.
5. Spores may be unicellular (uredo-and aeciospores) or bicellular (teleutospores) and stalked or sessile.	Chlamydospores are unicellular and sessile.
6. Spores are usually formed at the apex of the hyphae.	Chlamydospores are intercalary.

10

The Ustilago

The Sub-division Basidiomycotina includes most advanced fungi. They are represented by about 550 genera and 25,000 species, mostly parasitic but some are saprophytes. Puccinia (rusts) and Ustilago (smuts) are two important parasitic taxa of this group which are responsible for destructive diseases in cereals and other grasses, causing enormous economic losses. Saprophytic members of this group are popularly known as mushrooms, puff balls, pore fungi, toad stools, stinkhorn, bird's nest fungi, bracket fungi, etc. Agaricus and Polyporus are saprophytic genera.

General Characters of Basidiomycotina

(1) The mycelium is well developed, branched and septate. The septa are simple (*e.g.*, Uredinales and Ustilaginales) or dolipore (*e.g.*, Agaricales).

(2) In the life-cycle of these fungi, the mycelium is found in two or three different stages:

(a) *Primary Mycelium:* It is formed by the germination of basidiospores and it has a haploid nucleus in every cell. Such mycelium is known as monokaryotic. It is usually ephemeral and soon becomes dikaryotic.

(b) *Secondary Mycelium :* It develops by dikaryotization of primary mycelium. It is formed by the fusion of two cells of the primary mycelia of two different strains, conjugation between two basidiospores (sporidia) or the fusion of the primary hypha of one strain with the spermatium (oidium) of another strain. A heterokaryotic dikaryon is present in each cell of the secondary mycelium.

(c) *Tertiary Mycelium :* In several members, the secondary mycelium is organised to form specialized fruiting body (basidiocarp). This mycelium is known as tertiary mycelium and it is also dikaryotic.

(3) The cell wall is composed of chitin and glucan.

(4) The vegetative reproduction takes place by fragmentation or budding.

(5) The asexual reproduction takes place by oidia, conidia or chlamydospore formation. Out of the several types of spores developed in the life cycle of rusts and smuts, some (*e.g.*, uredospore) function as asexual spores. The higher taxa of this group lack asexual reproduction.

(6) No specialized sex organs develop in Basidiomycotina and sexual reproduction takes place by conjugation of nuclei of two different strains.

(7) During sexual reproduction, the dikaryotic cell is formed by spermatization, somatogamy, clamp connection or Buller phenomenon.

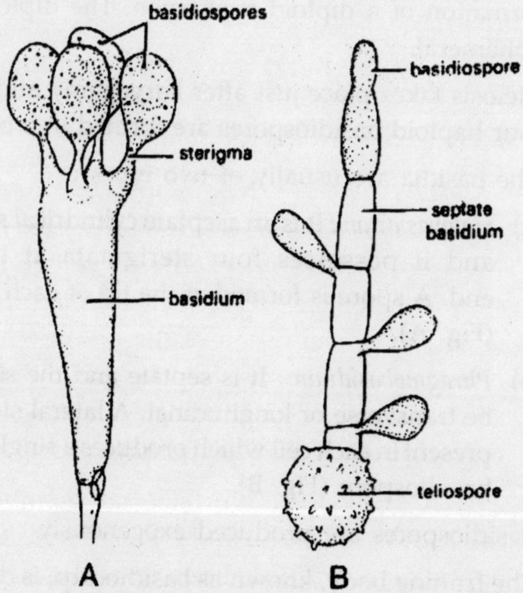

A-B. Basidia : A. Holobasidium, B. Phragmobasidium.

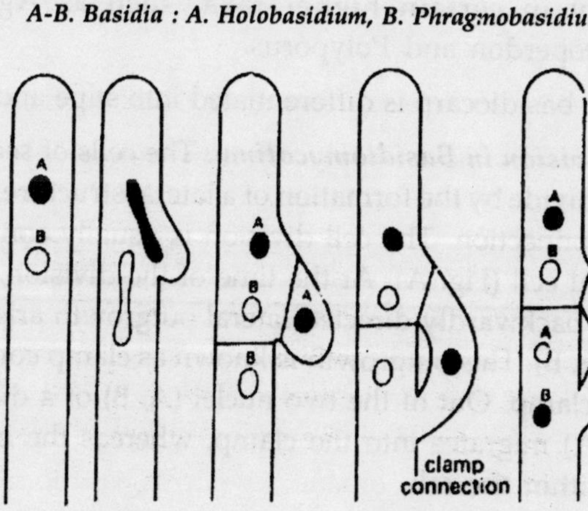

A-E. Cell division by the formation of clamp connections.

(8) Karyogamy does not occur just after plasmogamy. The fusion of nuclei of two different strains occurs within the basidium mother cell, and this results in the

formation of a diploid synkaryon. The diplophase is ephemeral.

(9) Meiosis takes place just after karyogamy and as such four haploid basidiospores are formed in a basidium.

(10) The basidia are usually of two types :

(a) *Holobasidium* : It is an aseptate cylindrical structure, and it possesses four sterigmata at its distal end. A spore is formed at the tip of each sterigma (Fig. A).

(b) *Phragmobasidium* : It is septate and the septa may be transverse or longitudinal. A lateral sterigma is present in each cell which produces a single haploid basidiospore (Fig. B).

(11) Basidiospores are produced exogenously.

(12) The fruiting body, known as basidiocarp, is developed only in certain higher taxa, such as Agaricus, Lycoperdon and Polyporus.

(13) The basidiocarp is differentiated into stipe and pileus.

Cell Division in Basidiomycotina : The cells of secondary mycelium divide by the formation of a lateral structure, known as clamp connection. The cell division is usually confined to the terminal cell (Fig. A). At the time of the division, a short pouch-like backwardly directed lateral outgrowth arises from the cell (Fig. B). This outgrowth is known as clamp connection or simply clamp. Out of the two nuclei (A, B) of a dikaryotic cell, one (A) migrates into the clamp, whereas the other (B) remains within the cell.

Thereafter, both the nuclei divide simultaneously (conjugate division). Out of the two daughter nuclei formed by the division of the nucleus (A) present in the clamp, one migrates to the terminal cell and the other remains within the clamp (Fig. C). Thereafter, two septa are laid down in such a way that one septum separates the clamp and the other develops just above the level of the clamp. This results in the formation of three cells, a terminal dikaryotic cell, a sub-terminal uninucleate cell and a uninucleate clamp cell. The wall between the clamp and subterminal cell dissolves and nucleus from the clamp migrates into this cell. Consequently the subterminal cell also becomes dikaryotic (Fig. D, E). The new cells are continuously formed by repetition of this process.

Ainsworth (1973) divided subdivision Basidiomycotina into three classes-Teliomycetes, Hymneomycetes and Gasteromycetes. However, on the basis of the ultrastructure of septa and basidia, Khan and Kimbrough (1982) divided Basidiomycotina into four classes - Teliomycetes, Hemi-basidiomycetes, Phragmobasidiomycetes and Holo-basidiomycetes.

Class Teliomycetes

The members of class Teliomycetes are parasitic on many vascular plants. The mycelium is well developed and septate; the septa are of simple type, *i.e.*, they lack dolipores. Basidiocarps are lacking and they possess thick-walled, binucleate resting spores, called teliospores.

The class Teliomycetes is usually divided into two orders - Ustilaginales and Uredinales.

Order Ustilaginales

The order Ustilaginales includes about 40 genera and over 1100 species. The members of the order are commonly known as smut fungi, as they form black dusty masses of spores on or within the tissues of the host. These masses of spores resemble with soot or smut. Most of the members, are parasite on grasses and sedges, and cause smut disease. They are not strict obligate parasites as many genera can be grown on artificial culture media. The mycelium is intercellular without any specialized haustoria. The primary mycelium is ephemeral and the secondary dikaryotic mycelium occurs in most phases of the life cycle. The secondary mycelium develops chlamydospores which are also known as brand spores, resting spores, smut spores, or teleutospores. These spores occur in free state (*e.g.*, Ustilago) or in groups, known as spore balls (*e.g.*, Tolyposporium). Sex organs or fruiting bodies are, however, lacking in majority of the members. Plasmogamy takes place by the fusion of two compatible hyphae or conidia, forming dikaryotic mycelium. Basidiospores are usually formed in large numbers in the basidium.

On the basis of the mode of germination of teliospores, the order is divided into two families, *viz.*, Ustilaginaceae and Tilletiaceae. Ustilago, a member of the family Ustilaginaceae, is described here in details.

Family Ustilaginaceae

(1) The chlamydospore gives rise to a septate and 4-celled promycelium (basidium) on germination.

(2) The number of basidiospores (sporidia) formed in a cell is indefinite.

Ustilago (Smut fungi)

Systematic Position

Kingdom	—	Mycota
Division	—	Eumycota
Sub-division	—	Basidiomycotina
Class	—	Teliomycetes
Order	—	Ustilaginales
Family	—	Ustilaginaceae

The genus Ustilago includes more than 300 species, commonly known as smut fungi. The species of Ustilago occur parasitically on several cereal crops, causing very destructive smut diseases. They are not strictly obligate parasites as they can be grown on artificial culture media. The infected tissues of the host are completely destroyed and the mycelium converts into a mass of spores. The species of Ustilago are autoecious, *i.e.*, they complete their life-cycle on a single host.

Loose and Covered Smut

The smut diseases are of two types, loose smut and covered smut. In loose smut, the chlamydospores remain covered by a very thin membranous structure, which bursts at the time of flowering and then the spores are gradually disseminated by the wind. On the other hand, in covered smut, the spores remain enclosed in a membranous covering of the grain and are not directly exposed to air. They are liberated usually at the time of threshing of the ear due to breaking of the wall of the grain.

Some cereal crops, such as oat and barley are infected by both types of smut diseases, whereas others, like wheat and maize are infected by only one type of smut disease. The loose smut as well as the covered smut do not affect the vegetative growth of the host significantly.

Smut Diseases of Some Important Cereal Crops

The following are the smut diseases of some important cereal crops.

Fungus	*Host*
A. Loose smuts	
1. Ustilago nuda var. tritici	Wheat
2. U. nuda var. hordei	Barley
3. U. nuda var. maydis	Maize
4. U. avenae	Oat
5. U. cynodontis	Doob grass

Contd...

Fungus	*Host*
B. Covered smuts	
6. U. hordei	Barley
7. U. kolleri	Oat
8. U. scitamineae	Sugarcane

Vegetative Structure : The mycelium is branched, septate and intercellular. It has two distinct stages-primary and secondary in the life-cycle of Ustilago. The primary mycelium is produced by the germination of the basidiospore. It consists of slender, hyaline and septate hyphae; each cell of the hypha contains a single haploid nucleus. Hence, it is also known as monokaryotic or haplomycelium. Depending on the strain of basidiospore, the mycelium may be of (+) or (-) strain. In majority of species, the primary mycelium, after growing for sometime, is converted into secondary mycelium.

The secondary mycelium is septate and branched. It contains two haploid nuclei in each cell. There is a simple pore in each septum which establishes continuity of cytoplasm of the adjacent cells. During the vegetative phase, the secondary mycelium becomes extensively branched and spreads into intercellular spaces in the various parts of the host. It absorbs nutrients from the host cells with the help of haustoria. The mycelium remains associated with almost every part of the host plant throughout the life (systemic) or restricted to certain parts of the host (localised).

Dikaryotization of Mycelium

The process by which primary haploid mycelium is converted into secondary dikaryotic mycelium is called dikaryotization. Dikaryotization may take place by various methods. Some important ones are described below :

By Fusion between Primary Hyphae : This type of dikaryotization occurs in U. maydis. The basidiospore, on germination, forms primary mycelium on the host surface. This mycelium enters the intercellular spaces by penetrating the host epidermis. Somatogamy takes place between the hyphae of two opposite strains, resulting in the establishment of dikaryotic phase. The dikaryotic cell divides several times by the process of clamp connections and forms secondary mycelium.

By Fusion between Germ Tubes of Two Basidiospores : In some species of Ustilago, *e.g.*, U. hordei the germ tubes of two different strains come in contact during basidiospore germination, their walls are dissolved at the contact point and the nucleus of one germ tube migrates into the other. This act establishes dikaryophase. The germ tube develops into secondary mycelium.

By Conjugation between the Basidiospores : In species like U. violacea and U. recepticularum, propagation of basidiospores takes place by budding, and a number of secondary spores, known as sporidia, are formed. A dikaryotic spore is formed by the fusion of two sporidia of different strains. The dikaryotic spore germinates to form secondary mycelium.

By Union of Infection Threads : In U. tritici the basidium is septate and it does not produce sporidia. Each haploid cell of the basidium produces a minute tubular filament, known as infection thread. Two adjacent infection threads of different strains fuse to form a dikaryotic hypha which divides and forms dikaryotic (secondary) mycelium.

Reproduction : The reproduction takes place by the formation of chlamydospores and basidiospores. In the life-cycle of Ustilago, first chlamydospores are formed and they give rise to basidia and basidiospores on germination.

Formation of Chlamydospores : Chlamydospores are

formed on dikaryotic mycelium. During the blossom period of the host, the mycelium becomes more active and infects the embryonic spikelets on reaching the inflorescence region. The mycelium then aggregates in the young ovaries and forms dense masses of hyphae. These hyphae, known as sporogenous hyphae, divide into dikaryotic cells. Each dikaryotic cell forms a binucleate chlamydospore. These binucleate chlamydospores are also known as teleutospores, brand spores or smut spores, and the spore mass, smut ball or sorus.

All species of Ustilago show the same mode of chlamydospore formation, but in loose smuts the sporogenous hyphae grow in between the anticlinal walls of the epidermis of the infected organ and as such the epidermis is destroyed. Thus in loose smuts, there is no envelope over the sorus. On the contrary, in covered smuts, the sporogenous hyphae do not destroy the host epidermis and the sorus in enveloped by the host epidermis.

Structure of Chlamydospores : The chlamydospores are globose or elliptical, light brown and measure 5-9 microns in diameter. The spore wall is differentiated into an outer thick exine and an inner thin intine. The exine may be smooth (U. hordei) or ornamented (U. maydis, U. tritici). The chlamydospores are binucleate as they are formed on dikaryotic secondary mycelium.

Germination of Chlamydospores and Development of Basidium : The chlamydospores are readily disseminated by wind. They remain viable up to 5 or 6 months. The mature spores of U. nuda var. tritici germinate soon after dissemination and infect the healthy plants. But in most of the species of Ustilago the chlamydospores undergo a period of rest and they germinate in the next season if conditions are favourable.

Before germination, the two nuclei of the binucleate chlamydospore fuse to form a diploid nucleus. During germination, the spore swells by absorbing water, the exine bursts and the intine comes out in the form of a germ tube, known as promycelium or basidium. At this stage, the diploid nucleus of the chlamydospore undergoes a reduction division and forms four haploid nuclei, two of '+' strain and two of '-' strain.

After reduction division, the basidium becomes 4- celled by the formation of septa. Each cell of the basidium has a single haploid nucleus. The latter divides to form two daughter nuclei, one of these nuclei remains in the germ cell (*i.e.*, in the basidial cell) and the other passes into basidiospore through sterigmata. Of the four basidiospores formed in a basidium, two are of '+' strain and two of '-' strain.

The remaining nucleus in the germ cell may undergo repeated divisions and as such each cell of the basidium may form numerous basidiospores (sporidia). The basidiospores are thin walled, uninucleate and globose or oval structures and may also multiply by budding.

With the exception of U. nuda var. tritici, all species of Ustilago form basidiospores. In U. nuda var. tritici the cells of the promycelium directly produce germ tubes, also known as infection threads.

Germination of Basidiospores : The haploid basidiospores germinate to form monokaryotic mycelium. This mycelium may be of (+) or (-) strain, depending on the strain of the spore. Species of Ustilago show different modes of infection. In covered smuts, the chlamydospores are usually present inside the grain and when such seeds are sown, the mycelium is formed inside the host tissue by the germination of spores. Such mode of infection is known as seedling infection.

In loose smuts, the infection occurs at flowering time, when the mycelium enters the young tissues at the base of the ovary. Secondary infection occurs by chlamydospores, blown away by wind. This mode of infection is known as blossom infection.

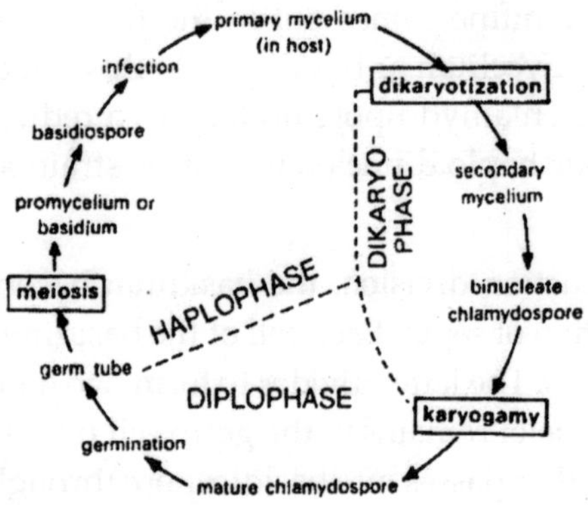

***Ustilago* : Schematic representation of the life-cycle.**

11

The Pezizales

Order Pezizales

The order Pezizales is represented by about 60 genera and 500 species. The members of the order are saprophytes, growing on decaying wood, humus rich soil or dung (coprophilous). The mycelium is usually underground, growing a few centimeters deep in the soil and consists of branched and septate hyphae.

The ascocarp is a fleshy or leathery apothecium. As the apothecium is cup-shaped, hence the members of the order are known as 'cup fungi'. Asexual reproduction is very uncommon but a few species multiply by conidia or chlamydospores. There is a distinct degeneration of sexuality. Normal antheridia and ascogonia develop only in few genera, such as Pyronema and Ascobolus; whereas in others like Peziza and Morchella sex organs do not develop at all. The latter reproduce by the process of somatogamy. Asci are cylindrical to clavate and operculate and are arranged in a hymenium along with paraphyses. Ascospores are hyaline or brown and smooth or variously sculptured. The ascospores are released from the asci by producing a hissing sound.

Alexopoulos and Mims (1979) recognised five families within the order - Sarcoscyphaceae, Pezizaceae, Ascobolaceae, Morchellaceae and Helvellaceae, whereas Korf (1973) included seven families Sacrcosmataceae, Sacrcoscyphaceae, Ascobolaceae, Pezizaceae, Morchellaceae, Helvellaceae and Pyronemataceae. Peziza (Pezizaceae), Morchella (Morchellaceae) and Ascobolus (Ascobolaceae) are described here in detail.

Family Pezizaceae

(1) Most of the members of Pezizaceae are saprophytes, growing on decayed wood or bark or humus rich soil and some species (*e.g.*, species of Coprobia) are coprophilous fungi, growing on dung.

(2) Conidial stage is absent with a few exceptions, such as Peziza repanda and P. vesiculosa.

(3) The fruiting body (ascocarp) is cup-or disc shaped apothecium, invariably epigeal.

(4) Ascospores are uninucleate and usually hyaline, and are arranged in the ascus in a single row. The ascus opens by a lid.

Peziza (Cup fungi)

Systematic Position

Kingdom	—	Mycota
Division	—	Eumycota
Sub-division	—	Ascomycotina
Class	—	Discomycetes
Order	—	Pezizales
Family	—	Pezizaceae

There are about 100 species of Peziza (cup fungi) and most of these are saprophytes, growing on rotting logs, bark of trees, humus rich soil, etc. Some species grow on burnt wood, whereas others are coprophilous. The genus is represented in India by about 25 species; the common ones are P. vesiculosa, P. pustulata, P. catinus, P. echinospora and P. mutiguttulata.

Vegetative Structure : The mycelium is usually perennial, well-developed, profusely branched and septate with uninucleate cells. The hyphae ramify within the substratum. They form a complex system and absorb nutrients from the substratum. The apothecia, however, develop above the substratum.

Reproduction : Peziza mostly reproduces sexually but asexual reproduction also takes place in a few species.

Asexual Reproduction : Asexual reproduction is absent in most of the species of Peziza. But a few species (*e.g.*, P. fuckeliana, P. ostracoderma, P. repanda, P. vesiculosa) produce conidia and chlamydospores. The conidial state of P. pustulata belongs to the form genus Oedocaphilum.

At the time of formation of conidia, some hyphae grow upwards and act as conidiophores. The conidiophores are long, cylindrical, erect and septate. The tip of the conidiophore swells to form a vesicle. Conidia are formed on the surface of the vesicle. They are hyaline or light coloured, elliptical and thin walled structures. They germinate on a suitable substratum and produce new mycelia.

Chlamydospores are thick walled resting cells which are formed in intercalary positions on the hyphae, either singly or in chains. In favourable conditions they germinate and form germ tubes which eventually grow into new mycelia.

Sexual Reproduction : Definite sex organs, *i.e.*, antheridium and ascogonium, are not formed in Peziza. Sexual reproduction takes place by somatogamous copulation. The hyphae grow in all directions and form a pseudoparenchymatous mass. Some hyphal branches in this parenchymatous mass have dense protoplasmic contents and their cells are uni-or multinucleate. Usually two adjacent cells of these hyphae fuse to form a dikaryotic cell, but sometimes two nuclei of the same cell form a dikaryon. The latter process is known as autogamous pairing. The dikaryotic cells give rise to ascogenous hyphae.

The terminal cell of the ascogenous hypha usually functions as the ascus mother cell. But in P. vesiculosa the dikaryotic cell directly acts as ascus mother cell and it elongates and forms a cylindrical or club-shaped ascus. The diploid nucleus undergoes a reduction division, followed by a mitotic division, resulting in eight haploid nuclei.

Each of these nuclei secretes a wall and organises into an ascospore. In this way, eight ascospores are formed in an ascus. The uninucleate ascospores are hyaline, oval or elliptical and have smooth or coarsely reticulate wall. The ascospores are placed obliquely in a row in the ascus.

Apothecium : The mature apothecium is a cup shaped, sessile or sub-sessile structure, 1-10 cm in diameter. They may be whitish yellow (P. coccinea), orange (P. aurantia) or bright red or grey in colour (P. vesiculosus). The asei are arranged in parallel rows within the apothecium. A longitudinal section of the apothecium shows the following regions.

(a) *Hymenium :* It is the innermost region of the apothecium. It is composed of asci and paraphyses which are placed at right angle to the surface of the apothecium.

(b) *Sub-hymenium :* The middle region of apothecium is known as sub-hymenium or hypothecium. It is composed of thin walled and light-coloured hyphae, which run parallel to the surface of the hymenium. The asci and paraphyses arise from these hyphae.

(c) *Excipulum* : It is the outermost region of the apothecium, composed of sterile hyphae. Ascospores are released from the apothecium generally in moist environmental conditions. On germination, the ascospore gives rise to a germ tube which develops into a new mycelium.

Family Helvellaceae

(1) The members of the family are saprophytes which grow commonly on humus soil, decaying wood and leaves and other similar substrates.

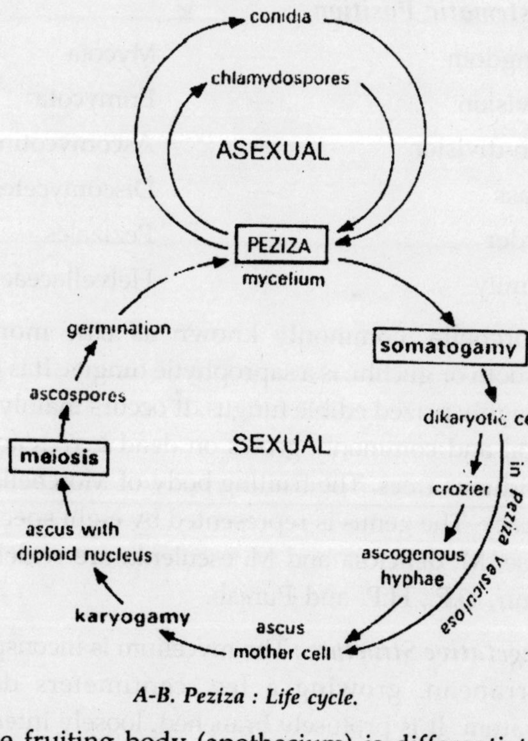

A-B. Peziza : Life cycle.

(2) The fruiting body (apothecium) is differentiated into a stipe and a pileus.

(3) The surface of the pileus is irregular and as such the hymenium becomes folded.

(4) Asexual reproduction is absent.

(5) Sex organs are absent and sexual reproduction takes place by somatogamy.

(6) The ascus has a terminal lid through which ascospores are liberated.

(7) Ascospores are hyaline and multinucleate with 20-60 nuclei.

Morchella (Guchhi)

Systematic Position

Kingdom	—	Mycota
Division	—	Eumycota
Sub-division	—	Ascomycotina
Class	—	Discomycetes
Order	—	Pezizales
Family	—	Helvellaceae

Morchella, commonly known as true morel, sponge mushroom or guchhi, is a saprophytic fungus. It is perhaps the most highly prized edible fungus. It occurs mainly on humus-rich soil and commonly grows on dead leaves and branches of deciduous trees. The fruiting body of Morchella is used as a delicacy. The genus is represented by eight species in India; of these, M. deliciosa and M. esculenta are widely grown in Kashmir, U.P., H.P. and Punjab.

Vegetative Structure : The mycelium is inconspicuous and subterranean, growing a few centimeters deep in the substratum. It is profusely branched, loosely interwoven and septate. The hyphal cells are multinucleate (Fig. A, B). The subterranean mycelium absorbs nutrients from the substratum. The aerial fruiting body (ascocarp) develops on the surface of the substratum after rainy season.

Development of the Fruiting Body (Ascocarp)

The mycelium grows rapidly under favourable conditions when sufficient food material is available. The hyphal mass aggregates under the soil surface and forms a compact structure, known, as hyphal knot. The hyphal knot develops into a stalked fruiting body under favourable growth conditions.

The young fruiting body is cup-shaped and asci are arranged in it vertically. Due to irregular growth of the hymenial surface, the fruiting body develops many ridges and depressions and thus becomes folded. The hymenium appears like a sponge at this state. The folds in the hymenial layer increase the fertile surface of the fruiting body.

Structure of the Fruiting Body

External Structure : The mature ascocarp is 3-13 cm in length and it is differentiated into a stalk-like structure, the stipe and a hollow cap, the pileus (Fig. A). The stipe and the pileus are nearly of the same length. The stipe is a thick, fleshy and hollow structure of light yellow colour, measuring 1-2 cm in diameter. The pileus is a brown or black conical structure and constitutes the fertile part of the ascocarp.

Internal Structure : The stipe is made up of pesudoparenchymatous tissue with a hollow cavity in the centre.

The pileus is differentiated into the following three regions (Fig. A, C, D).

1. *Trama* : The innermost region of the pileus is pseudoparenchymatous and is composed of sterile hyphae.
2. *Sub-hymenium (Hypothecium)* : This region is present outside the trama and is composed of many closely packed interwoven hyphae.

3. *Hymenium :* The hymenium is the fertile region of the pileus. It is confined only to the grooves. In this region, cylindrical, elongated and club-shaped hyphae are arranged perpendicular to the surface of the pileus and they form a palisade-like layer of asci and paraphysis. There are eight ascospores in each ascus. The ascospores are placed obliquely in a single row. They are oval, hyaline and uninucleate.

The paraphysis help in the dispersal of ascospores. The tips of the mature asci project beyond the hymenial layer and grow towards light. The turgor pressure of asci increases as they absorb water and ascospores are released through the apical pores.

Reproduction : Asexual reproduction is not known in Morchella and it reproduces only by the sexual method.

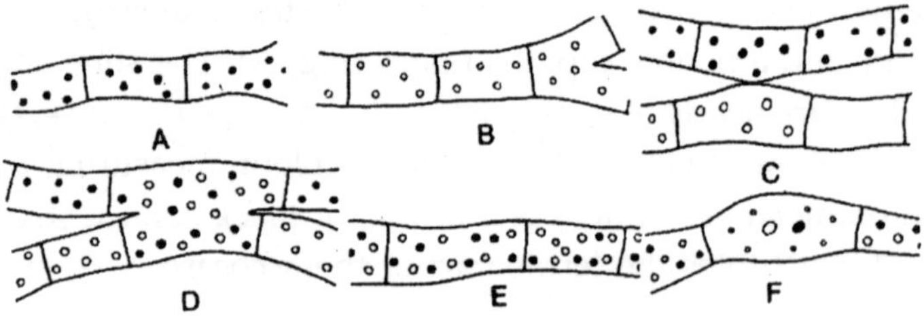

A-F. Morchella : Stages in somatogamy.

Apparent sex organs are not formed in Morchella and the fruiting body (ascocarp) develops by the fusion of two somatic cells (somatogamy) or by the fusion of two nuclei of the same cell (autogamy). In both, somatogamy and autogamy, the fusion of nuclei takes place along the following two steps.

Plasmogamy : In plasmogamy, the protoplasts of the two fusing cells mix with each other and all nuclei are transferred to one cell. These nuclei do not fuse and establish dikaryophase. There are two ways by which dikaryophase can be established in Morchella.

Somatogamous Copulation : In somatogamous copulation, two somatic hyphae of the sub-hymenium region come in contact with each other (Fig. A, C). The intervening walls between these cells dissolve and the protoplasts of both the cells mix freely (Fig. D). Two nuclei, one from each cell, pair to form a dikaryon (Fig. E, F). The remaining nuclei degenerate. The dikaryotic cells give rise to ascogenous hyphae. Each ascogenous hypha contains a pair of nuclei, formed by the division of the dikaryon. The terminal dikaryotic cell of the ascogenous hypha functions as ascus mother cell.

Autogamous Pairing : In autogamous pairing, dikaryon is formed by pairing of the two nuclei of the same somatic cell. The remaining nuclei of the cell degenerate. The dikaryotic cell gives rise to many ascogenous hyphae (Fig. A). The terminal dikaryotic cell of the ascogenous hypha functions as ascus mother cell.

Karyogamy : The two nuclei of the dikaryon present in the ascus mother cell fuse to form a diploid nucleus, the synkaryon (Fig. B).

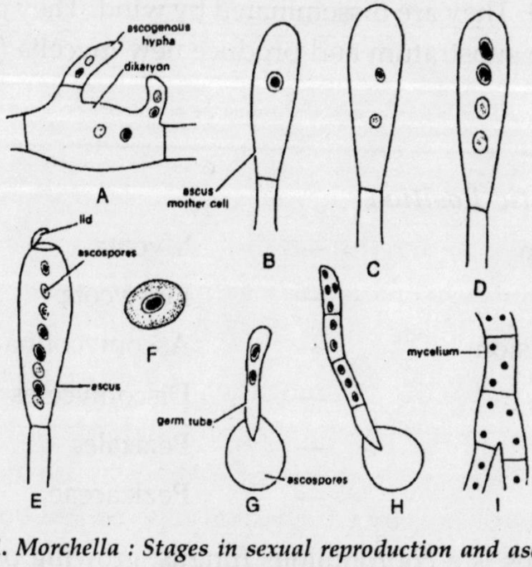

A-1. Morchella : Stages in sexual reproduction and ascus development; A. Autogamous pairing, B. Synkaryon, C-E. Formation of ascus. F. Ascospore, G-1. Germination of ascospore.

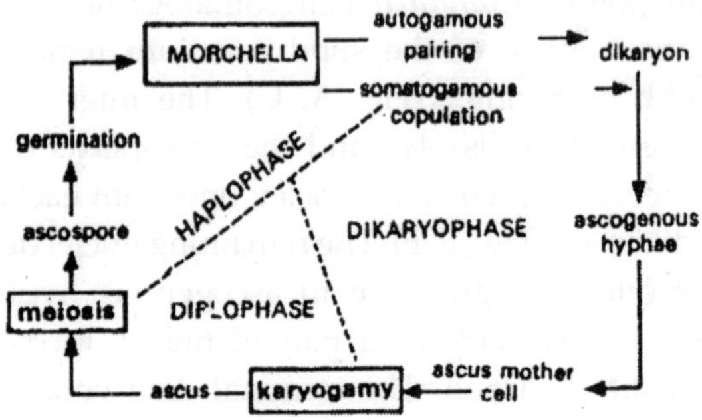

Morchella : Life-cycle.

Formation of Ascus and Ascospores

After the fusion of the nuclei, the ascus mother cell elongates to form a cylindrical structure, known as ascus. The diploid nucleus of the ascus divides by a meiotic division, followed by a mitotic division. Thus eight haploid nuclei are formed (Fig. C, E). Each nucleus secretes a thin wall around itself and develops into an ascospore (Fig. F).

Ascospores are released from the ascus by the removal of the apical lid. They are disseminated by wind. They germinate on a suitable substratum and produce new mycelia (Fig. G, I).

Ascobolus

Systematic Position

Kingdom	—	Mycota
Division	—	Eumycota
Sub-division	—	Ascomycotina
Class	—	Discomycetes
Order	—	Pezizales
Family	—	Pezizaceae

Ascobolus is a coprophilous fungus, growing on dung of cow and other herbivorous animals. Some species (*e.g.*, A.

carbonarius) grow on old bonfire sites. In India, about a dozen species of Ascobolus occur mostly on dung of herbivores. A. denudatus, A. magnificus, A. viridis and A. lignata are some-common Indian species.

Vegetative Structure : The mycelium is usually perennial, well developed, profusely branched and septate. The hyphal cells are short and multinucleate. The hyphae ramify within the substratum. They form a complex system and absorb nutrients from the substratum. The apothecia, however, develop above the substratum (Fig.).

Reproduction : Ascobolus mostly reproduces sexually and asexual reproduction by means of spores, is generally absent. However, some species (*e.g.*, A. furfuraceus) produce oidia or arthrospores in chains on small hyphal branches. They may germinate to form new mycelium. The oidia may also play some role in sexual reproduction.

Sexual Reproduction : Sexual reproduction is oogamous. Species may be homothallic (*e.g.*, A. crenulatus) or heterothallic (*e.g.*, A. carbonarius, A. immersus, A. magnificus). There is much variability in sexual reproduction in various species. In A. magnificus, well developed sex organs are produced (Fig. A, F). Multinucleate ascogonia and antheridia are developed when mycelia of different mating types are grown together. The male branch is cylindrical and clavate and remains erect. However, the female branch elongates and cuts off terminally a long trichogyne which by septation becomes a 5-8 celled structure, and simultaneously ascogonium swells to become a globose structure.

At maturity, the trichogyne coils around the body and reaches the antheridial tip. The walls at the point of contact dissolve and a pore is formed. The male nuclei pass into the apical cell of trichogyne and finally reach the ascogonium. At this stage, female nuclei also divide and form pairs with male nuclei. Soon the ascogenous hyphae are budded

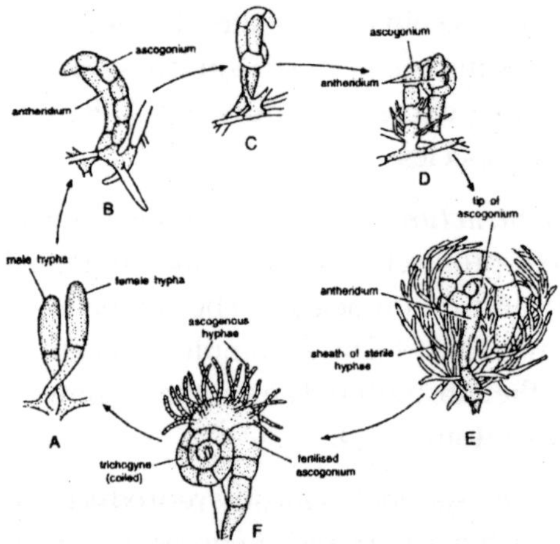

A-R Ascobolus : Stages in sexual reproduction.

out from the fertilised ascogonium. Ultimately apothecia are developed containing asci and paraphysis in the usual way. In A. carbonarius, instead of an antheridium, a conidium- like body is formed, while in A. citrinus, the antheridium is completely lacking and the female branch develops apothecium parthenogenetically without copulation.

In A. furfuraceus the oidia take part in the process of sexual reproduction (Fig.). Here, the oidium of one strain comes in contact with the mycelium of another strain and apothecia develop.

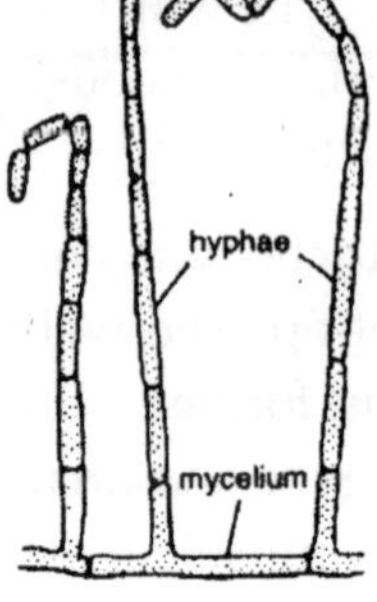

Ascobolus : Oidia formation.

Ascocarp Development : In Ascobolus two types of ascocarp development occur. In gymnohymenial type (as in A. magnificus) the hymenium is exposed in an earlier stage, whereas, in cleistohymenial type (as in A. furfuraceus, A. immersus) the hymenium is enclosed in earlier stage and at maturity, asci are exposed.

The apothecia are cup-shaped and yellowish with purplish dots structures. Asci are long, cylindrical structures intermixed with paraphyses. Each ascus contains eight ascospores and has an operculum at its apex. The ascospores are one-celled, ellipsoidal or spherical and dark brown in colour. The ascospore germinates by germ tube to form a new mycelium.

12

Erysiphales

Order Erysiphales

The order Erysiphales includes about 60 genera and 100 species which occur as ecto-obligate parasites on leaves and stems of a large number of flowering plants. These fungi appear as white powdery coating on the surface of the host, hence they are popularly known as powdery mildews.

The mycelium is superficial, well developed, branched and septate. They absorb food material from the host through globular haustoria. Asexual reproduction takes place by conidia, produced on conidiophores in basipetal succession. Conidiophores develop singly at right angle to the host surface. Sexual reproduction takes place by well defined antheridia and ascogonia. The fruiting body is a globose cleistothecium in which asci are arranged parallel to each other. Mature cleistothecia bear characteristic appendages.

The order Erysiphales though generally included in the class Plectomycetes, some mycologists place it in Pyrenomycetes as asci form a basal hymenium layer. The opinion is sharply divided with regard to the number of families within the order. Thrisseu and Sydow (1917) recognised five families (Erysiphaceae, Meliolaceae, Englerulaceae, Capnodiaceae and Trichothyriaceae); Yarwood (1973, 1978) two families (Perisporiaceae, dark mildews and Erysiphaceae, powdery or white mildews) and Alexopoulos and Mims (1979) only one family (Erysiphaceae) within the order.

Family Erysiphaceae

The characteristic features of the family are as follows:

(1) The members of the family are biotrophic parasites of flowering plants, causing the powdery mildew disease.

(2) The mycelium is superficial and appears as white cottony growth on the surface of the host. The simple or branched haustoria absorb food material from the host.

(3) Asexual reproduction takes place by unicellular, hyaline barrel-shaped conidia.

(4) The fruiting body (cleistothecium) develops on the surface of the host and it bears four types of appendages : (i) myceloid (*e.g.*, Erysiphe;), (ii) circinoid (*e.g.*, Uncinula;), (iii) dichotomously branched tips (*e.g.*, Microsphaera;), and (iv) bulbous (*e.g.*, Phyllactinia;).

(5) There are 2,8 ascospores in an ascus.

The family includes eight genera Sphaerotheca, Podosphaera, Erysiphe, Microsphaera, Uncinula, Leveillula, Phyllactinia and Acrosporium (Oidium). These genera can be recognised on the basis of the structure of ascoearp and appendages.

Three genera of the family, Erysiphe, Sphaerotheca and Uncinula are described here:

Erysiphe

Systematic Position

Kingdom	—	Mycota
Division	—	Eumycota
Sub-division	—	Ascomycotina
Class	—	Plectomycetes
Order	—	Erysiphales
Family	—	Erysiphaceae

The genus Erysiphe is represented by some 10 species, cosmopolitan in distribution. They occur as ecto-obligate parasites on several herbaceous flowering plants and cause the powdery mildew disease. Some common pathogenic species of the genus are as follows:

(1) E. polygoni — powdery mildew of peas.

(2) E. communis (syn. E. cichoracearum) — powdery mildew of cucurbits.

(3) E. graminis — This species infects the members of the family Poaceae. It has several physiological races, for example E. graminis-tritici infects only wheat plants. whereas E. graminis-hordei only barley.

Symptoms : The disease appears in the form of greyish-white powdery growth on the leaf surface. These symptoms later also appear on the floral parts. The infected plants have stunted growth and the size and number of leaves is reduced. In case of severe infection, leaves are twisted and deformed. Infected inflorescences wilt and seed setting is checked.

Vegetative Structure : The mycelium is superficial (ectophyte), well branched and septate. It forms a white cottony growth on the host surface. It is composed of short hyphae, made up of uninucleate cells. The food material is absorbed from the host with the help of haustoria that penetrate the epidermal cells of the host. The haustoria may be globular (*e.g.*, E. polygoni, E. communis) or digitate (*e.g.*, E. graminis). They are differentiated into a tubular neck and a swollen uninucleate body. The body is enclosed in a sheath which is invaginated plasmalemma of the host. The neck of the haustorium is

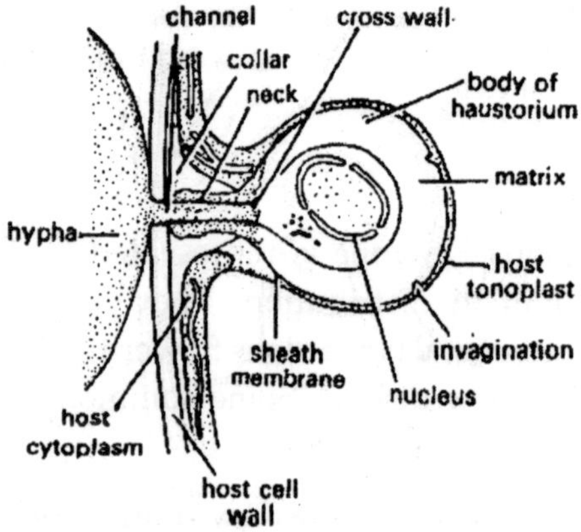

Erysiphe : Ultrastructure of haustorium.

surrounded by a thick collar, deposited on the host cell wall. Immediately outside the haustorium wall, there is a wide sheath matrix, bounded by sheath membrane. The haustorium is uninucleate and contains mitochondria, ER and unidentified vesicles.

Reproduction : Erysiphe reproduces both by asexual and sexual methods.

Asexual Reproduction : Asexual reproduction takes place by conidia. Numerous erect aerial hyphal branches arise from the fully established mycelium over the host epidermis. These hyphae grow at right angle to the host surface and they are called conidiophores. The conidiophore is differentiated into a long basal stalk cell and a comparatively small terminal generative cell. In some species (*e.g.*, E. graminis hordei) the stalk cell is swollen. The generative cell cuts off a chain of conidia in basipetal succession by abstraction process.

Formation of conidia is favoured by cool and humid conditions. The mass of conidia developed on the leaf surface appears as white to pale brown pustules.

Mature conidia are hyaline, unicellular and uninucleate. They are oval, barrel-shaped or cylindrical, depending on the species. They have enough quantity of water and have a waxy or lipoidal covering on their surface and hence are resistant to desiccation. They contain lipid granules and fibrosin bodies. They are dispersed by air. The conidium germinates on a suitable substratum by producing one or more germ tubes. These germ tubes form new mycelia.

Sexual Reproduction : Sex organs usually develop during the late summer when the process of conidia formation slows down. Majority of the species of Erysiphe are heterothallic; male (antheridium) and female (ascogonium) sex organs are borne on different but closely nearby hyphae.

Antheridium : The nucleus of the terminal cell of the male hypha divides to form two nuclei which are separated by a transverse septum. The terminal uninucleate cell functions as antheridium while the basal cell acts as stalk cell.

Ascogonium : The terminal cell of a female hypha swells to form a club-shaped ascogonium and a basal stalk cell in a similar fashion as described for the antheridium.

Plasmogamy : The antheridium comes in contact with the ascogonium .and the both more or less twist around each other. The walls, at the point of contact, dissolve and a pore is formed through which the male nucleus passes into the ascogonium. Thus a dikaryon is formed. It is generally believed that both the nuclei of a dikaryon fuses after the formation of ascus mother cell.

After plasmogamy the ascogonium elongates to form a tubular structure which divides by transverse walls to form a 4-5 celled structure. The penultimate cell of this tubular structure is a dikaryon, and ascogenous hyphae develop from this binucleate cell. The intercalary binucleate cells of the ascogenous hyphae function as ascus mother cells.

The two nuclei of the ascus mother cell fuse to form synkaryon. The ascus mother cell with a synkaryon increases in size and its diploid nucleus first divides meiotically and then mitotically to form eight haploid nuclei. Each nucleus secretes a wall around itself to form an ascospore. The structure that contains ascospores is now known as ascus.

Cleistothecium : Simultaneously with the development of ascus, several sterile hyphae arise from the basal stalk cell of the ascogonium. These hyphae form a protective covering around the ascogonium, which is known as peridium. Initially the peridium is only 1-2-layered but later it becomes 6-10-layered.

The fruiting body thus formed is called cleistothecium. It is a globose, dark brown or black and entirely closed structure. Cleistothecia develop mainly on basal leaves or on stems, nestling in a dense mass of mycelium. Some superficial cells of peridium develop into long, unbranched, hypha-like appendages.

The number of asci in a cleistothecium varies in different species of Erysiphe; for example, there are 2-8 asci in a cleistothecium in E. polygoni and up to 25 asci in E. graminis.

Dehiscence of Cleistothecium : Mature cleistothecia remain attached with the host surface for quite sometime. They are detached and dispersed either accidentally or by wind. They remain in dormant state throughout the winter season.

The thick peridium protects them. On the onset of spring season, the osmotic pressure of asci increases due to some physiological changes. The asci swell by absorbing water and this causes irregular cracks in the upper part of the peridium. This releases asci. Thereafter, ascospores are forcibly ejected

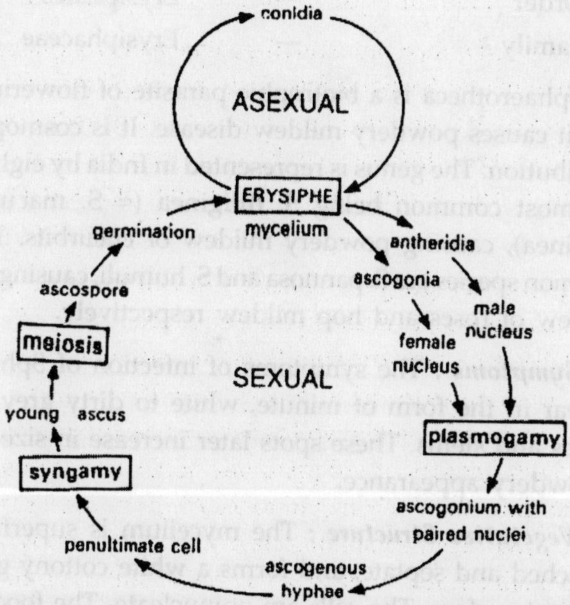

Erysiphe : Schematic representation of the life-cycle.

from asci. Usually an ascus contains eight ascospores, but in E. communis there are only two or rarely three ascospores in an ascus. In those species where the number of ascospores is less than eight, some nuclei in the ascus degenerate.

Ascospores may remain viable up to 13 years. When they fall on a suitable host, they germinate immediately. The ascospore produces a germ tube which develops into a new mycelium.

Sphaerotheca

Systematic Position

Kingdom	—	Mycota
Division	—	Eumycota
Sub-division	—	Ascomycotina
Class	—	Plectomycetes

Order — Erysiphales

Family — Erysiphaceae

Sphaerotheca is a biotrophic parasite of flowering plants and it causes powdery mildew disease. It is cosmopolitan in distribution. The genus is represented in India by eight species; the most common being S. fuliginea (= S. macularis vat. fuliginea), causing powdery mildew of cucurbits. The other common species are S. pannosa and S. humuli, causing powdery mildew of roses and hop mildew respectively.

Symptoms : The symptoms of infection of Sphaerotheca appear in the form of minute, white to dirty grey spots on leaves and stems. These spots later increase in size and give a powdery appearance.

Vegetative Structure : The mycelium is superficial, well branched and septate, and forms a white cottony growth on the host surface. The cells are uninucleate. The food material is derived from the host with the help of globular or lobed haustoria.

Reproduction : Sphaerotheca reproduces both by asexual and sexual means.

Asexual Reproduction : Asexual reproduction takes place by conidia, borne on short erect conidiophores, arising from the mycelium. Conidia are formed at the tips of conidiophores in long basipetal chains. They are ellipsoid to barrel-shaped, hyaline and with distinct fibrosin bodies. The fibrosin bodies are conical, disc or rod-shaped structures. Conidia detach easily at maturity and are dispersed by wind. After falling on a suitable host, the conidium germinates by producing a forked germ tube.

Sexual Reproduction : The sex organs (antheridia and ascogonia) usually develop in the late summer when conidia formation slows down. They are formed on hyphal tips of neighbouring branches (Fig. B). Both, antheridia and ascogonia are uninucleate.

Antheridium : The terminal cell of the male hypha divides to form a slender antheridium and a basal stalk cell.

Ascogonium : The terminal cell of the female hypha divides to form a club-shaped ascogonium and a basal stalk cell.

Plasmogamy : Both, the antheridium and ascogonium grow side by side, closely appressed. The walls, at the point of their contact, dissolve and a pore is formed through which the male nucleus passes into the ascogonium. After the migration of male nucleus, the antheridium collapses. The fertilised ascogonium develops into a fruiting body, known as cleistothecium.

Cleistothecium : Mature cleistothecia are globose and dark-brown in colour. They have a variable number of simple myceloid appendages. These appendages are usually as long as the diameter of the cleistothecium and they are tortuous and interwoven with the mycelium. Only one ascus develops in a cleistothecium.

The obovoid broadly elliptic to subglobose ascus has eight ascospores, which are ellipsoid to nearly spherical. The peridium is 2-3 layered. The innermost layer of peridium is made up of thin-walled cells. These cells, sometimes called as nurse cells, provide nourishment to the developing ascus. The outer layer of peridium has thicker, melanised and darker cells. It performs the function of protection.

Dehiscence of Cleistothecium : The ascus is released by the disintegration of cleistothecium wall. The ascospore germinates on a suitable host and forms a new mycelium. It is believed that the fertilized ascogonium does not play any functional role in the development of ascus. Thus the ascogonium is designated as pseudoascogonium and accordingly the male sex organs as pseudoantheridium.

Uncinula

Systematic Position

Kingdom	—	Mycota
Division	—	Eumycota
Sub-division	—	Ascomycotina
Class	—	Plectomycetes
Order	—	Erysiphales
Family	—	Erysiphaceae

Uncinula is an obligate parasitic fungus causing powdery mildew of grape-vines all over the world.

The most common species of the genus is U. necator and it sometimes causes complete destruction of grape crop. In India this species is common in Maharashtra, Gujarat, Madras and Andhra Pradesh, causing heavy losses to grape crop.

Symptoms : The symptoms appear first on leaves in the form of white patches on both the surfaces. The leaves may be discoloured and malformed and in case of severe infections the stems become stunted and turn dark brown.

These symptoms also appear on young blossoms and berries. The young fruits, when infected, turn dark, assume irregular form, remain undersized and develop cracks.

In case of early infection, fruits may not develop at all. The diseased grape vines have a wilted and dwarfed appearance.

Vegetative Structure : The well-branched and septate mycelium is superficial and spreads over the host surface. The hypha consists of uninucleate cells.

The mycelium remains attached to the host by means of appressoria, which gives out haustoria. The haustoria penetrate the epidermal cells of the host and derive food material.

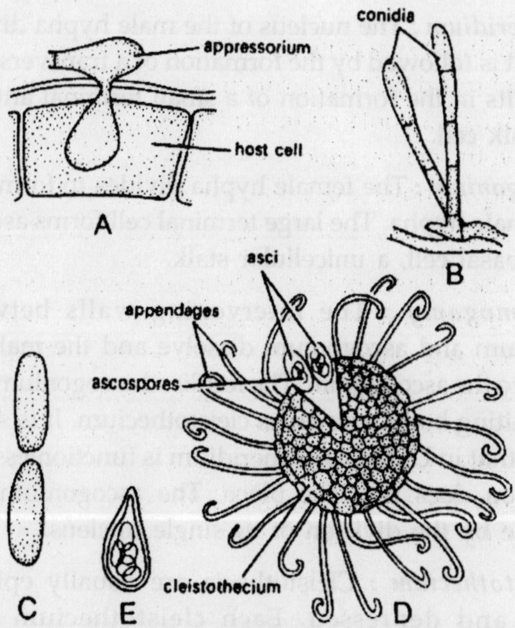

A-E Uncinula : A. V.S. of host leaf showing appressorium, B. Conidiophore and conidia, C. Conidia, D. Ruptured cleistothecium, E. Ascus.

Reproduction : Reproduction takes place by both, asexual and sexual means.

Asexual Reproduction : Asexual reproduction takes place with the help of conidia which develop on erect and unbranched conidiophores. The coni lia are ellipsoid to almost cylindric and are formed in long basipetal chains. The formation of conidia is favoured by warm dry weather, low humidity and cloudiness. They are disseminated by wind and on germination give rise to new hyphae.

Sexual Reproduction : Sexual reproduction takes place usually in late summer. Both, antheridia and ascogonia are small and uninucleate acid are produced on the neighbouring hyphae of a mycelium. They are closely appressed to each other.

Antheridium : The nucleus of the male hypha divides into two and it is followed by the formation of a transverse septum. This results in the formation of a small terminal antheridium and a stalk cell.

Ascogonium : The female hypha divides to form two cells like the male hypha. The large terminal cell forms ascogonium and the basal cell, a unicellular stalk.

Plasmogamy : The intervening walls between the antheridium and ascogonium dissolve and the male nucleus passes into the ascogonium. The fertilised ascogonium develops into a fruiting body, known as cleistothecium. It is sometimes believed that in Uncinula antheridium is functionless and thus fertilization does not take place. The ascogonium becomes binucleate by the division of its single nucleus.

Cleistothecium : Cleistothecia are usually epiphyllous, globose and depressed. Each cleistothecium has 5-30 equatorially inserted circinoid appendages. These appendages are one to six times longer than the diameter of the ascocarp. They are septate, thin walled, flexuous and flaccid. Their apices are inure or less helicoid when mature. There are 4-6 (rarely 6-9) broadly ovate asci in each cleistothecium. An ascus contains 4-7 ovate to ellipsoid ascospores. The mature cleistothecia can be easily seen by naked eye as tiny black balls, interspersed with white mycelium.

Dehiscence of Cleistothecium : Ascospores are released on rupture of asci after disintegration of the peridium. They are disseminated by wind and when they reach a suitable host they germinate and give rise to new mycelia.

13

Sphaeriales

Order Sphaeriales

The order Sphaeriales is included in the class Pyrenomycetes of the Sub-division Ascomycotina. Most of the members of the order are saprophytes, but a few are parasites (*e.g.*, Phyllacora) or coprophilous (*e.g.*, Sordaria). Asexual reproduction takes place by means of conidia which are formed on conidiophore, but some taxa are incapable of producing conidia. One of the characteristic feature of the members of the order is that either they do not produce antheridia or if at all formed, they are non-functional. Some species produce spermatia or microconidia. Sexual reproduction takes place by spermatia which fuse with the trichogyne of ascogonia. The fruiting body is usually flask-shaped or spherical perithecium. There are usually eight ascospores (occasionally two or four) in an ascus. The ascospores are unicellular or septate and hyaline or coloured.

Muller and Von Arx (1973) included 15 families (Ophiostomataceae, Melanosporaceae, Sphaeriaceae,

Hypocreaceae, Polystigamataceae, Coryneliaceae, Sordariaceae, Diaporthaceae, Halosphaeriaceae, Diatrypaceae, Amphisphaeriaceae, Xylariaceae, Verrucariaceae, Clavicipilaceae and Hypomycetaceae) in the order Sphaeriales. Representatives of the two families, Sordariaceae and Clavicipitaceae are described here in some details.

Family Sordariaceae

(1) The members of this family are saprophytes, occurring on decaying plants or dung.

(2) The fruiting bodies (perithecia) are dark-coloured and are usually stomatous.

(3) The asci are thin walled with a thickened apex.

(4) The ascospores are black and have either a mucilagenous exosporium or mucilagenous appendages.

The family contains 25 genera. Neurospora is described here as representative of the family.

Neurospora

Systematic Position

Kingdom	—	Mycota
Division	—	Eumycota
Sub-division	—	Ascomycotina
Class	—	Pyrenomycetes
Order	—	Sphaeriales
Family	—	Sordariaceae

Neurospora is a well known taxon and has been of considerable interest to mycologists, geneticists and biochemists. It is sometimes referred to as Drosophila of plant kingdom. The genus is represented by 12 species, which occur mainly on fruits of papaya, on soil or on burnt ground and on charred vegetation. Some species also colonise warm humid places, such as kilns and bakeries, and as such N. sitophila is commonly known as bakery mold or red bread mold, Some species of Neurospora, like N. tetrasperma, N. crassa and N. sitophila, have proved to be very useful material for genetical and biochemical researches.

Vegetative Structure : The mycelium is well branched, multicellular and septate. The cells are multinucleate and contain pigments of various colours. The hyphae usually grow superficially on the substratum.

Reproduction : Neurospora reproduces both by asexual and sexual means.

Asexual Reproduction : The formation of conidia is a common asexual method of rapid multiplication of Neurospora. The two types of conidia formed are macroconidia and microconidia.

Macroconidia : These conidia are large, oval, multinucleate and pink in colour. They develop on erect aerial branches of the mycelium, known as macroconidiophores. The branched conidiophores produce conidia in beaded chains at the tips of their branches. The terminal conidium of a chain may further produce more conidia by budding. Conidia are dispersed by wind. The macroconidia belong to the form-genus Monilia.

Microconidia : The microconidia are also formed on erect aerial branches, called microconidiophores. These conidia are uninucleate, sticky and comparatively small than the macroconidia. They are borne in groups in terminal or lateral positions on conidiophores.

Both, macro-and microconidia form new mycelium on germination.

Sexual Reproduction : Species of Neurospora may be

homothallic (*e.g.*, N. terricola, N. dodgei) or heterothallic (*e.g.*, N. crassa, N. sitophila).

Sex Organs : The female sex organs, known as ascogonia, develop as lateral outgrowths on the vegetative hyphae. The young ascogonium is a coiled, multinucleate and aseptate structure but septa are formed in the later stages of development. Some hyphae develop from the base of the ascogonium and they form a pseudoparenchymatous ball-like structure around the ascogonium. The upper cells of the ascogonium give rise to long tapering trichogyne. The female sex organs of Neurospora re also known as protoperithecia or bulbils.

Antheridia are absent in Neurospora, but both macro- and microconidia may act as spermatia.

Plasmogamy : At the time of fertilization, micro- or macroconidia (which act as spermatia) get adhered to the trichogyne. In heterothallic species, conidia of one strain unite with the trichogyne of the ascogonium of another strain. The walls between the trichogyne and the fusing conidium dissolve at the point of their contact and one or more nuclei from the conidium enter the ascogonium through the trichogyne. In some species of Neurospora, plasmogamy occurs through somatogamous copulation.

Development of Ascus : Shortly after plasmogamy the ascogonium develops one or more ascogenous hyphae. Each cell of the ascogenous hypha contains a dikaryon. The terminal cell of the ascogenous hypha is curved and forms a hook-like structure, known as crozier. Both the nuclei of crozier show conjugate division (*i.e.*, both nuclei of a dikaryon divide simultaneously) to form four daughter nuclei. Thereafter, by the septation of crozier, three cells are formed in which the nuclei are so distributed that curved terminal cell has one, penultimate cell two (dikaryon) and the basal cell one daughter nuclei. The two nuclei of the penultimate cell fuse to form a diploid nucleus. This cell acts as ascus mother cell and it elongates to form a club-shaped or cylindrical ascus. The diploid nucleus of the ascus first divides meiotically and then mitotically, resulting in the formation of eight haploid nuclei. Each nucleus secretes a wall around itself and metamorphoses into an ascospore. In N. tetrasperma, only four ascospores are formed in an ascus. The ascospores are arranged in a single row in the ascus. They are dark brown or black in colour and have characteristically ribbed wall. The globose envelope of sterile hyphae encloses the maturing ascogonium from all sides, resulting in the formation of young perithecium.

The mature perithecium is a dark-coloured, globose, flask-shaped and beaked structure, which remains surrounded by dark pseudoparenchymatous peridium. The canal of the neck is lined by paraphysis. Paraphyses are also present between the asci in mature perithecium. The asci are cylindrical, long and stalked.

Ascospores germinate to produce new mycelium. They remain viable for many years in N. crassa. Ascospores can be readily germinated by chemical (*e.g.*, furfural) or high temperature (60° C for 20 min.) treatment.

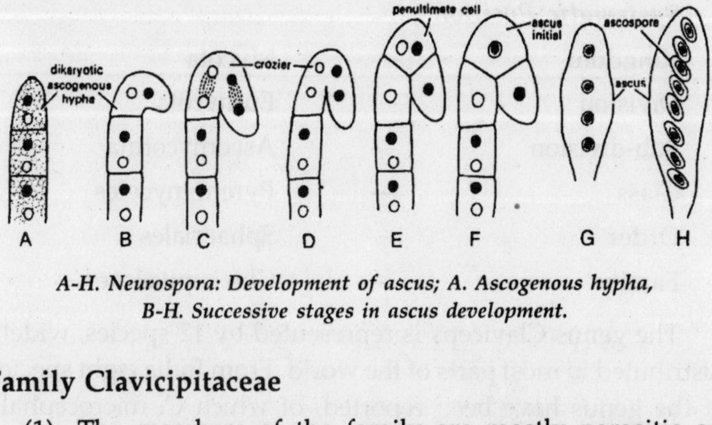

A-H. Neurospora: Development of ascus; A. Ascogenous hypha, B-H. Successive stages in ascus development.

Family Clavicipitaceae

(1) The members of the family are mostly parasitic on grasses (*e.g.* Claviceps) or on insects (*e.g.*, Cordyceps).

(2) Somatic hyphae are aggregated to form pseudoparenchymatous stroma.

(3) Fruiting bodies, known as perithecia, develop on well-developed fleshy stroma, and are ostiolate.

(4) Asci are long, narrow and cylindrical with distinct thick apical caps. The ascospores escape through a narrow pore in the ascus cap.

(5) Paraphyses develop on the inner walls of perithecia and do not occur in between asci.

(6) There are eight long, narrow and thread-like ascospores in each ascus. Sometimes ascospores are fragmented into short segments and each segment functions as a spore.

The family Clavicipitaceae has been included in the order Hypocreales (Bessey, 1950), in an order of its own, Clavicipitales (Gaumann, 1964; Rogerson, 1970; Alexopoulos and Mims, 1979) or in the order Sphaeriales (Miller, 1949; Luttrell, 1951; and Muller and Von Arx, 1973).

Claviceps

Systematic Position

Kingdom	—	Mycota
Division	—	Eumycota
Sub-division	—	Ascomycotina
Class	—	Pyrenomycetes
Order	—	Sphaeriales
Family	—	Clavicipitaceae

The genus Claviceps is represented by 12 species, widely distributed in most parts of the world. From India eight species of the genus have been reported, of which C. microcephala (ergot of pearl millet), C. gigantia (ergot of corn) and C. purpurea (ergot of rye) are most common. C. purpurea grows on a wide range of grasses, including Sorghum and sugarcane.

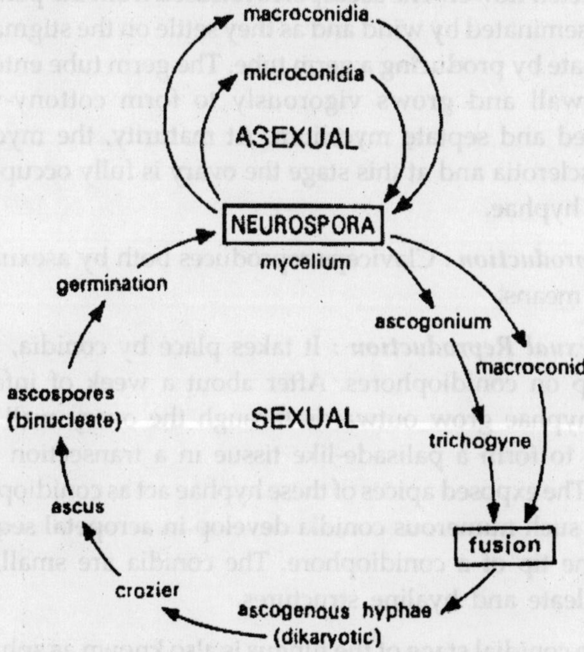

Neurospora : Schematic representation of the life-cycle.

The species of Claviceps infect several members of the family Poaceae, causing ergot disease. Hence, they are also known as ergot fungi. The fungus infects mostly the gynoecium of the host plant, which soon turns black due to accumulation of compact masses of fungal hyphae, called sclerotia. The sclerotia of Claviceps are called ergot'. The grains infected by ergot, when consumed by human beings or domestic animals, cause a serious disease, said to be ergotism.

The sclerotia of C. purpurea contain a number of toxic alkaloids, such as ergotamine, ergometrin and ergonovin, which affect blood vessels and nervous system of human beings. Sclerotia of Claviceps are also used medicinally to hasten uterine contraction during child birth.

Vegetative Structure : The mycelium of Claviceps is much branched, septate and hyaline. It develops within the ovary of the infected flower. The ascospores released from the perithecia ace disseminated by wind and as they settle on the stigma, they germinate by producing a germ tube. The germ tube enters the ovary wall and grows vigorously to form cottony-white, branched and septate mycelium. At maturity, the mycelium forms sclerotia and at this stage the ovary is fully occupied by fungal hyphae.

Reproduction : Claviceps reproduces both by asexual and sexual means.

Asexual Reproduction : It takes place by conidia, which develop on conidiophores. After about a week of infection, some hyphae grow outwards through the ovary wall. They appear to form a palisade-like tissue in a transection of the ovary. The exposed apices of these hyphae act as conidiophores, and as such numerous conidia develop in acropetal sequence from the tip of a conidiophore. The conidia are small, oval, uninucleate and hyaline structures.

The conidial stage of the fungus is also known as sphacelial stage, as conidia of Claviceps were earlier described under Sphacelia segetum. The ovary is enlarged and becomes spongy, and the infected flower secretes a sticky sugary substance, known as honey dew. The conidia stucked in honey dew are transferred to healthy flowers through insects. They germinate there and produce new mycelia.

Formation of Sclerotium : In the last stages of conidial formation, the mycelium present in the ovary (known as sphacelial mycelium) matures. The whole mycelium eventually becomes very hard and changes into a purplish or black pseudoparenchymatous body, called Sclerotium or ergot grain. As the Sclerotium matures, the ovary is completely destroyed and the Sclerotium is held in between the glumes. The size of the Sclerotium is slightly greater than the grain.

The mature Sclerotium is protected by a violet or black covering. Sclerotia fall on the soil and remain viable for a long period due to the presence of a resistant covering.

Germination of Sclerotium : Dispersal of Sclerotia takes place during winter season, but they remain dormant until next spring. In spring, when conditions are favourable, they germinate to produce a number of small pink or dark purple, drum-stick like structures, called perithecial stromata. These stromata are 1-2 cm in length, positively phototropic and are differentiated into stalk and capitate head. There are many minute cavities in the stromatal head. They are known as perithecial cavities, and are surrounded by pseudo-parenchymatous tissue.

Sexual Reproduction : Both, male (antheridium) and female (ascogonium) sex organs are present in a perithecial cavity. The sex organs develop from the terminal cells of the hyphae present at the base of the cavity. The cytoplasm of the fertile hypha is comparatively dense than that of the somatic hyphae. Both, antheridium and ascogonium are multinucleate at maturity. The ascogonium is usually stouter and broader than the slender and elongated antheridium.

As the ascogonium matures, a small papilla-like outgrowth appears on its lateral side. This outgrowth comes in contact of the neighbouring antheridium. The walls of the ascogonium and antheridium dissolve at the point of their contact and plasmogamy takes place. In the process, male nuclei from the antheridium migrate into the ascogonium.

Development of the Ascocarp : Following plasmogamy, numerous ascogenous hyphae are produced from the base of the ascogonium. The penultimate dikaryotic cell of an ascogenous hypha functions as ascus mother cell. The two nuclei of this cell fuse to form a diploid nucleus. It becomes narrow and tubular by elongation and forms an ascus. The diploid nucleus of the ascus divides first meiotically and then mitotically to form eight haploid nuclei. Each nucleus transforms into an ascospore by accumulating some cytoplasm and secreting a wall around itself.

The ascospores are elongated, hyaline and thread-like structures. All the eight ascospores lie parallel to one another in an ascus. A conspicuous cap is present at the tip of the ascus. The perithecial walls grow around the developing asci within the stromatal heads. The whole structure thus formed is called fruiting body or ascocarp.

Fruiting Body or Ascocarp : The fruiting body or the ascocarp of Claviceps is the perithecium which is a flask-shaped structure embedded in the apical part of stroma. The wall of the perithecium is 1-2-layered and the perithecium opens at the surface of the stroma by a narrow opening, known as ostiole. A perithecium contains many asci which arise in tufts from its base. Many paraphyses arise from the sides of the inner wall of the perithecium. They surround the asci but are not present in between the asci.

Dispersal of Ascospores : Ascospores are discharged forcibly due to hygroscopic pressure and are ejected one by one from the ascus. Soon after their release, they are disseminated by wind. They are carried to the flowers of healthy host plants, where they germinate and infect the ovaries.

Family Xylariaceae

(1) Ascocarps embedded in stroma formed entirely of fungal tissue.

(2) Asci with a crown at the apex and a characteristics amyloid ascal plug.

(3) Ascospores dark, unequilateral and with a germ slit.

(4) Mostly saprobes or weak parasites.

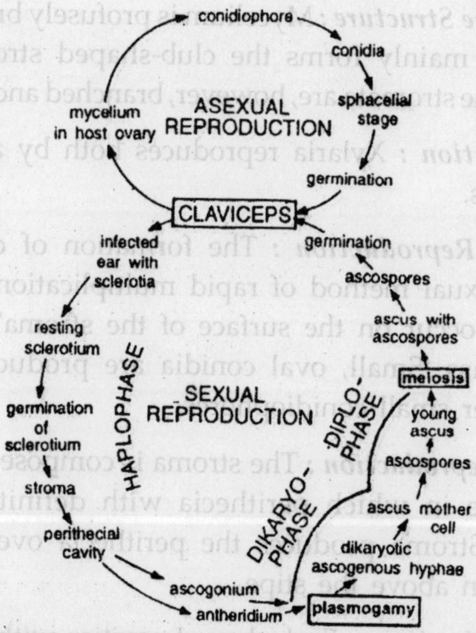

Claviceps : Schematic representation of the life-cycle.

Xylaria is described here as representative of the family.

Xylaria

Systematic Position

Kingdom	—	Mycota
Division	—	Eumycota
Sub-division	—	Ascomycotina
Class	—	Pyrenomycetes
Order	—	Sphaeriales
Family	—	Xylariaceae

The species of Xylaria are mostly saprobic or weak parasites on woody plants, causing white - rot. In India, about 10 species of Xylaria are found. X. hypoxylon (candle snuff fungus) and X. polymorpha (dead man's fingers) are the two most common species. X. mali causes black rot of apple trees.

Vegetative Structure : Mycelium is profusely branched and septate, and mainly forms the club-shaped stroma. In X. hypoxylon, the stromata are, however, branched and cylindrical.

Reproduction : Xylaria reproduces both by asexual and sexual means.

Asexual Reproduction : The formation of conidia is a common asexual method of rapid multiplication of Xylaria. The conidia occur on the surface of the stroma' as a white powdery mass. Small, oval conidia are produced in large numbers over small conidiophores.

Sexual Reproduction : The stroma is composed entirely of fungal tissue in which perithecia with definite walls are embedded. Stroma produces the perithecia over the entire fertile portion above the stipe.

The perithecia are flask-shaped cavities with their necks slightly protruding, out of the stroma. Sex organs are present in the perithecial cavity and develop from the terminal cells of the hyphae present at the base of the cavity. Asci are cylindrical and clavate, usually eight-spored, with a narrow pore at the apex and they develop from the ascogenous hyphae that are seen among the paraphyses to form a continuous hymenium. The ascospores are ellipsoidal, dark brown or black in colour and unequilateral with a germ slit.

14

The Eurotiales

The members of the order Eurotiales (Aspergillales) are mostly saprophytes, but a few species occur as parasites in plants as well as in animals. Some parasitic species cause skin diseases like dermatomycoses, in animals. Asexual reproduction takes place by conidia, produced on conidiophores in basipetal chains. The conidia give a characteristic tint to the fungus. The fruiting body (ascocarp) is a cleistothecium which is small, spherical, generally sessile and without ostiole, paraphyses and appendages. Asci are distributed irregularly and there are no definite hymenial layers. An ascus contains eight ascospores which are released by the decay of the ascus and the cleistothecial wall. The ascospores are unicellular and variously ornamented, and are without germ pores or germ slits.

Some members of the order, such as Aspergillus and Penicillium, are of great economic value.

Fennel (1973) divided the order into nine families, Amorphothecaceae, Gymnoascaceae, Onygenaceae, Monascaceae, Thermoascaceae, Trichomataceae, Eurotiaceae, Cephalotheaceae and Pseudoeurotiaceae. However,

Hawksworth et al. (1983) recognised only three families - Cephalothecaceae, Pseudoeurotiaceae and Eurotiaceae (Trichocomaceae) within the order. A detailed account of Aspergillus and Penicillium, both belonging to the family Eurotiaceae, is given here.

Family Eurotiaceae (Aspergillaceae)

The members of this family are commonly known as blue-green or black mould. The main characteristic features of the family are as follows:

(1) Most species are saprophytes, but a few occur as parasites on plants, animals and human beings.

(2) The mycelium is septate and profusely branched. The cells are multinucleate.

(3) Asexual reproduction takes place by conidia which develop on specialised cells (phialides) present at the tips of the conidiophores.

(4) The fruiting body is a cleistothecium, which varies in its structure in different taxa.

(5) The asci are globose or pyriform.

The nomenclature of the members of Eurotiaceae has been a vexing problem. In the past, separate generic names were given to the perfect (sexual) and imperfect (asexual) states of the same fungus. Perfect states of Aspergillus have been assigned to many different ascomycete genera, such as Edyuillia, Eurotium, Emericella, Sartorya and Hemicarpenteles. For example, the name Aspergillus nidulans was given to the imperfect state of a fungus whose perfect state is known as Emericella nidulans; similarly, A. fumigatus is the imperfect state of Sartorya fumigatus.

Moreover, perfect states of many species of Aspergillus and Penicillium have not been discovered so far. Thus, these fungi are being classified in Fungi Imperfecti till their perfect states are discovered. Some mycologists favour the use of conidial names (*e.g.*, Aspergillus, Penicillium) irrespective of the presence or absence of sexual state, whereas others advocate the use of only 'perfect states.

Aspergillus

Systematic Position

Kingdom	—	Mycota
Division	—	Eumycota
Sub-division	—	Ascomycotina
Class	—	Plectomycetes
Order	—	Eurotiales (Aspergillales)
Family	—	Eurotiaceae (Aspergillacene)

The genus Aspergillus, represented by about 132 species, is a saprophytic fungus, occurring over a wide range of habitats. Over 30 species of Aspergillus occur in India. Species like A. glaucus, A. niger, A. repens, A. flavus, A. candidus and A. fumigatus commonly grow on fruits, vegetables, pickles, jams, jellies and other food stuffs. They are also found on paper, leather, textiles, soil and even ill air. Some species of Aspergillus are found as parasites on plants, causing crown rot of groundnut and boll rot of cotton. Species like A. niger, A. fumigatus, A. flavus and A. terreus cause aspergillosis in animals and human beings. A. flavus contaminates groundnuts and other dry food stuffs and it produces some toxic substances, the aflatoxins, which are carcinogenic in nature.

Several species of Aspergillus are used in various industries, specially in the production of organic acids, enzymes, fats, vitamins and antibiotics. A. niger is used in the study of many biological processes and can detect even very minute quantities of trace elements (Zn, Cu, Mn, Mo, etc.) in an unknown sample. Species like A. fumigates possess antifungal properties.

Vegetative Structure : The mycelium is well developed, profusely branched, and septate and consists of interwoven mass of hyphae. The cells of the hyphae are multinucleate and mostly only one type of nuclei are-present, *i.e.*, the hyphae are homokaryotic. But sometimes they may become heterokaryotic by the anastomosis of hyphae of two different strains (+ or -). The species of Aspergillus mostly have some characteristic pigments in hyphae, conidiophores and conidia. The cytoplasm contains mitochondria, ER and ribosomes. The reserve food material is in the form of oil globules. A simple pore present in the centre of each septum permits the flow of cytoplasmic contents between adjacent cells.

Reproduction : Aspergillus reproduces by vegetative, asexual and sexual means.

Vegetative Reproduction : Vegetative reproduction takes place by fragmentation. Each detached fragment is capable to develop into a new mycelium.

Asexual Reproduction : It takes place by conidia, produced on the characteristic conidiophores. Some cells of the somatic hyphae grow more rapidly and develop a rigid wall. These cells are known as foot cells. An erect branch, known as conidiophore, arises from each foot cell. The conidiophore is usually unbranched and aseptate. The tip of the conidiophore swells considerably, accumulates cytoplasm and forms a vesicle. The shape of the vesicle varies in different species of Aspergillus; for example, the vesicle may be globose (A. niger), hemispherical or crescent-shaped (A. nidulans) or elongated club-shaped (A. clavatus). Several finger-shaped projections develop on the surface of the vesicle and they are known as phialides or sterigmata. There may be up to three layers of sterigmata on the vesicle.

The first formed layer is known as primary sterigmata, and the subsequent layers as secondary and tertiary sterigmata. During the formation of sterigmata many thin areas develop in the vesicle wall due to the dissolution of wall materials. The vesicle cytoplasm is then pushed in these areas synchronously and thus finger-like outgrowths are formed.

Formation of Conidia : Conidia develop in long basipetal chains on the sterigmata. There are 10-12 conidia in each chain. During the formation of a conidium, the nucleus of the sterigmata divides into two. Synchronously a globular protuberance develops at the tip of the sterigmata and it expands to form a conidium. One of the two daughter nuclei present in the sterigmata pass into the conidium. As the conidium expands, the wall of the sterigmata ruptures near its tip.

The remnants of the broken wall persist as a cap around the first formed conidium. The remaining nucleus in the sterigmata continues to divide in the same manner during the formation of successive conidia. A long chain of conidia is formed in this fashion. In early stages, the cytoplasm of the two successive conidia remains continuous due to the presence of a cylindrical isthmus. But later, due to the formation of an inner conidial wall layer, the cytoplasm of the two successive conidia become separated. The isthmus, which remains in between two conidia, is also called connective.

The conidium is a small, globose, echinulate, and uninucleate structure. It contains various pigments, such as yellow, green, brown or black, in different species. Conidia are dispersed by wind. They germinate on suitable substratum, each producing a new mycelium.

Sexual Reproduction : Sexual reproduction is uncommon in Aspergillus. It has been studied in details in A. herbariorum and A. repens. Most species of Aspergillus are homothallic, but a few (*e.g.*, A. fischeri and A. heterothallicus) are heterothallic. The sexual reproduction may be isogamous (*e.g.*, A. nidulans) or anisogamous (*e.g.*, A. repens). The male and female sex organs are known as antheridia and ascogonia (archicarp) respectively. In homothallic species, the antheridium and ascogonium develop on the same hypha or on two adjacent hyphae of the same mycelium.

Development of the Ascogonium : The hyphal branch which forms ascogonium soon becomes septate and loosely coiled. As ascogonium develops the coil becomes dense and looks as a spring-like structure. It is now known as archicarp. It is differentiated into the following three regions.

(1) The terminal unicellular and multinucleate part is known as trichogyne. It acts as the receptive region.

(2) The middle part is also unicellular and uninucleate. It functions as female gametangium and is known as ascogonium.

(3) The basal part is multicellular and multinucleate. It forms the stalk of the ascogonium.

Development of the Antheridium : The antheridium develops close to the ascogonium on the same or nearby hypha. It begins to initiate just before or during the septation of the ascogonial hypha. The antheridial branch, also known as pollinodium, becomes 2-celled by the formation of a septum. The upper cell forms the antheridium proper and the lower one, the stalk. Both the antheridium and stalk are multinucleate and remain unicellular even at maturity.

There is a gradual elimination of antheridia in the various species of Aspergillus. The following stages of progressive degeneration can be seen.

(1) In species like A. herbariorum antheridia are well developed and functional.

(2) In A. repens and some other related species though antheridia are well developed, their cytoplasmic contents are not transferred to the ascogonium.

(3) In many species of Aspergillus though antheridia are formed, the male nuclei soon degenerate and there are no functional nuclei in mature antheridia.

(4) In A flavus, A. fischeri and A. fumigatus antheridia do not develop at all.

Fertilization : Before fertilization, the stalk of the antheridium elongates, and coils and grows upwards in close contact with the ascogonium until the antheridium touches the trichogyne. This is known as gametangial contact. The intervening walls at the point of contact of the antheridium with the trichogyne dissolve and the contents of the, antheridium are transferred into the ascogonium through the trichogyne. The contents of the antheridium and the ascogonium readily mix but the fusion of male and female nuclei do not take place at once. The male and female nuclei arrange themselves in pairs within the ascogonium and this is known as dikaryon.

In some species of Aspergillus the process of fertilization differs from the one described above. In these species the role of antheridium in sexual process in not definitely known. For instance, in A. repens although the antheridium is present the male nuclei are not transferred to the ascogonium. In some species, such as A. flavus, A. fumigatus and A. fischeri, though antheridia do not develop at all, fruiting bodies are formed. There is pairing between the nuclei of the ascogonium itself.

Development of the Ascus

As described earlier, after fertilisation a dikaryon is formed in the ascogonium. This is followed by the septation of the ascogonium. Each cell thus formed contains a dikaryon and develops into a multicellular ascogenous hypha. In turn each cell of the ascogenous hypha also has a dikaryon. The terminal cell of the ascogenous hypha curves to form a hook-like structure, known as crozier. The two nuclei of the dikaryon present in this cell divide by a conjugate division and thus four daughter nuclei are formed. Now septa are formed in the crozier in such a way that the penultimate cell has two daughter nuclei, and the terminal and basal cells one daughter nucleus each. The two nuclei present in the penultimate cell fuse to form a diploid nucleus. This cell now functions as the ascus mother cell and elongates to form an ascus. The diploid nucleus of the ascus first undergoes a meiotic division, followed by mitotic division. Eight haploid nuclei are thus formed in an ascus and each of these ultimately transforms into an ascospore. In this way eight globose or pyriform ascospores are formed in each ascus. The wall of the ascospore is differentiated into an outer thick and sculpturous epispore and an inner thin endospore.

Development of Ascocarp

Simultaneously with the development of ascogenous hyphae, many sterile hyphal branches also develop from the cells lying below the ascogonium. These branches form a pseudo parenchymatous structure, known as peridium. The peridium, which is two layered structure, encloses ascogenous hyphae. The inner layer of the peridium is consumed by the developing asci and ascospores, whereas its outer layer forms a protective covering. The entire structure which appears like a hollow ball of the size of a pinhead is known as the fruiting body or ascocarp. The ascocarp is a completely closed cleistothecium.

Ascospores are released by the breakdown of the cleistothecial wall (peridium). They are disseminated by wind. They germinate when fall on a suitable substratum and produce new mycelia.

Perfect States of Aspergillus

The nomenclature of Aspergillus, like other members of the family, has been a vexing problem. Separate names have been given for the perfect (sexual) and imperfect (asexual) states of the same fungus. There are many instances where perfect states of Aspergillus have been placed in separate genera.

On the other hand, in many instances different conidial (imperfect) states have their perfect states in one genus or their perfect states are not known so far. Raper and Thom (1949) and Raper and Fennel (1965) favour the use of only conidial names irrespective of the presence or absence of sexual state, whereas Benjamin (1955) and Subramanian (1972) advocated the use of only perfect names irrespective of the presence of similar conidial (imperfect) states.

Subramanian (1972) assigned the perfect states of Aspergillus into nine different genera. Some important ones are as follows.

Imperfect state	*Perfect state*
Aspergillus alliaceus	Hemicarpenteles spp.
A. athecius	Edyuillia spp.
A. fumigatus	Sartorya fumigatus
A. glaucus	Eurotium repens
A. heterothallicus	Emericella heterothallica
A. nidulans	Emericella nidulans

Some distinguishing features of the important perfect states of Aspergillus *viz.*, Eurotium, Emericella, Sartorya, Hemicarpenteles and Edyuillia are enumerated below.

Eurotium : The best known species is E. repens. In Eurotium, the vesicle of the conidiophore is club-shaped, bearing directly on its surface a cluster of bottle-shaped phialides on which conidia develop in chains. The sex organs- the antheridia and ascogonia - are multinucleate elongated structures and are produced close to each other. The antheridium may be functional or non-functional, but a pairing of nuclei takes place in the ascogonium. If antheridium is functional, the male nuclei pair with the ascogonial nuclei, otherwise the ascogonial nuclei themselves approach each other in pairs. After the pairing of nuclei, the ascogonium produces a number of ascogenous hyphae. The peridium is single layered and yellow in colour. The asci are evanescent and are globose, ovoid or pear-shaped. The ascospores are hyaline, broadly lenticular without obvious surface ornamentations.

Emericella **:** In Emericella phialides are not borne directly on the vesicles, but on a series of cylindrical cells, called metulae.

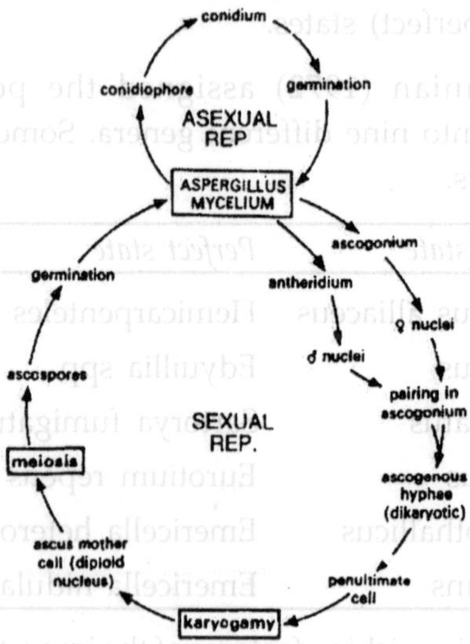

Aspergllus : life-cycle.

No sex organs are formed. The ascocarp originates from a loose hyphal coil. The peridium is multilayered and consists of chains of very thick-walled cells, the Hulle cells. The ascospores are red and provided with a double equatorial flange. The best known species are E. nidulans and E. heterothallica.

Sartorya **:** In Sartorya the ascocarp originates from the ascogonium alone; antheridium is not formed. The peridium consists of several layers of interwoven hyphae and is enveloped at maturity by a loose network of sterile hyphae which gives the ascocarp the appearance of a cottony ball. The best known species is S. fumigatus.

Hemicarpenteles : In Hemicarpenteles cleistothecia are formed within a sclerotial stroma.

Edyuillia : It is perfect state of Aspergillus athecius. In Edyulla asci develop in naked clusters and are not surrounded by investing hyphae.

Penicillium

Systematic Position

Kingdom	—	Mycota
Division	—	Eumycota
Sub-division	—	Ascomycotina
Class	—	Plectomycetes
Order	—	Eurotiales (Aspergillales)
Family	—	Eurotiaceae (Aspergillaceae)

Penicillium, commonly known as blue or green mold, has over 100 species which are distributed all over the world. The genus derives its name from the Latin word penicillus, means artist's brush, as the structure of its conidiophore to that of a artist's brush.

The species of Penicillium are saprophytes and are found growing on rotting fruits, vegetables, meat, and, other food stuffs, and on textiles, leather, paper, etc. Most of the species are harmful, *e.g.*, P. divaricatum causes decay of wood whereas P. chrysogenum, P. rubrum and P. purpurogenum are responsible for the spoilage of paper and paper products. P. citrinum is one of the common causes of fungus 'fouling' of optical instruments. P. digitatum, P. italicum and P. chrysogenum are the source of the antibiotic penicillin, the first 'wonder drug' of the world; the antibiotic griseofulvin is obtained from P. griseofulvum. P. rocqueforti and P. camembertile are used in the hydrolysis of fats and for flavouring of cheese. P. chrysogenum and P. vitale are used as a source of the enzyme glucose oxidase, which is used in ferment technology and food industry.

Vegetative Structure : The mycelium is profusely branched, and septate and is composed of thin-walled, uni- to multinucleate hyphal cells. There is a central pore in each septum through which the cytoplasm of adjacent cells remains continuous. The mycelium usually grows superficially, but some hyphae may penetrate deeper into the substratum to absorb food materials. The hyphae are hyaline and colourless in vegetative phase but their walls develop pigments in the reproductive phase. The reserve food is in the form of oil globules.

In some species of Penicillium, the mycelium becomes heterokaryotic as a result of anastomosis between two mycelia. In others, compact, discrete, thick-walled pseudo-parenchymatous structures, known as sclerotia, are formed.

Reproduction : Penicillium reproduces by vegetative, asexual and sexual means.

Vegetative Reproduction : It takes place by fragmentation. The mycelium breaks into small fragments by injury or any other reason. Each fragment is capable to develop into a new mycelium.

Asexual Reproduction : Asexual reproduction takes place by the formation of non-motile spores, known as conidia. The conidia are formed in long basipetal chains on specialised branches, called conidiophores.

The conidiophore arises as an erect branch from any cell of the mycelium. Rarely, conidiophores are unbranched as in P. spinulosum and P. thomii. But usually they divide one, two or more times forming the branches of different orders (primary, secondary, tertiary, etc.). The ultimate branches are known as metulae. A cluster of flask-shaped sterigmata (phialides) develops at the tip of each metulae. But when conidiophores are unbranched, sterigmata develop at the tips of the conidiophores.

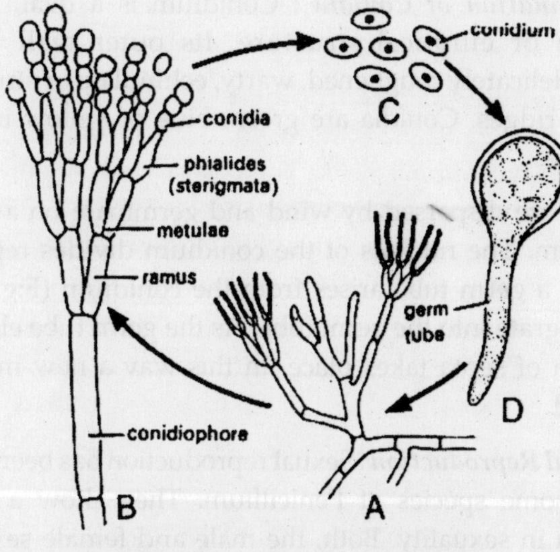

A-D. Penicillium : Asexual reproduction.

In some species of Penicillium (*e.g.*, P. expansum, the metulae are also formed on the small branches of conidiophores which are known as rami (singular = ramus). In P. claviforme, many conidiophores aggregate to form a compound club-shaped fructification, called coremium. The conidia formed on the coremium are known as coremiospores.

Development of Conidia : The tip of the sterigma swells a little and at the same time its nucleus divides to form two daughter nuclei. One of the daughter nuclei migrates into the swollen tip of the sterigma and then this portion separates from rest of the sterigma by the formation of a septum. In this way a uninucleate conidium is formed. The tip of the sterigma below the first formed conidium elongates, the nucleus divides and the second conidium is formed in the same way as the first. This process of conidium formation is repeated several times and as such a chain of conidia is formed. The youngest conidium is at the base of the chain and the oldest at the tip. This basipetal arrangement helps in the dissemination of conidia and provides proper nutrition to the developing conidia.

Germination of Conidia : Conidium is a oval, globose, pyriform or elliptical structure. Its outer wall may be smooth, delicately roughened, warty, echinulate or striate with irregular ridges. Conidia are green, blue or yellow in colour or hyaline.

They are dispersed by wind and germinate on a suitable substratum. The nucleus of the conidium divides repeatedly and then a germ tube arises from the conidium (Fig. D). All nuclei migrate into the germ tube. As the germ tube elongates, formation of septa takes place. In this way a new mycelium is formed.

Sexual Reproduction : Sexual reproduction has been studied only in some species of Penicillium. They show a gradual reduction in sexuality. Both, the male and female sex organs are functional in most of the species.

But in some (*e.g.*, P. vermiculatum) although both the sex organs are present, the antheridium does not take part in sexual process. The nuclei of the ascogonium only fuse in pair to form ascogenous hyphae. In others (*e.g.*, P. brefeldianum, P. stivitatum), sex organs do not develop at all and these species show the phenomenon of somatogamy, where fusion of morphologically similar hyphae takes place.

The following account deals with the process of sexual reproduction in P. vermiculatum. The male and female sex organs are known as antheridium and ascogonium respectively. Both develop on the same hypha or on two adjacent hyphae of the same mycelium.

(a) *Ascogonium* : Ascogonium develops as an erect unicellular branch from any cell of the mycelium. The young ascogonium is uninucleate, but the nucleus undergoes repeated mitotic divisions as the ascogonium matures. Thus, a mature ascogonium has 32 or 64 daughter nuclei and it remains unicellular.

(b) *Antheridium* : Following the initiation of ascogonium, an antheridial branch also develops from the same hypha on which ascogonium is initiated. As the antheridium grows, it coils spirally around the ascogonium.

(c) *Fertilization* : The tip of the antheridial branch touches the ascogonium and the intervening walls at the point of contact dissolve. The nucleus does not migrate into the ascogonium, probably it degenerates. But there is mixing of cytoplasm of the antheridium and the ascogonium. This is followed by the segmentation of the ascogonium. The binucleate cells thus formed show uniseriate arrangement. Some of these cells give rise to ascogenous hyphae which are multicellular with dikaryotic cells.

The terminal cells of ascogenous hyphae function as ascus mother cells. The two nuclei of the dikaryon present in the ascus mother cell fuse to form a diploid nucleus. The diploid nucleus of the young ascus first undergoes a meiotic division, followed by a mitotic division. This results in the formation of eight haploid nuclei. Each nucleus gets surrounded by cytoplasm and develops into an ascospore. Ascospores are globose, elliptical or lenticular in shape. The outer wall of the ascospore may be smooth, pitted, echinulate or banded. Out of the eight ascospores formed in an ascus, four are of one strain and the four of the another strain.

Development of Ascocarp

Simultaneously with the development of ascogonium and ascogenous hyphae, some sterile hyphae develop around the ascogonium and they form a multilayered protective covering, the peridium. The ascogonium and ascogenous hyphae surrounded by the peridium form a fruiting body, known as ascocarp. The fruiting body is a completely closed structure of cleistothecium type. There are many asci which are irregularly placed within the ascocarp.

At maturity, the ascus wall dissolves and ascospores float in the nutritive fluid present in the cleistothecium. Ascospores are liberated by the decay of the cleistothecial wall. They germinate on a suitable substratum and form new mycelia.

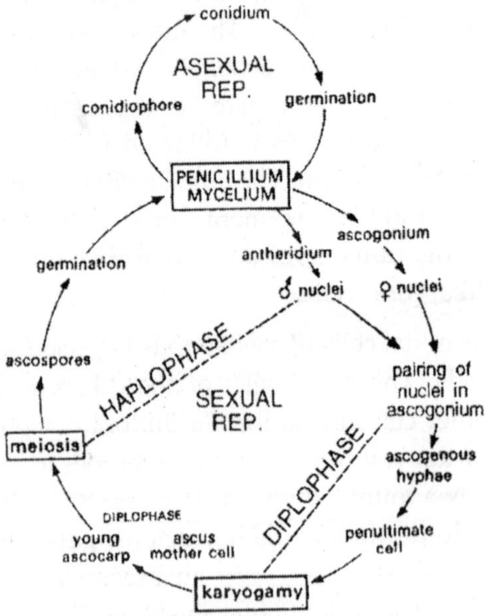

Penicillium : Life-cycle.

Perfect States of Penicillium

The nomenclature of Penicillium has been debated as ascocarps are rarely found. The perfect (sexual) state of Penicillium has been assigned to two different genera, *viz.*, Talaromyces and Eupenicillium. Some distinguishing features of the perfect states of Penicillium are as follows.

Eupenicillium

(1) Conidiophores are relatively simple.

(2) The ascocarp grows to a definite size.

(3) The cleistothecial wall is sclerotoid and is made up of thick walled pseudoparenchyma.

(4) Asci usually develop on short branches of ascogenous hyphae.

The important species are Eupenicillium javanicum and E. brefeldianum.

Talaromyces

(1) Conidiophores have tapering phialides.

(2) The ascocarp grows indefinitely.

(3) The cleistothecial wall is loose and soft and is made up of thin walled interwoven hyphae.

(4) Asci usually develop from croziers.

The important species are Talaromyces venniculatus (Penicillium venniculatum) and T. stipitatus (P. stipitatum).

15

The Saccharomyces

Ascomycotina, the largest sub-division of higher fungi, is represented by 1,700 genera and 15,000 species. They are popularly known as sac-fungi as the name of the group suggests (Greek Ascos = bladder, mykes = fungi). They are mostly terrestrial saprophytes or parasites, but some species are coprophilous. This group includes those fungi where zygospores are absent and the perfect state spores are ascospores.

Most commonly known Ascomycotina are yeasts, black or green molds, powdery mildews and morels. The yeast is commercially used in brewing and bakery. Several parasitic taxa cause destructive diseases in plants, such as ear rot, foot rot, brown rot, leaf curl and powdery mildew. Species of Aspergillus cause aspergillosis in human beings. Several species of Penicillium are' important source of penicillin and fruiting bodies of Morchella are used as delicacy.

General Characters of Ascomycotina : The following are important characteristic features of the Sub-division Ascomycotina.

(1) The members vary in their form and structure. Yeasts are unicellular, but mostly they have a well developed, profusely branched and septate mycelium with uni-or multinucleate cells and perforated septa.

(2) In unicellular forms, the cell wall is composed of glucans and mannans, whereas in septate forms it consists of chitin and glucans.

(3) Asexual reproduction takes place by various types of non-motile spores, such as oidia, chlamydospores and conidia. In unicellular forms, fission, fragmentation and budding are the most common methods of vegetative propagation.

(4) They are homothallic or heterothallic. In some heterothallic species, though male (antheridium) and female (ascogonium) sex organs develop on the thallus of the same strain they are self incompatible., In these species male gamete of one mating type fertilises ascogonium of other mating type. This process is known as physiological heterothallism.

(5) Sexual reproduction takes place by gametangial copulation (*e.g.*, yeast), gametangial contact (*e.g.*, Aspergillus, Penicillium, Erysiphe), somatogamy (*e.g.*, Morchella) or spermatization (*e.g.*, Polystigma).

(6) They show the phenomenon of heterokaryosis *i.e.*, the nuclei of two different genotypes are present in the same mycelium.

(7) The fruiting bodies are known as ascocarp. The following four types of fruiting bodies are found:

(i) *Apothecium* : It is a broad plate-like fruiting body in which asci develop on the surface of the hymenium and remain exposed at maturity, *e.g.*, Peziza.

(ii) *Perithecium* : It is a flask- shaped structure in which asci develop from the inner surface of the fruiting body. It has an apical pore, known as ostiole. Hair like sterile paraphysis are present on the neck and mouth of the fruiting body, *e.g.*, Claviceps.

(iii) *Cleistothecium* : It is globose and completely closed fruiting body in which asci are scattered, *e.g.*, Phyllactinia, Aspergillus, Penicillium.

(iv) *Ascostroma* : It is a pseudoparenchymatous structure composed of somatic hyphae. The fruiting body lacks a distinct wall and asci are usually bitunicate.

(8) The ascospores are endogenous in origin.

Degeneration of Sex in Ascomycotina : The higher members of Ascomycotina show gradual degeneration of sex. Like other living organisms, sexual reproduction in Ascomycotina aims at the union of two compatible nuclei. These nuclei are brought together in the same cell by one of the many methods developed by the Ascomycotina during their evolution. In Ascomycotina, the two nuclei remain in close association and undergo successive divisions which usually result in a number of dikaryotic cells. Nuclear fusion eventually takes place in the ascus mother cell, which develops into the ascus.

In lower Ascomycotina there is a functional and morphological differentiation among sex organs. The male and female sex organs are respectively called as antheridium and ascogonium. In Eurotium, the sex organs, antheridia and ascogonia, are produced close to each other on somatic hyphae. Both are multinucleate, elongated structures. In Sartorya, antheridia are not formed and the ascocarp is initiated by a coiled ascogonium and in Emericella, sex organs are completely lacking and the ascocarp originates from a loose hyphal coil. In Penicillium too, sexual reproduction appears to take place by functional gametangia in some species, but in others (*e.g.*, Talaromyces) the antheridia no longer seem to function.

In such species the ascogonial nuclei pair and pass into the ascogenous hyphae. Some of the higher Ascomycotina produce ascogonia but no antheridia. Instead in them the conidia- like male sex cells called spermatia are developed in flask-shaped cavities known as the spermagonia. In Neurospora, conidia produced at the tips of the conidiophores act as spermatia. When mature the spermatia are detached and carried by some external agency to the ascogonia where they get attached to the receptive organs such as the trichogyne or other receptive hypha.

However, in higher Ascomycotina (*e.g.*, Helvella, Morchella, Peziza, etc.), definite sex organs, *i.e.*, antheridium and ascogonium, are not formed. The sexual reproduction takes place by somatogamous copulation, *i.e.*, fusion between somatic hyphae of two compatible mycelia takes place.

Classification

On the basis of the presence or absence of fruiting bodies the group is usually divided into the following two sub-classes:

Sub-class 1 : Hemiascomycetes. Ascocarps are absent and asci are naked. This sub-class has been divided into three orders, Protomycetales, Endomycetales and Taphrinales.

Sub-class 2 : Euascomycetes. Asci develop in the fruiting bodies. This sub-class has been divided into the following five series.

Series 1 : Plectomycetes. Asci evanescent, scattered within cleistothecium; ascospores aseptate. Two orders, Eurotiales and Erysiphales are included in this series.

Series 2 : Pyrenomycetes. Ascocarp is typically a perithecium; asci inoperculate with an apical pore. It includes two orders, Hypocreales and Sphaeriales.

Series 3 : Loculascomycetes. Asci bitunicate; ascocarp an ascostroma. It includes four orders, Myriangiales, Dothidiales, Hemisphaeriales and Hysteriales.

Series 4 : Laboulbenimycetes. Ascocarp a perithecium, asci inoperculate; thallus reduced; exoparasites of arthropods. It includes a single order, Laboulbeniales.

Series 5 : Discomycetes. Ascocarp is an apothecium. Three orders, Pezizales, Helotiales and Tibberales, are included in this series.

However, Ainsworth (1973) divided the Sub-division Ascomycotina into six classes - Hemiascomycetes, Loculascomycetes, Plectomycetes, Laboulbeniomycetes, Pyrenomycetes and Discomycetes.

Class Hemiascomycetes

Majority of the members are saprophytes, whereas a few are parasites on vascular plants, human-beings and animals, causing diseases. The mycelium is either poorly developed or completely absent, and the majority of the members are yeast-like. Asexual reproduction takes place by budding or fission. Sexual reproduction occurs by gametangial copulation. Ascogenous hyphae and ascocarps absent, *i.e.*, asci are naked.

Martin (1961) included three orders—Protomycetales, Endomycetales and Taphrinales in Hemiascomycetes.

Order Endomycetales

The members of this order are mostly unicellular, but a few like Eremascus are mycelial. The cell wall is composed of glucans and mannans. The asexual reproduction takes place by fission, budding and arthrospore formation. Budding in succession in some yeasts results in the formation of pseudomycelium. The sexual reproduction takes place by the fusion of two somatic cells (which behave as gametangia) or two ascospores. Dikaryotic phase is absent. Ascocarps are also absent and thus asci are naked.

Kreger-van Rij (1973) included four families in the Endomycetales - (i) Ascoideaceae, (ii) Endomycetaceae, (iii) Spermophthoraceae, and (iv) Saccharomyctaccae.

Family Saccharomycetaceae

(1) The members are mostly unicellular.

(2) Vegetative reproduction takes place by budding or fission.

(3) Ascospores are globose, oval, ellipsoid or sickle-shaped, but never needle or spindle-shaped.

Saccharomyces

Systematic Position

Kingdom	—	Mycota
Division	—	Eumycota
Sub-division	—	Ascomycotina
Class	—	Hemiascomycetes
Order	—	Endomycetales
Family	—	Saccharomycetaceae

Saccharomyces, commonly known as yeasts, is represented by some 40 saprophytic species found ubiquitously. They are mostly found on media rich in sugar or an organic matter of vegetable origin.

Yeasts are of considerable economic importance. S. cerevisiae (commonly known as baker's yeast or brewer yeast) is widely used in brewing and baking industries. Besides, yeast cakes are used as food on account of their protein and vitamin contents. S. cerevisiae is also used for the extraction of the enzymes invertase and zymase. The high fermentative property of yeasts which make them useful also make them destructive. They easily destroy soft cheese and similar foods, for they impart repulsive yeasty flavour. A number of asporogenous yeasts are pathogenic to man, causing effective infection.

Vegetative Structure : The plant body is unicellular but sometimes cells remain attached in short chains, forming a pseudomycelium. The cells are extremely variable in shape being globose, oval, elliptical, elongated or rectangular. The size of the cells ranges between 6 and 8 μm x 5 and 6 μm. Individually yeast cells appear hyaline, but in colonies they appear white, cream coloured or tinged with brownish pigments. Sometimes, the colonies are brilliantly coloured.

The yeast cell is surrounded by a distinct thin and firm cell wall, chiefly composed of chitin. Besides chitin, the cell wall also contains polysaccharides (glycogen and mannan), lipids, proteins and phosphates; but cellulose is absent. Electron microscopic studies reveal that the cell wall consists of two layers; the outer electron dense layer of about 0.2 μm thick and the inner microfibrillar layer of about 0.2μm thick. But in some species (*e.g.*, S, cerevisiae) the cell wall is relatively thick, consisting of three layers - the outer layer mainly composed of mannan-protein and some chitin; the middle layer of glucan and the innermost layer of protein-glucan.

Inner to the cell wall a plasma membrane (plasmalemma) is present which has a series of shallow, elongated pits or invaginations. The cell is rich in granular cytoplasm. The vacuolated cytoplasm contains oil drops, glycogen bodies, volutin granules, besides several organelles like endoplasmic reticulum, ribosomes, mitochondria, and Golgi apparatus. The centre of the mature cell is occupied by a large vacuole, limited by a single vacuolar membrane, the tonoplast. The vacuole contains a watery substance and granules of polymetaphosphate and lipid. The vacuolar sap has several hydrolytic enzymes like proteases, ribonucleases, esterases, etc. A minute nucleus is present which is attached to the vacuole.

There is much controversy regarding the structure of the nucleus. According to Wager and Peniston (1910), the central vacuole is in reality a vacuolated nucleus and the body at its one side is the nucleolus. The nuclear vacuole contains a net work of chromatin threads. But Alexopoulos (1958) was of the opinion that in the centre of the vacuolated nucleus (nuclear vacuole) lies a small nucleolus. The structure connected to vacuole is centrosome, which remains surrounded by centrioles and heterochromatin. However, electron microscopic studies reveal that the nucleus and vacuole are two separate entities. The nucleus is surrounded by a definite nuclear membrane and it has no contact with the vacuole.

The mitotic division is intranuclear, *i.e.*, the nuclear membrane remains intact during the division. At the time of division, a pair of spindle-pole bodies develop on opposite sides. Between these two spindle pole bodies there is an intranuclear spindle made up of microtubules. The spindle pole bodies separate at opposite poles and develop into two daughter nuclei. The nuclear mass, however, remains surrounded by the nuclear envelope.

Reproduction : Yeast reproduces both by vegetative and sexual means.

Vegetative Reproduction : Vegetative reproduction takes place by the following methods.

By Budding : It is the most common method of reproduction under favourable conditions. The protoplast of yeast cell bulges out in the form of a bud. The nucleus divides into two daughter nuclei and of these one migrates into the enlarging bud. The bud grows and is eventually separated from the mother cell.

As the bud separates a scar with a convex- surface is left on the parent cell. This scar is called bud scar. A concave scar is also retained by the newly formed daughter cell.

It is called birth scar. This process of the formation of new buds from the parent cell continues and each time a bud scar is left. Sometimes, a large number of buds develop without being detached from one another and as the result chain of cells is formed. Such chains may be branched or unbranched and give the appearance of a false mycelium, called pseudomycelium. The cells of the pseudomycelium are loosely attached and they finally detach.

By Fission : Fission is simple splitting of' a cell into two daughter cells by the constriction and formation of a transverse wall. The parent cell elongates and its nucleus divides by intranuclear mitosis. Then a constriction appears somewhere near the middle of the mother cell, followed by a transverse septum, and as such two uninucleate daughter cells are formed.

By Endospore Formation : During unfavourable conditions, thick walled endospores are formed. The protoplast usually divides into four parts; each part is surrounded by a thick wall. These structures are known as endospores. The endospore may survive adverse conditions and on the return of favourable conditions it germinates by budding and produces a chain of cells.

Sexual Reproduction : Sexual reproduction usually takes place during unfavourable conditions when the food supply is exhausted. Sex organs are entirely absent and as such the sexual reproduction is a simple process. In this process either two somatic cells or two ascospores are involved which assume the function of copulating gametangia. The zygote formed by their fusion eventually develops into an ascus which contains 4 or 8 ascospores depending upon the number of nuclear divisions.

Three types of life cycle patterns have been recognised in yeasts.

Haplobiontic Type : This type of life cycle is found in S. octosporus. Here the haplophase is very elaborate and the diplophase is very short, confined to the zygote cell only. The somatic cells are haploid and function as a potential gametangia. At the time of sexual reproduction, two somatic cells meet in pairs, each sending a small protuberance towards the other. Both the protuberances come in contact and the wall at the point of contact dissolves to form a common passage, called conjugation tube. Thereafter, the nuclei of both the gametangia move into the conjugation tube; they fuse there and form a diploid zygotic nucleus. The zygote functions as an ascus. It undergoes meiosis immediately after karyogamy and thus eight haploid nuclei are formed. These nuclei organize themselves into ascospores. The ascospores are liberated by breaking of the ascus wall. The ascospore behaves like a somatic cell.

Diplobiontic Type : This type of life cycle is represented by S. ludwigii. In this type, the diploid somatic stage is long and the haplophase, represented by ascospores, is very short. Only four ascospores are formed in an ascus, and they are not liberated from the ascus. The ascospores behave as gametangia and they copulate in pairs and each pair by fusion produces

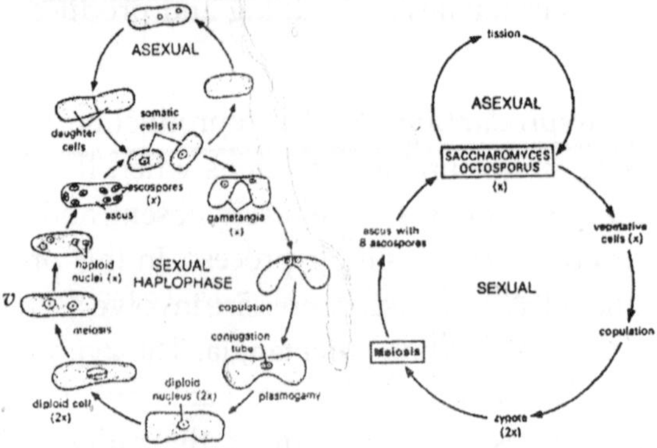

Saccharomyces octosporus : Life cycle.

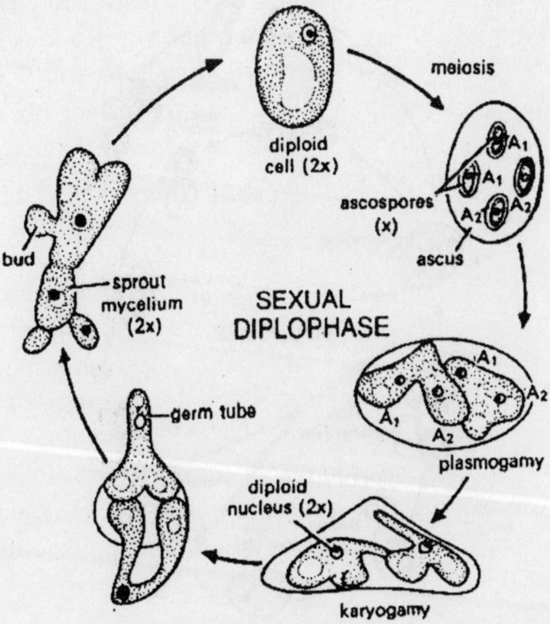

Saccharomyces ludwigii : Life-cycle.

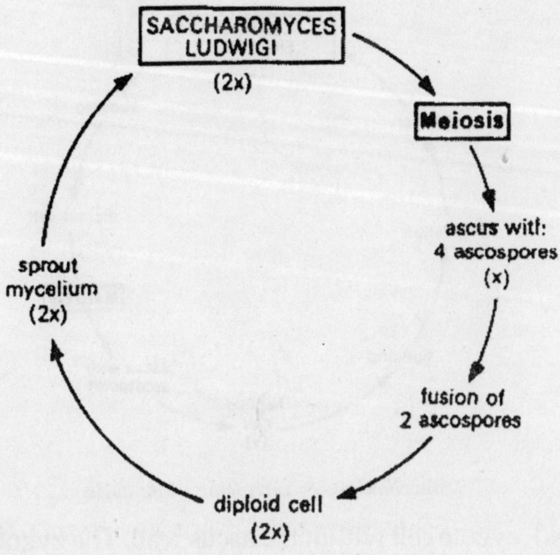

Saccharomyces: cerevisiac : Life-cycle.

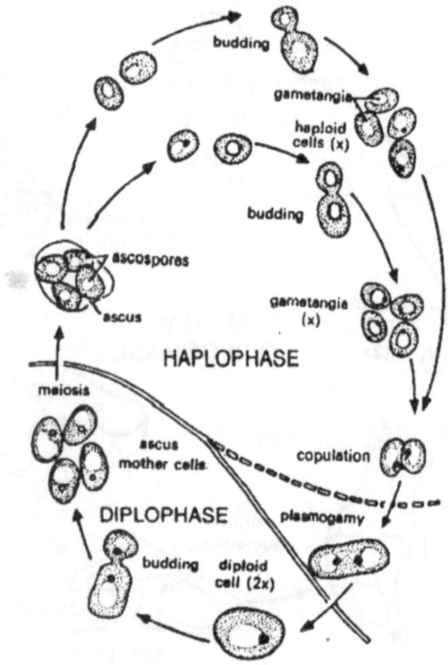

Saccharomyces ludwigii : Life-cycle.

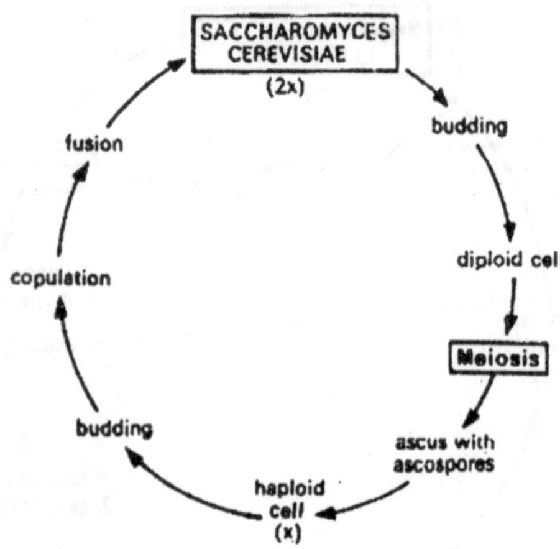

Saccharomyces cerevisiae : Life-cycle.

a diploid zygote cell within the ascus wall. The zygotic diploid cell produces a germ tube that pushes through the ascus wall.

The germ tube becomes multicellular and functions as a diploid sprout mycelium. The new diploid yeast cells are budded from it. The diploid buds are detached from the parent sprout mycelium and function as diploid sprout cells. Under favourable conditions, these cells function as asci; each producing four ascospores as the result of a reduction division.

Haplo-diplobiontic Type : This type of life cycle occurs in S. cerevisiae. Both, the haploid and diploid phases are of pretty long duration representing somewhat an alternation of generations. Out of the four ascospores forned in an ascus, two are of one mating type and two of the other mating type. Normally the haploid cells of different mating types multiply by budding. But under certain environmental conditions, two somatic cells of opposite mating types behave as gametangia. The zygote formed by the fusion of two gametangia propagates by budding, producing a large number of diploid cells. When there is scarcity of food, the diploid cells function as ascus. Four haploid ascospores are formed in each ascus by a reduction division. On liberation the ascospore forms many haploid somatic cells by vigorous budding.

Zygomycetes Mucorales

The class Zygomycetes includes those members in which the resting spore (zygospore) develops by the fusion of two gametangia. They do not have motile cells (zoospores) in any stage of their life-cycle, Some important features of the class are as follows:

(1) The members are saprobes or weak parasites on plants to specialized parasites on animals. A few occur on dung, thus coprophilous in nature.

(2) The plant body consists of well developed, branched, filamentous and coenocytic mycelium.

(3) The cell wall is mainly composed of chitosan-chitin.

(4) Asexual reproduction takes place usually by means of non-motile sporangiospores, called aplanospores, but some members also reproduce by chlamydospores or by oidia formation.

(5) Sexual reproduction takes place by means of gametangial copulation, resulting in the formation of thick-walled zygospore.

(6) The zygospore germinates by producing a germ sporangiophore that terminally bears a germ sporangium.

Hesselting and Ellis (1973) divided Zygomycetes into three orders — Mucorales, Entomophthorales and Zoopagales. However, Alexopoulos and Mims (1979) recognised six orders —Mucorales, Dimargaritales, Kickxellales, Endogonales, Entomophthorales and Zoopagales.

Benjamin (1979) added another order Harpellales to the list of the orders proposed by Alexopoulos and Mims (1979). Webster (1980) recognised only two orders, Mucorales and Entomophthorales in the class.

Order Mucorales

The order Mucorales is represented by about 45 genera and 250 species, which are saprophytes, growing on dead decaying organic matter.

Some species are coprophilous and still others are weak parasites on higher plants and animals. A few members of the order are also obligate parasites on other Mucorales and mushrooms. The mycelium consists of well developed, coenocytic hyphae.

Septa, however, develop during the formation of reproductive structures. The cell wall is composed of chitin and chitosan. Asexual reproduction takes place by non-motile sporangiospores (aplanospores) produced in very large number in sporangia. Some Mucorales also produce chlamydospores and oidia. Sexual reproduction takes place by the fusion of two multinucleate gametangia of usually equal size. The gametangial fusion results in the formation of a thick-walled resting spore, called zygospore. The wall of the zygospore is heavily pigmented and is from light yellow to black in colour. The zygospore germinates by producing a germ sporangiophore which terminally bears a sporangium. In the absence of gametangial copulation, zygospores develop.

Martin (1961) recognised nine families within the order Mucorales - Mucoraceae, Pilobolaceae, Thamnidiaceae, Choanephoraceae, Piptocephalidaceae, Kickxellaceae, Cunninghamellaceae, Mortierellaceae and Endogonaceae. Hesseltine and Ellis (1973), however, included as many as 14 families within the order, adding Dimargaritaceae, Helicocephalidaceae, Radiomycetaceae, Saksenaceae, and Syncephalastraceae to the Maftin's list.

Family Mucoraceae

Some important features of the family Mucoraceae are as follows:

(1) Most of the members of the family are saprobic but a few are weak parasites or coprophilous.

(2) The mycelium is well developed, branched and coenocytic.

(3) The cell wall is largely composed of chitin-chitosan, besides some polysaccharides like glucosamine and galactose and proteins and lipids.

(4) Asexual reproduction takes place with the help of non-motile spores.

(5) Sporangia are globular or pyriform and columellate.

(6) Sexual reproduction occurs by gametangial copulation.

Rhizopus

Systematic Position

Kingdom	—	Mycota
Division	—	Eumycota
Sub-division	—	Zygomycotina
Class	—	Zygomycetes
Order	—	Mucorales
Family	—	Mucoraceae

The genus Rhizopus is represented by some 50 species which grow as saprophytes on a wide variety of organic substrates like bread, jellies, syrups, leather, etc. Rhizopus stolonifer, commonly known as bread mould, is the most widely known species. Some species are weak parasites; for instance, R. artocarpi causes fruit drop of jack fruit (Artocarpus heterophyllus) and R. arrhizus causes fruit rot of apples. Sometimes, R. stolonifer also acts as parasite, causing soft rot of sweet potato and 'leak' disease of strawberry and tomato. Besides, some species of Rhizopus cause 'mucormycosis' in domestic animals and still others attack the internal nervous system in human beings with fatal consequences.

Several species of Rhizopus are capable of synthesizing some important industrial products- R. stolonifer is used for the manufacture of fumaric acid and cortisone, R. oryzae in alcoholic fermentation and R. nodosus in lactic acid fermentation. Some species are also used in the production of some popular oriental foods like Tempeh, made from soyabeans.

Vegetative Structure : The mycelium, which appears as white cottony growth, is coenocytic and profusely branched. Three types of hyphae recognised in the mycelium are : stolons, rhizoids and sporangiophores. Stolons are those hyphae which grow horizontally above the substratum and are comparatively stout and less branched. The stolons develop tufts of rhizoids at certain points of contact with the substratum. Rhizoids are repeatedly branched hyphae which penetrate the substratum and absorb water and nourishment from it. Sporangiophores are erect, aerial and unbranched hyphae which grow upwards in tufts from the stolons exactly opposite to the rhizoids. The sporangiophores bear terminal sporangia and are thus reproductive in function.

The hyphae are tubular structures. The hyphal wall is microfibrillar and complex and consists of chitin-chitosan. Besides, some other polysaccharides such as glucosamine and galactose, proteins, purines, lipids, magnesium and calcium are also present in the cell wall. The cell wall is lined internally by a thin plasma membrane. The protoplast is granular and contains many nuclei, ribosomes, mitochondria, endoplasmic reticulum, vacuoles, oil droplets and glycogen bodies. However, in older hyphae the smaller vacuoles coalesce to form a large central vacuole. The cytoplasm often shows rapid streaming. The nuclei are irregular in shape and divide by constriction.

Reproduction : Rhizopus reproduces by vegetative, asexual and sexual methods.

Vegetative Reproduction : Vegetative reproduction takes place by fragmentation. The vegetative hyphae (stolons) may break up into smaller fragments. Each fragment is capable of developing into new mycelium.

Asexual Reproduction : Asexual reproduction takes place by the formation of aplanospores (sporangiospores) or by chlamydospores.

Aplanospore Formation : During favourable conditions, it is the most common method of asexual reproduction. The multinucleate non-motile spores, known as aplanospores or sporangiospores, are formed inside the round black sporangia, which occur singly at the tips of sporangiophores. Sporangiophores develop in tufts from the stolon opposite the rhizoids. At the time of sporangium formation, the tip of the sporangiophore swells into a knob-like vesicle. The cytoplasm, alongwith many nuclei, flows from the sporangiophore into the swollen vesicle. Ultimately, the swollen tip develops into a large globose structure, the young sporangium.

The contents of the young sporangium soon differentiate into a peripheral dense multinucleate region and a central region with small flattened vacuoles and fewer nuclei. Thereafter, a cleft is formed between the two regions by the coalescence of vacuoles. Finally, a dome-shaped septum is laid down, separating the two regions completely. The central vacuolated region is sterile and is known as columella, whereas the outer fertile sporangiferous region forms spores. The columella remains in continuity with the protoplast of the sporangiophore. Cleavage in the peripheral sporangiferous zone results in the formation of several multinucleate segments. Each segment ultimately transforms into a globose, multinucleate, non-motile aplanospore.

At maturity, the sporangium wall dries and the columella collapses like an inverted cup-like dish. Ultimately, the sporangial wall breaks, liberating spores to the atmosphere. The remnants of the sporangium wall can be seen as a frill at the base of the columella.

When the spore falls on a suitable substratum, it germinates by producing a germ tube. The germ tube ultimately develops into a fluffy and profusely branched mycelium.

Chlamydospore Formation : During unfavourable conditions, asexual reproduction takes place by means of chlamydospores. At the time of chlamydospore formation, the mature mycelium becomes septate and the protoplast of each cell forms a rounded and thick walled structure, known as chlamydospore. The chlamydospores are perennating bodies and they can pass unfavourable conditions. They germinate and form new mycelia.

Sexual Reproduction : Sexual reproduction takes place by the copulation of two morphologically similar multinucleate gametangia. Most of the species of Rhizopus are heterothallic, but a few (*e.g.*, R. sexualis) are homothallic.

In heterothallic species, zygospores are formed only when two mycelia of compatible strains come in contact with each other. But in homothallic species, zygospores may develop by the mycelia derived from a single spore.

During sexual reproduction, gametangia are produced as terminal swellings on the tips of two compatible hyphae or hyphal branches. These special hyphae are known as progametangia. The progametangia of opposite strains adhere by their tips and enlarge by accumulating cytoplasm. A septum is then formed a little below the tip of the progametangium, separating the terminal gametangium from a proximal suspensor cell.

The gametangium has densely granular multinucleate protoplast, whereas the suspensor has a highly vacuolated protoplast. The walls of the two gametangia at the point of their contact dissolve and the protoplasts of both gametangia unite to form a zygospore. The nuclei of opposite strains pair and fuse to form diploid nuclei in the combined protoplast. The nuclei, which fail to fuse in pairs, ultimately degenerate. Soon, the young zygospore enlarges and secretes several layered thick wall around it.

The zygospore germinates after a long period of rest. During germination, the zygospore cracks open and a germ sporangiophore emerges which develops a germ sporangium at its tip. Reduction division occurs during the formation of germ sporangium and the sporangium contains numerous haploid spores. Each spore, after liberation, germinates to form a new mycelium (Fig. F, H).

Occasionally, copulation does not take place between

gametangia and these gametangia are surrounded by a many layered wall and then develop in zygospores.

Another member of the family Mucoraceae, Mucor, resembles with Rhizopus except for some minor differences. These differences are listed below:

Differences between Rhizopus and Mucor

Rhizopus	*Mucor*
1. Mycelium is differentiated into three kinds of hyphae, stolons, rhizoids and sporanglophores.	Mycelium is undifferentiated, *i.e.*, it has only one kind of hyphae.
2. Rhizoids present.	Rhizoids absent.
3. Food material is absorbed mainly by rhizoids.	Food is absorbed by the entire mycelial surface.
4. Sporangiophores occur in tufts from the stolons opposite the rhizoids.	Sporangiophores usually occur singly from any point on the mycelium.
5. Spores easily disseminated by wind.	Spores remain adhered to columella and thus are not easily disseminated.
6. Reduction division occurs during zygospore germination after a period of rest.	Reduction division occurs soon after karyogamy before the zygospore undergoes a period of rest.

16

The Peronosporales

Order Peronosporales

The order Peronosporales includes 13 genera and about 325 species. The members of this order are more advanced than those of Saprolegniales. They are mostly aquatic or terrestrial obligate parasites, but some genera like Pythium are facultative parasites. They cause serious diseases in higher plants, such as white rust, downy mildew, late blight, fruit rot, damping off and seedling blights.

Some genera can also survive saprophytically on dead and decaying plant parts and a few in soil. The mycelium is delicate, well branched, aseptate, coenocytic and inter- or intracellular.

It absorbs food materials from the host cells with the help of knob-like or rhizoid-like haustoria. The cell wall is fibrillar and is composed of polysaccharide, lipid and protein. The members reproduce asexually by zoospore or conidia formation. The zoospores are kidney-shaped, monoplanetic and laterally biflagellate. The posterior flagellum is of whiplash type and the anterior of tinsel type. Sexual reproduction is oogamous. The oogonium characteristically shows differentiation into ooplasm and periplasm and contains a single egg.

The antheridium is club-shaped and usually develops just below the oogonium. Plasmogamy is by gametangial contact. The oospore germinates by the formation of a germ tube or it produces a large number of zoospores, whereas Alexopoulos and Mims (1979) recognised four families :

Pythiaceae, Albuginaceae, Peronosporaceae and Peronophythoraceae on the basis ot sporangiophore characters.

Waterhouse (1973) divided the order Peronosporales into three families, Pythiaceae, Peronosporaceae and Albuginaceae.

Phytophthora and Pythium (Pythiaceae) and Albugo (Albuginaceae) are described here In detail as the representatives of the order.

Family Pythiaceae

(1) The members are aquatic, amphibious or terrestrial. They are mostly parasites, but a few facultative parasites occur as saprophytes.

(2) The mycelium is branched, aseptate and coenocytic, and it may be inter- or intracellular. The intercellular mycelium has haustoria.

(3) Asexual reproduction takes place by zoospores, produced in conidiosporangium.

(4) Sporangiophores or conidiophores are usually undifferentiated from the mycelium. They are branched indeterminate structures and resume growth after the production of sporangium or conidium.

(5) Sexual reproduction is oogamous; oogonial periplasm is either absent or forms only a thin layer.

Phytophthora

Systematic Position

Kingdom	—	Mycota
Division	—	Eumycota
Sub-division	—	Mastigomycotina
Class	—	Oomycetes
Order	—	Peronosporales
Family	—	Pythiaceae

The genus Phytophthora is represented by some 40 species which are worldwide in distribution. The majority of the species are parasitic, but a few are facultative saprophytes. It is one of the most destructive genus, hence named Phytophthora (Greek : phyton = plant, phthora = destruction). The most well known species of the genus is P. infestans, which causes late blight disease of potato. Low temperature and humid atmosphere favour the spread of the disease. The disease is widely spread in the hilly areas of India during rainy season. Some other important pathogenic species of Phytophthora and the diseases caused by them are as follows:

(1)	Phytophthora himalayensis	—	late blight of catechu.
(2)	P. parasitica	—	seedling blight of castor.
(3)	P. palmivora	—	bud rot of palms.
(4)	P. colocasiae	—	blight of colocasia.
(5)	P. arecae	—	koleroga or mahli disease of areca palms.
(6)	P. cactorum	—	damping off, fruit rot and root rot of angiospermic plants.
(7)	P. citrophthora	—	gummosis of Citrus.
(8)	P. erythroseptica	—	pink rot of potato.
(9)	P. nicotianae	—	black shank of tobacco.

Symptoms : The symptoms of the disease caused by Phytophthora infection can be seen on any part of the plant, *viz.*, leaves, petioles, stems and tubers. On leaves the symptoms appear in the form of dark brown or olive green, oval or irregular water soaked areas. In the early stages, the symptoms develop at the tips or the margins of old leaves.

The infection spreads vigorously when temperature is low and atmosphere is humid, and soon appears in the form of blight. On the ventral surface of the leaf, white cottony structures develop due to the growth of conidiophores. In case of severe infection, practically all parts of the host plant become brown and then degenerate.

P. infestans infects the stem, leaves and tubers of potato. When infection is severe, aerial parts of the host die. The mycelium then infects tubers and this turns their skin brown. If conditions are humid, the cells of the tuber become soft and dark brown. This symptom is known as wet rot. In dry atmosphere, however, the potato pulp does not rot but its interior part becomes black. This condition is called dry rot.

Vegetative Structure : The mycelium is hyaline, branched, aseptate and coenocytic. But septa are formed on older parts of the mycelium or at the base of the sex organs. In most species, hyphae are intracellular but in a few they are intercellular. In latter, haustoria are formed for the absorption of food material from the host. The hyphae are 4-8 μm in diameter and their wall is about 0.1 μm thick.

The cell wall of hyphae mainly consists of glucan; chitin is however, lacking. The cytoplasm contains numerous nuclei, dictyosomes, endoplasmic reticulum, ribosomes, mitochondria and vacuoles. Haustoria develop as lateral outgrowths from intercellular hyphae. Each of these outgrowths enlarges into a club-shaped head with a narrow constricted region, called stalk. The young haustorium invaginates the host cell. It remains surrounded by an extra-haustorial sheath, an extra-haustorial membrane and the cytoplasm of the host cell.

In P. infestans, haustoria are finger-like protuberances surrounded partly by thickenings of the host wall material. The cytoplasm of the haustorium contains ribosomes, mitochondria, lomasomes, well developed endoplasmic reticulum and conspicuous nuclei.

Reproduction : Phytophthora reproduces by asexual and sexual means.

Asexual Reproduction : Asexual reproduction takes place by zoospores formed in the conidiosporangium. The most favourable climatic conditions for the formation of zoospores are a temperature of 20°-22°C and 100% atmospheric relative humidity. Under favourable conditions, the hyphal branches from the intercellular mycelium emerge out through stomata or by piercing epidermal cells. These hyphal branches are hyaline, freely branched, indeterminate and they act as conidiophores or sporangiophores. A sporangium develops at the tip of the sporangiophore.

The first sporangium on a sporangiophore is terminal, but soon it comes to lie in a lateral position as the sporangiophore pushes the previously formed sporangium to one side to produce another sporangium. The sporangiophore is, therefore, sympodially branched.

The sporangium is a thin walled, hyaline, pear-shaped or oval structure with a terminal papilla or beak like projection. It measures 22-25 μm in length and 16-24 μm in breadth. It is separated from the sporangiophore by a transverse septum formed at its base.

The mature sporangia are detached from the sporangiophore and are dispersed by wind, rain splashes or by contact with other leaves. They remain viable only for a few hours and germinate on a suitable substratum. For their germination, a temperature of 12°-14°C and 100% humidity is necessary. The sporangia of P. infestans may germinate directly or indirectly. In indirect germination, each sporangium acts as zoosporangium in which numerous zoospores are formed.

In direct germination, the sporangium acts as a conidium. The conidium produces a germ tube which ultimately grows into a new mycelium. The germination of sporangium is governed by temperature and moisture; low temperature (12°-15°C) and wet conditions favour indirect germination and high temperature (20°-23°C) and dry conditions direct germination.

Indirect Germination of Sporangium : If temperature is below 15°C and weather is humid then the sporangium functions as zoosporangium. The multinucleate protoplast of the sporangium divides into 5-10 uninucleate segments. Each segment then metamorphoses into a biflagellate reniform zoospore. The flagella of the zoospore are lateral, one of them is of whiplash type and the other of tinsel type.

The zoospores are released by breaking of apical papilla. After liberation, they swim in a thin film of water for sometime, then withdraw their flagella and secrete a thick wall. These cysts germinate in favourable conditions by producing germ tube. The germ tube penetrates the host, generally through a stoma, and forms an intercellular mycelium.

Direct Germination of Sporangium : In high temperatures (20°- 23°C) and dry conditions, the sporangium behaves like a conidium. A new wall, known as germination wall, develops between the sporangial wall and the plasma membrane. The germination wall is more prominent towards the apical papilla. When the sporangium comes in contact of a suitable host, the germination wall protrudes in the form of a germ tube from the central part of the apical papilla. It forms a continuous layer alongwith the germ tube wall. The germ tube ultimately develops into a well branched mycelium.

In some species of Phytophthora (*e.g.*, P. cinnamomi, P. palmivora) thick walled, spherical, terminal or intercalary chlamydospores are formed. They germinate to form new mycelia.

Sexual Reproduction : Sexual reproduction is oogamous. The male and female reproductive structures are known as antheridium and oogonium respectively. Some species of Phytophthora (*e.g.*, P. himalayensis) are homothallic, whereas the others (*e.g.*, P. palmivora, P. infestans) heterothallic.

On the basis of the development of gametangia, species of Phytophthora may be classified as paragynous or amphigynous. In paragynous species (*e.g.*, P. cactorum), first an oogonium is formed and then from the same or nearby hypha an antheridium develops. In such species, therefore, antheridia are attached laterally to the oogonium.

On the other hand, in amphigynous species (*e.g.*, P. erythroseptica; P. himalayensis, P. infestans, P. colocasiae) first an antheridium is formed. When the antheridium is still in the developing stage, a nearby hypha emerges out by piercing the antheridium. The tip of this hypha inflates to form a spherical oogonium. Thus, in these species the antheridium appears as a collar above the base of the oogonium. The following account pertains to the sex organs of P. infestans.

Antheridium : The antheridium is formed at the tip or on the lateral side of a hypha. The region of the hypha destined to form antheridium becomes club-shaped and the cytoplasm of this region has only one or two nuclei. The young antheridium is separated from rest of the hypha by the formation of a transverse septum. The formation of oogonium stimulates divisions of the antheridial nucleus/nuclei. Of the 8-12 nuclei thus formed, only one remains functional in the mature antheridium and the others degenerate.

Oogonium : The oogonium develops laterally on a hypha adjacent to the antheridium. This hypha pierces the antheridium and its tip inflates to form a globose oogonium above the antheridium. Just after the penetration of the oogonial initial hypha, the antheridium secretes a thick wall. The protoplast of the oogonium is multinucleate and these nuclei divide repeatedly to form about 40 nuclei. The cytoplasm of the mature oogonium is differentiated into a peripheral periplasm and a central ooplasm. Both, the periplasm and the ooplasm are multinucleate in early stages, but gradually all nuclei except one move to the periplasm from the ooplasm. Ultimately all nuclei of the periplasm degenerate and the single active nucleus of the ooplasm functions as egg nucleus. The mature oogonium is a uninucleate, pyriform or globose, red or yellow and smooth walled structure.

Fertilization : At the time of fertilization, the common wall at the contact point of the antheridium and oogonium dissolves and the male nucleus from the antheridium enters into the oogonium. In paragynous species (*e.g.*, P. cactorum), the antheridium develops a fertilization tube which penetrates the oogonium and delivers the male nucleus to the oogonium. The fusion of egg and male nucleus results in the formation of a diploid oospore. Karyogamy does not occur just after fertilization and in some species (*e.g.*, P. himalayensis, P. erythroseptica) it is delayed until the oospore wall is mature.

The oospore is enveloped by a double layered wall. The outer exospore is comparatively thin and composed of pectic substances, whereas the inner endospore is thicker and made up of cellulose, protein and reserve substances.

Parthenogenetic development of oospore, in the absence of antheridia, has also been observed in P. infestans.

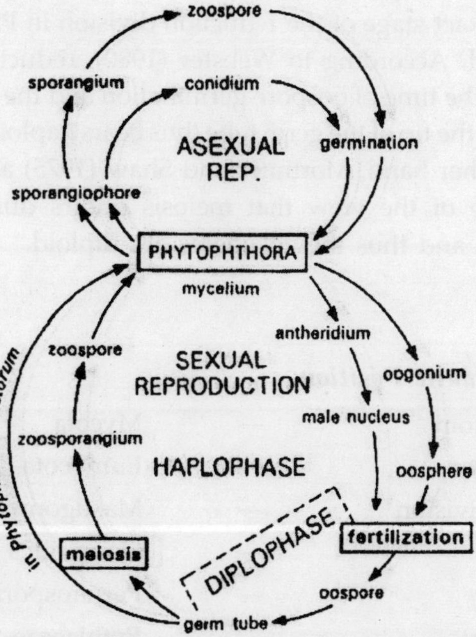

Phytophthora : Schematic representation of the life-cycle.

Germination of Oospore

The oospore undergoes a period of maturation which extends from several weeks to months. Before germination, the diploid nucleus of oospore divides mitotically to form a number of nuclei. Then the oospore produces a germ tube and the tip of the germ tube ultimately develops into a sporangium.

The sporangium is cut off from rest of the germ tube by a transverse septum. The nuclei present in the sporangium undergo first a meiotic division, followed by many mitotic divisions. Each nucleus ultimately metamorphoses into a kidney-shaped biflagellate zoospore. The zoospores remain motile only for sometime and then they are encysted. The cyst germinates to form a branched, coenocytic, inter- or intracellular mycelium.

The exact stage of the reduction division in Phytophthora is debated. According to Webster (1980), reduction division occurs at the time of oospore germination and the sporangium present at the tip of the germ tube thus bears haploid zoospores. On the other hand, Mortimer and Shaw (1975) and Sansome (1976) are of the view that meiosis occurs during gamete formation and thus the zoospores are diploid.

Pythium

Systematic Position

Kingdom	—	Mycota
Division	—	Eumycota
Sub-division	—	Mastigomycotina
Class	—	Oomycetes
Order	—	Peronosporales
Family	—	Pythlaceae

The genus Pythium is represented by some 92 species which are worldwide in distribution. Many species occur as saprophytes in water, in soil or on decaying organic matter. Others are parasites on plants and few on animals. From India about 28 species have been reported; the common species being P. debaryanum, P. aphanidennatum and P. ultimum, all causing damping off of seedlings. Some important pathogenic species of Pythium and the diseases caused by them are given below

(1) P. debaryanum — causes damping off of seedlings of tobacco.

(2) P. aphanidermatum — causes fruit rot disease of sugar beets and cucurbits.

(3) P. graminicolum — causes foot rot of wheat.

(4) P. myretonianum — causes foot rot of ginger.

Symptoms : P. debaryanum is a well known species causing damping off of seedlings in tobacco, tomato, mustard and chillies. Damping off occurs in two following stages : (i) Pre-emergence phase, and (ii) post-emergence phase. In pre-emergence phase of the disease, the young seedlings are killed before reaching the soil surface.

Since this happens under the surface of the soil, the disease is often not recognised. Initial infection appears as a slightly darkened water-soaked spot and the infected area enlarges rapidly, whereas, the post-emergence phase is characterized by the toppling over of infected seedlings often after emergence from the soil surface. The infected tissue appears soft and water-soaked. The stems become constricted at the base as the disease advances and finally the plant collapses.

Vegetative Structure : The mycelium is hyaline, branched, aseptate and coenocytic. But septa are formed on the old parts of the mycelium or at the base of the sex organs. In most of the species, the hyphae are intracellular but in a few they are intercellular. In the latter, haustoria are formed for the absorption of food materials from the host.

Reproduction : Pythium reproduces by asexual and sexual means.

Asexual Reproduction : The asexual reproduction takes place by zoospores which are formed in the sporangia. The sporangia are either terminal on intercalary on the vegetative hyphae. In P. gracile, P. monospermum and P. papillatum, the sporangia are simple and scarcely distinguishable from the vegetative hyphae, whereas, it is globose as in P. debaryanum or lobulated as in P. aphanidermatum and P. graminicola.

Each sporangium has an apical papilla, through which sporangial contents are discharged into a vesicle. In the vesicle are formed many biflagellate zoospores. The zoospores are kidney - shaped and have two lateral flagella attached on the concave side. They are liberated by the bursting of the vesicular wall. After a period of swarming the spore comes to rest, encysts, and germinates by a germ tube.

Sexual Reproduction : The sexual reproduction is oogamous. The male and female reproductive structures are known as antheridium and oogonium respectively. Majority of the species are homothallic, whereas some (*e.g.*, P. heterothallicum, P. splendens, P. sylvaticum) are heterothallic. Oogonia and antheridia are developed in close proximity, usually with the antheridium just below the oogonium.

Antheridium : The antheridium is formed at the tip or on the lateral side of a hypha. They arise either as branches of the stalk of oogonia (monoclinous) or on separate hyphae (diclinous). The region of the hypha destined to form antheridium becomes club-shaped. The young antheridium is separated from rest of the hypha by the formation of a transverse septum. The young antheridium is somewhat elongated or club-shaped and multinucleate structure. But, at maturity, all except one nucleus degenerate.

Oogonium : The oogonium is generally terminal, ovoid or spherical with smooth or papillate wall. The cytoplasm of the mature oogonium is differentiated into a peripheral periplasm and a central ooplasm.

Both, the periplasm and the ooplasm are multinucleate in early stages, but gradually all the nuclei except one from the ooplasm move to the periplasm. Ultimately all the nuclei of periplasm degenerate and the single active nucleus of ooplasm functions as egg nucleus.

Fertilization : At maturity, the antheridium gives out a fertilization tube through which the male nucleus passes into the oogonium and unites with its nucleus thus affecting fertilization. The oospore develops a thick and ornamented wall.

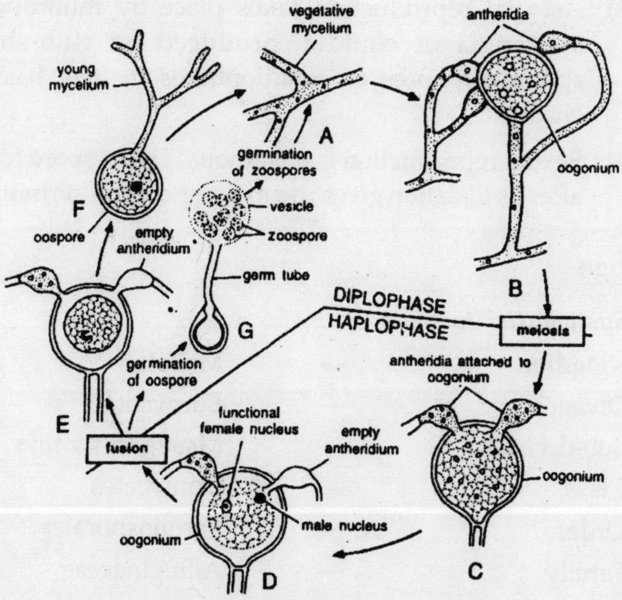

Pythium : Stages in sexual reproduction.

Germination of Oospore : The oospores undergo a period of rest of several weeks before they germinate. When the temperature is very high, it germinates by germ tube which develops into a mycelium. However, at lower temperature, the germ tube stops growing and forms a vesicle at its tip. Inside the vesicle, zoospores develop which on liberation develop into a new mycelium. Reduction division occurs in the first division of the zygote nucleus.

Family Albuginaceae

(1) This is a monotypic family represented by the genus Albugo (Cystopus), which occurs as obligate parasite on herbaceous plants, causing white rust disease.

(2) The mycelium is branched, coenocytic and intercellular. It has knob-like haustoria for the absorption of food materials from the host tissues.

(3) Asexual reproduction takes place by multinucleate sporangia or conidia, produced on club-shaped sporangiophores or conidiophores in long basipetal chains.

(4) Sexual reproduction is oogamous. The oospore formed after fertilization gives rise to zoospores on germination.

Albugo

Systematic Position

Kingdom	—	Mycota
Division	—	Eumycota
Sub-division	—	Mastigomycotina
Class	—	Oomycetes
Order	—	Peronosporales
Family	—	Albuginaceae

The genus Albugo is represented by about 30 species, distributed in the most parts of the world. They are obligate parasites occurring as endoparasites in higher plants. They usually infect the members of the families Cruciferae, Convolvulaceae, Compositae, Amaranthaceae, etc.

About 18 species of Albugo have been reported from India; the common ones are A. bliti, A. candida, A. evolvulii, A. portulacae, A. ipomoeae panduranae, A. platensis and A. trapogonosis. A. candida (= Cystopus candidus) is the most common species of the genus and it parasitises the members of Cruciferae, like radish, turnip, cabbage and mustard and causes white rust or blisters of crucifers.

Symptoms : The fungus infects mainly the aerial parts of the plant, such as petioles, leaves, stems, flowers and fruits. The early symptom of the disease is the formation of rust-like white, shiny, irregular pustules on the lower surface of leaves which may also spread on the upper surface of the leaves and stems. In later stages, several pustules coalesce to form large irregular erupted pustules. The host epidermis is ruptured and white powdery mass comes out. The infected parts of the host become abnormal and the abnormality may be in the form of hypertrophy, fleshyness or distortion. The infected leaves become thick and fleshy and the stem is malformed. The floral parts are also malformed beyond recognition. The sepals, petals, stamens and carpels are converted into thick fleshy structures. The infected flowers thus become sterile. All fleshy tissues of the host are hypertrophied and the growth of the entire plant is stunted.

Vegetative Structure : The mycelium of Albugo is well-branched, aseptate and coenocytic. The hyphal wall is composed of chitin and the cytoplasm is granular and contains many nuclei, vacuoles, oil globules and glycogen. A system of unit membranes and tubules (lomasomes) develops from the plasma membrane towards the cytoplasm. The fungus grows in intercellular spaces of the host tissue and absorbs food material with the help of small knob-like haustoria.

The haustorium is differentiated into a haustorial head (about 4 μm in diameter) and a slender neck-like stalk (about 0.5 μm in width). In the haustorial head; the cytoplasm is dense and packed with mitochondria, ribosomes, endoplasmic reticulum and lipid inclusions, but nuclei are, however, -absent. The base of the haustorium is surrounded by a collar-like sheath which is an extension of the host cell wall. The haustorium is separated from the host plasma membrane by an encapsulation made up of extra-haustorial matrix.

Reproduction : Albugo reproduces both by asexual and sexual means.

Asexual Reproduction : Asexual reproduction takes place by multinucleate structures, known as conidia, sporangia or zoosporangia. These structures develop in long basipetal chains on short, erect, club-shaped structures, called conidiophores or sporangiophores.

Formation of Conidiophore or Sporangiophore : After establishing in the host tissue, the intercellular mycelium collects beneath the host epidermis. The tips of these hyphae develop into short, erect, thick-walled and clubshaped structures, called conidiophores or sporangiophores. They develop as a palisade-like layer beneath the host epidermis and lie perpendicular to the host surface. The apical end of the conidiophore is multinucleate, thin walled and densely cytoplasmic.

Formation of Conidia or Sporangia or Zoosporangia : Conidia or sporangia develop in long basipetal chains at the distal end of the conidiophore or sporangiophore. During the formation of the conidium, a deep constriction appears below the swollen distal end of the conidiophore, which gradually increases from the periphery to the centre. Ultimately this constriction develops in the form of a transverse wall and a small bud-like structure develops into a spherical, multinucleate (usually 5-8 nuclei) conidium or sporangium. This process is repeated several times and each time a new conidium is formed on the conidiophore, below the previously formed conidium. Thus a chain of conidia is formed on each conidiophore. Conidia develop in basipetal succession, *i.e.*, the youngest conidium is at the base and the oldest at the top of the chain. An intercalary mucilagenous disc, known as disjunctor, develops between the two successive conidia (Fig. E, F). The disjunctor is formed by the common walls of the adjacent conidia.

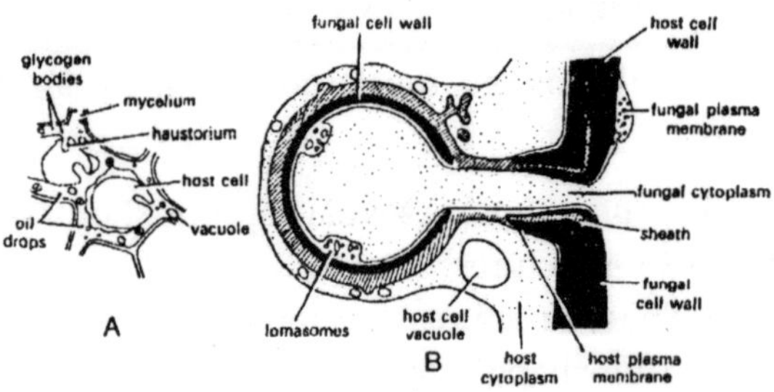

A-B. Albugo : A. Intercellular mycenum showing knoo-like haustoria, B. Ultrastructure of haustorium.

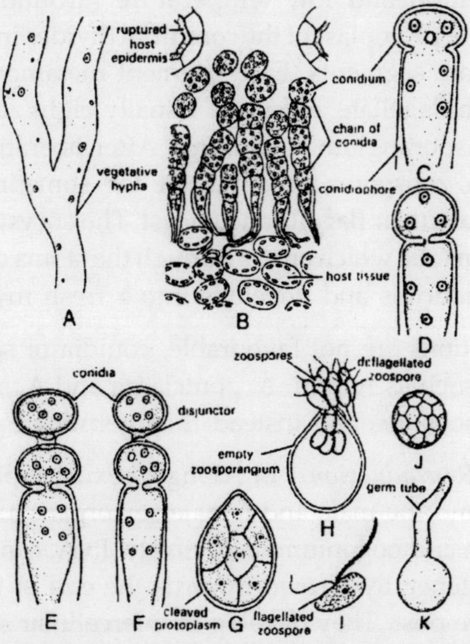

A-K. Albugo : Asexual reproduction; A. Hyphae, B. A part of vertical section of infected leaf, C-F. Stages in the formation of conidia, G. A conidium with cleaved protoplasm, H. Liberation of zoospores from sporangium, I. A zoospore, J-K. Germination of zoospore.

Due to the pressure exerted by the developing conidial chains, host epidermis is raised and finally ruptured. Conidia are now visible on the host surface as white powdery mass. Hence, the disease caused by this fungus is known as white rust. Conidia are smooth, hyaline, multinucleate and spherical or shortly ellipsoid-cylindrical structures, measuring 13-18μm in diameter. On the basis of the form of conidia, Waterhouse (1973) divided the genus Albugo into two groups : (i) Aequales-conidia with an uniformly thin wall, and (ii) Annulati - conidia with an internal equatorial thickening on the wall.

Germination of Conidia : Conidia are dispersed from the pustules by shrinking and drying of the disjunctor. They are blown by wind or washed away by rain splash. They germinate after 2-3 hours of their dispersal on suitable substratum under moist condition and low temperature (around 10°C). The multinucleate protoplast of the conidium divides into a number of uninucleate segments. Each segment metamorphoses into a reniform biflagellate zoospore. Usually eight zoospores are formed per sporangium (conidium). After liberation from the sporangium, zoospores remain active for sometime and then they withdraw their flagella and encyst. The encysted zoospore forms a germ tube which enters through the stoma or penetrates the host epidermis and develops into a fresh mycelium.

If conditions are not favourable, conidia of some species of Albugo, such as A. bliti, A. pontclacae and A. spinulosa do not form zoospores and instead they germinate directly.

Sexual Reproduction : In Albugo, sexual reproduction is oogamous and the male and female sex organs are known as antheridium and oogonium respectively. In A. candida the sex organs are generally formed towards the end of the growing season of the post. They develop in intercellular spaces, quite deep into the host tissues. Their presence is externally indicated by hypertrophy and deformation of the organ. Antheridia and oogonia develop side by side.

Oogonium : During oogonial development, the apex of the hyphal branch swells considerably and becomes globular by accumulating cytoplasm and nuclei. The swollen apex, with 6-12 nuclei, is separated from the main hypha by the formation of a transverse septum. The nuclei in the young oogonium undergo many mitotic divisions forming about 100-300 nuclei. The cytoplasm of the young oogonium is uniformly vacuolated and the nuclei are evenly distributed throughout. But as the oogonium matures, the protoplast is differentiated into an outer more vacuolated periplasm and a central densely cytoplasmic ooplasm. The periplasm and ooplasm are separated by a plasma membrane. The ooplasm represents the egg. There are many nuclei in the egg when it is first delimited from the periplasm, but later all nuclei except one migrate to the periplasm. However, some workers believe that all nuclei in the ooplasm except one disintegrate. Thus the mature female gamete or egg has a single nucleus.

Antheridium : Antheridia are elongated, club-shaped and multinucleate structures. They develop on male hyphae which are placed very close to the oogonium. The tip of the male hypha accumulates cytoplasm and nuclei and the swollen antheridial tip is soon cut off by a transverse septum. There are 6-12 nuclei in a young antheridium but the mature antheridium has only a single functional nucleus as the other nuclei degenerate.

Fertilization : Plasmogamy takes place by gametangial contact. The walls of the antheridium and oogonium become thin at the region of their contact. The antheridium forms a slender fertilization tube that penetrates the oogonial wall and the periplasm and reaches the ooplasm. Prior to fertilization, the granular cytoplasm of the oogonium forms a mass of protoplast, known as coenocentrum. The single functional nucleus of the oogonium remains attached with the coenocentrum. The fertilization tube bursts near the coenocentrum, releasing the male nucleus which fuses with the egg nucleus. The coenocentrum disappears after fertilization.

The fertilized egg secretes a thick wall around itself and is now known as oospore. The wall of the oospore is differentiated into two layers. The outermost layer, exine, is thick, warty or tuberculate and the inner, intine, is thin. Sometimes, a third layer is also present between the exine and intine. The diploid nucleus of the oospore divides first by a meiotic division, followed by several mitotic divisions resulting in the formation of about 32 haploid nuclei. Sansome and Sansome (1974), however, reported that meiosis occurs within the gametangia and the zygotic nucleus divides mitotically and the vegetative mycelium is thus diploid in nature. Before germination, the oospore undergoes resting period, extending to several months. During this period, the host tissue first shows hypertrophy and then decays setting oospores free.

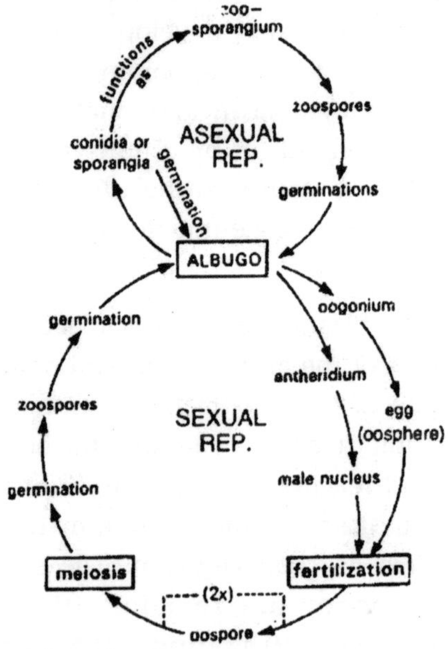

Albugo : Schematic representation of the life- cycle

Germination of Oospore

The oospore germinates in the following spring. The nuclei divide further mitotically and the protoplast of the oospore is cleaved into numerous uninucleate segments. Each segment metamorphoses into a uninucleate, reniform and biflagellate zoospore. The exine of the oospore bursts by absorbing water and the intine is protruded as a thin spherical vesicle. The zoospores are then transferred into the vesicle and later released by dissolution of the vesicle. The zoospore germinates on the suitable host by producing a germ tube. The germ tube enters through the stoma and extends into the intercellular spaces of the host.

17

Oomycetes Saprolegniales

Class Oomycetes

The members of the class Oomycetes are characterised by oogamous type of sexual reproduction and the presence of biflagellate zoospores which lack cell wall. The gametes are non-flagellate. The primitive members of the class are aquatic. Some important features of the class are as follows:

(1) Most of the members are aquatic but some grow on soil and others attack the shoots of terrestrial plants.

(2) The primitive forms are unicellular, whereas the advanced forms have a well developed branched, filamentous and coenocytic mycelium.

(3) The cell wall is mainly composed of cellulose and β-glucan. Chitin is altogether absent in most species.

(4) Asexual reproduction takes place by means of pyriform or reniform biflagellated zoospores which are devoid of cell wall. In pyriform zoospores, the flagella are attached anteriorly and in reniform laterally. One

flagellum is of whiplash type and the other of tinsel type.

(5) In primitive aquatic forms, zoosporangia are not well differentiated from somatic hyphae, whereas in advanced forms distinct zoosporangia are found.

(6) Some forms also produce non-motile asexual spores, known as conidia.

(7) Most Oomycetes are eucarpic, but the members of the order Lagenidiales are holocarpic.

(8) Sexual reproduction is of oogamous type. The male and female sex organs are known as antheridia and oogonia respectively. The sex organs are either terminal or intercalary.

(9) The gametes are non-flagellate. The fusion of gametes results in the formation of thick-walled oospores.

Martin (1961) has divided the class Oomycetes into four orders : (1) Saprolegniales, (2) Leptomitales, (3) Lagenidiales, and (4) Peronosporales whereas Sparrow (1976) recognised six orders : (1) Eurychasmales, (2) Saprolegniales, (3) Lagenidiales, (4) Peronosporales, (5) Tiraustochytriales, and (6) Labyrinthulales.

Order Saprolegniales

The order Saprolegniales includes 25 genera and about 250 species. They are the best known aquatic fungi, commonly known as water moulds. Most of the species are fresh water saprophytes, but a few species of Saprolegnia and Achlya grow as parasites on fishes and their eggs. Some members parasitize algae and fungi. The mycelium is coarse, stiff, sparsely branched and coenocytic. Asexual reproduction takes place by biflagellate zoospores. The zoospores develop in terminal zoosporangia which are not well differentiated from somatic hyphae. Sexual reproduction is oogamous and sex organs develop terminally on hyphae. The oogonium is usually spherical and contains one to many oospheres (eggs). Plasmogamy takes place by gametangial contact; only one male nucleus enters the egg.

Dick (1973) divided the order Saprolegniales into five families, Saprolegniaceae, Thraustochytriaceae, Ectrogellaceae, Haliphthoraceae and Leptolegniellaceae.

Family Saprolegniaceae

(1) Aquatic forms of the family are parasite on algae, higher aquatic plants and fishes, whereas terrestrial forms are saprophytes.

(2) The mycelium is well developed, profusely branched, aseptate and coenocytic. Septa, however, develop in the hyphae or at the base of sex organs.

(3) Asexual reproduction takes place by biflagellate zoospores which develop in long, terminal and cylindrical zoosporangia. Most of the members show the phenomenon of diplanetism.

(4) Sexual reproduction is oogamous and takes place by gametangial contact.

(5) The oogonium is not differentiated into periplasm and ooplasm. It contains one or more eggs. The male gamete enters the oogonium through a fertilisation tube.

Two members of the class Oomycetes, Achlya and Saprolegnia, belonging to the family Saprolegniaceae of the order Saprolegniales are described here.

Achlya

Systematic Position

Kingdom — Mycota

Division — Mastigomycotina

Sub-division	—	Eumycota
Class	—	Oomycetes
Order	—	Saprolegniales
Family	—	Saprolegniaceae

Achlya, commonly known as water mould, is the largest genus of the family and is represented by about 47 species, distributed all over the world. The genus is represented by some 28 species in India. The common ones are Achlya bisexualis, A. ambisexualis. A. racemosa, A. flagellata and A. caroliniana. Most of the species are aquatic saprophytes, commonly found in ponds, pools, ditches and streams or on damp soil. The fungus can be grown in laboratory on dead flies or on soaked seeds of hemp plant.

Vegetative Structure : The mycelium of Achlya is well branched, tubular, coenocytic and usually white in colour. Some minute hyphal branches penetrate into substratum to form a rhizoidal system which acts as an anchoring and absorbing organ. The hyphae are aseptate in the vegetative phase but septa are formed at the time of the formation of sex organs. The cell wall is composed of cellulose and glucan. There are many nuclei, arranged peripherally in the cytoplasm around the central vacuole.

Reproduction : Reproduction takes place by vegetative, asexual and sexual means.

Vegetative Reproduction : It takes place by fragmentation or by chlamydospore formation.

1. *By Fragmentation :* The mycelium breaks into small fragments and each fragment is capable to grow into a new mycelium.
2. *By Chlamydospore (Gemmae) Formation :* Chlamydospores are formed during unfavourable conditions. At some terminal or intercalary portions of the hyphae, nuclei

aggregate and the cytoplasm becomes dense. These dense cytoplasmic coenocytic parts are separated from rest of the mycelium by septa. These structures, known as chlamydospores or gemmae, are of various shapes and occur singly or in chains. The spores perennate during unfavourable periods. On the onset of favourable conditions, they germinate and give rise to new mycelia or zoosporangia.

Asexual Reproduction : Asexual reproduction takes place by zoospores. Achlya is diplanetic, *i.e.*, it forms two types of zoospores - primary and secondary.

Zoosporangia are formed singly at the tips of somatic hyphae. These hyphal tips have dense cytoplasm with many nuclei around the central vacuole. The tip swells and becomes club-shaped and rounded. It is separated from the vegetative hyphae by the formation of a septum which at 'first is convex with respect to the sporangial tip. There are numerous nuclei around the central vacuole in the young sporangium. The central vacuole soon disappears and the cytoplasm becomes homogeneous. This stage is known as homogeneous stage. Then the cytoplasm divides into a number of uninucleate parts, called spore initials. Each spore initial differentiates into a pear shaped primary zoospore. The primary zoospore lacks flagella in terrestrial species, but in aquatic species it is a biflagellate structure.

The apical papillae of the mature zoosporangium disappears and as such zoospores are discharged through the mouth of the zoosporangium. The zoospores withdraw their flagella and secrete a thick wall to form primary cysts. After 2-3 hours of rest period the cysts germinate. They now release another kind of zoospores, known as secondary zoospores. The secondary zoospores are bean-shaped with two lateral flagella. They swim for sometime and then encyst to form secondary cysts. The secondary cyst germinates giving rise to a germ tube and forms a new mycelium. The phenomenon of the formation of two types of zoospores, primary and secondary, is known as diplanetism.

Some species of Achlya show the phenomenon of polyplanetism. In these species, secondary cysts, on germination, form reniform, biflagellate tertiary zoospores. These spores encyst. There may be repeated emergence of secondary zoospores which ultimately germinate and form new mycelia.

After the maturation of the zoosporangium, the growth of sterile hyphae is usually renewed laterally by outpushing a new hyphal apex just beneath the first zoosporangium.

Sexual Reproduction : The sex organs are usually formed in unfavourable conditions and the sexual reproduction is of oogamous type. Some species of Achlya (*e.g.*, A. racemosa, A. colorata) are monoecious or homothallic, whereas the others (*e.g.*, A. bisexualis, A. ambisexualis) are dioecious or heterothallic. In the monoecious forms, the male (antheridium) and female (oogonium) sex organs are formed on the same or different branches of a hypha, whereas in the dioecious forms, antheridia and oogonia develop on the hyphae of two different strains.

Oogonium : The oogonia are terminal of intercalary on the main hypha or on their short lateral branches. The mature oogonium is a globose or pyriform structure, surrounded by a thin, smooth and hyaline wall. The terminal or intercalary region of the hypha destined to form oogonium swells by aggregation of cytoplasm and nuclei. The swollen portion eventually becomes spherical. A large vacuole appears in the developing oogonium in continuation with the vacuole of the hypha. Finally, the oogonium is cut off from the hypha and its contents metamorphose into one or more globose uninucleate oospheres or eggs. It is generally believed that in the developing oogonium most of the nuclei degenerate and the protoplast divides into 1-8 uninucleate segments, depending on the number of the surviving nuclei. Each segment metamorphoses into an egg and all the eggs in an oogonium are placed in the centre. Each, egg has a distinct cell wall.

Antheridium **:** The monoecious species may be monoclinous (antheridium and oogonium developing on the same hypha) or diclinous (antheridium and oogonium developing on different neighbouring hyphae). The tip of the hypha destined to form antheridium enlarges by accumulating cytoplasm and numerous nuclei. It is then cut off from the main hypha by transverse septum. The mature antheridium is club shaped and multinucleate.

Hormonal Control of Sex-organs in Heterothallic Species : Hormonal control of sexual reproduction has been studied in detail in two heterothallic species of Achlya, A. bisexualis and A. ambisexualis. In these species sex organs do not develop if the male and female thalli are growing separately and instead zoosporangia are formed. However, if the male and female thalli are growing together then they develop antheridia and oogonia respectively. On the basis of his extensive studies, Raper has shown that male and female thalli of heterothallic species secrete some hormones which stimulate the development of oogonia and antheridia. This process is controlled by four hormones (A, B, C and D). The effect of these hormones on the sexual system of A. ambisexualis is shown in Table.

Effect of Sex Hormones in Achlya Ambiserualis

Hormone	*Source*	*Affected structures*	*Specific action*
1. A-complex*			
A, A_2,	female (vegetative) hyphae	male (vegetative) hyphae	induces the development of antheridial branches.
A_1, A_3	male (vegetative) hyphae	male (vegetative) hyphae	A_1 promotes the effect of other hormones of A-complex (A, A_2 A_3), whereas A_3 reduces the response towards A, A_1 and A_2.

Contd...

Hormone	*Source*	*Affected structures*	*Specific action*
2. B	antheridial branches	female (vegetative) hyphae	initiates formation of oogonial initials.
3. C	oogonial initials	antheridial branches	(i) attracts antheridial branches. (ii) induces thigmotropic response and delimitation of anthendia.
4. D	antheridia	oogonial initials	induces delimitation of oogonia by the formation of basal septum.

* It is a mixture of four A hormones (A, Al, A2, A3).

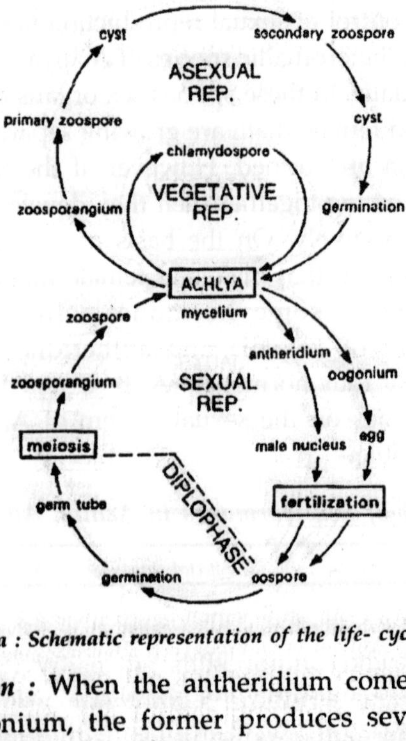

***Achlya* : Schematic representation of the life- cycle.**

Fertilization : When the antheridium comes in contact with the oogonium, the former produces several minute filamentous outgrowths. These outgrowths, which act as fertilization tubes, penetrate the oogonium. Sometimes fertilization tubes may branch. The male nuclei pass into the oogonium through the fertilization tube. Several nuclei may penetrate an egg but only one of them fuses with the egg nucleus and the rest degenerate. The fertilized egg secretes a thick wall. This diploid structure is known as oospore. There may be one or more oospores in an oogonium depending on the number of eggs.

Oospores are released by disintegration of the oogonial wall. They do not germinate immediately and undergo a period of rest. In favourable conditions, the oospore germinates by the formation of a germ tube. There is reduction division of the oospore nucleus in the germ tube. The germ tube develops into a terminal zoosporangium. The zoospores formed in a sporangium swim freely for sometime and then encyst. In favourable periods, they germinate to produce new mycelia. Sometimes young germ tubes rupture the oogonial wall and directly develop into new mycelia.

Saprolegnia

Systematic Position

Kingdom	—	Mycota
Division	—	Eumycota
Sub-division	—	Mastigomycotina
Class	—	Oomycetes
Order	—	Saprolegniales
Family	—	Saprolegniaceae

The genus Saprolegnia is represented by about 30 species. Most of the species are aquatic saprophytes, found on dead decaying organic material or on the bodies of floating dead insects. Some species like S. parasitica and S. ferax are parasitic on fishes or on their eggs and they cause salmon disease. In India, the genus is represented by seven species; of these, S. monoeca, S. ferax, S. litoralis and S. parasitica are more common.

Saprolegnia can be grown easily in the laboratory by keeping a dead fly in pond water in a test tube for few days.

Vegetative Structure : The mycelium of Saprolegnia is profusely branched and coenocytic. Two types of hyphae can be recognised during vegetative phase.

1. *Intra-material or Rhizoidal Hyphae* : These hyphae are short and are capable to absorb food materials from the substratum or the host.
2. *Extra-material or External Hyphae* : These hyphae grow on the surface of the host and spread in all directions in the water. They form visible part of the fungus and develop sex organs during the reproductive phase. They are profusely branched, coenocytic and aseptate, but septa are formed at the bases of sex organs.

The cell wall is made up of cellulose and glucans. There are many nuclei in the vacuolated cytoplasm. The reserve food material is in the form of oil globules and glycogen. The thallus is eucarpic, *i.e.,* only a part of the thallus is utilised in the formation of fruiting bodies.

Reproduction : The reproduction takes place by vegetative, asexual and sexual means.

Vegetative Reproduction : It takes place by fragmentation or by chlamydospore formation.

1. *Fragmentation* : Vegetative hyphae break up into small segments and each segment is capable to grow into a new mycelium.
2. *Chlamydospore (Gemmae) Formation* : Chlamydospores are thick-walled, ovoid, globose or irregular bodies. They develop singly or in chains and may be terminal or intercalary in position. During the formation of chlamydospores, the tips of the hyphae become swollen. The swollen portion accumulates large amount of food

material and many nuclei. Subsequently, it cuts off from the rest of the mycelium by the formation of a transverse septum. The swollen portion ultimately transforms into a chlamydospore. On maturation, chlamydospores are detached from the hyphae; they germinate on a suitable substratum and produce new mycelia.

Asexual Reproduction : Asexual reproduction takes place by zoospores which are produced in zoosporangium. The zoosporangium is an elongated cylindrical structure, formed at the tip of the hypha. Intercalary zoosporangia are, however, found in S. torulosa.

Zoosporangia are formed only on extra-material hyphae. The tip of the hypha becomes swollen, accumulates numerous nuclei and the cytoplasm becomes dense in this region. This structure is now separated from the hypha by a basal septum. The septum is initially convex and bulges into the young sporangium. The mature sporangium is club-shaped. The multinucleate protoplast of the sporangium is cleaved into uninucleate portions. Each uninucleate portion differentiates into a biflagellate pear- shaped primary zoospore.

One of the flagellum of the zoospore is of whiplash type and the other of tinsel type. Zoospores are released through an apical pore in the zoosporangium. They swim freely for some time and then come to rest, withdraw their flagella and each secretes a thick wall. These resting zoospores are now known as primary cysts. During favourable conditions, the cysts germinate and produce biflagellate kidney-shaped secondary zoospores. The secondary zoospores have lateral flagella : They swim for sometime and then germinate, producing new mycelia.

The phenomenon of the production of two types of zoospores (primary and secondary) is known as diplanetism and these zoospores are called diplanetic.

One of the characteristic features of Saprolegnia is the proliferation of zoosporangium. Following the discharge of zoospores from the zoosporangium, growth is renewed from the septum at the base of the sporangium and a new apex is formed inside the old zoosporangium. This apex develops into a new zoosporangium inside the old one. The zoospores of this new zoosporangium are discharged through the pore in the old zoosporangium. In some species of Saprolegnia this process is repeated several times and as such one can find several empty zoosporangia one inside the other.

Sexual Reproduction : Sexual reproduction usually takes place during unfavourable conditions and is of typically oogamous type. Male and female sex organs, known as antheridia and oogonia respectively, usually develop terminally on the lateral branches of the mycelium (Fig. A, B). But in S. litoralis sex organs are intercalary in position. Most species of Saprolegnia are homothallic or monoecious but a few are heterothallic or dioecious.

1. *Oogonium :* Oogonia are usually formed singly or in chains at the tips of the main hypha or its lateral branches. In some species they are intercalary (Fig. B). The tip of the hyphal branch that develops into an oogonium swells considerably and eventually becomes spherical or globose. Many nuclei alongwith cytoplasmic contents migrate into the swollen portion and then it is separated from the rest of the mycelium by a septum. As the oogonium increases in size, a large vacuole develops in its centre and the multinucleate protoplast forms a thin peripheral layer. The nuclei undergo some mitotic divisions and the protoplast is cleaved into as many segments as the number of nuclei. Each segment transforms into an egg. Initially the egg is multinucleate, but later on it becomes uninucleate with the degeneration of all nuclei but one.

The mature oogonium is a thick-walled globose structure with usually 4-10 naked uninucleate spherical eggs. Occasionally, there are up to 32 eggs in an oogonium.

2. *Antheridium :* Antheridia are also formed at the tips of the main hyphae or their lateral branches. The tips which develop into antheridia are relatively broad than the supporting hyphae and they contain dense cytoplasm and many nuclei. These swollen tips are cut off from the hyphae by septa. The hyphae which bear antheridia are known as sporangiophores.

The mature antheridium is a club-shaped structure with dense cytoplasmic contents and many nuclei.

3. *Fertilization :* As antheridia mature, sporangiophores elongate and extend towards oogonia and finally one or more antheridia become closely attached to an oogonium. Thereafter, each antheridium gives off many delicate protuberances, known as fertilization tubes,

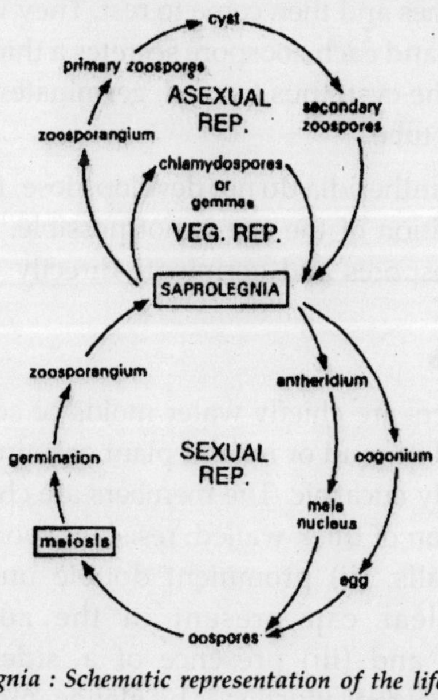

Saprolegnia : Schematic representation of the life- cycle.

which penetrate the oogonium. In some species of Saprolegnia fertilization tubes are branched. The male nuclei pass into the oogonium through the fertilization tubes and come in contact with the eggs. Only one male nucleus fuses with an egg. The fertilised egg secretes a thick wall and this diploid structure is known as oospore.

4. *Germination of Oospore* : Oospores undergo a prolonged resting period. In favourable conditions, the oospore germinates by producing an unbranched germ tube. The diploid nucleus of the oospore undergoes a meiotic division, followed by several mitotic divisions. Thus many haploid nuclei are formed. The germ tube develops into a new mycelium. Sometimes the tip of the germ tube develops into a zoosporangium with many biflagellate zoospores (equal to the number of the haploid nuclei). The zoospores are released by the rupture of the wall of the sporangium. They swim for sometimes and then come to rest. They withdraw their flagella and each zoospore secretes a thick wall around itself. The cyst, thus formed, germinates by producing a germ tube.

In S. ferax, antheridia do not develop close, to oogonia and as such fertilization of the egg is not possible. Thus the eggs act as parthenospores and germinate directly.

Blastocladiales

The members are chiefly water molds or soil inhabitants. Some are found in mud or inhabit plant or animal debris. The thallus is usually eucarpic. The members are characterized by the (i) production of thick-walled, resistant sporangia, usually with pitted walls, (ii) prominent double unit membrane-bounded nuclear cap present in the zoospores and planogametes, and (iii) presence of a 'side-body' in the zoospores. Sexual reproduction is by planogametic copulation.

The members differ from Chytridiales in having a true mycelium and in differentiation of male and female gametes and gametangia; and from Monoblepharidales in producing resistant sporangia.

Sparrow (1960,73) and Alexopoulos and Mims (1979) divided the order into three families, Coelomomycetaceae, Catenariaceae, and Blastocladiaceae. Of these, the last family is the most typical and well known.

Family Blastocladiaceae

(1) Plant body of varying complexity, typically with a prominent basal part which bears one or more reproductive structures above and a rhizoidal system below.

(2) Hyphae are non-septate, but pseudosepta, in the form of thickened rings, are present in some species.

(3) Sexual reproduction takes place by planogametic copulation.

(4) Life cycle is diplobiontic type.

The family is represented by five genera, *viz.* Allomyces, Blastocladiella, Blastocladia, Blastocladiopsis and Microallomyces. Of these, Allomyces is best known and thus described here as representative of the family.

Allomyces

Systematic Position

Kingdom	—	Mycota
Division	—	Eumycota
Sub-division	—	Mastig omycotina
Class	—	Chytridiomycetes
Order	—	Blastocladiales
Family	—	Blastocladiaceae

The genus Allomyces is represented by five species, which are worldwide in distribution, chiefly in tropical countries. From India, the genus was first reported by E. J. Butler in 1911. In India, the genus is represented by eight species; of these A. arbusculus, A. macrogynous and A. javanicus are most common. All the species are saprophytic, soil dwellers and are found in wet soil. It is common to find the species in places which are periodically wetted and dried, in rice fields, ponds and sluggish rivers.

Species can be easily cultivated in the laboratory. Soil samples are placed in water to which baits of boiled hemp seeds are added. Baits become colonised by zoospores.

Vegetative Structure : The mycelium remains differentiated into a trunk-like portion which remains attached to the substratum by a tuft of delicate but well formed, branched rhizoidal hyphae constituting the rhizoidal system. The trunk undergoes several successive dichotomous branching to form the main part of the mycelium on which the reproductive organs are borne terminally. The hyphae possess pseudosepta at fairly regular intervals. Cell wall is composed of chitin with appreciable quantities, of β–glucan, ash and proteins. Besides the nuclei there are mitochondria and roughly unique organelles termed the concentric granules. The concentric granules move about in a slow random fashion and probably act as plugs for septal pores and thus prevent protoplasmic loss following damage.

Reproduction : There is a great variation in the life-cycle pattern in various species of Allomyces. Hence, life-cycle pattern of A. arbusculus is exemplified in the following pages and a brief account of other species is given afterwards.

In A. arbusculus, the vegetative plant body is of two types, gametothalli and sporothalli. The former is haploid and concerned with sexual reproduction, while the latter is diploid and concerned with asexual reproduction.

Asexual Reproduction : On reaching a certain stage of maturity the sporothalli produce two types of sporangia (i) the colourless thin-walled zoosporangia (mitosporangia), and (ii) reddish-brown, thick-walled resistant sporangia (meiosporangia). Both types of sporangia are borne on the same thallus singly or in chains at the tips of ultimate dichotomies. The mitosporangia are papillate multinucleate structures. Within these sporangia, the nuclei undergo meiosis and the cytoplasmic contents become cleaved around the nuclei to form diploid colourless zoospores- the mitospores. Each zoospore is a naked uninucleate tiny mass of protoplast with a single whiplash type of flagellum inserted at the posterior end.

After liberating from the mitosporangia the zoospores swim about for a while, come to rest, round up and secrete a wall. The encysted zoospores germinate immediately to form a new generation of diploid asexual plants resembling the parents.

The mature meiosporangia (resting sporangia) are dark-brown, thick-walled and pitted structures, formed at the tips of branches. The multinucleate diploid sporangial contents of them undergo a resting period of 2-6 weeks or even more prior to the cleavage of the protoplast into zoospores. Meiotic divisions within them result in the formation of haploid meiospores. About 48 posteriorly uniflagellate zoospores are formed in a sporangium. At maturity, the wall of the mature sporangium cracks open and the zoospores escape in water. They swim for a brief period, encyst and germinate to form gametothalli (sexual plants).

Sexual Reproduction : Sexual reproduction takes place by planogametic copulation, and the gametangia are developed on the tips of the branches of the gametothalli. The gametothalli are homothallic; the male and female gametangia are developed in pairs, one above the other. The male gametangia are comparatively smaller in size, organge-red in colour. In A. arbusculus, it is subterminal or hypogynous, beneath the terminal female gametangium, whereas in A. macrogynous, the positions are reversed and the male gametangium is terminal above the subterminal female gametangium. On the contrary, the female gametangium is a grey or hyaline structure. Both male and female gametangia form posteriorly uniflagellate gametes which are released through 1-3 discharge papillae. The male gametes are comparatively smaller, more active orange-coloured whereas the female gametes are colourless and much bigger in size.

Before the release of the female gametes from the gametangium, the gametes secrete a hormonesirenin which diffuses outside in the water. The male gametes surround the female gametangium under the influence of sirenin. The number of male gametes surrounding a female gametangium depends upon the concentration of sirenin. As the female gamete is released from the gametangium, it fuses with the male gametes. Thus, a diploid motile zygote is formed by the fusion of the two gametes. Female gametes which fail to pair with male gametes can give rise to haploid gametothalli parthenogenetically.

The zygote swims for sometime, then comes to rest, becomes round and encysts. It soon germinates to produce a sporothallus that resembles the sexual plants in habit.

Life-cycle Patterns

Various species of Allomyces can be distinguished into three types on the basis of life-cycle pattern. Accordingly, the genus Allomyces is usually divided into three sub-genera, *viz.* Eu-Allomyces Cystogenes and Brachy-Allomyces.

In Eu-Allomyces (e.g, A. arbusculus, A macrogynous, A. javanicus), life cycle exhibits isomorphic alternation of generations (Fig. A). In these forms, there are two types of thalli-haploid gametothalli (sexual plants) and diploid sporothalli (asexual plants). On the gametothalli, gametes are produced inside the gametangia. The zygote formed by the fusion of male and female gametes develop into diploid sporothalli. On the sporothalli, two types of sporangia- the mitosporangia and meiosporangia are formed. In the former diploid zoospores are produced while in the latter by reduction divisions haploid zoomeiospores are produced. The zoomeiospores then develop into the gametothalli.

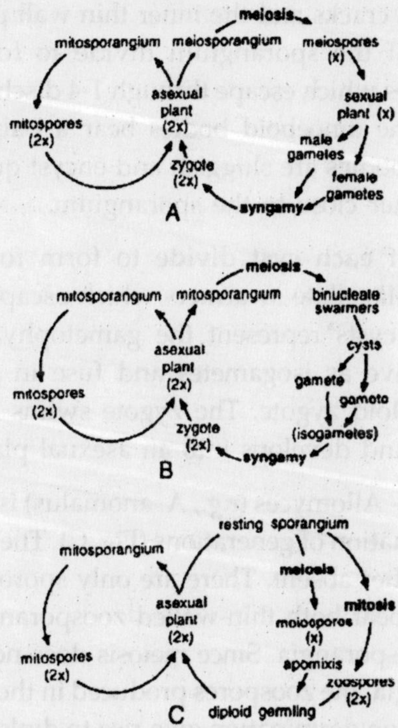

A-C. Life-cycle patterns in Allomyces : A. Eu-Allomyces B. Cystogenes; C. Branchy-Allomyces.

In Cystogenes (*e.g.*, A moniliformis, A neomoniliformis), the gametophytic generation is suppressed and is represented by a cyst only (Fig.B). The sporothalli (asexual plants) resemble those of Eu-Allomyces bearing both thin-walled mitosporangia and thick-walled meiosporangia. The zoospores produced in the mitosporangia germinate to form sporothalli again. However, the behaviour of meiosporangia (resting sporangia) is quite different. The meiosporangia are contained in a sac-like structure.

The wall of the sac ruptures irregularly and the sporangium slips out. The sporangium then germinates. During germination, the sporangium assumes a spherical shape and its outer thick pigmented wall cracks and the inner thin wall protrudes out. The contents of the sporangium divide to form up to 30 amoeboid bodies which escape through 1-4 discharge papillae. Occasionally, the amoeboid bodies bear a pair of posterior flagella. These bodies are sluggish and encyst quickly, and lie clumped together close to the sporangium.

Contents of each cyst divide to form four colourless posteriorly uniflagellate swarmers which escape through the papilla. These cysts represent the gametophytic stage. The swarmers behave as isogametes and fuse in pairs to form biflagellate diploid zygote. The zygote swims for sometime, comes to rest and develops into an asexual plant.

The Brachy- Allomyces (*e.g.*, A. anomalus) is characterized by lack of alternation of generations (Fig. C). The gametophytic body is altogether absent. There are only sporothalli (asexual plants) which bear both thin-walled zoosporangia and thick-walled resting sporangia. Since meiosis does not occur in the resting sporangia, the zoospores produced in those are diploid. The zoospores on germination give rise to diploid sporothalli.

18

The Synchytrium

Class Chytridiomycetes

The members of the class Chytridiomycetes are mostly aquatic, but a few species occur on the soil as saprophytes and some as parasites on many land plants. The important features of this class are :

(1) Most of the members are unicellular, but some advanced taxa form short chains of cells which are attached to the substratum with the help of the rhizoids. Some forms possess undeveloped mycelium.

(2) Chitin and glucan are the main constituents of the cell wall.

(3) In unicellular forms the thallus is holocarpic, whereas in filamentous forms it is eucarpic. The members of the class may be epibiotic (reproductive bodies present on the host's surface) or endobiotic (live completely within the cells of the host) and monocentric (having only a single reproductive structure) or polycentric (having more than one reproductive structures).

(4) The thallus is coenocytic but sex organs are separated from vegetative part by a septum.

(5) Asexual reproduction takes place with the help of zoospores which are posteriorly uniflagellate. The flagellum is of whiplash type. Zoosporangia are spherical or pear-shaped and inoperculate or operculate.

(6) Planogametes are also posteriorly uniflagellate.

(7) The zygote is formed by the fusion of planogametes and it is transformed into a resting spore which produces zoospores on germination.

Sparrow (1973) divided the class Chytridiomycetes into four orders : Chytridiales, Harpochytriales, Blastocladiales and Monoblepharidiales.

Order Chytridiales

The members of this order are aquatic parasites or saprophytes. The thallus is microscopic, unicellular, naked or with a distinct cell wall. The cell wall, when present, is made up of chitin or of chitin and cellulose. The primitive forms are holocarpic, whereas the advanced forms are eucarpic. Asexual reproduction takes place with the help of posteriorly uniflagellate zoospores. Sexual reproduction is isogamous or anisogamous. The zygote develops into a thick walled resting sporangium.

Sparrow (1973) divided the order Chytridiales into nine families, Olpidiaceae, Achlyogetonaceae, Synchytriaceae, Phlyctidiaceae, Rhizidiaceae, Cladochytriaceae, Physodermataceae, Chytridiaceae and Megachytriaceae.

Family Synchytriaceae

(1) The members of the family are microscopic, unicellular and amoeboid. The cell is initially naked but later may develop a thin wall.

(2) They are endobiotic, holocarpic, monocentric or polycentric.

(3) At the time of reproduction the thallus is directly converted into a group of sporangia (sorus) or into a prosorus.

(4) Asexual reproduction takes place by uniflagellate zoospores, formed in inoperculate sporangia.

(5) Sexual reproduction is of isogamous type. Synchytriaceae consist of three genera, Synchytrium, Endodesmidium and Micromyces; of these, Synchytrium, the largest genus is described here as the representative of the family.

Synchytrium

Systematic Position

Kingdom	—	Mycota
Division	—	Eumycota
Sub-division	—	Mastigomycotina
Class	—	Chytridiomycetes
Order	—	Chytridiales
Family	—	Synchytriaceae

The genus Synchytrium is represented by more than 200 species which occur as obligate parasites on flowering plants, ferns, mosses and algae. In India, the genus is represented by about 80 species. S. endobioticum, causing black wart disease of potato, is a well known species; it occurs in almost all potato growing regions of the world and causes enormous loss to the crop. Some other common species of the genus are S. fuegens (parasitic on Oenothera sp.), S. rytzii (on the members of the family Lamiaceae), S. trichosanthoides (on many cucurbits), S. sisamicola (on Sesamum indicum) and S. taraxaci (on Taraxacum officinale).

Symptoms : Synchytrium endobioticum causes the most serious disease of potato, called black wart disease. The disease symptoms can be seen in all underground parts of the host. Diseased potato tubers bear dark-brown, warty and cauliflower-like excrescences. These warts may be simple or highly branched and in several cases they are larger than the size of the tuber. Sometimes galls are also formed on aerial parts. The outgrowths appear as green convoluted leafy structures on shoots.

Vegetative Structure : The thallus is unicellular and holocarpic. It forms a mass of naked, uninucleate, amoeba-like protoplast. It is endobiotic and occurs as parasite in the epidermal cells of the host. Shortly after infection, the naked protoplast gets enveloped by a thin membrane.

Reproduction : Synchytrium reproduces both by asexual and sexual means. At the time of reproduction, whole vegetative structure transforms into a reproductive unit, *i.e.*, it is holocarpic monocentric.

Asexual Reproduction : Asexual reproduction usually takes place under favourable conditions. During spring season numerous microscopic, naked, uninucleate and uniflagellate zoospores are released from the infected tubers. These zoospores swim actively for about 2-3 hours in the film of soil water. As soon as they come in contact with the host cell, they rest and withdraw their flagella. At the site of contact a minute pore is formed in the epidermis of the host and the naked protoplast of the zoospore enters the host epidermis by amoeboid movement.

The zoospore increases in size considerably by absorption of food from the host cell and it assumes spherical form. The host cell is also stimulated and enlarges abnormally. The cells around the infected cell also enlarge and as such a rosette of hypertrophied cells surrounds a central infected cell. After sometime the zoospore becomes pear-shaped and is surrounded by a golden brown thick wall. This structure is known as summer spore. The encysted summer spore contains a prominent lipid body and a crescent-shaped nucleus on one side.

It occupies the lower half of the host cell. The cells surrounding the infected cell show abnormal cell divisions (hyperplasia) and this results in the formation of a wart or tumour-like structure. The infected host cell dies ultimately.

The summer spore germinates within the host cell. The nucleus of the spore increases in size just before germination. The inner wall of the spore protrudes out through a pore in the outer wall and forms vesicle that fills the upper half of the host cell. The contents of the summer spore, including the nucleus, are transferred to the vesicle. The nucleus undergoes repeated mitotic divisions to form 32 nuclei. This multinucleate protoplast is known as prosorus. The prosorus is divided into 4-9 multinucleate chambers with the help of thin hyaline walls. This structure is known as zoosporangium or summer sporangium, and this stage of the prosorus is now known as sores.

The nuclei of the zoosporangium continue to divide further and as many as 200-300 nuclei are formed. Now, the protoplast of the zoosporangium divides into as many uninucleate segments as the number of nuclei.

As the sporangium matures, it swells by absorbing water and the host cell which contains sporangium bursts open by the pressure. Zoospores are released through a small slit in the sporangial wall. They are naked, uninucleate and uniflagellate. They swim actively in the water film for sometime and then infect the new host. When there is scarcity of water, zoospores behave as gametes.

Sexual Reproduction : During unfavourable conditions, multinucleate segments of prosorus function as gametangia instead of zoosporangia. In gametangium, motile gametes, slightly smaller in size than the zoospores, are formed. Two planogametes from different gametangia fuse in the water film present on the host surface or in the soil. Plasmogamy is followed by the fusion of two nuclei (karyogamy), resulting in the formation of a diploid biflagellate zygote.

The zygote swims actively for sometime and then withdraws its flagella. Now, it infects the host cell like zoospores. Following the infection, the epidermal cells of the host become hypertrophied and divide repeatedly (hyperplasia) and as such the infected cell is buried deeply within the host tissue.

The diploid zygote enlarges and is enclosed by a thick, 2-3-layered wall. It is now called resting sporangium. It is also known as winter sporangium as it remains dormant throughout the winter season.

The surface of the resting sporangium contains dense fibrillar structures. It has a large centrally placed nucleus and the cytoplasm contains many membrane-bound lipid and osmiophilic bodies.

The resting sporangia are released by the decay of host tissue. With the onset of spring season, the resting sporangia become active. The nucleus of the sporangium becomes active and divides repeatedly. The first division is meiotic, followed by many mitotic divisions. Several haploid nuclei are thus formed within the sporangium. The multinucleate protoplast of the resting sporangium undergoes cleavage to form many uninucleate daughter protoplasts. Each daughter protoplast metamorphoses into a uniflagellate zoospore. The wall of the sporangium bursts open as it absorbs water and zoospores are thus released. The zoospores are oval, haploid and uniflagellate. They are comparatively large than the gametes. They infect healthy host cells.

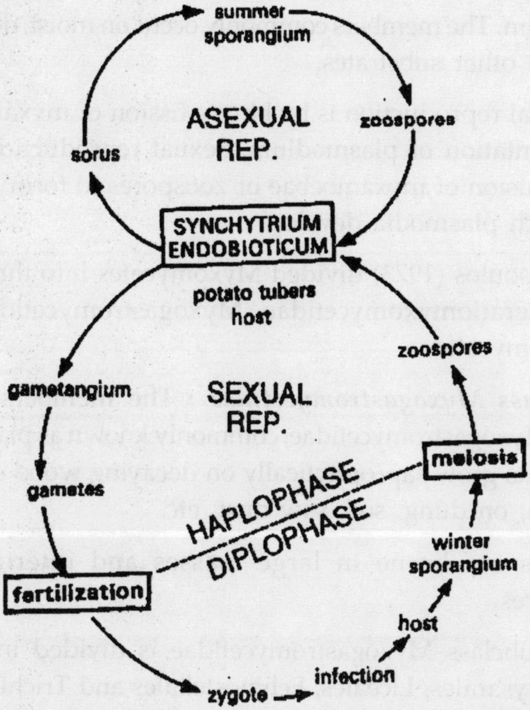

Synchytrium endobioticum : Schematic representation of the life-cycle.

Myxomycetes Physarales : The members of the division Myxomycota (commonly known as slime molds) are characterised by possessing a plasmodium or pseudoplasmodium like plant body. They are highly intriguing organisms which share characters of both animals and fungi. Their nutrition is holozoic (animal-like) but the fruiting stage shows close relationship with fungi.

Ainsworth (1973) classified Myxomycota into four classes, Acrasiomycetes, Hydromyxomycetes, Myxomycetes and Plasmodiophoromycetes.

Class Myxomycetes

The members of class Myxomycetes are commonly called true slime molds. The plant body is a free-living true plasmodium. The members commonly occur on moist, decaying wood and other substrates.

Asexual reproduction is by binary fission of myxamoebae or fragmentation of plasmodium. Sexual reproduction takes place by fusion of myxamoebae or zoospores to form zygotes from which plasmodia develop.

Alexopoulos (1973) divided Myxomycetes into three subclasses, Ceratiomyxomycetidae, Myxogastromycetidae and Stemonitomycetidae.

Subclass Myxogastromycetidae : The members of the subclass Myxogastromycetidae, commonly known as plasmodia slime molds grow saprophytically on decaying wood or other vegetation, on dung, soil, saw dust, etc.

Spores are borne in large masses and internally in sporophores.

The subclass Myxogastromycetidae is divided into four orders, Physarales, Liceales, Echinosteliales and Trichiales. Of these Physarales is most important.

Order Physarales

The members of the order Physarales are characterized by the presence of abundant lime on the sporangium and purplish brown spores. The assimilative stage is of phaneroplasmodia type. The order is represented by about 142 species and 12 genera placed in two families, Physaraceae and Didymataceae. Of these Physaraceae with the type genus Physarum is well known.

Family Physaraceae

(1) Plasmodial slime molds.

(2) Granular lime present on both the peridium and the capillitium.

Physarum

Systematic Position

Kingdom	—	Mycota
Division	—	Myxomycota
Class	—	Myxomycetes
Subclass	—	Myxogastromycetidae
Order	—	Physarales
Family	—	Physaraceae

The genus Physarum with over 100 species, is the largest and the best studied slime mold in class Myxomycetes. The species commonly occur on dead and decaying fallen rotting leaves, twigs and logs of wood. Species occur in all parts of the world, but are quite common and active in the humid tropical forests. A few species occur on the alpine region. P. cinereum forms large bluish patches on city lawns. P. viride, P. bivalve and P. compressum are some of the most widely distributed species. P. polycephalum has been used extensively for genetic, physiological and biochemical studies.

Life Cycle

The life cycle of a slime mold consists of two distinct phases, (i) vegetative (somatic) phase represented by the diploid plasmodium, and (ii) reproductive phase.

Vegetative (Somatic) Phase : The vegetative phase is represented by a free living, irregularly shaped mass of slimy protoplasm with several diploid nuclei embedded in it. This is called the plasmodium. The plasmodium is acellular and is formed by repeated divisions of the diploid nucleus of the zygote. The plasmodium feeds phagocytically on bacteria, fungal cells, myxamoebae, etc. The plasmodium increases in size and ultimately organised into fan-shaped reticulate masses, yellow in colour.

Within the veins, the protoplasm shows forward and backward streaming. Thus, in Physarum, the plasmodia are phaneroplasmodium type. Within the plasma membrane, the plasmodial protoplasm is differentiated into an outer gelatinous and an inner plasmosol zone. The outer zone consists of firm cytoplasm of less liquid consistency and is devoid of nuclei and other cell organelles, whereas the inner zone contains cytoplasm of more fluid consistency and contains numerous nuclei, pigment granules, numerous vacuoles and cell organelles, like mitochondria, endoplasmic reticulum and ribosomes. Besides, the endoplasm shows a complex network of micro-channels. Micro-channels are fine tubular structures that arise as invaginations of the plasma membrane from the exterior into the cytoplasm.

Under favourable conditions, the phaneroplasmodia is converted into fruiting body and starts sporulation, whereas, under unfavourable conditions, plasmodia instead of sporulating may give rise to dark thick-walled resting structures, known as sclerotia. During sclerotial development, the plasmodium is cleaved into numerous bits, known as spherules and finally a thick, two-layered wall is formed around each bit. The sclerotium represents the only diploid cellular stage in its life-cycle. It remains dormant for a long period under conditions unfavourable for growth. With return of favourable conditions the sclerotium grows into a plasmodium again. Thus, the sclerotia and spherules primarily serve as means of perennation.

Reproductive Phase : Under favourable conditions, the mature phaneroplasmodium passes into the reproductive phase. During this phase, the entire plasmodium is converted into one or more fruiting bodies, called sporophore or sporangium. The sporangia are seated mostly on a membranous layer, the hypothallus persisting at the base. Each sporangium is enclosed in a tough, slimy, non-cellular layer, called peridium. Numerous tiny, rounded spores occur in a mature sporangium that remain closely packed in between fine tube-like structures constituting the capillitium. The capillitical tubes may be simple or branched, possess rigid walls and open at the peridial surface. These capillitial openings look funnel-like cavities on the peridial surface. The capillitium serves (i) as an endoskeleton providing mechanical support to the delicate sporangium, (ii) as a conducting system, and (iii) assistance in the dispersal of spores.

The spores are haploid, unicellular, usually globose with smooth, spiny or reticulately thickened walls. They are dispersed by air currents, rain and mites. The peridium disintegrates exposing the capillitium and mass of mature spores.

The spores remain viable for a considerable period. On favourable conditions, spores germinate by the cracking of the spore wall, or a tiny pore appears in it. The contents usually escape in the form of spindle-shaped structures through it. These structures are known as swarm cells or myxamoebae. Each swarm cell has two flagella inserted at its anterior end. One of them is shorter than the other but both are of whiplash type, Sometimes, spores germinate producing amoeboid structures, which are known as myxamoebae. Normally, abundant moisture favours the formation of swarm cells and comparatively drier conditions induce the formation of myxamoebae.

The swarm cells or myxamoebae behave as gametes and fuse in pairs to form a diploid zygote. The species may be homothallic (P. compressum, P. gyrosum, P. pusillum) or heterothallic (P. cinereum). Karyogamy occurs soon after plasmogamy. The diploid zygote initiates the formation of plasmodium. The zygote keeps on moving on a suitable substratum for sometime. Its diploid nucleus undergoes repeated mitotic divisions resulting into a multinucleate amoeboid structure, the plasmodium.

19

The Classification

Various Methods

Traditionally the living organisms are grouped into plant and animal kingdoms, and fungi are placed in the plant kingdom alongwith other plants due to the presence of a definite cell wall and the mode of their nutrition. However, in the second half of the 19th century, the opinion begin to be revised with regard to the status of fungi-should they be retained in the plant kingdom? E. H. Haeckel (1866), a German biologist, suggested that the living organisms, instead of two, should be divided into three kingdoms - Animalia, Plantae and Protists. He placed fungi alongwith algae, bacteria and protozoa in Protists. Later, Copeland (1956) suggested a four-kingdom system of classification of living organisms. He placed bacteria and blue-green algae in the fourth kingdom, Monera, and fungi in the third kingdom, Protista. Subsequently, Otto and Towel (1956), Wesiz (1967), Barkley (1968), Stafleu (1969), etc. also supported the view that fungi should be placed in a separate kingdom, Protista.

Whittaker (1969) proposed a five kingdom system of classification of living organisms (Plantae, Fungi, Animalia, Protista and Monera); and favoured the treatment of fungi as a discrete major taxonomic unit distinct from plants and animals. This view is also shared by Martin (1968, 69) and Cooke (1977). Cytochromic studies of different organisms, made by Nolan and Margoliash (1968), also suggested that fungi form a phylogenetic line distinct from animal and plant kingdoms. Ainsworth (1966,71,73) treated fungi as a separate kingdom Mycota. Alexopoulos and Mims (1979) also included fungi, in a separate kingdom Myceteae of the super kingdom Eukaryonta.

In his review in the status of fungi, Gregory (1981) commented that "if all living organisms had to be classified into either animals or plants, I would be inclined to put the fungi with the animals." According to him, the following unique characteristics of fungi make them distinct from plants and animals.

(1) Their ability to exploit the environment by absorbing organic materials, whether from soil, compost or living animal or plant.

(2) The remarkable hyphal anastomosing and the phenomenon of protoplasmic migration.

(3) The active or passive launching of nucleated cytoplasmic masses from the hyphal system in the form of spores able to colonise fresh substrate.

The classification of fungi is not in a settled state and many systems of classification have been proposed from time to time by various mycologists. With the advancement of our knowledge, newer systems of fungal classification are being proposed and the older ones are discarded. Most of the earlier systems of fungal classification were based on the following criteria (i) thallus organization and cell wall composition,

(ii) kinds of asexual spores-their colour, shape, septation and flagellation, (iii) presence or absence of sexual reproduction, and (iv) nature of the life cycle. However, in more recent systems of classification, besides morphological criteria, host specialization, physiological, cytological, genetical, serological and biochemical characters have been taken into consideration.

The first reference of fungal classification is found in Bauhin's Pinax Theatri Botanici (1623), where he described some 100 species of fungi. Tournefort (1694), in his book 'Elements de Botanique' divided fungi into six classes, whereas Linnaeus (1753), in Species Plantarum, placed all fungi in his 24th class 'Cryptogamia'. However, it was Elias Fries who gave the first detailed classification of fungi in his three volume work :Systema Mycologicum, published between 1821 and 1832. He recognised four classes of fungi (i) Coniomycetes, (ii) Hyphomycetes, (iii) Gastromycetes, and (iv) Hymenomycetes.

Saccardo (1882) in 'Sylloge Fungorum' divided fungi into the following six classes (i) Schizomycetes, (ii) Myxomycetes, (iii) Phycomycetes, (iv) Ascomycetes, (v) Basidiomycetes, and (vi) Deuteromycetes (Fungi Imperfecti).

Classification Proposed by Gwynne-Vaughan and Barnes : H.C.I. Gwynne-Vaughan and B. Barnes (1926) recognised the following four classes in fungi on the basis of the structure of somatic mycelium and characteristics of spores.

1. *Phycomycetes :* Mycelium branched, aseptate or poorly developed; sexual spores are oospores (Oomycetes) or zygospores (Zygomycetes).
2. *Ascomycetes :* Mycelium branched, and septate; main spores are endogenously produced ascospores.
3. *Basidiomycetes :* Mycelium branched and septate; main spores are exogenously produced basidiospores.

4. *Deuteromycetes or Fungi Imperfecti* : Mycelium branched and septate; sexual spores absent.

Classification Proposed by Gaumann and Dodge : E.A. Gaumann and C.W. Dodge (1928) classified fungi on the basis of haploid and diploid phases of their life cycle and the structure of the thallus and fructification. They recognised the following four classes:

1. *Archimycetes* : Thallus naked; diploid stage represented only by the zygote.
2. *Phycomycetes* : Thallus with distinct cell wall; diploid stage represented only by the zygote.
3. *Ascomycetes* : Thallus advanced, with distinct cell wall; during sexual process dikaryotic mycelium is formed by plasmogamy; meiosis takes place in specific sporangia, known as asci.
4. *Basidiomycetes* : Thallus structure is similar to that of Ascomycetes; dikaryotic mycelium is also formed; fruiting body is known as basidiocarp, where basidiospores develop exogenously on basidia.

Classification Proposed by Tippo : O. Tippo (1942) placed both algae and fungi in the division Thallophyta. This division was divided into 10 phyla and fungi and bacteria were included in the last three phyla.

Schizomycophyta : Bacteria were included in this phylum.

Myxomycophyta : Slime molds were included in this phylum.

Eumycophyta : True fungi were included in this phylum and it was further sub-divided into four classes (i) Phycomycetes, (ii) Ascomycetes, (iii) Basidiomycetes, and (iv) Deuteromycetes.

Classification Proposed by Bessey : E.A. Bessey (1950) gave an exhaustive system of classification of fungi. His system was based largely on reproductive characters. The following is an outline of his system of classification:

Lower Fungi **:** Fruiting bodies are not formed.

1. Class Phycomycetes
   (i) *Filamentous Fungi* : Oomycetes, Zygomycetes.
   (ii) *Non-filamentous Fungi* : Plasmodiophorales, Chytridiales.

Higher Fungi—Phylum Capromyceteae **:** Fruiting bodies are formed.

1. *Class Ascomyceteae* : Fruiting bodies are known as asci; ascospores are formed endogenously.
2. *Class Basidiomyceteae* : Fruiting bodies are known as basidia; basidiospores are formed exogenously.
3. *Class Fungi Imperfecti* : Sexual stage is lacking.

Classification Proposed by Martin **:** G.W. Martin (1961) proposed a system of classification of fungi which was based on septation of mycelium and characters of spores. His classification is as follows:

Myxomycetes **:** Plant body in the form of naked protoplast.

Eumycetes **:** Plant body unicellular or filamentous.

1. *Class Phycomycetes* : Mycelium aseptate.
2. *Class Ascomycetes* : Mycelium septate; main spores ascospores, produced endogenously.
3. *Class Basidiomycetes* : Mycelium septate; main spores basidiospores, produced exogenously.
4. *Class Deuteromycetes* : Mycelium septate; propagation takes place only by asexual spores (conidia).

Classification Proposed by Alexopoulos **:** C. J. Alexopoulos (1962, 68) placed all fungi (including slime molds) in a separate division Mycota, and it was divided into two sub-divisions—

Myxomycotina and Eumycotina on the basis of the absence and presence of cell wall respectively. Slime molds were included in the Sub-division Myxomycotina and the true fungi in Eumycotina. His classification is as follows:

Division Mycota : Thallus microscopic, unicellular or filamentous; nucleus with a distinct nuclear membrane and nucleolus; cell wall chitinous or cellulosic; reproduction by asexual and sexual means.

A. *Sub-division Myxomycotina* : Plant body is in the form of naked protoplast, known as plasmodium.

   1. *Class Myxomycetes* : Vegetative phase is represented by a solitary large multinucleate naked protoplast (plasmodium); reproduction by minute multinucleate walled spores.

B. *Sub-division Eumycotina* : Vegetative phase is represented by unicellular or branched siphonaceous mycelium, cell possesses distinct cell wall; hyphae are aseptate and multinucleate or septate; cells are uni-, bi-, or multinucleate; reproduction by spores or gametes.

The Sub-division Eumycotina was divided into eight classes:

1. *Class Chytridiomycetes* : Motile cells with solitary posterior whiplash flagellum.
2. *Class Hyphochytridiomycetes.* : Motile cells with solitary anterior tinsel flagellum; includes aquatic fungi.
3. *Class Oomycetes* : Mycelium well developed and multinucleate during vegetative phase; motile cells biflagellate (one flagellum whiplash and the other tinsel type); flagella are arranged in opposite directions.
4. *Class Plasmodiophoromycetes* : Parasitic fungi; cell wall lacking; multinucleate thallus remains inside the host tissue; motile cells with two unequal anterior tinsel flagella.

5. *Class Zygomycetes :* Parasitic or saprophytic fungi; mycelium well developed and multinucleate; motile structures absent.
6. *Class Trichomycetes :* Thallus simple or branched and multinucleate; often parasitic on anthropods.
7. *Class Ascomycetes :* Hyphae septate; ascospores produced endogenously in specialised sporangium, known as ascus.
8. *Class Basidiomycetes :* Hyphae septate; basidiospores exogenously on basidium.

Form Class Deuteromycetes : Hyphae septate; reproduces only by asexual spores; sexual phase lacking.

Classification Proposed by Alexopoulos and Mims : Constantine J. Alexopoulos and Charles W. Mims (1979) placed all fungi, including the slim molds, into a kingdom of their own, Myceteae, of the superkingdom Eukaryonta. They divided the kingdom Myceteae into 3 divisions, 8 sub-divisions, 11 classes and 1 form class.

Kingdom Myceteae- Achlorophyllous; usually surrounded by cell wall that characteristically consists of chitin.

Division I. 1. Gymnomycota- Phagotropic; thallus lacks cell wall.

Sub-division 1. Acrasiogymnomycotina — includes one class Acrasiomycetes.

Sub-division 2. Plasmodiogymnomycotina — includes two classes, Protosteliomycetes and Myxomycetes.

Division II. Mastigomycota — Nutrition absorptive; motile structures present.

Sub-division 1. Haplomastigomycotina — includes three classes, Chytridiomycetes, Hyphochytridiomycetes and Plasmodiophoromycetes.

Sub-division 2. Diplomastigomycotina — includes a single class Oomycetes.

Division III. Amastigomycota — Nutrition absorptive; motile structures absent.

Sub-division 1. Zygomycotina — divide into two classes, Zygomycetes and Trichomycetes.

Sub-division 2. Ascomycotina — includes a single class Ascomycetes.

Sub-division 3. Basidiomycotina — includes a single class Basidiomycetes.

Sub-division 4. Deuteromycotina — includes a single form-class Deuteromycetes.

Classification Proposed by Ainsworth : A more natural system of classification of fungi was proposed by G. C. Ainsworth (1966,71,73). He included all fungi in the kingdom Mycota. An outline of his classification is as follows.

Kingdom Mycota (fungal kingdom)

Division I. Myxomycota (slime molds)

Class. 1. Acrasiomycetes (cellular slime molds) — assimilative phase free-living amoebae that aggregate to form a pseudoplasmodium before reproduction.

2. Hydromyxomycetes — plasmodium forms a slimy net work; mostly parasitic on marine plants.

3. Myxomycetes — a true plasmodium; free-living, saprophytic.

4. Plasmodiophoromycetes — plasmodium parasitic within the cells of the host plants. Division 11. Eumycota (true fungi)

Sub-division 1. Mastigomycotina

Class. 1. Chytridiomycetes — often unicellular, zoospores with single, posterior, whiplash flagellum

2. Hyphochytridiomycetes — often unicellular, zoospores with single, anterior, tinsel flagellum.

3. Oomycetes- usually mycelial; zoospores biflagellate with posterior flagellum whiplash and anterior flagellum tinsel type.

Sub-division 2. Zygomycotina

Class. 1. Zygomycetes- mycelium immersed in host tissue; usually saprophytic, parasitic or predacious.

2. Trichomycetes- mycelium not immersed in host tissue; often parasitic on anthropods.

Sub-division 3. Ascomycotina

Class 1. Hemiascomycetes- asci naked; no ascocarps and ascogenous hyphae.

2. Loculoascomycetes-asci bitunicate; ascocarp an ascostroma.

3. Piectomycetes-asci unitunicate, evanescent; fruiting body cleistothecium type.

4. Laboulbeniomycetes — fruiting body perithecium type, asci unitunicate, inoperculate; exoparasites of anthropods.

5. Pyrenomycetes- fruiting body perithecium type, asci unitunicate, inoperculate with apical pore or slit.

6. Discomycetes — fruiting body apothecium type; asci unitunicate, operculate with apical pore.

Sub-division 4. Basidiomycotina

Class. 1. Teliomycetes — basidiocarp lacking; teliospores grouped in sori or scattered within the host tissue.

2. Hymenomycetes — basidiocarp present; basidia arranged in hymenium; completely or partly exposed at maturity.

3. Gasteromycetes — basidiocarp present; basidia arranged in hymenium, enclosed within the basidiocarp, basidia aseptate.

Sub-division 5. Deuteromycotina (Fungi Imperfecti)

Class 1. Blastomycetes- true mycelium lacking; budding cells with or without promycelium.

2. Hyphomycetes — mycelial, sterile or asexual spores producing directly on hyphae or conidiophores.

3. Coelomycetes — mycelial; asexual spores on pycnidium or acervuli.

General Character

The word 'fungus' has been taken from Latin, which means mushroom. The fungi are achlorophyllous and heterotrophic thallophytes. Their plant body consists mostly of a thallus, made up of hyphae, which together constitute mycelium. The study of fungi is known as mycology (mykes = mushroom; logos = study) and the scientist who is concerned with fungi is called mycologist.

Several large species of fungi like mushrooms, morels, smuts, rots and rusts, were known to man since the time immemorial, but their nature, origin and development were unknown at that time. The fossil record shows their presence in the Devonian and Pre-Cambrian periods. There are evidences in the ancient literature of human culture that these plants were used as a food by human-beings.

At present, about 5,100 genera and more than 50,000 species of fungi are known. They grow in a variety of habitats and show much diversity in their from, structure of plant body, physiological processes and methods of reproduction.

Characteristics of Fungi

(1) They are cosmopolitan in distribution and can occur in any habitat where life is possible. Some fungi occur in fresh or marine water, others are terrestrial and still

others are air borne. Many species are parasitic, infecting plants, animals and human-beings.

(2) The plant body of fungi typically consists of branched and filamentous hyphae, which form a net-like structure, known as mycelium.

(3) The hyphae are aseptate and coenocytic (as in Phycomycetes) or septate and uni-, bi- or multinucleate (as in Ascomycetes, Basidiomycetes and Deuteromycetes).

(4) The septa usually have simple pores but in Basidiomycetes septa with dolipore occur. On both the sides of the dolipore, a double membranous structure - septal pore cap or parenthosome occurs. These septal pores form a valve like structure.

(5) With the exception of slime molds, the protoplast remains surrounded by a distinct cell well. The main component of the cell wall is the fungal cellulose, also known as chitin. In some lower fungi (*e.g.*, members of Oomycetes), the cell wall is composed of cellulose and glucan.

(6) The hyphae contain colourless and vacuolated cytoplasm. Besides endoplasmic reticulum, mitochondria and Golgi bodies, many non-living substances, like reserve food also occur in the cytoplasm.

(7) Fungi are entirely devoid of chlorophyll, but carotenoids are normally present.

(8) The hyphae are homo-or heterokaryotic, haploid, dikaryotic or diploid (diploid phase is usually ephemeral).

(9) The higher fungi (Ascomycetes, Basidiomycetes and Deuteromycetes) are immotile, *i.e.*, motile cells do not occur in their life-cycle at any stage. But in lower fungi

(Phycomycetes) the reproductive cells (gametes and asexual zoospores) are uni- or biflagellate.

(10) The flagella are of two types : (i) whiplash (acronematic) flagella are smooth with 9 + 2 organisation, and (ii) tinsel (pantonematic) flagella with numerous minute hair-like mastigonemes on their surface; the mastigonemes originate from the axial filament of flagella.

(11) In majority of fungi the hyphae form a felt-like structure, but in some advanced fungi they may undergo certain modifications in response to functional needs. The hyphal structure is variously modified into :

(a) *Prosenchyma* : When the component hyphae lie more or less parallel to one-another, they unite to form a rather loosely interwoven structure where their individuality is not lost.

(b) *Pseudo-parenchyma* : When the hyphae become closely intertwined, forming a tissue which consists of hollow tubes running in all directions. The hyphae loose their individuality and are not distinguishable from each other. They resemble with the parenchymatous tissue of higher plants in cross section.

(c) *Rhizomorph* : It is a thick strand or root-like aggregation of somatic hyphae.

The hyphae lose their individuality and the whole mass behaves as an organised unit. It is believed that rhizomorph has a higher infection capacity than individual hyphae (Fig. A, B).

(d) *Sclerotium* : It is a compact globose structure, formed by the aggregation and adhesion of hyphae. It may survive for long periods, sometimes for several years, and thus represents the resting stage of the fungus. They accumulate reserve food and help in vegetative propagation. (Fig. C, D).

(e) *Appressorium* : It is a terminal simple or lobed swollen structure of germ tubes or infecting hyphae, found in many parasitic fungi, such as rusts and powdery mildews. It adheres to the surface of the host and helps in the penetration of infection hyphae (Fig. E).

(f) *Haustoria* : They are intracellular absorbing structures of obligate parasitic fungi, meant for absorbing food material from the host. They vary in shape and may be knob like, elongated, finger like or branched. They secrete some specific enzymes which hydrolyse the proteins and carbohydrates of the host cell (Fig. F).

(g) *Hyphal traps (Snares)* : The predacious fungi develop sticky hyphae or network of hyphal loops, known as hyphal traps. They help in capturing nematodes.

(h) *Stromata* : These are compact somatic structures much like mattresses. Fructifications are usually formed on/in them.

(12) They are heterotrophic, *i.e.*, they do not manufacture their own food material. They are usually grouped into the following three categories on the basis of their modes of nutrition:

(a) *Parasites* : Those fungi which obtain their nutrition from other living plants or animals are known as parasites. Some of these fungi can live only on living protoplasm and they are known as obligate parasites (*e.g.*, Ravenelia, Puccinia), whereas others (*e.g.*, Ustilago, Sphacelotheca, Tolyposporium) can also grow on the dead organic material in the absence of living host and then they are known as facultative saprophytes.

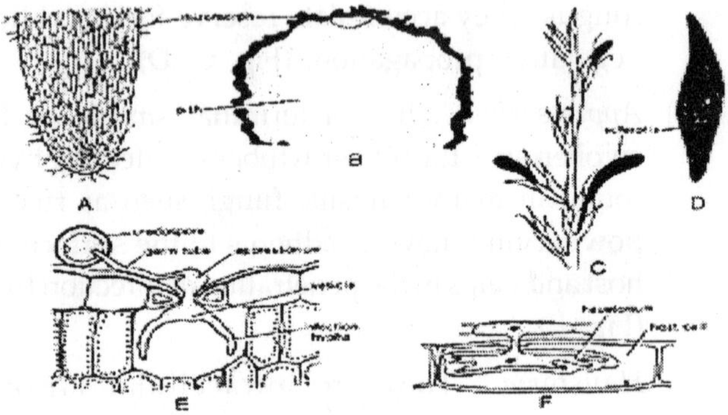

A-F. *Modifications of hyphal structures. : A-B. Rhizomorph; C-D. Sclerotia; E. Appressorium; F Haustorium; F Haustorium.*

(b) *Saprophytes* : These fungi obtain their nutrition from the dead decaying organic matter. Some saprophytes, such as Mucor mucedo, can obtain their nutrition only from the dead organic material and they do not have the capacity to infect living plants or animals. They are known as obligate saprophytes. On the other hand, those saprophytic fungi (*e.g.*, Fusarium, Pythium) which have the capacity to infect living organisms, are known as facultative parasites.

(c) *Symbionts* : Some fungi grow on other living organisms and both are mutually benefited. Such an association is known as symbiosis. Lichens and mycorrhiza are common examples of symbionts.

(13) In most of the fungi only a part of the vegetative mycelium forms reproductive unit and the rest remains vegetative. Such fungi are known as eucarpic. However, in unicellular fungi whole vegetative cell is transformed into a reproductive unit, and as such they are known as holocarpic.

(14) They reproduce by vegetative, asexual and sexual means.

(15) The vegetative reproduction may take place by fragmentation (*e.g.*, Rhizopus, Mucor, Aspergillus, Altemaria), fission (*e.g.*, yeast) and budding (*e.g.*, yeast, Ustilago).

(16) About 20% fungi propagate only by asexual means. Asexual reproduction takes place during favourable conditions by the formation of a variety of conidia or spores. The spores may be unicellular (*e.g.*, Aspergillus, Penicillium) or multicellular (*e.g.*, Altemaria, Cercospora). They may be endogenous, when developed in pycnidia or sporangia (*e.g.*, Mucor, Rhizopus), or exogenous, when developed on sporophores or conidiophores (*e.g.*, Aspergillus, Penicillium).

(17) Some common asexual spores in fungi are zoospores (*e.g.*, Achlya, Pythium, Phytophthora); chlamydospores (*e.g.*, Ustilago), oidia (*e.g.*, Collybia, Coprinus) and aplanospores (*e.g.*, Mucor, Rhizopus, Pilobolus).

(18) With the exception of the class Deuteromycetes, sexual reproduction occurs in all the groups of fungi. The process of sexual reproduction is completed in the following three distinct phases :

(a) *Plasmogamy* : In the first step fusion of the protoplasts of two compatible gametes or sex cells takes place, and the two compatible nuclei come close to each other.

(b) *Karyogamy* : In the second step, fusion of the two nuclei from the two fusing gametes takes place and a diploid zygotic nucleus is formed. In Phycomycetes karyogamy occurs just after plasmogamy, but in Ascomycetes and Basidiomycetes karyogamy is much delayed. In the latter groups, the nuclei of the opposite strains get themselves arranged in pairs (dikaryon). This phase in the life cycle is known as dikaryophase and the process by which this stage is accomplished is called dikaryotization. The two nuclei of a dikaryon fuse to form a diploid nucleus.

(c) *Meiosis* : After karyogamy, reduction division takes place in the diploid nucleus, and haploid stage is reestablished.

(19) The compatible nuclei are brought together by the following sexual processes :

(a) *Planogametic Copulation* : This involves fusion of two naked motile gametes (planogametes). The planogametes copulation is of three types, depending on the nature and structure of the fusing gametes.

(i) *Isogamy* : The fusing gametes are morphologically similar but physiologically different, and are formed on different hyphae (*e.g.*, Synchytrium, Catenaria).

(ii) *Anisogamy* : The fusing gametes are both morphologically and physiologically different; the male gamete is smaller and more active than the female gamete (*e.g.* Allomyces).

(iii) *Oogamy* : The female gamete (egg) is non-motile and the male gamete (antherozoid) is motile; they are formed in specialised gametangia, known as oogonium and antheridium respectively.

(b) *Gametangial Contact* : Here gametes are never released from gametangia, instead the male and female gametangia come in close contact with the help of a fertilization tube. Then one or more male nuclei migrate to the female gametangium. The gametangia never fuse or lose their identity during the sexual act. The male and female gametangia are known as antheridium and oogonium (ascogonium in Ascomycetes) respectively. Albugo, Aspergillus, Pythium and Phytophthora are some common examples of gametangial contact..

(c) *Gametangial Copulation* : This process involves fusion of the entire contents of two compatible gametangia. The gametangia come in close contact, their walls at the point of contact dissolve and their contents mix. The two gametangia ultimately fuse, resulting in karyogamy (*e.g.*, Mucor, Rhizopus).

(d) *Spermatization* : In some advanced genera (*e.g.*, Puccinia), sex organs are completely absent and the sexual process is accomplished by minute spore-like spermatia (male gametes) and specialized receptive hyphae (female gametes). The spermatia are carried by air, water or insects to the receptive hyphae, and the contents of the spermatium enter the receptive hyphae through a pore.

(e) *Somatogamy* : Here sex organs are not at all formed, but two vegetative cells or two vegetative hyphae take over the sexual function and fuse together. Morchella, Peziza and Agaricus are some common examples of somatogamy.

(20) There are five basic types of life cycles in fungi, *i.e.*, asexual, haploid, haploid dikaryotic, haploid-diploid and diploid.

Differences between Algae and Fungi

	Algae	*Fungi*
1.	They are chlorophyllous autotrophic plants.	They are non-chlorophyllous heterotrophic plants.
2.	The plant body is filamentous or true parenchymatous.	The plant body is filamentous or pseudoparenchymatous.

Contd...

	Algae	*Fungi*
3.	The cell wall is made up of cellulose.	The cell wall is made up of chiefly fungal cellulose or chitin.
4.	The reserve food material is usually a carbohydrate.	The reserve food material is in the form of glycogen or fat globules.
5.	They are autotrophic and light is necessary for their growth.	They are heterotrophic and light is not required for their growth and development.
6.	They show progressive complexity in sexual reproduction, *i.e.*, sexual reproduction is simple in the primitive forms and reproduction, *i.e.*, sexual reproduction is simple in the primitive forms and complex in the advanced forms.	They show progressive simplicity in sexual reproduction, *i.e.*, sexual reproduction is complex in the primitive forms and simple in the advanced forms.
7.	They are mostly aquatics found in fresh and marine waters.	They occur in all types of habitats.

PART-II

Applied Microbiology

20

Practical Application

The man lives in an environment that contains innumerable micro-organisms, which are so small to be seen only under high power of a microscope. These organisms have a profound influence, directly or indirectly, on our daily life. Some of these micro-organisms contaminate various food and other useful products. While several others provide many beneficial products by their vital activities. Beneficial activities of microbes are described in the following pages.

Dairy Industry

Milk is one of the best complete foods for man and it has high nutritive value. Because of its richness in various nutritional elements, microbes find it a suitable medium for their growth. There are various sources which contaminate milk. The teat openings of cattle are usually contaminated with microbes and the first stream of milk drawn from the udder contains more micro-organisms than the later streams of milk. Subsequent contamination of milk is through air-borne microbes setting in the utensils and milk, through water, drainage, etc.

Various dairy products, such as curd, cheese, butter milk, butter, etc. are obtained by the activity of lactic acid bacteria. These bacteria convert the lactose sugar present in the milk into lactic acid. As a result, the milk becomes sour and the casein protein is separated. Some microbes which ferment milk are Lactobacillus acidophilus, L. helviticus, L. plantarum, Streptococcus lactis, S. thermophilus and Leuconostoc citrovorum.

Butter is also an important milk product, prepared by churning sour milk. In the process, fat globules are separated and butter milk (skim) is obtained as a by-product. In order to get desired flavour and aroma, a starter culture of Streptococcus cremoris, Leuconostoc citrovorum or L. dextranicum is added to the milk. Mostly pasteurized milk is used for manufacturing butter to ensure that there are no pathogenic micro-organisms.

Cheese making is also an important microbiological process in dairy industry. The first step known as curdling, involves separation of milk protein (casein) from whey. The second step involves ripening of solid curd by the action of bacteria and fungi. Some important microbes associated with cheese manufacturing are Streptococcus lactis, S. cremoris, Leuconostoc citrovorum and L. dextranicum. Certain bacteria and yeasts grow on the surface of cheese where they produce enzymes which are responsible for the flavour and aroma of the cheese. For instance, rocquefort and combert cheese are ripened by the blue-moulds, Penicillium roqueforti and P. camemberti respectively.

Alcoholic Fermentation

Large number of micro-organisms; both prokaryotes (bacteria) and eukaryotes (fungi) are used in alcohol fermentation. Of these Saccharomyces cerevisiae (brewer's yeast) is the most important. Beer, rum, whisky, wine and several other alcoholic beverages are all products of yeast fermentations. The nature of the beverage depends on the material fermented and the strains of yeast used in fermentation. For example, beer is made by yeast fermentation of grains to ethanol and carbon dioxide. Rum is a product of blackstrap molasses fermentation and whisky is a product of corn and rye fermentation. For the production of wine, grapes are fermented by yeast and as such soluble sugars (glucose and fructose) are converted into carbon dioxide and ethyl alcohol. In order to get the desired quality of wine, ethyl alcohol is further fermented by using other microbes like Pediococcus, Leuconostoc and Lactobacillus.

Wine is also prepared by fermenting rice grains by Aspergillus oryzae.

Butyl alcohol is obtained on fermenting molasses by anaerobic bacteria like Clostridium acetobutylicum and C. butyricum.

Production of Organic Acids

Several microbes are used in the commercial production of a variety of organic acids.

Citric acid is manufactured by fermenting sucrose and molasses by the species of Penicillium and Aspergillus (A. niger, A. wentii).

Itaconic acid is obtained on fermenting glucose or sucrose solution by Asspergillus itaconicum and A. terreus.

Gluconic acid is prepared by fermenting sugary solution by the species of Penicillium like P. chrysogenum and P. perpurogenum.

Kojic acid is produced when Aspergillus oryzae is grown in a medium with sugar plus minerals. The fungus is allowed to grow aerobically at 30-35°C for about 12 days.

Acetic acid is produced by fermenting sugary solutions. It involves : (i) alcoholic fermentation of sugar, and (ii) oxidation of the alcohol to acetic acid. Usually yeast fermentation is used for the production of alcohol. Thereafter, alcohol is exposed to the action of acetic acid bacteria (Acetobacter spp.).

For the production of lactic acid sugars are fermented by Lactobacillus and Streptococcus lactis.

Fumaric acid is produced when sugar to which other nutrient salts and calcium carbonate are added is fermented by Rhizopus stolonifer.

Production of Enzymes

Several microbes synthesize large quantities of enzymes which can be obtained on a commercial scale. Amylase, zymase, invertase, glucose oxidase, pectinase and protease are some of these enzymes which are synthesized by microbes.

Amylase is obtained from Aspergillus oryzae, A. niger, Rhizopus delamar and Bacillus subtilis. The enzyme is used to hydrolyse starch to dextrin and sugars and in preparing sizes and adhesives, desizing textiles, clarifying fruit juices, manufacturing pharmaceuticals and several other purposes.

Zymase is obtained from Saccharomyces cerevisiae and it is used in the production of ethyl alcohol by fermenting carbohydrates.

Invertase is also obtained from Saccharomyces cerevisiae. It hydrolyzes sucrose to a mixture of glucose and fructose and is widely used in the production of candy and non-crystallizable syrups from sucrose.

Pectinase is obtained from the species of Penicillium. It is used to clarify fruit juices and also to hydrolyse pectins in the stems of flax hemp and jute plants, thus releasing cellulose fibres which are used in the manufacture of linen and burlap, respectively.

Proteases and peptidases are obtained from Aspergillus oryzae and Bacillus subtilis. Proteases are largely used for baiting and other leather-processing steps, degumming of silks and cleaning in laundries as an adjunct to soap.

Glucose oxidase is produced from Aspergillus niger and it is used to remove glucose from egg white and to oxidise glucose to obtain gluconic acid.

Production of Vitamins

Several vitamins are produced by microbes as primary growth products or as bye-products in certain microbial reactions. Vitamin B_{12} is commercially produced by Streptomyces and Bacillus when they are grown in nutritive medium containing sugarcane molasses. Riboflavin is produced as a growth product of the mold Achlya and Clostridium acetobutylicum. Yeast cells are a very good source of vitamin B and D. Vitamin A is produced by Rhodotorula gracilis.

Production of Antibiotics and Medicines : Micro-organisms are also used in the manufacture of several important medicines. Ergot, prepared from a fungus, Claviceps purpurea, is used to induce uterine contraction for abortions, menstrual disorders and to check haemorrhages. e-Ephedrine is synthesized from benzaldehyde by the activity of yeast; and it is used for the treatment of asthama and nasal troubles. Cortisones are prepared by fermentation of glycosides of plants by moulds like Rhizopus nigricans, Aspergillus niger and species of Fusarium, Cunninghamella and Neurospora. Cortisones are used in the treatment of rheumatoid arthritis.

Besides, a variety of microbiological antigens (vaccines) are prepared on commercial scale for the control of infectious diseases through immunization. Development of effective immunizing antigens constitutes major programs in the pharmaceutical industry.

Baking Industry : Saccharomyces cerevisiae is commonly used to leaven the bread, to produce desired flavour and for conditioning the bread. Yeast produces CO_2 and alcohol by fermentation of carbohydrates. The carbon dioxide produced during the fermentation is responsible for leavening of dough. The quality of the product depends on the proper selection of yeasts, incubation conditions and choice of raw materials. Some bacteria like Streptococcus lactis and Lactobacillus casei are used to improve the flavour of dough and quality of the bread.

As Food : A number of micro-organisms are consumed by human beings. Chlorella, a green alga rich in proteins, is used as food. Several fungi like yeast, Rhizopus stolonifer, Aspergillus oryzae, Penicillium notatum and Torulopsis utilis are also rich sources of proteins. Fats and lipids are synthesized in large quantities by some microbes and they can be used for industrial manufacture of fat at the time of emergency. Torulopsis pulcherrima, T. lipofera, Endomyces vemalis, Rhodotorula glutinis, Aspergillus fischeri, Penicillium javanicum and Geotrichum candidum are some important fungi which are rich sources of fats. Molasses, waste cellulose material, grain wastes, lactose in whey, etc. can be used as substrates for growing these fungi. A food article, Indonesian tempeh, is made by fermenting soaked soybeans with Rhizopus spp. Similarly, shoyu is prepared by fermenting soybean paste, rice, wheat, etc. by Aspergillus oryzae, Saccharomyces rouxii and Lactobacillus.

In Agriculture : Microbes play an important role in agriculture in many ways, such as in biological nitrogen fixation, sewage treatment and waste recycling, bioconversion of cellulose and lignocellulose, in mobilization of insoluble phosphorus through ecto- and endomycorrhiza and preparation of new agricultural pesticides. Some bacteria (*e.g.*, Azotobacter, Rhizobium) and cyanobacteria have the capacity to fix atmospheric nitrogen into nitrogenous compounds, thus increasing soil fertility. Gene splicing techniques are used to transfer genes from one organism to another in order to produce new strains capable of better nitrogen fixation. Biotechnology has made significant contribution in isolation of plasmids carrying nitrogen-fixing genes from N_2-fixing bacteria like Azotobacter and Rhizobium, and to transfer them to some non-nitrogen fixing strains of Klebsiella, etc. which thereafter started fixing nitrogen. Some actinomycetes (*e.g.*, Frankia) produce root nodules in some non-leguminous plants like Alnus and Casuarina, thus enabling them to fix atmospheric nitrogen.

In developing countries, there is a growing trend of using organic manures alongwith nitrogenous fertilizers. The application of mass grown algae (blue-green algae) and bacteria (*e.g.*, Rhizobium, Azotobacter and Azospirillum) to crop fields have shown significant increase in yield of cereals and pulses. Certain microbes (*e.g.*, Bacillus, Aspergillus, Penicillium and Pseudomonas) solubilize soil phosphates and make them available for plant growth. Mycorrhizae enable plants to grow in nutrient deficient and moisture limited soils.

Micro-organisms have contributed greatly to the development of a novel type of chemical called microbial insecticide. This insecticide is prepared commercially from Bacillus thuringiensis, and is very effective on a large number of lepidopterous insect pests of crops. The use of such insecticides does not pollute the environment. Besides some antagonistic microbes are used for biological control of some soil-borne diseases.

In Production of Bio-gas : Due to oil crisis there is increased interest in use of power alcohol and bio-gas as a supplement to conventional fuels. Gasohol (mixture of 90% unleaded gasoline and 10% alcohol) is used in several countries to combat the energy shortages. However, widespread usage of this fuel mixture depends on the cost of producing alcohol. Large quantities of suitable low-cost raw materials are available for fermentation. Microbial fermentation of sugarcane juice to ethyl alcohol can partially meet energy requirement. Ethyl alcohol can also be produced by bioconversion of lignocellulose. Livestock waste can be used for generation of bio-gas (methane) through microbial fermentation.

The methane generation is dependent upon microbial conversion of cellulose and other materials to methane under anaerobic conditions. The residue is used as organic fertilizer for crops. Microbes synthesize hydrocarbons and oxygenated compounds, which may serve as liquid fuel. Many marine and fresh water red, brown and green algae, diatoms and phytoplanktons produce heneicosahexane (C_{21}) hydrocarbons. Pentadecane (C_{15}) predominates in brown algae and heptadecane (C_{17}) in red algae.

Hydrogen is a valuable fuel. Several algae such as Chlorella, Chlamydomonas, Scenedesmus, Dunaliella and Oscillatoria have the capability of producing hydrogen from water in sunlight. This process is known as biophotolysis and is mediated by the enzyme hydrogenase. However, in cyanobacteria and phototrophic eubacteria, the chief hydrogen-producing enzyme is nitrogenase. This enzyme reduces dinitrogen to ammonia with the use of electrons, protons and ATP. Some of the reduced protons are evolved as hydrogen gas. In cyanobacteria, hydrogen evolution depends upon light which provides electron donors. The pretreatment of Nostoc muscorum with sulphide induces a 3- fold stimulation of nitrogenasecatalyzed hydrogen production.

In Petroleum Industry : Micro-organisms are also involved in such diverse areas as petroleum formation, exploration for petroleum, clean-up of oil- spills and deterioration of petroleum products. Many microbes can utilize hydrocarbons as sole sources of carbon and energy and can degrade petroleum hydrocarbons both in marine and fresh water ecosystems. Some important microbes capable of degrading petroleum hydrocarbons are the species of Pseudomonas, Achromobacter, Micrococcus, Nocardia, Rhodotorula, Cunninghainella, Penicillium, Themomicrobium, Oscillatoria, Anabaena, Chlorella, Dunaliella and Porphyridium. Clostridium perfringens, Vibrio desulphuricans and Beggiatoa alba are some important microbes which are associated with petroleum formation.

In Waste Treatment and Recycling : Microbes play, an important role in the treatment and recycling of sewage, organic and municipal wastes, and other wastes arising from various industries. Recycled organic wastes can be used as animal feed, fertilizer, fermentable substrates, or soil conditioners. Sewage consists of domestic and industrial water- borne wastes. The application of algal-bacterial systems is of much significance in sewage disposal. The algae release oxygen during photosynthesis and synthesize the bacterial degradation products into new protein-rich algal biomass. Water treatment may be carried out in facultative or integrated ponds. These ponds are suitable for treating municipal sewage and other domestic wastes. Certain algae, such as Spirulina, Scenedesmus and Chlorella, are suitable for this purpose.

The mass cultures of these algae in the system are used as animal feeds. Concentrated organic wastes are best treated in anaerobic lagoons which have a diverse variety of micro-organisms that generate methane and carbon dioxide. Besides, some protozoans (*e.g.*, Opercularia microdiscum, Vorticella convallaria, Chilodonella uncinata and Trachelophyllum pusillum play an important role in aerobic sewage treatment. They occur in plenty in both percolating filters and in activated sludge plants.

In Mining : The traditional method of processing ores is a major cause of air pollution. In recent years, micro-organisms have been used in the recovery of minerals from ores without any pollution. Some autotrophic aerobic bacteria like Thiobacillus ferro-oxidans and T. thiomidans when grown in the presence of copper ores produce acid and affect oxidation of the ore with subsequent precipitation of the metal. The technique improves the recovery of the metal from the ore and there is no pollution of the atmosphere.

Microbial Immobilization : Immobilized microbial cells of Escherichia coli, Pseudomonas, Streptococcus, Saccharomyces, Aspergillus, Micrococcus, Streptomyces, Enterobacter, Nocardia, Rhodotorula, etc. have various applications. They are used to produce amino acids, malate, glutathione, fructose, glucose, ATP, NADP, coenzyme A, vitamin B_{12} and menthol. Anabaena, when adsorbed on glass beads, breaks up water in the presence of light, producing hydrogen gas. Immobilized yeast cells produce ethanol from molasses.

In Bioassay and Biological Warfare : Micro-organisms are found to be useful in several biological studies. Many genetical and biochemical studies are being conducted on micro-organisms. The studies on fine structure of genes and their synthesis were possible only on microbes. Some micro-organisms can be used to assay the potency of chemical substances, such as drugs, vitamins, antibiotics and amino acids.

Thus they have practical applications in pharmaceutical industries. Microbiological assays are highly specific and unusually sensitive. For instance, as little as 0.1 nanogram (0.1^{-9} g) per ml of vitamin biotin can be detected by using Lactobacillus casei. Similarly, Neurospora crassa is proved useful for the detection of vitamin B in a sample. Besides, several microbes like Aspergillus niger are used in the detection of traces of zinc, nickel, copper, etc. in a sample. Azotobacter is used as test organism to detect the presence of potassium and phosphorus in the soil.

Some microbes act as good pollution indicators. For instance, Clostridium perfringens acts as a water quality indicator, especially for recreational water quality and as a monitoring agent for examining chlorinated drinking water in distribution system.

Now-a-days, many harmful and pathogenic microbes (*e.g.*, Coccidioidomyces) are used in biological warfare to cause disease and death in enemy camp.

In recent times, microbes have found a prominent place in space research. They have been found to be the most useful and the cheapest forerunners of man to explore the space and bring back the required information. Recent studies have demonstrated the capability of bacteria, fungi and viruses to survive fairly long periods of residence in space. Fungi have been sent into space aboard balloons, earth satellites and rockets. Certain yeasts (Saccharomyces cerevisiae and Rhodotorula rubra) and Neurospora crassa showed mutational and physiological damage in space.

21

The Antibiotics

Antibiotics are metabolic products of one organism that is detrimental or inhibitory to other micro-organisms in very small amounts. The study of antibiotics began in 1928, when Alexander Fleming noticed that a fungal colony (Penicillium notatum) inhibited the growth of staphylococci.

The compound excreted by this fungus could diffuse in the agar and was named penicillin. Since then a great many compounds with antibiotic actions have been isolated.

The ability to produce antibiotics has been found mainly in fungi of the group Aspergillales, in actinomycetes, and in a few other bacteria.

The streptomycetes are remarkable for the chemical diversity of antibiotics that they produce. More than 2,000 antibiotics have been characterised so far, but only about 50 or so are used therapeutically. But the few effective substances have radically changed the entire medical practice of treating infectious diseases.

Qualities of an Ideal Antibiotic

An ideal antibiotic should possess the following qualities:

(1) It should have the ability to destroy or inhibit specific pathogenic microorganisms without injuring host cells.

(2) It should not cause the development of resistant forms of the parasites.

(3) It should not produce any undesirable side effects in the host, such as allergic reactions, nerve damage, or irritation of the gastro-intestinal tract.

(4) It should not eliminate the normal microbial flora of the host.

(5) It should be able to be given orally without inactivation by gastric juices, or by injection without binding to blood proteins.

(6) It should have a high level of solubility in body fluids.

(7) It must contact with the parasite by penetrating the cells of the host in effective concentrations.

(8) It should be a broad spectrum antibiotic.

Mode of Action of Antibiotics

The action of antibiotics depends mainly on their chemical nature, concentration of the preparation, the particular species of the organism and conditions under which the microorganism functions. The biological mechanism of antibiotics is studied to determine the disorders provoked in the microbial cell by antibiotics, to determine the sites of their main attack in the chain of metabolic reactions, the molecular principles underlying the action of antibiotics, and also the cause of inefficiency of antibiotics against forms of microbes and macro-organisms resistant to it.

In general, the mechanism of biological action of the

antibiotics are quite specific. They may exert their antimicrobial activity by inhibiting the synthesis of the cell wall (penicillins, bacitracin, cephalosporin), by upsetting the function of the membrane (gramicidins, nystatin), by inhibiting synthesis of nucleic acids selectively (actinomycin, griseofulvin, neomycin, novobiocin), by inhibiting synthesis of purines and pyrimidines (sarcomycin), by inhibiting synthesis of protein (bacitracin, neomycin, tetracyclines, chloramphenicol, erythromycin), by inhibiting respiration (actinomycins) or by inhibiting oxidative phosphorylation (gramicidins).

Some Therapeutically Important Antibiotics

Penicillins : The first of the modern antibiotics, and still among the most useful and widely used are the penicillins. Structurally, all penicillins have a common core, a fused β–lactam thiazolidine ring (β–aminopenicillanic acid); different side chains give each penicillin its unique properties.

Natural penicillins (*e.g.*, penicillin G and penicillin V) are produced during the growth and metabolism of certain fungi, namely Penicillium notatum and P. chrysogenum. Several different kinds of penicillin molecules are produced by the same culture. They can be prepared as salts of sodium, potassium, and other bases and are freely soluble in water.

In recent times, several semisynthetic penicillins (*e.g.*, phenethicillin, methicillin, ampicillin, etc.) have been produced by developing culturing techniques to obtain core compound in quantity and then, by chemical reactions, different side chains are added on.

The phenethicillin is more readily absorbed than penicillin V and just as effective as penicillin G. The methicillin is comparatively more resistant to penicillinase, whereas ampicillin is strongly bactericidal and lacks toxicity.

Penicillins mainly act upon actively growing bacteria. They

inhibit bacterial cell wall formation by preventing the incorporation of N-acetyl muramic acid, produced within the cell, into the mucopeptide structure that normally constitutes the rigid bacterial cell wall. Penicillin sensitive bacterial cells grown in presence of this antibiotic are abnormally large and have unusual shapes.

Penicillin is the most valuable and powerful means to treat diseases caused by cocci and some anaerobic bacilli. It is most effective against gram '+ve' bacteria, particularly pneumococci, -hemolytic streptococci, and some staphylococci, some gram '-ve' bacteria (meningococci and gonococci); and the spirochete that causes syphilis.

Penicillin is practically non-toxic to humans, except for a small percentage who develop allergic reactions. The high curative properties of penicillin and its extremely low toxicity account for its wide use in medicine. Penicillin is widely used in surgery; it is very helpful in treatment of osteomyelitis, carbuncles, infected wounds, and other diseases. Penicillin is very effective in peritonitis and pneumonia.

Cephalosporins : Cephalosporins are again β–lactam group of antibiotics, derived from strain Cephalosporium acremonium having therapeutic activity against typhoid. In chemical structure, cephalosporin is close to penicillins. As for instance cephalosporin N resembles penicillin G, while cephalosporin C resembles penicillin N. Several semisynthetic cephalosporins (*e.g.*, cephalothin, cephaloridin, cephalexin) are produced by substitution of T-aminocephalosporamic acid, obtained from cephalosporin, with various side chains.

The cephalosporins, like the penicillins, exert their antibacterial effect by inhibiting the synthesis of the bacterial cell wall. They are not destroyed by penicillinase, and some are stable at acid pH.

Cephalosporins are active against many gram '+ve' and gram '-ve' bacteria and are most effective in urinary and respiratory tracts infections.

Griseofulvin : Griseofulvin is commercially produced from a mutant strain of Penicillium patulum using stirred aerated fermenters.

Griseofulvin has no antibacterial activity but is highly active against fungi. It interferes with cell wall chitin biosynthesis in growing hyphae. Besides, it also interferes with synthesis of nucleic acid, protein, maleic acid and causes destruction of organelle membrane.

Griseofulvin is used in the treatment of many superficial fungus infections of the skin and body surfaces, such an ring worm. It is also effective in the treatment of some systemic mycoses.

Streptomycin : Streptomycin belongs to the chemical class of antibiotics known as aminoglycosides. It was first isolated by Waksman and his associates (1944) from the culture medium of Streptomyces griseus. Dihydrostreptomycin, spectinomycin, neomycin, kanamycin, gentamicin and tobramycin are some other important antibiotics belonging to this group.

Streptomycin exerts its antimicrobial action through combination with and distortion of ribosome subunits, thus interfering with protein synthesis.

Streptomycin achieved its clinical success because it was effective against a series of acid-fast and gram '-ve' bacteria that were not sensitive to penicillin. It can give rise to quite marked allergic side reactions in humans.

Spectinomycin is more suitable for the treatment of gonorrhoea in persons who are allergic to penicillin. Gentamicin is especially active against some strains of Pseudomonas. Neomycin is usually used in the form of lotions and ointments for local application against skin and eye infections.

Kanamycin is used to treat tuberculosis; it is also helpful against staphylococcal diseases and to control anthrax, gonorrhoea, and other infections, against which other antibiotics are ineffective. Besides, streptomycin is also used in veterinary medicine and against plant infections.

Tetracyclines : The tetracyclines are excreted by a number of streptomycetes, including Streptomyces aureofaciens. The best known are chlorotetracycline (aurcomycin), oxytetracycline (terramycin), and tetracycline. They are also closely related chemically, being derived from a naphthacene skeleton.

The tetracyclines act by blocking the binding of RNA (aminoacyl t-RNA) to a specific site on the ribosome during peptide chain elongation, thus inhibiting the protein synthesis.

The tetracyclines are well tolerated by humans. They are regarded as broad-spectrum antibiotics and are effectively used for the treatment of infections caused by many gram '-ve' and some gram '+ve' bacteria. Chlorotetracycline is successfully used to treat bacterial pneumonia, pertussis, scarlet fever, anthrax, and other bacterial diseases.

Chloromycetin (Chloramphenicol) : Chloromycetin was first discovered in cultures of Streptomyces venezuelae, but now, it is also produced synthetically. It is very stable and a broad-spectrum antibiotic effective against many gram '-ve' and gram '+ve' bacteria, spirochaetes, rickettsiae and actinomycetes.

Chloromycetin exerts its antimicrobial action by combining with ribosome subunits, interfering with protein synthesis.

The antibiotic is a specific preparation io treat typhoid and paratyphoid fevers, dysentery, toxic dyspepsia, trachoma, and other diseases.

Erythromycin : Erythromycin was first derived by Selman Waksman (1952) from Streptomyces erythreus, isolated from a sample of Philippine soil. It belongs to macrolide group of antibiotics. They are characterised by the presence of a macrocyclic lactone ring connected with one or several sugar residues. Erythromycin is active against most gram '+ve' bacteria, some gram `-ve' bacteria (Neissena spp., Bordetella pertussis, etc.), and pathogenic spirochaetes.

Erythromycin shows its antimicrobial activity by interacting with ribosome subunits to prevent the normal sequence of reactions for protein synthesis.

Erythromycin is specially used to treat many infections caused by staphylococci, streptococci, and pneumococci. It is effective against pneumonia, tonsillitis, sepsis, wound infections and diphtheria.

Polymyxin : The antibiotic polymyxin is obtained by a bacterium Bacillus polymyxa. Chemically, it is a polypeptide antibiotic consisting of a ring of seven amino acids, with a side chain attached by a peptide bond. They are active against many gram '-ve' bacteria including Pseudomonas aeruginosa frequently involved in infections of the urinary tract or persons with extensive skin burns.

The antibiotic exerts its antibiotic activity by damaging the cell wall structure.

Polymyxins are used to treat meningitis, infections of air passages and the urinary system. Besides, polymyxin ointments are used to treat some forms of eczema, furuncles, and other skin diseases.

Bacitracin : Bacitracin is produced by Bacillus subtilis. It is also a polypeptide antibiotic, which is highly active against gram '+ve' bacteria and is almost inefficient against gram '-ve' ones. It is effective against many penicillin resistant microbes. In combination with other antibiotics, *e.g.*, with penicillins, streptomycin, chlorotetracycline, the bacitracins have a synergic effect. It is highly toxic, and hence its use is restricted to external application.

Like polymyxin, it inhibits the synthesis of the bacterial cell wall structure and affects the integrity of the cytoplasmic membrane.

The bacitracins are mostly used for prophylaxis and treatment of surgical infections. Sometimes they are used to treat skin diseases, pneumonia, bacillary dysentery, etc.

Actinomycin : Actinomycin is one of the first actinomyces antibiotics that is isolated from Streptomyces antibioticus by Waksman and Woodruff (1940) but now is obtained by several other species of Streptomyces.

It is actually a mixture of several compounds, all of which contain a phenoxazone chromophore, but all are substituted with different polypeptide chains.

It is active against gram '+ve' but less active against gram '-ve' bacteria. Some actinomycins have anticancer properties, but their high toxicity is a limiting factor.

Non-medical Uses of Antibiotics Antibiotics are now widely used not only in medicine but also in agriculture and in food industry.

Antibiotics in Agriculture : The use of antibiotics for plant disease control has been studied intensely during the past two decades. Diseases of plants are caused by various phytopathogenic agents, such as viruses, bacteria, fungi, protozoa, etc. Antibiotics are used to control various agents and to prevent propagation of infection.

The method by which antibiotics are applied depends on the disease, the stage of plant growth, the plant size, the area of cultivation and the planting method. All techinques are based on the inhibition of growth or destruction of phytopathogenic organisms on the plant surface or inside plant tissues by the antibiotic applied to the surface of the plants or added to the soil.

Streptomycin, tetracyclines, griscofulvin, trichothecin, etc. are some important antibiotics used to control plant pathogens.

Antibiotics in Nutrition and Veterinary : Antibiotics are now widely used as growth stimulants in poultry and livestock feeds.

Commercially the addition of aureomycin, terramycin or penicillin to poultry feed at the rate of 5 to 20 g/ton of feed increases the rate of young ones by at least 10 per cent.

Antibiotics in Food Preservation : Antibiotics were first reported to be used in canning industry in 1943. Subtilin, nisin, and some other antibiotics are generally used for this purpose. Besides, antibiotics are also used to preserve fresh foods like meat, etc. For preventing quick spoilage of meat, usually two methods are in use;

(i) antemortem feeding of animal with food containing an antibiotic, and

(ii) postmortem injection of the antibiotic into the blood system (immediately after slaughtering and releasing the blood). This prolongs the time of safe storage of meat to three days and improves its quality.

Similarly, antibiotics are very effective in preservation of fish spoilage. Films and other packaging materials containing antibiotics are sometimes used for packing perishables.

This prolongs the time of their storage. Antibiotics are also used for preserving milk and other dairy products. As for example, nisin is used in the manufacture and storage of cheeses.

Despite the apparent advantages of using antibiotics for nutrition of animals and preservation of foods, they should be used with great precautions.

Getting inside man with food, antibiotics in small concentrations may stimulate formation of antibiotic-resistant micro-organisms.

Some Important Antibiotics and their Source, Range of Action and Mode of Action

	Antibiotic	*Source*	*Range of action*	*Mode of action*
A.	**Antibiotics obtained from actinomycetes**			
1.	Streptomycin	Streptomyces griseus	Gram '+ve' and Gram '-ve' bacteria, tuberculosis bacteria.	Induces abnormal protein synthesis by inhibiting 30 s ribosome function.
2.	Neomycin	Streptomyces fradiae	Gram '-ve' bacteria (inhibits most of the intestinal bacteria)	Induces abnormal protein synthesis by inhibiting 30 s ribosome function.
3.	Viomycin	Streptomyces puniceus	Gram '-ve' bacteria, including tuberculosis bacteria	Interferes with protein synthesis.
4.	Chlorotetra-cycline (Aureomycin)	Streptomyces aureofaciens	Gram '+ve' and Gram '-ve' bacteria, Treponema, rickettsia	Interferes with protein synthesis.
5.	Oxytetracycline (Terramycin)	Streptomyces rimosus	Gram '+ve' and Gram '-ve' bacteria, rickettsia	Interferes with protein synthesis.
6.	Novobiocin	Streptomyces niveus	Proteus and Diplococcus	Interferes with protein synthesis.
7.	Chloramphenicol (Chloromycetin)	Streptomyces venezuelae	Gram '+ve' and Gram '-ve' bacteria, rickettsia	Inhibits protein synthesis by interfering with the ribosome function.
8.	Erythromycin	Streptomyces erythreus	Gram '+ve' bacteria	Inhibits protein synthesis by interfering with 50 s ribosome function.

Contd...

	Antibiotic	*Source*	*Range of action*	*Mode of action*
9.	Linomycin	Streptomyces spp.	Gram '+ve' bacteria	Interferes with protein synthesis
10.	Rifamycin	Streptomyces mediterranei	Tuberculosis bacteria	Interferes with protein synthesis
11.	Nystatin (Mycostatin)	Streptomyces noursei	Fungal infections due to Candida	Damages cell membrane.
12.	Kanamycin	Streptomyces kanomyceticus	Most Gram '—ve bacteria except Pseudomonas	Induces abnormal protein synthesis.
13.	Gentamicin	Micromonospora purpurea	Gram'+'ve and Gram '-ve' bacteria	Induces abnormal protein synthesis.
14.	Vancomycin	Streptomyces orientalis	Gram '+' ve bacteria	Interferes with protein synthesis.
15.	Amphotericin	Streptomyces nodosus	Deep-seated mycotic infections	Interferes with membrane function.
B. Antibiotics obtained from bacteria				
16.	Bacitracin	Bacillus subtilis	Gram '+' ve bacteria	Inhibits synthesis of bacterial cell wail.
17.	Polymyxin B	Bacillus polymyxa	Gram'-' ve bacteria	Destroys cytoplasmic membrane.
18.	Colistin	Bacillus colistinus	Gram '-' ve bacteria	Destroys cell membrane. (Polymyxin E)
19.	Tyrothrycin	Bacillus brevis	Gram -' ve and cocci	Inhibits synthesis of bacterial cell wall.
C. Antibiotics obtained from fungi				
20.	Penicillin	Penicillium notatum, P. chrysogenum	Gram '+'ve bacteria	Inhibits cell wall synthesis.

Contd...

	Antibiotic	*Source*	*Range of action*	*Mode of action*
21.	Proliferin	Aspergillus proliferans	Tuberculosis	Inhibits growth.
22.	Isopenicillin N	Penicillium chrysogenum	Bacteria	Inhibits growth.
23.	L-sarcin	Aspergillus giganteus	Tumour	Inhibits growth.
24.	Ramycin	Mucor remannianus	Bacteria	Inhibits growth.
25.	Campestrin	Psalliota campestris	Bacteria	Inhibits growth.
26.	Ustilagic acid	Ustilago maydis	Fungi	Upsets normal growth.
27.	Chaetomin	Chaetomium cochloides		
28.	Ergotine	Claviceps purpurea		
29.	Citrinin	Penicillium citrinum		
30.	Jawaharene	Aspergillus niger	Bacteria, tumours	Inhibits growth.
31.	Baccatin A	Gibberella baccata	Bacteria, fungi	Upsets normal growth.
32.	Cephalosporin	Cephalosporium spp.	Gram '+'ve and Gram '-'ve bacteria	Inhibits cell wall synthesis.
33.	Fumagillin	Aspergillus fumigatus	Amoeba, staphylococcal bacteriophage	Inhibits growth and multiplication.
34.	Griseofulvin	Penicillium nigricans, P. urticae, P. griseofulvum.	Fungi	Upsets normal growth.
35.	Trichothecin	Trichothecium roseum	Fungi, virus	Arrests growth.
36.	Gliotoxin	Trichoderma spp.		
37.	Puberulic acid	Penicillium puberulum		

22

Infection and Immunity

Micro-organisms are responsible for a wide range of diseases in plants and animals. The disease is disorders developed in an organism by harmful physiological changes. The causative agent of a disease is known as pathogen, and the biological process of the establishment of a pathogen inside the organism is known as infection. The infection results in 'the development of visible or latent disease in an organism. Broadly speaking, infection is a process of inter-struggle between organisms (pathogen and host) inhabiting two different environmental conditions. The potential capacity of infection of a pathogen is known as its pathogenicity. The pathogenicity of a pathogen is its specific character, and it depends on the parasitic adaptation and capacity to life-struggle of the pathogen.

Infection

The term infection (Latin, infectio = to infect) signifies the sum of biological processes which take place in the body of the host after the penetration of the pathogenic micro-organism, independent of the fact whether the microbe will cause a disease or not. Thus, infection is the establishment of a pathogenic micro-organism within the host following its entrance. The following are the requirements for the origin and development of infection.

Entrance of Pathogen into the Host : The place of entry of the pathogen in the body of the host is important for the occurrence and kind of disease. Certain microbes cause disease only if they enter the body of the host through a specific path. For instance, when dysentery bacilli enter the body through wound, they do not cause any harm, but if swallowed they might prove to be fatal. Similarly, typhoid bacteria are harmful if they enter the body through mouth. Gonococci and some other microbes like spirochetes enter through abrasions in the skin and set up local infections, such as boils caused by staphylococci.

Growth and Multiplication of Pathogen in the Host : After having gained access to the body of the host successfully, the pathogen secretes several chemical substances like enzymes and toxin. These substances cause physiological, morphological and anatomical abnormalities in the body of the host. Some pathogens also secrete such substances which protect them from the lytic enzymes produced by the host. The enzymes and toxins secreted by the pathogen kill and drive away leucocytes of the host. Thereafter, the pathogen gets an opportunity to grow in the host.

Toxicity of Microbes : The pathogens secrete some poisonous substances which are harmful to the host, but helpful in the disease development. Such substances are called toxins. The toxins do not break the structural entity of the host but cause a minor change in its metabolism. The potency of the toxins is an important factor in the establishment of the disease. The toxins may be exotoxins or endotoxins.

Exotoxins are proteinaceous, highly toxic and diffusable substances which microbes secrete in the host tissue, circulatory system, etc. They lose their toxicity when heated or treated chemically and form toxoids.

Endotoxins are less toxic and consist of glucoside lipoprotein. They are relatively more heat stable and do not form toxoids. They are usually located in the cell wall and are liberated only after the cell disintegrates. They generally play a contributory rather than a primary role in causing disease and are mainly responsible for many of the symptoms. Many pathogens, particularly Gram '-ve' bacteria produce endotoxins.

Some important characteristics of exotoxins and endotoxins are summarized in the Table.

Important Characteristics of Toxins

Feature	*Exotoxins*	*Endotoxins*
1. Bacterial source	Mainly Gram '+ve' bacteria.	Mainly the cell wall of lysed Gram '-ve' bacteria.
2. Chemical nature	Proteinaceous.	Composed of glucoside lipoprotein complexes and polysaccharide specific complexes.
3. Heat tolerance	Inactivated easily by treating at 60-100°C for 30 min.	Withstand autoclaving.
A. Diffusibility	Easily diffuse from the cell into the surrounding medium.	Firmly bound within the bacterial cell.
5. Toxicity	Highlytoxic, characterized by selective affection of certain organs and tissues.	Less toxic, selective action poorly expressed.
6. Immunity	Can be converted to toxoids and readily neutralized by antitoxin. Produce highly active antibodies, the antitoxins on parental infection.	Cannot form toxoids and neutralization with antitoxin is not easily possible. Produce precipitins, lysins, opsonins agglutinins and complement using antibodies on parental infection.
7. Biological effect	Specific for a particular type of cell function.	Show various effects especially the symptoms of hypersensitivity.
8. Lethal dose	Minute amounts.	Comparatively more.

Besides toxins, virulence of microbes also depends on extracellular enzymes produced by them. Several types of cocci, clostridia and some other bacteria produce the enzyme hyaluronidase which helps the pathogen to penetrate the tissues of the host by hydrolyzing hyaluronic acid. Lecithinase, produced by Clostridium perfringens, destroys various tissue cells and is especially active in the lysis of red blood corpuscles. Collagenase, also produced by Clostridium perfringens, destroys collagen tissue found in muscles, bones and cartilage. Some staphylococci and streptococci secrete leucocidin enzyme which kills white blood cells. The enzyme coagulase, produced by some virulent staphylococci (Staphylococcus aureus), clots plasma. The virulence of pathogenic bacteria also depends on the presence of capsule and pili. The increased virulence of capsulated strains is due to the ability of capsular polysaccharide in preventing phagocytosis by host leucocytes. Presence of pili helps the pathogen to adhere to the surface of host cells and tissues, thus enhancing the virulence of the pathogen (*e.g.*, Neisseria gonorrhoea, Escherichia colt).

***Dosage* :** This factor in establishing infection is a quantitative one. Under normal conditions, the larger is the dose of the infective microbe the greater the chances of infection.

***Defence Mechanisms of Host* :** All organisms show resistance towards pathogens and as such the host resists the effects of pathogen by active or passive activities. The resistance of the organism towards a specific pathogen depends upon the environmental conditions and genetic constitution of the host. The defensive mechanism of the host determines the extent of infection and severity of the disease.

According to manifestations, infection may be acute or chronic, obvious or latent, mixed or secondary. Acute infections are characterized by a sudden onset and a comparatively short course. Influenza and measles are the excellent examples of acute infections. On the other hand chronic infections have a long span. Malaria and tuberculosis are chronic infection. Typical symptoms do not appear in latent infection. In some cases infection causes a weakening of the body of the host which then becomes susceptible to other diseases. This is known as secondary infection. For instance after measles or influenza the body becomes susceptible to pneumonia.

Immunity

The resistance of the host towards a pathogenic microbe and its toxin is known as immunity and the study of immunity to infectious diseases is called immunology. The degree of resistance depends on the defense mechanism of the host and the virulence and dosage of the pathogen immunity can be classified into the following two groups.

Natural, Heritable or Non-specific Immunity : This type of immunity is hereditary and is passed from one generation to the next. For instance, human beings are immune to cattle plague, and cattle are immune to diseases like cholera and measles. The natural immunity may occur at racial or individual level. Most racial immunities are due to nonspecific factors related to people's way of life. Such immunity may reflect the evolution of resistant humans. For instance, African black people are immune to malaria, whereas majority of the white people are susceptible to this disease.At individual level, the resistance depends on several factors, such as age, nutrient status, sex, occupation, health, and personal hygiene. Young ones and aged are more susceptible to diseases. Infants are protected from many diseases from birth to six months owing to the presence of maternal antibodies in the blood. In the aged person there is a natural decline in the immune response to infection and that is why they become more susceptible to some diseases such as pneumonia.

Acquired or Specific Immunity : The acquired immunity may be of following two types.

Natural Acquired Immunity : The host acquires this type of immunity naturally during the course of its life time. The natural acquired immunity may be :

(a) *Natural Active Immunity :* This is the resistance which is developed as a result of some disease or through recovery from a disease; for instance, the host develops immunity for the diseases like small pox and diphtheria after subjecting them.

(b) *Natural Passive Immunity :* This type of resistance is found in the newly borne child, acquired from the mother through the placenta. For instance the child borne by mother immune to tetanus also shows resistance towards the disease.

Artificial Acquired Immunity : The host acquires this type of immunity during the course of life time by artificial means. It is classified into the following two types.

(a) *Artificial Active Immunity :* This type of immunity is developed by vaccination, made by dead organisms pathogens. The toxin thus entering the body stimulates the production of immunity.

(b) *Artificial Passive Immunity :* This type of immunity is developed through injection of immune serum present in already immunized organisms.

Antigen

Antigen is a substance that, when introduced into a vertebrate host, provokes an immune response leading to acquired immunity. The immune response results in the formation of specific antibodies. The antibodies are circulated in the blood stream or they stimulate the increase in the number of lymphocytes. The lymphocytes have the capability to destroy other cells. The ability of the antigen to stimulate antibody formation is called its antigenicity.

Antigens are organic compounds of high molecular weight. The antigens of proteinaceous nature are known as complete antigens. They can produce antibodies. Some antigens which consist of lipids and carbohydrates, and can not produce antibodies on their own are known as partial antigens. The partial antigen stimulates antibody formation by combining with proteins. Bacteria, viruses and other microbes, pollen grains, and blood cells of other organisms may act as antigens. Besides, some important antigens (ABO, MN, S, P, Rh) are also found in red blood cells of human being. The bacterial antigen is usually secreted in the form of exotoxins and enzymes. Vaccines (suspensions of killed, living or attenuated microbes) are also used as antigens to produce immunity against infection of specific pathogens. Toxoids are made by destroying the poisonous portions of toxins. In the process, the properties of antigen is altered. Toxoids are used specifically for protecting individuals against diphtheria, tetanus, etc.

Antibody

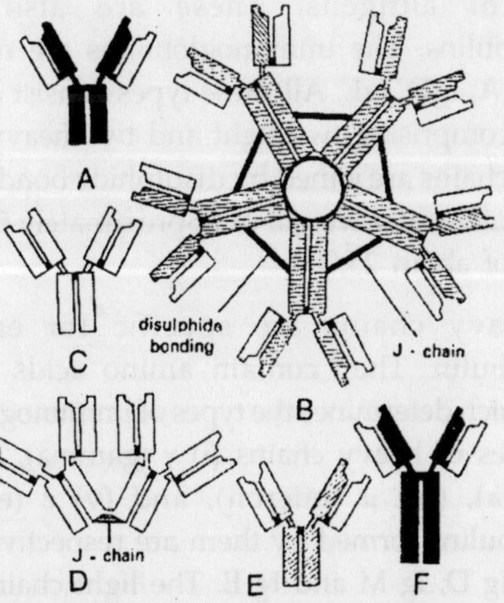

A-F. Structure of different types of immunoglobulin molecules. : A. IgG, B. Ig M, C. Ig A (monomeric), D. Ig A (dimeric), E Ig D, F. Ig E.

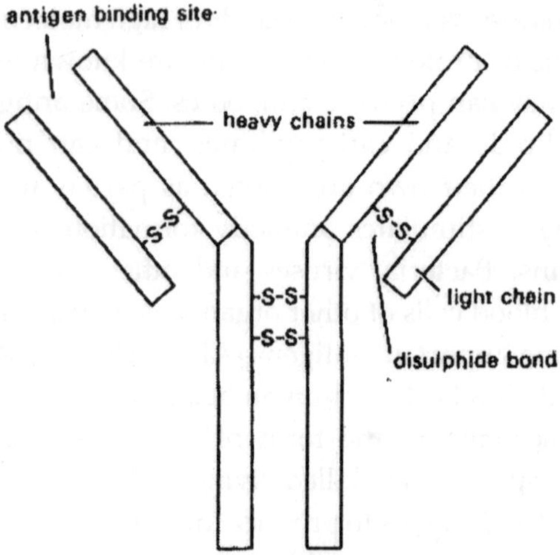

Monomers of the immunoglobulin molecule joined by two heavy and two light polypeptide chains.

Antibodies are specific substances produced in the body of the host on introduction of antigens. All antibody molecules are globulin proteins which have been altered under the influence of antigens. These are also known as immunoglobulins. The immunoglobulins are of five types : IgG, IgM, IgA, IgD, IgE. All these types consist of monomeric units; each comprising two light and two heavy polypeptide chains. The chains are joined by disulphide bonds. Each heavy chain has a molecular weight of approximately 55,000 and the light chain of about 25,000.

The heavy chains are specific for each type of immunoglobulin. They contain amino acids in a specific sequence which determines the types of immunoglobulin. There are five types of heavy chains (i) γ (gamma), (ii) α (alpha), (iii) δ (delta), (iv) μ (micron), and (v) ε (epsilon). The immunoglobulins formed by them are respectively known as Ig G, Ig A, Ig D, Ig M and Ig E. The light chains are mainly of two types : (i) κ (kappa), and (ii) λ (lambda). Both can occur in an immunoglobulin. The terminal portions of both heavy and light chains of each monomeric unit show considerable variation, whereas the remaining portions of the chains are relatively constant in amino acid structure.

Each immunoglobulin is specific in its function. Approximately 70% of normal human serum is Ig G. It is the most common type of antibody and it passes from the mother to the foetus through placenta. It provides immunity to the new born. On the basis of antigenic differences, lg G may be classified into IgG_1, IgG_2, IgG_3, and IgG_4. Each IgM molecule is composed of five monomeric units. There is an additional peptide, called J-chain, attached to one or both the heavy chains. About 6% of the human sera is Ig M.

It usually appears following induction by an antigen and is very effective against viral and bacterial antigens. Ig A constitutes about, 10% of the total serum. Its basic structure is similar to that of Ig G. Besides blood serum, it also occurs in almost all body secretions like tear, saliva, seminal fluid, urine and colostrum. Ig A is an essential part of defensive mechanism of the host. Only 1% of the serum is Ig D. Little is known about this serum, but it is understood that it regulates the synthesis of other immunoglobulins. About 0.002% of total immunoglobulins present in the serum constitutes Ig E. When combined with antigens, it shows allergic reactions.

Antibodies react against a specific pathogen or its toxic products. Hence, antibodies are important in the treatment of infection caused by specific micro-organisms.

Serological Reactions

An antibody is an immunoglobulin molecule secreted into the tissue fluids when an organism is exposed to a foreign substance the antigen. The antibody can combine only with an antigen which is identical or nearly identical to the inducing antigen. In an antigen - antibody reaction, first, union of antigen and antibody occurs, which is usually not visible. The second stage is the visible result of antigen - antibody union, which is exemplified by agglutination, precipitation, opsonization, lysis, etc. Such different manifestations are not necessarily due to various types of antibodies. A particular antibody may be involved in any type of reaction, depending upon the kind and size of antigen, the physical conditions of the suspending fluids, the presence of electolytes, and several other factors. These reactions can be used to detect antigen using a known antibody, or to detect antibody with a known antigen. Such reactions are thus the basis for several in vitro serological tests. Serology is the study of blood serum containing antibodies. Some important serological reactions are as follows.

Agglutination : Agglutination is the visible stage as a result of mixing a particulate or cellular antigen with a homologous antiserum. Agglutinins are antibodies that cause clumping of the bacterial cells for which they are specific. This test is performed by mixing bacterial cells with the serum prepared from animal's blood. When cells interact with the appropriate antibody they clump together and form flocculent masses. This serological reaction is useful in the identification of bacteria. It is easy to perform and simple to interpret. Agglutination tests are classified as macroscopic when the test is carried out in small test tubes and microscopic when antigen and antiserum are mixed on a slide and examined under microscope. When agglutinated cells are erythrocytes the phenomenon is called haemoglutination.

Precipitation : Precipitation occurs when the antigen is soluble instead of cellular. Precipitins are those antibodies that cause precipitation or flocculation of extracts of bacterial cells or other soluble antigens. In precipitin test a reaction takes place between a soluble antigen and its homologous antibody. The reaction is manifested by the formation of a visible precipitate at the interface of the reactants. This reaction is used in medico-legal work in the identification of blood stains on clothes and weapons used in murder cases.

Opsonization : Opsonins are antibodies that make microorganisms more susceptible to ingestion by phagocytes. The effect is known as opsonization (Gr. opsonin = to prepare food for). Antigens that have bound antibody are far more easily phagocytized than the antigen alone. Complement-dependent opsonization is important in limiting bacterial infections, particularly in capsulated bacteria.

Lysins : Lysins cause lysis or breakdown of bacterial or other cells which are specifically sensitive to their action. This test is based on the presence of complement-fixing antibodies in the serum. In the presence of complement these antibodies cause lysis of the specific cells. The test determines whether specific antibodies are present in the serum or not. The test is widely used in laboratory diagnosis of many infectious diseases, It is especially useful when the test antigen and antibody combination does not give a visible reaction such as that occurring in agglutination and precipitation.

Acquired Immunodeficiency Syndrome : Acquired immunodeficiency syndrome or AIDS is a fatal disease comprising serious clinical condition of various manifestations characterized by underlying cellular immunodeficiency. The disease probably originated in southern parts of Africa, but it was first detected in June, 1981 at 'Centre for Disease Control,' Georgia in America.

At first when it surfaced in USA, it appeared to be a disease of the male homosexuals and male prostitutes, but nevertheless huge number of others also suffer. AIDS has in recent years grown exponentially into a frightening worldwide epidemic. It is the first great pandemic of 20th century which has claimed thousands of lives throughout the world. WHO estimates that more than 10 million people are currently infected with human immune deficiency virus (HIV). In India, first report of AIDS case came in Jan. 1986 from Madras and now, in some areas, it is spreading alarmingly.

Cause of AIDS

AIDS is caused by a retrovirus, HIV (human immune deficiency virus). Previously, it was called HTLV-III (Human T-lymphocyte Virus III) by Robert Gallo of National Cancer Institute, USA or LAV (lymphadenopathy associated virus) by Luc. Montagnier of Pasteur Institute, Paris. The virus destroys part of the body's immune system, leaving victims unable to defend themselves against infections and certain cancers.

Structure of AIDS Virus

The structural model of HIV shows that virus is a sphere containing RNA as genetic material and an enzyme reverse transcriptase. The lentivirus nature is reflected by the capsid which is in the form of an elongated protein core. The major inner protein, called P_{24} has a molecular weight of 24,000 whereas another protein, called P_{18} covers the inner surface of the envelope. The envelope is a two-layered lipid or fatty coat. The surface of the virus particle is studded with knob-like structures. The full genome of HIV codes for its various components. The gene 'Gag' codes for the RNA containing core; gene 'pol' is responsible for the production of enzymes reverse transcriptase and endonuclease, whereas gene 'env' produces the envelope proteins. The newly discovered genes, tat, art, sor arid 3'orf encode small protein that help to regulate gene expression.

Mechanism of Action

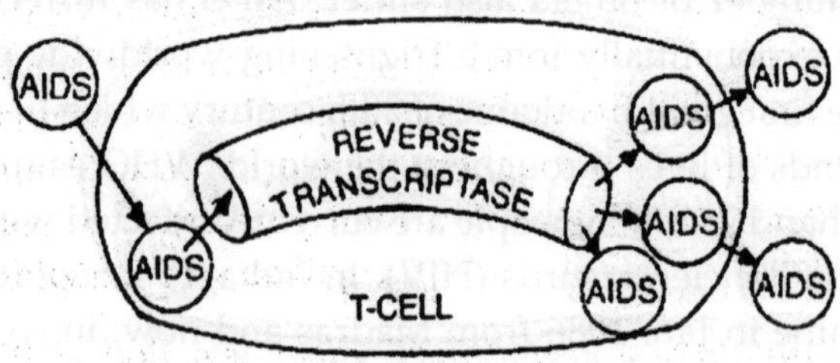

Invasion and multiplication of HIV in the human body.

The primary target of HIV is T_4 lymphocytes which play a major role in maintaining the body's response to infection. Once the virus enters the blood stream, it attacks T_4 lymphocytes and the viral genes are integrated into these cells. There, they can apparently remain dormant for indefinite period without causing any ill effects. This is called incubation period which ranges from six months to five years or more. The virus converts T-cells from a lymphocyte to HIV factory, thus damaged T-cells are unable to stimulate β–cells. As a result, β–cells could not multiply and produce antibodies for the invading virus. The resulting immunodeficiency is so severe that the host is unable to cope with even those infectious agents which pose no problem to individuals with intact immune mechanisms. Besides, HIV also attacks monocytes, macrophages and' even glial cells of the brain and CNS. Since HIV comprises many variants, its infection may be variable.

Symptoms of AIDS

Early symptoms of AIDS may be vague and ill-defined. They include weight loss, fever, diarrhoea, oral thrush and enlargement of the lymph glands. Not all sufferers of early symptoms will go on to develop the full-blown illness. A number of individuals develops a mild condition, known as ARC (AIDS-related complex). In severe condition, the patients may develop pneumonia, kaposi's sarcoma (a skin cancer) and lymphoma (cancer of the lymph system). Although AIDS has a long incubation period (possibly five years), once it does develop there is a rapid decline in health. For an individual who develops AIDS, the likelihood of death within a year, is very high.

Transmission of HIV

HIV can be transmitted from an infected person to a healthy one by the following means.

Sexual Intercourse : HIV is found in semen and vaginal fluids, thus sexual transmission can occur from man to woman, woman to man or man to man.

Blood and Blood Products : Transfusion of HIV-contaminated blood can infect the recipient.

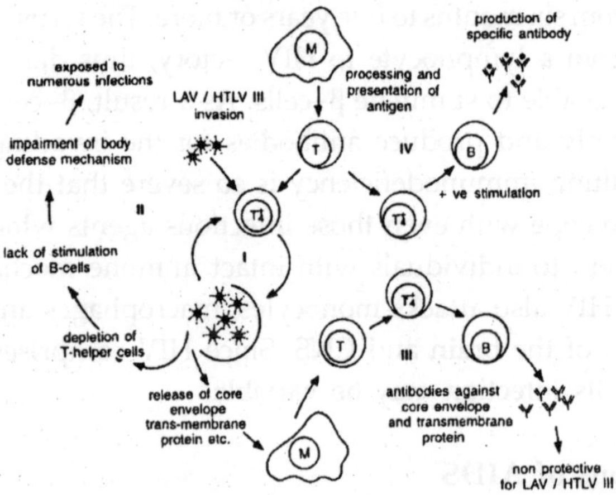

Mechanism of action of HIV.

Shared Needles : Users of intravenous drugs are a major risk group because many of them share needles and syringes without proper cleaning. However, any unsterilised, skin-piercing instrument- including ear- piercing or tattooing needles can spread the disease from one person to another.

Mother-to-child : A woman infected with HIV may spread the disease to her child during pregnancy, during birth or shortly after birth. It is also possible that an infected mother could transmit the virus through breast-feeding.

HIV is not spread through casual contact in school, on the job, in the swimming pool, or at the market. It is also not spread by toilet seats, hand shakes, hugs, casual kissing, eating from the same dish, drinking from the same glass, or by food handlers in restaurants. It also does not spread by mosquitoes or other insects.

AIDS Diagnosis

The individuals suffering from HIV are usually diagnosed by the following two methods:

Clinician make Diagnosis of AIDS by History and Medical Examination for Symptoms : Doctors usually diagnose AIDS - infected persons from typical history and symptoms shown by the patient. The symptoms of full blown AIDS disease or AIDS-related complex (ARC) are clearly different from the symptoms of other known diseases. Moreover if the patient is of young age group and in high risk group (intravenous drug user, transfusion recipient, hemophilic, or polygamic), one can say with certainty that person is suffering from full blown AIDS or ARC. Besides, the bood sample analysis shows the sign of anaemia, leukopenia, lymphopenia (low T-helper cell count), reversal of T4:T8 ratio (0.9), elevated immunoglobulin level and B-cell abnormalities.

Screening Tests

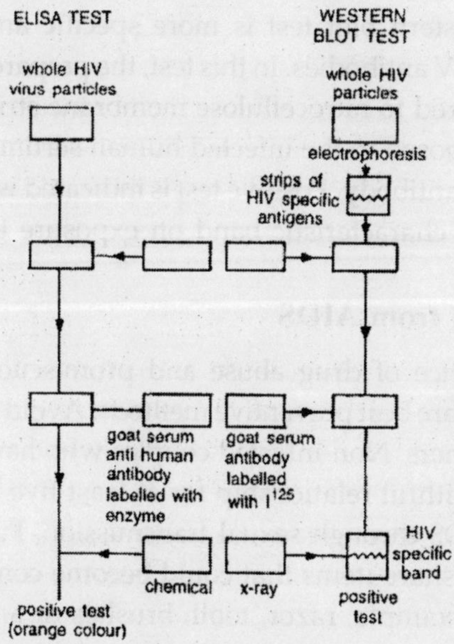

Screening tests for AIDS

Two types of screening tests - elisa test and western blot test are done to confirm whether the patient is suffering from AIDS. In these tests the individual's serum is used to detect HIV antibodies.

The enzyme - linked immunosorbent assay (ELISA) for antibody to human immune deficiency virus (HIV) has proved to be a valuable tool in examining the association between exposure to HIV and AIDS and its related complex. ELISA test is simple and convenient to perform.

It is a colour reaction test in which antigen (HIV particles) binds antibodies to HIV (in infected human serum). This complex fixes the added goat antihuman antibody labelled with a chemical enzyme. It gives a colour reaction with a particular chemical, *e.g.*, orange colour with peroxidase labelled goat antihuman antibody. However, ELISA test is not 100% reliable because it is not completely specific for HIV antibodies but also reacts with other related viral antibodies.

The western blot test is more specific and confirmatory assay for HIV antibodies. In this test, the prepared HIV particles are transferred to nitrocellulose membrane strips. These strips are then exposed to the infected human serum and to the goat antihuman antibody. Positive test is indicated when these strips display the characteristic band on exposure to X-ray films.

Prevention from AIDS

Avoidance of drug abuse and promiscuous homosexual intercourse are best preventive methods. Avoid having multiple sexual partners. Non-infected couples who have maintained a mutually faithful relationship for at least five years are not at risk for AIDS through sexual transmission. Furthermore, one should not share items that could become contaminated with blood, for example, razor, tooth brushes or any skin-piercing instrument.

Treatment of AIDS

The most distressing thing is that at present there is no cure for AIDS patients. So, fatality rate is 100%. There is no effective drug or vaccine available to restore the collapsed immune system of the patient and to control infections permanently and effectively. A recently introduced drug, N-butyl deoxynojorimycin, however, holds some promise. Currently, AZT (azidothymidine), which prevents reverse transcription, is used for the treatment of AIDS.

Patients receiving AZT showed increased numbers of circulating helper T-cells, increased immunological capacity, a virostatic effect, weight gain and improved general well being. Ribavirin is an another drug which is tried intensively in AIDS patients. The drug stops the progression of infection associated with AIDS.

23

The Cyanobacteria

Cyanobacteria, also known as blue-green bacteria, comprise about 1,500 species of photosynthetic procaryotes that live in a great variety of habitats. They occur in soil, in fresh and sea water and even in the extreme conditions like glacier ice and hot springs. One reason for their wide distribution is their great independence. Their photosynthetic pigments are chlorophyll a and phycobiliproteins. They differ from two other major groups of photosynthetic bacteria - purple and green bacteria - in the nature of their photosynthetic pigments and in their capacity to perform oxygenic photosynthesis. Thus cyanobacteria show physiological resemblances to aerobic photoautotrophs.

Since their recognition as a biological group early in the 19th century, Cyanobacteria have been treated as a class or division of algae, commonly known as blue-green algae or Cyanophyta. This taxonomic treatment, primarily based on their plant-like photosynthetic activities, became untenable around 1960 when fundamental differences in cellular organization of procaryotes and eucaryotes were fully elucidated. It became evident that the cell of blue-green algae, long known to be unlike those of any other algal group, is typically procaryotec.

Many phycologists (Fritsch, 1945; Prescott, 1969; Morris, 1973; Round, 1973), however, consider that recognition of the procaryotec organisms as bacteria is an intrusion of bacteriologists into their traditional field. Phycologists consider that an organism with chlorophyll a and the plant body (thallus) not differentiated into roots, stems and leaves is an alga. However, the blue-green algae share the following characters with bacteria.

(1) Both, the blue-green algae and bacteria have a procaryotec cell structure.

(2) Nuclear membrane, nucleolus and membrane bound plastids are absent in both the groups.

(3) The blue-green algae have a mucilagenous sheath outside the cell wall as in bacteria.

(4) Structures like hormogonia (a characteristic of blue-green algae) are present in some bacteria like Thiothrix.

(5) Beggiotoa, a sulphur bacterium and Oscillatoria show resemblance in shape and movement.

(6) The blue - green algae are sensitive to antibiotics like bacteria.

(7) The blue - green algae resemble with bacteria in many biochemical processes, such as sulphur and nitrogen metabolism.

(8) As in bacteria, genetic recombination is also known to occur in many blue-green algae.

Cell Structure

Cells of cyanobacteria may occur singly, enclosed within a mucilagenous envelope (*e.g.*, Gleocapsa, Microcystis) or form chains called trichomes (*e.g.*, Oscillatoria, Gloeotricha) or as a sheet (*e.g.*, Dermocarpa). The filament consists of one or more trichomes enveloped in a gelatinous sheath.

The cells show a typical procaryotic structure. There are no membrane-limited organelles within the cells although a variety of vesicles and inclusions may be present. The cytoplasm is surrounded by a plasma membrane, followed by a firm cell wall and a mucilage sheath.

Mucilage Sheath : The cell wall of cyanobacteria is surrounded by a thin (*e.g.*, Anacystis montana) or thick (*e.g.*, Anabaena) mucilage sheath, also known as slime layer or capsule. In unicellular forms, each individual has a distinct sheath of its own, whereas in filamentous forms, there is a common gelatinous sheath.

The sheath is composed of a complex mixture of mucopolysaccharides and glucoronic acid. It is made up of fibres, held together in the mucilage. The fibres arise from the cell wall, but are not firmly attached to it and as such the trichome may turn within the sheath. The orientation of fibres varies in different species. The mucilage sheath probably helps in protecting the cells from desiccation, particularly in those species which do not produce resistant spores. It also helps in gliding movements.

Cell Wall : The cell wall of cyanobacteria is thin and firm and is basically similar to that of Gram-negative bacteria. It is differentiated into four layers which are designated as L_I, L_{II} L_{III} and L_{IV}; L_I being the inner most lying next to plasma membrane and L_{IV} the outer most. The cell wall is made up of peptidoglycan together with carbohydrates, amino acid and fatty acids.

Cytoplasmic Membrane : It forms the limiting membrane of the cytoplasm and is made up of protein-lipid-protein layers. It has many infoldings which penetrate the cytoplasm. It has been suggested that these infoldings give rise to new pigment containing thylakoids.

Chromatoplasm : It is the peripheral pigmented portion of the protoplast and has a reticulate, alveolar or homogeneous structure. The chromatoplasm is characterized by the presence of flattened vesicular structures, called thylakoids or photosynthetic lamellae. A lamella is composed of two membranes joined at the ends and enclosing an intrathylakoid space of 70- 80Å width.

The lamellae thus form closed flattened sacs which contain chlorophyll a, carotenoids and three phycobiliproteins : c-phycocyanin, allophycocyanin and c- phycoerythrin. The thylakoids are arranged in parallel rows close to the periphery of the cell or they are distributed irregularly throughout the cell. Thus the pigments are not contained in true membrane-limited plastids.

The photosynthetic lamellae of Cyanobacteria are also considered as the sites of cellular respiration and as such they are preferably called photosynthetic respiratory membranes. The cells of Cyanobacteria are devoid of Golgi apparatus, mitochondria, endoplasmic reticulum and other membrane bound organelles. In addition to thylakoids, chromatoplasm contains protein granules and gas and oil vesicles. Cyanophycin granules, also known as structured granules, are irregular or polyhedral in shape and they store protein in the form of polypeptides. Polyhedral bodies are angular in shape and are usually associated with the DNA in the central part of the cell.

They are considered to be the carboxysomes which store the carbon dioxide-fixing enzyme ribulose-1,5-bisphosphate carboxylase. Polyphosphate bodies, also known as volutin granules, are small spherical structures, located near the centre of the cell. They store phosphate and are found abundantly in mature cells grown in the medium with high phosphate contents. Gas vesicles are small, tubular structures, composed of protein ribs or spirals exclusively. These vesicles which contain gas have quite rigid wall but they are collapsed when pressure is applied. The gas vacuoles shield light and also provide buoyancy.

Centroplasm : The central colourless region of the cell is known as centroplasm or nucleoplasm. It consists of several naked DNA which are not associated with histones. However, in some Cyanobacteria a histone-like protein binds nonspecifically with DNA.

The nuclear material is not surrounded by a nuclear membrane and nucleoli are also absent. The nucleoplasm is not sharply differentiated and usually intrudes into the peripheral chromatoplasm. Although this region stains like that of the nucleus of eucaryotes, it cannot be regarded as true nucleus and is called incipient nucleus.

Reproduction : Gametes or flagellated zoospores are not found in Cyanobacteria. The most common method of multiplication is binary fission. The rate of cell division is very high and in certain coccoid forms (*e.g.*, Anacystis nidulans) there are as many as 12 doublings in a day.

Such a rapid growth results in the formation of dense water blooms. Many bloom-forming species are buoyant and may rise to the surface, forming a scum or mat. Multiplication also takes place by some specialized structures, such as hormogonia, akinetes, endospores and heterocysts. These structures are discussed in details in unit Algae.

Ecological Importance of Cyanobacteria : Cyanobacteria occur in almost every moist environment, in the sea, in fresh water, and on the land. As all Cyanobacteria are photosynthetic, they live near the surface, where they can obtain light. They form a considerable part of the phytoplankton on the sea and fresh water. They produce oxygen as a by-product of photosynthesis; and some species provide food for heterotrophs. The unpleasant aspect of Cyanobacteria is that they cause water pollution as many forms are toxic and slimy. They thrive well in water that gets industrial and sewage waste and form smelly blooms. These blooms cause a decline in the fish population. Cyanobacteria, as they die, provide food for an enormous array of aerobic bacteria, which use all oxygen in the water and suffocate the fish.

Fish cannot control populations of cyanobacteria by consuming them because toxins and gelatinous sheaths of Cyanobacteria make them inedible. Pretreatment and controlled discharge of industrial and sewage waste in lakes, ponds and rivers can help in reducing water pollution by cyanobacteria.

Actinomycetes

Actinomycetes are unicellular organisms placed in the order Actinomycetales or the class Schizomycetes. They occur abundantly in soil, water, mud, manure, milk and other food products. Most of the actinomycetes are saprophytes but some are parasites. The latter cause some serious diseases in plants as well as animals.

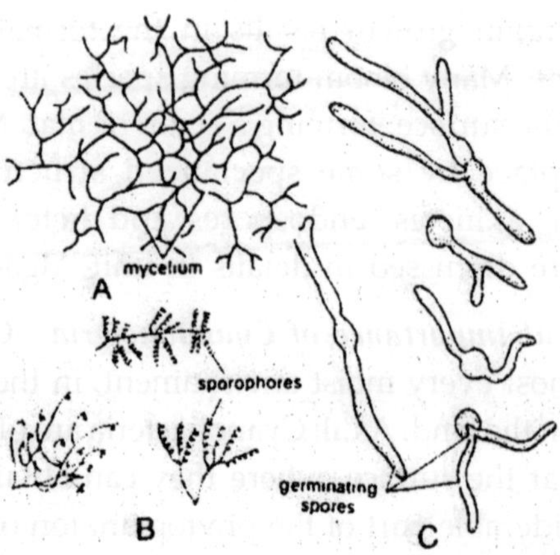

A-C. Actinomycetes : A. Mycelium, B. Sporophores, C. Spore.

Structure : Actinomycetes (actis = ray, mykes = fungus) are fungus like bacteria with cylindrical cells which are usually united to form filaments resembling the mycelium of a true fungus. The mycelium is branched, non-septate and thin (0.2-1.2 μm in width). In some species the mycelium breaks up into small bacteria-like cells. In young mycelium the cytoplasm is homogeneous, but at maturity many vacuoles, fat droplets, granules and few rod-shaped bodies develop in the cytoplasm. In actinomycetes, as in true bacteria, there are no well differentiated nuclei but many chromatin granules are present. At maturity, the cell wall of the mycelium becomes fragile and breaks up easily. Most of the actinomycetes are Gram-positive. They are non-motile but in Actinoplanes small flagella are present.

Reproduction : They generally multiply by means of fragmentation. The mycelium breaks up into small hyphae and each hypha grows into a new mycelium. Some species of actinomycetes also reproduce by means of spores and conidia, which develop on sporophores and conidiophores, singly or in long chains.

Classification of Actinomycetes : The order Actinomycetales consists of the following four families.

Actinomycetaceae : The members of this family have a tendency to form branched filaments which are very slender and nonseptate and about a micron in diameter. They form spores and conidia in long chains.

Example : Actinomyces (commonly known as,ray fungus').

Mycobacteriaceae : These actinomycetes form a rudimentary mycelium, which is unstable and is fragmented into immotile rods. These bacteria are acid-fast.

Examples : Mycobacterium, Corynebacterium, Nocardia, Mycoplan.

Micromonosporaceae : The vegetative mycelium of these fungi does not fragment in rod-shaped or coccid forms. It produces a compact colony with a moist and smooth surface. The terminal hyphae of loose aerial mycelia, which arise over the surface of the colony, bear spores either in chains (*e.g.*, Streptomyces) or singly (*e.g.*, Micromonospora).

Examples : Micromonospora, Streptornyces.

Proactinomycetaceae : In these forms the mycelial development is transitory and often limited. They do not produce specialized spores and reproduce by fragment of mycelium into short rod-shaped cells.

Examples : Proactinomyces, Connistreptothrix.

Economic Importance of Actinomycetes

Useful Activities : Economically, the members of the family Micromonosporaceae of the Actinomycetes are very useful as they are capable of synthesizing antibiotic substances. Some important antibiotics synthesised by them are listed in the Table.

Harmful Activities : Many actinomycetes cause serious diseases in plants as well as in animals and human beings. Some common diseases caused by actinomycetes are as follows:

		Pathogen
1.	Human diseases	
	Tuberculosis	Mycobacterium tuberculosis
	Leprosy	Mycobacterium leprae
	Diphtheria	Corynebacterium diptheriae
	Actinomycosis	Actinomyces Israeli
	Madura foot	Nocardia madurae
2.	Animal diseases	
	Actinomycosis or	Actinomyces bovis Lumpy jaw
	Tuberculosis of cattle	Mycobacterium bovis
	Tuberculosis of domestic fowls and other birds	Mycobacterium avium
3.	Plant diseases	
	Tondu of wheat	Corynebacterium tritici
	Scab of potato	Actinomyces scabies

Some Important Antibiotics and their Source, Range and Mode of Action

Antibiotic	*Source*	*Range of action*	*Mode of action*
Streptomycin	Streptomyces griseus	Gram'+'ve and Gram'-'ve bacteria, tuberculosis bacteria	Induce abnormal protein synthesis by inhibiting 30s ribosome function
Neomycin	Streptomyces fradiae	Gram'-'ve bacteria inhibits (most of the intestinal bacteria)	Induces abnormal protein synthesis by inhibiting 30s ribosome function
Viomycin	Streptomyces puniceus	Gram'-'ve bacteria, including tuberculosis bacteria	Interferes with protein synthesis
Chlorotetra-cycline or Aureomycin	Streptomyces aureofaciens	Gram'+'ve and Gram '-'ve bacteria, rickettsia	Interferes with protein synthesis
Oxytetracycline or Terramycin	Streptomyces rimosus	Gram'+'ve and Gram '-'ve bacteria, rickettsia	Interferes with protein synthesis
Novobiocin	Streptomyces niveus	Proteus and Diplococcus protein synthesis	Interferes with
Chloramphenicol or	Streptomyces venezuelae	Gram'+'ve, Gram '-' ve bacteria,	Inhibits protein synthesis by interfering with 50s ribosome function

Contd....

Antibiotic	*Source*	*Range of action*	*Mode of action*
Chloromycetin		rickettsia	
Erythromycin	Streptomyces erythreus	Gram'+'ve bacteria	Inhibits protein synthesis
Linomycin	Streptomyces lincolnesis	Gram'+'ve bacteria	Interferes with protein synthesis
Rifamycin	Streptomyces mediterranei	Active against tuberculosis bacteria	Interferes with protein synthesis
Nystatin	Streptomyces noursei	Fungal infections due to Candida	Damages cell membrane

Mycoplasmas

Mycoplasmas are the smallest known aerobic procaryotes without a cell wall. These microorganisms were first discovered by Pasteur (1843) while studying the causative agent of pleuropneumonia in cattle (bovine pleuropneumonia). They were designated as PPLO (pleuropneumonia-like organisms). However, Pasteur could not isolate them in pure cultures on standard nutrient media or observe them under the light microscope. In 1898, two French microbiologists, Nocard and Roux, were successful in obtaining pure cultures of these microorganisms in media containing serum. They observed that these organisms could produce disease when inoculated in healthy cattle.

Mycoplasmas are frequent contaminants in tissue cultures rich in organic matter. They have also been found in hot-water springs and other thermal environments. They occur in soil, sewage water, different substrates and in humans, animals and plants. These pleuromorphic microorganisms were named as Asterococcus mycoides by Borrel et al. (1910). Nowak (1929) placed these organisms in the genus Mycoplasma, which belongs to the class Mollicutes of the order Mycoplasmatales.

Classification of Mycoplasmas

The following three genera of mycoplasmas have been recognised on the basis of their nutritional requirements.

Mycoplasma : These mycoplasmas require cholesterol for their growth which is incorporated in their cell membrane. They are parasitic and infect mucous membranes and joints of humans and animals.

Acholeplasma : These mycoplasmas do not require cholesterol for their growth, but will incorporate it into the membrane if it is furnished in the medium. They are found in vertebrates and sometimes in plants, and also occur in soil and sewage water.

Thermoplasma : These mycoplasmas also do not require cholesterol for their growth. They are strictly aerobic, acid loving microorganisms, growing well at pH 0.96-3.0. The optimum temperature for their growth is 59°C.

Structure of Mycoplasmas

The absence of a true cell wall makes these organisms highly plastic and readily deformable, hence mycoplasmas are irregular and variable in shape. The cells may be coccoid, granular, pear-shaped, cluster-like, ring-like or filamentous. The filaments may be branched or unbranched. There is so much variability in shape that no two forms are alike. The cells are small, ranging in diameter between 0.3 and 0.9μm. Since mycoplasmas pass through many filters (due to their plastic nature) and grow on media which do not contain live tissue, they are considered to be intermediate between bacteria and viruses.

These organisms are covered with a unit lipoprotein cytoplasmic membrane, 7.5-10 υm thick. The cytoplasm contains ribosomes and nucleoplasm-like structure. Though genetic material is composed of both DNA and RNA, it is less than half that usually occurs in other procaryotes, and is perhaps the lowest limit required for a cellular organism. The amount of DNA is up to 4 per cent and RNA about 8 per cent. The G+C content in DNA ranges between 23 and 40 per cent. Mycoplasmas may be the simplest form of life capable of independent growth and metabolism.

Mycoplasmas are Gram-negative and stain slowly on long exposure to dyes. They are usually non-motile; some forms, however, show gliding movements. They reproduce by budding or binary fission.

They are sensitive to oxytetracycline, streptomycin, erythromycin and chloramphenicol.

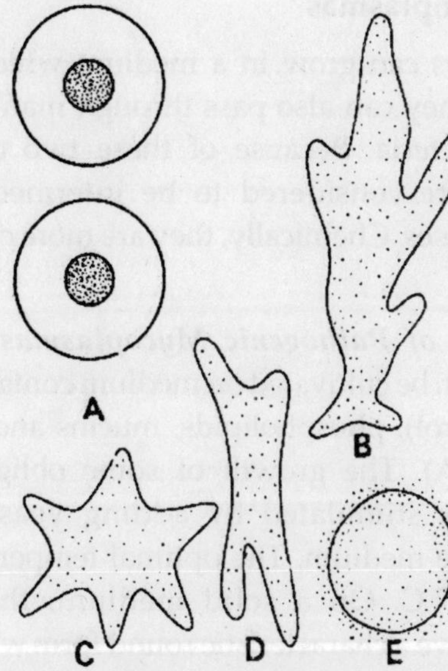

A-E. Mycoplasma : Morphology of colony and cell shape; A. colony, B-E. Cells of different shapes.

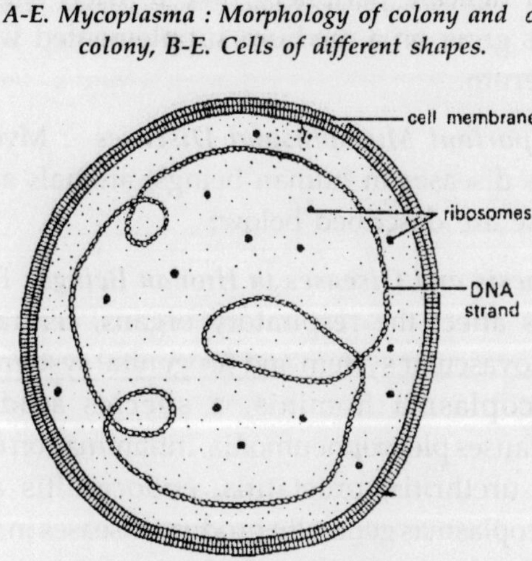

Mycoplasma : Electron micrograph of a cell.

They do not have cell wall with peptidoglycans (like that of bacteria) and are insensitive to penicillin and some other antibiotics such as ampicillin and methicillin, which specifically affect peptidoglycan synthesis.

Nature of Mycoplasmas

Mycoplasmas can grow in a medium which contains no living tissues. They can also pass through many filters which cannot pass bacteria. Because of these two characteristics, mycoplasmas are considered to be intermediate between bacteria and viruses. Chemically, they are more close to bacteria than viruses.

Cultivation of Pathogenic Mycoplasmas : Pathogenic mycoplasmas can be cultivated on medium containing proteins, sterols (cholesterol), phospholipids, mucins and nucleic acids (DNA and RNA). The growth of some obligate anaerobic mycoplasmas is stimulated by adding yeast extract and cholesterol to the medium. The optimal temperature for their growth is 36-37°C. On a solid medium, the colonies of mycoplasmas have a characteristic round form with a thickened centre and a delicate periphery, like a 'fried egg'. Aerobic mycoplasmas grow on a medium supplemented with 20 per cent horse serum.

Some Important Mycoplasmal Diseases : Mycoplasmas cause serious diseases in human beings, animals and plants. Some of these are described below:

Pathogenesis and Diseases in Human Beings : Pathogenic mycoplasmas affect the respiratory organs, central nervous system, cardiovascular system and urogenital system in human beings. Mycoplasma hominis, a species studied most extensively, causes pleuropneumonia, inflammation of genitals, non-specific urethritis, prostatitis, endocarditis and other diseases. Mycoplasmas generally produce diseases in man when his body resistance is significantly low.

Mycoplasma pneumoniae causes primary atypical pneumonia (PAP), haemorrhagic laryngitis and vesicular inflammation of tympanic membrane. Mycoplasma hominis and M. fermentants are known to cause infertility in man. M. orale and M. salivarium also affect human beings.

Pathogenicity for Animals : Mycoplasma mycoides causes pleuropneumonia in cattle. Some strains of this mycoplasma produce pleuropneumonia in goat. Mycoplasma bovigenitalium causes inflammation of genitals in animals. Another species, Mycoplasma agalactia, produces agalactia of sheep and goat.

Mycoplasmal Diseases of Plants : In plants, mycoplasmas produce diseases usually by upsetting the hormonal balance. This imbalance is expressed in the form of symptoms such as dwarfness, loss of apical dominance, enlargement of floral buds and bunchy growth of newly formed sprouts. In plants the mycoplasmal diseases are transmitted by insect vectors, belonging to the class Cicadellidae and Psyllidae. Multiplication of mycoplasma within insect body has been demonstrated in many experiments. Insects fed on mycoplasma infected plants become infectious only after a specific latent period. Some common mycoplasmal diseases of plants are : (i) Little leaf disease of brinjal, (ii) Clover dwarf, (iii) Clover phyllody, (iv) Clover virescence, (v) Cotton virescence, (vi) Bunchy top of papaya, (vii) Witches broom of legumes, (viii) Stripe disease of sugar cane, (ix) Big bud of tomato, and (x) Yellow dwarf of tobacco.

Rickettsias

Rickettsias are the smallest procaryotes, 0.3-0.7 pin wide and 1- 2μm long. They are larger than any of the viruses and can be seen with any ordinary microscope at the magnification around 1000 X. Their shape shows considerable variation; they may be cigarette-or rod-shaped, spherical or ovoid and often pleomorphic. They multiply by transverse binary fission, like bacteria. They form pairs or chains after fission. They are non-motile and do not produce spores. Unlike most bacteria, it is difficult to stain rickettsias with ordinary basic aniline dyes. They can, however, be stained with Giemsa's stain, and if they take Gram's stain, they are gram '+'ve.

Rickettsias are intracellular obligate parasites which can grow only in the presence of another living cell. In laboratory, they are easily cultivated in 'live chick embryos'. In nature, they exhibit close relationship with blood-sucking arthropods like fleas, lice and ticks.

Structure : Like true bacteria, the cell wall of rickettsias contains mucopolysaccharide as the main supporting component, along with the amir sugar and muramic acid. The cell structure is of typical procaryotic type. Ribosomes and nucleoids occur in the cytoplasm. Invaginations of the cell membrane suggest the presence of mesosomes or a primitive endoplasmic reticulum.

Rickettsias show autonomous, but incomplete metabolic activity. Though they do not possess their own independent systems of digestive and synthetic enzymes, they are markedly susceptible to antibiotics that interfere with energy-mediating enzymes.

Transmission : Rickettsias are primarily parasites of insects, and appear only secondarily in man and other animals. They inhabit the cells lining the intestines and other tissues of arthropods, such as ticks, fleas, lice, bed bugs, spiders and mosquitoes. Some species of pathogen rickettsias are found in the salivary glands of biting insects, from where they may be transmitted to man. Other species occur in the intestinal contents of sanguisugent (blood sucking) arthropods and appear in the faeces. Transmission of rickettsias in animal hosts may thus be through bites or by rubbing or scratching the fecal deposits of insect vectors into the skin.

Pathogenic Rickettsias : Rickettsias are the causative agents of several important diseases like typhus fever, rocky mountain spotted fever, scrub typhus and Q fever. Some important pathogenic rickettsias, their vectors and diseases caused by them are presented in the Table.

Some Important Pathogenic Rickettsias

	Disease	*Pathogenic rickettsia*	*Insect vector*	*Interrelationship of insect-pathogen -human*
1.	Rocky mountain spotted fever	Rickettsia rickettsii	Ticks (Dermacentor, Amblyomma, Orinithodoros, etc.)	Pathogen multiplies in wall of tick's midgut; congenitally transferred in tick, humans infected through bite.
2.	Scrub typhus	R. trutsugamushi	Red mites (Trombicula)	Pathogen multiplies in gut of mite; congeni tally transmitted in mite; humans, infected through bite of larval mite.
3.	Rickettsiae	R. akari	Mouse mite (Allodermonyssus sanguineus)	Pathogen multiplies in gut of mite; humans infected by bite of mite.
4.	Classical typhus fever	R. prowazekii	Body louse (Pediculus humanus)	Pathogen multiplies in epithelium of louse's midgut humans infected by bite, faeces, or crushing of louse on skin.
5.	Trench fever	R. quintana	Body louse (Pediculus humanus)	Pathogen multiplies in midgut of louse; humans infected by faeces or crushing of louse on skin.
6.	Murine typhus fever	R. typhi	Fleas (Aenopsylla cheopis, etc.)	Pathogen multiplies in epithelium of midgut of flea; humans infected through bite.

24

Viruses and Viroids

The name virus (Latin virus = venom of poisonous fluid) was given by Pasteur to the causative agents of infectious diseases. Adolph Mayer (1885), a Dutch scientist, and D.J. Ivanowsky (1892), a Russian scientist, recognized certain microbes as causative agent of mosaic disease of tobacco.

The basic criterion used to differentiate these agents from other familiar microbial agents of diseases was their ability to pass through bacteria-proof filters. These agents were thus designated as filterable viruses (now it has been established that many microbes of bacterial nature are also capable of passing through such filters, therefore, the word 'filterable' was subsequently dropped).

M.W. Beijerinck, a Dutch bacteriologist, demonstrated in 1898 that viruses differ from other cellular organisms. He discovered that the virus of tobacco mosaic disease could be precipitated from a suspension of alcohol without losing its infectious power and the fluid was capable of diffusing through agar gel. These characteristics are not possessed by bacteria or any other living organism. This led Beijerinck to believe that the fluid itself was living and he put forward the principle of "Contagium vivum fluidum" (infectious living fluid). In the same year, Loeffer and Frosch demonstrated that foot-and-mouth disease of cattle is caused by a filtrate apparently free of any bacteria. They concluded that if the agent of this disease is particulate, it must be smaller than the diameter of the smallest known bacteria. Nearly 40 years after these onservations, the structure of virus was studied by Wendell M. Stanley, an organic chemist, in 1935. He showed that the infectious principle of virus could be crystallized and that the crystals consisted largely of proteins.

For many years it was thought that virus is simply a protein molecule, but later it was discovered that virus contains a small but constant amount of RNA or DNA in addition to protein. A virus is, therefore, not simply a protein but a nucleoprotein and its infectious principle is the nucleic acid rather than protein. Stanley was awarded Nobel prize in 1946 for this discovery.

Subsequently, many already known diseases like mumps, chicken pox and hog cholera, as well as many newly identified diseases, were found to be caused by filterable agents, *i.e.*, viruses. An English scientist, F.W. Twort (1915), and a Canadian, Felix d'Herelle (1917), independently discovered that a virus was capable of dissolving or lysing bacterial cells. This virus was designated as bacteriophage, *i.e.*, eater of bacteria.

Luria (1953) described viruses as submicroscopic entities capable of being introduced into specific living cells and reproducing inside such cells only. According to Andre Lwoff (1966), a Nobel Laureate French virologist, the most appropriate definition of viruses is "viruses are viruses". Some biologists think that viruses are descendants of cellular organisms that have become highly specialized as parasites, others consider that viruses were originally fragments of DNA or RNA broken off from the nucleic acids of cellular organisms.

However, it is now well established that viruses are obligately parasitic, self-replicating non-cellular organism and essentially composed of a protein covering surrounding a central nucleic acid molecule (either DNA or RNA). Most of the modern virologists prefer not to define viruses by any specific definition but describe them by their various physical, chemical, biological and clinical properties.

General Characteristics of Viruses

Viruses are sub-microscopic particles which have been studied in detail by electron microscope using selective techniques. A simple virus particle, often designated as virion, consists of a nucleic acid core of genetic material, enclosed within a protein coat. The amount of protein in different viruses varies from 60-95 per cent and the rest is nucleic acid.

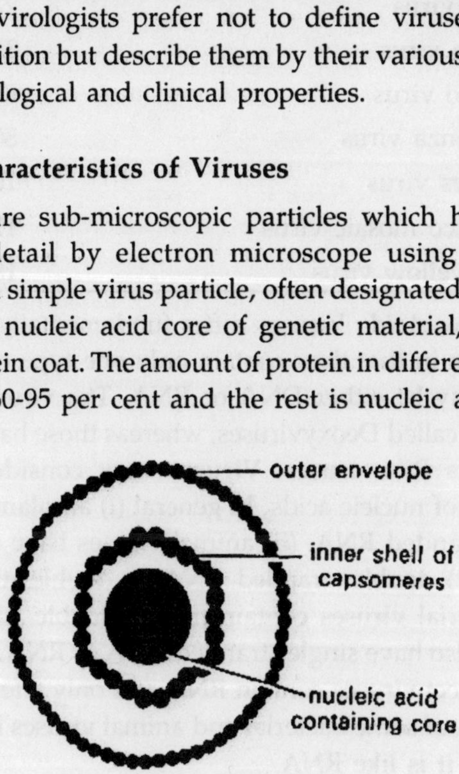

Virus : General structure.

Size : Earlier the size of viruses was measured by using the technique of filtration through collodion membranes of known porosity. But now techniques like ultracentrifugation and electron microscopy are employed. Viruses are very small in size, varying over a wide range from 20-350 nm. The dimensions of some common viruses are shown in the Table.

Size of some Well known Viruses

Virus	*Size (nm)*
Turnip yellow mosaic virus	28
ϕ X-174 virus	22
Tomato bushy stunt virus	30
Polio virus	27-30
Alpha virus	35-80
Adeno virus	60-90
Influenza virus	80-120
Herpes virus	180-200
Tobacco mosaic virus	17.5 X 300
Beet yellow virus	10 X 1250

Nucleic Acid : Viruses differ fundamentally from cellular organisms in that they contain only one type of nucleic acid which may be either DNA or RNA. The viruses containing DNA are called Deoxyviruses, whereas those having RNA are known as Riboviruses. Viruses vary considerably in the structure of nucleic acids. In general (i) all plant viruses have single stranded RNA, (ii) animal viruses have either single - or (rarely) double-stranded RNA or double-stranded DNA, (iii) bacterial viruses contain mostly double- stranded DNA but can also have single stranded DNA or RNA, and (iv) most of the insect viruses contain RNA and only a few have DNA. The DNA of some bacterial and animal viruses is circular, but in others it is like RNA.

Careful extraction of nucleic acids from viruses has shown that a virion contains only a single molecule of nucleic acid. The number of nucleotide pairs in a molecule varies from 1000-250,000 pairs. But the number of pairs in a specific virion is always constant. The amount of nucleic acid depends on the size of virion, usually larger the size of virion greater is the amount of nucleic acid.

Characteristics of nucleic acids of some better known plant and animal viruses are given in the Table.

Characteristics of Nucleic Acids of some Viruses

Virus	*Nucleic acid type*	*Strands pairs*	*Nucleotide*	*Molecular weight in dalton*
Polyoma	DNA	Double	4,500	3×10^6
Adenovirus	DNA	Double	35,000	23×10^6
Coliphage T4	DNA	Double	200,000	130×10^6
Coliphage T2	DNA	Double	60,000	40×10^6
Tobacco mosaic virus	RNA	Single	7,500	2.5×10^6
Tobacco necrosis virus	RNA	Single	1,500	0.5×10^6
Bean mosaic virus	RNA	Single	3,000	1×10^6
Polio virus	RNA	Single	6,000	2×10^6

The Protein Coat : The nucleic acid core of the virus is protected by a protein coat, called capsid. The capsid consists of several identical protein subunits, known as capsomeres. These subunits are usually arranged in the helical or polyhedral geometric forms which are specific for a particular virus.

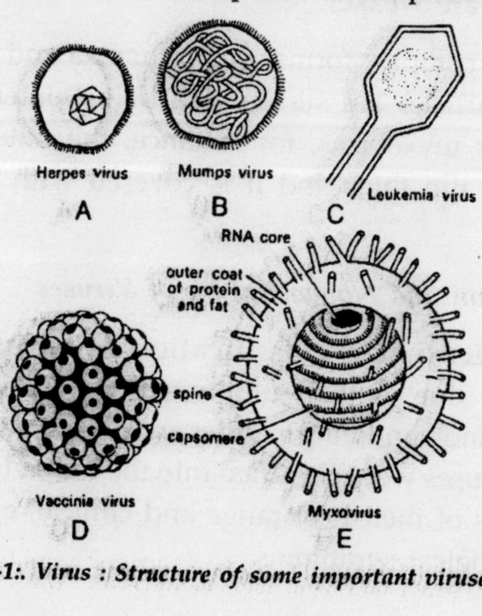

A-1:. Virus : Structure of some important viruses.

Helical Viruses : These viruses are cylindrical or rod-like in form and the central nucleic acid strand is coiled like a helical spring. The protein subunits (capsomeres) are helically arranged around the helical spring. The common examples of helical viruses are New Castle virus, Mumps virus and Tobacco mosaic virus.

Polyhedral Viruses : In these viruses the nucleic acid is packed in an unknown manner within a hollow polyhedral head. They have been classified further into tetrahedral, octohedral and icosahedral forms depending upon the number of faces. Of these, icosahedral form is considered to be the most efficient shape because of packing and bending of capsomeres in a near spherical form. An icosahedron has 12 corners, 20 triangular faces and 30 edges.

The capsomeres, forming the capsid (protein coat) of a virion, are of two types : pentamer - made up of five identical monomers and hexamer - having six monomers. Each monomer is connected with the neighbouring monomers on either side with the help of bonds. Likewise, the capsomeres are also connected with each other, but the bonds between the capsomeres are weaker.

In some complex forms (*e.g.*, influenza and herpes virus), the virus particles are surrounded by an outer envelope. In influenza or myxovirus, for instance, the outer envelope is about 10-15 nm thick and it is covered with projections or spikes.

Classification and Nomenclature of Viruses

Classification : Classification of viruses has been problematic due to their ultramicroscopic size, presence of both living and non-living features and absence of fossil records. Initially, viruses were classified into the following four groups on the basis of their host range and clinical, epidemiological and pathological symptoms.

1. *Plant Viruses :* These viruses infect only plants and depending upon the host they have been subdivided into bacterial viruses, algal viruses, fungal viruses, etc.
2. *Invertebrate Viruses :* Viruses infecting invertebrates have been included in this group.
3. *Vertebrate Viruses :* This group includes viruses infecting vertebrate animals.
4. *Dual-host Viruses :* This group includes those viruses which infect two different hosts mentioned above.

Holmes (1948) included all viruses in a single order Virales which was divided into three suborders.

1. *Phagineae :* This suborder includes viruses infecting bacteria, *i.e.,* bacteriophage.
2. *Phytophagineae :* It includes viruses infecting plants.
3. *Zoophagineae :* It includes viruses infecting animals.

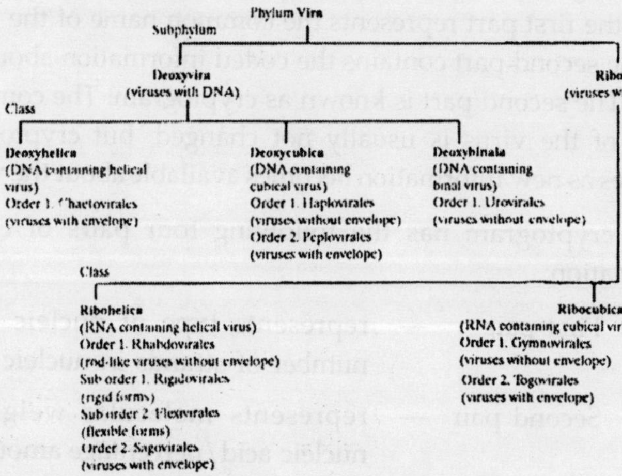

Based on information obtained from ultracentrifuge and electron microscopic studies, Lwoff, Horne and Tournier (1962) proposed a system of classification of viruses (known as LHT system) which was accepted by International Committee for Virus Nomenclature (1966). Their classification is based upon the following characters : (i) type of nucleic acid, (ii) molecular weight of virus, (iii) shape and size of virus, (iv) symmetry of virus, (v) number of protein subunits in a capsid, (vi) diameter of nucleic acid coil, (vii) presence of outer envelope, (viii) intercellular multiplication, (ix) temperature inactivation of virus, (x) method of viral transmission, and (xi) symptoms of virus on the host plant.

The following is an outline of Lwoff, Horne and Tournier's system of classification. Sherwood Gasjens and Jonathan King (1975) proposed the following classification on the basis of genome structure, symmetry, site of assembly and presence and absence of envelope.

Nomenclature : As the binomial system of nomenclature is not suitable for viruses, International Committee for Virus Nomenclature put forward a new system for naming viruses. According to this system, the name of a virus consists of two parts, the first part represents the common name of the virus and the second part contains the coded information about the virus. The second part is known as cryptogram. The common name of the virus is usually not changed, but cryptogram changes as new information becomes available about the virus.

A cryptogram has the following four pairs of coded information:

(1) First pair — represents type of nucleic acid-number of strands in nucleic acid.

(2) Second pair — represents molecular weight of nucleic acid/percentage amount of nucleic acid in the virus.

(3) Third pair — denotes the size of virus/size of nucleoprotein.

(4) Fourth pair — denotes the type of host/ carrier used in transmission of virus.

The cryptograms of two viruses are explained below.

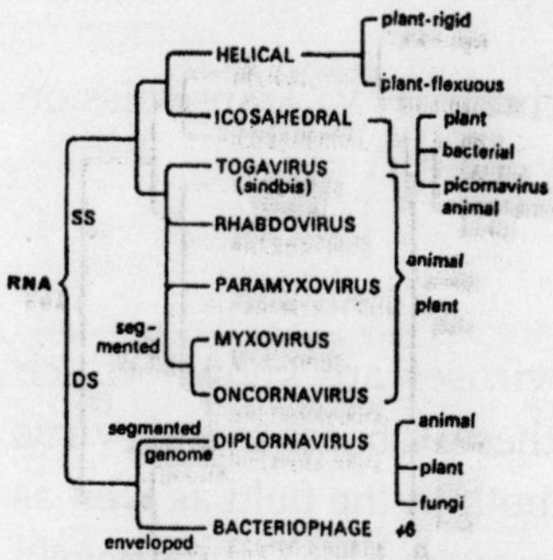

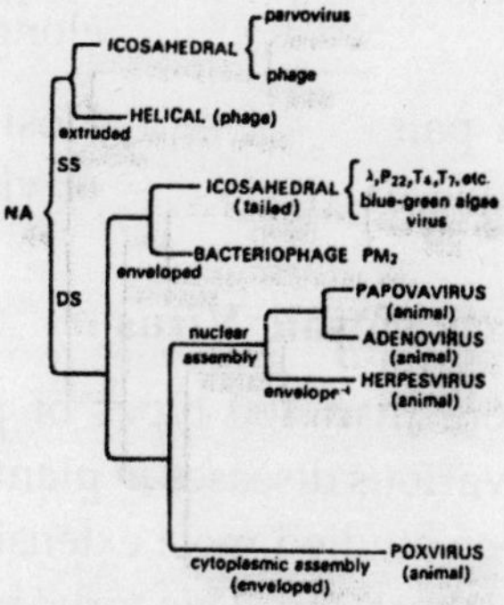

Cryptogram of Tobacco Mosaic Virus (TMV) : The cryptogram of TMV is R/1 : 2/5 : E/E : S/A.

Explanation

First pair — Nucleic acid RNA = (R)/Nucleic acid single stranded = 1

Second pair — Molecular weight of nucleic acid =2 (in hundred thousands)/Amount of nucleic acid 5%

Third pair — Virus elongated = E/Nucleoprotein elongated = E

Fourth pair — Most seeded plants (Spermatophytes) = S/Transmission of virus by air = A

Influenza Virus : The cryptogram of this virus is R/1 : (2- 3)/10: S/E : V/A.

Explanation

First pair — Nucleic acid RNA = (R)/ Nucleic acid single stranded =1

Second pair — Molecular weight of nucleic acid = 2 or 3 (hundred thousands)/Amount of nucleic acid = 10%

Third pair — Virus spherical = S/Nucleoproteins elongated=E

Fourth pair — Host vertebrate = V/Trans-mission of virus by air = A

Tobacco Mosaic Virus

More than 100 types of plant viruses are known which cause various diseases in plants. Of these, tobacco mosaic virus has been studied most extensively, both in the field as well as in the laboratory. This virus was discovered by D. lwanowski in 1892, but its isolation from infected plants and crystallization was done by W.M. Stanley in 1935.

Structure : TMV particles appear as bundle of rods or needles under electron microscope. Each rod is approximately 3000 λ in length and 170 λ in diameter and has a molecular weight of approximately 400,000. The electron microscopic and X-ray crystallographic investigations have revealed that these particles have two constituents, a protein coat and nucleic acid. The protein coat (capsid) is made up of approximately 2130 identical protein subunits called capsomeres. Each

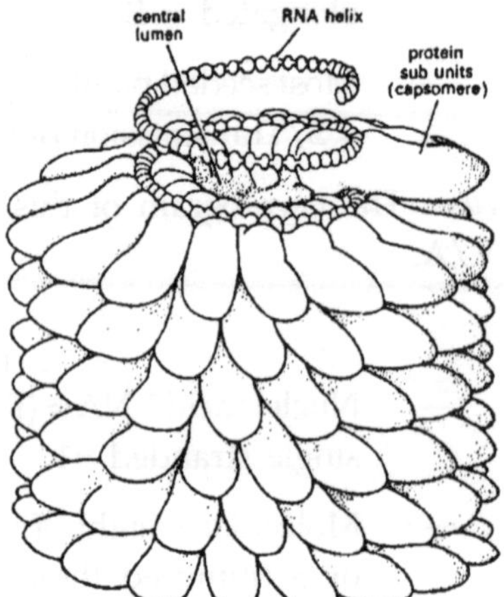

Tobacco mosaic virus : Structure

capsomere consists of a long chain of 158 amino acids and its molecular weight is 18,000. The capsomeres are helically arranged around a central single stranded RNA molecule. The latter consists of some 6,000 nucleotide pairs. The protein and nucleic acid ratio in these particles is 94.4:5.6.

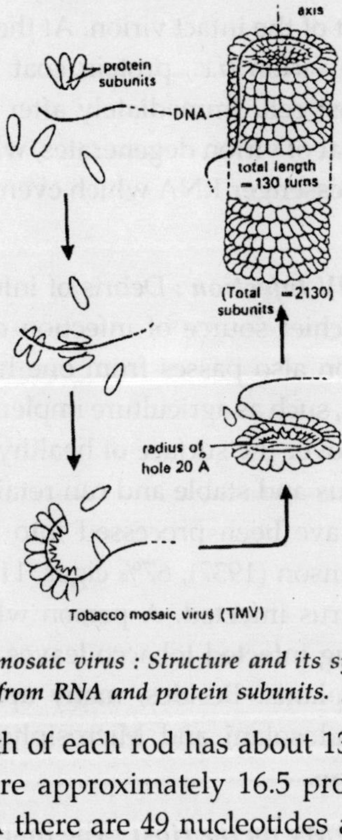

Tobacco mosaic virus : Structure and its synthesis from RNA and protein subunits.

The total length of each rod has about 130 helicals and in each helix there are approximately 16.5 protein subunits. In each helix of RNA there are 49 nucleotides and the helix has a pitch of 23λ. The genetic information necessary for the formation of a complete TMV particle is contained in its RNA. This information determines the replication of RNA and sequence of amino acids in the protein subunits of the capsid. As all subunits are identical, only one coded gene is required for the formation of all capsomeres. It is possible to remove the protein coat from RNA. The naked RNA is capable of infecting tobacco plant; once inside the host cell, the virus RNA directs the protein synthesizing apparatus of the host cell to synthesize its own proteins. Thus RNA has two functions : (i) self replication, and (ii) synthesis of virus specific proteins for which it takes the raw material from the host cell.

Although the naked RNA is capable of infecting the host cell, the efficiency of infection is usually one thousandth to one- millionth that of the intact virion. At the time of infection, usually the entire virion (*i.e.*, protein coat and nucleic acid) enters into the host cell. Immediately after entering the host cell, the protein coat of virion degenerates, whereas RNA starts the synthesis of messenger RNA which eventually synthesizes protein subunits.

Sources of TMV Infection : Debris of infected plants lying in the soil is the chief source of infection of tobacco mosaic virus. The infection also passes from one host to another by mechanical means, such as agriculture implements and rubbing of infected leaves over the surface of healthy leaves. The virus is highly contagious and stable and can retain infectivity even after the leaves have been processed into smoking tobacco. According to Johanson (1937), 67% cigar, 11% cigarettes, 62% pipe tobacco is virus infected. A person who has handled a cigarette containing infected tobacco leaves may transmit the virus to healthy plants. Besides, many aphids (*e.g.*, Myzus persicae, M. pseudosolani and Microsiphon solanifolii) are vectors of this virus.

Symptoms of TMV on the Host : The first visible symptoms of TMV infection in tobacco plant appear in the form of downward curling and distortion of young apical leaves. Large spots of dark green colour appear on the upper surface of the leaf which eventually develop into irregular blistered areas. Other areas of the leaf show typical yellow and green mottling or mosaic symptoms. The plant becomes stunted.

Common Symptoms of Viral Infection : Viral infection produces various symptoms on the host which are widespread (systemic) or are localized to a particular part. In most of the plants these symptoms are highly specific and diagnostic, but sometimes they are non-specific. Occasionally, the symptoms of viral infection are latent and in such cases the presence of virus can be determined by electron microscopic examinations.

Symptoms of viral infection can be grouped into the following three categories:

(1) External symptoms

(2) Internal symptoms

(3) Physiological symptoms

External Symptoms : These symptoms appear externally on various parts of the plant, such as leaf, stem, fruit, etc.

General External Symptoms : Some general symptoms common to most viral infections are (i) general reduction in the growth of the host, (ii) reduction in the yield, and (iii) reduction in the life-span of the infected plant.

External Symptoms on Leaves : The external symptoms on leaves may be of the following types:

(a) *Mosaic :* Spots of white or yellow colour appear on, the leaf surface as a result of viral infection. These spots are known as mosaic. Sometimes bands of green colour occur only on the leaf veins and rest of the area becomes chloretic (*i.e.*, yellow or white). This is known as vein banding. Sometimes the parenchymatous cells surrounding veins become chloretic and due to this veins become clearly visible. This symptom is called vein clearing.

(b) *Yellowing :* It is uniform yellowing of the leaf and is different from mosaic.

(c) *Morphological Changes :* An important symptom of viral infection is distortion of leaves. Rolling and curling of leaves is very common in tomato and papaya. In 'little leaf of brinjal', leaves become unusually small and closely placed, forming a rosette-like structure. In some other plants the diseased leaves become leathery.

External Symptoms on Stems : The stem of infected plants shows many abnormalities : (i) Internodes become short and there is proliferation of lateral buds. It results in the formation of densely packed broom-like structure. This symptom is known as witches broom. (ii) In some cases the stem swells at certain points due to excessive growth of xylem in localized areas. (iii) In infected plants of Citrus the stem becomes pitted. (iv) Galls are formed on the stem of infected sugarcane.

External Symptoms on Flowers : The flowers of infected plants develop the following symptoms: (i) They become distorted or variegated and their size is reduced. (ii) The infection of 'big bud virus of tomato' leads to the enlargement of buds. (iii) Sometimes, as in 'peach yellows' disease, flowers bloom early. There is also formation of pseudo-inflorescences where flowers are sterile or wither before fruiting. (iv) Virescenee is a very common symptom in which flowers become green.

External Symptoms on Fruits : Symptoms of viral infection which commonly appear on fruits are : (i) Fruits usually become small and pitted. (ii) The colour, shape and flavour of fruits is changed. (iii) In 'bushy stunt disease of tomato', flowers become enlarged and mottled, whereas in 'peach yellows' fruits ripe quite early.

Internal Symptoms : The virus infected plants also show many abnormalities in tissues and cells. For example, in 'tobacco mosaic disease', 'dahlia mosaic', 'sugar cane mosaic', 'tomato bushy stunt', and many other viral diseases intracellular inclusion bodies are formed. These bodies are lamellar or amoeboid. Besides, hyperplasia and necrosis of cells and tissues is common in diseased plants.

Physiological Symptoms : The functional system of plants becomes affected directly or indirectly due to virus infection. It leads to the accumulation of many metabolic products in cells.

The concentration of oxidised phenolic compounds in cells is responsible for the development of necrotic areas. Virus infected cells possibly synthesize certain new proteins which affect their permeability and enzyme system.

The chlorophyll contents of diseased leaves are low hence the photosynthetic rate of such leaves is decreased. It ultimately results in lower yield. Viruses accelerate the activity of oxidising enzymes like cytochrome oxidase, peroxidase, etc., and as such the rate of respiration is usually high in diseased leaves.

Diagnostic Symptoms of Some Common Viral Diseases : The diagnostic symptoms of some common viral diseases are as follows:

Tobacco Mosaic Disease

(1) Apical leaves show downward curling.

(2) There is distortion, dwarfing and blistering of leaves.

(3) Unusually large mottled light-green yellow patches develop on the upper surface of leaves.

(4) Epidermal cells and hairs show lamellar or amoeboid bodies.

(5) Plants show stunted growth.

(6) There is distortion and discoloration of flowers.

Sugarcane Mosaic

(1) Mottling of young leaves.

(2) Light green and yellow areas (mosaic) develop in alternate sequence.

(3) Stunting of young plants.

Bunchy Top of Banana

(1) In infected plants leaves usually do not come out of the pseudo-stem.

(2) Leaves form dense rosette and remain stunted.

(3) Flowering is inhibited.

Yellow Vein Mosaic of Bhindi

(1) Stunted growth of leaves.

(2) Yellowing and vein clearing of leaves. In case of severe infection, chlorosis may involve interveinal areas, resulting in yellowing of the entire leaf.

(3) Flowering is reduced and fruits become malformed, distorted and yellow green in colour.

Clump Disease of Groundnut

(1) Infected plants become bushy and sterile.

(2) Leaves become smaller and show mosaic character.

(3) Petunia and Catharanthus are alternate hosts of this virus.

Leaf Curl of Papaya

(1) Curling of young leaves.

(2) Vein clearing and mottling of leaves.

(3) Flowers become sterile.

(4) Reduced growth of the plant.

Leaf Mosaic of Cucurbits

(1) Mottling, distortion and wrinkling of young leaves.

(2) Downward curling of leaf margins.

(3) Leaves form a rosette-like clump near the ground, which gives bushy appearance to the plant.

(4) Fruits develop mosaic symptoms.

Transmission of Virus

The virus moves from one host to another for its survival and in search of a suitable host. For its transmission the virus is dependent on external carriers. Some common modes of transmission of plant viruses are described below:

Transmission by Vegetative Propagation : The plants, which are propagated vegetatively, once infected with a virus disease, transmit the pathogen from one generation to the next. Eventually the entire population of a given clonal variety may become infected with the same pathogen. Viruses may be transmitted by tubers, corms, bulbs, grafting and cutting. In grafting operations, there is contact between stock and scion and if either of these carries a virus, it may be transmitted to the other host. Transmission of 'potato spindle tuber viroid' (PSTV) takes place through tubers and that of 'Katie disease virus' through bulbs.

Transmission by Friction and Rubbing : In nature, when leaves of closely growing infected and healthy plants rub together, the healthy plants get infected. Even a gentle rubbing with no obvious damage is enough for transmission of the virus. Viruses are also transmitted by contact of infected leaves with clothes of men and bodies of animals. Potato virus X (PVX), tobacco mosaic virus (TMV) and many others are transmitted by this method.

Transmission through Alternate Hosts : Viruses causing 'yellow vein mosaic of bhindi' and 'cucurbit mosaic' are transmitted through alternate or collateral hosts.

Transmission through Soil and Seeds : Viruses which spread through underground natural methods are called soil-borne viruses. The tobacco mosaic virus remains in the field after harvest on debris and it infects the crop in the next season. The virus enters the host through roots by mechanical or biological methods. Although more than 100 seed borne viruses are known, all virus infected seeds are not equally efficient in transmitting the disease. Bean mosaic virus, barley stripe mosaic virus, curly top virus of beet sugar and ring spot virus of soyabean are transmitted through seeds.

Transmission through Pollen Grains : If flowers of healthy plants are pollinated with the pollens from infected plants, they produce seeds which have virus infection. Such seeds, on germination, give rise to infected plants. 'Stone fruit ring spot virus' and 'bean mosaic virus' are commonly transmitted through pollens.

Transmission through Fungi : Tobacco necrosis virus is transmitted by a root infecting fungus, Olpidium brassicae. Besides, Synchytrium brassicae and Polymyxa graminis also act as agents for transmission of potato virus.

Transmission through Insects : Insects are the most common and potential agents of virus transmission. The insects transmitting virus diseases are called vectors. Insects like white fly, leaf hopper, plant hopper, aphids, mealy bugs, scale insects and flee-beetles are the potent vectors. 'Yellow vein mosaic of bhindi', 'leaf curl of chilli', 'yellow mosaic of mung', and many other viral diseases are transmitted by white flies, such as Bemisia and Aleyrodes. Aphids such as Rhopalosiphum, Myzus and Brachycaudus are vectors for 'chirke virus', 'maize mosaic virus' and maize dwarf mosaic virus' respectively. When insects suck the juice from the infected plants, viruses enter into their proboscis and salivary glands. Such insects transmit the virus to healthy plants when they visit the latter.

Transmission by Infested Agricultural Tools : It is a common method of virus transmission. Viruses usually get stuck to implements used for cutting, pruning and weeding of infected crops. When virus infested tools are used for healthy plants, the virus is transmitted to the latter. Potato virus X (PVX), 'tobacco mosaic virus' and `leaf mosaic of cucurbits' are some common viruses which are transmitted by this method.

Nature of Viruses

The evolution and nature of viruses have been the subjects of numerous investigations and theoretical discussions. Some virologists consider viruses to be acellular forms of a live parasitic system that are functionally closely related with the host cell but develop independently and are genetically free from it.

Others consider them cellular genetical factors capable of synthesizing the protein cell wall which protects them from the ill-effects of environment and allows them to penetrate into host cells. Still others consider viruses as non-living transmissive nucleoproteins which have pathogenic properties that constantly arise from cellular substances.

Stanley (1935) successfully crystallized tobacco mosaic virus (TMV) and the crystals retained the property to induce mosaic disease in tobacco. Frankel Conart and Takahasi (1956) and Cochrane et al. (1963) were able to reconstitute a typical tobacco mosaic virus from biologically inactive components. It produced a characteristic disease in plants. Later, it was confirmed that infectioe properties of viruses are due to nucleic acid and the cell wall does not take part in this process.

Thus, despite the simplicity in physical and chemical composition, all attributes of life, such as self-reproduction, variation and transmission of genetic information are inherent in viruses as in other living systems. The main characteristics of viruses are listed below.

Characters Specific to Viruses

(1) They are granular and can be resolved with electron microscope only.

(2) Most viruses can not be separated by bacterial filters.

(3) They can not be grown in artificial medium.

(4) They produce characteristic symptoms on specific hosts.

(5) They can be inactivated by chemotherapy and thermotherapy.

(6) They show response to change in temperature and humidity.

(7) They are obligate intracellular parasites. In host tissue they show characters of living organisms.

(8) They lack functional autonomy, but develop independently and are genetically free from the host cell.

Differences between Viruses and Other Organisms : Viruses differ from other organisms in the following characters.

(1) Viruses possess only one kind of nucleic acid (either DNA or RNA), whereas other organisms have both the nucleic acids.

(2) Viruses are devoid of polysaccharides.

(3) In viruses there is no cell division. They show replication of genetic material only (for which they are dependent upon the machinery of the host cell).

(4) Viruses lack cellular organelles like ribosomes, mitochondria, etc. They make use of the ribosomes of host cells.

Viruses possess both living and non-living properties.

Living Properties of Viruses

(1) The genetic material of viruses replicates.

(2) Mutant forms of certain viruses are known, and mutation is essentially a characteristic of living organisms.

(3) They show sensitivity of many stimulants, such as temperature, chemical substances and radiation.

(4) They possess antigenic property.

(5) They multiply in living host cells.

(6) They show host specificity.

Inanimate Properties of Viruses

(1) They can be crystallized.

(2) They are inert outside their specific host cells.

(3) They are autocatalytic and do not have functional autonomy.

(4) They are devoid of cell membrane and cell wall.

(5) They do not respire.

Bacteriophage

Viruses, which infect bacterial cells, are known as bacteriophage or viruses of bacteria. There is hardly any species of bacteria which does not serve as host to one or more viruses. Bacteriophage was discovered by Frederick W. Twort (1915) and Felix d'Herelle (1917) independently while investigating certain types of soil bacteria. They observed that if few drops of highly concentrated bacterial viruses are introduced into a dish with nutrient medium seeded with a culture, then there was no growth of bacteria at the point of the

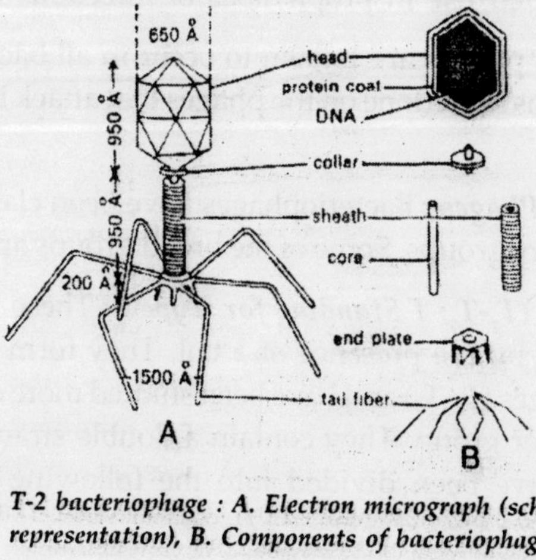

A-B. T-2 bacteriophage : A. Electron micrograph (schematic representation), B. Components of bacteriophage.

introduction of the virus. This interaction was not restricted only to the solid medium. They observed that when bacteriophage is added in a suspension of Staphylococcus albus the bacterial cell wall breaks open (lysis). This suspension also retained the capability of destroying bacteria. This phenomenon is said to be Twort-d'Herelle phenomenon. They called the viruses which destroy bacteria as bacteriophage, literally means eaters of bacteria' (Greek phagein = to eat). They are also known as phages.

Bacteriophages are obligate parasites and are found in all such habitats where bacteria can survive. They are abundant in soil, sewage water, fruits, vegetables, milk and nodules of legumes. Specific phages have also been found in the intestine of birds and animals. In human beings phages can be found in intestinal contents, urine, blood, sputum, saliva, pus and nasal exudate. It is easy to isolate the phage during the period of convalescence. Besides, some viruses occur as parasites in actinomycetes (Actinophages), yeast cells (Zymophages) and blue-green algae (Cyanophages). The discovery of bacterial viruses led to great excitement, for it was hoped that they would serve as an effective and simple way to combat bacterial diseases. But due to introduction of antibiotics, phage therapy has only limited use in prophylaxis of infectious diseases.

Although phages are known to occur in all bacteria, most of the work has been done on the phages that attack Escherichia coli.

Types of Phages : Bacteriophages have been classified into many arbitrary groups. Some of the broad groups are as under.

T-phages (T_1-T_7; T Standing for 'Type') : These phages are characterised by the presence of a tail. They form the largest group of phages (T_1-T_7) and have been studied more extensively than any other group. They contain a double stranded DNA. T- phages have been divided into the following three sub-groups.

(a) *T-even Phages (T_2, T_4, T_6)* : These phages are genetically as well as serologically closely related. They have an angular head and a contractile tail. Their multiplication in the bacterial cell is independent of the bacterial DNA, the latter is rapidly destroyed following infection. The DNA of T-even phages contains a unique base, 5-hydroxymethyl cytosine, in place of cytosine.

(b) *T-odd Phages (T_1, T_3, T_7)* : These phages have different genetic and serological features. They have an angular head and a short non-contractile tail. The DNA of these phages is characterised by the presence of cystosine.

(c) *T5 Phages* : These phages contain an angular head and a long non-contractile tail. Like T-odd phages, these phages also have cytosine in their DNA.

Virulent and Temperate Phages : Depending on the interaction of phages with the bacterial cell, they have been distinguished into virulent or lytic and avirulent or temperate phages. According to Lwoff (1953), Bacteriophages exist in three clear-cut states, namely extracellular virion, vegetative phage and prophage. Extracellular virions are complete phage particles with both nucleic acid core and protein coat. They infect bacterial cell. Vegetative phage and prophage, on the other hand, are intracellular, represented by nucleic acid only.

As a vegetative phage, the nucleic acid is capable of independent replication, while as a prophage it never replicates independently and is first inserted in the bacterial (host) DNA and then replicates along with bacterial DNA. Bacteria containing prophages are called lysogenic bacteria, and those viruses whose nucleic acid can become prophage (*i.e.*, gets incorporated in bacterial DNA) are known as lysogonic, temperate or avirulent phages (*e.g.*, F_2, M_{12}). On the other hand, those viruses which always multiply when they enter the host cell are called virulent or lytic phages (*e.g.*, T_2, T_4 phages).

Important characteristics of some Bacteriophages are listed in the Table.

Important Characteristics of Some Bacteriophages.

Phage	*Head*	*Tail*	*Nucleic acid type*
T-even	Polygonal	Contractile	Double stranded DNA
T_3, T_7	Hexagonal	Short non-contractile	Double stranded DNA
T_1, T_5	Hexagonal	Short non-contractile	Double stranded DNA
fdm, M_{13}	Filamentous	No differentiation into head & tail	Single stranded DNA
ϕ x 174	Hexagonal	Tail absent	Single stranded DNA
f_2, R_{17}, fr	Hexagonal	Tail absent	Single stranded RNA

Structure of Bacteriophage

Morphology : Bacteriophages are very small particles and they cannot be separated even by bacterial filters. The structural details of Bacteriophages are known through electron microscopic studies of some larger particles of T-even group which infect the bacterium Escherichia coli. In their appearance the phage resembles a tadpole or spermatozoid; it is differentiated into a head and a tail.

In most phages (T, T_2, T_6) the head is prismoid but in T_3 and T_7 phages it is hexagonal. There are some phages (*e.g.*, fl, fd, M_{13}) which are filamentous and do not show differentiation into head and tail.

The size of the head of T_2 phages is approximately 950 λ x 650λ. The extended part between the head and the tail is called collar. The tail is almost equal to the length of the head (950 λ) and has a diameter of 80 λ. At the proximal end of the tail a hexagonal tail plate or end plate is present. The end plate is approximately 200 Å thick. It has six tail pins or fibres on its under surface, each about 1500λ long. The tail pins have two main functions :

(i) they help in the adsorption of phage particle on the surface of the bacterium, and (ii) the enzymes secreted by these pins are helpful in the lysis of bacterial cell wall.

Chemical Composition : These phage particles are made up of protein (about 50-60%) and nucleic acids (40-50%). They also contain a small proportion of lipids in the form of neutral fats. The wall of the head is composed of some 2000 similar subunits of proteins. The nucleic acid of phages is either double-stranded DNA, single stranded DNA or single stranded RNA (DNA and RNA are never present together). Except for coliphage, $\phi \times 174$ and few others, most phages have double stranded DNA. The phage DNA differs from bacterial DNA chemically. The molecular weight of phage DNA is 2,500,000, and in each phage particle the amount of nucleic acid is approximately 6×10^{-3} mg. DNA is the genetic material of the phage particle, it carries infection and induces the host cell to synthesize more and more phage particles.

Replication of bacteriophage

The process of replication of bacteriophage involves the following four steps:

(1) Infection.

(2) Synthesis of phage components in the host cells.

(3) Assembly of new phage particle.

(4) Liberation of phage particles from the host cells.

Infection : The two important events of this step are adsorption of the phage on the host bacterium by means of its tail fibres, and transfer of phage nucleic- acid into the host cell. Adsorption of phage on the host 'bacterium depends on the mutual affinity of the phage and bacterium. Only a specific phage can infect a particular bacterium. Some of the coliphages have so much specificity that they are adsorbed only at a specific site on the host surface. These sites are known as receptor sites. In addition, factors such as composition and viscosity of the medium and temperature also affect adsorption.

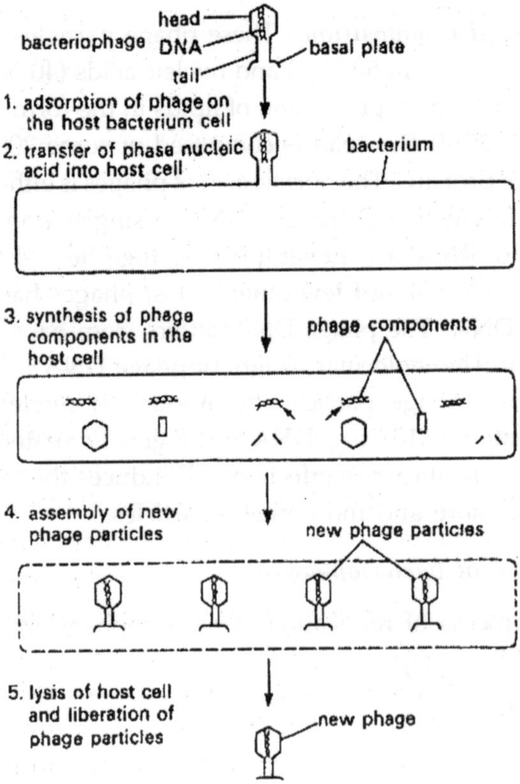

Bacteriophage : Various stages of replication.

Adsorption of phage on the surface of host is facilitated by interaction of amino groups of proteins present at the margins of the phage tail, and negatively charged carboxyl groups on the surface of the host. Thus the host specificity of the phage particle is due to its protein and not the nucleic acid. As such the host range of free viral nucleic acid is much wider than that of the intact virion. After adsorption, the phage particle secretes a special enzyme of lysozyme type which hydrolyses muramic acid-peptide complex of the bacterial cell wall. Consequently, a minute pore is formed through which the nucleic acid of the phage enters into the host cell. The infection is accomplished through the following steps.

(1) Immediately after landing on the host surface the tail fibres bent, bringing the end plate of the phage in contact with the bacterial cell wall.

(2) The tail sheath then contracts, pushing the central tubular part of the tail into the host cell wall just like an injection needle.

(3) The nucleic acid of the phage flows into the host cell through the hollow centre of the tubular needle.

The protein shell of phage remains attached to the host cell even after the transfer of nucleic acid. Such empty protein shells are called ghosts. A host cell once infected by a phage particle becomes immune against the infection of the phage of the same type. A phage particle loses its ability of infection after its nucleic acid has been released into the host cell.

Synthesis of Phage Components in the Host Cell : The phage nucleic acid, once inside the bacterial cell, takes over the protein 'synthesis machinery of the cell. It suppresses the synthesis of bacterial protein and directs the metabolism of the cell to synthesize the proteins of the phage particle. This is accomplished by the synthesis of viral specific m-RNA (using the host RNA polymerase).

The replication of phage DNA follows the semi-conservative mechanism. Most of the phage DNA serves as a template in its own synthesis and the rest is used as a template for the synthesis of viral specific m-RNA. The latter directs the host cell to synthesize proteins which are used as sub-units (capsomeres) of the protein coat of the phage particle. These proteins are called late proteins. Towards the end of replication of phage nucleic acid, a late protein, called phage lysozyme, is synthesized.

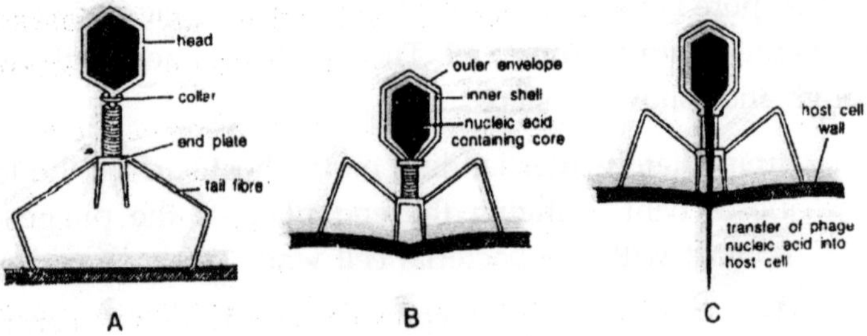

A-C. Bacteriophage : Mechanism of infection; A. Adsorption of phage on the host surface with the help of tail fibres, B. Contact of end plate with the host surface, C. Transfer of phage nucleic acid into the host cell.

Assembly of New Phage Particle : Assembly of nucleic acids and proteins (late proteins) into new phage particles is called maturation. This process is controlled by a viral genome. The first step in maturation is the condensation of nucleic acid molecule in crystalline form.

The protein sub-units then aggregate around DNA to form the head of the phage. Meanwhile, assembly of the tail starts. It is initiated with the attachment of core tube with the tail plate. The sheath around the core tube is formed afterwards. At this stage the tail gets attached to the base of the head. The tail fibres are attached to the end plate in the last.

Liberation of Phage Particles from the Host Cells : The entire cycle of phage development is completed in 30-90 minutes. In an infected bacterium 7-8 phage particles are formed per minute and a total of about 200 phages are formed in a bacterium. Lysis of the host cell wall is essential for the liberation of phage particles. It is facilitated by the enzyme lysozyme secreted by the phage DNA in the host cell. The host cell ruptures as a result of lysis and the phage particles are liberated.

Economic Importance of Phage Particles

Phages have been used in prophylaxis and treatment against dysentery, enteric fever, cholera, plague, and many other pathogenic bacterial diseases. However currently, due to introduction of antibiotics, phage therapy and prophylaxis of infectious diseases are used only to a limited extent. Phages are also used in the diagnosis of certain infections. The causative agents of plague, cholera, etc. can be determined with the help of special phages. Phages are also helpful in the lysis of bacteria present in the polluted water. Hence, . they can also be used as scavangers.

In space microbiology lysogenic cultures are used as radiation detector. They were used by Russians in the space ship, Vostok-2.

Temperate phages help in transduction of genetic material from one bacterial cell to another. They are also used widely as models in genetic research.

Phages are often very harmful as they kill beneficial microorganisms by the lysogenic activity during the process of manufacture of antibiotics and milk products.

Viroids

Viruses are no longer considered the simplest form of life. In 1971, T.O. Diener discovered new infectious agents which were even smaller than viruses. He introduced the term viroid for these subviral pathogens. The first viroid came to light in attempts to isolate and characterize the agent of the potato spindle tuber disease, which was assumed to be caused by a virus. The infectious agent of this disease was found to be a RNA strand devoid of nucleoprotein coat. This infectious RNA has a very low molecular weight unlike conventional viruses. Since then many plant diseases (*e.g.*, citrus exocortis, chrysanthemum stunt, cucumber pale fruit and chrysanthemum eblorotic mottle) are known to be caused by viroids.

However, the search for viroids of animal disease has been relatively slow and we know only few animal diseases (*e.g.*, scrapie disease of sheep and Alzheimer's disease of human) of viroid origin. Presumably, certain infectious diseases of obscure etiology are caused by agents resembling viroids.

Structure of Viroids

A viroid consists of a very short strand of RNA without any protective coat. Electron microscopic studies of purified 'potato spindle tuber viroid' (PSTV) revealed that it has a single stranded RNA molecule containing 250-350 nucleotides. The adenine : uracil (A : U = 21.'7:20.9) and guanine: cytosine (G : C = 28.9 28.3) ratios are close to unity. Viroids do not possess capsid (protein coat) around the RNA molecule.

Besides linear RNA, some circular molecules of PSTV have also been reported. The RNA fingerprinting, however, has shown that the circular and linear molecules of PSTV structures are probably not two distinct RNA species. The two structures, more likely, represent two stages of maturity of PSTV. The molecular weight of viroids is in the range of 115,000 - 130,000 daltons.

Although the precise mechanism of viroid replication is not known as yet, the following two hypothesis have been put forward:

1. *RNA-directed Replication* : According to this view, in uninfected plants there exists a replicase enzyme which accepts a wide variety of RNA species as templates. The presence of RNA-directed RNA polymerase has been shown in healthy plants of chinese cabbage and tobacco.
2. *DNA-directed Replication* : This hypothesis suggests that viroids might be replicated on DNA templates which are either already present in repressed form in uninfected hosts or are synthesized as a result of viroid infection.

Viroids and Plant Diseases : Viroid diseases are persistent infections, *i.e.*, there is no recovery of infected plants and viroids can be isolated from the diseased plants as long as it lives. They do not differ significantly from the viral diseases in their symptomatology.

The symptoms observed in viral diseases also occur with viroid diseases. These include stunting, epinasty, veinal discoloration, leaf' distortion, vein clearing, localized chloretic or necrotic spots, mottling of leaves and death of the whole plant. Precise mode of viroid interference with their host metabolism to produce characteristic symptoms is not known. In some viroid infected plants certain host proteins occur in larger amounts than in healthy plants. It is presumably due to viroid-induced metabolic aberrations in the host protein synthesis.

All known viroids are transmitted from one plant to another (*i.e.*, horizontally) by mechanical means. Besides, vertical transmission through the seed and pollen of the infected plants has been shown at least in PSTV. No insect vectors have been identified so far with any viroids.

Bibliography

Adil, A. : *Manual of Practical Algae*, Mohit Pub., New Delhi, 2003.

Aggarwal, S.K. : *Vistas in Mycology and Plant Pathology*, APH, New Delhi, 2000.

Ahuja, B. : *Plant Anatomy*, Rawat Pub., New Delhi, 2000.

Antony, T. : *Quarry : Closing in on the Missing Link*, Free Press, New York, 1993.

Arora, J.K.: *Biopesticides and Pest Management*, Ankit Pub., New Delhi, 2003.

Aslam, M.S. : *Genetics-Behavioural and Molecular*, Campus Books, New Delhi, 2003.

Bacon, Mark : *Water use Efficiency in Plant Biology*, Blackwell, London, 2004.

Bahadur, Rai K. Ranga : *A Manual of Botany for India*, Atlantic Publishers, New Delhi, 1994.

Bayleybolfour, I. :*Plant Geography Upon A Physiological Basis*, ABS, London, 2004.

Behnke, H.D. : *Sieve Elements : Comparative Structure, Introduction and Development*, Springer, London, 1990.

Bentley, Robert : *Study of Structural Morphological and Physiological Botany*, Atlantic Publishers, New Delhi, 1994.

Bilgrami, C : *Fundamentals of Botany,* Vikas Publishing House, New Delhi, 2003.

Bilgrami, K.S. : *Physiology of Fungi,* Vikas Publishing House, New Delhi, 2003.

Boodle, L.A. and Sinnott, E.W. : *Student Handbook of Botany and Plant Science,* Atlantic Publishers, New Delhi, 1995.

Bose, J.C. : *Irritabiligy of Plants,* Discovery Publishing House, New Delhi, 1999.

Brooks, F.T., and Smith, M. : *Dictionary of Fungus Diseases of Plants,* Campus Books, New Delhi, 2003.

Busch, L. : *Plants, Power and Profit : Social, Economic, Lawrence Busch, and Ethical Consequences of the New Biotechnologies,* Blackwell, London, 1992.

Chrispeels, M.J. : *Plants, Genes and Agriculture,* Jones & Bart., London, 1994.

Coulter, John Gaylord : *Fundamentals of Plant Science,* Atlantic Publishers, New Delhi, 1994.

Dayal, R. : *Predaceous Fungi,* Campus Books, New Delhi, 2000.

———— : *Predaceous Fungi,* Commonwealth Publishers, New Delhi, 2000.

Dhaliwal, G.S. and Singh, Balwinder : *Pesticides and Environment,* Campus Books, New Delhi, 2000,

Dube, H.C. : *Introduction to Fungi,* Vikas Publishing House, New Delhi, 2003.

Dyson, Freeman : *Origin of Life,* Oxford, New York, 1999.

Frank, S. Smythe : *Flowering Plants of Himalaya,* Commonwealth Publishers, New Delhi, 1994.

Fuller, Harry J. : *University Botany,* Atlantic Publishers, New Delhi, 1994.

Gaston, Kevin J. and spice, John I. : *Biodiversity : An Introduction,* Blackwell, 2004.

Grewal, R.C. : *Medicinal Plants,* Campus Books, New Delhi, 2003.

——— : *Dictionary of Botany,* Arihant Pub., New Delhi, 2003.

Hall, Marion and Halliday, Tim : *Behaviour and Evolution,* Springer, London, 1998.

Hanan, James S. : *The Algorithmic Beauty of Plants,* Springer, London, 1990.

Hardie, D.G. : *Protein Phosphorylation : A Practical Approach,* Oxford, New York, 1999.

Johri, Lata and Tyagi : *A Textbook of Bryophyta,* Dominant Publishers, New Delhi, 2004.

——— : *A Textbook of Fungi,* Dominant Publishers, New Delhi, 2004.

——— : *A Textbook of Gymnosperm,* Dominant Publishers, New Delhi, 2005.

Kains, M.G. and Mcquesten, L.M. : *Fundamentals of Plant Science,* Campus Books, New Delhi, 2002.

——— : *An Introduction to Plant Science,* Atlantic Publishers, New Delhi, 1995.

Kumar, H.D. : *General Ecology,* Vikas Publishing House, New Delhi, 2003.

Kumar, Suresh : *A Textbook of Plant Taxonomy,* Campus Books, New Delhi, 2004.

——— : *Economic Botany,* Campus Books, New Delhi, 2002.

Lang, W.H. : *A Textbook of Botany,* Atlantic Publishers, New Delhi, 1994.

Leung, D.Y.M. : *Superantigens : Molecular Biology, Immunology, and Relevance to Human Disease,* Dekker, London, 1997.

Lozzo, Renato V. : *Proteoglycans : Structure, Biology, and Molecular Interactions,* Dekker, London, 2000.

Madigan, M.T. : *Brock : Biology of Microorganisms,* Prentice Hall, New York, 2000.

Magner, Lois N. : *A History of the Life Sciences,* Dekker, London, 1994.

Magurran, Anne E. : *Measuring Biological Diversity,* Blackwell, London, 2003.

Mishra, S.R. : *Bacterial Plant Diseases,* Discovery Publishing House, New Delhi, 2003.

———— : *Morphology of Plants,* Discovery Publishing House, New Delhi, 2004.

Nairne, Alexander : *Scientific Classification of Flowering Plants,* Atlantic Publishers, New Delhi, 1994.

Narasaiah, M.L. : *Biodiversity and Environment,* Discovery Publishing House, New Delhi, 2004.

Nover, L. : *Plant Promoters and Transcription Factors Results and Problems in Cell Differentiation,* Springer, London, 1994.

Oliver, Daniel : *Study of Indian Botany,* Atlantic Publishers, New Delhi, 1994.

Panday, S.N. : *Economic Botany,* Vikas Publishing House, New Delhi, 2003.

Panday, S.N. and Sinha, B.K. : *Plant Physiology,* Vikas Publishing House, New Delhi, 2003.

Panday, S.N. and Trivedi, P.S. : *A Textbook of Botany,* Vikas Publishing House, New Delhi, 2003.

Pfleiderer, I. : *The Life of Indian Plants,* Atlantic Publishers, New Delhi, 1992.

Prasad, S.N. : *Environmental Biology,* Campus Books, New Delhi, 2003.

———— : *Cell Biology,* Campus Books, New Delhi, 2003.

Pugnaire, Ed. F.I. : *Handbook of functional Plant Ecology,* Dekker, London, 1999.

Rai, Mahendra and Mares, Donatella : *Plant-Derived Antimycotics : Current Trends and Future Prospects,* Blackwell, London, 2003.

Rashid, A. : *An Introduction to Pteridophyta,* Vikas Publishing House, New Delhi, 2003.

Rashid, A., Kumar, Santosh and Tembhre, Manju : *Anatomy and Physiology of Fishes,* Vikas Publishing House, New Delhi, 2003.

Roy, E. Biles, Seymour, E.L.D. : *Encyclopaedia of Horticulture,* Commonwealth Publishers, New Delhi, 1994.

Sandhu, G.S. : *Textbook of Cytology,* Campus Books, New Delhi, 2002.

Sandhu, G.S. and Kumar, Suresh : *Cyto-Genetics,* Campus Books, New Delhi, 2003.

Sarkar, Dilip De : *The Silkworm: Biology, Genetics and Breeding,* Vikas Publishing House, New Delhi, 2003.

Sharma Rajni : *An Introduction to Plant Morphology,* Campus Books, New Delhi, 2004.

———— : *Plant Pathology,* Campus Books, New Delhi, 2004.

———— : *Plant Tissue Culture,* Campus Books, New Delhi, 2003.

Shukla, S.K. and Srivastava, P.R. : *Encyclopaedia of Environmental Pollution,* Commonwealth Publishers, New Delhi, 1994.

Singh, Mahinder : *A Textbook of Biochemistry,* Dominant Publishers, New Delhi, 2005.

Singh, S.K. : *Plant Physiology,* Campus Books, New Delhi, 2000.

Sinha, Asha *Microbes and Plants,* Campus Books, New Delhi, 2001.

Sinha, R.K. :*Biodiversity–Global Concerns,* Commonwealth Publishers, New Delhi, 1996.

Sinnott, E.W. : *Botany and Plant Science,* Commonwealth Publishers, New Delhi, 2000.

Srivastava, M.B.: *Introduction to Forestry,* Vikas Publishing House, New Delhi, 2003.

Stiles, Walter :*Principles of Plant Physiology,* Atlantic Publishers, New Delhi, 1995.

Sundara, Rajan, S. : *Introduction to Cell Biology,* Vikas Publishing House, New Delhi, 2003.

Talbot, Nicholas J. : *Plant-Pathogen Interactions,* Blackwell, London, 2001.

Tassy, Pascal : *The Message of Fossils,* McGraw-Hill, London, 1993.

Transeau, E.N. : *General Botany,* Atlantic Publishers, New Delhi, 1995.

Tripathi, G. and Tripathi, Y.C. : *Biological and Biotechnological Resource,* Campus Books, New Delhi, 2002.

———— : *Bioresource and Environment,* Campus Books, New Delhi, 2002.

Waisel, Yoav : *Plant Roots : The Hidden Half,* Dekker, London, 1996.

Warren, L. : *Bound Carbohydrates in Nature,* Oxford, New York, 1994.

Wrigglesworth, J. : *Energy and Life : Modules in Life Sciences,* T & F, New York, 1997.

Zaidi, N. : *An Introduction to Algae,* Ankit Pub., New Delhi, 2002.

ENCYCLOPAEDIA OF
BOTANY

ENCYCLOPAEDIA
OF
BOTANY

(Vol. 3)

Prof. Harmeet Kour

ANMOL PUBLICATIONS PVT. LTD.

NEW DELHI - 110 002 (INDIA)

ANMOL PUBLICATIONS PVT. LTD.
H.O.: 4374/4B, Ansari Road, Darya Ganj,
New Delhi-110 002 (India)
Ph.: 23261597, 23278000
B.O.: No. 1015, Ist Main Road, BSK IIIrd Stage
IIIrd Phase, IIIrd Block,
Bangalore - 560 085 (India)
Visit us at: www.anmolpublications.com

Encyclopaedia of Botany

First Published, 2006

ISBN 81-261-2673-6 (Set)

[Responsibility for the facts stated, opinions expressed, conclusions reached and plagiarism, if any, in these volumes is entirely that of the Editor. The Publishers bear no responsibility for them, whatsoever.]

PRINTED IN INDIA

Published by J.L. Kumar for Anmol Publications Pvt. Ltd., New Delhi - 110 002 and Printed at Mehra Offset Press, Delhi.

Contents

PART-II
Gymnosperms

(xvi)

Preface

Biology comprises mainly two branches of study, namely, Zoology and Botany. Botany pertains to the study of plants, which have always been of great import since the formation of the planet 'Earth'. Plants play a crucial part in the life of living creatures as well as in balancing the environment. In other words, they are as much necessary for life as air or water is.

Apart from numerous benefits, obtained from plants, they have also been a source of medication since days immemorial. Taking all these dimensions of plants into consideration, they have been classified into different groups and every one of them has been given a unique Botanical name. A plant may have different names in different dialects of different parts of the world, but its botanical name remains one and the same the world over.

This multi-volume aims at furnishing a detailed account of plants from every possible dimension. A ready reference book for scholars and students, this encyclopaedia is an authority on the subject without an iota of doubt.

Editor

PART-I

Pteridophyta

1

The Telomes

It is now widely accepted that the present day plants with green leaves and branches have evolved from the axial sporophyte of the earliest and most primitive land plants. Devonian Psilophytales (e.g., Rhynia, Homeophyton, Psilophyton, Asteroxylon), which are considered to be the earliest vascular land plants, had a simple dichotomously branched axis with leaves and roots. Some of their terminal branches bore sporangia while others were sterile. Discovery of these primitive plants prompted many workers to believe that from these simple vascular plants, by the process of elaboration, branches with green leaves developed. One such view that has attracted maximum attention is the telome theory put forward by Zimmermann (1930).

Definition

The name telome has been given to the simple ultimate terminal portions of a dichotomously branched axis. These axes are undifferentiated and single nerved. Two telomes of a dichotomising axis are united below the point of dichotomy to form a fused structure, called mesome (Fig. A). In fact, each mesome is basically a telome but during ontogeny its apex was formed and it was transformed into a mesome. The following two types of telomes have been recognised on the basis of their function.

1. *Vegetative or Sterile Telomes :* These telomes are without sporangia and they are also called phylloids (Fig. B).

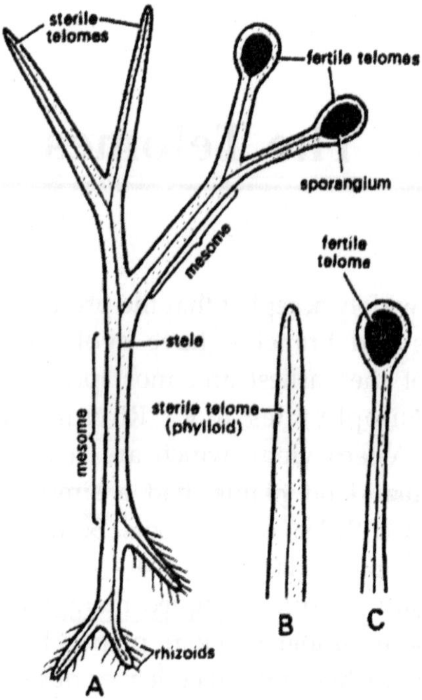

A-C. Telome : A. Diagrammatic representation of a hypothetical primitive vascular plant, B. A sterile telome, C. A fertile telome.

2. *Fertile Telomes :* Those telomes which bore terminal sporangia are called fertile telomes (Fig.C).

During the course of evolutionary development, telomes became grouped together to form a more complex structure, called syntelome or telome truss. A syntelome is called phylloid truss if it consisted of only sterile telomes, a fertile truss if it consisted only fertile telomes and a mixed telome truss or mixed syntelome when it consisted of both sterile and fertile telomes.

Telome and the origin of higher land plants according to Zimmermann, the primitive vascular land plants have originated from unicellular green algae. Repeated divisions in unicellular forms resulted in the development of large undifferentiated forms with parenchymatous thalli. Further development and differentiation led to the establishment of meristematic tissue which gave rise to erect branches. It was followed by the appearance of heteromorphic alternation of generation. The branches of the sporophyte were dichotomously branched and had a central mechanical strand. Zimmermann visualized such green algae as the immediate ancestors of the primitive vascular cryptogams (e.g., Rhynia, Homeophyton, Psilophyton, etc.) which evolved during the Upper Silurian and Lower and Middle Devonian periods. The sporophyte of these primitive vascular plants was relatively undifferentiated; the aerial telomes developed stomata and the basal telomes rhizoids. Internally, the axis was undifferentiated with an axial, typically protostelic vascular strand.

Zimmermann suggested that the following elementary processes were responsible for the development of higher vascular plants from the early vascular cryptogams.

(1) Overtopping

(2) Plantation

(3) Syngenesis

(4) Reduction

(5) Curvation

***Overtopping* :** In this process, one of the two dichotomising branches of an axis became larger, stronger and grew vertically upward as the main axis. The shorter dichotomy was displaced laterally and it served as precursor of megaphylls (Fig. A-B).

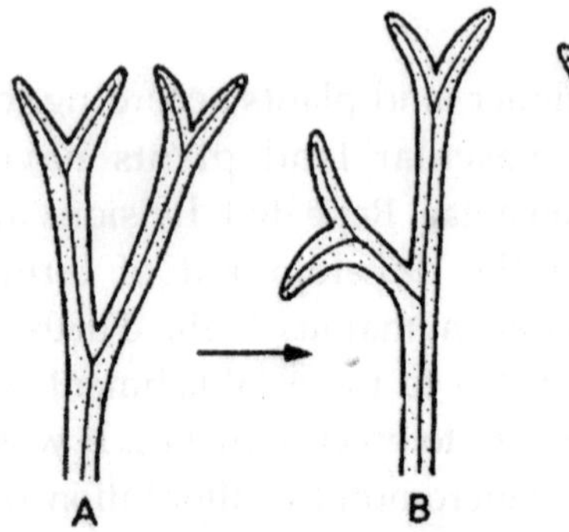

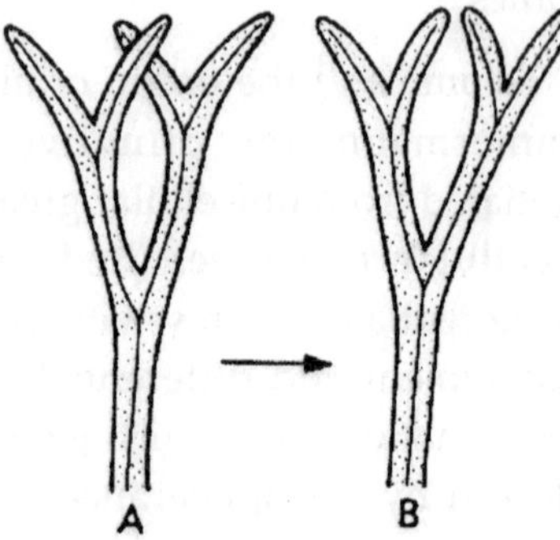

Overtopping of telomes. *Planation of telomes.*

Thus in this process the weaker branch was overtopped by the stronger branch. It resulted in the development of monopodial branches from equal dichotomies. Evolutionary studies have also shown that dichotomous branching is frequent in primitive pteridophytes and monopodial in advanced forms.

Planation : The equal dichotomies of a system, which were in more than one plane (called cruciate dichotomy), come to lie in a single plane, (fan-shaped dichotomy; Fig.). This process, known as planation, helps in the interpretation of the development of organs of bilateral symmetry from those of radial symmetry. Thus, planation must have led to the evolution of leaf.

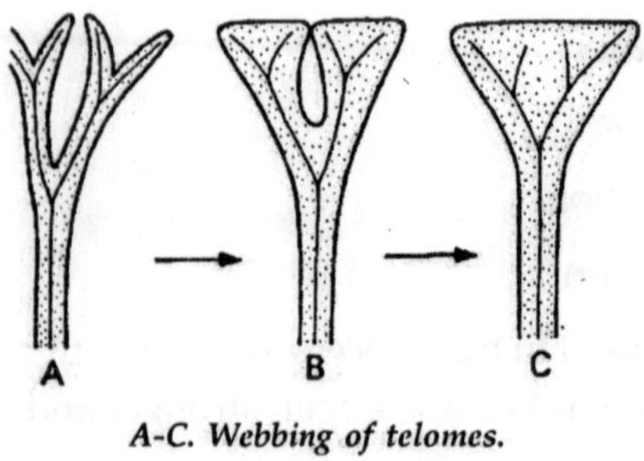

A-C. Webbing of telomes.

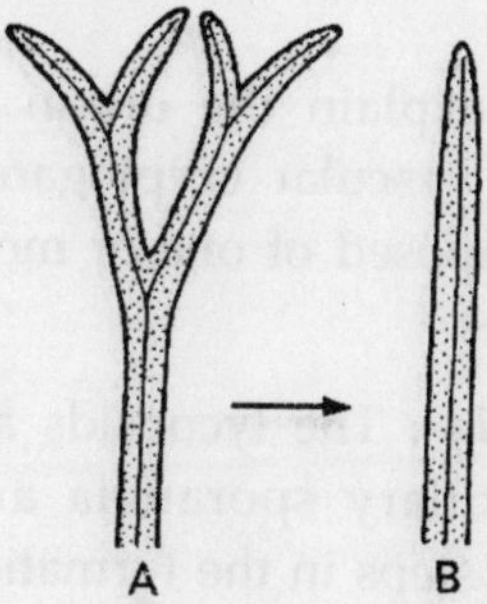

A-B. Reduction of telomes.

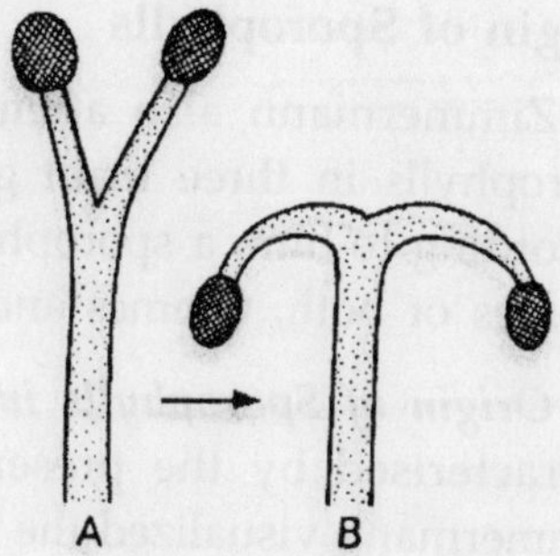

A-B. Recurvation of telomes.

Syngenesis : In this process, telomes and mesomes came to lie within a common parenchymatous tissue. This process is also known as fusion or webbing (Fig. A-C). Syngenesis also involves the fusion of vascular strands of telomes. It is an important process which explains the formation of :

(i) leaves with open dichotomous, pinnatified and reticulate venation, and

(ii) polystelic condition found in Selaginella and some fossil members of the Devonian and Lower Carboniferous period (e.g., Cladoxylon, Steloxylon).

Reduction : It involved transformation of a syntelome into a single needle- like leaf (Fig. A-B). This process thus accounts for the evolution of simple microphyllous leaves of the lycopods.

Curvation : It is brought about by unequal growth of tissues on two opposite flanks of the telome. The following two types of curvation processes have been recognized.

1. *Recurvation* : When telomes bent downwards, it is called recurvation (Fig. A-B). It is believed that recurved position of sporangia in the Sphenopsida is the result of this process.

2. *Incurvation* : This process accounts for the shifting of sporangia to the ventral surface of the leaf in ferns.

Origin of Sporophylls

Zimmermann also attempted to explain the origin of sporophylls in three main groups of vascular cryptogams. According to him, a sporophyll is composed of one or more telomes or both, telomes and mesomes.

Origin of Sporophylls in Lycopsida : The lycopsids are characterised by the presence of axillary sporangia and Zimmermann visualized the following steps in the formation of axillary sporangia.

(1) Aggregation of sterile and fertile telomes.

(2) Reduction in the number of telomes and mesomes.

(3) Reduction in the number of sporangia. Transitional stages of the origin of axillary sporangia by the above process are shown in figure A-D.

Origin of Sporophylls in Sphenopsida : In the Sphenopsida, sporangia occur on the underside of a stalked peltate disc, called sporangiophore. The following steps are involved in the origin of such a structure (Fig. A-D).

(1) There was downward bending (recurvation) of fertile telomes.

(2) The recurved telomes then come to lie within a common parenchymatous web (webbing). This resulted in the formation of a peltate disc.

Origin of Sporophylls in Pteropsida : The Pteropsida usually have large pinnately divided sporophylls. According to Zimmermann, elementary processes, such as overtopping, reduction, syngenesis and incurvation have resulted in the development of such sporophylls. He visualized the following steps in the development of sporophylls of the Pteropsida (Fig. A-B).

(1) Overtopping of sterile and fertile telomes has led to the development of pinnate sporophylls.

(2) Lateral fusion (webbing) of mesomes has resulted in the development of pinnate sporophylls with marginal sporangia.

(3) Incurvation resulted in the shifting of sporangia from marginal to lower superficial position.

Origin of Leaves

The ultimate branches of the dichotomous axis of the primitive sporophyte either terminated into sporangia (fertile telomes) or remained as sterile telomes (phylloids). According to Zimmermann (1930), microphylls originated by gradual reduction of the sterile telomes and megaphylls by a process involving planation and webbing.

Merits of Telome Theory : Telome theory provides an excellent interpretation of the origin and evolution of sporophyte of land plants. It is based on phyletic relationships between the various groups of plants, both living and fossils. The elementary processes proposed by Zimmermann (such as overtopping, planation, webbing, reduction and curvation) provide a basis of interpretation which remove outstanding morphological difficulties in the lower vascular plants. According to Eames (1936), though the theory is built upon structure in the lowest known vascular plants, higher plants can also be safely interpreted in this way. Bierhorst (1971) is of the view that the theory is too simple and too easily applicable but unfortunately its excessive use has greatly diminished its value.

Demerits of Telome Theory : The following are the drawbacks of the telome theory.

(1) An important drawback of the telome concept is that Zimmermann has taken telome as a ready-made structural unit, without explaining as to how such a structure really came into existence. Although this problem was later realized by Zimmermann himself

and he put forward several other elementary processes, but they do not satisfy plant morphologists.

(2) Many fossil plants of much greater complexity than Rhynia have been discovered in the beds of the same age. They bore whorled (e.g., Psilophyton princeps) or lateral (e.g., Zosterophyllum) sporangia instead of terminal sporangia. The telome theory does not explain such arrangements of sporangia.

(3) According to the telome theory, all leaves in plants are telomic in nature. But Bower (1935) considered that microphyllous leaves are simply outgrowths of the stem, i.e., they are not the end product of reduction.

(4) According to the telome theory, the polystelic condition, such as found in the stem of Selaginella, has developed due to syngenesis (parenchymatous webbing) and siphonostelic and actinostelic conditions are supposed to be the products of tangential or radial fusion of vascular systems of polystelic axis. Such an explanation is opposed to the widely-accepted concept of stelar theory.

In support of the telome theory-Zimmermann (1965) states that as the theory has dealt with all the 'test cases' satisfactorily and there is no alternative theory, the universality of this theory is unquestioned.

2

Apospory and Apogamy

Life-cycle of ferns is essentially similar and it comprises two alternating phases - the sporophytic and the gametophytic generations. The sporophytic generation is large, leafy and diploid, whereas the gametophytic generation is relatively small, simple and haploid. The gametophyte gives rise to the sporophyte by gametic union and the sporophyte forms haploid spores which on germination give rise to gametophyte. However, sometimes certain stages may become interpolated in the normal cycle or some essential phases are entirely omitted. Two important processes which involve the elimination of essential events in the life- cycle are apospory and apogamy. These processes are briefly discussed here.

Apospory

Development of gametophyte directly from the vegetative cells of the sporophyte without the formation of spores is known as apospory. Thus, apospory involves the omission of the event of spore production from the life-cycle of the fern or it implies the absence of meiosis in the life-cycle of the organism. Therefore, aposporous gametophytes have the same number of chromosomes as the sporophyte, i.e., they are diploid. Recent experimental studies have, however, shown that in ferns apospory also leads to the development of polyploid gametophytes and sporophytes. Triploid, tetraploid and even higher polyploids have been obtained by inducing apospory in Osmunda regalis, Aspidium marginale and Woodwardia virginica. An aposporous prothallus develops from the cells of an abortive sporangium or directly from the cells of the leaf.

Apospory was first demonstrated by Druery (1884) as a natural phenomenon in Athyrium filix femina var. clarissima. In this fern, the prothalli are produced on pinnules in place of sporangia, a condition described as soral asospory (Fig.). Since then apospory has been reported from many diverse families of ferns as a pathway of natural polyploidisation in pteridophytes. Soral apospory, as reported by Druery, is also found in cultivated plants of Trichomanes kaulfussi and Polystichum angulare var. pulcherrimum where leaves bore no sori and instead gametophytes developed on the margins (Bower, 1888; Fig.). In some varieties of Polystichum angulare prothalli developed at the tips of the pinnules in their direct continuation (apical apospory; Bower, 1923; Fig.). Aposporous gametophytes have also been observed on the young leaves of Osmunda regalis (Lang, 1924) and Osmunda javanica (Sarbadhikari, 1936). Antheridia borne on such gametophytes produced normal antherozoids but archegonia were absent. Sporophyte was formed on these gametophytes as direct growth of the prothallus.

Besides naturally occurring apospory, apospory has also been induced in many ferns. Different parts of sporophytes when grown in a medium with little or no carbohydrate, develop into aposporous gametophytes. Such experimental studies on apospory have been made by Steil (1944; on Tactaria trifoliata), Bristwo (1962; on Pteris cretica), Takahashi (1962;

on Phebodium aureum), Munore and Bell (1970; on Pteridium aquilinum), Hirsch (1975, 1976; on Microgramma vaccinifolia), and Whittier (1976, 1978; on Botrychium dissectum), etc.

Factors Regulating Apospory

Aposporous development of garnetophyte is influenced by a number of factors.

Nutrition : Nutritional and biochemical status of medium determines the nature of callus. Sucrose, besides providing nutrition, also has an important role in triggering different gene blocks into action that ultimately determine the nature of the regenerated tissue. Complete absence of carbohydrate in growth medium (called starvation medium) induces development of aposporous gametophytes. It has been found that in growth medium with 1% carbohydrate, callus formed structures which were partly gametophytes and partly sporophytes (i.e., gametosporophytes), while in growth medium with 2% carbohydrate callus formed only sporophytes.

Light Intensity : Low light intensity favours development of aposporous gametophyte. Experiments have shown that under high light intensity leaves of many ferns, such as Drynaria rididula, Polypodium aurenum, Platycerium bifurcatum and P. grande develop sporophytic buds, but they formed prothalli under low light intensity.

Age of Sporophytic Cells : The nature of tissue regenerated by a sporophytic cell is greatly influenced by its age. Young cells, like those of shoot apex or leaf tip, form gametophytes, whereas those of the basal part of the shoot or leaf regenerate sporophytic tissue.

Apogamy

Development of sporophyte from the gametophytic tissue without involving fusion of gametes is known as apogamy.

Such sporophytes grow vegetatively by budding from the gametophyte, hence they have the same level of ploidy as gametophytes. Though this process was first observed by Farlow (1874) in Pteris cretica, the term apogamy was introduced by de Bary in 1878 to include all cases of elimination of sexual function.

Amongst ferns apogamy is more widespread than apospory. There are more than 50 species of ferns in which apogamous sporophytes are known to occur naturally due to the absence of or nonfunctional nature of one or both sex organs. Such apogamy is known as obligate apogamy.

Besides, gametophytes having functional sex organs can be induced to bear apogamous sporophytes. This type of apogamy is known as induced or facultative apogamy.

Species of Dryopteris, Adiantum, Diplazium, Pteris, Polystichum, Osmunda and Asplenium are some common ferns where obligate apogamy has been observed. The prothalli of most of these ferns have functional antheridia but archegonia are either non-functional or absent.

Ferns of dry habitats usually do not get sufficient water for fertilisation hence fail to reproduce sexually. In such forms, obligate apogamy is an adaptation to dry conditions.

The following are some important facts regarding apogamy.

(1) Apogamously produced sporophytes may have their inception (i) in a single cell or in a group of cells of the gametophyte, (ii) in one or more cells adjacent to the base of the archegonium, or (iii) in the ovum in an unopened archegonium.

(2) The transition from gametophytic organization (of the prothallus) to sporophytic tissue is gradual and as such in apogamous sporophytes it is usually difficult

to determine the boundaries of gametophyte and sporophyte.

(3) The apogamous sporophyte may be represented either only by a sporangium with spores (e.g., Phyllitis scolopendrium) or by a complete structure with root, stem and leaf (e.g., Doodia caudata).

(4) Apogamous sporophyte and gametophyte have the same level of ploidy as the former develops vegetatively from the latter. A characteristic of obligate apogamous species is that in them both generations have diploid chromosome number. Diploid gametophytes are obtained by viable diploid spores which develop due to certain modifications in the process of spore production.

(5) The organogenic development of apogamous sporophyte resembles that of a normal embryo. It has been very well demonstrated in Pteris cretica, Nephrodium molle and Marsilea drummondii. In each instance, nutrition of the young sporophyte is drawn from the cells of the prothallus. Thus, the young sporophyte, whether produced apogamously or after normal fertilisation, manifests a specific organization from an early stage. The development of organs is as such determined by genetic factor.

Induction of Apogamy

Lang (1896) was the first to induce apogamy successfully in many varieties of ferns. His work, as well as subsequent studies, have established that three factors play an important role in the induction of apogamy :

(i) ability of prothallial cell to become meristematic,

(ii) prevention of fertilisation, and

(iii) exposure of prothallus to direct illumination.

Lang (1898, in Phyllitis scolopendrium), Brown (1923, in Pheyopteris polypodiodes), Whittier and Steeves (1960, in Osmunda and Adiantum), Whittier (1964, in Pteridium aquilinum), and many others have been successful in inducing apogamy in ferns.

Factors Regulating Apogamy : Development of apogamous sporophytes in ferns is determined by a number of factors. Some important factors are as follows :

Nutritional Factors : Experiments have shown that maximum number of apogamous sporophytes are obtained if the gametophytes are placed in 4% sucrose solution at the 'initiative phase'. Further development can, however, proceed normally even if sucrose concentration of all mineral nutrients is necessary and deficiency or absence of any one of them inhibits apogamy.

Light Quality : Far-red light (705 nm) has been found to be the most effective in inducing apogamous development of sporophyte, followed by blue (445 nm) and white light. Experiments of Whittier and Pratt (1971) on Pteridium aquilinum have shown that pattern of vegetative growth of gametophyte (which is maximum in white light) is not related to the phenomenon of apogamy.

Hormones : Presence of hormones like 2, 4-D (2, 4 dichlorophenoxy acetic acid), gibberellic acid (GA3), indole 3-acetic acid (IAA) and tryptophan stimulates gametophytes to produce apogamous sporophytes even at very low concentration (0.5%) of carbohydrates. Besides, formation of apogamous plants is increased in prothalli exposed to ethylene.

Chromosome Number : Apogamy is more prevalent in polyploid gametophytes than in normal haploid prothalli. Thus, there seems to be a correlation between ploidy of the gametophyte and apogamy.

Heterospory and Seed Habit

Heterospory : Most of the pteridophytes produce only one kind of spores, i.e., they are homosporous. But some produce two kinds of spores, differing in size. The smaller spores are called microspores and the larger ones megaspores. They are formed in microsporangia and megasporangia respectively. The difference in the size of the spores is related to their function. The microspore on germination gives rise to male gametophyte and the megaspore to female gametophyte. Thus, the difference in the size of the spores is related to the sex of the gametophyte. The phenomenon where two types of spores differing in size, structure and function are formed on the same plant is known as heterospory.

Heterospory is known in nine genera of pteridophytes. Selaginella, Isoetes, Stylites, Marsilea, Pilularia, Regnellidum, Salvinia, Azolla and Platyzoma. Some of these genera also nearly approach to seed habit.

Origin of Heterospory

A detailed study of heterosporous forms has revealed that heterospory originated due to reduction in the number of spores within sporangia. When all the spores in a sporangium are functional there is greater competition for nutrition and as such a spore receives a limited food supply and its size remains smaller, i.e., microspores are formed. On the other hand, if some of the spore mother cells in a sporangium disintegrate during development, the remaining ones get sufficient nutrition for their development and consequently their size is increased, i.e., megaspores are formed.

The above interpretation of the origin of heterospory received support from palaeobotanical, developmental and experimental studies.

Palaeobotanical Evidences of Heterospory : Although most of the earliest vascular plants were homosporous, many fossils of the late Devonian and early Carboniferous period (e.g., Lepidocarpon, Lepidostrobus, Mazocarpon, Sigillariostrobus, Calamocarpon and Calamostachys) were heterosporous. According to Scott (1894), an indication of heterospory can be seen in two species of Calamostachys viz., C. binneyana and C. casheana. In C. binneyana most of the sporangia were with large number of small spores in tetrads but in some sporangia spores were larger. However, in C. casheana two distinct types of sporangia- micro- and megasporangium occurred. There were some aborted spores in the megasporangia indicating that megaspores were formed at the cost of abortion of some of the spores in the sporangium. A similar abortion of spores was also observed in certain species of Lepidocarpon, Calamocarpon and Stauropteris. For example, a mature megasporangium of Lepidocarpon and Calamocarpon had only a single functional megaspore as other spores aborted during development and in Stauropteris a megasporangium had two functional megaspores and two aborted spores.

The above examples indicate that (i) heterospory has not evolved in living forms but was also present in fossil plants, and (ii) it originated due to disintegration of some spores in a sporangium.

Evidences from Developmental Studies : Developmental studies in pteridophytes, particularly the events that take place during the formation of sporocytes, meiosis and maturation of spores, provide a real insight in the understanding of heterospory.

In Selaginella, the development of the micro and megasporangium is similar till the differentiation of sporocytes. Subsequently, in the microsporangium, all the sporocytes undergo meiosis, with the result a large number of microspores are formed. On the other hand, in the megasporangium, all sporocytes, except one, abort and the surviving sporocyte undergoes meiosis forming four large functional megaspores. In some species of Selaginella, there is variation in the size of megaspores; for instance, in S. stenophylla two megaspores of a tetrad are slightly larger than the other two and in S. molliceps one megaspore is larger and the remaining three are smaller. An extreme case of reduction is found in S. erythropus where a megasporangium has only a single large megaspore with no traces of aborted spores. Rashid (1976) observed some larger spores in microsporangia of Selaginella sp. Such sporangia represent initial stages in the origin of heterospory.

In Isoetes, the microsporangia and megasporangia are identical till the differentiation of sporogenous tissue. In the microsporangium almost entire sporogenous tissue forms sporocytes, which, after meiosis, give rise to a very large number (1,50,000–1,000,000) of microspores. In the megasporangium, however, a part of the sporogenous tissue and so also many sporocytes degenerate and they provide nutrition to the growing sporocytes (megaspores). There are only 50-300 megaspores in a megasporangium.

In Marsilea, differences in the development of microsporangium and megasporangium do not become evident until after meiosis. In a microsporangium, all the 64 microspores formed after reduction division are functional, but in the megasporangium also though 64 spores are formed, only one is functional and the rest disintegrate. A similar condition also exists in Salvinia and Azolla.

Developmental studies have thus shown that determining process of heterospory becomes operative either before (e.g., Selaginella, Isoetes) or after meiosis (e.g., Marsilea, Salvinia, Azolla).

Evidences from Experimental Studies : Experimental studies on Selaginella and Marsilea have shown that heterospory originated due to nutritional factors. It was observed that if the photosynthetic activity of Selaginella was slowed down by keeping it in low light intensity, then only microsporangia developed. Due to low photosynthetic activity, nutrition became a limiting factor and spores could not grow in size. Thus, under such conditions only microspores were produced. Similar results were obtained in experiments done on Marsilea.

Importance of Heterospory

(1) Heterospory expresses sex determining capability of the plant. In homosporous species, differentiation of sex takes place at the gametophytic stages, whereas, in heterosporous species, differences in size of the spores is related to the sex of the gametophyte; a microspore always gives rise to male gametophyte and a megaspore to female gametophyte. Thus, in heterosporous forms the sex of the gametophyte can be predicted at the spore stage.

(2) The biological significance of heterospory is that in heterosporous forms development of gametophyte is endosporic and the nutrition for the developing gametophyte is derived from the sporophyte. Hence, the development of gametophyte is not affected by ecological factors as in case of independently growing gametophytes.

(3) An important characteristic of seed habit is that the megaspore is retained by the parent even after fertilisation. This ensures nutrition for the developing embryo.

Seed Habit

In order to evaluate seed habit in vascular cryptogams, let us first recapitulate the situation in seed bearing plants (spermatophytes). The spermatophytes produce two types of spores - microspores (pollen grains) and megaspores. In angiosperms, there is only one functional megaspore which is not liberated from the megasporangium (ovule). The male and female gametes fuse to form zygote or oospore, which eventually develops into embryo inside the ovule. The embryo and the rest of the gametophytic tissue (integuments and ovary wall) form the seed. After shedding, the seed germinates and forms a new plant.

Thus, the requirements for the formation of seed are as follows :

(1) Formation of two types of spores-microspores and megaspores (heterospory).

(2) Reduction in the number of functional megaspores to one per megasporangium.

(3) Retention of megaspore in the megasporangium until embryo development.

(4) Elaboration of the apical part of mega sporangium to receive microspores or pollen grains.

(5) Availability of sufficient nutrition for the development of embryo.

The evolution of seed habit in heterosporous pteridophytes can be evaluated on the basis of above mentioned requirements. Selaginella, the most common genus of the heterosporous pteridophytes, provides the best example. Most of the species of Selaginella are heterosporous and they have only one functional megaspore mother cell which gives rise to four megaspores after meiosis. Only a single functional megaspore in a sporangium is present in Selaginella rupestris, S. monospora and S. erythropus like spermatophytes. The development of female gametophyte, fertilisation and embryo development takes place within the megasporangium. Thus, evolution of seed habit took place in such species of Selaginella. However, the seeds developed in these species cannot be called true seeds because (i) the megasporangium is not covered with integuments, and (ii) there is no resting stage after embryo development; the development of embryo is accompanied with the development of shoot and rhizophore.

Stelar System in Pteridophytes : The central vascular cylinder of the primary axis of pteridophytes is usually referred to as stele (Greek stele = column). Besides xylem and phloem, it includes pith (if present) and is delimited from the cortex by the pericycle. The concept that stele is the fundamental unit of vascular system was put forward by Van Tieghem and Douliot (1886). They proposed the stelar theory, according to which the root and stem have the same basic structure consisting of two fundamental units - the cortex and the central cylinder. Van Tieghem considered 'endodermis (the innermost layer of the cortex) to be the anatomical boundary between these two fundamental units. Although a distinct endodermis is present between the cortex and stele - in vascular cryptogams, higher plants usually lack a distinct endodermis and pericycle. Esau (1953) and Foster and Gifford (1963), therefore, opined that in relation to higher plants, the term vascular cylinder is better suited than stele.

In lower vascular cryptogams like Lycopodium and Selaginella, the stele of the axis and the leaf traces develop independently, the stele being mainly cauline and the leaf traces are attached only to its surface. Contrary to this, the stele of higher vascular cryptogams (e.g., Pteridium, Pteris) is a composite structure consisting of the vascular cylinder of the stem and leaf traces.

Types of Steles in Pteridophytes : Schmid (1982) recognized the two principal types of steles in pteridophytes.— (1) protostele, (2) siphonostele.

Protostele : It is a non-medullated stele consisting of a central core of xylem, surrounded by a band of phloem. There is a single or multiple layer of pericycle outside the phloem which is delimited externally by a continuous sheath of endodermis. The protostele is considered to be the most primitive, both phylogenetically as well as ontogenetically. Fossil Psilophytales (e.g., Rhynia, Homeophyton), as well as many living primitive vascular plants (e.g., Psilotum, Tmesipteris), characteristically show this type of stele. In Filicales (which are otherwise advanced vascular cryptogams) protostele is seen generally in the sporelings, but is also permanently maintained in the adult stems of some relatively primitive Filicales.

The protostele gives off leaf traces without any break in the continuity of the enveloping endodermis. With the departure of leaf trace there is no marked change in the solid xylem core. Each leaf trace is surrounded by its own endodermis.

The following four types of protosteles are recognized in pteridophytes.

1. *Haplostele* : It is the simplest and most primitive type of protostele. It consists of a solid xylem core with smooth circular outline, which is surrounded by a ring of phloem. Haplostele is found in fossil (e.g., Rhynia, Homeophyton, Cooksonia) as well as many living pteridophytes (e.g., Psilotum, Selaginella, Lycopodium).
2. *Actinostele* : In this type of protostele, the xylem is star-shaped with many radiating arms. The phloem, instead of forming a continuous ring, is present in small patches in between the radiating arms of the xylem (Fig.). Actinostele is characteristic of many living (e.g., Psilotum, Lycopodium serratum) and fossil forms (e.g., Asteroxylon, Sphenophyllum).
3. *Plectostele* : A protostele in which the central xylem core breaks into more or less parallel plates is known as plectostele. Each xylem plate is surrounded by phloem. Aerial shoots and cone axis of Lycopodium clavatum and L. volubile have this type of stele.

4. *Mixed Protostele :* Sometimes the solid xylem core of the protostele is broken into small groups of tracheids which remain embedded in the phloem (Fig.). Such a protostele is known as mixed protostele and is found in the stems of Lycopodium cemuum. A mixed protostele is different from a protostele with mixed pith. The latter possesses groups of tracheids intermixed with parenchyma cells (e.g., Osmunda regalis).

Siphonostele : A simple protostele is sufficient to meet the requirements of a small stem or a few larger ones under certain circumstances. However, the requirement of increasing diameter of the stem is met with by certain internal modifications in the stele. The first step in this direction is the appearance of definite parenchymatous pith or medulla in the centre of the protostele. Such a modified protostele with a central pith is known as siphonostele or medullated stele. It is an advancement over protostele.

Jeffery (1910), on the basis of their association with leaf and branch traces, recognised two types of siphonosteles - cladosiphonic and phyllosiphonic. A cladosiphonic stele is characterised by the absence of leaf traces. Microphyllous lycopsids are the main example of this group. Phyllosiphonic stele, on the other hand, has both leaf and branch traces and is found in the members of the order Filicales.

Siphonosteles of Equisetum, Osmunda, Schizaea, etc., have a single phloem ring external to the xylem. Such a siphonostele is called ectophloic siphonostele (Fig.). Contrary to this, the siphonosteles of Adiantum, Dryopteris and Marsilea have a ring of phloem each external and internal to the xylem. This type of siphonostele is called amphiphloic siphonostele (Fig.). Such steles characteristically have two endodermal layers : outer endodermis - that lies outside the outer phloem and inner endodermis - that lies inner to the inner phloem, outside the pith.

1. *Origin of Siphonostele* : As mentioned earlier, siphonostele has originated by the development of pith in the centre of the protostele. The process of medullation of protostele has been debated. The following two views have been put forward regarding the origin of pith in protostele.

   (a) *Intra-stelar Origin of Pith* : According to this view, pith originated as a result of transformation of tracheary elements of the central xylem core into parenchyma. Thus the pith is wholly intra-stelar. Examples of Botrychium ternatum, B. virginianum, Osmunda regalis, Lunularia, etc., are cited in support of this view where tracheids are scattered throughout the pith (i.e., mixed pith). This view was supported by Boodle (1901), Gwynne-Vaughani (1908), Bower (1911) and Fahn (1960).

   (b) *Extra-stelar Origin of Pith* : According to this view, protostele has transformed into a siphonostele due to the migration of cortical cells into the stelar axis (Jeffery, 1902, 1910, 1917). Openings, such as leaf and branch gaps, probably provided passage for the invasion of parenchyma. This view derives support from amphiphloic siphonostele that has two endodermal layers, one delimits the stele from the cortex and the other from the pith. Since endodermis is a structure peculiar to the cortex, pith is considered as cortical in origin.

2. *Modifications of Siphonostele* : Depending on the presence of non-overlapping or overlapping gaps, the following three types of siphonosteles are recognised.

   (a) *Solenostele* : A siphonostele with non-overlapping leaf gaps is known as solenostele. The solenostele may be ectophloic or amphiphloic. Usually in the lower part of the stem where leaves are sparsely placed, the stele is solenostelic.

(b) *Dictyostele* : Many ferns like Dryopteris, Pteris, Ophioglossum, etc., have a very small rhizome with crowded leaves. Consequently, the leaf gaps overlap with each other (Fig.). A siphonostele with overlaping leaf gaps is known as dictyostele. A dictyostele has many scattered vascular strands, each of these strands is called a meristele (Fig.).

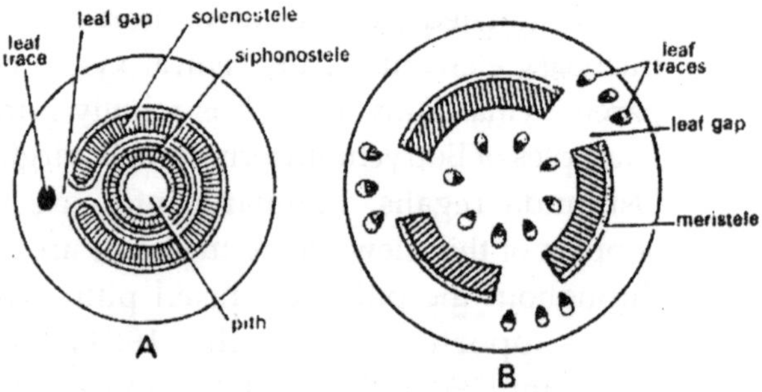

A-B. Polycyclic stele : A. A polycyclic solenostele, B. A polycyclic dictyostele.

(c) *Polycyclic Stele* : Stelar structure of certain pteridophytes is still more complex. They possess two or more concentric rings of vascular tissue and are called polycyclic (Fig.). For instance, there are two rings of vascular tissue in Pteridium aquilinum, three in Montania pectinata and four in Pteris podophylla and Platycerium aethiopicum.

If in a polycyclic stele the outer cylinder is solenostelic, it is called polycyclic solenostele (Fig.) and if the outer cylinder is dictyostelic, it is known as polycyclic dictyostele (Fig.).

3

The Salviniales

The order Salviniales includes two genera of water ferns-Azolla and Salvinia. The sporophytic plant body of these ferns is soft and herbaceous and floats on the surface of permanent water bodies forming dense mats. The stem is represented by a horizontal rhizome. It is covered with two (Azolla) or four (Salvinia) rows of leaves. True roots are present in Azolla, but they are absent in Salvinia and instead one of the ventral leaves becomes dissected and appears root-like.

The members are heterosporous and sporangia are borne within special fruiting bodies, called sporocarps. The sporocarp bears either microsporangia (microsporocarp) or megasporangia (megasporocarp). The spores alongwith the cytoplasm of degenerating tapetal cells form small units, called massulae.

Reimers (1954) and Sermolli (1977) included two families-Azollaceae (Azolla) and Salviniaceae (Salvinia) in the order Salviniales. The life cycles of Azolla and Salvinia are described here as representatives of the order.

Azolla

Systematic Position

Sub-division	—	Pteropsida
Class	—	Leptosporangiatae
Order	—	Salviniales
Family	—	Azollaceae

The genus Azolla is represented by six species, which are confined to the tropical and subtropical regions of the world. They are small moss-like floating plants, forming a dense covering on the surface of permanent ponds. Mature plants of Azolla impart a characteristic reddish colour to the pond. The genus is represented in India by three species; of these A. pinnata and A. filiculoides are widely distributed throughout the country and A. imbricata is confined to the Eastern Himalayas.

Sporophyte

External Morphology : The sporophytic plant body is small and herbaceous and consists of a horizontal rhizome which bears leaves and roots.

The rhizome is thin and delicate and floats horizontally on the water surface. It is pinnately branched; the branches are extra-axillary. On the lower side of the rhizome, long adventitious roots are present. The roots are usually close to the point of branching and are pendent in water. The rhizome as well as its branches are covered with minute (0.5 mm or less) leaves. They are borne in two alternate overlapping rows on the dorsal surface of the rhizome. Each leaf is divided into two almost equal lobes. The upper (aerial) lobe is several cells thick, green and photosynthetic. It is somewhat obliquely placed and touches water only on one edge. The lower (submerged) lobe is one cell thick and nearly colourless; it is probably concerned with the absorption of water.

Internal Structure

1. *Rhizome* : A transverse section of the rhizome shows a single layered epidermis, a cortex consisting of 5-8 layers of large parenchymatous cells, and a central stele. The stele is surrounded by a single layer of endodermis, followed by a layer of pericycle. The vascular tissue is greatly reduced. In young stems, the stele is essentially protostelic but mature stems have ectophloic siphonostele with typical leaf and branch traces. The leaf and branch gaps are of non-overlapping type. The xylem consists of only few tracheids and is separated from the phloem by a thin layer of parenchyma. The leaf trace forks dichotomously before entering into the leaf; one branch enters the upper lobe and other the lower lobe.

2. *Root* : The epidermis of the root is composed of a single layer of small cells. The cortex is made up of two layers of large thin walled parenchymatous cells; the outer layer has nine cells and the inner only six cells. Inner to the cortex, there is a single layer of endodermis, followed by a layer of pericycle. The endodermal and pericycle layers consist of six cells each. The xylem is represented by two centrally placed metaxylem

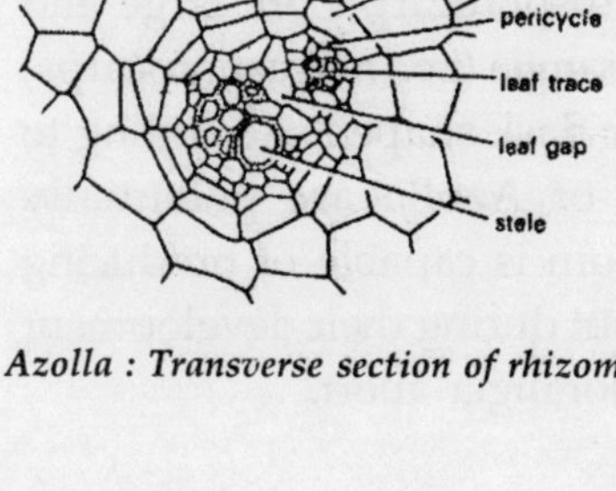

Azolla : Transverse section of rhizome.

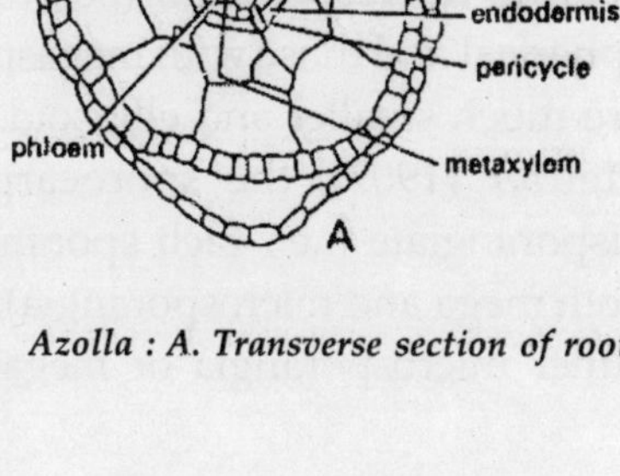

Azolla : A. Transverse section of root

tracheids surrounded by four small peripheral groups of protoxylem elements. The phloem consists of only few cells present on either side of the metaxylem elements.

3. *Leaf :* The upper green photosynthetic lobe of the leaf is many cells thick in the central region. The cells of the upper epidermis give rise to a tuft of unicellular or bi-celled hairs which trap considerable amount of air. This prevents leaves from wetting. Stomata are present in both the upper and lower epidermal layers. The mesophyll is composed of columnar palisade-like photosynthetic cells with intercellular spaces in between them. The upper lobe has a large cavity at its base, opening externally by a circular pore. The cavity-contains Anabaena azollae, a blue-green nitrogen fixing alga.

The lower subriterged lobe of the leaf is only one cell thick, colourless and non-photosynthetic.

Reproduction : Azolla reproduces vegetatively and by spores.

Vegetative Reproduction : Vegetative propagation takes place by fragmentation. The lateral branch separates from the main axis due to the formation of an abscission layer at its base. It grows into an independent plant.

Reproduction by Spores : Azolla is a heterosporous fern and spores are borne within specialized fruiting bodies, called sporocarps. The sporocarps of Azolla are dimorphic; those bearing microsporangia (i.e., microsporocarps) are large and spherical and those with megasporangia (i.e., megasporocarps) are much smaller and ellipsoid or flask-shaped. According to Pfeiffer (1907), the sporocarps of Azolla are potentially bisporangiate (i.e., each sporangium is capable of producing both mega and microsporangia), but during their development either microsporangia or megasporangia abort.

1. *Position of Sporocarps :* Sporocarps are borne on the lowermost leaf of a lateral branch. Development of the lower submerged lobe of the fertile leaf stops when its primordium is 2-4-celled. Subsequently, each cell of the leaf primordium develops into a sporocarp. Thus each lateral branch bears 2-4 sporocarps depending upon the number of cells in the lower lobes of the leaf primordium. The upper lobe of the fertile leaf forms a hood-like covering around the sporocarps.

2. *Development of Mega and Microsporocarps :* In both types of sporocarps, megasporangium is always the first to be formed. Microsporangial initials arise from the stalk of the megasporangium. If the latter matures, the microsporangial initials degenerate, and a megasporocarp containing a single megasporangium is formed. But if the megasporangium degenerates, then the microsporangial initial develops further to form microsporocarp.

The fertile leaf initial divides anticlinally. The upper lobe initial undergoes normal anticlinal and periclinal divisions, while the lower lobe initial, after a few divisions, directly gives rise to the megasporangial initial. The involucre, which covers the sporocarp, develops from the upper leaf lobe. The megasporangium develops from a single initial cell, thus the development is of leptosporangiate type. First few divisions in the sporangial initial are oblique and intersecting and as a result a dome-shaped apical cell is formed. The lateral derivatives of the apical cell give rise to the stalk. After a few cells of the stalk have been cut off, a distal periclinal division occurs in the apical cell forming a central cell surrounded by a single layered wall.

Meanwhile, the stalk cells of the sporangium, by divisions and enlargement, form a short; annular protuberance around the base of the apical cell. This protuberance gradually grows, becomes 2-layered and forms the indusium. It surrounds the sporangium completely, except for a small opening at the tip. The upper part of the indusium of the megasporocarp becomes quite hard due to lignification and deposition of tannins. After dehiscence, the lower portion of the indusium decays while the upper hard portion persists as a cap till it is pushed aside by the developing embryo. The central cell cuts off a layer of tapetum by periclinal divisions and functions as sporogenous cell. The number of cells in the tapetal layer increases by repeated anticlinal divisions.

The sporogenous cell divides transversely and then by two vertical divisions at right angles to each other give rise to 8 megaspore mother cells. By this time, the wall of the densely cytoplasmic tapetal cells dissolves and their cytoplasm intermingles to form multinucleate periplasmodium. Later, the walls of the megaspore mother cells also dissolve, and the rounded cells become surrounded by periplasmodium. Each megaspore mother cell undergoes meiosis giving rise to four megaspores, thus there are 32 megaspores in a megasporangium. Of these, only one is functional while the rest degenerate at various stages of development.

The vacuolated periplasmodium, surrounding the degenerating megaspores, gradually hardens to form a honeycomb-like structure, called float corpuscles or massulae, at the distal end of the megaspore. It is believed by some workers that the massulae serve as swimming apparatus and provide buoyancy to the large megaspore. But as the megaspore remains at the bottom of the pond until the development of female gametophyte, this view does not seem to be convincing.

The microsporangial initials arise from the stalk of the megasporangium. Its development is dependent upon complete degeneration of all megaspores. Once the development of microsporangial initials starts, additional initials appear in basipetal succession and thus at maturity, as many as 120 or more microsporangia may develop in a sporocarp.

The microsporangial initial undergoes two oblique intersecting divisions, differentiating an apical cell with three cutting faces. The cells cut laterally function as stalk cells. A number of divisions occur in the stalk cells resulting in the formation of an elongated filament which is two cells broad. The apical cell undergoes two oblique and one periclinal division, giving rise to a single layered sporangial wall and a centrally situated cell.

The cells of the sporangial wall divide anticlinally to keep pace with the developing sporogenous tissue. The central cell divides periclinally separating a parietal tapetal layer and a sporogenous cell. The latter by four successive divisions forms 16 microspore mother cells. Initially, the tapetal layer divides anticlinally forming many small uninucleate tapetal cells but gradually their radial walls dissolve to give rise to a multinucleate periplasmodium. Meanwhile, the spore mother cells undergo meiosis, each forming a microspore tetrad. Thus in a microsporangium, there are 64 microspores derived from 16 microspore mother cells.

As the microsporangium matures, the microspores move towards the periphery and the periplasmodium organises into four or more massulae each with many peripherally situated glochidia, which are hardened cytoplasmic simple or branched strands. The massulae, when set free in water, get themselves attached to the rugged surface of the megaspore with the help of glochidia.

Both, the mega- and microsporocarps sink to the bottom of the pond. The massulae are set free after the death and decay of the sporocarp wall.

Gametophyte

Female Gametophyte : The megaspore is the mother cell

of the female gametophyte. The first division of the megaspore is asymmetrical establishing a small lenticular cell and a large basal cell. Although the basal cell becomes multinucleate after repeated nuclear divisions and is full of starch grains, it does not take part in the formation of the gametophyte. The upper cell divides repeatedly and forms a hemispherical cushion-shaped non-chloronhyllous female gametophyte. Archegonia develop from archegonial initials which differentiate near the centre on the surface of the gametophyte. The archegonial initial divides periclinally into a central cell and a primary cover cell.

The latter divides by two vertical walls at right angles to each other forming four neck initials. These cells by further transverse divisions give rise to 3-4 cells high neck of the archegonium. Meanwhile, the central cell divides transversely into a small primary neck canal cell and a large primary venter cell. The primary neck canal cell usually does not divide further and directly functions as neck canal cell. The primary venter cell divides transversely into a small venter canal cell and a large egg. Mature archegonium thus has a short neck and a swollen venter. The neck has four vertical rows of cells with 3-4 cells in each row. The narrow neck canal is occupied by a single (or rarely two) neck canal cell. The venter has a venter canal cell and a large egg.

Male Gametophyte : The microspore is the mother cell of the male gametophyte. It germinates within the massula at the bottom or on the surface of the pond. The germinating spore absorbs water due to which outer spore wall ruptures along the triradiate ridge and the inner spore wall protrudes in the form of a papilla. The papilla is cut off near its base by a transverse wall. The outer cell divides by two periclinal walls and thus three cells are established in the outer protruded papilla. Meanwhile, the basal cell also divides asymmetrically forming a small prothallial cell. Of the three cells formed in the papilla, the apical and basal cells do not divide further and eventually function as jacket cells. The central cell, however, by two periclinal divisions forms two more jacket cells.

One of these jacket cells divides further. Thus five jacket cells are formed which surround the central cell. The latter divides repeatedly forming eight spermatocytes. Each spermatocyte eventually metamorphoses into an antherozoids. Thus mature male gametophyte consists of eight antherozoids surrounded by five jacket cells. The gametophyte remains embedded in the outer part of massula and is released when massula softens.

Embryo : The zygote, formed by the fusion of male and female gametes, enlarges considerably within the venter of the archegonium. It divides transversely, followed by a vertical division, thus a quadrant is formed. The derivatives of the two epibasal cells give rise to the leaf and stem and those of the hypobasal cells form primary root and foot. The leaf and stem develop more rapidly than the primary root.

Salvinia

Systematic Position

Sub-division	—	Pteropsida
Class	—	Leptosporangiate
Order	—	Salviniales
Family	—	Salviniaceae

Salvinia, primarily a tropical genus, has about 12 species, mostly native of Africa. The genus is represented in India by four species - S. auriculata, S. cucullata, S. modesta and S. natans. These are free floating perennial aquatic ferns which occur in still waters of ponds and lakes and back water of sluggish streams. The plants multiply fast by fragmentation and grow rapidly and as such cover water bodies.

Sporophyte

External Morphology : Salvinia has a horizontal, floating, hairy and branched rhizome about 10 cm long, differentiated into nodes and internodes. The nodes bear leaves as well as give off lateral branches but these branches are not axillary in position. There are no roots.

The sessile or shortly stalked leaves are arranged in whorls of three at the nodes. The leaves of successive nodes alternate, thus giving rise to six rows of leaves. Salvinia shows dimorphism, i.e., it has two types of leaves-floating and submerged. The floating leaves are entire, cup or boat-shaped, approximately 2 cm long, laminate and with reticulate venation. The upper surface of these leaves is covered with peculiar unwettable stiff and papillose hairs that can trap considerable amount of air. Thus they act as floating organs. The submerged leaves are divided into fine linear segments; they are hairy and have striking resemblance to roots, but it is doubtful whether they perform the functions of roots. Most probably they act as balancers. The hairs on the submerged leaves are composed of three cells each, a foot cell and one long and one short horizontally oriented cell.

Internal Structure

1. *Rhizome :* The internal structure of the rhizome shows characteristic aquatic features. It has a distinct epidermis, a broad cortex and a central vascular cylinder.

   The epidermis is a continuous layer, composed of thin walled cells that are covered with a thin layer of cuticle. It does not possess stomata.

   The cortex is broad and possesses large number of air chambers separated from each other by partition walls.

   The stele is surrounded by a distinct layer each of endodermis and pericycle. The xylem is highly reduced and is represented by a few patches of annular, spiral or scalariform tracheids. The xylem is surrounded by the phloem. In the centre, there is a small parenchymatous pith which has few tracheidal cells.

The stele of Salvinia can be regarded as ectophloic siphonostele as it is interrupted by leaf gaps. However, in S. auriculata the stele is amphiphloic siphonostele in the internodal region and ectophloic siphonostele at the node.

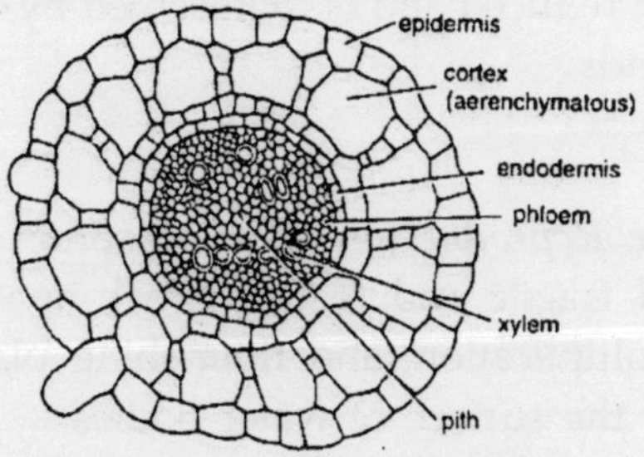

Salvinia : Internal structure of rhizome.

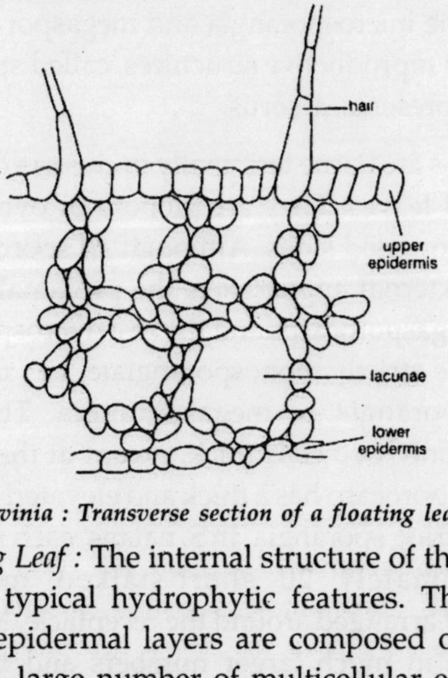

Salvinia : Transverse section of a floating leaf.

2. *Floating Leaf :* The internal structure of the floating leaf shows typical hydrophytic features. The upper and lower epidermal layers are composed of thin walled cells. A large number of multicellular egg-beater like hairs are present on the upper epidermis. The

mesophyll, which is not differentiated into palisade and spongy parenchyma, has large air chambers. The vascular strands are represented by few xylem tracheids which are encircled by phloem.

3. *Submerged Leaf*: The vascular cylinder of the submerged leaf is surrounded by a single layered endodermis and pericycle. At the base of the leaf there is present an amphiphloic siphonostele but as it traverses upward, it breaks into many ectophloic steles. The xylem is highly reduced and is represented by only one or two tracheids.

Reproduction

Vegetative Reproduction : The rhizomes of Salvinia are very soft and fragile and readily break apart. This secures vegetative multiplication, and flourishing plants of Salvinia quickly cover the surface of water bodies

Reproduction by Spores : Salvinia is also a heterosporous water fern. The microsporangia and megasporangia are borne within special reproductive structures, called sporocarps. Each Sporocarp represents a sorus.

Sporocarps are borne terminally in clusters on the segments of submerged leaves. They are globose or ovoid with ridged or smoothly rounded sides. Although all sporocarps are alike in size and external appearance, the proximal 2 to 5 in each cluster are megasporocarps and the rest microsporocarps. Thus sporocarps are strictly monosporangiate, i.e., they bear either only microsporangia or megasporangia. The wall of the sporocarp is only two cells thick, except at the apex where it is lobed. The sporocarp has a thick and elevated sporangiferous receptacle bearing sporangia. In S. natans, each megasporocarp has approximately 20 short-stalked megasporangia symmetrically arranged around the receptacle. Microsporangia are produced in much larger numbers and their stalks are longer, thinner and occasionally branched.

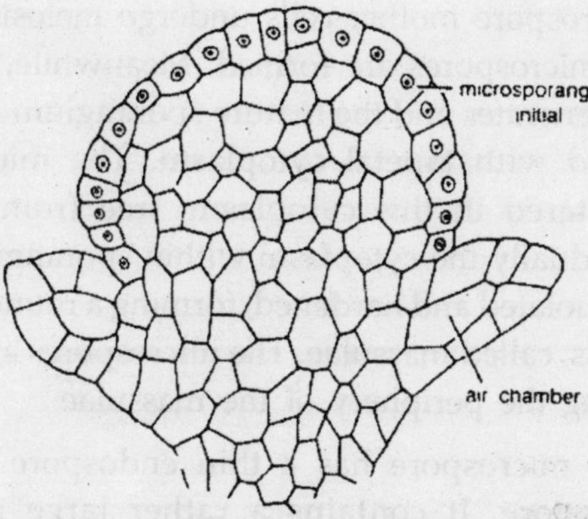

Salvinia* : *Longitudinal section of a young Sporocarp showing initials of microsporangia.

1. *Development of Microsporangium* : Microsporangium initials are formed on the surface of a very young Sporocarp. Each initial cell undergoes a transverse division forming a stalk cell and an apical cell. The stalk cell divides only transversely forming a several celled long stalk of the sporangium. Besides, the derivatives of the stalk cell also form a part of the sporangium wall attached to the stalk.

   The apical cell divides by a transverse and two more or less periclinal walls, thus cutting off three primary wall cells and a central archesporial cell. The primary wall cells divide only anticlinally and form a thin delicate single layered wall of the sporangium. The archesporial cell divides by periclinal walls, forming three primary tapetal cells and a central cell. The central cell undergoes a series of divisions as a result of which 16 microspore mother cells are formed. Meanwhile, the three primary tapetal cells also divide by radial walls and form a single layered tapetum.

   The full grown microspore mother cell has denser cytoplasmic contents and a much larger nucleus than the surrounding cells of the sporangium. The microspore mother cells undergo meiosis, as a result 64 microspores are formed. Meanwhile, the tapetum degenerates and the mature sporangium is completely filled with tapetal cytoplasm. The microspores lie scattered in this cytoplasm, free from each other. Gradually the cytoplasm within sporangium becomes vacuolated and hardened, forming a rounded alveolate mass, called massulae. The microspores are embedded along the periphery of the massulae.

The microspore has a thin endospore and a thick exospore. It contains a rather large nucleus and abundant starch grains.

2. *Development of the Male Gametophyte* : The microsporangium does not open and the microspore germinates in the massulae within the sporangium. The microspore divides by two almost transverse walls, and the lowest cell of the 3-celled prothallus cuts off a small root cell. The prothallus thus now has 4 cells- a root cell, a prothallus cell adjacent to the root cell, an inner cell and an outer cell. At this stage, the exospore splits along the triradiate ridge on the surface of the microspore.

   The upper most cell (outer cell) of the 4-celled prothallus divides by two intersecting oblique cross walls forming an apical cell and an antheridial mother cell. The inner cell also divides transversely, giving rise to a sterile cell above and an antheridial mother cell below. Subsequently, each antheridial mother cell cuts off a small sterile wall cell by an oblique wall. Thus at this stage, the male gametophyte has two young antheridia, each made up of a wall cell and an antheridial cell. The latter divides into four androcytes. Each androcyte forms a spirally coiled multiflagellate antherozoid. It has a starch filled vesicle at its posterior end which is detached sooner or later from its body.

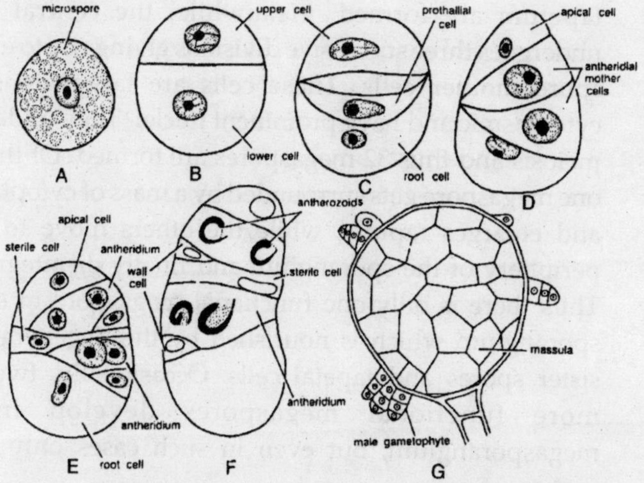

Salvinia* : *Development of male gametophyte; A- E. Stages in the development of male gametophyte; F. Mature male gametophyte, G. Male gametophytes breaking through massula,

Figure shows the structure of a fully developed male gametophyte, consisting of a root cell, a prothallial cell, an apical cell and two antheridia which are separated by a sterile cell. The antheridia protrude out of the massula.

3. *Development of Megasporangium* : The initials of megasporangia appear at the apex of the young megasporocarp. They are distinguishable from the neighbouring cells by their denser cytoplasm and larger nuclei. Each initial cuts off 2-3 cells at its base by oblique walls. Thus a short stalk and a somewhat conical apical cell is formed. The latter divides by three intersecting periclinal walls, forming an archesporial cell and three primary wall cells. Further periclinal divisions in the archesporial cell results in the formation of three primary cells of the tapetum and a central cell. Both the primary wall-cells as well as primary tapetal cells undergo repeated radial divisions and consequently a one cell thick wall of the sporangium and a layer of

tapetum are formed. Meanwhile, the central cell undergoes three successive divisions giving rise to eight spore mother cells. These cells are large, densely cytoplasmic and have prominent nuclei. They undergo meiosis and thus 32 megaspores are formed. Of these, one megaspore gets surrounded by a mass of cytoplasm and enlarges rapidly, while the others move to the periphery of the sporangium and finally disintegrate. Thus there is only one functional megaspore in each sporangium which is nourished by the degenerating sister spores and tapetal cells. Occasionally, two or more functional megaspores develop in a megasporangium, but even in such cases only one matures finally.

The mature megaspore is surrounded by three membranes. The outer most membrane, called epispore or perispore, is formed by the vacuolization and hardening of the cytoplasm derived from the tapetum and degenerating sister spores. This membrane is thicker at the apex of the spore, where it is divided into three folds and forms a cavity like the pollen-chamber of a gymnospermic ovule. The middle membrane, known as exospore, is homogeneous in structure and almost uniform in thickness. It is probably derived from the protoplasmic mass surrounding the megaspore. The innermost membrane, the endospore, is very thin and lines the inner surface of the exospore.

The mature sporocarps sink to the bottom of the pond. There is no mechanism for the dehiscence of the sporocarp and spores are liberated only after the decay of its wall.

4. *Development of Female Gametophyte* : Germination of the megaspore takes place on the surface of the water after the wall of the sporocarp has decayed. The megaspore

maintains a definite orientation while floating and becomes stabilized in some way. At the time of germination, the cytoplasm in the upper part of the spore becomes very dense. It divides forming a small apical lens-shaped cell and a considerably larger lower cell.

The entire prothallus tissue is formed by further divisions of the upper small cell. The nucleus of the lower large cell multiplies by free nuclear divisions, and many nuclei thus formed are uniformly distributed, throughout the parietal layer of the cytoplasm. There is no wall formation in the lower cell and this cell serves as a food reservoir. The apical cell forms a cushion of cells. Before the prothallus pushes out of the spore cavity, a large cell appears on the upper surface of the prothallus. This cell becomes the initial cell of the archegonium.

5. *Development of Archegonium :* The archegonial initial divides by a transverse wall into an outer primary cover cell and an inner cell. The primary cover cell divides by two vertical walls at right angles to each other into four neck initials. All the neck initials undergo transverse divisions forming 2-4 cells high neck of the archegonium.

Almost simultaneously, the inner cell divides unequally by a curved wall into a smaller neck canal cell and a larger central cell. The latter divides again to form an outer small venter canal cell and an inner large egg cell. Although the nucleus of the neck canal cell divides into two there is no wall formation to separate the nuclei. More archegonia may develop subsequently on both sides of the first formed archegonium.

Fertilisation : When the egg cell becomes mature, the neck of the archegonium opens widely. A slimy substance formed by the disintegration of the neck canal cell and the venter canal cell oozes out of the neck and attracts the antherozoids. A large number of antherozoids enter the neck of the archegonium and reach the exposed surface of the egg cell. Normally, only one antherozoid penetrates the cytoplasm of the egg and fertilisation occurs.

Embryo : The first division of the zygote is parallel to the long axis of the archegonium and the two cells thus formed are somewhat unequal in size. Both these embryonic cells divide transversely forming a quadrant. Each cell of the quadrant divides by a vertical wall and thus an octant is formed. The four lower (hypobasal) cells form the foot, which does not enlarge much and is mainly haustorial in function. The two outer cells of the epibasal half become the first leaf. Of the remaining two cells (of the epibasal half), one becomes the stem apex and the other one is apparently functionless.

There is no root in Salvinia. However, some workers consider that an embryonic root is formed but its development ceases at an early stage and later its tissue cannot be distinguished from that of the root.

Soon after the formation of octant, its epibasal and hypobasal cells form a plate of eight cells at the centre by periclinal divisions. The cells of the central plate divide rapidly and form a column-like structure which connects the root with the leaf and stem tip. Further elongation of the column pushes the stem tip and leaf out of the gametophytic tissue. At this stage, the stem apex is inverted, i.e., it faces the gametophyte but subsequently the arched tip of the column straightens. The stem apex grows and forms the rhizome which bears leaves at nodes.

4

The Marsileales

The order Marsileales is a smaller group of heterosporous leptosporangiates. These are mostly aquatic or semiaquatic plants and are commonly known as 'water ferns'. The sporophyte has a branched creeping rhizome which bears roots and upright leaves at intervals. Young leaves show circinate vernation. The plants are heterosporous and both microsporangia and megasporangia occur within a specialized structure, called sporocarp. Marsileales include a single family Marsileaceae with three genera Marsilea, Pilularia and Regnellidium. Here life-history of Marsilea, the most widely distributed genus of the family, is described in detail.

Marsilea

Systematic Position

Sub-division	—	Pteropsida
Class	—	Leptosporangiatae
Order	—	Marsileales
Family	—	Marsileaceae

Marsilea, commonly known as water fern, is represented by some 53 living and 10 fossil species. The living species occur in all parts of the world, but are more common in the tropical Africa and Australia. They are aquatic or amphibious; the aquatic species usually grow in shallow ponds and pools, but fruiting bodies (sporocarps) are formed only in terrestrial habitats. The amphibious species grow in water-logged soil, partly submerged. The same species may grow in both aquatic and terrestrial habitats. Marsilea condensata and M. rajasthanensis are near xerophytic and M. hirsuta, an Australian species, can also withstand long dry spells.

In India, the genus is represented by nine living species; of these, M. minuta is the most widely distributed. Marsilea aegyptiaca, M. braehypus, M. brachyearpa, M. condensata, M. rajasthanensis and M. quadrifolia are the other common Indian species.

Sporophyte

External Morphology : The sporophyte of Marsilea is differentiated into rhizome or stem, roots and leaves (Fig.).

1. *Rhizome (Stem) :* The stem of Marsilea is modified into a long and slender rhizome, which creeps on the surface of the soil or is subterranean. The rhizome grows indefinitely and sometimes it may attain a length of up to 25 meters. It is profusely branched and often forms a mat. The branches arise in the axils of the leaves (axillary) or are lateral to the leaves. It is differentiated into nodes and internodes; the internodes are longer in aquatic species than in semiaquatic or terrestrial species (e.g., M. hirsuta). Underground tubers are formed on rhizomes.
2. *Roots :* The primary root is ephemeral and one or more adventitious roots arise at the nodes on the underside of the rhizome. Sometimes roots may also arise at

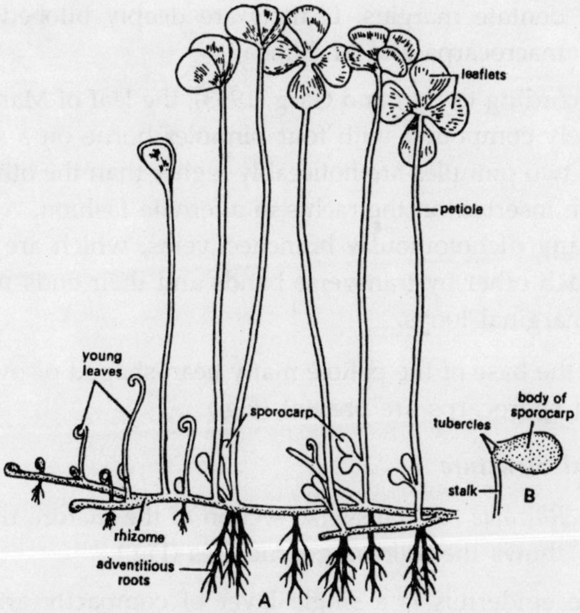

A-B Marsilea : A. External morphology, B. A sporocarp.

internodes (M. aegyptiaca) or laterally (M. minuta). They are sparingly branched and develop in an acropetal sequence, i.e., the youngest root is towards the apex of the rhizome. The number of roots at a node and their size vary considerably. For example, in M. aegyptiaca roots are about 2 cm long, whereas in M. minuta they may attain a length of up to 13 cm.

3. *Leaves* : The leaves are long petioled and compound. They are borne at the nodes and are arranged in two alternate rows on the upperside of the rhizome. The young leaves are strongly circinate. In submerged aquatic forms, the petiole is usually long, soft and flexible, whereas in those forms which grow in muddy habitats it is relatively small and stout. The lamina is usually divided into four leaflets, but occasionally the number of leaflets may vary from 3-8. The leaflet is obovate, elliptical or wedge-shaped with entire or

dentate margins. Leaflets are deeply bilobed in M. macrocarpa and M. biloba.

According to Puri and Garg (1953), the leaf of Marsilea is pinnately compound with four pinnules borne on a slender rachis; two pinnules are noticeably higher than the other two and are inserted on the rachis in alternate fashion. A leaflet has many dichotomously branched veins, which are joined with each other by transverse bands and their ends unite to form marginal loops.

At the base of the petiole many bean-shaped or oval and stalked sporocarps are present (Fig.).

Internal Structure

1. *Rhizome :* A transverse section of the mature rhizome shows the following structures (Fig.).

The epidermis is a single layer of compactly arranged thick walled cells. It does not have stomata.

The cortex is differentiated into an outer, a middle and an inner region. The outer cortex is composed of parenchymatous cells. It is characterized by the presence of aerenchyma with many radially elongated air chambers arranged in a single ring. The air chambers are separated from each other by a single layered septum, called trabeculae. The middle cortex consists of few layers of sclerenchyma, whereas the inner cortex is made up of several layers of compactly arranged parenchymatous cells. Some cells of the inner cortex are filled with tannins.

The vascular cylinder is an amphiphloic siphonostele. Xylem is surrounded both externally and internally by phloem, called outer phloem and inner phloem respectively. Vessels have been reported in the rhizomes of two Australian species of Marsilea, via., M. drummondii and M. elata. The outer endodermis and pericycle surround the stele externally and the inner endodermis and pericycle internally. The outer and inner endodermis as well as pericycle are single-layered. The central part of the stele is occupied by a sclerenchymatous pith.

Environmental conditions affect the internal structure of rhizome. For instance, in xerophytic species (e.g., M. aegyptiaca), pith is sclerotic and air chambers are not present in the outer cortex.

2. *Petiole* : The petiole is also differentiated into epidermis, cortex and stele (Fig.).

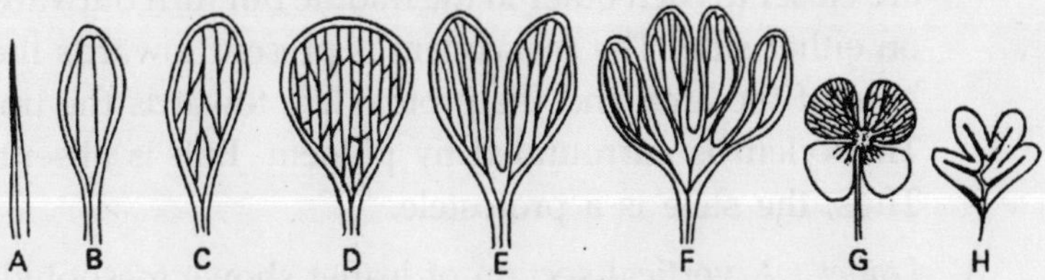

A-H. Marsilea : Venation in Juvenile leaves.

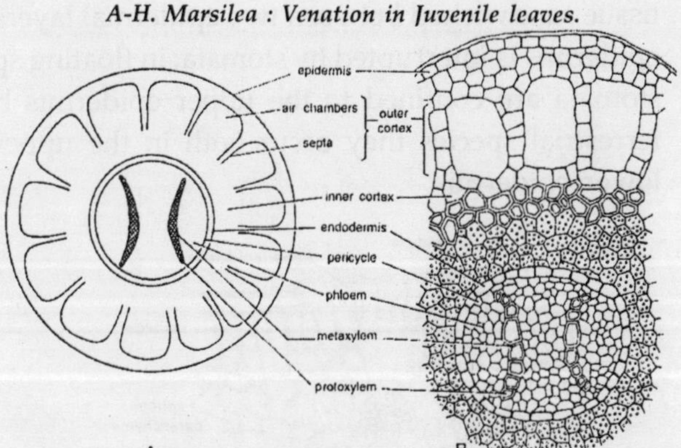

Marsilea : Transverse section of rhizome; A. Diagrammatic representation, B. A part cellular.

Marsilea : Transverse section of petiole; A. Diagrammatic representation, B. A part cellular.

The epidermis is composed of a single layer of rectangular cells.

The cortex is differentiated into an outer aerenchymatous and an inner parenchymatous region. A few outer layers of parenchyma develop thickenings. The ratio of aerenchyma and parenchyma differs in aquatic and terrestrial forms; the former have larger proportion of aerenchyma. The cells of the inner cortex contain starch grains. Besides, dark brown tannin is also present in some cells.

The stele is surrounded by a layer each of endodermis and pericycle. 'The endodermal cells are characterized by the presence of casparian bands. The xylem consists of two strands, each forming a curved arch. The arches are closer to each other in the middle but turn outward on either side. The metaxylem is present towards the base of the arch and the protoxylem towards the tip. The xylem is surrounded by phloem. Pith is absent. Thus, the stele is a protostele.

3. *Leaflet* : A vertical section of leaflet shows mesophyll tissue sandwiched between two epidermal layers. The epidermis is interrupted by stomata; in floating species stomata are confined to the upper epidermis but in terrestrial species they occur both in the upper and lower epidermis.

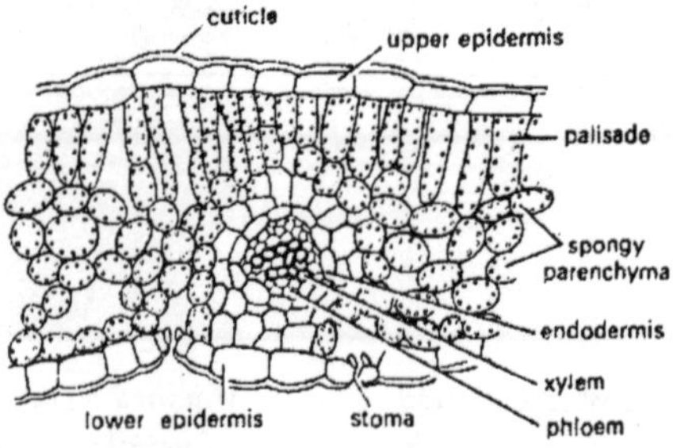

Marsilea : Transverse section of lamina.

The mesophyll is differentiated into palisade and spongy parenchyma. One or two layers of compactly arranged palisade cells are present just below the upper epidermis. They are followed by loosely arranged cells of spongy parenchyma with large intercellular spaces. In submerged species, however, the mesophyll is not differentiated into palisade and spongy parenchyma.

The vascular bundles of the leaflet are concentric with a central xylem core surrounded by phloem. Each vascular bundle is surrounded by a layer of endodermis.

4. *Root* : A transverse section of the root shows three distinct regions - epidermis, cortex and stele. The single layered epidermis is composed of compactly arranged cells.

The cortex is differentiated into an outer aerenchymatous, a middle parenchymatous and an inner sclerenchymatous region. The outer cortex has large air chambers, arranged in a single row. Adjacent air chambers are separated by a septum made up of a single file of parenchymatous cells. The middle cortex is a wide zone of thin walled parenchymatous cells. The inner cortex is represented by a narrow zone of sclerotic cells.

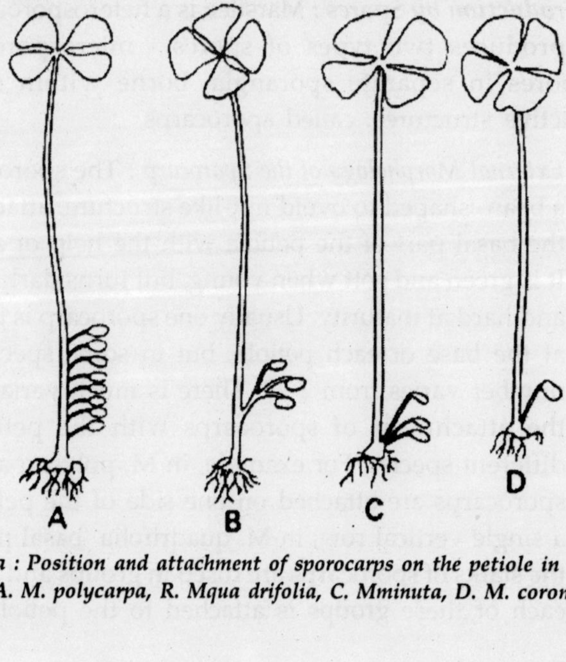

Marsilea : Position and attachment of sporocarps on the petiole in different species; A. M. polycarpa, R. Mqua drifolia, C. Mminuta, D. M. coromandelica

The stele is surrounded by a layer each of endodermis and pericycle. The xylem is diarch and exarch. It has two metaxylem elements in the centre and protoxylem elements towards the periphery. Vessels with scalariform to simple perforation plates have been reported in the roots of some species of Marsilea. Two phloem bands are present on either side of the xylem plate.

Reproduction

Marsilea reproduces vegetatively and by spores.

Vegetative Reproduction : In species like M. minuta, M. hirsuta and M. erosa, some subterranean branches of the rhizome form tubers during unfavourable conditions. The tubers have reserve food in the form of oil globules. They are capable of tiding over unfavourable period and thus serve as perennating bodies. On the return of favourable conditions, they germinate and form new plants.

Reproduction by Spores : Marsilea is a heterosporous fern, which produces two types of spores - microspores and megaspores in separate sporangia, borne within special reproductive structures, called sporocarps.

1. *External Morphology of the Sporocarp :* The sporocarp is a bean- shaped to ovoid nut-like structure, attached to the basal part of the petiole with the help of a stalk. It is green and soft when young, but turns dark brown and hard at maturity. Usually one sporocarp is present at the base of each petiole, but in some species the number varies from 2-20. There is much variation in the attachment of sporocarps with the petiole in different species. For example, in M. polycarpa many sporocarps are attached on one side of the petiole in a single vertical row; in M. quadrifolia 'basal parts of the stalks of sporocarps are fused in groups and in turn each of these groups is attached to the petiole by a

common stalk and in M. minuta the stalks of all sporocarps though free are attached to the petiole at a single point.

The sporocarp is differentiated into a stalk and a body. The stalk is fused laterally to the back of the body of the sporocarp, generally forming a distinct ridge, called raphe. In some species the distal end of the raphe is marked by one or two teeth-like projections, known as tubercles. The lower tubercle (present at the end of raphe) is usually stouter and prominent.

2. *Internal Structure of the Sporocarp :* The internal structure of the sporocarp can be studied under the following headings :

   (a) *Sporocarp Wall :* The wall of the sporocarp is very hard, thick and highly resistant to mechanical injury. It is differentiated into an outer epidermis, a middle hypodermis and an inner parenchymatous zone. The single layered epidermis is made up of cuboidal cells, covered with a thick layer of cuticle. Some epidermal cells elongate to form multicellular hairs. A large number of sunken stomata are present in the epidermis. The hypodermis consists of two layers of radially elongated palisade-like cells which are compactly arranged without any intercellular spaces between them. It is followed by a 2-3 layered parenchymatous zone. The inner cells of this zone form a gelatinous ring inside the sporocarp wall.

   (b) *Vascular System of the Sporocarp :* The sporocarp receives its vascular supply from the vascular cylinder of the petiole. A single strand enters into the stalk, extends through the raphe and then passes along the dorsal side of the sporocarp. This strand, called dorsal bundle, gives out a number of traces on both the faces of the sporocarp and these traces extend downward just like ribs. They are called lateral bundles or commissural bundles. The number and position of lateral bundles roughly correspond to the number and position of sori. Each lateral bundle is formed dichotomously. The adjacent dichotomies anastomose and form loops on the ventral side of the sporocarp. At the point of dichotomy a small branch arises that enters the receptacle bearing sporangia. This branch is called placental bundle.

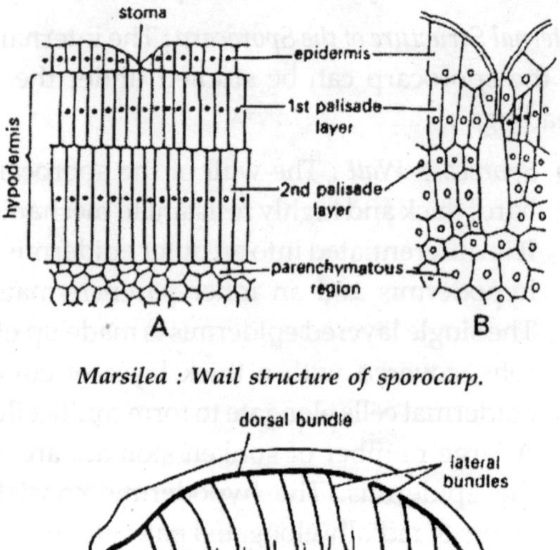

***Marsilea* : Wail structure of sporocarp.**

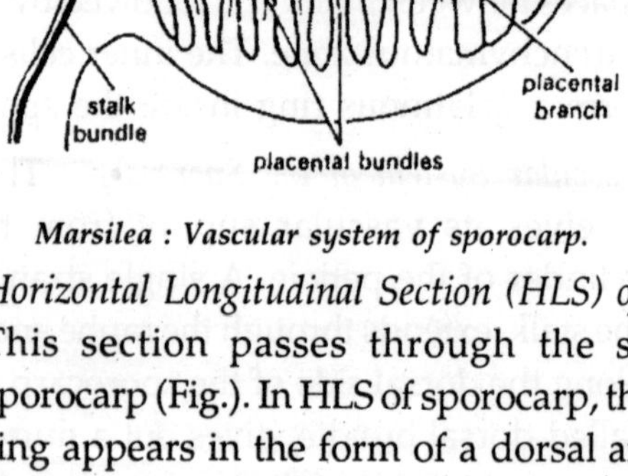

***Marsilea* : Vascular system of sporocarp.**

(c) *Horizontal Longitudinal Section (HLS) of Sporocarp :* This section passes through the stalk of the sporocarp (Fig.). In HLS of sporocarp, the gelatinous ring appears in the form of a dorsal and a ventral mass; the dorsal mass is more prominent than the ventral mass. In this plane both micro and megasporangia are visible in sori. The sori are seen in two alternate rows. Each sorus has a receptacle which contains one terminal megasporangium and two microsporangia on the lateral sides. It is surrounded by an indusium.

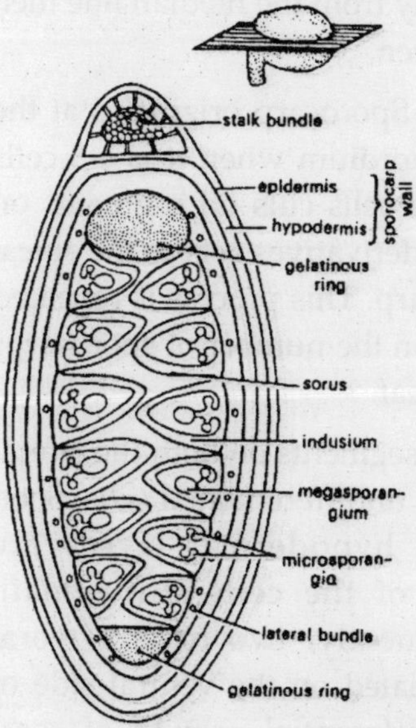

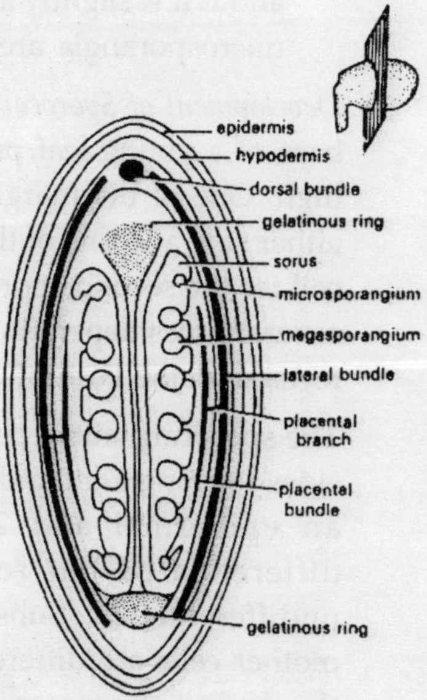

Marsilea* : *Horizontal longitudinal section (HLS) of sporocarp.

Marsilea* : *Vertical transverse section (VTS) of sporocarp.

(d) *Vertical Transverse Section (VTS) of the Sporocarp* : This plane of section does not pass through the stalk. The gelatinous ring appears in the form of a dorsal and a ventral mass as in HLS. The dorsal mass is more prominent than the ventral mass. In a particular plane in VTS the sorus can be seen either with only microsporangia or megasporangia. The lateral bundle extending from the dorsal to the ventral side of the sporocarp can be seen in this view. The placental bundle, which arises from the lateral bundle, is also clearly visible.

(e) *Vertical Longitudinal Section (VLS) of Sporocarp* : In VLS, the gelatinous ring appears as a complete ring around the sori which are arranged in many vertical rows. In this plane of section either microsporangia or megasporangia are visible. If the section is perfectly median then megasporangia are visible and if it is slightly away from the median line then microsporangia are seen.

3. *Development of Sporocarp* : Sporocarp originates at the base of a young leaf primordium when it is 6-8 cells high. One of the marginal cells cuts off segments on either side and one of the derivatives acts as the apical cell for the second sporocarp. This process is repeated several times depending on the number of sporocarps formed at the base of a leaf.

   The sporocarp initial cuts segments on both the lateral sides forming a mound of undifferentiated cells. Soon an epidermal and 2-3 hypodermal layers are differentiated but rest of the cells still remain undifferentiated. Subsequently, two rows of soral mother cells are differentiated on the ventral side of the young sporocarp. Differential growth of soral mother cells and their derivatives result in the formation of involutions on the upper surface of the sorus. These involutions grow into arc- shaped depression, called soral canals. Each soral canal is lined on its inner face by a tissue called indusium.

   The soral mother cells divide to form receptacular surfaces in two alternating rows. The sporangial initials develop on the receptacular surface in acropetal succession. The first sporangial initial is formed at the apex of the receptacle and it forms megasporangium. Subsequently, two initials are formed on the lateral sides of the receptacle. They function as microsporangium initials and each of these forms a microsporangium.

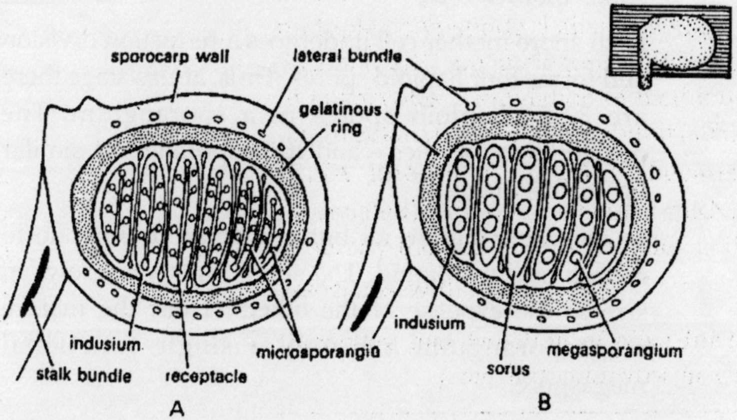

Marsilea : Vertical longitudinal sections (VLS) of sporocarp; A. Section passing through microsporangia, B. Section passing through megasporangia.

4. *Development of Sporangia* : The development of microsporangium as well as megasporangium is almost similar. The sporangial initial divides transversely and an outer and an inner cell is established. The entire sporangium develops from the outer cell and the inner cell does not take part in the development of sporangium. The outer cell divides by three successive diagonal walls and as a result a tetrahedral apical cell is formed. The latter cuts derivatives from the three faces. Initially, the apical cell cuts off two segments from each face and the six cells thus formed constitute the stalk of the sporangium. Thereafter, the apical cell divides periclinally and as such an outer jacket initial and an inner tetrahedral archesporial cell is formed. The jacket initial divides only anticlinally and forms a single layered sporangial wall. The tetrahedral archesporial cell divides by periclinal walls into an outer primary tapetal cell. The primary tapetal cell usually divides only anticlinally and forms one or two

layered tapetum. The centrally situated primary sporogenous cell divides in all planes and forms 8-16 spore mother cells.

Each spore mother cell undergoes a reduction division and forms four haploid spores. Thus, at this stage there are 32-64 haploid spores in a sporangium. The development of micro- and megasporangia is similar up to this stage.

In megasporangium all but one spore degenerate to form a nutritive liquid. The surviving spore grows in size at the expense of the others. Thus, the mature megasporangium has only a single functional megaspore.

Almost all spores in a microsporangium are functional. They remain much smaller in size than the megaspore due to nutritional factors. Thus, there are 32-64 microspores in a mature microsporangium.

5. *Dehiscence of Sporocarp* : There is no special mechanism for the liberation of sporangia from the sporocarp. They are set free only when the wall of the sporocarp splits open. The mature sporocarp is hard and nut-like, and may remain viable for 20-40 years. Spores from even 50 years old sporocarps could be made to germinate by soaking them in water. The internal tissues of the ripe sporocarp, including the gelatinous ring, swell up by absorbing water. The expanding gelatinous ring frees the sori from the wall of the sporocarp by abscission along the receptacle. Expansion of the internal gelatinous material results in the splitting of sporocarp wall along with its ventral margins. In the meantime, the gelatinous ring breaks at one point and comes out through the split wall of the sporocarp dragging sori alongwith it. The gelatinous ring bears two rows of sori and it appears as a worm-like structure.

One end of the gelatinous ring, however, remains attached to the wall of the empty sporocarp.

Following the release of sori from the sporocarp, the indusium (i.e., wall of the soral sac) disorganises liberating spores which function as mother cells of the gametophytic generation.

Morphological Nature of the Sporocarp : The morphological nature of the sporocarp of Marsilea is debated and several interpretations have been put forward from time to time by different workers. These views can be grouped into the following three categories :

(1) Leaf-segment or laminar hypothesis.

(2) Petiolar or whole-leaf hypothesis.

(3) A phylogenetically synthesized new organ.

Leaf-segment or Laminar Hypothesis : According to the laminar hypothesis, the sporocarp is the lateral fertile part of a leaf. This view was supported by Russow (1872), Campbell (1892), Bower (1926), Goebel (1930), Eames (1936), Smith (1938), Puri and Garg (1953) and Gupta (1957). Russow (1872) and Busgen (1890) consider that the sporocarp is made up of two leaflets with their ventral surfaces facing each other.

Campbell (1892) interpreted the sporocarp as a modification of a pinnately compound leaf, in which leaflets (pinnae) fuse to form the sporocarp. Bower (1926) considered the sporocarp as a modified unipinnately compound leaf, with rachis bearing two rows of pinnae. The lateral fusion of these pinnae have given rise to sporocarps of Marsilea. This interpretation receives support from the vascular supply of the sporocarp.

Eames (1936) states that the sporocarp is morphologically comparable to the tip of the leaf with its four leaflets. The two distal leaflets form the body of the sporocarp and the two proximal are represented by two tubercles present on the raphe on the dorsal surface of sporocarp. Evidences in support of this interpretation are derived from (i) the course of vascular bundles in the sporocarp and their similarity with the vasculature of the leaf, and (ii) the structure of abnormal leaves. In their colour and texture some abnormal leaves were found to be intermediate between normal leaves and sporocarps. Besides, in many cases, the proximal leaflets are very small, whereas the distal ones are large and valve-like.

Smith (1938, 55) interpreted the sporocarp in terms of a modified pinna which has a midrib with 15-20 lateral veins on either side in alternate positions. The lateral veins are forked, and at the point of forking, a linear receptacle develops on the abaxial side of the pinna. In the axil of each receptacle there is a flap-like outgrowth, known as indusium. The indusium projects forward to the pinna margin and thus covering the sorus. The infolding of the pinna, along the midvein, results in the formation of the sporocarp.

According to Puri and Garg (1953), the sporocarp is a modified leaflet in which the number of fertile pinnules corresponds to the number of commissural (lateral) bundles. The commissural bundles fuse to form the midrib of the pinnule. Each pinna splits into two up to half of its length and each lobe receives a branch of the commissural bundle. The sori are restricted to the inner fertile margins of the pinnules (Fig. A). This hypothesis is based on the anatomical and ontogenetic studies of the sporocarp. The vascular supply of the sporocarp is similar to that of the leaflets and this provides support to the laminar nature of the sporocarp.

Petiolar or Whole-leaf Hypothesis : This hypothesis, propounded by Johnson (1898), postulates that the sporocarp of Marsilea is homologous to a sterile leaf where the marginal cells of the leaf develop into sporangia instead of leaflets. Thus, according to this hypothesis, the sporocarp is the modification of the whole leaf and not just of a pinna.

A Phylogenetically Synthesized New Organ : This interpretation of the sporocarp was put forward by Bierhorst (1971). According to him, the early ontogeny of sporocarps of Marsilea polycarpa contradicts the leaf segment interpretation. Here, a number of sporocarps are borne on one side of the petiole but the ontogeny shows that the entire group of sporocarps originates after the vegetative pinnae are formed and they cannot be traced through the marginal meristems of the vegetative pinnae. Instead, each sporocarp initiates from a single cell with three cutting faces, a condition unknown for any fern pinna. The marginal series of sporocarp initials in Marsilea are comparable to those of filicalean sporangial stalk initials. Thus, Bierhorst (1971) states that "these are organs which truly changed their morphological nature in mid-ontogeny. That segment of the genome governing sporangial ontogeny is expressed in early stages and that concern with pinna ontogeny in later stages." Therefore, he concluded that the sporocarp of Marsilea is a phylogenetically synthesized new organ.

Gametophyte : Marsilea is a heterosporous fern, producing two types of spores - microspores and megaspores. The former gives rise to the male gametophyte and the latter to the female gametophyte.

Microspores and Development of Male Gametophyte : Microspores are minute spherical structures, ranging from 0.060 - 0.075 mm in diameter. The spore wall is differentiated into an outer exine and an inner intine. The exine is covered by a thin layer, called perispore. The microspore has a haploid nucleus and its cytoplasm is rich in starch grains.

The microspore germinates immediately on shedding. The first division produces a small biconvex prothallial cell and a large apical cell. In some species (e.g., M. diffusa, M. elata), two prothallial cells are formed instead of one. The apical cell divides transversely into two nearly equal antheridial initial cells. Each antheridial initial cell divides periclinally forming an outer first jacket cell and a larger wedge-shaped cell. The latter divides further by a periclinal wall and thus a small second jacket cell and an outer large cell are distinguished. The outer cell again divides periclinally forming an outer third jacket cell and an inner primary androgonial cell. Thus, at this stage, the male gametophyte consists of two primary androgonial cells, each surrounded by three jacket cells. Each primary androgonial cell divides repeatedly forming a group of 16 antherozoid mother cells or androcytes. During the development of androcytes the prothallial cell and the jacket cells of both the antheridia degenerate and as such the two groups of androcytes lie free within the spore wall (or male gametophyte;). Each androcyte metamorphoses into a cork screw-shaped multiflagellate antherozoid, characterized by the presence of a prominent terminal vesicle.

The wall of the male gametophyte breaks when antherozoids mature and thus setting them free. The development of male gametophyte is extremely rapid. For instance, the microspores of M. crenata form antherozoids within 5 hours and that of M. quadrifolia within 7-9 hours at room temperature.

Megaspore and Development of Female Gametophyte : The megaspore is an oval or elliptic structure with an apical papilla. It varies in size from 0.15-0. 40 mm and is large enough to be clearly seen with naked eyes. The mature megaspore has a single haploid nucleus and a large quantity of stored food material. It germinates immediately after shedding from the megasporangium.

Before germination, the nucleus of the megaspore along with the cytoplasm moves into the apical papilla. The first division is transverse, forming a small nipple-shaped apical cell and a large basal prothallial or nutritive cell. The apical cell forms the female gametophyte and the prothallial cell is sterile and provides nutrition to the developing female gametophyte.

The apical cell divides by three intersecting vertical divisions, establishing an axial cell surrounded by three lateral cells. The axial cell functions as archegonial initial.

The archegonial initial divides by a periclinal wall forming an outer primary cover cell and an inner central cell. The primary cover cell divides by two successive anticlinal walls and gives rise to four cells which function as neck initials. Each neck initial divides by an oblique wall and the eight cells thus resulted form a small neck of the archegonium. The neck cells are arranged in two tiers of four each. The central cell divides transversely thus forming an upper primary canal cell and a lower primary venter cell. The former does not divide further and functions as neck canal cell, whereas the latter divides again by a transverse wall forming a venter canal cell and an egg. Thus, the mature archegonium has an 8-celled neck with a single neck canal cell, a small venter canal cell and a large egg.

The female gametophyte is surrounded by a gelatinous sheath. The sheath extends to the apex of the spore as a thin, cone-shaped mass, leaving a small opening.

Fertilisation

Just before fertilisation, the neck canal cell and venter canal cell degenerate to form a mucilagenous mass, which attracts sperms chemotactically. At this stage, the apical mucilagenous cone may be seen with a number of sperms which are directed toward the archegonium. One of the sperms fuses with the egg and forms diploid zygote.

Embryo Development

The diploid zygote is the mother cell of the next sporophyte generation. The development of embryo is extremely rapid and a young sporophyte is established within 2-4 days after fertilisation.

The first division of the zygote is longitudinal, followed by a transverse division. Thus, a four-celled embryo is established. The upper two cells of the embryo eventually give rise to the leaf and root, whereas the lower two cells form the stem and foot. The embryo of Marsilea is composed of relatively few cells and grows entirely within the apical papilla of the femalc gametophyte. As the embryo develops, the cells of the venter wall are stimulated and divide periclinally, forming a calyptra all around the young sporophyte. The sporophyte remains completely covered by the calyptra till the differentiation of the first leaf. Afterwards, the calyptra splits at the tip and the young sporophyte comes out of it.

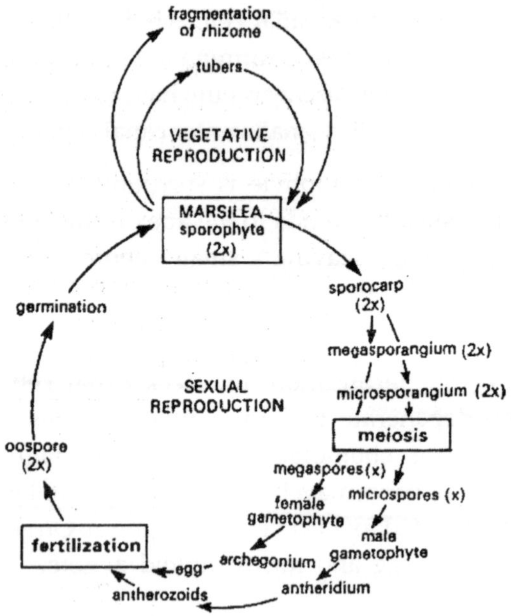

Marsilea : Life-cycle.

5

Leptosporangiatae Filicales

The Leptosporangiatae are a large assemblage of mostly living ferns, particularly abundant in tropical rain forests. Most of these ferns are perennial mesophytes or epiphytes but some also grow on exposed rocky slopes in mountains. They are characterised by the following features.

(1) Sporangium develops from a single superficial cell.

(2) Sporangial jacket is always single layered.

(3) Position and structure of annulus and stomium is definite. The latter is usually composed of four sister cells.

(4) The number of spores in a sporangium is definite (spore output of leptosporangiate ferns is less than that of eusporangiate ferns).

(5) They are either homosporous (e.g., Pteridium, Adiantum, Pteris) or heterosporous (e.g., Marsilea, Azolla, Salvinia).

Order Filicales

The order Filicales with about 300 genera and 9000 species is the largest group of Pteridophyta. They include most of the common ferns, used extensively by florists in bouquets and in green houses. Most of these ferns are terrestrial, making up a substantial part of the ground flora in moist forests. Some of these ferns are epiphytes on tree trunks and still others grow on exposed rocks and rock crevices. The leaves are circinately coiled when young. The sporangia are marginal or superficial and are grouped together in sori. The sporangium develops from a single initial (i.e., leptosporangiate). They are homosporous and the spore gives rise to a monoecious gametophyte on germination. The antheridia usually project from the surface of the gametophyte.

The antheridial jacket consists of only few cells (thus there is reduction in the number of sterile cells). The archegonium has a short neck which is curved backward because of the unequal growth of neck cells. Each cell of the quadrant embryo is related to a definite organ of the young sporophyte.

Pteridium, Adiantum, Nephrolepis and Polypodium are described here as representatives of the order.

Pteridium

Systematic Position

Sub-division	—	Pleropsida
Class	—	Leptosporangiatae
Order	—	Filicales
Family	—	Polypodiaceae

Pteridium is a monotypic genus represented by P. aquilinum. This species is widely distributed throughout the world. It grows in a variety of habitats, ranging from thick humid forests to dry exposed areas. In India, it is abundant along the entire Himalayan tract at altitudes between 1000 and 3000 meters. The subterranean rhizome has rapid and indefinite growth and as such it can spread in a very wide area in a short time.

Sporophyte

External Morphology : The main plant body, which is sporophytic, is differentiated into root, rhizome and leaves (Fig.).

1. *Root :* The primary root is ephemeral and is soon replaced by a large number of adventitious roots which develop throughout the length of the rhizome in acropetal succession. These roots are usually small, branched and endogenous.

2. *Rhizome :* The rhizome or stem is long, slender,' dichotomously branched and subterranean, growing parallel to the surface of the soil. It is differentiated into nodes and internodes. It is capable of indefinite growth and accounts for the rapid vegetative growth of the fern. The entire surface of the rhizome is covered with brown multi-cellular hairs.

   According to Watt (1940), the rhizome of Pteridium aquilinum forrns a sympodial system which gives rise to the following three types of branches in alternate sequence.

   (a) *Long Shoots :* They have an average internode length of 30-40 cm and do not bear fronds (leaves).

   (b) *Short Shoots :* These are with very short internodes (0.5-2 cm) and they bear fronds.

   (c) *Intermediate Shoots :* These are intermediate between the long and short shoots. They vary in length from 2 - 30 cm. Initially, these branches are erect and grow upward but soon they become horizontal. The intermediate shoots do not bear fronds.

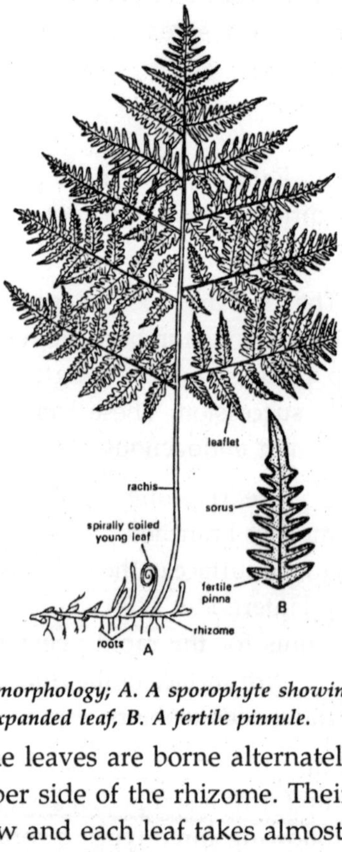

***Pteridium* : External morphology; A. A sporophyte showing a young and a expanded leaf, B. A fertile pinnule.**

3. *Leaves :* The leaves are borne alternately at the nodes on the upper side of the rhizome. Their development is very slow and each leaf takes almost three years to expand fully. When young, they are spirally coiled and show typical circinate vernation. They are exstipulate and tripinnately compound, varying from 0.5 - 3 meters in length. The upper part of the petiole that extends into the laminar region is known as rachis. A young rachis is usually covered with multicellular brown hairs called ramenta.

A fully expanded leaf assumes a roughly conical shape. Each pinna has a distinct vein which gives out lateral branches

(Fig.); a branch enters into each pinnule. A nectary is present at the base of each pinna (Fig.).

Internal Structure

1. *Root :* A vertical section of the leaf shows three distinct regions- epiblema, cortex and vascular cylinder (Fig.).

   The epiblema is the outermost layer of thin walled cells. Some of the cells of epiblema elongate to form long unicellular root hairs.

   The cortex is differentiated into an outer parenchymatous and an inner sclerenchymatous region. The sclerenchymatous cortex, which completely surrounds the stele, provides mechanical support to the root. The cortex is followed by an endodermis; the radial walls of endodermal cells possess casparian thickenings. Next to endodermis, there are one or two layers of parenchymatous pericycle.

   The vascular cylinder is diarch and exarch. The xylem plate consists of two central metaxylem tracheids with small groups of protoxylem elements on either side. Phloem is present on both sides of the xylem plate.

2. *Rhizome :* Internally, the rhizome is differentiated into epidermis, hypodermis, ground tissue and stele. The latter consists of many meristeles (Fig.).

   The epidermis is the outermost protective layer of compactly arranged cells. It is covered with a thick layer of cuticle.

   The hypodermis is continuous or interrupted on the lateral sides by parenchymatous cells. The hypodermal cells are heavily sclerotic with pitted walls. The hypodermis is mechanically effective and probably helps in meeting soil pressures. It is followed by parenchymatous ground tissue. The cells of the ground tissue are rich in starch grains and thus function as storage parenchyma.

The structure of stele varies considerably within the same rhizome. In a week-old sporeling, which bears one or two leaves, the rhizome is protostelic. But as the rhizome matures, there is progressive elaboration of the stele. For example, in a 5-6 week old rhizome, bearing 2-5 leaves, the vascular cylinder is ectophloic siphonostelic. But in a mature rhizome with many leaves, the vascular cylinder becomes dictyostelic as it gets dissected into a number of strands. Each of these strands is called a meristele. The meristeles are arranged in two rings, separated by two sclerenchymatous bands.

Each meristele is surrounded by its own endodermis. Next to endodermis, there are one or two layers of parenchymatous pericycle. It is followed by phloem; sieve tube elements close to the pericycle form protophloem and the inner elements form metaphloem. The central part of the stele is occupied by xylem. The protoxylem tracheids have reticulate thickenings, whereas metaxylem tracheids show predominantly scalariform pittings. True vessels with scalariform perforation plates have also been reported from vegetative organs of Pteridium.

3. *Rachis :* The rachis or petiole is somewhat hemispherical in a transverse section. The epidermis consists of a single layer of compactly arranged cells. The outer walls of these cells have a thick deposition of cuticle.

   The epidermis is followed by 2-3 layered sclerenchymatous hypodermis, which provides mechanical strength to the frond.

   Next to hypodermis is a broad zone of parenchymatous ground tissue. The cells of the ground tissue are compactly arranged and do not have intercellular spaces. The ground tissue of a young rachis consists of a U-shaped stele but as the rachis matures the stele gets dissected into two or more meristeles. The structure of meristele is like that of the rhizome.

4. *Pinnule* : In internal structure the pinnule resembles a dicotyledonous leaf (Fig.). The upper and lower epidermis are single layered.

The mesophyll is differentiated into palisade and spongy parenchyma. The palisade consists of compactly arranged columnar cells with many discoid chloroplasts. The spongy tissue is composed of thin-walled parenchymatous cells with numerous chloroplasts. Intercellular spaces in between these cells communicate with the outer environment through stomata present in the lower epidermis.

The mid-vascular region of the pinnule is composed of compactly-arranged parenchymatous cells. The central part of this parenchymatous zone is occupied by a concentric vascular bundle. The vascular bundle is surrounded by a distinct bundle sheath.

Reproduction : Pteridium reproduces vegetatively and by spores.

Vegetative Propagation : Death and decay of the underground rhizome is the common method of vegetative propagation. The rhizome is dichotomously branched and grows indefinitely. When the death and decay of rhizome reaches up to the point of dichotomy, both the branches separate and each grows into a new plant.

Reproduction by Spores : Pteridium is a homosporous fern. The sporangia develop in large numbers on the abaxial side of pinnules. Thus the fronds also function as sporophylls. The sporangia 'form a continuous linear sorus along the margins. Such a sorus is known as coenosorus. Thus in Pteridium individuality of sori is lost.

1. *Development of Sorus :* The sorus originates on the margins of the frond. The marginal initials of the frond form the receptacle (or placenta) on which sporangia are borne. The abaxial and adaxial flaps of the indusium appear as superficial growths lateral to a very young receptacle.
2. *Development of Sporangium :* Development of sporangium is strictly leptosporangiate, i.e., it develops from a single initial cell. A marginal cell, which functions as sporangial initial, slightly projects above the receptacular surface in the form of a papilla. It divides periclinally into an outer and an inner cell. The inner cell may or may not divide and does not participate in further development of the sporangium; its derivatives become indistinguishable from the other cells of the receptacle. The outer cell divides by three intersecting diagonal divisions as a result of which a pyramidal apical cell is distinguished. The apical cell cuts off four or five segments from each of its three lateral cutting faces. The upper three segments contribute to the proximal part of the sporangial jacket, whereas the lower segments form the stalk of the sporangium (Fig.).

   The pyramidal apical cell divides periclinally into a flat outer cell, the jacket initial, and an inner tetrahedral archesporial cell (Fig.). The latter functions as internal apical cell and cuts off tubular tapetal initials from each of its four cutting faces, leaving a central tetrahedral primary sporogenous cell (Fig.). The tapetal initials divide anticlinally as well as periclinally and eventually form a double-layered tapetum. The primary sporogenous cell undergoes four successive divisions forming 16 spore mother cells (Fig.). Meanwhile, the tapetal cells break down and form a nutritive fluid in which the spore mother cells float. Each spore mother cell undergoes meiosis and forms four haploid spores, but sometimes one or more spore mother cell may abort or fail to divide further. As such a sporangium does not always contain 64 spores.

The distal part of the sporangial jacket is formed by repeated anticlinal division of the jacket initial. It is always single-layered. As the jacket develops, a vertical row of 13 strongly thickened cells, the annulus, becomes differentiated. It encircles about 3/4th of the sporangium from the base. A thin area of tubular cells present between the annulus and stalk is known as stomium. Both annulus and stomium help in the dehiscence of the sporangium.

3. *Mature Sporangium :* The coenosorus is covered by two indusial flaps (Fig.). Both the indusial flaps have fringed margins, but the outer flap is more developed than the inner flap. Mature sporangium is differentiated into a stalk and a capsule. The stalk is long, slender and composed of three vertical rows of cells. The capsule is oval or biconvex. The sporangial jacket is single-layered. An annulus consisting of a vertical strip of 13 thick-walled cells extends from the base to almost three-fourth of the capsule. At its distal end the annulus joins a strip of thin walled cells, the stomium. There are 32-64 spores in a capsule (Fig.).

4. *Dehiscence of Sporangium :* Dehiscence of sporangium takes place due to changes in the cells of the annulus. The radial and inner tangential walls of these cells are thickened, whereas the outer tangential walls are thin. In dry weather water evaporates from the thin outer walls and thus there is loss of turgidity. Consequently, the outer wall being thin, bends inwards and due to this structural change the annulus tears the sporangial wall from the stomium and turns backward. Some of

the spores are blown away by wind through the ruptured capsule but most part of the capsule in which spores are contained is attached to the distal end of the annulus (Fig.). Eventually, the cells of the annulus become still drier due to further loss of water; there is sudden release of tension and the annulus snaps back to its original position ejecting the spores forcibly (Fig.).

Gametophyte

Spore : In Pteridium all spores are alike; they are nearly spherical (0.03 mm in diameter), haploid and uninucleate. The spore wall is differentiated into an outer dark and relatively thick exine and an inner thin intine.

Development of Prothallus : Haploid spore is the mother cell of the gametophytic generation. It germinates immediately after dispersal if moisture and suitable temperature is available. Viability of spores, however, gradually decreases if germination is delayed. The spore absorbs water and swells. Consequently, the exine ruptures at the triradiate mark and the intine protrudes out in the form of a germ tube. The germ tube usually grows into a short multi-cellular chlorophyllous filament. The basal cell of the filament gives rise to colourless rhizoids which attach the developing prothallus to the soil (Fig.). The cells of the filamentous prothallus divide repeatedly and eventually form a flat green cordate (heart-shaped) prothallus (Fig.). Young prothallus is essentially a single-layered sheet. The peripheral part remains single-layered even at maturity but the central region, below the emarginate apex, becomes massive cushion-like due to cell divisions. The mature prothallus is 3-8 mm in diameter. It is capable of independent physiological existence, nourishing itself by absorption from the soil and by photosynthesis.

Sex Organs : Both, male (antheridia) and female (archegonia) sex organs are borne on the same prothallus. Antlteridia which usually appear first are confined to the basal central or lateral regions and archegonia develop near the apical notch.

1. *Antheridium* : An antheridium develops from a superficial cell on the ventral surface of the prothallus. The antheridial initial grows as a papillate outgrowth which is cut off from the vegetative cell by a transverse wall (Fig.). The initial cell now divides transversely into a lower first ring cell and an upper daughter cell. Due to higher turgor pressure in the upper daughter cell, the wall between these two cells is pressed downward and as a result the upper cell becomes dome-shaped (Fig.). The upper cell divides by an arched periclinal wall into an outer dome cell and an inner primary androgonial cell.

   (Fig.) Dome cell now divides transversely into a cover cell and a second ring cell (Fig.). The cover cell and the first and second ring cells eventually form a single-layered jacket of the antheridium by anticlinal divisions (Fig.). The primary androgonial cell divides repeatedly and gives rise to 30-40 androcytes. Each androcyte metamorphoses into a multi-flagellate antherozoid (Fig.).

2. *Archegonium* : Archegonia develop on the ventral surface of the prothallus near the apical notch. The archegonial initial divides transversely into an upper primary cover cell and a lower central cell (Fig.). The primary cover cell divides by two vertical walls at right angles to each other, as a result four neck initials are formed. The neck initials, by repeated transverse divisions, give rise to 5-7 celled high neck of the archegonium. The central cell divides transversely and eventually forms three cells— a neck canal cell, a venter canal cell and an egg (Fig.). The nucleus of the neck canal cell may divide further but it is not accompanied by wall formation.

Thus a mature archegonium is differentiated into a 5-7 cells high neck composed of four vertical rows of cells and a venter. The neck canal is occupied by a neck canal cell and a venter canal cell. The venter contains a single large egg. The neck of the archegonium is curved towards the posterior side of the prothallus as the two anterior neck canal cells grow faster than the posterior cells. The neck projects above the surface of the prothallus, whereas the venter remains embedded in it.

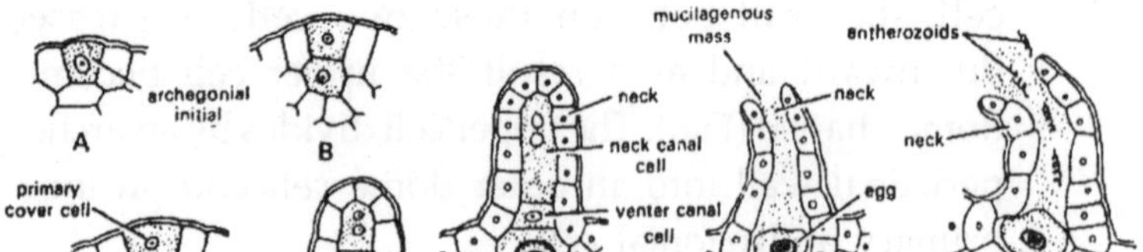

Pteridium : A-E. Stages in the development of archegonium, F-G. Archegonium just before fertilisation.

Fertilisation

Water is essential for fertilisation. At maturity the antheridium absorbs water and as such there is an increase in the pressure within the antheridium. Consequently, the cover cell of the antheridium splits apart and antherozoids swim freely in a thin film of water present on the surface of the prothallus. Almost at the same time the neck canal cell and venter canal cell of the archegonium disintegrate and form a viscous mucilagenous mass (Fig.). The mucilage swells due to absorption of water and as a result neck opens. The mucilagenous mass probably contains chemical substances like malic acid which attract antherozoids to the neck of the archegonium chemotactically (Fig.). Although many antherozoids may enter the neck, only one of them fuses with the egg and forms zygote.

Embryo : The zygote, formed by the fusion of antherozoid

and egg, is the starting point for the next diploid (sporophytic) generation (Fig.). It increases in size considerably and almost completely occupies the cavity of the venter. The mucilagenous mass, present in the archegonium, dries up immediately after fertilisation, thus the opening of the archegonium is completely sealed. The first division of the zygote is vertical (i.e., parallel to the axis of the archegonium), followed by a second vertical division, at right angles to the first one.

The four cells thus formed divide by a transverse wall, giving rise to an octant (Fig.). The octant is composed of an epibasal and a hypobasal tier, each consisting of four cells. The four epibasal cells form shoot apex and cotyledon initial, whereas the hypobasal cells give rise to root and foot (Fig.). The foot absorbs nutrition for the developing embryo from the prothallus. The venter of the archegonium forms a protective covering called calyptra around the developing embryo. The root and cotyledon grow more rapidly than the shoot and rupture the surrounding calyptra (Fig.). The shoot grows slowly and a young sporophyte with 4-5 leaves is formed in about a year. The foot and prothallus gradually decay once the young sporophyte is established.

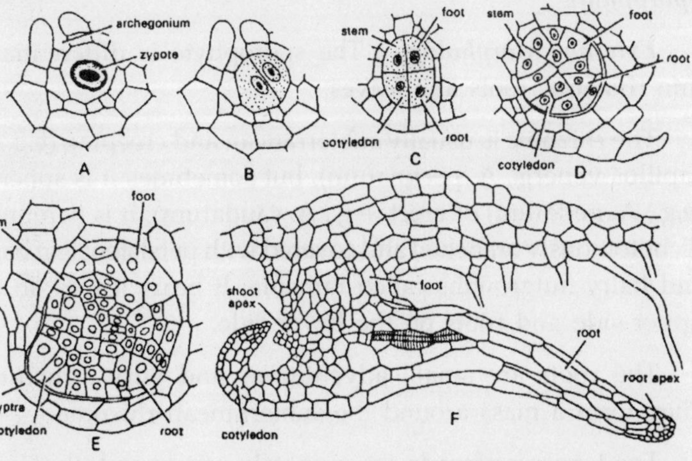

A-F. Pteridium : A. Zygote, B-F. Stages in the development of embryo.

Adiantum

Systematic Position

Sub-division	—	Pteropsida
Class	—	Leptosporangiatae
Order	—	Filicales
Family	—	Adiantaceae

It is a large genus comprising of some 200 species, widely distributed in both tropical and temperate regions of the world. Species of Adiantum occur in a wide variety of habitats, but they are most abundant in moist and shady places. They occur on wet walls and in shady rock crevices. The common name 'maiden hair fern' is given to the species of Adiantum because of their bright black petioles which resemble with black hair of ladies. Several species of Adiantum are commonly grown in gardens as ornamentals.

The genus is represented in India by 10 species; of these, Adiantum caudatum (Fig.), A. capillus-veneris (Fig.), A. venustum, A. lunulatum and A. athiopicum are fairly common.

Sporophyte

External Morphology : The sporophyte is differentiated into rhizome, roots and leaves.

The rhizome is usually subterranean and creeping (e.g., A. capillus-veneris, A. pectinatum), but sometimes it is suberect (e.g., A. pedatum) or erect (e.g., A. caudatum). It is perennial, dichotomously branched and covered with persistent leaf bases and hairy outgrowths called ramenta. It bears leaves on the upper side and roots on the lower side.

The roots are small, adventitious and wiry structures. They form a mass around the subterranean rhizome.

The leaves or fronds are pinnately compound; they may

be simply unipinnate (e.g., A caudatum, Fig.), bipinnate (e.g., A. capillus-veneris; Fig.) or 3-4 pinnate (e.g., Avenustum). The pinnae are alternate or opposite and nearly sessile (e.g., A. caudatum), shortly (e.g., A. venustum) or long petiolulate (e.g., A. lunulatum). The rachis is smooth and naked or villose and its length varies from 3 to 40 cm in different species. In A. caudatum, the terminal part of the rachis has only sparsely placed pinnae and produces roots at the apex. The veins are repeatedly forked and the venules are free, parallel and continue into the indusium.

The leaves bear marginal sori which are covered by reflexed marginal flaps of the leaf (Fig.).

Internal Structure

1. *Root :* Internally, the root consists of a single layer of epidermis, a wide cortex, and a central cylinder (Fig.). Unicellular root hairs are present in the epidermis. The cortex is differentiated into an outer parenchymatous region consisting of 4-6 layers of cells and an inner relatively narrow sclerenchymatous region. The central cylinder is surrounded by endodermis and pericycle. The stele consists of central metaxylem tracheids with two radiating arms of protoxylem. A group of phloem cells is present on either side of the xylem plate.

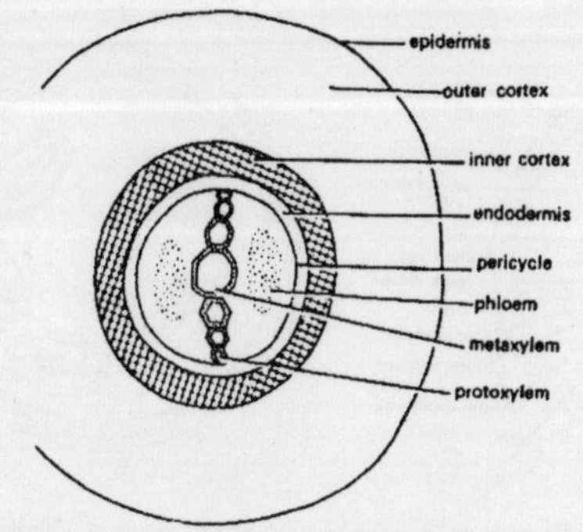

Adiantum : Transverse section of root.

2. *Rhizome* : The rhizome shows differentiation into epidermis, hypodermis and ground tissue with many meristeles (Fig.).

   The single-layered epidermis is covered by a thick layer of cuticle. Some epidermal cells give rise to long multicellular hairs.

   The epidermis is followed by 4-6 layers of sclerenchymatous hypodermis, which provides mechanical strength to the rhizome. The hypodermal cells are, however, not thickened in the young rhizome.

   Next to the hypodermis is parenchymatous ground tissue. The structure of the stele not only varies in different species but also within the same species at different stages of development. There is an amphiphloic siphonostele with a central parenchymatous pith in a young rhizome. As the rhizome becomes older, this stele transforms into a solenostele or dictyostele. Species with long rhizome (e.g., A. pedatum) characteristically have a solenostele (as leaf gaps do not overlap), but in rhizomes with short internodes the stele is essentially dictyostelic as leaf gaps overlap each other. In such cases several meristeles are found embedded in the ground tissue.

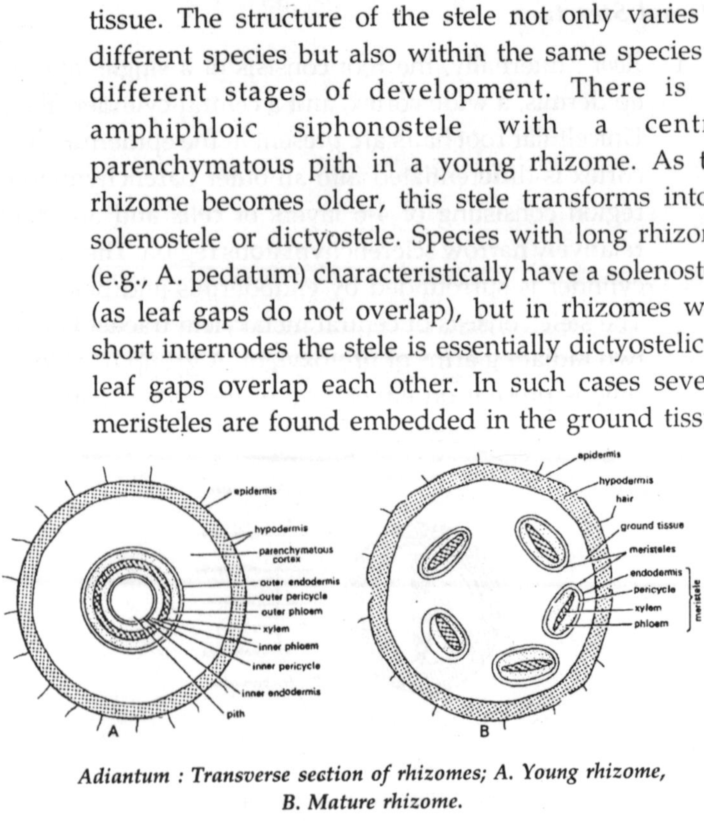

Adiantum : Transverse section of rhizomes; A. Young rhizome, B. Mature rhizome.

3. *Petiole* : A transverse section of the petiole shows the following structures (Fig.). The epidermis is single layered and is covered by a thick cuticle. In some species (e.g., A. caudatum) epidermal cells form multi-cellular hairs.

   The epidermis is followed by a few layers of sclerenchymatous hypodermis, which provides mechanical support to the elongated rachis. Inner to the hypodermic is parenchymatous ground tissue. Some cells of the ground tissue contain dark tannin.

   The central part of the ground tissue is occupied by either a single large horse-shoe shaped (e.g., A. phillipense, A. caudatum) or two arc-shaped vascular strands facing each other. In the latter, each strand is surrounded by a layer of endodermis and pericycle. The central part of the strand Js composed of xylem surrounded by phloem.

4. *Pinna* : The epidermal cells of pinna are irregular in outline and contain chloroplasts. Stomata are mostly confined to the lower epidermis (Fig.).

   The mesophyll is not differentiated into palisade and spongy parenchyma. But the cells adjacent to the upper epidermis are somewhat columnar and those away from it are irregularly lobed.

   The vascular bundles are embedded in the mesophyll tissue. Each bundle is surrounded by a sclerenchymatous bundle sheath. The larger veins are concentric, whereas the smaller veins are collateral.

Reproduction

Spore-bearing Organs : In Adiantum, reproduction takes place mainly by spores which are produced in large numbers within sporangia. The sporangia form sori close to the margins on the dorsal side of the mature pinnae (the leaves bearing sporangia/sori are known as sporophylls). The sori though appear to be marginal are in fact superficial and they are covered by reflex margins of the pinna. These protective margins are membranous and form indusium-like structure, called false indusium (Fig.).

The development of sporangium is leptosporangiate as in Pteridium. Sporangia in different stages of development can be seen in a sorus without any definite sequence. The sorus is thus of mixed type.

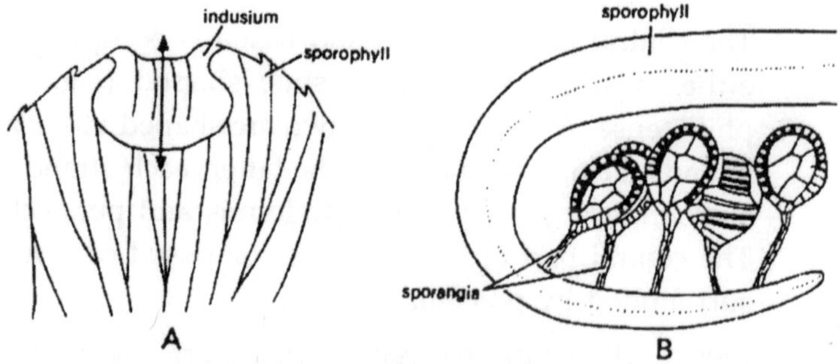

Adiantum : A. Leaflet showing reflexed indusium, B. Longitudinal section of leaflet through indusium.

Mature Sporangium : The mature sporangium has a 2-3 cells thick stalk and a spherical or elliptic head or capsule. The capsule wall is only one cell thick. A vertical row of 10-20 cells in the capsule wall develops thick walls and they form annulus. The stomium is represented by 2-6 thin-walled cells present at the distal end of the annulus. This arrangement of cells permits an easy cleavage when sporangium dehisces.

There are many spore tetrads within the capsule. All spores are alike, i.e., Adiantum is homosporous. The spores are haploid, uninucleate and bilateral. The spore wall is differentiated into an outer thick and ornamented exine and an inner thin intine.

Gametophyte : On germination the haploid spore gives rise to a heart-shaped prothallus with a distinct apical notch.

The prothalli of Adiantum are monoecious and bear both antheridia and archegonia. The development of sex organs is similar to that of Pteridium.

Embryo : The development of zygote is uniform in all leptosporangiate ferns. The first division of zygote is transverse and this results in the formation of an epibasal and a hypobasal cell. It is followed by two successive divisions at right angles to each other and as such an 8-celled embryo is established. The four cells of this embryo, which form epibasal tier, give rise to the axis and leaf of the sporeling, and the remaining four cells, forming the hypobasal tier, give rise to the first root and foot.

Polypodium

Systematic Position

Sub-division	—	Pteropsida
Class	—	Leptosporangintae
Order	—	Filicales
Family	—	Polypodiaceae

Polypodium has about 50 species, cosmopolitan in distribution. Most of the species are perennial and epiphytic, but some species also grow on exposed and shady places as well (e.g., P. parvulum). There are 25 species of Polypodium in India; the common ones are P. parasiticum (Fig.), P. virginianum, P. parvulum, P. obliquatum (Fig.), P. wallii, P. khasyanum, P. decorum and P. nigro-carpum (Fig.).

Sporophyte

External Morphology : The sporophytic plant body is differentiated into rhizome, roots and leaves.

The rhizome is branched and creeping or erect. The creeping rhizome in different species shows varying degrees

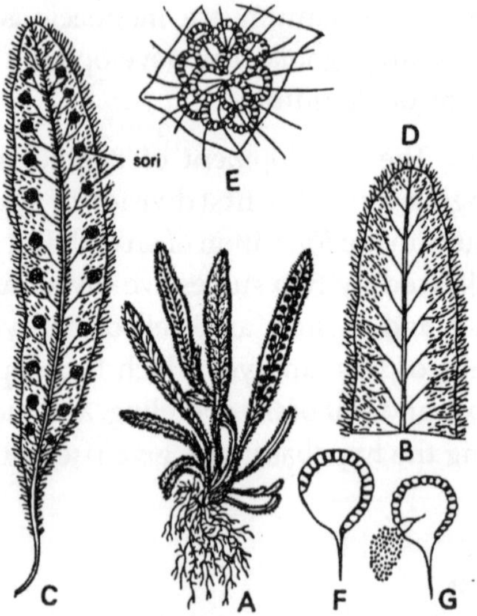

A-G. Polypodium parasiticum : A. Habit, B. A part of vegetative leaf, C. A sporophyll, D. Scale, E. A sorus, F-G. Sporangium.

of dorsiventrality, determined by the gravity. The dorsiventral-creeping rhizome bears leaves on its upper surface, usually inserted alternately to right and left. Species with erect rhizomes, however, show spiral phyllotaxy. The entire surface of the thizome is covered with multi-cellular hairs- called ramenta. In P. lingua and some other species, the ramenta are stellate (Fig.), but in P. incanum they form specialized flattened protective scales, known as ramentum (Fig.). Such scales provide permanent protective covering against drought. The rhizome is also covered by persistent leaf bases.

The roots are adventitious and they arise throughout the undersurface of the rhizome in acropetal succession. They are branched and often form a closely-packed mass around the rhizome.

The leaves vary considerably in their structure. They are simple and linear-lanceolate (e.g., P. parasiticum, Fig. wallii) or pinnatifid (e.g., P. obliquatum, Fig. P. paludosum; P.

cucullatum). The length of frond ranges from 2 (P. hirtellum) to 280 cm (P. ornatum, P. rugulosum). The petiole as well as lamina are clothed with delicate black hairs. The vein and veinlets are thick and simple or forked. The mature leaves (or pinnae) bear sori in rows one on either side of the vein. Each sorus is present at the tip of the upper veinlet (or vein in simple leaves).

The leaves of Polypodium have characteristic water glands (Fig.). These glands are composed of a group of thin-walled epidermal cells, sunk in a shallow depression immediately above a vein ending. Thus they are in direct communication of the vascular system. The water glands are uniformly scattered over the leaf surface or are placed in a series along the leaf margins.

Internal Structure : The internal structure of rhizome, root and leaf is essentially similar to that of Adiantum. The rhizome is differentiated into a single-layered epidermis, a 4-6 layered sclerenchymatous hypodermis and a parenchymatous ground tissue. The stele is dictyostelic, consisting of many small meristeles embedded in the ground tissue. Each meristele is surrounded by its own endodermis and pericycle.

The root is differentiated into epidermis, cortex and central cylinder (Fig.). Unicellular root hairs are present in the epidermis. The cortex is differentiated into an outer wide parenchymatous and an inner relatively narrow sclerenchymatous region. The vascular cylinder is surrounded by a layer each of endodermis and pericycle. It is protostelic with exarch xylem.

The leaves are isobilateral. The upper and lower epidermal cells often possess soft multicellular hairs and stomata. The mesophyll is undifferentiated and consists of spongy parenchyma only. The vascular bundles are concentric or collateral. The petiole has a large horseshoe-shaped vascular strand.

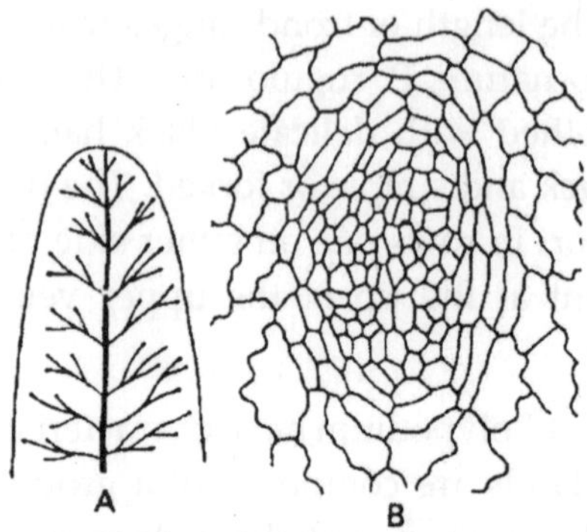

Polypodium : Water glands; A. As seen on dorsal surface, B. A water gland enlarged.

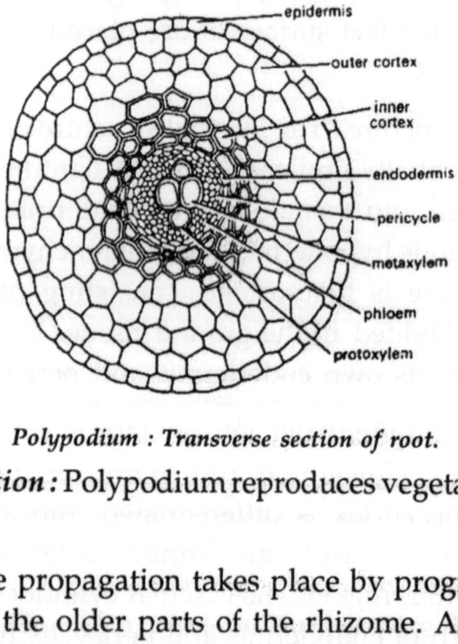

Polypodium : Transverse section of root.

Reproduction : Polypodium reproduces vegetatively as well as by spores.

Vegetative propagation takes place by progressive death and decay of the older parts of the rhizome. As the process of decay reaches up to the point of dichotomy, the two branches separate and each of these grows into an independent plant.

Spores are, however, the main reproducing bodies and they occur within the sporangium. The sporangia form definite sori on the under surface of the mature leaves (called sporophylls). Thus in a mature plant the foliage leaves perform the function of photosynthesis as well as spore production.

The sori are superficial; they are terminal (e.g., P. paludosum) or median (e.g., P. nigro-carpum) on the veinlet. The sporangia develop on a flat receptacle and form a mixed sorus, i.e., sporangia of different ages are aggregated in a sorus without any definite sequence. The sori of Polypodium are characterised by the absence of indusium, i.e., they are naked (Fig.).

Dehiscence of sporangium, structure and development of gametophyte and embryo is essentially similar to that of Pteridium.

Nephrolepis

Systematic Position

Sub-division	—	Pteropsida
Class	—	Leptosporangiatne
Order	—	Filicales
Family	—	Dennstaedtiaceae

Nephrolepis is also a tropical fern with about 30 species. Most of the species are terrestrial but N. volubilis and N. ramosa are epiphytic, climbing up to 15 meters over trees. Several species of Nephrolepis are grown as ornamentals. The genus is represented in India by 5 species, of these N. acuta and N. ruberosa are common.

Sporophyte

External Morphology : The sporophytic plant body is differentiated into rhizome, roots and leaves (Fig.).

1. *Rhizome :* The rhizome is short and erect (e.g., N. acuta) or sub-erect (e.g., N. tuberosa, N. exaltata), producing elongated slender stolons. N. ramosa and N. volubilis have creeping rhizome with adpressed scales. The rhizome of N. tuberosa bears tubers which act as

reservoirs for carbohydrates and water. The rhizome is covered with peltate scales.

2. *Roots* : The roots arise from the rhizome and stolons in acropetal succession. They are adventitious and branched.

3. *Leaves* : The leaves are long, narrow and herbaceous or sub-coriaceous. They are unipinnately compound and their length varies from 40 (N. tuberosa, N. exaltata) to 70 cm or more (N. acuta). The pinnae are sessile (N. tuberosa), sub-sessile (N. exaltata) or shortly petioled (N. acuta). They have usually rounded or cordate base. The veins are prominent and the veinlets are branched with open ends. The tips of veinlets are gland dotted and they extend up to the margins (Fig.).

The petiole, rachis and pinnae are covered with multicellular hairs or scales.

Internal Structure : Internal structure of the rhizome is similar to that of Pteridium. The vascular cylinder is dictyostelic. The stolon, however, has a typical protostele. In N. tuberosa, where stolons possess tubers, the protostele expands at the base of the tuber and splits into cylindrical network, then again contracts at the apex of the tuber. A transverse section passing through the middle of the tuber would thus show a ring of meristeles, each surrounded by its own endodermis (Fig.).

The structure of the root, petiole and pinna is essentially similar to that of Pteridium.

Reproduction : Sori, formed on the underside of the mature pinnae, are arranged in two rows, one on either side of the vein. They are superficial, arising at the tip of the veinlet (Fig. A-B). They are distinct and maintain their individuality. Some species of Nephrolepis, however, show a tendency towards the loss of individuality of sorus. For example, in N. dicksonioides the adjacent sori tend to fuse and a near complete fusion occurs in N. acutifolia where coenosori or fusion sori may extend along the whole margin. The sori are indusiate; the indusium is reniform (kidney-shaped), roundish or sub-orbicular (Fig.).

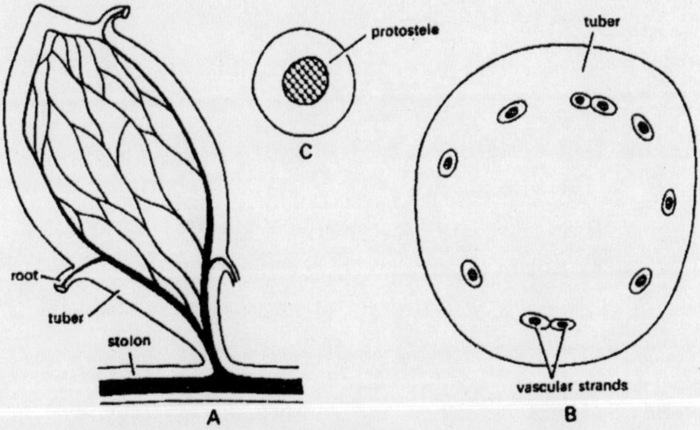

***Nephrolepis tuberosa* : Anatomy of tuber; A. Longitudinal section of tuber, B. Transverse section of tuber, C. A meristele (protostelic).**

The structure of the sporangium, gametophyte and embryo is essentially similar to that of Pteridium.

6

The Osmundales

The osmundaceous ferns are intermediate between Eusporangiate and Leptosporangiate in many respects. In the structure of petiole, presence of stipule-like expansions at the base of the petiole, development of sporangium and structure and development of antheridium and archegonium, they show resemblance with eusporangiate ferns. On the other hand, they show affinity with leptosporangiate ferns in the development of tapetum from the archesporial cells, single layered sporangial wall, projected antheridia and archegonia, and development of embryo. Hirmer (1927) thus created a new class - Protoleptosporangiate for these ferns. Reimers (1954) also treated them as a distinct group and placed in a separate sub-group, Osmundidae of Pteropsida.

The sub-group Osmundidae has a single order Osmundales, which is represented by the family Osmundaceae.

The members of Osmundaceae perhaps originated in the early Permian times. The family is represented by three living (Osmunda, Leptopteris and Todea) and three extinct genera

(Zalesskya, Thamnopteris, Osmundacaulis). The living members of the family are often called 'living fossils'.

The life-history of Osmunda is described here as the representative of Osmundales.

Osmunda

Systematic Position

Subdivision	—	Pteropsida
Order	—	Osmundales
Family	—	Osmundaceae

Osmunda is widely distributed both in the tropical and temperate regions of the world. The genus is represented by over 12 species; 'of these five species, viz., O. regalis, O. claytoniana, O. japonica, O. javanica and O. cinnamomea occur in India. The plants prefer cool and shaded habitat, and are abundant in north and north-western Himalayas up to an altitude of 3000 metres.

Sporophyte

External Morphology : Osmunda is a medium-sized fern with a short upright or semiprostrate rhizomatous stem, clothed with hard persistent bases of old leaves and numerous dark coloured wiry fibrous roots. These increase the diameter of the stem and provide it mechanical strength. The branching of the stem, which is very infrequent, is dichotomous.

The roots develop endogenously and help to fix the plant to the substratum. Normally, there are two roots at the base of each leaf. As the leaf bases are very close to each other, the roots are flattened at the point of their origin. The roots branch freely and form dark dense masses around the stumpy stem.

The leaves are arranged in close spiral phyllotaxy at the apex of the stem. They are circinately coiled when young and sometimes take up to 3-4 years to unfold fully. The young leaves are covered with simple uniseriate hairs, but the mature leaves are, however, glabrous. The leaves are often large, attaining a length of about lm but in O. cinnamomea they are up to 3m long. They are once (e.g., O. claytoniana) or twice pinnate (e.g., O. regalis). The margins of the pinnae are variously incised. The venation is open dichotomous. The leaves die at the end of the growing season leaving behind hard persistent bases on the stem.

Some species of Osmunda (like O. cinnamomea and O. japonica) are dimorphic, i.e., they possess two types of leaves-sterile and fertile. The fertile leaves appear before the sterile ones; they are non-green and bear only sporangia directly on the branches of rachis. In the absence of lamina, these sporangia cannot be called superficial and are described as marginal. The fertile leaves shrivel soon after the spores are shed. In most of the species all leaves are fertile and dimorphism is shown only by pinnae. For example, in O. claytoniana the pinnae at the middle of the leaf bear sporangia, while in O. regalis the pinnae at the distal region produce sporangia (Fig.).

Internal Structure

1. *Stem :* The stem is very irregular in outline due to the presence of persistent leaf bases and roots. The epidermis is indistinct.

   The bulk of the stem is represented by cortex. It is differentiated into two distinct regions, the outer cortex is dark brown and has highly sclerified cells, whereas the inner cortex is relatively narrow and made up of thin-walled parenchymatous cells, often packed with starch grains. Many crescent-shaped leaf traces traverse both the outer as well as inner cortex. Each leaf trace is surrounded by an endodermis and a 2-3-layered pericycle. The central part of the leaf trace has a C-shaped xylem with convex side facing outside. The xylem, made up of scalariform tracheids, is surrounded by phloem.

In young sporophytes of Osmunda, the stem is protostelic, but pith develops after the formation of the first few leaves. Thus the mature stem exhibits a typical dictyostelic condition. The stele is delimited from the cortex by a well defined endodermis with characteristic casparian- strips. An unusual feature of the endodermis is that it is not interrupted by leaf gaps. The endodermis is followed by a 2-4 layered pericycle. A continuous thin band of phloem is present internal to the pericycle. This band becomes thicker in between the bundles. The phloem is made up of sieve cells of a fairly large diameter with sieve areas on their lateral walls. Occasionally, an internal phloem and internal endodermis are also present as in O. cinnamomea.

The phloem is followed by a reticulate cylinder of xylem strands. In a transverse section these strands mostly appear slightly elongate in radial direction and are solitary or in close pairs. The former are terete, U-, O-, or S-shaped. The xylem strands are of variable sizes and are separated from each other by narrow parenchymatous rays that pass outwards from the pith (Fig.). They actually represent leaf gaps. The xylem strand shows mesarch condition with protoxylem lying in the sinus formed by the two arms of the metaxylem. The metaxylem tracheids are scalariform, while those of protoxylem show annular or spiral thickenings.

In mature stems, the central part of the stem is occupied by a large pith. The pith may be entirely parenchymatous (e.g., O. claytoniana) or have few strands of sclerenchyma (e.g., O. cinnamomea). Some species of Osmunda (e.g., O. regalis, O. javanica), however, have a mixed pith, i.e., tracheids are intermingled in the parenchymatous pith cells.

2. *Roots* : The roots of Osmunda are larger than most other ferns. They are tough and resistant to decay. In a transverse section, the root appears more or less circular in outline (Fig.). The epidermis is composed of a single layer of radially elongated cells.

   The cortex is broad and differentiated into outer sclerenchymatous and inner parenchymatous region. In species like O. cinnamomea, however, a few layers of sclerenchyma also occur around the stele. The cortex and stele are delimited by a well-defined endodermis and pericycle.

   The stele is usually diarch or triarch. The surface contours of the root frequently conform with those of the xylem mass. The metaxylem, which occupies the central part of the stele, has large scalariform tracheids. The protoxylem, present at an exarch position, has annular or spiral tracheids. Phloem bundles, which alternate with xylem strands, are separated from the latter by 3- 4 layers of parenchymatous cells.

3. *Leaf* : The leaf primordia are massive and develop at a distance from the stem apex. The young leaf primordium has an apical cell with three cutting faces. In O. cinnamomea, although the leaf arises from a group of cells of the apical meristem, subsequently a single pyramidal apical cell is set off which is responsible for the apical growth of the leaf.

   (a) *Petiole* : The petiole is surrounded by a distinct epidermis which is interrupted by stomata and simple multicellular hairs.

   The epidermis is followed by a broad zone of ground tissue. A few outer thick walled layers of the ground tissue form a distinct hypodermis, while the rest of the ground tissue is made up of thin-walled parenchymatous cells. A single large crescent-shaped concentric vascular strand is present in the centre of the ground tissue which is surrounded by an inconspicuous endodermis. The central xylem core of the strand has an almost continuous band of metaxylem with numerous protoxylem groups along its concave side. Some mucilage cells alternate with the protoxylem groups. The xylem band is completely surrounded by phloem.

(b) *Pinna :* The pinnae are bifacial with anatomically distinct upper and lower surfaces. The upper epidermis is continuous while the lower epidermis is interrupted by stomata. The mesophyll is undifferentiated, but the cells immediately below the upper epidermis are closely packed, whereas those towards the lower epidermis are loosely arranged enclosing large air spaces. A single concentric bundle is present in each pinna.

Reproduction : Osmunda is a homosporous fern and the sporangia do not form sori and lack indusia. As mentioned earlier, the sporangia are borne on the non-laminated pinna forming tassels; either the entire leaf (e.g., O. cinnamomea) or only a part of it is fertile (e.g., O. regalis, O. claytoniana). The sporangia are clustered and disposed radially on the ultimate units of the fertile pinnae.

The sporangia of Osmunda are large, pyriform with a short stout stalk. The sporangial jacket is single layered and the annulus is represented by a patch of thick walled cells on one side of the apex (Fig.). Occasionally, the annulus is in the form of a transverse band that extends to a considerable distance around the circumference of the sporangium. A row of vertically elongated thin walled cells is present perpendicular to the annulus. This row extends over the top of the sporangium down to the other end (Fig.). Each sporangium produces 128-572 spores.

Development of Sporangium : The development of sporangium in Osmunda is intermediate between the eusporangiate and leptosporangiate ferns. The sporangium first appears as small protuberance on the pinna or pinnule (Fig.). The protuberance has one large cell as in leptosporangiate ferns, but it differs from them in that the whole sporangium does not develop from this initial alone. The massive stalk of the sporangium of Osmunda is formed entirely or in part from the cells adjoining the initial cell.

The sporangial initial divides periclinally into an outer jacket initial (or primary wall cell) and an inner archesporial cell (Fig.). The shape of the archesporial cell is variable; it may be tetrahedral (as in Leptosporangiate) of cubical (as in Eusporangiate). The jacket initial and its daughter cells divide only anticlinally, as a result a one cell thick wall of the sporangium is formed (Fig.). The archesporial cell functions as internal apical cell and cuts off a flat cell from each of its four faces. Bower (1889) designated these cells as primary tapetal cells, as they were considered similar to the tapetal initials of the Leptosporangiate. They divide anticlinally and form a layer of cells inner to the sporangial wall. This layer however, cannot be interpreted as tapetum layer because its cells do not disintegrate even after the maturation of spores. These cells instead become appressed to the sporangial wall so that the jacket appears 2-layered at this stage (Fig.).

The archesporial cell, after cutting flat tapetal cells, functions as the primary sporogenous cell. It divides repeatedly and forms a large number of sporogenous cells (Fig.). In O. claytoniana, the outermost layer of the sporogenous cells functions as tapetum and the rest of the sporogenous cells function as spore mother cells and form 256-512 spores after meiosis. All the sporangia in a region mature together (Fig.).

Dehiscence of Sporangium : Dehiscence of the sporangium initiates at a point slightly above the annulus. The wall of the sporangium splits over the apex and down the opposite side along the line of thin walled cells and the sporangium opens widely (Fig.). The cells on the flanks of the sporangium shrink along with the cells of the annulus and facilitate opening.

Gametophyte

Spores : Haploid spores (n = 22) are the first cells of the gametophytic generation. Spores of Osmunda are large (55-60 × 60-75 um) and globose with a prominent triradiate ridge. The outer spore wall (exine) is granulose, whereas the inner one (intine) is thin and smooth. The mature spores contain abundant chloroplasts and, therefore, they remain viable for a very short time. Normally, germination occurs within 24 hours after the spores have reached a suitable substrate.

Development of Gametophyte : The spore germinates when adequate moisture is available. During germination, the exine ruptures along the triradiate ridge and the intine protrudes out along with the contents of the spore. The first division in spore occurs quite early, usually at the time of the protrusion of the intine. It is generally at right angles to the long (polar) axis and divides the spore into a small first rhizoidal cell and the first prothallial cell (Fig.). The rhizoidal cell, unlike those of other ferns, contains chloroplasts, but these chloroplasts disappear as the rhizoids push its way down into the soil.

The prothallial cell contains numerous chloroplasts and grows in a direction opposite to the rhizoidal cell (thus the germination of the spore is bipolar). It divides transversely to form a 3-4- celled filament. Subsequently, the terminal cell of the filament, by two intersecting diagonal divisions, differentiates an apical cell which gives rise to the prothallus (Fig.).

The mature gametophyte is large (upto 5 cm long), fleshy, thalloid, surface living and green (Fig.). It has a characteristic very distinct midrib. The margins of the prothallus often become deeply cut with the age and at certain points the prothallus is represented by the mid-rib only. As such some parts of the gametophyte may appear as axial, but in fact they develop from a linear meristem as do thalloid gametophytes in general. The rhizoids develop along the midrib on the ventral surface of the thallus. They are generally unicellular but in mature prothalli they may become septate.

Reproductive Structures : The prothallus of Osmunda is monoecious and protandrous. The young prothalli bear only antheridia, which are borne either marginally or terminally. Later, towards the maturity, the same prothalli bear archegonia on the ventral side, on or lateral to the midrib. The smaller and weaker prothalli are usually male.

The mature antheridia project from the surface of the prothallus as in Leptosporangiate ferns. They are relatively large, comparable in size to those of Psilotaceae. Each antheridium rests on a short and stout stalk (Fig.). The antheridial wall is highly complex; it consists of 8-12 chlorophyllous cells. The antheridium opens through a distinct large opercular cell situated either at the apex or laterally each antheridium produces 100 or more spirally coiled antherozoids each with a tuft of flagella at its apex (Fig.).

The archegonium has a long projecting neck, made up of four vertical rows of cells (Fig.). In each row there are upto 8 neck cells. The neck is occupied by a single binucleate neck canal cell, a venter canal cell and an egg embedded in the venter.

Embryo Development

The diploid zygote, formed after the fertilisation of the egg, is the mother cell of the sporophytic generation. The first division of the zygote is vertical, i.e., parallel to the archegonium axis (Fig.). The two daughter cells thus formed also divide vertically, but in a place at right angles to the first division.

Then all the four cells divide transversely forming eight cells which are arranged in two tiers of four each (Fig.). The young embryo differs from other leptosporangiate ferns in two respects; firstly, the various parts of the embryo, i.e., cotyledon, stem, primary root and foot are not referable to specific cells of the embryo, and secondly, the many celled embryo, which is soon built up, retains its more or less spherical shape for periods longer than in the typical leptosporangiate ferns.

In Osmunda, the cotyledon and stem are derived from the outer quadrant and the root and foot from the inner quadrant (Fig.). The stem and leaf develop by the activity of superficial apical cells. The apical cell of the root, however, does not arise at the surface (Fig.). Compared to other ferns, the foot in Osmunda is large, often extending deep into the prothallial tissue and has an irregular haustorial surface.

The young embryo remains within the archegonial wall till the initials of the various parts are differentiated. A calyptra (protective sheath), formed by the gametophytec tissue adjoining archegonium, also encloses the young embryo. Soon the root and cotyledon burst through the calyptra. The cotyledon grows out laterally from under the prothallus, instead of the apical end as in leptosporangiate ferns.

Brown (1920) has reported apogamy in Osmunda. The sporophytes develop from the spores directly without the intervention of sexual process. Likewise, gametophytes also develop aposporously from the sporophytec tissue without undergoing meiosis. Aposporously-produced gametophytes are naturally diploid and apogamously-produced sporophytes are haploid.

Ophioglossales : Ophioglossum : The order Ophioglossales, a small group of Eusporangiatae, is represented by three living genera— Ophioglossum, Botrychium and Helminthostachys. They do not have any definitely known fossil progenitor. Though they are generally called ferns, their relationship to ferns is extremely obscure. The sporophyte is herbaceous and lacks sclerenchyma. The rhizome is fleshy, naked and upright or dorsiventral. The leaves are non-circinate, simple or pinnate with sheathing stipules. The venation is open dichotomous (e.g., Botrychium, Helminthostachys) or netted with blind ends (e.g., Ophioglossum). The fertile spike projects from the leaf near the juncture of the blade and petiole; it is simple or pinnately branched. The sporangial wall has several layers; the sporangia are homosporous. The gametophyte is tuberous, colourless, mycorrhizic and subterranean. It bears numerous sex organs.

All the three genera of the order are included in a single family, Ophioglossaceae. The genus Ophioglossum is described in detail.

Ophioglossum

Systematic Position

Sub-division	—	Pteropsida
Class	—	Eusporangiatae
Order	—	Ophioglossales
Family	—	Ophioglossaceae

The genus Ophioglossum is represented by 30-50 species, widely distributed in tropical and temperate regions of the world. These are perennial herbs of moderate size, growing luxuriantly in soils rich in organic matter. Ophioglossum pendulum is the largest species, attaining a length of 1.5-2.7 meters. It is often grown as ornamental plant in tropics. Ophioglossum palmatum and O. pendulum (Fig.) are the two epiphytic species of the genus.

The genus is represented in India by about 10 species; of these, O. vulgatum, O. petiolatum, O. reticulatum and O. pendunculosum are common.

Sporophyte

External Morphology : The sporophyte is differentiated into stem, leaves and roots.

1. *Stem :* The stem is represented by a small upright subterranean rhizome. But in epiphytic species the rhizome is creeping and dorsiventral. Sometimes (e.g., O. aitchisoni, O. lucitanicum), the rhizome is covered by the old leaf bases and sheathing stipules. Branches usually develop from the buds present in the axil of leaves, but occasionally branching is due to dichotomy of the stem apex.
2. *Leaves :* The leaves are borne in an irregular spiral succession on the stem. Most of the species are monophyllous, producing only one leaf in a year, but several tropical species are known to produce up to five leaves in a year (e.g., O. brevips;). The leaves are simple and petiolate; the lamina may be palmately lobed (O. palmatum), simply lanceolate (O. vulgatum; Fig.) or linear (O. pendulum; Fig.). The venation is reticulate. There is no distinct midrib but in some species veins near the central axis become more conspicuous. Each leaf, when young, is enclosed by a thin sheathing stipule which remains attached to the stem even after the leaf is dead. In Ophioglossum (as also in other members of Ophioglossaceae) the development of a leaf takes about 4-5 years. It remains in bud condition for about three years and unfolds in the fourth year. Circinate vernation (spiral coiling of young leaves), a characteristic feature of ferns, is absent in Ophioglossum.
3. *Roots :* The rhizome bears many thick, fleshy adventitious roots. They arise singly below each leaf scar of the previous season and the leaf (or leaves) of the current season. The roots are sparingly branched;

when branched, the branching is monopodial or dichotomous. They are heavily infested with mycorrhiza. Root hairs are absent.

Apical growth takes place by means of a single pyramidal apical cell with three cutting faces.

Internal Structure

1. *Rhizome* : A transverse section of rhizome is somewhat irregular in outline due to the presence of persistent leaf bases and roots. It has a distinct epidermis, a parenchymatous cortex and a central stele (Fig.).

   The epidermis is single layered and is fairly distinct in the younger parts of the rhizome. There are, however, no stomata in the epidermis.

   The cortex is exclusively composed of thin-walled parenchymatous cells and as such the rhizome is soft and fleshy. In older parts of the rhizome a periderm is present in the outer region of the cortex but there is no functional cork cambium.

   The stele shows considerable variations not only in different species but also within the same plant. In the young basal part of the rhizome there is a protostele delimited by an endodermis. However, in the upper mature part pith develops gradually in the centre and as such the stele becomes siphonostelic. This condition is achieved through a transitional stele with mixed pith (parenchymatous pith with some tracheary elements).

   Since leaves receive their vascular supply from the stele of the rhizome, in that region the siphonostele may have one leaf trace (a condition known as solenostelic) or many leaf traces, overlapping each other (Fig. a condition known as dictyostelic). In the latter case, the siphonostele is broken into many small or large crescent-shaped meristeles without any endodermis around them (Fig.). In a meristele the protoxylem faces inwards. The protoxylem tracheids have annular thickenings, whereas the metaxylem tracheids have scalariform-bordered pits.

2. *Leaf* :

   (a) *Lamina* : The leaf is isobilateral and consists of spongy mesophyll cells with large intercellular spaces (Fig.). A large number of vascular bundles are seen in a vertical section of the leaf, and each bundle is surrounded by a bundle sheath. Stomata are uniformly distributed on both the upper and lower surfaces in erect leaves (e.g., O.vulgatum), but horizontal leaves have more stomata on the lower surface (e.g., O.polyphyllum).

   (b) *Petiole* : The epidermis of the petiole is composed of a single layer of moderately thick- walled cells. There are no stomata in the epidermis.

   The ground tissue is parenchymatous. The cells of the outer layers of ground tissue contain chloroplasts. The leaf receives a single vascular strand which branches repeatedly after entering the petiole. Thus a transverse section of the basal part of the petiole shows a single median vascular bundle, but there are many bundles in the upper part of the petiole, arranged in a C-shaped arc (Fig.). Some of these bundles enter the lamina and others supply the fertile spike.

3. *Root* : A transverse section of root shows a single layer of epidermis, a broad cortex and a central stele (Fig.).

The epidermis is made up of compactly-arranged cells with suberized outer walls. As the roots are mycorrhizic, epidermal cells do-not form root hairs. In the older roots of some species (O. costatum) epidermis is replaced by an exodermis which develops from the outer layers of the cortex.

The cortex consists of parenchymatous cells without or with small intercellular spaces. Some of the cells of the outer layers of the cortex contain endophytic mycorrhiza. The cells in the inner region of the cortex contain starch grains. A distinct endodermis with characteristic casparian bands on radial walls is present outside the stele.

The stele shows considerable variation; it may be monarch (O. vulgatum; Fig.), diarch (O. fibrosum; Fig.), triarch or tetrarch (O. aitchisoni; Fig.). In monarch steles the xylem is in the form of a shallow plate with protoxylem at each pole and phloem in between the arms of the xylem. But in di-, tri- and tetrarch steles the xylem groups are separated from each other by intervening phloem cells.

Reproduction : Ophioglossum reproduces by means of spores. Some species also multiply vegetatively by adventitious buds formed on the roots.

Spore-producing Organs : Spores are formed within sporangia. The latter are borne on a fertile segment which projects from the leaf near the junction of the 'petiole and lamina. This sporangiferous fertile segment is known as fertile spike. Usually a single spike is associated with a leaf but in O. palmatum there are several spikes on a leaf (Fig.). In the latter condition, many spikes emerge from the same point. Almost every leaf of a mature flourishing plant bears fertile spikes. The spike is 'simple or occasionally equally or unequally dichotomously branched. The sporangia are fused laterally into elongate synangium. But each sporangium receives independent vascular supply (Fig.).

Development of Fertile Spike and Sporangium : The fertile spike is initiated quite early in the ontogeny of the leaf. It differentiates from a cell on the adaxial side of the apical cell of the vegetative apex. This cell acts as the apical cell and forms a conical hump of meristematic tissue which grows independent of the vegetative apex.

When the primordium of the spike is very young, two vertical strips of cells in the epidermal layer are differentiated on its lateral sides. Each strip is 2-3 cells broad and several cells in length, and is known as sporangiogenic band (Fig.). Some hypodermal cells of these bands differentiate into alternate groups of fertile archesporial cells and sterile cells (Fig.).

Each group of archesporial cells represents a future sporangium. Thus the development of sporangium in Ophioglossum is of eusporangiate type. The archesporial cells of each group divide repeatedly and form a mass of sporogenous cells (Fig.). Meanwhile, the cells outside the sporogenous cells divide anticlinally as well as periclinally and, eventually form multi-layered sterile jacket of the sporangium (Fig.). The innermost layer of the jacket differentiates into a poorly defined tapetum (some workers believe that the tapetum is derived from the outermost layer of the sporogenous cells). The tapetal cells ultimately breakdown and spores float in the cytoplasm of degenerated tapetal cells (Fig.). Stomata are present in the outermost jacket layer of the sporangium.

The archesporial cells divide repeatedly and form a mass of sporogenous cells, which function as spore mother cells (Fig.). Each spore mother cell divides meiotically and forms a tetrad of four haploid spores (Fig.). The spore output per sporangium varies from 1,500 to 15,000. Dehiscence of sporangium takes place by means of a transverse slit. There is, however, no structure like annulus, a characteristic of other ferns. The intersporangial sterile tissue keeps pace with the sporangial growth and as such the massive sporangia are not raised above the marginal surface.

Gametophyte

Spores : The haploid spore is the mother cell of the gametophytic generation. It is small, with an outer- sculptured exine and an inner thin intine. It is almost hyaline but appears pale yellow when in a mass. In some species (e.g., O. vulgatum), spores are pale green as they contain chloroplasts.

Development of Gametophyte : Usually spores germinate within a few days after shedding, but sometimes their germination may be delayed for months or even for years. Before germination the spore enlarges considerably by absorbing water and consequently the exine ruptures along the triradiate ridge (Fig.). Then it divides transversely into an outer upper cell and inner lower cell (Fig.). The lower cell divides vertically and forms a three-celled gametophyte. Further development of the gametophyte takes place only if it gets infected with mycorrhizal fungus, which always enters through the lower cell (Fig.). Failing mycorrhizal infection, the young gametophyte aborts. The fungus forms endophytic mycorrhiza and enables gametophyte to continue its further growth.

The gametophyte is subterranean, achlorophyllous and has endophytic mycorrhiza. Occasionally superficial prothalli have also been observed; such gametophytes contain chloroplasts, thus are green. The gametophyte is oriented almost vertically in the soil buried at 1.5 mm to 13 mm depth. It is ovoid, linear, stellate or cylindrical; the latter being the most common form. It is unbranched (e.g., O. vulgatum) or profusely branched (e.g., O. pendulum). There is considerable variation in the size of the gametophyte and it ranges from few mm to 6 cm in length and 0.5 mm to 2.5 mm in diameter. For the, purpose of description, the gametophytic body can be divided into three distinct but intergrading regions : (i) basal extended tubercle, (ii) cylindrical main body, and (iii) narrow apical region (Fig.). The entire gametophytic body consists of thin-walled parenchymatous cells. The tissue of the cylindrical main body is typically differentiated into two zones—a central core of elongated lignified cells and a peripheral zone of more or less isodiametric cells. These two zones, although not sharply delimited, are recognizable almost to the apex of the gametophyte.

The gametophyte bears slender, hyaline, unicellular rhizoids, which are unbranched and usually slightly expanded at the base. They are profuse at or near the base of the gametophyte and become sparse towards the apex. They are points of connection between the endophytic fungus and the substrate as almost every rhizoid contains at least one hypha.

Sex Organ : The gametophyte of Ophioglossum is homothallic (monoecious), i.e., both male (antheridia) and female (archegonia) sex organs are formed on the same prothallus. In erect cylindrical prothalli sex organs are usually distributed all over the surface, whereas in flattened forms they are confined to the upper surface only.

1. *Antheridia :*

(a) *Development :* The antheridium develops from a single superficial cell, called antheridial initial, which is differentiated near the apex of the prothallus (Fig.). The antheridial initial divides by a periclinal wall, establishing an outer jacket initial (also called primary cover cell) and an inner primary androgonial cell (Fig.). The jacket initial divides anticlinally, forming two daughter cells of almost equal size. One of these divides by an anticlinal wall in a plane diagonal to the first anticlinal division. Consequently, one of the two cells thus formed is triangular that forms derivatives in all the three planes just like an apical cell. Thereafter, it is called opercular cell (Fig.).

The primary androgonial cell divides usually first by an anticlinal wall, followed by a periclinal wall. As a result four androgonial cells are formed which are arranged in two tiers of two each. These cells divide repeatedly and form a large mass of androgonial cells

(Fig.). The last generation of these cells functions as androcyte mother cells. An androcyte mother cell divides to form two androcytes; each androcyte metamorphoses into an antherozoid.

(b) *Mature Antheridium :* The mature antheridium is a massive oblong or spherical structure. It is almost completely embedded in the tissue of the gametophyte (Fig.). The antheridial jacket is usually single layered but in some species (e.g., O. palmatum, O. vulgatum, O. costatum) it is partly or wholly double layered. The cavity of the antheridium is packed with antherozoids. Each antherozoid has a large basal vesicle and a terminal tuft of flagella (Fig.). The number of antherozoids in an antheridium may be upto five thousand.

As the antheridium matures, the terminal opercular cell of the antheridial jacket becomes dark brown or black and then disintegrates. The antherozoids are liberated through the pore formed by the disintegration of opercular cell.

2. *Archegonia :*

(a) *Development :* The archegonium also differentiates near the apex of the prothallus or its branches. One of the superficial cells of the prothallus functions as archegonial initial (Fig.), which divides periclinally and forms an outer primary cover cell and an inner cell (Fig.). The primary cover cell divides by two vertical walls at right angles to each other, forming four quadrantly arranged neck initials (Fig.). Further diagonal transverse division of the neck initials results in the formation of neck of the archegonium. The neck is usually composed of eight cells, arranged in two tiers of four each. Sometimes, these cells may divide further and then the neck of the archegonium becomes three or more cells in height.

Meanwhile, the inner cell divides transversely into a basal cell and a central cell (Fig.).The latter divides transversely into a primary canal cell and a primary venter cell (Fig.). The primary canal cell usually does not divide further and directly functions as neck canal cell. The primary venter cell, on the other hand, divides further by a transverse wall and forms a venter canal cell (that lies adjacent to the neck canal cell) and a large egg. The basal cell undergoes one or more vertical divisions and forms a sterile pad at the base of the archegonium.

(b) *Mature Archegonium :* Archegonia are usually sunken in the tissue of the prothallus. The narrow neck is occupied by an elongated neck canal cell. The basal swollen venter contains a venter canal cell and a large egg (Fig.). In mature archegonia the four terminal neck cells separate. It is followed by gradual lateral radiation of the four rows of the neck cells, down to about four cells from the distal end.

Young Sporophyte

The fertilized egg (zygote) is the mother cell of the next sporophytic generation. The first division of the zygote is usually transverse to the archegonial axis and as such an epibasal and a hypobasal cell is established (Fig.). Both these cells are embryonic in nature and usually no suspensor is formed. But in O. nudicaule, the epibasal cell forms suspensor and the embryo develops from the hypobasal cell. The second division of the zygote is vertical and the four cells thus formed divide irregularly to form a mass of cells (Fig.). Although at this stage it is impossible to distinguish between the derivatives of the epibasal and hypobasal cells, apparently the cells of the outer half (epibasal part) form root, stem and leaf and those of the inner half (hypobasal part) give rise to a foot (Fig.). Thus the embryogeny is exoscopic.

In Ophioglossum the development of root is much rapid than stem and leaf. The root is usually more than a centimeter long by the time stem and leaf initiate (Fig.). But in some species (e.g., O. moluccanum) the stem and root develop almost simultaneously.

The embryo enlarges considerably within the archegonium: Meanwhile, the cells of prothallus surrounding the archegonium divide and form a prominent cushion over the young sporophyte. In some species the young sporophyte remains enclosed within the prothallus for years.

Eusporangiatae Marattiales : Angiopterzs : The class Eusporangiatae is a small group of primitive ferns, represented by some 17 genera and 30 species. The members of this class show the following distinguishing features.

(1) Sporangia are borne on fertile spikes (e.g., Botrychium, Ophioglossum) or on the abaxial side of the lamina (e.g, Angiopteris, Marattia). They are free or laterally adherent to form a synangium.

(2) The sporangium develops from a group of initial cells, i.e., the development is eusporangiate.

(3) The sporangial jacket is more than one cell in thickness.

(4) They produce only one type of spores, i.e., they are homosporous.

(5) Antheridia are embedded in the gametophyte and there are usually many antherozoids in an antheridium.

The class Eusporangiatae includes two orders—Marattiales and Ophioglossales.

Order Marattiales

The order Marattiales includes those eusporangiate ferns which bear sporangia on the abaxial side of lamina. They mostly occur in tropical rain forests. There are seven living genera; of these Archangiopteris, Macroglossum, Protomarattia and Christensenia are very small and Angiopteris, Marattia and Danaea are conspicuous because of their massive fleshy axis and large fronds. The young leaves are circinately coiled and each leaf has characteristically two stipules at its base. Sporangia form simple superficial sori on the abaxial side of the lamina and the sporangia in a sorus are free or show various degrees of fusion. Typical indusium is absent and instead a few hairs or scales are present about the margins of the sorus (e.g., Angiopteris) or the entire synangium (fused sporangia) is enclosed within a bivalved spoiocarp-like structure (e.g., Marattia). The wall of the sporangium is multilayered. The sporangium opens by slits on either side or by a terminal pore. However, a rudimentary annulus is present in Angiopteris and Macroglossum. The gametophyte is large, thalloid, green and perennial; it is mycorrhizic and monoecious.

Riemers (1954) recognised five families in the order Marattiales. The life history of Angiopteris (Angiopteridaceae) is described here as representative of the order.

Angiopteris

Systematic Position

Sub-division	—	Pteropsida
Class	—	Eusporangiatae
Order	—	Marattiales
Family	—	Angtopteridaceae

Angiopteris has some 100 species, distributed in Japan, Tropical Australia, New Caledonia, Madagascar and Polynesia. A. evecta occurs throughout India upto an elevation of 2500 meters.

Sporophyte

External Morphology : Sporophytic plant body consists of a stem which bears roots and leaves (Fig.).

The stem is large, globose, upright and 60 cm or more in height. It may attain a diameter of upto 60 cm. It is shielded by fleshy stipules. The leaves are spirally arranged and form a crown at the apex of the stem. They are pinnately compound, often large, 2-6 meters long. The pinnae are herbaceous or sub-coriaceous and glabrous, varying in length from 20-60 cm (Fig.). The pinnae are sessile or shortly stalked. A conspicuous swelling, called pulvinus, is present at the base of each pinna (Fig.). The venation is open dichotomous and veinlets are usually sub-parallel. Each leaf bears a pair of fleshy stipules at its base, which enclose the next younger leaf. The young leaves show typical circinate vernation.

Numerous adventitious roots arise from the basal part of the stem. They are somewhat fleshy, branched and mycorrhizic. They arise endogenously and are inserted below the leaves. They pierce the fleshy stipules in their outward course. The young stem grows by means of a single narrow apical cell, which is replaced by group of initials at maturity.

Internal Structure

1. *Stem :* The anatomy of the stem is complex and it varies with the age of the plant. The parenchymatous cortex has numerous mucilagenous ducts. The fleshy nature of the stem is due to the absence of thick-walled cells, and mechanical support to the stem is provided by thick leaf bases.

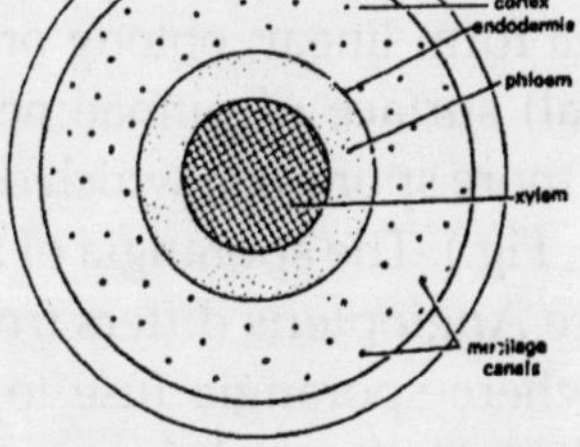

Angiopteris: Transverse section of a young stem.

The young stem is protostelic. The stele is surrounded by a distinct layer of endodermis (Fig.). As the stem grows older, the stele becomes amphiphloic siphonostele with distinct leaf gaps. In a mature stem, however, the stele breaks into many meristeles due to overlapping leaf gaps (Fig.). It appears to be polycyclic as several curved sheets of vascular strands are arranged in concentric rings. Root traces, however, arise only from the meristeles of the inner rings .

2. *Root* : The internal structure of the root is relatively simple (Fig.). The epidermal cells of the root form root hairs; some root hairs are multicellular, a distinctive feature of Marattiales. There is a large parenchymatous cortex with many mucilagenous canals. The latter extend throughout the plant body. The stele is surrounded by a single layer of endodermis and one or two layers of pericycle. It is polyarch with 4-10 (or more) long and thin xylem arms. Small groups of phloem alternate with the xylem arms. A small parenchymatous pith is present in the centre of the stele.

3. *Pinna* : The internal structure of a pinna is similar to that of a dicotyledonous leaf. Both, the lower and upper epidermis are single layered. Stomata are confined to the lower epidermis. The mesophyll is differentiated into 1-2 adaxial layers of slightly elongated palisade-like cells followed by spongy parenchyma.

Spore-producing Organs

The sporangia form linear, oblong or boat-shaped sori on the lower (abaxial) surface of pinnae near the margin. Each sorus has 8-14 or more sporangia which develop over a veinlet (and not at its tip; Fig.). The sporangia of a sorus are free (Fig.) and in this feature Angiopteris differs from other members of the Marattiales where sporangia fuse to form synangia. The sori are non-indusiate, instead few multicellular hairs are present along the margin of the sorus (Fig.).

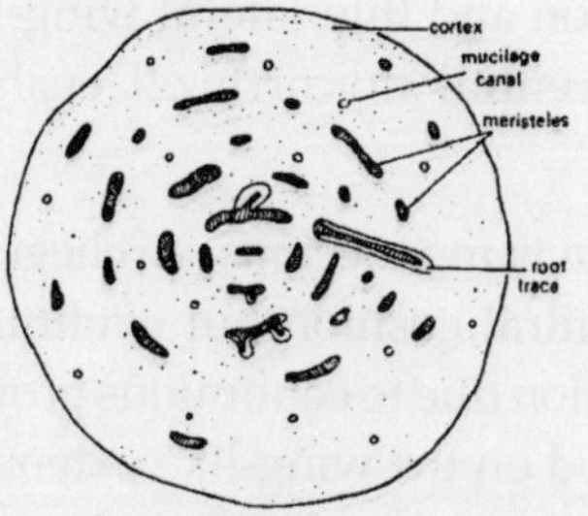

Angiopteris: Transverse section of a mature stem.

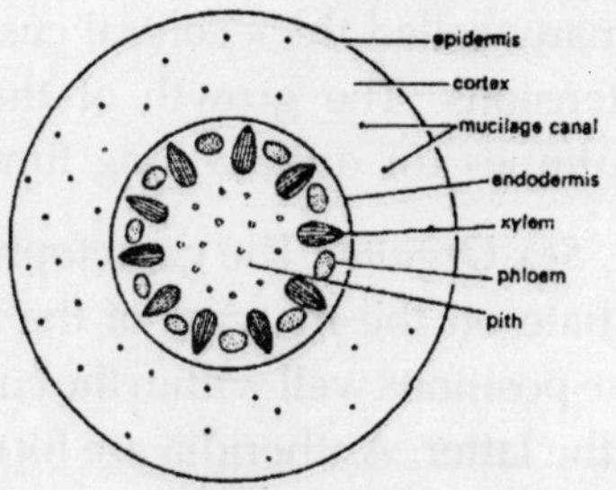

Angiopteris: Transverse section of root.

The development of sporangium is of eusporangiate type. In Angiopteris, sporangia of a sorus do not initiate individually, instead the entire sorus arises as a surface receptacular meristem. All surface cells of this meristem divide periclinally and then groups of hypodermal cells differentiate as sporangial initials. Adjacent groups of sporangial initials are separated by sterile intersporangial areas. Each group of sporangial initial eventually forms a sorus. The inter-sporangial cells, however, do not keep pace with the dividing sporangial initials, and as such the sporangia of a sorus do not fuse with each other.

The mature sporangium is large, massive and sessile. The proximal part of sporangium is directly attached to the receptacle. The wall of the sporangium is several layered; the outermost layer has stomata and the innermost layer, lining the sporogenous tissue, forms tapetum. An arch of slightly thick walled cells, the annulus, is differentiated in the sporangial wall. It extends vertically from one end to the other across the top of the sporangium. Contractions in the cells of annulus result in the splitting of sporangium along the line of dehiscence. The spore output për sporangium is very large (1400-7500) in comparison to leptosporangiate ferns.

Gametophyte : Germination of spores takes place within a few days after their shedding and they form large, perennial prothalli. The prothallus is dark-green, mycorrhizic and thalloid, resembling very much with thallose liverworts (Fig.). It contains a many-celled thick central cushion and thin lateral wing-like extensions. The growth of the central mycorrhizal cushion continues for quite a long time.

Sex Organs : The gametophyte is monoecious. Archegonia initiate on the margins of the central cushion but eventually take positions well within the cushion due to continuous growth of the latter. Antheridia are formed on the wing-like extension of the prothallus and they are more or less scattered.

1. *Antheridium :* The mature antheridium is massive and sunken in the tissue of the gametophyte. It is surrounded by a single layered antheridial jacket (Fig.). It produces a large number of coiled multiflagellate antherozoids and opens on the surface of the prothallus through a triangular opercular cell.
2. *Archegonium :* The archegonium consists of a 3-4 cells long neck and a venter. The neck canal is occupied by a single binucleate neck canal cell and in the venter a venter canal cell and a large egg is present (Fig.). As the archegonium matures, it undergoes considerable stretching due to the growth of the cushion.

Embryo

Zygote is formed by the fusion of egg and antherozoid. It enlarges and then divides transversely establishing an epibasal and a hypobasal cell (Fig.). The epibasal cell, lying close to the neck of the archegonium, forms a short single-celled suspensor. The hypobasal cell, called embryonic cell, divides to form a small spherical mass of cells. The polarity of the embryo becomes apparent at quadrant or octant stage. The cells close to the suspensor (i.e., those towards the neck of the archegonium) form foot and the inner cells form stem, root and the first leaf (Fig.). Thus the development of the embryo in Angiopteris is endoscopic.

7

The Equisetales

The order Equisetales comprises a single family Equisetaceae. It includes two genera— Equisetum and Equisetites. Of these, Equisetum is living, whereas Equisetites is extinct. The sporophyte shows differentiation into stem, leaves and roots. The stem has distinct nodes and internodes. The internodes are longitudinally ribbed and the ribs of successive internodes alternate with each other. The stem shows both hydrophytic and xerophytic features. The leaves are small, scaly and form a whorl around each node. The number of leaves at a node corresponds to the number of ridges. Both members of the order are homosporous. The sporangia are borne on the underside of a peltate sporangiophore. The sporangiophores form compact terminal cones. The spores possess characteristic elaters which develop from the spore wall itself. The gametophyte is monoecious. The antherozoids are multiflagellate. The development of embryo is exoscopic.

Equisetum is described here in detail as representative of Equisetales.

Equisetum (Horse-tail)

Systematic Position

Sub-division	—	Sphenopsida
Order	—	Equisetales
Family	—	Equisetaceae

Equisetum is represented by some 30 species, commonly known as horse-tail or scouring rushes. They are distributed all over the world, except Australia and New Zealand. They grow in a variety of habitats. Some species prefer damp and shady places (e.g., E. pratense), while others grow in marshy places, ponds or along the sandy river banks (e.g., E. palustre). E. arvense is found in xerophytic habitats, such as open grasslands, along roadsides and railway tracks. Some common Indian species of the genus are E. arvense, E. debile, E. diffusum, E. ramosissimum and E. elongatum. Of these, E. debile is found in plains along the river banks and E. arvense is common in the western Himalayan ranges.

Species of Equisetum are used as indicators for minerals in the soil. Some species accumulate gold from the soil. There is deposition of silica on the outer wall of the epidermal cells due to which plants become rough (just like sand paper) in texture and are used as abrasive. Silica also provides a protective covering to these plants against predators and pathogens. Some species of Equisetum are also used in ayurvedic medicines as diuretic.

Sporophyte

Habit : The sporophyte of Equisetum has a perennial, branched and creeping underground rhizome, which gives annual erect aerial branches (Fig.). The aerial branches are herbaceous and usually grow to a height of 10-60 cm, but they attain a height of 2-3 meters in E. telmateia, and about 4 meters in E. ramosissimum. E. giganteum, the largest species of

Equisetum, which grows in the tropics of South America, has stem upto 13 meters long, but since the stem is relatively thin (0.5 to 2.0 cm), it is vine-like in habit and climbs over adjacent trees. The diameter of the stem varies from few millimeters to 10 cm (e.g., E. schaffneri).

External Morphology : The sporophyte of Equisetum is differentiated into stem, root and leaves.

1. *Stem :* The main stem is underground creeping and perennial rhizome. It often lies more than a meter below the surface of the soil. It is differentiated into nodes and internodes. There is a whorl of small scaly leaves at each node. The leaves of a whorl are joined at their bases, forming a sheath around the node. A branch bud is formed alternate to each scale leaf at the node and these buds develop into erect aerial branches. In species like E. arvense, E. telmateia some branch buds instead of developing into aerial branches form small subterranean tubers. The tuber is a perennating organ and it develops into a new plant on the return of favourable conditions. The rhizome bears two types of aerial shoots, vegetative shoots and fertile shoots. Both are differentiated into nodes and internodes. The vegetative shoots are usually branched (unbranched in E. hyemale), green and photosynthetic. The fertile shoots are unbranched and achlorophyllous, and each terminates into a cone or strobilus. In E. palustre some fertile shoots become branched, green and photosynthetic after the dehiscence of strobili. In some species (e.g., E. debile, E. diffusum) there is no distinction between sterile and fertile shoots and all aerial shoots bear strobili.

   A whorl of lateral branches is present at each node of the vegetative shoot. Their number corresponds to the number of leaves at the node. These branches are of the following two types.

(a) *Branches of Unlimited Growth* : They are further branched like the main vegetative shoot and are differentiated into nodes and internodes.

(b) *Branches of Limited Growth* : They are also differentiated into nodes and intemodes like vegetative shoots but remain unbranched.

The rhizome as well as the aerial shoots of Equisetum present articulated or jointed appearance due to the presence of distinct nodes and internodes. The internodes are hollow and have longitudinal ridges and grooves. The ridges of the successive internodes alternate with each other (Fig.). The number of ridges is equal to the number of leaves at the node.

2. *Leaves* : The leaves are minute, scaly and isophyllous. They arise in whorls at the nodes of both the underground rhizome and aerial branch. They are fused at their bases to form a distinct sheath and their distal-free ends give a frill-like appearance (Fig.). The degree of fusion between the leaves and the number of leaves at a node varies considerably not only among different species of Equisetum but also within the same species. The species with narrow stems usually have few leaves (e.g., 2-3 in E. scirpoides) and those with thick stems have many leaves (up to 40 in E. schaffneri) at each node. The leaves at the successive nodes alternate with each other.

The leaves of Equisetum are nonphotosynthetic and their main function is to protect the branch buds at the node.

Internal Structure

Aerial Shoot : The anatomy of the stem of Equisetum is interesting as it shows a combination of hydrophytic and xerophytic characters. The nodes and internodes have different internal structures.

Internode : A transverse section of the internode is wavy in outline due to the presence of ridges and grooves (Fig.). It has a distinct epidermis, a well-developed cortex, a stele with a ring of vascular bundles and a large central pith cavity.

The epidermis is composed of a single layer of cells with a heavy incrustation of silica on their outer and radial walls. Silica provides mechanical strength to the shoot. The epidermis is interrupted by stomata, which are usually confined to the grooves. A stomatal apparatus is composed of four cells arranged in two tiers of two cells each (Fig.). The cells of the lower tier are guard cells and those of the upper tier, subsidiary or accessory cells. Crescent-shaped accessory cells completely cover the guard cells. A series of transverse thickening bands of siliceous matter develops on the lower surface of the accessory cells and they appear to radiate from the stomatal openings (Fig.). The stomata are usually sunken but in species like E. pratense and E. palustre they are at the level of the epidermis. The cortex is differentiated into an outer and an inner region (Fig.). The outer cortex consists of sclerenchymatous and chlorenchymatous cells. The sclerenchyma lies just below the epidermis and provides mechanical strength to the shoot. It is many layered below the ridges but in the region of the grooves it is only 1-2 layered. However, there is no sclerenchyma beneath the stomata. The sclerenchyma is followed by a band of chlorenchyma, which forms photosynthetic region of the shoot. The chlorenchyma is thicker below the ridges. Conspicuous intercellular spaces are present in chlorenchyma just beneath the stomata. The inner cortex is composed of thin walled parenchymatous cells. It has large schizolysigenous canals known as vallecular canals below the furrows. These canals extend throughout the length of the internode and form a distinct aerating system.

The cortex is delimited from the stele by an endodermis, which varies in its position in different species : (i) in species like E. arvense, E. palustre, E. scirpoides and E. telmateia there is a single layer of endodermis outside the stele (Fig.), (ii) in E. debile, E. hyemale, etc., there are two endodermal layers, one on the outside and the other on the inner side of the stele (Fig.), and (iii) in species such as E. limnosum and E. giganteum there is a separate endodermal layer around each vascular bundle (Fig.). On the basis of fossil evidences, the first type is considered to be the most primitive and the last, the most advanced.

The endodermis is followed by a single layer of parenchymatous pericycle. The vascular cylinder is a siphonostele where vascular bundles are arranged in a ring around the large pith. The vascular bundles lie opposite to the ridges in positions alternate to the vallecular canals. Thus the number of vascular bundles corresponds with the number of ridges. The vascular bundles are conjoint, collateral and endarch. The xylem of a bundle is in the form of two lateral and a median group of tracheids. In a very young vascular bundle the protoxylem is represented by tracheids with annular or spiral thickenings. But in the mature bundle the protoxylem elements disintegrate to form a conspicuous protoxylem lacuna, called carinal canal. Some protoxylem elements, however, can be seen along the periphery of the carinal canal. The metaxylem tracheids have scalariform, reticulate or pitted thickenings. In several species of Equisetum vessels with simple and reticulate perforation plates have been reported.

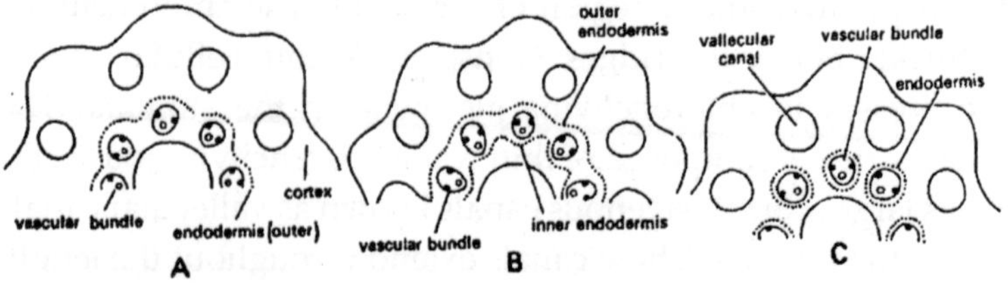

Equisetum : Different positions of endodermis; A. Transverse section of aerial shoot of E. palustre (note a single-layered endodermis). B. Transverse section of rhizome of E. hyemale (note the inner and outer endodermal layers), C. Transverse section of rhizome of E. limosum (note the presence of individual endodermal layers around each vascular bundle).

The phloem lies outside the xylem and in radial alignment with the carinal canal. It is composed of phloem parenchyma and sieve tubes with sieve areas on radial walls.

The carinal canals are filled with water and help in conduction of water. Thus they are different from the vallecular canals which contain air only. The central part of the internode of the aerial shoot is occupied by a large pith cavity. In some species (e.g., E. scirpoides) pith is absent. The internode shows the following xerophytic and hydrophytic features.

Xerophytic Features

(1) Presence of ridges and grooves.

(2) The outer and radial walls of the epidermal cells are impregnated with silica.

(3) Presence of sunken stomata.

(4) Presence of well-developed sclerenchymatous hypodermis.

(5) Presence of reduced and scaly leaves

(6) Presence of photosynthetic tissue (chlorenchyma) in the stem.

(7) Presence of well-developed vascular cylinder.

Hydrophytic Features

(1) Presence of well-developed aerating system (vallecular canals and central pith cavity).

(2) Presence of reduced xylem, represented by few protoxylem and metaxylem tracheids only.

Node : The internal structure of the node differs from the internode in the following features.

(1) The pith is not hollow at the node and instead a solid diaphragm, called nodal diaphragm, is present (Fig.).

(2) Vallecular canals are usually not present at the node.

(3) The vascular bundles fuse together and form a vascular cylinder around the nodal diaphragm.

(4) Carinal canals are absent in the nodal region.

(5) Leaf and branch traces arise from the vascular cylinder of the node (Fig.).

Rhizome : The internal structure of the rhizome differs from the aerial shoot in the following features (Fig.).

(1) The epidermis of rhizome is without stomata.

(2) The outer cortex of rhizome is less sclerenchymatous.

(3) Chlorcphyllous cells are not found in the cortex.

(4) The vallecular, carinal and central canals are narrow in comparison to those in the aerial shoots.

Root : A transverse section of the root shows epiblema, cortex, and stele (Fig.).

The single layered epiblema forms the outer-most protective covering of the root. Some cells of epiblema elongate to form root hairs.

The epiblema is followed by a wide cortex, usually differentiated into an outer sclerenchymatous and an inner parenchymatous region. The latter has intercellular spaces. The sclerenchymatous outer cortex is also known as exodermis.

A characteristic feature of the root of Equisetum is the presence of a two-layered endodermis. The cells of the outer endodermal layer are larger and have casparian bands but those of the inner endodermis are smaller and without casparian bands. Pericycle is absent and lateral roots arise from the inner endodermis.

A protostele is present in the centre of the root with diarch to tetrarch xylem. The number of xylem strands usually increases with the diameter of the root. A large metaxylem tracheid is present in the centre of the stele and the protoxylem strands lie around it. The phloem is present in between the protoxylem strands.

Apical Growth

The rhizome and aerial shoot of Equisetum have a broad pyramidal apical cell with three cutting faces (Fig.). It cuts cells on three lateral sides successively. Each derivative of the apical cell divides anticlinally forming an upper and a lower daughter cell. The upper daughter cell forms the nodal region while the lower cell gives rise to the internodal region of the stem. The initials of the lateral branches develop from the cells of the nodal region.

Reproduction : Equisetum reproduces by vegetative and sexual methods.

Vegetative Reproduction : The underground rhizomes of some species of Equisetum (e.g., E. arvense, E. telmateia) form tubers, which help in vegetative propagation. The tubers develop due to irregular growth of some branch buds at the nodes of the rhizome. They are oval (E. arvense) or pear-shaped (E. telmateia). The tuber has a central parenchymatous region, rich in starch grains, and it is surrounded by 2-3 layers of sclerenchyma. Several collateral vascular bundles are present in the central region. Each bundle is individually surrounded by its endodermal layer. After detachment from the parent plant, the tubers remain in the soil, and on the return of favourable conditions they grow into new plants.

Reproduction by Spores : Equisetum is a homosporous pteridophyte. The spores develop within sporangia borne on sporangiophores. The sporangiophores are aggregated into a compact cone or strobilus.

1. *Position and Structure of Strobilus :* In most species strobili are borne terminally on the vegetative shoots. But in

some species (e.g., E. arvense) they are borne on special fertile shoots, which are short-lived and die soon after spores are shed.

The strobilus has a central massive axis, called strobilus axis, and a large number of sporangiophores. The sporangiophores project at right angles in successive alternate whorls from the strobilus axis. Each sporangiophore is a stalked structure with a hexagonal peltate disc at its distal end. On the underside of the sporangiophore disc 5-10 sac-like sporangia are borne near its periphery in a ring. The free round apices of sporangia face the strobilus axis.

In some species a whorl of scale-like outgrowths, called annulus, is present at the base of the strobilus. The morphological nature of annulus is debated. It has been interpreted as the last whorl of scale leaves of the fertile shoot by some, while others consider it of sporangiophorec nature as it bears small sporangia occasionally.

2. *Development of Sporangium* : The apical growth of the shoot, which is destined to bear strobilus, is slowed down and it assumes a conical shape. This conical apex now functions as strobilus axis. The primordia of sporangiophores are formed on this axis in an acropetal succession, in positions of 'leaf primordia. Initially, the primordium is hemispherical but soon it becomes peltate due to the formation of a constriction near its base. Sporangia are initiated on the apical dome of the sporangiophore.

   The development of sporangium is eusporangiate, i.e., a sporangium develops from a group of initials. Sporangia are initiated near the centre of the apical dome of the sporangiophore. But later the dome expands laterally (due to rapid growth of the central part of the dome) and as such the sporangia are pushed to the margins and finally to the underside of the peltate disc.

The first division of the sporangial initial is periclinal, establishing an outer and an inner daughter cell. The inner daughter cell forms sporogenous tissue by repeated divisions in all possible planes, whereas the outer daughter cell divides both anticlinally and periclinally and forms 2-7 layered sporangial wall. The innermost layer of the sporangial wall functions as tapetum. The cells of the tapetum are irregular, thick-walled and nutritive in function. The cells of the outer wall layer develop thin bands of helicoid secondary wall thickening, which presumably help in the dehiscence of the sporangium.

As the sporangium matures, all out the two outermost wall layers disintegrate and form a periplasmodial liquid. At this stage, the sporogenous cells separate from each other and function as spore mother cells. Before meiosis, about 30% spore mother cells also disintegrate and form plasmodial fluid. The remaining spore mother cells remain active and float freely in the periplasmodial fluid. Each spore mother cell undergoes a reduction division and forms four spores. The spores soon become spherical.

3. *Structure of the Mature Sporangium* : The mature sporangia are sac-like structure attached to the underside of the peltate disc of the sporangiophore. The wall of the mature sporangium is only two layered. All spores in a sporangium are alike.

4. *Dehiscence of Sporangium* : As the sporangia mature, the strobilus axis elongates. Consequently, the compactly arranged sporangiophores separate from each other and the sporangia are exposed: As the sporangium

dries, the helicoid thickening bands present in the outer wall layer shrink and the sporangium ruptures.

5. *Spores* : The spores of Equisetum are spherical, uninucleate and green; they contain numerous chloroplasts. The spore wall is differentiated into four concentric layers, the outermost perispore, the second middle layer, the third exospore and the innermost endospore. The perispore, also known as epispore, splits into four strips or bands. Although these bands are separate from one another, they are attached to a common point on the spore. Initially, these bands are wrapped around the spherical spore but as the spore dries the bands are uncoiled. These bands are known as elaters and they have expanded spoon-like tips. The elaters are hygroscopic and respond to change in humidity. Although their definite function is not known, they probably help in the dehiscence of the sporangium by expanding and contracting. More often they seem to keep spore groups entangled. They also act as parachute and help spores to float in air.

Gametophyte

Germination of Spore and Development of the Gametophyte: The spores of Equisetum germinate shortly after shedding. They are short-lived and remain viable for 1-48 hours after shedding. Such a short duration of viability of Equisetum spores is perhaps due to their high rate of respiration. The spores germinate on a suitable substratum. They swell by absorbing water and shed the outer coat. The first divison of the spore is asymmetrical; it produces a small rhizoidal cell and a larger cell. The former develops into the first rhizoid, whereas the latter eventually gives rise to prothallus. The shape and size of prothallus depends on prevailing conditions; if a large number of spores germinate within a limited space, the prothalli formed are usually filamentous whereas sparsely germinating spores usually form relatively thick and cushion-shaped prothalli. The cells in the upper part of the prothallus are chlorophyllous and those in the lower portion are colourless. The basal cells in the lower portion give rise to unicellular rhizoids, which help in the attachment of prothallus to the substratum. The upper marginal cells of the prothallus are meristematic. If the growth is uniform along the entire margin, it results in the formation of a circular gametophyte and if growth is more active at certain points then lobed gametophyte is formed. Mature field grown gametophytes of Equisetum are very small, seldom exceeding a few millimeter.

Several studies have shown that prothalli of Equisetum are basically monoecious and heterothallism (i.e., dioecious nature) develops only under the influence of adverse environmental conditions.

Kashyap (1917) observed that if the spores of E. debile germinate densely then they form very small prothalli, which bear either antheridia or archegonia. But when they germinate sparsely, relatively large prothalli are formed, which first bear archegonia and then antheridia. Thus the sex of the prothallus depends upon the conditions available for the germination of spores. Though E. arvense is a monoecious species, under adverse conditions prothalli remain unisexual.

Gametophytes of Equisetum are rarely seen in nature because of two factors. Firstly, the young gametophytes appear as filamentous algae or moss protonema to the naked eyes and often get obscured by other small plants growing nearby. Secondly, they often die while quite young as a result of drying and in a given season only very few gametophytes attain maturity.

Sex Organs : Both, antheridia and archegonia develop from the superficial meristematic cells of the prothallus. Archegonia are confined to the basal cushion region of the prothallus and antheridia on the lobes of the upright branches. Usually archegonia develop before antheridia.

1. *Development of Archegonium :* Any superficial cell from the margin of the prothallus may function as archegonial initial. It divides periclinally into an outer primary cover cell and a central cell. The primary cover cell divides by two vertical divisions at right angles to each other, forming four primary neck cells. These cells divide further and form 3-4-celled high neck of the archegonium. In the meantime, the central cell divides transversely into a primary neck canal cell and a venter cell. The primary neck canal cell, by a further transverse division, gives rise to two neck canal cells. But in some species (e.g., E. debile) the primary neck canal cell does not divide and directly functions as neck canal cell, and in E. hyemale and E. ramosissimum it divides vertically into two boot- shaped cells, which are attached at the base of the neck canal. In the meantime, the venter cell forms two unequal cells by a transverse divison; the larger basal cell forms egg and the smaller upper cell functions as venter canal cell.
2. *Mature Archegonium :* The mature archegonium has a small neck and a swollen base. The neck is made up of 3-4 tiers of neck cells; each tier usually has four cells. The neck canal has one or two neck canal cells. The swollen basal part of the archegonium has a venter canal cell and an egg. The neck of the archegonium usually projects above the surface of the prothallus.
3. *Development of Antheridium :* In monoecious species antheridia develop after archegonia. There are usually two types of antheridia : (i) projecting antheridia, and (ii) embedded antheridia. The projecting antheridia develop on the margins of the prothallus or on the

lobes of the upright branches, whereas the embedded antheridia develop in the massive cushion-shaped part of the prothallus.

The superficial antheridial initial divides periclinally into an outer jacket initial and an inner androgonial cell (Fig.). The jacket initial divides only anticlinally and gives rise to a single layered antheridial jacket. A large triangular jacket cell at the apex of the antheridium differentiates into a cover cell or opercular cell (Fig.). The nucleus of the androcyte progressively changes in shape to become helicoid (Fig.). Then each androcyte metamorphoses into an antherozoid (Fig.). The anterior part of the antherozoid is spirally coiled, and has numerous flagella, whereas the posterior part is somewhat expanded.

Fertilisation

Before fertilisation, the neck canal cells and venter canal cell disintegrate and form a passage for the entry of antherozoids. The disintegrated mass of these cells contains malic acid, which attracts freely swimming antherozoids towards the neck of the archegonium. Although many antherozoids move towards an archegonium, only one of them swims into it and fuses with the egg. Thus zygote is formed. Usually more than one archegonia are fertilized in a prothallus. In E. debile as many as 15 young embryos have been observed in a single prothallus.

Embryo Development

The diploid zygote, formed by the fusion of the male and female gametes, is the mother cell of the next sporophyte generation. It develops into an embryo, which forms a new sporophyte.

The embryogeny of Equisetum shows some variations in different species. In most of the species, the first division of the zygote is transverse (i.e., at right angles to the neck of the archegonium), establishing an epibasal and a hypobasal cell. Both of these cells contribute to embryo. Unlike Selaginella and Lycopodium, no suspensor is formed in Equisetum. The second division of the zygote is vertical and as the result a four-celled embryo is established (Fig.). All the four cells of the embryo are of different shape and size. In E. arvense, the 4-celled embryo divides by a second vertical division at right angles to the first one, thus forming eight cells arranged in two tiers of four each. The largest of the four epihasal cells differentiates as apical cell of the shoot and the remaining three cells form leaf initials. The largest of the hypobasal cells functions as root initial and the remaining three cells give rise to the foot (Fig.).

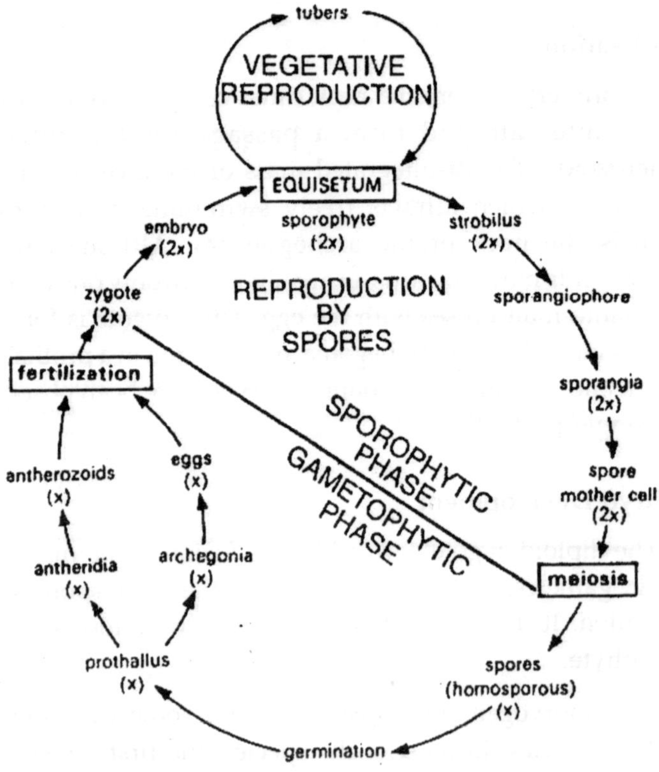

Equisetum : Life-cycle (diagrammatic representation).

Active divisions in the apical cell of the shoot and the root initial results in the establishment of the shoot and root. The embryonic root penetrates the prothallus and enters the soil (Fig.). The first shoot, called primary stem, bursts through the neck of the archegonium and grows upright. It is differentiated into nodes and internodes. Each node of the primary stem has only three leaves.

The primary stem shows only limited growth. Its apical growth ceases after 10-15 internodes have been formed. Then a secondary branch develops at the base of the primary stem from a bud. It has 1-5 leaves at each node. The secondary branch also has limited growth and is replaced by another upright branch arising from its base. This process may be repeated until 3-4 shoots have appeared and finally, a shoot, instead of growing in usual upright position, grows horizontally and penetrates the ground forming the first underground rhizome. It is perennial and gives out new aerial shoot regularly.

The Sphenophyllales

The order Sphenophyllales includes herbaceous fossil sphenopsids. These plants first appeared in the Upper Devonian period and flourished until the lower Triassic. The sporophyte had longitudinally ridged stems, differentiated into nodes and internodes. At each node, a whorl of wedge-shaped leaves was present. The primary vascular cylinder of the stem was an actinostele. Secondary vascular tissues were also formed.

The strobilus was made up of whorls of fertile appendages, each of which was differentiated into a sporangiophore and a bract-like portion. Most of the Sphenophyllales were homosporous. The order includes two families - Sphenophyllaceae and Cheirostrobaceae. Here only Sphenophyllaceae is described as representative of the order.

Some common genera of the Sphenophyllaceae are Sphenophyllum, Sphenophyllostachys (= Bowmanites) and Sphenostrobus. In India, the family is represented by Sphenophyllum.

Sphenophyllum

Systematic Position

Sub-division	—	Sphenopsida
Order	—	Sphenophyllales
Family	—	Sphenophyllaceae

The generic name Sphenophyllum is applied to the impression-compression specimens of stems, roots, leaves, and some strobili (that have been found in organic attachment) of the Carboniferous period. S. speciosum is the most common Indian species, reported from Barker stage in Bokaro and other coal-fields of Jharkhand, Bengal, Satpura basin and Sikkim.

External Structure : The plants were small and probably herbaceous with a climbing or erect stem (Fig.). The stem was joined with non-alternating ribs. It was very delicate and despite secondary thickening, seldom exceeded 1 cm in diameter. It was irregularly branched, the lateral branches arising at nodes in positions alternating with the leaves. Sometimes, the branches were dichotomous. The roots were adventitious and were borne along the internodes. They were rarely produced on lateral branches.

A whorl of 6-18 leaves was present on each node. The leaf was constricted at the base and expanded distally. It was supplied by a single trace that dichotomised within the lamina (Fig.). The leaves were less than 2 cm in length and showed a wide range of form. In S. emarginatum and S. majus, they were entire and wedge-shaped; in S. myriophyllum, they were dichotomously branched into several equal lobes (Fig.) and

S. cuneifolium was heterophyllous bearing both entire, and wedge-shaped and deeply dissected leaves (Fig.). The anatomy of the leaf suggested mesomorphic to xeromorphic nature.

Internal Structure : The stems were protostelic with a solid, exarch primary xylem, which was 3-angled in a cross section (Fig.). At the tip of each arm protoxylem groups were present. Protoxylem canals were observed in species like S. insigne and S. plurifoliatum. The primary xylem consisted entirely of tracheids. The secondary xylem formed a thick sheath around this primary core. It was dimorphic in the sense that the tracheids opposite the protoxylem poles were comparatively smaller than the tracheids on the intermediate radii. It consisted of pitted tracheids and intervening parenchyma. Usually the pitting was of multiseriate bordered type and the pits were present on the lateral walls. The rays were conspicuous and closely placed. However, in S. plurifoliatum, the rays were unusual consisting of horizontally aligned cells abutting against vertically extending parenchyma at the corners of tracheids.

The primary phloem consisted of thin-walled sieve elements. The sieve cells were axially elongated with horizontal to slightly tapered end walls. The secondary phloem consisted of vertical series of sieve elements with horizontal end walls. The sieve elements were of relatively small diameter in the areas opposite the protoxylem arms.

A massive cortex was present outside the stele that was in part sclerenchymatous. The stem- produced typical periderm which was deep-seated in origin and thus resulted in the eventual loss of all external primary tissues. A vascular cambium originated first in the bays between the xylem arms and then it became continuous in older stems. The structure of leaf shows mesomorphic to xeromorphic features, although these plants occurred within coal swamps. The chlorenchyma was reduced and the dermal and hypodermal tissues were sclerenchymatous (Fig.).

Sphenophyllum is an important taxon of the palaeozoic flora. It shows affinities with the Equisetales in their joined ridged stem, whorled leaves, and the arrangement of sporangia. But it differs from them in having broader, more complex and 3-merous plan of leaves, simple protostelic central cylinder and more complex sporophylls. In the structure of sporophyll and leaf and dichotomous venation, however, it resembles with the Hyeniales.

Sphenophyllostachys

Systematic Position

Subdivision	—	Sphenopsida
Order	—	Sphenophyllales
Family	—	Sphenophyllaceae

The form-genus Sphenophyllostachys refers to the cones of Sphenophyllum. Bourean (1964) reported some 31 species of Sphenophyllostachys; of which only three were found in organic connection with Sphenophyllum. Other species have been placed in Sphenophyllostachys on the basis of their general similarity. All members have whorls of free or laterally-fused bracts and whorls of sporangiophores that were either free' or partly fused with bracts. Some common species of Sphenophyllostachys are described here briefly.

Sphenophyllostachys Dawsoni : These cones, found on the stems, were up to 12 cm long and 2 cm in diameter. A whorl of 14 to 20 bracts was present on the cone and their bases were fused into a cup but their distal ends were free (Fig.). In the axil of each bract there was a sporangiophore which was partially fused with the bract. The sporangiophores were branched, each branch terminated in a recurved sporangium (Fig.).

Sphenophyllostachys Fertilis : These cones were not found in organic connection with the stem but because of the triarch wood in the axis, they were presumed to belong to Sphenophyllales. These were .complex cones upto 6 cm long and 2.5 cm in diameter. Each cone consisted of many superimposed whorls of sporangiophores (Fig.). In a whorl there were six sporangiophores, each subtended by a single bract bifurcated at its base. The sporangium was a compound structure which terminated into a cluster of 16 branches, each bearing a pair of recurved sporangia.

Sphenopsida Calamitales : Calamites

Order Calamitales : Members of the order Calamitales originated in the upper Devonian. They were most abundant during the Carboniferous period and became extinct early in the Triassic. They include perhaps both herbaceous and arborescent forms but most of these were trees. They were rhizomatous and the rhizome was very much the tree-trunk. The stem bore whorls of leaves at nodes. The fertile regions of the branches bore superimposed whorls of peltate sporangiophores, each bearing four reflexed sporangia.

Calamitales is the best known group of the fossil sphenopsids and has many interesting features. Most of the described specimens are fragmentary and were described under following generic names.

Asterophyllites	small
Annularia	stems and foliage
Dicalamophylluin	(as impressions and compressions)
Calamites stems	(impressions and pith casts)
Asteromycelon	roots
Palaeostachya	
Calamostachya	sporangiate
Cingularia	fructifications

Calamites is being described as a natural genus hereunder.

Calamites

Systematic Position

Sub-division	—	Sphenopsida
Order	—	Calamitales
Family	—	Calamitaceae

The generic name Calamites was initially given to pith casts of stems and branches of palaeozoic times, but subsequently the entire arborescent sphenopsids were referred as Calamites.

External Structure : Calamites was a large and highly branched tree. Though it is difficult to make an estimation of its height because of the fragmentary nature of the fossils, it appears that it attained a height of 20 to 30 m. The internal diameter of the stem was 40 cm. The plant was differentiated into underground rhizome, roots and aerial shoot (Fig.).

The underground rhizome was horizontal and differentiated into nodes and internodes. At each node whorls of adventitious roots were present. The arborescent aerial shoot of the plant had a central axis bearing lateral branches. The central axis developed from the buds present at the nodes of the horizontally growing rhizome. The aerial shoot had a conspicuous articulated structure with nodes and internodes. The internodes were longitudinally ribbed and in many specimens the ribs of the successive internodes were alternate as in the living genus Equisetum. In some specimens the ribs, however, continued straight across the nodes, i.e., without alternating at successive internodes.

Calamitean plants showed some variations in their branching pattern and on this basis three sub-genera - Eucalamites, Calamitina and Stylocalamites have been recognized. In Eucalamites, branches were present at almost all nodes and they were usually in pairs. In Calamitina, nodes with whorls of branches were separated by a series of branchless nodes and internodes. In Stylocalamites, there were usually no branches or only few irregularly scattered branches were present. The leaves were borne in a whorl at each node of the stem. The whorls at successive nodes were alternate. Each whorl had 4-60 leaves. They were usually free at the base but in some specimens they showed some degree of fusion into a sheath. The leaves of Calamites were placed in two form genera- Annularia and Asterophyllites, on the basis of their shape. Annularia was upto several centimeters long, spathulate or deltoid (Fig.) while Asterophyllites was very small and linear (Fig.). In the latter, stomata were confined to the adaxial surface, suggesting that the branches bearing leaves were perhaps pendulous. The leaves in both form genera had only a single midvein.

Internal Structure : Pith casts of Calamites showed ridges and grooves, corresponding in number with the protoxylem strands. A central pith cavity was present in internodes but the pith was solid at nodes forming a diaphragm. A ring of primary collateral vascular bundles was present around the pith. A carinal canal, representing the protoxylem, was present in each vascular bundle (Fig.). The metaxylem was centrifugal and consisted of scalariform and pitted tracheids. The secondary xylem was composed of radial rows of tracheids with scalariform pittings or with multiseriate circular bordered pits on the radial walls. Interfascicular rays traversed the secondary xylem. A number of leaf traces diverged from the main vascular cylinder. The cortex was composed of thin- walled parenchymatous cells. The older branches were surrounded by a thick and smooth periderm.

Fructifications

Several types of strobili have been found on the preserved remains of the calamitean type of plant body. Calamostachys, Palaeostachys and Cingularia are some common cone genera of Calamites. They are briefly described below.

The calamite cone, Calamostachys, was compact, measuring 3-4 cm in length and up to 7.5 mm in diameter. It was composed of alternating whorls of sterile bracts and sporangiophores (Fig.). There were 12 sterile bracts in a whorl which were partly fused at the base, whereas there were only 6 sporangiophores in a whorl (Fig.). The sporangiophores were placed at right angles to the cone axis and inserted midway between the two whorls of sterile bracts. The distal part of the sporangiophore was extended into a peltate disc which bore four sporangia on its undersurface. Most of the species of Calamostachys (e.g., C. casheana, C. americana) were heterosporous with microsporangia in one part of the cone and megasporangia in the other part. A few cones were distinctly homosporous.

Palaeostachys was similar to Calamostachys in general appearance but its sporangiophores were placed at an angle of about 45° and thus were axillary to the bracts (Fig.). The bracts in a whorl were usually twice the number of sporangiophores in the whorl. The most interesting feature of these cones was the course of the vascular bundle supplying the sporangiophore; it traversed the cortex of the cone axis almost midway between the two successive whorls of bracts and then turned downward forming a loop before entering the stalk of the sporangiophore.

The structure of Cingularia was markedly different from those of Calamostachys and Palaeostachys. The sporangiophores of Cingularia were associated with the whorl of bracts above and not with the whorl below as in Calamostachys and Palaeostachys. There were four pendent sporangia on each sporangiophore. The two other cone genera of these fructifications, viz., Metacalamostachys and Mazostachys also had an almost similar structure, except that in Metacalamostachys one and in Mazostachys two sporangia were present on each sporangiophore.

8

The Selaginellales

The order Selaginellales includes a single family Selaginellaceae, represented by only two genera—Selaginella and Selaginellites. Of these, the former is living and the later is a fossil. The sporophyte' is herbaceous and usually lacks secondary thickenings. The leaves are microphyllous and ligulate. In creeping forms, roots are borne at the distal end of leafless branches, called rhizophores. Sporophylls aggregate to form terminal strobili. They are heterosporous; the mega- and microsporangia are borne on the same strobilus. The development of male and female gametophytes begins while spores are still within the sporangium.

Selaginella is described here in detail.

Selaginella

Systematic Position

Sub-division	—	Lycopsida
Order	—	Selaginellales
Family	—	Selaginellacene

There are about 700 species of Selaginella, cosmopolitan in distribution. They are commonly known as spike moss or small club moss. Most of the species inhabit damp and shaded forests of tropics, but some (e.g., S. densa, S. rupestris, S. lepidophylla) grow in xerophytic habitats, such as exposed rock surfaces. S. oregana is an epiphyte that grows on tree trunks in tropical rain forests. Several species of Selaginella are grown in gardens as ornamentals. Some xerophytic species of Selaginella (e.g.,S. lepidophylla, S. pilifera) show cespitose habit; they curl and become ball-like during dry season and again become green and fresh when moisture is available. These are called resurrection plants.

The genus is represented in India by more than 70 species. Among these species, Selaginella kraussiana, S. monospora, S. biformis, S. rupestris, S. megaphylla, S. bryopteris, S. ciliaris, S. chrysorhizos and S. pentagona are common.

Sporophyte

External Morphology : The plant body of Selaginella is sporophytec. There is considerable variations in shape-, symmetry and morphology of plants in various species. They are nearly flat and creeping (e.g., S. kraussiana, S. apoda; Fig.), sub-erect with upright moss-like branches (e.g., S. trachyphylla, S. rupestris) or upright with several centimeters long branches (e.g., S. erythropus, S. selaginoides, S. viridangula; Fig.). Species like S. alligans are climbers and they possess special discoid pads on their rhizophores for climbing. Most species are perennial but some like S. pygmaea and S. gracilina are annual. The sporophytes vary from few centimeters (S. selaginoides) to 20 meters (S. willdenovii) in length.

Selaginella has been divided into the following two sub-genera on the basis of the structure of sporophytic plant body.

(a) *Homoeophyllum* : The species belonging to this sub-genus generally have erect and dichotomously branched stem. The leaves are isophyllous (i.e., of the same size) and spirally arranged. This sub-genus includes about 50 species, which include S. rupestris, S. pygmaea, S. selaginoides and S. oregana.

(b) *Heterophyllum :* Most of the species of Selaginella belong to this sub-genus. They usually have prostrate and dorsiventral stem with erect branches; the branching is always irregular. The foliage leaves are arranged in four vertical rows on the stem-two dorsal rows have small leaves and the two ventral rows, large leaves (Fig.). S. kraussiana, S. lepidophylla and S. chrysocaulos are some common representatives of this sub-genus.

The sporophytic plant body of Selaginella is differentiated into root, stem and leaves. Besides, some species also have rhizophores.

1. *Roots :* Primary roots are ephemeral and the adult plant has adventitious roots. The adventitious roots usually have specific locations in relation to stem dichotomies. In most of the creeping species with dorsiventral stems (e.g., S. kraussiana, S. laevigata) roots arise at or close to the point of dichotomy; in species like S. rupestris and S.wallichii they arise at the point of dichotomy as well as at other positions and in S.selaginoides and S. spinulosa they arise from knot-like swellings present at the basal portion of the stem.

   The roots arise endogenously and are dichotomously branched. The dichotomies are at right angles to each other.

2. *Rhizophore :* In some species of Selaginella many long, cylindrical, unbranched and leafless structures arise from the lower side of the stem at the point of dichotomy. These grow vertically downward and bear tuft of adventitious roots at their distal end. They are

known as rhizophores. The rhizophore may develop into a typical leafy shoot under certain conditions.

3. *Stem :* The stem is erect and dichotomously branched in the sub-genus Homoeophyllum, and prostrate or sub-erect with lateral branching in the sub-genus Heterophyllum.

   The stem apex usually has a single well-defined apical cell, but in S. oregana a group of meristematic cells has been observed.

4. *Leaves :* The leaves of Selaginella are microphyllous, sessile and simple. Their shape varies from ovate to lanceolate. The leaf has a single midvein that remains unbranched throughout its course. Most of the species have thin and soft leaves, but in xerophytic species they are thick. The vegetative leaf as well as sporophyll has a small, membranous projection on its adaxial (upper) surface, close to the base. This projection is known as ligule (Fig.). The basal part of the ligule has a distinct hemispherical foot-like structure, called glossopodium. It is composed of highly vacuolated thin-walled tubular cells. The ligule is embedded at the base of the leaf in a pit-like structure known as ligular pit (Fig.). The projected part of the ligule is only one cell in thickness and is tongue-like (e.g., S.svogelli), lobed (e.g., S. caulescens) or fan-shaped (e.g., S. martensii). It develops precociously and matures long before its associated leaf. Although the definite function of ligule is not known, it has been suggested that in some way they are associated with water absorption and secretion and thus prevent desiccation of the shoot. Some consider that the ligules in Selaginella are concerned with upward movement of inorganic solutes and thus compensate for smaller and less effective leaf primordia.

In the sub-genus Homoeophyllum all leaves are alike and spirally arranged. But the species belonging to the sub-genus Heterophyllum have two types of leaves - two dorsal rows of small leaves (microphylls) and two ventral rows of large leaves (megaphylls). The leaves occur in pairs and the two leaves of a pair are always unequal.

Internal Structure

1. *Root* : A cross section of the root shows a simple structure (Fig.). The epidermis is made up of tangentially elongated cells. In exposed roots, the outer wall of epidermal cells is cutinized but in roots that penetrate the substratum, the epidermal cells are delicate and have root hairs.

   The cortex is usually homogeneous consisting of many layers of thin walled parenchymatous cells. But in some species the outer layers of the cortex become thick walled and form hypodermis. In S. selaginoides the parenchymatous cells of the cortex show mycorrhizal association. The innermost layer of the cortex forms endodermis. In species like S. densa and S. rubella endodermis is fairly distinct.

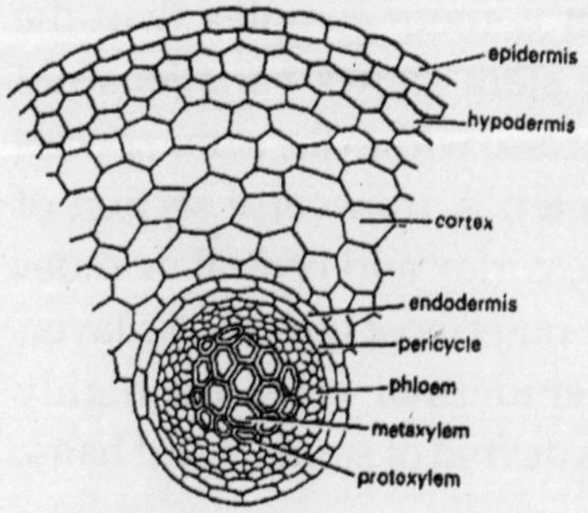

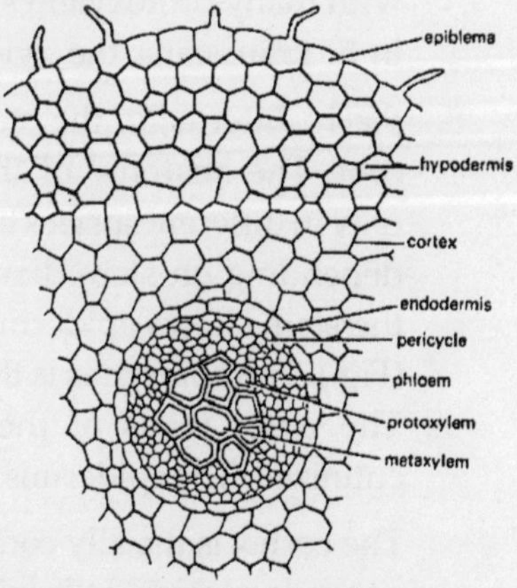

Selaginella : Transverse section of root. Selaginella : Transverse section of rhizophore.

The central part of the root is occupied by a protostele, surrounded by 1-3 layers of parenchymatous pericycle. The xylem, which forms the central solid core of the stele, is monarch to tetrach and exarch. The phloem occurs in the form of a ring around the xylem.

2. *Rhizophore* : The structure of the rhizophore resembles the root (Fig.). Some variations in the internal organization are due to the fact that the rhizophore is an aerial structure, whereas the root is a subterranean organ. The epidermis is single layered and the outer wall of the epidermal cells is covered with a thick layer of cuticle. Root hairs, characteristic of roots, are absent on rhizophores.

   The cortex is differentiated into an outer sclerenchymatous and an inner relatively wide parenchymatous zone. The innermost layer of the cortex forms endodermis.

   The protostele of the rhizophore is surrounded by a parenchymatous pericycle. Usually the stele is monarch and exarch, but shows some variations. For example, in S. afro- viridis the metaxylem is crescent-shaped with many protoxylem strands on its concave side and in S. kraussiana the xylem is centrifugal.

3. *Stem* : Internally, the stem is more complex than the root. The anatomy of the stem shows variations not only in different species but also within the same species depending on stem diameter. A transverse section of the stem shows epidermis, cortex and central cylinder (Fig.). The epidermis is the outermost unistratose layer. The outer walls of the epidermal cells are highly cutinized. The epidermis is devoid of stomata and hairs.

   The cortex is usually composed of compactly arranged parenchymatous cells without intercellular spaces. But in mature stems of many species outer layers of cortex become partially sclerenchymatous, forming a tough hypodermis. In xerophytic species (e.g., S. rupestris, S. lepidophylla), most part of the cortex is composed of thick-walled cells.

A distinctive feature of Selaginella stem is the presence of radially elongated endodermal cells called trabeculae. They have characteristic casparian bands on their lateral walls. Due to the presence of trabeculae, the central stele is separated from the cortex by large air spaces. In fact, in a transverse section the stele appears suspended in an axial air column with the help of trabeculae. The air spaces develop due to more rapid expansion of the cortical region than the stele. This differential growth also results in radial stretching of some endodermal cells. Xerophytic species of Selaginella, however, do not have trabeculae.

The number of steles in the stem shows considerable variation in different species of Selaginella. For example, the stem is monostelic in S. spinulosa and S. flabellata, distelic in S. kraussiana and polystelic (with 12-16 steles) in S. laevigata. Besides, the number of steles may also vary within different parts of the same plant. For example, the creeping branches of S. braunii are distelic, whereas the erect branches are monostelic, and in S. lyalli the creeping branches are distelic and the erect branches are polystelic.

The stele is surrounded by a single-layered pericycle. The shape and structure of the stele is also variable. It is circular in S. kraussiana, and flat or ribbon-like in S. viridangula and S. vagelii. Most of the species have a protostele with a solid xylem core surrounded by phloem, but S. laevigata var. lyalli has a siphonostele. The xylem is usually monarch (S. kraussiana) or diarch (S. chrysocaulos, S. oregana) and exarch (mesarch in S. selagincides). It usually consists of only tracheids; the protoxylem tracheids have annular or helical thickenings; whereas the metaxylem tracheids show scalariform thickenings. In S. oregano, S. densa and S. rupestris, however, the xylem has true vessels with transverse perforation plates. Although secondary growth is absent, some secondary xylem elements have been found in the basal part of the stem of S.selaginoides.

4. *Leaf* : Both, the upper and lower epidermis of the leaf are unistratose (Fig.). The epidermal cells have chloroplasts. The leaves are mostly amphistomatic but sometimes they are hypostomatic as in S. martensii. Stomata are distributed mostly in the midrib region.

The mesophyll consists of loosely arranged thin walled cells with many small or large intercellular spaces. It is usually made up of only spongy parenchyma, but occasionally a distinct palisade layer may be present towards the morphological upper side. A mesophyll cell has 1-8 cup-shaped chloroplasts, which have many spindle-shaped pyrenoid-like bodies.

The leaf has a median vascular bundle surrounded by a distinct bundle sheath. The xylem, which occupies the central part of the bundle, consists of only tracheids with annular or spiral thickenings. It is surrounded by phloem.

Morphological Nature of Rhizophore : Morphological nature of the rhizophore is debated because of its unusual position and structure. It is considered as a root or a stem. Bower (1908, 1935), Goebel (1905), etc. however, consider it a structure intermediate between stem and root. They regard it as an organ sui-generis. Arguments put forward in support of the above mentioned views are presented below.

Rhizophore is a Root : Van Tieghem and Harvey-Gibson (1902), Uphof (1920) Webster and Steeves (1962) and Wochok and Sussex (1974) consider that rhizophore is a root, because of the following reasons.

(1) It is positively geotropic.

(2) It does not bear roots.

(3) The internal structure of rhizophore is similar to that of a root, i.e., it is monostelic like the root.

(4) A distinct root cap is shown to be present at the apex of the rhizophore (Webster and Steeves, 1967).

(5) Auxin transport in rhizophore is similar to roots (Wochok and Sussex, 1974).

Rhizophore is a Stem : Bruchmann (1905), Worsdell (1910) and Cusick (1954) consider that rhizophore is more like a stem than root due to the following facts.

(1) Root cap and root hairs are absent in the rhizophore.

(2) The rhizophore arises exogenously, whereas roots arise endogenously.

(3) It grows by means of an angle meristem, and according to Cusick (1954), angle meristem is an embryonic shoot.

(4) Under controlled experimental conditions, the rhizophore can be induced to grow into a shoot (Foster and Gifford, 1959).

It has been now conclusively established by Steeves (1967) and Sussex (1974) that rhizophore is of root nature. It probably functions as additional root and provides mechanical support to the weak stem.

Reproduction

The sporophyte of Selaginella reproduces vegetatively and by spores.

Vegetative Reproduction : Vegetative propagation in Selaginella takes place by tubers, bulbils, dormant buds and by fragmentation.

In S. rupestris, prostrate branches produce roots during

favourable conditions. These root- bearing prostrate branches separate from the parent plant and grow into new sporophytes.

Species like S. chrysorhizos and S. chrysocaulos propagate with the help of tubers and bulbils. The tubers may be aerial, developing at the apices of aerial branches (e.g., S. chrysocaulos) or subterranean (e.g., S. chrysorhizos, Fig.). During favourable conditions the tuber germinates into a new sporophyte. Aerial branches of S. chrysocaulos also bear some dormant (resting) buds which grow into new plants during favourable conditions (Fig.).

Reproduction by Spores : Selaginella is a heterosporous pteridophyte ; it produces two types of spores - megaspores and microspores. The megaspores form female gametophytes on germination and the microspores give rise to male gametophytes. The sporangia are strictly dimorphic, i.e., micro- and megaspores are formed in separate sporangia. The sporangia-bearing microspores are called microsporangia and those bearing megaspores as megasporangia. There are many microspores in a microsporangium but a megasporangium usually has 1-4 (or rarely more) megaspores. The megaspores are much larger than the microspores:

The sporangia are borne singly in the axils of sporophylls. The sporophyll-bearing microsporangium is called microsporophyll and the one with megasporangium is known as megasporophyll (Fig.). The sporophylls are spirally arranged around a central axis.

1. *Strobilus or Cone :* In most of the species of Selaginella sporophylls are aggregated at the apex of the main stem and its branches in definite loose or compact cones, called strobili (singular = strobilus; Fig.). The size of the strobilus varies from 5 μm to 6-7 cm. It is often inconspicuous due to its small size and similarity between sporophylls and vegetative leaves. Usually a branch terminates in strobilus, but in species like

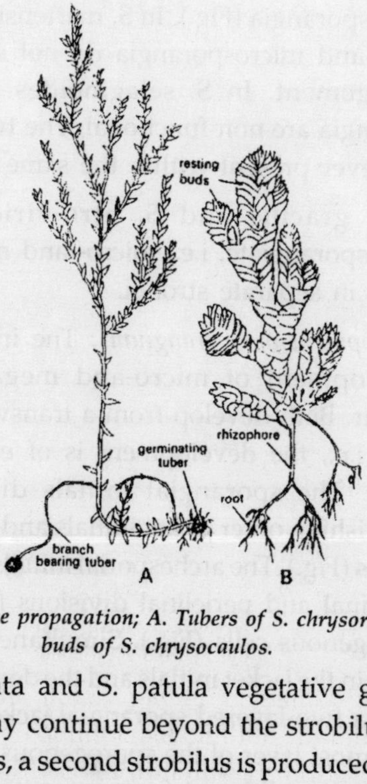

Selaginella : Vegetative propagation; A. Tubers of S. chrysorhizos, B. Resting buds of S. chrysocaulos.

S. cuspidata and S. patula vegetative growth of the branch may continue beyond the strobilus (Fig.). In S. erythropus, a second strobilus is produced on the fertile branch after an intervening vegetative region. Thus, in this species sporophylls and vegetative leaves occur in alternate segments.

2. *Distribution of Micro and Megasporangia in Strobilus* : In most of the species of Selaginella both micro-and megasporangia are found within the same strobilus. Their distribution is specific. For example, in S. selaginoides, S. rupestris and S. helvetica mega-sporangia are present in the basal part and micro-sporangia in the upper part of the strobilus (Fig.), in S. kraussiana there is only a single megasporangium at the base of the strobilus and the rest are microsporangia (Fig.), and in S. inaequalifolia one side of the strobilus bears only megasporangia and the other

microsporangia (Fig.). In S. martensii and S.caulescens mega-and microsporangia do not show any definite arrangement. In S. selaginoides a series of basal sporangia are non-functional. The two types of spores are never present within the same sporangium.

In S. gracilis and S. atro-viridis, strobili are monosporangiate, i.e., micro- and megasporangia are borne in separate strobili.

3. *Development of Sporangium :* The initial stages of the development of micro-and megasporangium are similar. Both develop from a transverse row of initial cells, i.e., the development is of eusporangiate type (Fig.). The sporangial initials divide periclinally, establishing outer jacket initials and inner archesporial initials (Fig.). The archesporial initials undergo repeated anticlinal and periclinal divisions forming a mass of sporogenous cells (Fig.). Simultaneous divisions also occur in the jacket initials and the derivatives eventually form a two-layered sporangial jacket. The cells of the outermost layer of the sporogenous tissue (adjacent to the inner wall layer) form a nutritive layer, known as tapetum (Fig.). The tapetal layer disintegrates as spores mature.

   The last generation of sporogenous cells functions as spore mother cells. The micro- and megasporangium differ in subsequent development.

   (a) *Further Development of Microsporangium :* In microsporangium about 80-90% spore mother cells are functional and behave as microspore mother cells. The remaining spore mother cells degenerate and form a viscous nourishing fluid. The functional microspore mother cells undergo meiosis and form haploid microspores (Fig.), which are arranged in tetrahedral tetrads.

(b) *Further Development of Megasporangium* : In megasporangium all spore mother cells, but one, degenerate. The functional spore mother cell behaves as megaspore mother cell. It divides meiotically forming four tetrahedrally arranged haploid megaspores (Fig.). All the four megaspores derived from a megaspore mother cell may not always be functional. For example, in S. sulcata only one and in S. rupestris two megaspores are functional. Sometimes there are more than one megaspore mother cells in a megasporangium and in such cases the megasporangium has 8 or more megaspores. The megaspores are much larger than microspores (Fig.).

The expression of maleness or femaleness is not genetically determined, it appears to be influenced by the nutritional factor and the specific environment in which the sporangium develops.

4. *Mature Sporangium* : Mature sporangia are stalked structures with a two-layered sporangial jacket. The cells of the outer jacket layer are elongated and contain chloroplasts. The micro and megasporangia differ in shape, size and colour. The microsporangia are slightly elongated, yellow, red or orange in colour. The megasporangia are larger and paler and assume the shape dictated by the enlarging megaspores within.

The mature sporangium dehisces along the line of dehiscence present at its distal end and oriented transverse to the axis of the sporophyll. Structural modification of the surface cells along this line and at its flanks results in splitting of the distal part of the sporangium into two valves. The lower cup-shaped part of the sporangium shrinks on drying and throws out spores violently.

Gametophyte

The spore is the mother cell of the gametophytic generation. As mentioned earlier, Selaginella is heterosporous and produces two types of spores, the smaller microspores and the larger megaspores. This difference in the size of the spores is related to their fate and function; microspores develop into male gametophytes and megaspores into female gametophytes.

In Selaginella both microspores and megaspores begin to germinate while still inside the sporangium (i.e., they germinate in situ). Thus, spores are shed at multi-cellular stage.

Microspores and Development of Male Gametophyte

1. *Microspore :* The microspores are small spherical structures, ranging 0.015-0.06 mm in diameter. A microspore is surrounded by a thick ornamented exine and a relatively thin intine. The ornamentations in the exine may be papillate, echinulate or granulate (Fig.). The spore has a single haploid nucleus and granular cytoplasm, rich in fatty substances (about 35%). The fats probably provide food to the developing male gametophytes as spores contain no chlorophyll.
2. *Development of Male Gametophyte :* The microspores germinate inside the microsporangium and are shed at B-celled stage. The first division of the microspore (1-1) is asymmetrical and as a result a small lenticular prothallial cell and a large antheridial initial is established (Fig.). The prothallial cell does not divide further and the entire sporangium develops from the antheridial initial.

The first division of the antheridial initial (2-2) is nearly at right angles to the prothallial cell. It results in the formation of two antheridial cells of almost equal size (Fig.). Both these cells divide by a vertical wall (3-3, at right angles to the first division 2-2) to produce a group of four cells. Thus, at this stage the gametophyte consists of five cells (four cells derived from the antheridial initial and a prothallial cell; Fig.). The two basal cells, derived from the antheridial initial, do not divide further, whereas the upper two daughter cells divide repeatedly and form ten cells. At this stage the gametophyte has 13 cells (10 cells derived from the upper daughter cells of the antheridial initial, 2 basal daughter cells and 1 prothallial cell). Of these, four central cells function as primary androgonial cells and eight peripheral ones as jacket cells (Fig.). The male gametophyte is shed from the microsporangium at B-celled stage. It enters into partially opened megasporangium where further development of the male gametophyte takes place in close proximity of the developing female gametophyte. In some species, it is believed that further development of the male gametophyte takes place in the soil.

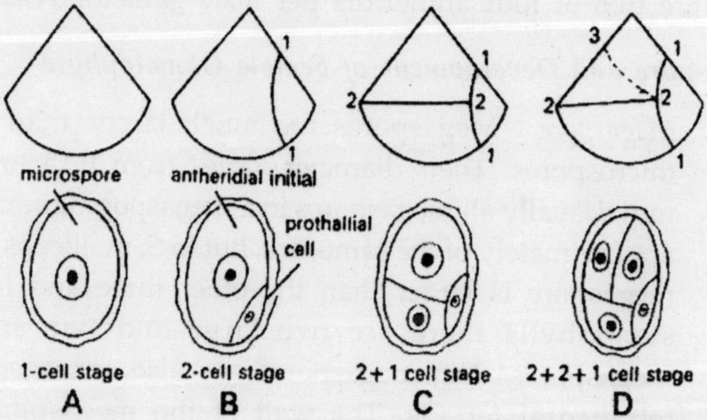

Selaginella* : *Stages in the development of male gametophyte.

The four central primary androgonial cells of the male gametophyte divide repeatedly forming a mass of 128-256 antherozoid mother cells or androcytes. Each androcyte metamorphoses into a spindle-shaped biflagellate antherozoid (Fig.). The antherozoids of Selaginella are perhaps the smallest amongst the vascular plant.

With the formation of antherozoids, the jacket cells decompose and form a mucilagenous substance. The antherozoids float in this substance. Until this stage the male gametophyte is completely enclosed within the wall of the microspore. Thus, it is entirely endosporic and extremely reduced structure. Unlike other pteridophytes, vegetative prothalli are not formed in Selaginella. The gametophyte is not set free and is dependent on the parent sporophyte for nutrition.

There appears to be some controversy regarding the number of antheridia per gametophyte. Belajeff, Pfeffer, Campbell, etc., believe that since each gametophyte has only one antheridial initial, there is only a single antheridium per gametangium. On the other hand, Bruchmann opined that as the antheridial initial divides to form two (e.g., S. cuspidata, S. caulescens) or four (e.g., S. kraussiana) antheridial cells, there are two or four antheridia per male gametophyte.

Megaspore and Development of Female Gametophyte

1. *Megaspore :* Megaspores are much larger than the microspores. Their diameter varies from 0.15 to 0.5 mm. Usually all megaspores in a megasporangium are approximately of the same size, but in S. molliceps one megaspore is larger than the other three and in S. stenophylla there are two large and two small megaspores. The megaspores are also arranged in tetrahedral tetrads. The wall of the megaspore is differentiated into an outer massive exine and an inner thin intine, but in S. rupestris and S. apus it is differentiated into three distinct layers the outer exospore, the middle mesospore and the inner endospore (Fig.). The megaspore has a single haploid nucleus, surrounded by granular cytoplasm, rich in fatty substances (about 48%).
2. *Development of Female Gametophyte :* Like male gametophyte, the development of the female gametophyte of Selaginella also begins while it is still

within the megasporangium. In S. kraussiana, the gametophyte is liberated from the megasporangium after the first archegonium is differentiated, whereas in S. rupestris and S. apus it is retained in the megasporangium even after the development of embryo has started. However, in S. spinulosa and S. helvetica the development of female gametophyte starts only after the megaspore is shed from the sporangium.

Immediately before the development of female gametophyte initiates, a large vacuole appears in the centre of the megaspore and as a result the cytoplasm is pushed along the spore wall in the form of a thin membrane (Fig.). There is considerable enlargement of the megaspore. The outer spore wall (exospore) grows more rapidly than the mesospore and endospore, consequently a large gap is formed in between the exospore and mesospore (Fig.). At this stage, the exospore is attached to the mesospore only at one point. The space between the exospore and mesospore is filled with a homogeneous liquid.

The haploid nucleus of the megaspore divides repeatedly without any wall formation. The free nuclei are unequally distributed in the peripheral cytoplasm; they are clustered beneath the triradiate ridge of the spore and sparsely distributed elsewhere. Now, wall formation begins in the apical region and a lens-shaped pad of small cells is formed at the apical end. It is separated from the rest of the female gametophyte by a distinct diaphragm (Fig.). The cytoplasmic layer becomes thicker gradually and pushes the mesospore outward. As a result the mesospore again comes in contact with the exospore. With the increase in the amount of cytoplasm, the central vacuole diminishes and eventually disappears. The part of the gametophyte below the diaphragm is multinucleate in early stages but becomes multicellular as wall formation proceeds inward. At this stage, the spore wall ruptures along the triradiate ridge exposing the apical cellular pad (Fig.). The exposed part of the female gametophyte may develop chloroplasts

But the photosynthetic ability of this part is of limited importance as food for the developing embryo is stored in the lower multicellular part of the gametophyte. Many rhizoids develop from the exposed part of the gametophyte. They attach the gametophyte to the substratum and also help in absorption of water.

3. *Development of Archegonia :* Archegonia develop from the apical tissue of the gametophyte. All superficial cells of this tissue have the potential of forming archegonia. The archegonial initial divides periclinally into a primary cover cell and a central cell (Fig.). The primary cover cell divides by two vertical divisions at right angles to each other and forms four neck initials (Fig.). The neck initials divide transversely so as to form eight neck cells, arranged in two tiers of four each. In the meantime, the central cell divides by a periclinal wall and an outer primary neck cell and an inner primary venter cell is established (Fig.). The former does not divide further and directly functions as neck canal cell, whereas the latter divides transversely into a venter canal cell and an egg.

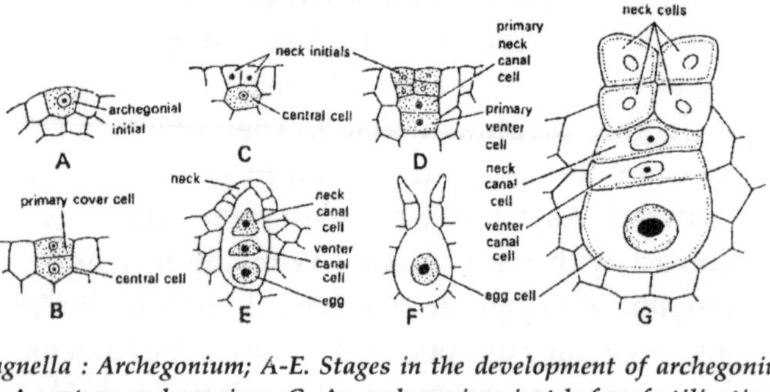

Selagnella : Archegonium; A-E. Stages in the development of archegonium, F. A mature archegonium, G. An archegonium just before fertilisation.

The mature archegonium of Selaginella has two cell long neck (consisting of eight cells in two tiers of four each), a neck canal cell, a venter canal cell and an egg (Fig.). The four terminal cells of the neck project beyond the surface of the gametophyte as asymmetric nipples. Rest of the archegonium remains embedded in the tissue of the gametophyte.

Fertilisation : Fertilisation usually takes place after the megasporangium has fallen on the soil, but in some species it may occur while the female gametophyte is still within the sporangium. Just before fertilisation, the neck cells of the archegonium separate from each other and form a passage for the entry of antherozoids (Fig.). After liberation from the male gametophyte, antherozoids swim in rain or dew water and reach the archegonia. Usually only one antherozoid enters into an archegonium and fuses with the egg to form a diploid zygote.

Some species of Selaginella (e.g.,S. rupestris, S. apoda) show seed habit. In these species, the sporangium has only a single megaspore and at maturity of the archegonium the spore wall ruptures, but the developing female gametophyte does not come out of the spore wall. The developing male gametophyte, when shed from the microsporangium (present in the distal part of the strobilus), lands on the partially open megasporangium. Thus, at this stage, both the male and female gametophytes lie within the megasporangium. As such fertilisation and embryo development take place inside the megasporangium. The sporangium is shed after the development of root and primary shoot of the new sporophyte. This feature is of considerable importance from the point of view of seed habit because when the megaspore with young sporophyte is shed, it has all typical characters of a seed.

Embryo Development

The diploid zygote is the mother cell of the sporophyte generation (Fig.). It divides transversely, establishing an epibasal (upper) suspensor cell and a hypobasal (lower) embryonic cell (Fig.). As development proceeds, the suspensor cell repeatedly divides to form a suspensor which pushes the developing embryo deep into the female gametophyte. The rest of the embryo develops from the embryonic cell. It divides by two vertical walls at right angles to each other and thus a four-celled embryo is formed (Fig.). One of the four cells of the embryo divides by an oblique vertical wall and thus an apical cell with three cutting faces is established (Fig.). This eventually functions as the apical cell of the embryonic shoot.

The remaining three cells of the 4-celled embryo and the sister cell of the apical cell (i.e., total four cells) divide transversely to form two tiers of four cells each. The cells of both the tiers divide irregularly forming a multicellular embryo. Usually the cells of the lower tier divide more rapidly than the upper tier and due to this differential growth the entire embryo apex rotates at 180° and emerges through the apical pad of the gametophyte. The derivatives of the lower tier form the foot. At first the foot grows on one side but eventually comes to lie opposite the suspensor (Fig.). The foot acts as a haustorial organ; its main function is to absorb nutrition for the developing sporophyte from the female gametophyte. At this stage, a superficial cell in each of the two diagonally opposed quadrants of the upper tier differentiates as the apical cell of a foliar appendage, which eventually forms a cotyledon. In the axil of each cotyledon a ligule develops.

The part of the embryo immediately posterior to cotyledons develops into hypocotyledonary part of the stem. The stem grows with the help of the apical cell of the embryo. After the formation of cotyledons and stem, the apical cell of the root differentiates on the lateral surface of the foot. The derivatives of this cell develop into a root-like structure, called rhizophore. Roots, infact, develop at the apex of the rhizophore. In early stages of development the young sporophyte is attached to the megaspore and derives its food from the female gametophyte with the help of its foot. But after the establishment of root and stem, the sporophyte becomes independent.

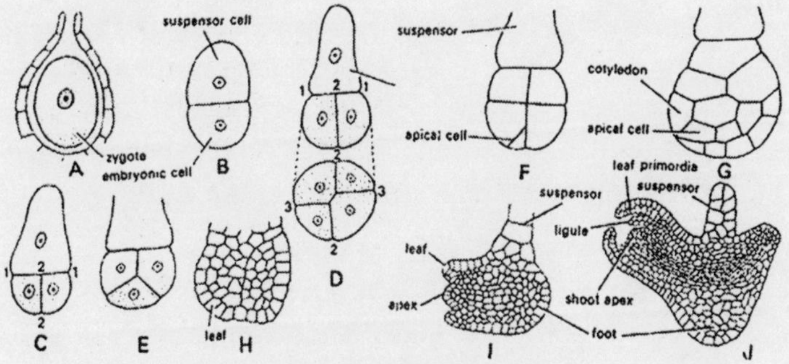

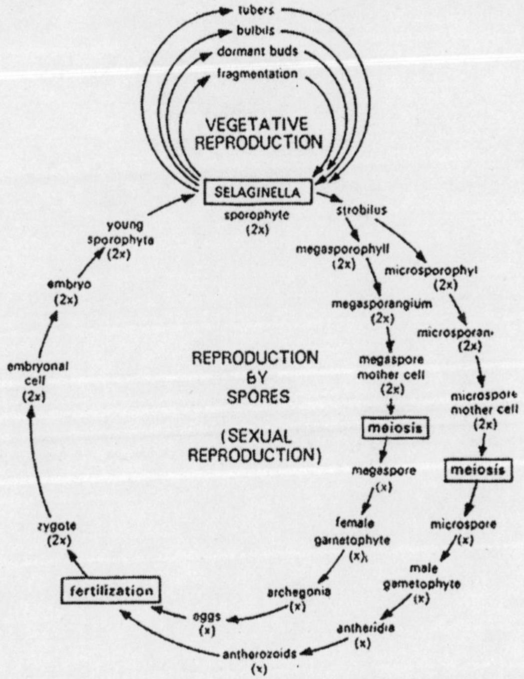

9

The Isoetales

The order Isoetales includes a single family Isoetaceae with two genera—Isoetes and Stylites. These are herbaceous lycopods, characterised by the presence of long awl-shaped microphyllous and ligulate leaves. The axis shows secondary growth which takes place by means of lateral and basal meristems. All leaves are potential sporophylls. Both genera are heterosporous and a sporophyll bears either a megasporangium or a microsporangium. The sporangium is completely embedded in the basal expanded part of the leaf. The sporophylls are not organized into cones. The development of gametophyte is endosporic.

The female gametophyte does not protrude out of the megaspore wall and lacks rhizoids. The embryo develops endoscopically ; the epibasal cells of the zygote form the foot but during ontogeny, due to unequal growth, it changes its orientation. In the following pages Isoetes is described as representative of the order.

Isoetes

Systematic Position

Sub-division	—	Lycopsida
Order	—	Isoetales
Family	—	Isoetaceae

Isoetes is represented by about 100 species, which are worldwide in distribution. They prefer a cool temperate climate and are less abundant in tropics. They are aquatic perennial herbs, occurring in shallow ponds, lakes and rivers either partly (e.g., I. engelmanni, I. coromandelina) or completely submerged (e.g., I. japonica). Some species are terrestrial (e.g., I. butleri). The genus is represented in India by nine species. I. coromandelina, an amphibious perennial, is the most common Indian species.

Sporophyte

External Morphology : The plant has general appearance of a monocot. It consists of a massive axis which bears a tuft of roots at the base and quill-like leaves on the upper part.

1. *Axis :* The axis is subterranean perennial part of the sporophyte. It is cylindrical when young, but it becomes two, three or four-lobed in the lower part as the plant matures. It is differentiated into an upper leaf bearing and a lower root bearing part. It is covered with dried leaf bases and roots of the previous year. The flakes of the dried tissue are sloughed off towards the end of the growing season. The axis is a peculiar organ of Isoetes and its morphological nature is debated. It has been variously interpreted as stock, stem, corm or rhizome. According to Eames (1936), the axis is made up of two parts : (i) stem—the upper leaf bearing part, and (ii) rhizomorph— the lower part which bears are

not evident externally (because of its condensed structure and thick cortex), external form gives no indication of the true morphological nature of the axis. Anatomical studies, particularly a three-dimensional view of stele indicates that the axis is differentiated into stem and rhizomorph.

2. *Leaves* : The leaves are quill-like, ranging from 2 to 60 cm (sometimes more) in length. They are borne in a close spiral phyllotaxy in an acropetal order. In young axis there are only 2 or 3 leaves around the apex (i.e., 1/2 or 1/3 phyllotaxy), but as the plant matures the phyllotaxy becomes 3/8, 5/13 or 8/21. The latter conditions are brought about by the condensation of more than one spirals. The leaf is ligulate, microphyllous, with a somewhat expanded base and a straight or slightly curved tapering apex. It is light green in colour but the lower portion usually lacks chlorophyll and is pink or white. A healthy plant may have 100 or more leaves. In terrestrial species the peripheral leaves die and decay at the end of the growing season leaving behind only a few immature leaves, whereas in aquatic species healthy leaves can be seen throughout the year.

The leaf of Isoetes is characterised by the presence of a triangular, cordate or bifid, and colourless ligule and a velum. The ligule is 2-15 mm in length. It is embedded in the tissue of the leaf just above the sporangium with the help of a lobed structure, called glossopodium. Any definite function of ligule is not known, but it is considered as an absorbing and secreting structure.

The velum is a small flap-like structure, present between the sporangium and the ligule. It shows much variation. It may have only a single lobe, which either extends downward and reaches nearly half the length of the sporangium (e.g., I. coromandelina) or stretches upward (e.g., I. malinvemiana, I. asiatica). It has two well developed lobes that cover the sporangium completely in I. sahyadrii and I. sampathkumarnii. Velum is absent in some species of Isoetes.

In Isoetes all leaves have the potential of bearing sporangia.

3. *Roots* : They arise on the lower lobed part of the axis in definite rows, radiating across each lobe from the central groove (in acropetal order). The younger roots are close to the central groove and the older ones are gradually pushed upward and outward. The roots are usually dichotomously branched and possess numerous root hairs. They remain attached to the axis for a long period or are sloughed off every year.

Internal Structure

1. *Axis* : The axis represents a condensed stem and rhizomorph. The internal structure of the mature axis is complex.

   (a) *Primary Structure* : Transverse section of a young axis shows a distinct epidermis, followed by a wide cortex. The outer layers of the cortex are collenchymatous, while the inner layers are parenchymatous with many intercellular spaces. The cortical cells are rich in starch grains. The central part of the axis is occupied by a protostele with a central core of xylem, surrounded by phloem. The xylem is not differentiated into protoxylem and metaxylem elements. The xylem tracheids have mostly spiral thickenings. A characteristic feature of the axis is the presence of a continuous parenchymatous sheath between the xylem and phloem.

The complete vascular cylinder of the axis can be seen in a longitudinal section. It is an anchor-shaped structure. The upper cylindrical part has 2, 3 or 4 lobes radiating from its base. Since the upper part gives off leaf traces, it represents the stem. The lower part gives out only root traces and hence it is rhizomorphic in nature.

(b) *Secondary Structure* : The secondary tissues are formed by a perennially active meristem, which extends over the entire surface of the stem except near the apical meristem. This meristem is abnormal in position as it develops outside the phloem (Fig.). It produces secondary parenchyma on the outer side and compactly arranged longitudinal cells on the inner side (thus abnormal in function). The secondary parenchyma constitutes the major bulk and it gets sloughed off every year as flakes of dead tissue. These cells eventually mature into elements of xylem, phloem and parenchyma. Such a mixed tissue is said to be prismatic tissue.

2. *Leaf* : In a transverse section the leaf appears quadrangular in outline with two expanded lateral wings on the lower side (Fig. A). The epidermis consists of a single layer of uniform cells covered by a thin layer of cuticle. Stomata are present in the epidermis, usually on the exposed parts of the leaf and are oriented along its long axis (Fig.).

Most of the interior of the leaf is occupied by four longitudinal air canals, separated from each other by transverse diaphragms. The mesophyll is undifferentiated and consists of thin walled parenchymatous cells with many chloroplasts and starch grains. There are only a few chloroplasts in the basal part of the leaf and hence this part is almost colourless.

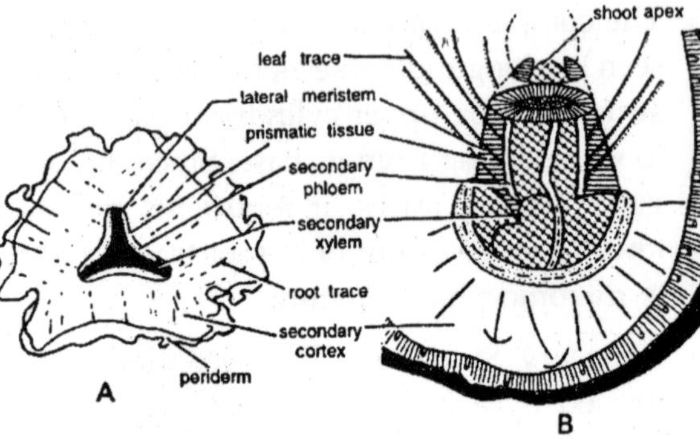

Isoetes* : *Secondary structure of axis: A. Transverse section of an old axis, B. A three -dimensional view of the longitudinal section of the axis.

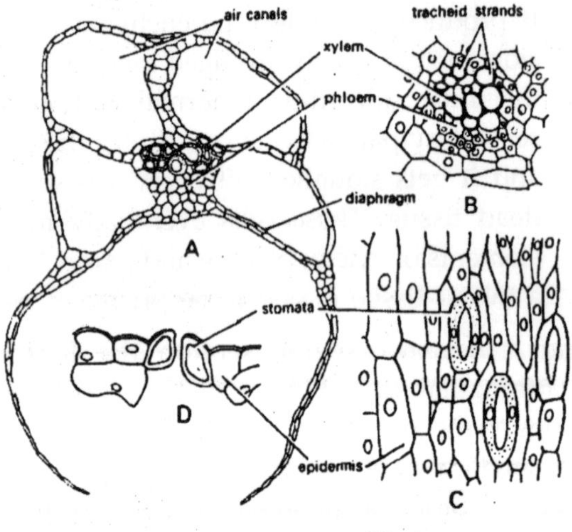

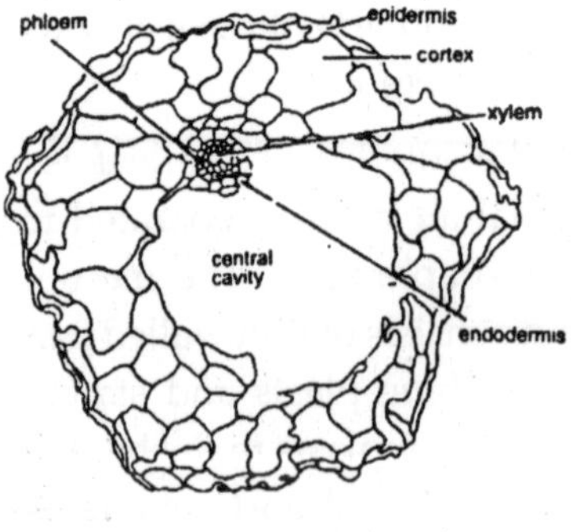

Isoetes* : *Transverse section of root.

A single collateral vascular bundle is present in the median part of the leaf. In the basal part xylem and phloem are compactly arranged but slightly upward (at the level of the sporangium) xylem tracheids break into small groups.

3. *Root :* The internal structure of the root is comparatively simple (Fig.). It has a single layer of persistent epidermis, consisting of compactly arranged thin walled cells. It is followed by 4-8 layers of a parenchymatous cortex. The cortical cells have conspicuous intercellular spaces. The centre of the root is occupied by a large central cavity, formed by the disintegration of cells in the inner part of the cortex. The stele is represented by a single vascular bundle, attached to the inner margins of the cortex at one point. The stele is surrounded by an endodermis. The xylem of the stele is on the side away from the central lacuna.

Growth

The shoot apical meristem of Isoetes is a dome-shaped protuberance, situated at the base of a conical depression. It is protected by a rosette of leaves. The shape of the apical protuberance varies at different stages of development. The cells of the superficial layer of the apex are columnar with dense cytoplasm. They divide anticlinally except at the site of leaf initiation where periclinal divisions occur. The inner homogeneous mass consists of vacuolated cells which divide in all planes.

Reproduction

Isoetes reproduces chiefly by spores. Occasionally, vegetative multiplication takes place by means of buds developing on the stem.

Spore Producing Organ

1. *Sporangia :* Isoetes is heterosporous, producing two types of spores—microspores and megaspores—in different sporangia. Sporangia are borne singly on a sporophyll (all leaves on a plant have the potential of bearing sporangia). There are two types of sporophylls— those bearing microsporangia are known as microsporophylls and those megasporangia as megasporophylls. The micro- and megasporophylls are irregularly distributed on the axis. In some species, however, megasporophylls are restricted to the periphery and the central part of the axis is occupied by microsporophylls. New sporophylls develop in every growing season and the sporophylls of the previous season fall off.

   The sporangium of Isoetes is the largest amongst living spore producing plants. It is oblong cake-like structures, measuring 3-7 mm in length. It is sunken on the adaxial (upper) face of the sporophyll in the expanded basal part just below the ligule. A small flap-like one or two-lobed velum is present in between the sporangium and the ligule, which covers the sporangium partly or completely. Internally, the sporangium is traversed by ribbon or plate-like trabeculae, but they do not subdivide the sporangium. Definite function of trabeculae is not known, but they probably help in extending the tapetal layer inside the sporogenous tissue thus making the tapetum more effective for nutrition.

2. *Development of Sporangium :* The early stages of development of micro- and megasporangium are similar. The sporangium develops very early in the ontogeny of leaf. The first sign of sporangial initiation is the differentiation of a group of embryonic epidermal

cells just below the velum. These cells are larger and densely cytoplasmic and function as sporangial initials. They divide periclinally forming an outer layer of jacket initials and an inner layer of archesporial cells (Fig.). The jacket initials divide both in anticlinal and periclinal planes and eventually give rise to 2-4-layered wail of the sporangium. The archesporial cells divide in all possible planes and the derivatives differentiate into spore mother cells (Fig.). Subsequently, some transverse bands of sporogenous tissue lose their protoplasm and cease to divide. These bands of sterile cells eventually form trabeculae (Fig.). The sterile cells bordering the trabeculae as well as those immediately inner to the wall layer form a nutritive layer, called tapetum.

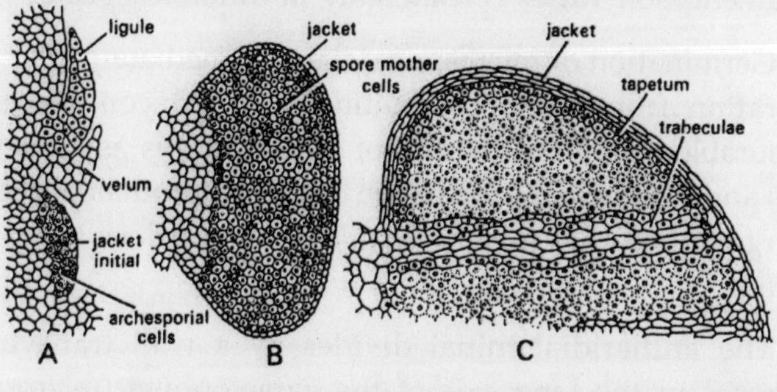

A-C. Isoetes : Vertical section of young leaf primordia showing development of sporangium.

In a sporangium, destined to form microspores, almost all sporogenous cells function as microspore mother cells. Each of these cells undergoes reduction division forming a tetrad of microspores. There is a very large number (1,30,000-2,50,000) of microspores in a sporangium. In the megasporangium most of the sporogenous cells degenerate and only a limited number of surviving sporogenous cells function as megaspore mother cells. They form 50-300 megaspores after reduction division. The degenerating cells help in the nutrition of the developing megaspores.

3. *Dehiscence of Sporangium* : In Isoetes sporangia do not have special mechanism for dehiscence. Spores are liberated only after the death and decay of the sporophyte.

Gametophyte

The microspores and megaspores on germination give rise to male and female gametophytes respectively. Both, the male and female gametophytes are endosporic, i.e., they develop inside the spore wall.

Microspores and Development of Male Gametophyte **:** The microspores are very small (30-45 μm in diameter), longitudinally ridged and light brown in colour. Their shape and ornamentation varies considerably in different species (Fig.).

Germination of microspores starts immediately after their liberation from the sporangium, provided conditions are favourable. The first division of microspore is asymmetrical (1-1) and results in the formation of a small prothallial cell and a large antheridial initial (Fig.). The former does not divide further.

The antheridial initial divides by a wall transversely diagonal to the long axis of the gametophyte (microspore, 2-2; Fig.). Of the two daughter cells thus formed, the one close to the prothallial cell becomes the first jacket cell of the antheridium, whereas the other cell divides in a plane perpendicular to the first transverse diagonal division (3-3;). The cell farthest from the prothallial cell becomes the second jacket cell and the other divides in the direction of the long axis of the gametophyte (4-4). The outer derivative of this cell forms the third jacket cell and the inner derivative divides periclinally (5-5), forming the fourth jacket cell and a central primary androgonial cell (Fig.). The latter by two successive divisions at right angles to each other forms four androcytes (Fig.).

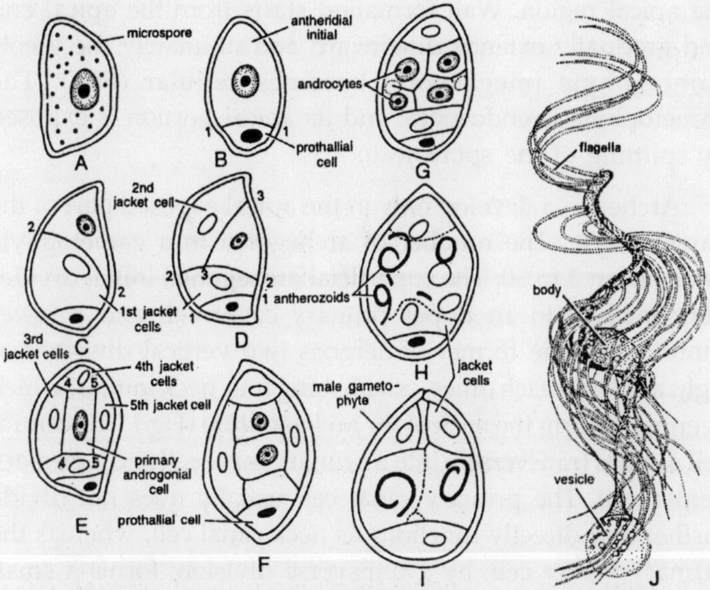

Isoetes* : *Development of male gametophyte; A-I. Successive stage in the development of male gametophyte, J. Antherozoid.

Thus the mature male gametophyte consists of nine cells—a prothallial cell, four jacket cells and four androcytes. Each androcyte metamorphoses into a multiflagellate antherozoid with a pronounced terminal vesicle (Fig.). The antherozoids are liberated through an irregular crack in the spore wall. The gametophyte is thus endosporic i.e., it develops within the spore wall.

Megaspore and Development of Female Gametophyte : The megaspores are white, grey or black and are much larger (250-900 µm in diameter) than the microspores. They are arranged in tetrahedral tetrads and each megaspore has a prominent triradiate ridge. The outer spore wall bears conspicuous spiculate projections.

On germination, the megaspore undergoes free nuclear divisions. About 30-50 nuclei thus formed are distributed along the periphery in the protoplast, but there are more nuclei in the apical region. Wall formation starts from the apical end and gradually extends downward and ultimately the whole gametophyte (megaspore) becomes cellular (Fig.). The gametophyte is endosporic and its apical portion is exposed by splitting of the spore wall.

Archegonia develop only in the apical exposed part of the gametophyte. The number of archegonia in a gametophyte varies from 2 to 30. The superficial archegonial initial divides periclinally into an upper primary cover cell and a lower central cell. The former undergoes two vertical divisions at right angles to each other, establishing four neck initials, which eventually form the neck of the archegonium (Fig.). The central cell divides transversely into a primary canal cell and a primary venter cell. The primary canal cell usually does not divide further and directly functions as neck canal cell, whereas the primary venter cell, by a transverse division, forms a small venter canal cell and an egg (Fig.).

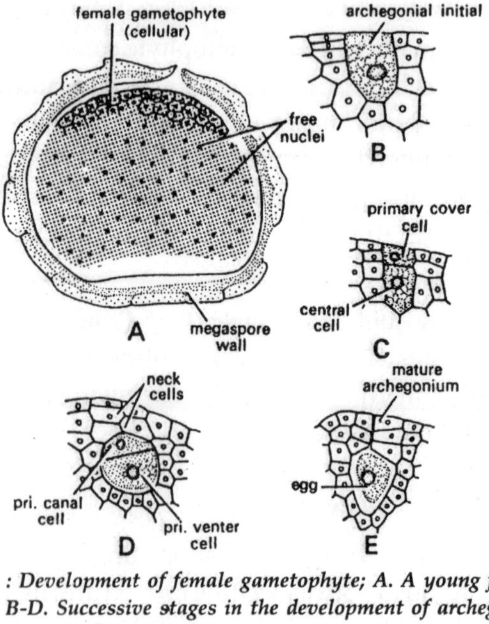

A-E. Isoetes : Development of female gametophyte; A. A young female gametophyte, B-D. Successive stages in the development of archegonium, E- A mature archegonium.

Thus, in the mature archegonium the neck is composed of four vertical rows of neck cells with four cells in each row. The neck canal is occupied by a single uni- or binucleate neck canal cell and a venter canal cell. The lower swollen part of the archegonium contains an egg. The archegonium is completely sunken and the adjoining cells of the gametophyte form a sterile jacket around it.

Embryo

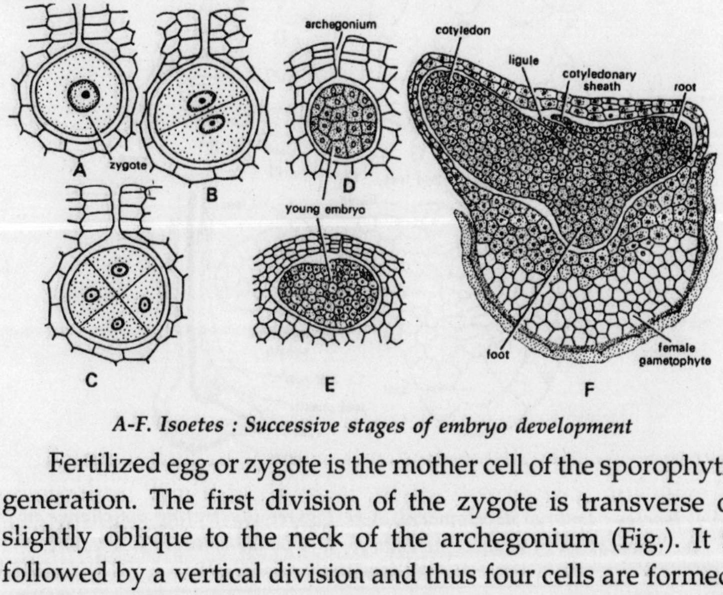

A-F. Isoetes : Successive stages of embryo development

Fertilized egg or zygote is the mother cell of the sporophytic generation. The first division of the zygote is transverse or slightly oblique to the neck of the archegonium (Fig.). It is followed by a vertical division and thus four cells are formed, which are arranged in two tiers of two each (Fig.). The cells of both the tiers contribute to the formation of embryo. There is no suspensor. The polarity of embryo is established at this stage; the two upper cells form the foot and one of the two basal cells forms the first embryonic leaf and the other basal cell, the root (Fig.). The apex of the embryo at this stage is thus directed inward (i.e., towards the base of the archegonium). This type of embryo development is known as endoscopic (Fig.). In subsequent development, the embryo grows unevenly and consequently the foot assumes an inner and the leaf and root an outer position (Fig.). As the growth of the embryo continues, the gametophytic tissue forms a cushion at its top. The young leaf, which grows much faster, comes out by piercing the gametophytic tissue. The latter forms a conspicuous sheath at the base of the leaf (Fig.). The root grows out through the gametophyte in the direction opposite to the leaf.

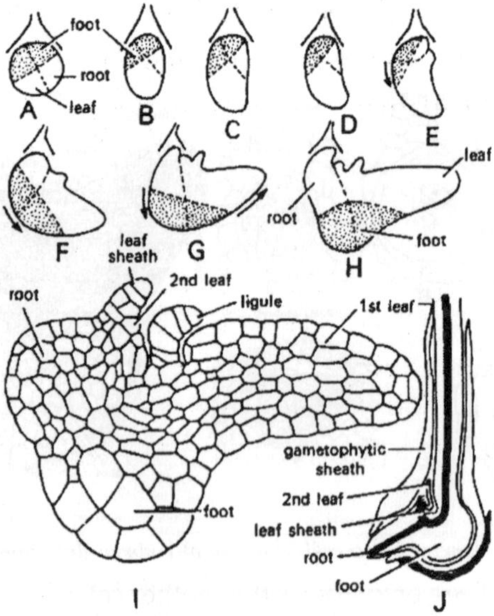

A-J. Isoetes : Embryo development; A-H. Figures illustrating the change in the orientation of endoscopic embryo, I. A young embryo, J. A young sporophyte.

The stem apex differentiates at last between the young leaf and the root.

Lepidodendrales : Lepidodendron

Order Lepidodendrales : Order Lepidodendrales includes heterosporous fossil lycopsids. These plants originated approximately 850 million years ago in the Upper Devonian period and reached their greatest development in the Upper Carboniferous Swamp Forests. The order, represented by about 200 species, is the best known amongst the fossil lycopods because of the abundance and excellent preservation of fossil remains in the form of casts and impressions.

Most of the lepidodendrids were huge trees, some attaining a height of more than 30 m and a diameter of possibly 2 m. The fossils of these plants are usually found as fragments and hence each part was described under a different generic name (i.e., form genera). Thus within the lepidodendrids there are generic names Stigmaria for underground axes, Lepidodendron for stems with leaf scars, Lepidophyllum for leaves, Lepidostrobus for strobili and Lepidocarpon for seeds.

Since the nomenclatural rules for fossil plants implied that the oldest published name that was applied to any part of the plant, becomes the name of the entire plant, the name Lepidodendron which was applied to stem with leaf scars becomes the generic name of the entire (reconstructed) plant. All the form genera attributed to Lepidodendron have been described here under the natural genus Lepidodendron.

Lepidodendron

Systematic Position

Sub-division	—	Lycopsida
Order	—	Lepidodendrales
Family	—	Lepidodendraceae

Lepidodendron was a giant tree of the Carboniferous period. It was about 30 m in height and 2 m in diameter near the base. The main trunk was unbranched. However, at the top the trunk was profusely branched forming a large crown. These branches showed repeated dichotomies and each successive branch was of reduced size.

The leaves, known as Lepidophyllum, were spirally arranged on the branches; they were simple, ligulate, acicular to linear and about 2.5 cm in length and only a few mm in width. In L. obovatum, however, the leaves were 4-5 cm long on smaller twigs and 60-80 cm long on larger branches. There was a single vein which traversed the whole length of the leaf.

The main trunk had a characteristic pattern of leaf scars and leaf cushions, which was helpful in distinguishing various genera. The cushions were somewhat vertically elongated, tapering at the base. They represented expanded decurrent leaf bases. The actual leaf scar was present above the mid-line of the leaf cushion. In the leaf scar the bundle scar mark was flanked by two parichnos strands. In some species two additional larger scars were also present at a lower level. These were once thought to be associated with parichnos system, but are now known to be formed by the shrinkage of loosely arranged parenchyma.

The underground system of the lepidodendrids is placed in the form genus Stigmaria. The base of the trunk bifurcated once and then immediately branch again to produce four massive horizontal axes (commonly known as stigmarian bases). The latter were further branched dichotomously many times in a horizontal plane. These branches bore spirally arranged appendages, known as stigmarian appendages or stigmarian rootlets. These were up to 1 cm in diameter and without root hairs. Occasionally, the position of rootlets was marked by circular scars which they left behind.

The morphological nature of the stigmarian base is debated. It has been compared with the rhizophore of Selaginella and the rhizomorph of Isoetes. Some palaeobotanists consider it as organ sui-generis.

Internal Structure

The mature trunk was covered with a massive periderm. Near the base of the trunk a protostele was present with a tiny primary xylem strand and well developed secondary xylem. Higher up, the primary xylem strand was medullated and became progressively larger. The secondary xylem was poorly developed in the upper part of the trunk. The steles of the branches had smaller primary xylem cylinders with poorly developed secondary xylem and no pith. The species of Lepidodendron showed much variations in their stelar structure. For instance, L. pettycurrensis and L. vasculare the stele was of intermediate type with mixed pith and in L. veltheimii there was a well defined pith. The primary xylem was exarch. The secondary xylem consisted of radial rows of elongated tracheids with scalariform secondary thickenings. The phloem was composed of small parenchymatous cells and large sieve cells. The rays were mostly one cell thick and many cells high, but in the central part of the secondary xylem they were two or more cells in thickness. The stele was surrounded by a massive cortical zone. The cortex was differentiated into three zones : (i) an inner cortex, composed of parenchymatous cells surrounded by a secretory zone, (ii) a large middle cortex of delicate cells, and (iii) a massive outer cortex of alternating radial patches of thin-walled and thick walled tissues or of uniformly thick walled tissue. The leaf traces as passed through the middle cortex, each acquired a strand of secretory cells which ran parallel with it. On entering the base of the leaf, the trace splitted into two parichnos strands.

Lepidocarpon, seeds of the natural genus Lepidodendron, are found in the form of compressions. The megasporangium was completely enveloped by a flap of sporophyll on either side, leaving an elongated slit like adaxial opening between the integumentary halves. This opening probably served as the crude micropyle. On the basis of the study of some early ontogenetic stages, Balbach (1962) concluded that the flaps were not new organs but simply outgrowths of the lateral margins of the pedicel. The most distinctive feature of Lepidocarpon is the presence of a single functional megaspore. It developed into a mature gametophyte without being shed from the strobilus. In some species three aborted spores of a spore tetrad and well developed archegonia were observed. Thus, the sporophyll of Lepidocarpon with its sporangium, megaspore and megagametophyte showed many similarities with the gymnospermous ovule. The megasporophyll enveloping the sporangium can be compared with the integument, and the sporangium is homologous with the megasporangium (nucleus) of gymnosperms. The megaspore wall consisted of loosely arranged strands of sporopollenin. Embryos could not be seen in Lepidocarpon.

Lepidocarpon has been referred to as a seed by many palaeobotanists, but some consider it a seed-like structure as true integument (which arises as a ring-like growth of the stalk of the megasporangium from below the nucleus) is not present here. Furthermore, in Lepidocarpon the sporangium actually dehisced, indicating that it can not be considered as a seed.

Nevertheless, Lepidocarpon illustrates the important evolutionary principle of convergent evolution. It is quite possible that such a sporangium with many megaspores has given rise to an integumented nucleus with a single functional spore.

10

Lycopsida Lycopodiales

The order Lycopodiales is represented by a single family, Lycopodiaceae, with two living genera - Lycopodium and Phylloglossum. The sporophyte is herbaceous with erect or creeping stem.

The stem has a single protostele. It is covered with microphyllous leaves without ligule. The leaf has a single vein. The sporophyll and sterile foliage leaf may be similar or dissimilar. The sporophylls may form compact cones or are loosely arranged. Each sporophyll has a single sporangium on the adaxial face near the base.

The development of sporangium is eusporangiate. Both genera are homosporous. The gametophyte is monoecious and exosporic, and it is subterranean or partially exposed. It is usually associated with mycorrhiza.

The life history of Lycopodium is described here as representative of the order.

Lycopodium

Systematic Position

Sub-division	—	Lycopsida
Order	—	Lycopodiales
Family	—	Lycopodiaceae

Lycopodium is represented by about 400 species which are worldwide in distribution, but most of the species grow in the tropical and sub-tropical regions. Some species also occur in the arctic and alpine regions of the world. The genus is represented in India by about 33 species which are mainly confined to the Eastern Himalayas and South Indian hills. Lycopodium hamiltonii, L. phyllanthum, L. clavatum, L. cernuum, L. wightianum, L. complanatum, L. serratum, L. squarrosum and L. phlegmaria are some of the common Indian species. They grow mostly in moist and shady places with enough humus and organic matter. Lycopodiums are commonly known as 'clubmoss', 'ground pine' or 'trailing evergreens'.

Sporophyte

Habit : Species of Lycopodium are small herbaceous sporophytes but vary greatly in their habit. Most of the tropical species grow on tree trunks and branches as pendulous epiphytes (e.g., L. phlegmaria, L. squarrosum, L. phlegmarioides, L. linifolium, L. corinatum;). The terrestrial species are either prostrate with a creeping stem (e.g., L. clavatum, L. cernuum, L. densum) or erect with aerial stems (e.g., L. complanatum; Fig.). Lycopodium volubile is a climber.

External Morphology : The sporophyte plant body of Lycopodium is differentiated into root, stem and leaves. As there is great variation in the form and habit of various species, Pritzel (1900) divided Lycopodium into two sub-genera —

Urostachya and Rhopalostachya. The sub-genus Urostachya is characterized by (i) the presence of erect or pendent (but never creeping) stems, (ii) dichotomous branching in the stem, (iii) roots arising from the basal part of the stem only, (iv) foliage leaves and sporophylls are approximately of the same size, (v) sporophylls are not organised into compact cones, and (vi) vegetative reproduction takes place by means of bulbils. Species like Lycopodium selago, L. subulatum, L. compactum and L. phlegmaria are included in this sub-genus. The other sub-genus, Rhopalostachya, is characterised by (i) the presence of prostrate stem bearing upright branches, (ii) dichotomous branching is confined to the first formed branches and those developing later are always monopodial, (iii) adventitious roots are borne along the entire length of the prostrate stem, (iv) sporophylls are smaller than the foliage leaves, (v) sporophylls are organised into strobili, and (vi) vegetative reproduction never takes place by means of bulbils. Species like Lycopodium cernuum, L. clavatum, L. annotinum, L. densum, L. obscurum and L. innundatum represent this sub-genus.

1. *Root* : The first formed root is ephemeral, hence older plants have only adventitious roots. In the sub-genus Urostachya, adventitious roots arise from the basal part of the axis, but in Rhopalostachya they develop acropetally along the underside of the stem either single or in groups. In some species (e.g., L. obscurum and L. lucidulum) branching is dichotomous, but in others dichotomy is obscure.

   Adventitious roots develop mainly from the cells of the stem pericycle, but sometimes they also originate from the endodermis. They do not grow out straight through the cortex, instead they take a vertical course before emerging out. As a result, in a transverse section of the stem, many roots are seen in the region of the cortex. Such roots are known as cortical roots.

2. *Stem* : The stem is weak, slender and rhizomatous. It is usually erect or pendent in the sub-genus Urostachya and creeping in Rhopalostachya. It is sparsely (e.g., L. annotinum) or profusely branched (e.g., L. obscurus). The branching in Lycopodium is basically dichotomous, but formation of unequal growth centers at the apex is not uncommon. In the latter case branching becomes monopodial. Unequal branching occurs mostly during the formation of reproductive branches such as bulbils and gemmae. The stem and branches are covered with leaves.

3. *Leaves* : The leaves are small, simple, eligulate and sessile with a narrow apex and slightly broad base. They are mostly entire but sometimes the margins are serrate (e.g., L. serratum). The leaf has a single unbranched midvein. The size of the leaf usually varies from 2-10 mm, but in L. mandioccanum leaves are 25-30 mm long (Fig.), whereas in L. densum they are in the form of minute scales. The leaves are arranged in closely spiral (L. clavatum), whorled (L. cernuum;) or opposite decussate (L. complanatum, L. alpinum;) phyllotaxis. But sometimes their arrangement is irregular. More than one type of leaf arrangements may be found on the same plant.

Some species of Lycopodium, especially those with flattened stems (e.g., L. complanatum and L. volubile), show dimorphism. The lateral branches bear four vertical rows of leaves; the two lateral rows have large wing-like leaves, the ventral row has small leaves and those of the dorsal row are of intermediate size.

Internal Structure

1. *Root* : A transverse section of the root shows three distinct regions - epidermis, cortex and stele.

The epidermis is single layered, made up of thin walled cells. In the apical region of the root, some epidermal cells form unicellular hairs. A characteristic feature of Lycopodium root is that the root hairs arise in pairs.

The cortex is composed of parenchymatous cells. In older roots, some outer layers of the cortex become thick walled and they provide mechanical support to the root.

The central part of the root is occupied by a protostele which is monarch in young roots but becomes diarch or triarch in older roots. Thus the same root may show all the above three conditions at different levels. The xylem is C-or U-shaped (crescent-shaped) and the phloem lies in between the arms of the xylem.

The internal organization of cortical roots is almost identical to that of aerial roots. In cortical roots epidermis is indistinct and root hairs are usually absent or represented by small protuberances.

2. *Stem :* Internally, the stem is also differentiated into three tissue systems, epidermis, cortex, and stele.

The epidermis forms the outermost protective layer. The outer walls of the epidermal cells are thick and cutinized. Stomata are also present in the epidermis of young stem.

The structure of cortex in Lycopodium varies depending on the species and stem diameter. It is homogeneous or shows distinct tissue zonation. In some species, cortex remains parenchymatous throughout. In large above ground stems, the cortex has three distinct zones — chlorenchymatous in the hypodermal region, sclerenchymatous around the stele and parenchymatous in between these two zones. In L. clavatum, the outer and inner cortex represented by narrow sclerenchymatous zones and the middle cortex has large parenchymatous cells with intercellular spaces. Sometimes large air spaces develop in the middle cortex due to disruption of parenchyma. In smaller stems sclerenchyma is reduced or absent, and often the entire cortex is composed of loosely packed chlorenchyma cells similar to leaf mesophyll. However, the cortex of subterranean stems is often sclerenchymatous throughout.

The innermost layer of the cortex forms endodermis. It is usually ill-defined in mature stems, but in younger stems it is often distinct with characteristic casparian bands.

Endodermis is followed by 2-6-layered pericycle, composed of thin walled cells.

Though all species of Lycopodium have a protostele, the arrangement of xylem, phloem and the related parenchyma within the stele varies not only in different species but even within the same species. In L. clavatum and L. complanatum, xylem and phloem occur in parallel rows, separated by stelar parenchyma. Such a protostele is called plectostele; Lycopodium serratum, L. phlegmaria and L. selago have an actinostele with star-shaped xylem, and phloem contained in between the arms of xylem. Lycopodium cernuum shows an extreme case of mixing of xylem and phloem. In this species, xylem and phloem are uniformly intermixed and such a stele is known as mixed protostele.

The xylem is differentiated into protoxylem and metaxylem elements. The protoxylem is situated at the outer edge of each xylem arm, band or patch (i.e., exarch) and the metaxylem in the centre. The metaxylem tracheids have scalariform or pitted thickenings, while the protoxylem tracheids have spiral or annular thickenings. The phloem is composed of sieve cells with relatively inconspicuous sieve areas on their lateral walls. There is no cambium hence secondary growth does not occur.

3. *Leaves* : The leaf is somewhat triangular in outline in a vertical section. The epidermis is covered by a thick layer of cuticle. Most of the species of Lycopodium are amphistomatic, i.e., stomata are uniformly distributed on both surfaces of the leaf, but some species (e.g., L. complanatum and L. volubile) are hypostomatic, where stomata are confined to the lower surface only.

   The mesophyll is undifferentiated and consists of chlorophyllous cells with large intercellular spaces. The leaf has a median concentric vascular bundle with central xylem surrounded by phloem. The tracheary elements of xylem have annular or spiral thickenings and the phloem has only sieve cells with inconspicuous sieve areas.

Reproduction

Lycopodium reproduces vegetatively and by spores.

Vegetative Reproduction : Propagation of sporophyte by vegetative methods is quite frequent in Lycopodium and takes place by the following methods.

1. *By Bulbils or Gemmae* : Species belonging to the sub-genus Urostachya bear bud-like bulbils or gemmae on the stem apex in the axils of leaves (Fig.). The bulbil has many thick fleshy leaves (which store food), arranged spirally around a small axis. When the bulbil falls off on the ground, it gives out many adventitious roots and grows into a new sporophyte (Fig.).

   The bulbil has been variously interpreted as a modified

lateral branch borne in the axil of leaf, a modified sporangium or a modified leaf. The latter interpretation receives support from the facts that bulbil resembles with leaf in position, dorsiventrality and vascular supply.

2. *By Adventitious Buds :* In some species (e.g., L. aloifolium, L. reflexum), adventitious buds occur at the base of the stem. These buds are capable of forming new plants when detached from the parent plant.

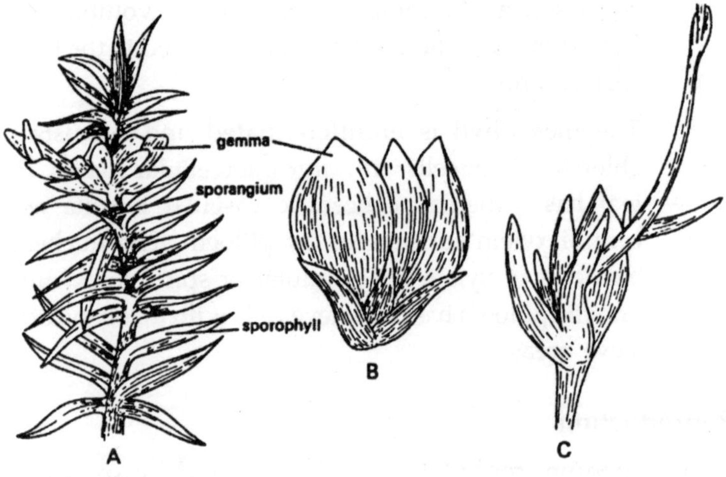

A-C Lycopodium : Vegetative propagation; A. A branch with gemmae, B. A gemma, C. A germinating gemma.

3. *By Fragmentation and Decay :* This method of propagation is common in creeping species of Lycopodium. In these species, when progressive death of older portions reaches to the point of dichotomy, the two branches separate and each grows into a new plant.

4. *By Root Tubers :* Adventitious roots of L. cernuum and L. ramulosum bear bud-like tubercles, which are formed by the proliferation of parenchymatous cells of the apical region of the root. These tubercles, on germination, give rise to new sporophytes.

Reproduction by Spores

1. *Spore Producing Organs :* Spores are formed within sporangia which originate singly on the adaxial (upper) side of specialized leaves, called sporophylls or fertile leaves. The sporangium develops very close to the leaf axil and remains so till maturity in most of the species. The sporophyll forms a protective covering around the sporangium (Fig.). In some species, there is a shifting in the position of sporangium during ontogeny. For example, the mature sporangium is abaxial and protected in L. cemuum (Fig.), subfoliar and exposed in L. squarrosum (Fig.) and axial and exposed in L. lucidulum (Fig.).

   Sporophylls vary in shape and size in different species. In most of the species belonging to the subgenus Urostachya, sporophylls are similar to foliage leaves, but are not organized into strobili or cones (Fig.). On the other hand, in the subgenus Rhopalostachya, sporophylls are smaller than the foliage leaves and usually have serrate margins. They are organized into definite strobili or cones at the apices of the main stem or its branches (Fig.). In some species (e.g., L. selago, L. drumondii, L. obscurum) the strobilar apex resumes growth and as a result alternating series of foliage leaves and sporophylls are formed.

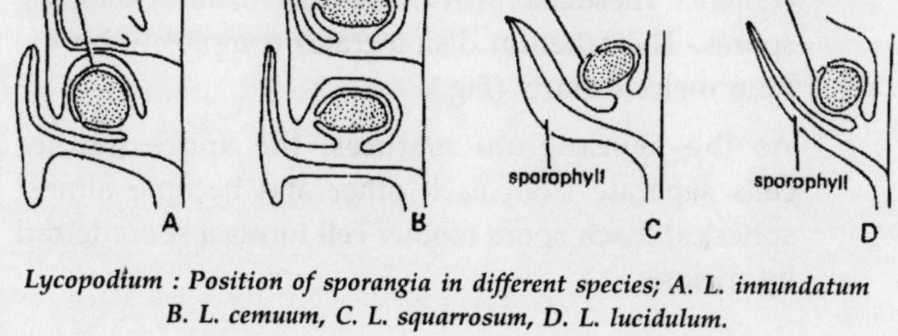

Lycopodium* : *Position of sporangia in different species; A. L. innundatum B. L. cemuum, C. L. squarrosum, D. L. lucidulum.

The sporangium is unilocular, sub-spherical or reniform at maturity. It is approximately 2 mm in length and 1-2.5 mm in diameter. It is oriented in such a way that its full length and curvature are revealed in surface view of the stem. The orientation of the line of dehiscence is always transverse to the sporophyll axis. The sporangia are homosporous and yellow or orange at maturity.

2. *Development of Sporangium :* The sporangium initiates on a very young sporophyll and all developmental stages can be seen in a longitudinal section of a young growing strobilus. The development of sporangium is of eusporangiate type. A group of 2-5 surface cells becomes differentiated as sporangial initials on the adaxial side of the sporophyll close to the axil (Fig.). These cells divide periclinally and differentiate into an outer and an inner layer. The cells of the outer layer, called primary wall cells, undergo periclinal and anticlinal divisions and eventually form three or more layered jacket of the sporangium. The cells of the inner layer which function as primary sporogenous cells, undergo repeated mitotic divisions and ultimately give rise to spore mother cells (Fig.).

   As the sporangium develops, the innermost wall layer, adjacent to the sporogenous mass, differentiates into a nutritive layer, called tapetum (Fig.). The cells of the tapetum stain deeply indicating their high protein contents. These cells provide nutrition to the developing spores. The tapetum disintegrates completely by the time meiosis starts (Fig.).

   As the sporangium matures, the spore mother cells separate from each other and become almost spherical. Each spore mother cell forms a spore tetrad by meiosis.

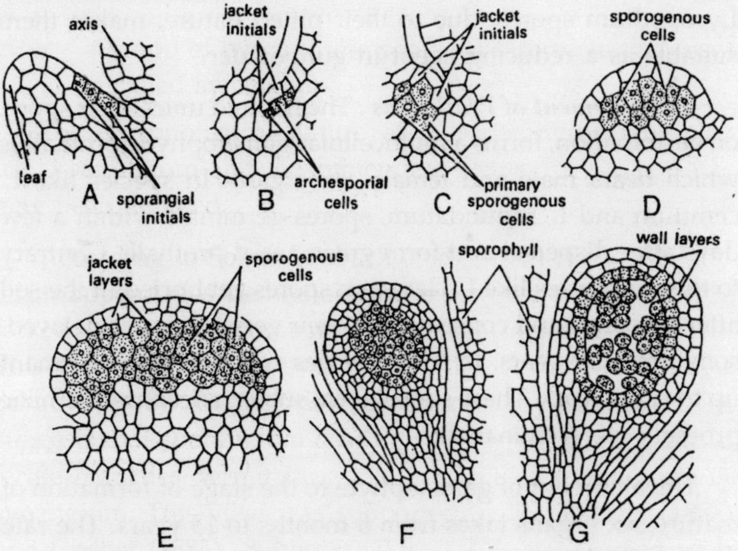

A-G. Lycopodium : Successive stages in sporangium development.

3. *Dehiscence of Sporangium :* With the maturation of the sporangium, all jacket layers, except the outermost, disintegrate. The cells of the outer wall layer elongate anticlinally and become thick walled. But a narrow transverse strip of cells across the apical portion of the sporangium fails to elongate and forms a line of dehiscence, called stomium. The thickenings on these cells are different from other jacket cells. The cells of the jacket shrink on drying and the sporangium ruptures along the stomium due to the nature of wall thickenings in this region.

Gametophyte

Spore : The haploid spore is the mother cell of the gametophytic generation. Lycopodium is homosporous and all spores are similar in shape and size. The diameter of the spores is in the range of 0.03 and 0.05 mm. The spore wall is differentiated into an outer lamellate layer with pitted surface and an inner granular layer. The large surface area of

Lycopodium spores, due to their pitted nature, makes them suitable as a reducing agent in gunpowder.

Development of Prothallus : The haploid unicellular spore, on germination, forms a multicellular gametophytic prothallus which bears male and female sex organs. In species like L. cemuum and L. innundatum, spores germinate within a few days after dispersal and form green aerial prothalli. Contrary to this, in species like L. clavatum spores get buried in the soil after dispersal and consequently their germination is delayed, sometimes by years. In some species spores remain dormant up to eight years. These spores form subterranean mycorrhizic prothalli on germination.

Development of gametophyte to the stage of formation of mature sex organs takes from 8 months to 15 years. The rate of growth of gametophyte is dependent upon the proportion of the photosynthetic tissue. Species with colourless gametophyte nourished by a mycorrhizal fungus, show slow rate of development.

Just before germination, the spore absorbs water and becomes almost spherical. At this stage, the outer spore wall (exine) ruptures along the triradiate ridge (Fig.). The first division of the spore is asymmetrical, resulting in the formation of a very small biconvex rhizoidal cell and a large cell. The rhizoidal cell disorganizes and does not participate in further development of the •prothallus. The large cell divides by a vertical wall into a basal cell (adjacent to the rhizoidal cell) and an outer cell. The outer cell divides by two oblique divisions, intersecting each other, forming an apical cell with two cutting faces. The gametophyte develops up to this stage with the help of the food stored in the spore. Further development proceeds only after the entry of symbiotic phycomycetous fungus into the basal cell of the gametophyte. This fungus forms endophytic mycorrhiza (Fig.). If there is no fungal infection, further development of gametophyte is checked.

The apical cell of the gametophyte becomes active after the establishment of endophytic mycorrhiza and forms a major part of the adult gametophyte (Fig.).

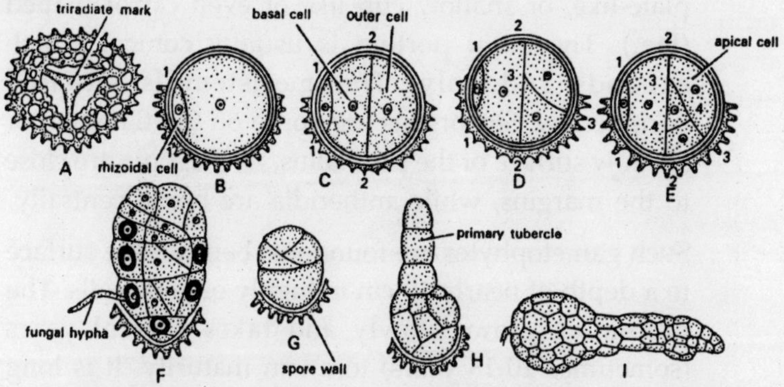

***Lycopodium* : Successive stages in the development of prothallus.**

***Structure of the Mature Ggametophyte* :** There is great diversity in the form and structure of the mature gametophyte. Primarily, three distinct types of prothalli are found in Lycopodium, but many intermediate types have also been reported which interconnect these three types.

1. *Cernuum Type :* This type of gametophyte, found in species like L. cemuum and L. innundatum, consists of a massive conical erect prothallus. The basal part of the prothallus is buried in the soil and the upper part is exposed above the ground. The prothallus is invariably green and may bear numerous irregular leaf-like lobes, which are very small and seldom exceed 3 mm in length. Sex organs appear between the lobes. The prothallus matures quickly, usually in one season, and is short lived. The leafy sporeling, when formed, is attached laterally.

   This type of prothallus is a self nourishing body but additional nourishment is provided by endophytic fungus present in the cortex.

2. *Clavatum Type* : This is a non-green, subterranean, tuberous, often much larger (1-2 cm in length) than the cernuum type. The shape is variable; it may be top-like, plate-like, or shallow cup like or even carrot-shaped (Fig.). The basal portion is usually conical which expands suddenly. The meristem is situated peripherally. Sex organs are borne on the flattened or shallow surface of the prothallus. Archegonia are close to the margins, while antheridia are borne centrally.

   Such gametophytes are found just beneath the surface to a depth of nearly 10 cm in highly organic soils. The prothallus grows slowly and takes several years (sometimes 10-15 years) to attain maturity. It is long lived and may remain functional for 25 years or even more. Thus it may bear several generations of sporophyte.

   The only method of nourishment in these prothalli is saprophytic with the help of endophytic fungus. The cortex has very well differentiated mycorrhizal region. The central part is, however, completely free from fungus and is meant for storage of food material.

   This type of prothallus is found in L. clavatum, L. complanatum, L. obscurum and many other creeping species.

3. *Phlegmaria Type* : This is a non-green prothallus found in epiphytic species of Lycopodium. It is small and repeatedly branched. The colourless cylindrical branches arise irregularly from the points distant to the main branch. These branches extend through the dead bark in which prothallus grows. They are provided with prominent absorbing hairs. The central core of the branches is traversed by mycorrhiza. The stored food material is in the form of oil. The branches show indefinite apical growth, the older parts die but the

apex continues to survive. Thus prothalli continue to propagate vegetatively by fragmentation. They also propagate by gemmae and surface budding.

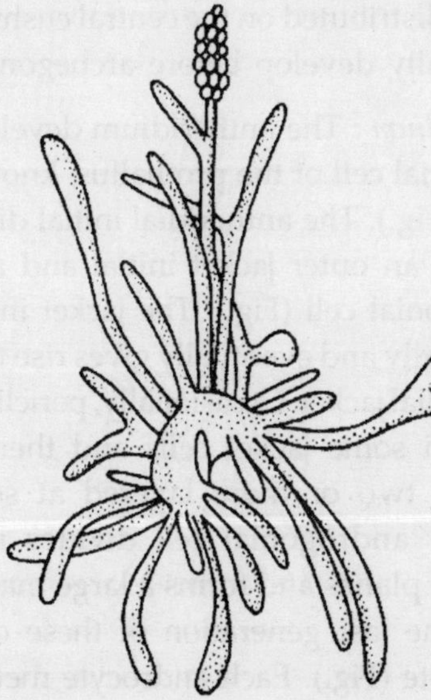

***Lycopodium complanatum* : Prothallus.**

Sex organs occur irregularly on branches and they are generally surrounded by glandular hairs, called paraphysis.

This type of prothallus is found in L. phlegmaria and many other epiphytic species. Recent studies have shown that there are also some intermediate forms in between these three types of prothalli; for instance, the prothallus of L. selago is intermediate between the cernuum and clavatum types. Besides, more than one type of prothalli may also occur in the same species. For example, in L. selago, if spores germinate immediately after their dispersal from the sporangium, they form cernuum type of prothalli and if their germination is delayed (as they get buried in the soil), they give rise to subterranean colourless prothalli.

Sex Organs : The prothalli of Lycopodium are monoecious, i.e., male (antheridia) and female (archegonia) sex organs are borne on the same prothallus. In subterranean prothalli, antheridia and archegonia occur in distinct patches on the whole surface, whereas in erect surface living prothalli they are irregularly distributed on the central cushion. In both types, antheridia usually develop before archegonia.

1. *Antheridium* : The antheridium develops from a single superficial cell of the prothallus, known as antheridial initial (Fig.). The antheridial initial divides periclinally to form an outer jacket initial and an inner primary androgonial cell (Fig.). The jacket initial divides only anticlinally and eventually gives rise to a single layered antheridial jacket. Occasionally, periclinal divisions may occur in some jacket cells and then the jacket may become two or more layered at some places. The primary androgonial cell divides repeatedly in all possible planes and forms a large mass of androgonial cells. The last generation of these cells functions as androcyte (Fig.). Each androcyte metamorphoses into a pear-shaped biflagellate antherozoid (Fig.).

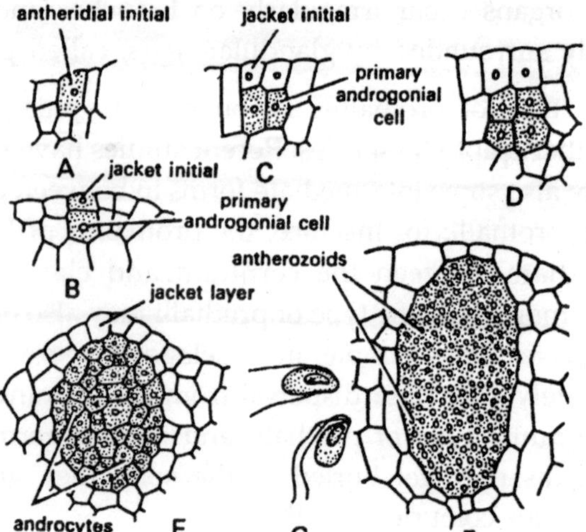

A-G Lycopodium : Antheridium : A-E. Successive stages in the development of antheridium, F. Mature antheridium, G. Antherozoid.

The antheridia are sunken in the tissue of the prothallus. Their position is demarcated by a slight bulge at the surface of the prothallus. The antheridium has an operculum at its apex, formed by the decomposition of a single jacket cell. As water enters into the antheridium through this pore, antheridial jacket is ruptured and antherozoids are liberated.

2. *Archegonium* : Like antheridium the archegonium also develops from a single superficial cell, called archegonial initial (Fig.). It divides by a periclinal wall into an outer primary cover cell and an inner cell (Fig.). The primary cover cell divides by two anticlinal (vertical) walls at right angles to each other thus forming four neck initials (Fig.). The neck initials divide transversely and form 3-4 cells high neck of the archegonium. In the meantime, the inner cell also divides by a transverse wall into a basal cell and a central cell. In some species, the inner cell directly functions as central cell and hence basal cell is not present in these species. The central cell divides transversely into an outer primary canal cell and a lower primary venter cell. The primary canal cell, by repeated divisions, forms 4-8 neck canal cells. However, there is only a single neck canal cell in L. cemuum and as many as 14 in L. complanatum. The primary venter cell divides only once and forms a small venter canal cell and a relatively large egg cell (Fig.).

   The neck of the mature archegonium usually projects above the surface of the prothallus but its basal part remains embedded in the tissue of the prothallus.

Fertilisation : As the archegonium matures, the tips of neck cells split apart and the neck canal cells and venter canal cell degenerate to form a mucilagenous substance (Fig.). This mucilagenous substance which contains citric acid diffuses from the archegonium and chemotactically attracts biflagellate pear-shaped antherozoids. Although many antherozoids may enter into an archegonium, only one of them fuses with the egg. Thus an oospore is formed.

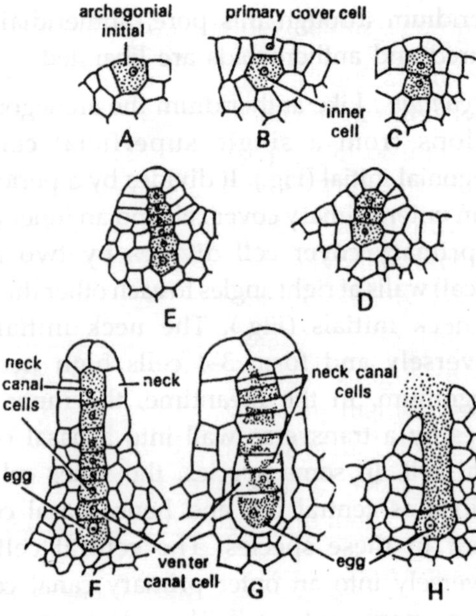

A-H. Lycopodium : Successive stages in the development of archegonium.

Embryo Development

The diploid oospore is the mother cell of the next sporophytic generation (Fig.). The first division of the zygote is transverse, i.e. at right angles to the long axis of the archegonium. Of the two cells thus formed, the outer functions as suspensor cell and the inner as embryonic cell (Fig.). The suspensor cell does not divide further but in some species of Lycopodium it undergoes considerable elongation and pushes the proembryo deeper into the gametophytic tissue. The embryonic cell divides by two vertical walls at right angles to each other forming four cells. Each of these four cells divides transversely, thus eight cells are formed which are arranged in two superimposed tiers of four cells each. The embryo has nine cells at this stage.— one suspensor cell and eight proembryonic cells.

In L. clavatum, L. complanatum, L. phlegmaria and many other species with subterranean prothalli, the upper four cells of the proembryo, adjacent to the suspensor cell, form a foot by repeated divisions (Fig.). The foot is massive and is the most prominent part of the embryo (Fig.). It acts as an intraprothallial haustorium and absorbs food for the developing embryo until it is established into an independent sporophyte. The lower four cells of the proembryo eventually give rise to stem and leaves.

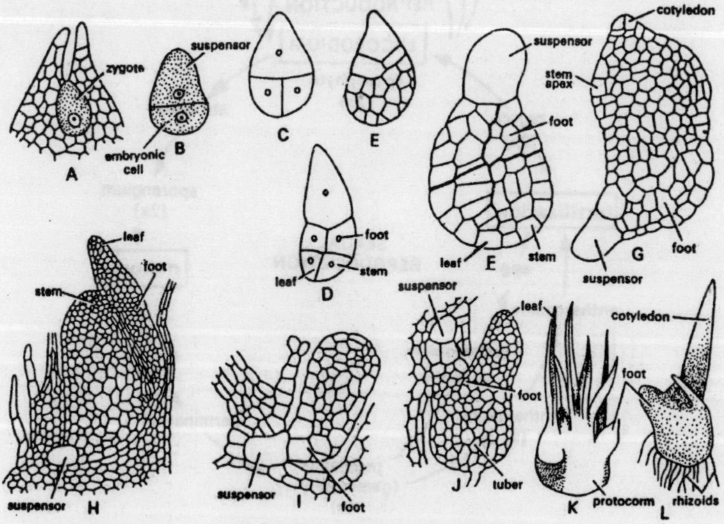

A-L. Lycopodium : Embryo development; A. Zygote, B-J. Stages in embryo development, K-L. Protocorm.

Development of the proembryo in species with aerial, green and photosynthetic prothallus (e.g., L. cemuum, L. innundatum) is similar to the subterranean prothallus till the 9-celled stage. Subsequently, the four cells of the upper tier form a foot, which is a relatively small part of the embryo

(Fig.). The four cells of the lower tier instead of forming stem and leaves, divide irregularly and give rise to a bulbous structure, called protocorm. The protocorm is a mass of parenchymatous cells, and it bears rhizoids on its lower surface and non-vascular leaf-like structures, called prophylls, on its upper surface. After a period of irregular growth of the protocorm, a meristem differentiates among the prophylls which gives rise to the axis of the young sporophyte. The first roots arise from the base of the axis.

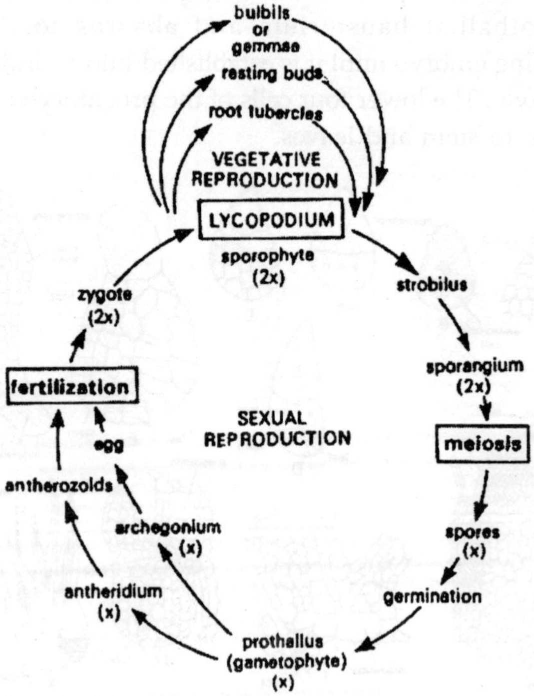

Lycopodium. Life-cycle.

11

Psilopsida Psilotales

Psilotales are primitive land plants which show resemblance with fossil vascular plants belonging to the order Psilophytales (Rhynia, Horneophyton, etc.) The sporophyte is rootless and consists of a subterranean rhizome with numerous rhizoids and the aerial shoot which is leafless (e.g., Psilotum) or bears foliage leaves (e.g., Tmesipteris). The sporangia, borne in groups of two or three (i.e., synangia), are homosporous. The gametophyte is exosporic and monoecious. The development of embryo is exoscopic.

The life-history of Psilotum is described here as representative of the order.

Psilotum

Systematic Position

Sub-division	—	Psilotopsida
Order	—	Psilotales
Family	—	Psilotaceae

The genus Psilotum consists of two species, P. nudum and P. flaccidum. P. nudum is fairly common in tropical and subtropical parts of both hemispheres, whereas P. flaccidum, a pendulous epiphyte on tree trunks, is restricted to Jamaica, Hawaii, Florida, Mexico and some Pacific Islands. In India, the genus is represented by P. nudum only. It has been collected from Shimla, Kulu (Himachal Pradesh), Panchmarhi (Madhya Pradesh) and Darjeeling hills (West Bengal). Tetraploid races of P. nudum are widely cultivated in gardens as botanical curiosities.

Sporophyte

External Morphology : The plant body is sporophytic and consists of a subterranean rhizome and dichotomously branched aerial shoots. The rhizome is cylindrical, prostrate and dichotomously branched, but sometimes the branching tends to become lateral or irregular due to injury in the meristematic region. The rhizome anchors the plant to the substratum. It is associated with an intracellular mycorrhiza; the fungal hyphae gain access to the cortex through the rhizoids. In the absence of true roots, rhizoids form the major absorbing surface of the subterranean system. In P. nudum, sometimes the tips of immature rhizoids proliferate to form gemmae, which are capable of regenerating into new plants.

The rhizome and aerial shoots form a continuous system. In fact, some branches of the rhizome turn upward and develop into aerial leafy shoots (or fronds). The aerial shoots of P. nudum are small and 15-20 cm long, but in P. flaccidum, they attain a height of up to 90 cm. In terrestrial forms the aerial axis is erect but in epiphytic forms it is pendent. The axis appears to be dichotomously branched but technically the branch is lateral as one of the branches arise in the axil of the leaf and the other (considered to be a dichotomy) is the continuation of the main axis. The basal part of the axis is usually cylindrical but the distal part is flattened (e.g.,

P. flaccidum) or longitudinally ribbed (e.g., P. nudum). The basal part of the aerial shoot is generally smooth and without any lateral appendages but in the distal part small, scaly, awl-shaped sterile appendages or 'leaves' are present. These appendages are arranged in definite spiral (1/3) phyllotaxis in P. nudum, whereas in P. flaccidum they are arranged in sub-opposite pairs along the edges of the flattened branches. The 'leaves' do not have any vascular trace or stoma. The aerial green shoots serve as photosynthetic organ.

'Leaves' are replaced by fertile appendages at the distal end of the aerial shoot.

Internal Structure

1. *Subterranean Rhizome :* The internal structure of the rhizome varies with its diameter. Rhizomes with less than 1 mm in diameter are avascular and made up of parenchymatous cells with endophytic mycorrhiza. But larger rhizomes (2-3 mm in diameter) show differentiation of tissues (Fig.). There is a superficial layer of uninterrupted epidermis composed of thin walled cells. The epidermis is followed by a broad parenchymatous cortex, differentiated into three distinct zones. The outer cortex is characterised by the presence of endophytic mycorrhiza, the middle cortex has large parenchymatous cells with starch grains, and the inner cortex is frequently dark brown in colour due to the presence of phlobaphene, an oxidation product of tannins, in the cells. Inner to the cortex is a well defined endodermis with conspicuous casparian strips on radial walls. It is followed by a single layered pericycle of parenchymatous cells which encircle the protostele. The stele of the young rhizome is represented by few tracheids (without differentiation into metaxylem and protoxylem), surrounded by phloem. In larger rhizomes, however, the xylem breaks into

small segments due to the differentiation of parenchyma between the tracheids.

2. *Aerial Shoot* : A transverse section of the aerial shoot is irregular in outline due to the presence of grooves and ridges. It is bounded by a single layered epidermis with heavily cutinised outer wall. The epidermis is interrupted by large number of stomata, mainly confined to grooves. Each stoma has two dumb-bell shaped guard cells with heavily cutinised walls like those of the epidermal cells (Fig.).

   The epidermis is followed by a well developed cortex, differentiated into three zones : (i) outer cortex is made up of 2-5 layers of loosely arranged chlorophyllous cells. The intercellular spaces in between these cells are in contact with the external atmosphere through stomata. It is the main photosynthetic region of the plant, (ii) middle cortex consisting of 4-5 layers of vertically elongated sclerenchymatous cells which provide mechanical support, (iii) inner cortex consists of compactly arranged thin walled parenchymatous cells, rich in starch.

   The stele, is surrounded by a distinct endodermal layer with characteristic casparian bands on the radial walls of its cells.

   The aerial shoot of Psilotum is characterised by the presence of a siphonostele. The central part of the stele is occupied by a sclerenchymatous pith. The xylem cylinder has radiating rays. In the apical part of the young branches there are only 2-3 radiating rays, but in the older branches there are as many as 10 rays. The protoxylem tracheids have annular or helical thickenings and are present at the tips of the rays, whereas the metaxylem tracheids with scalariform or pitted thickenings develop centripetally. The phloem, which is composed of only sieve elements, completely surrounds the xylem. A zone of thin walled cells, several layers in thickness, is also present between the endodermis and xylem. This zone though referred to as phloem, does not have typical sieve tubes, characteristic of phloem.

The sclerenchymatous pith gradually diminishes towards the basal part of the aerial shoot and consequently the stele becomes actinostele.

3. *Leaf* : The internal structure of the leaf is very simple. It is covered by a single layered epidermis with cutinised outer walls. There are no stomata in the epidermis. The mesophyll of the leaf is made up of chlorophyllous parenchymatous cells without any differentiation into palisade and spongy cells. The air spaces between the mesophyll cells are well developed in P. nudum and very narrow in P. flaccidum. There are no veins, but in P. flaccidum minute leaf traces originate from the stele but they fade out in the cortex without entering the leaf. Thus, in the absence of stomata and vascular traces, the leaves do not perform photosynthetic function.

Reproduction

Sporophyte of Psilotum reproduces vegetatively as well as by spores.

Vegetative Propagation : In P. nudum minute multicellular bodies, called gemmae, develop on the surface of the rhizome. They are mostly concentrated near the apex of the rhizome (Fig.). The gemma is a small oval structure, only one cell in thickness. It grows by means of a two-sided apical cell. It germinates into a new sporophyte while still attached to the rhizome or after falling on a suitable substratum. On germination, it gives rise to a subterranean shoot.

Gemmae are also formed on the surface of the gametophytic prothallus by the proliferation of the terminal cell of the rhizoid (Fig.). They are similar to those formed on the rhizome of the sporophyte. These gemmae, on germination, form prothalli which also bear some reproductive structures.

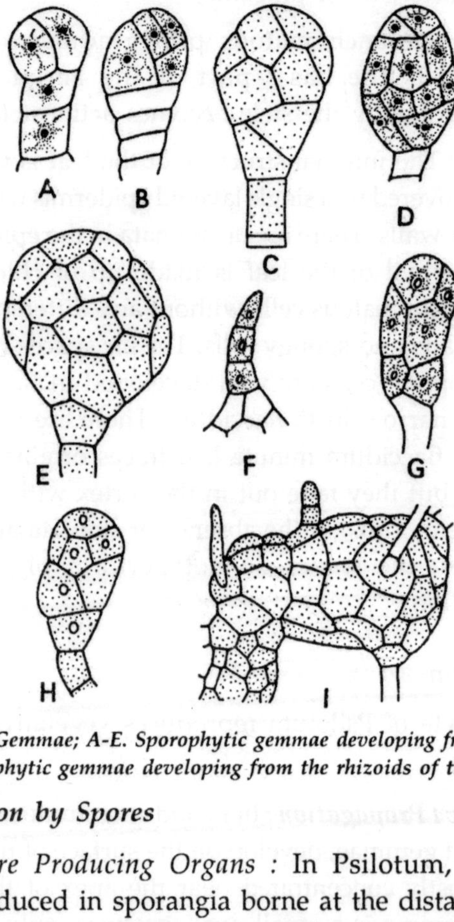

Psilotum : Gemmae; A-E. Sporophytic gemmae developing from rhizome, F-I. Gametophytic gemmae developing from the rhizoids of the prothallus.

Reproduction by Spores

1. *Spore Producing Organs :* In Psilotum, spores are produced in sporangia borne at the distal end of the dichotomously branched aerial shoot. The sporangium develops on a small appendage subtended by a forked bract. It is a trilobed structure, measuring 2-3 mm in diameter. Each lobe of the sporangium has a spore sac with numerous spores. Such a sporangium, formed by

a group of two or more spore sacs (three in this case) is called synangium.

The morphological nature of the sporangium complex (i.e., sporangium and the subtending fertile appendage) has long been debated.

According to Solms Laubach (1884), both the sterile and fertile appendages are homologous structures; the latter being forked and bearing sporangia and thus represents a sporophyll. This view was upheld by Bower (1908), Seward (1910) and Schoute (1938). Another view, propounded by Juranyi (1871), suggests that sporangium bearing structures are short lateral fertile branches, terminating into a trilocular sporangium or synangium. This view was supported by Strasburger (1873) and Goebel (1881). Bierhorst (1956, 1977) also agreed with Juranyi's view and postulated that phylogenetically a fertile appendage is a reduced branch which bears a terminal sporangium after a short period of apical growth.

The fertile appendage receives a single vascular trace from the stele of the aerial shoot. This trace divides into three just below the synangium, and these three branches fade out at the base of the sporangium.

2. *Development of Sporangium :* The development of sporangium is of eusporangiate type. It has been studied in detail in Psilotum nudum.. The primordium of the synangium differentiates quite early in the ontogeny of the fertile appendage. Each of the three sporangia develops from a single or a group of superficial cells of the sporangiophore. The first division of the sporangial initial is by a periclinal wall, as the result an outer jacket initial and an inner archesporial cell (also called primary sporogenous cell) are formed. Repeated periclinal and anticlinal divisions in jacket initials give rise to a 4-5 layered thick sporangial wall. The outermost wall layer differentiates into epidermis.

The cells of this layer elongate anticlinally except at the line of dehiscence that extends vertically from the base to the apex of the sporangium. The cells of the inner wall layers are thickened, primarily on their lateral walls.

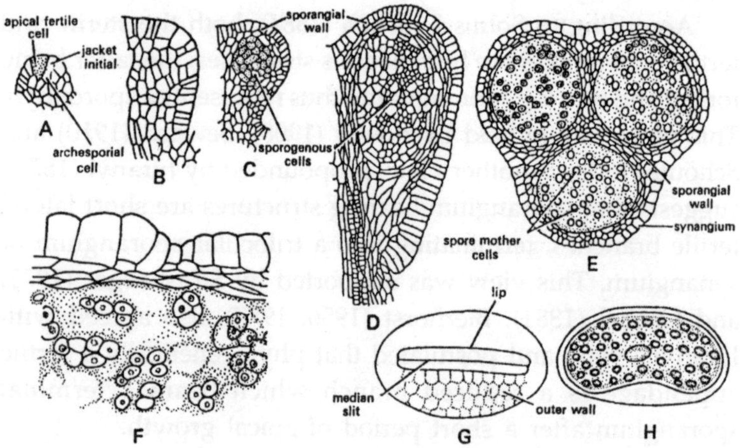

Psillotum : A-D. Stages in the development of synangium, E. Transverse section of young synangium, F. Transverse section of a mature synangium, G-H. Spores.

The archesporial cells divide repeatedly and produce a mass of cells that differentiate into sporogenous tissue (Fig.). As the sporangium matures, irregular groups of cells in the sporogenous tissue become densely cytoplasmic. These cells divide repeatedly and give rise to spore mother cells. The protoplast of the rest of the cells of the sporogenous tissue becomes watery (Fig.). These cells slowly degenerate into a plasmodial mass which provides nourishment to spore mother cells. Each spore mother cell undergoes meiosis, forming a spore tetrad (Fig.). Thus a large number of spore tetrads are formed in each sporangium. The spores are colourless and bean-shaped, each with a slit on the concave side surrounded by a smooth ridge of wall thickening, called lip (Fig.). All spores are of the same type (i.e., homosporous).

The dehiscence is initiated from the centre of the three-sporangiate synangium and the entire synangium is torn open at the centre.

Garnetophyte

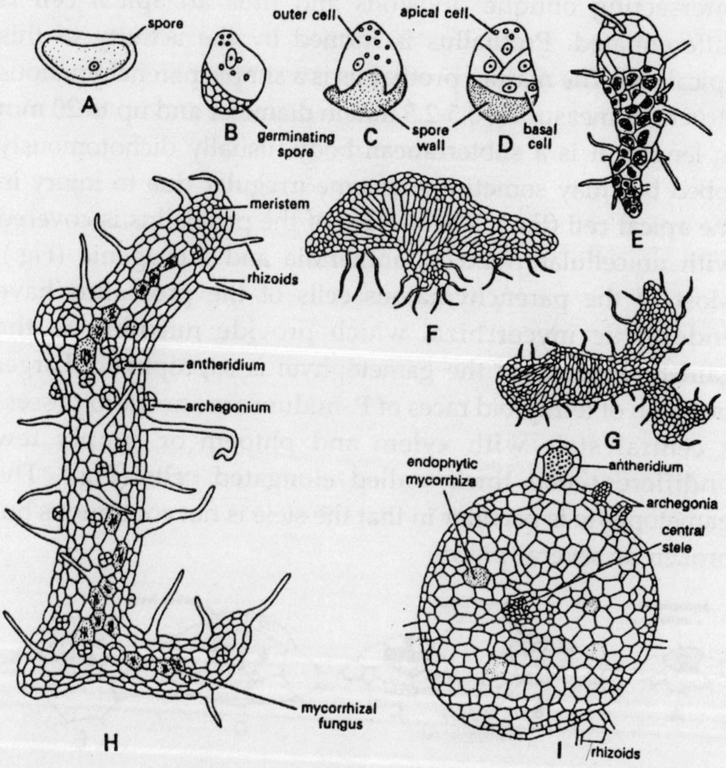

***Psillotum* : A-E. Stages in the development of prothallus, F-G. Irregular Prothalli, H. A cylindrical prothallus, I. Transverse section of prothallus.**

Development and Structure of Gametophyte

Spore is the mother cell of the gametophytic generation. On germination it produces a gametophytic prothallus. The spores of Psilotum usually germinate in 3-4 months after their liberation from the sporangium. They absorb water and swell up; consequently the outer spore wall (exine) opens along the median slit and the inner thin and smooth intine extrudes out in the form of a pouch along with spore contents (Fig.). A transverse wall cuts off this pouch (henceforth called outer cell) from the portion lying within the spore wall in the form of a large spherical cell. The outer cell divides by two intersecting oblique divisions and thus an apical cell is differentiated. Prothallus is formed by the activity of this apical cell. The mature prothallus is a simple parenchymatous structure, measuring 0.5-2.5 mm in diameter and up to 20 mm in length. It is a subterranean body, usually dichotomously lobed but may sometimes become irregular due to injury in the apical cell (Fig.). The surface of the prothallus is covered with unicellular rhizoids, antheridia and archegonia (Fig.). Most of the parenchymatous cells of the prothallus have endophytic mycorrhiza, which provide nutrition to the gametophyte. Thus the gametophyte is saprophytic. Larger prothalli of tetraploid races of P. nudum are known to possess a central stele with xylem and phloem or simply few undifferentiated thick walled elongated cells (Fig.). The gametophyte is peculiar in that the stele is not continuous but broken at several places.

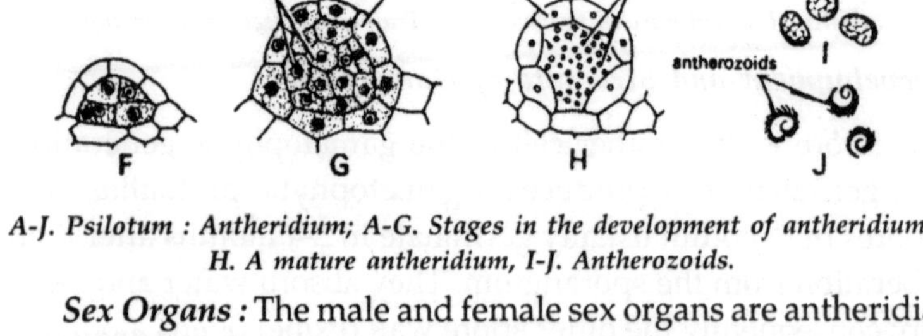

A-J. Psilotum : Antheridium; A-G. Stages in the development of antheridium, H. A mature antheridium, I-J. Antherozoids.

Sex Organs : The male and female sex organs are antheridia and archegonia respectively. They are scattered over the surface of the gametophyte. The gametophyte is monoecious and the number of antheridia is usually higher than archegonia.

1. *Antheridium:*

(a) *Development :* The antheridium develops from a single superficial cell of the gametophytic prothallus. This antheridial initial divides periclinally into an outer jacket initial and an inner primary androgonial cell (Fig.). The jacket initial undergoes repeated anticlinal divisions and forms a single layered jacket of the antheridium. The primary androgonial cell first divides by a vertical wall, followed by divisions in all possible planes. It results in the formation of a mass of 128-512 or more androgonial cells. The last generation of these cells functions as androcytes (Fig.).

(b) *Mature Antheridium :* The mature antheridium is a small spherical structure, partly projected above the surface of the gametophytic prothallus as a hemispherical dome. It has a single layered jacket of 10-12 sterile cells which enclose a mass of spirally coiled multiflagellate antherozoids (Fig.). The jacket has a 4-5 sided laterally situated opercular cell. The antherozoids are liberated by the disintegration of the opercular cell. Following the liberation of antherozoids, the jacket layer also disorganises.

2. *Archegonium:*

(a) *Development :* The archegonium also develops from a superficial cell of the prothallus. This cell, known as archegonial initial, divides periclinally to form an outer small primary cover cell and an inner large central cell (Fig.). The primary cover cell divides by two successive vertical (anticlinal) walls at right angles to each other, forming four quadrately arranged neck initials. The neck initials undergo transverse divisions and eventually form 4-7 cells high neck of the archegonium.

The central cell divides transversely and forms an upper primary neck canal cell and a lower primary venter cell (Fig.). The nucleus of the primary neck canal cell divides to form two neck canal nuclei, but this division is not accompanied by wall formation. Simultaneously, the primary venter cell also divides by a transverse wall into a large egg cell and a small venter-canal cell (Fig.).

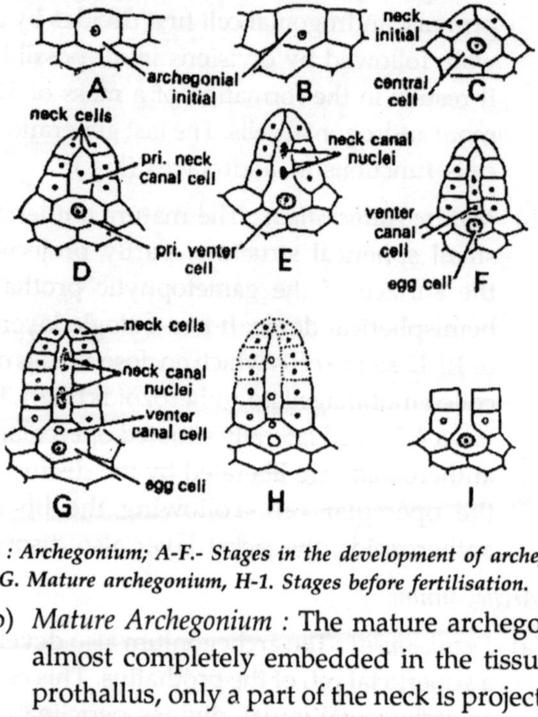

Psilotum* : *Archegonium; A-F.- Stages in the development of archegonium, G. Mature archegonium, H-1. Stages before fertilisation.

(b) *Mature Archegonium* : The mature archegonium is almost completely embedded in the tissue of the prothallus, only a part of the neck is projecting out. The neck is made up of four vertical rows of cells with 4-5 cells in each row. It has two neck canal nuclei, a venter canal cell and an egg cell (Fig.).

3. *Fertilisation* : As the archegonium matures, the cells of the lowermost tier of the neck become thick-walled

and cutinised. Then the distal tiers of the neck break away leaving 1-3 basal tiers only (Fig.). Thus a mature archegonium looks like a shallow plate. The neck canal and the ventral canal nuclei disintegrate and a free passage for the entrance of antherozoids is thus formed.

Young Sporophyte

The diploid zygote, formed by the fusion of the male (antherozoid) and the female (egg) gametes, is the mother cell of the sporophyte generation. It increases in size and eventually almost completely occupies the cavity of the venter (Fig.). It divides by a transverse wall into an upper epibasal and a lower hypobasal cell. This division establishes polarity in the embryo as the epibasal cell always develops into shoot and the hypobasal cell into foot (Fig.). This type of embryogeny, where the shoot forming cell is directed towards the neck of the archegonium, is described as exoscopic.

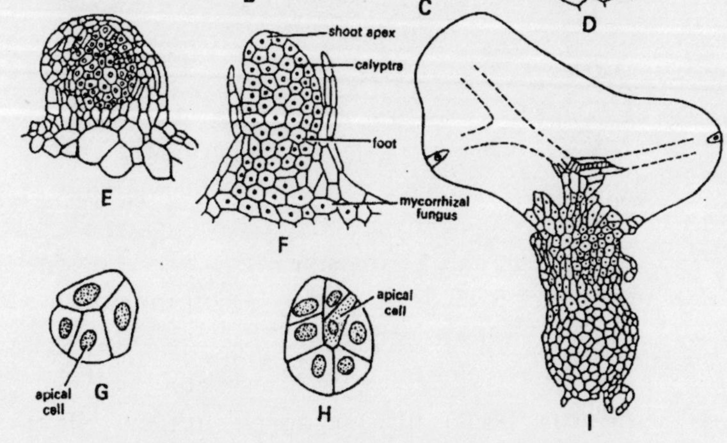

Psilotum : Stages in the development of embryo.

The hypobasal cell divides by a vertical division (at right angles to the first division) and the two daughter cells thus formed divide irregularly. Consequently, a cylindrical foot is formed (Fig.). Some individual cells or strands of cells extend from the surface of the foot into the tissue of the gametophyte. These cells are haustorial in nature and derive nourishment from the prothallus for the developing sporophyte.

The epibasal cell undergoes a vertical division, followed by transverse divisions, and soon one or rarely two apical cells are established (Fig.). Further growth of the shoot is due to the activity of the apical cell/cells. The cells of the prothallus adjoining the young embryo form a calyptra-like structure around it. The embryo grows through the calyptra, projecting beyond the prothallus. It grows horizontally to form a rhizome that develops several unicellular rhizoids. The rhizome soon becomes infected by mycorrhizal fungus. Rhizomes with mycorrhiza are capable of nourishing themselves. The rhizome branches dichotomously and some of the branches turn upward and develop into aerial shoots (Fig.). Once the sporophyte is established, it breaks off from the foot.

12

Psilophytopsida Psilophytales

Psilophytopsida is an assemblage of the simplest and the earliest land plants. Their rootless sporophyte was differentiated into rhizome and slender aerial branches that were more or less dichotomously branched. The aerial axes were either naked or had small spirally arranged appendages. Thick walled homosporous sporangia were terminal or lateral on aerial shoots. The vasculature was protostelic.

The first member of this group, discovered by Sir William Dawson, as early as 1859 from the rocks of the early Devonian period, was called Psilophyton due to its naked body (Psilos = naked, phyton = plant). The discovery of this simple plant did not fit into the preconceived notion of land plants. However, when some other plants with similar simple organization were collected by R. Kidston and W.H. Lang in 1917, it became evident that land plants with such a simple organization did existed.

Riemers (1954) placed these simplest fossil plants in the sub-division Psilophytopsida. It included only a single order Psilophytales and the following four families.

1. Rhyniaceae (e.g., Rhynia, Homeophyton, Cooksonia, Yarravia)
2. Zosterophyllaceae (e.g., Zosterophyllum)
3. Psilophytaceae (e.g., Psilophyton)
4. Asteroxylaceae (e.g., Asteroxylon)

Rhyniacea and Asteroxylaceae are described.

Rhyniaceae

The sporophytic plant body was rootless and differentiated into a cylindrical or lobed rhizome and dichotomously branched erect aerial shoots. The rhizome bore rhizoids in tufts. The aerial axes had appendages; true leaves were, however, absent. Sporangia were borne at the tips of the branches either singly or in groups (synangia). The sporangial wall was several layers thick. Rhynia, Homeophyton, Cooksonia and Yarravia are some common members of Rhyniaceae. Of these, Rhynia and Homeophyton are described here in detail.

Rhynia

Systematic Position

Sub-division	—	Psilophytopsida
Order	—	Psilophytales
Family	—	Rhyniaceae

The fossils of the genus Rhynia were discovered by Kidston and Lang in 1917 from the Rhynie locality of Aberdeenshire in Scotland. So far only two species of Rhynia, viz., R. major and R. gwynne-vaughani are known. Both these species are preserved as complete plants in hard siliceous rocks of the Lower Devonian age.

The remains of these plants are found in perfect petrified

state and their orientation within the rock is often upright. The available evidences suggest that they grow near volcanoes in habitats saturated with acid water from hot springs. The plant body of Rhynia was sporophytic.

Structure of Sporophyte : Both species of Rhynia were herbaceous. The plant body consisted of a subterranean, creeping, cylindrical and dichotomously branched rhizome that bore upright, simple dichotomously branched aerial leafless shoots. The aerial shoots of R. major were about 50 cm in height and 1.5- 6 mm in diameter (Fig. A), and those of R. gwynne-vaughani were comparatively small, measuring about 20 cm in height and 1-3 mm in diameter (Fig. B). Roots were absent but tufts of rhizoids were present on the underside of the rhizome. The rhizome and aerial shoots of R. gwynne vaughani had many adventitious branches which probably developed by proliferation of the cortical cells. These branches had no vascular connection with the main axis and they probably helped in vegetative propagation. The aerial branches end in tapering vegetative apices or bore pear-shaped terminal sporangia.

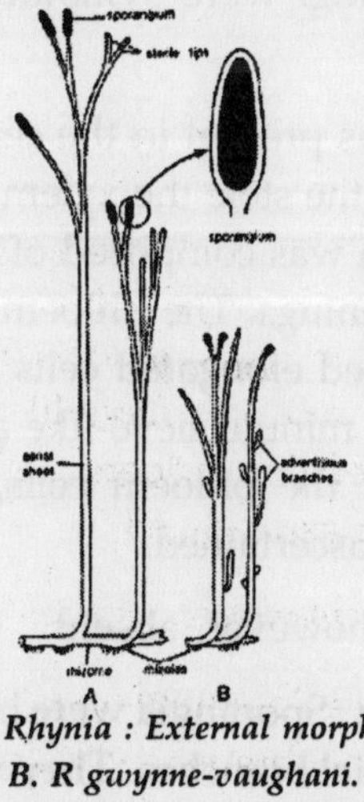

A-B. Rhynia : External morphology: B. R gwynne-vaughani. (after Kidston and Lang).

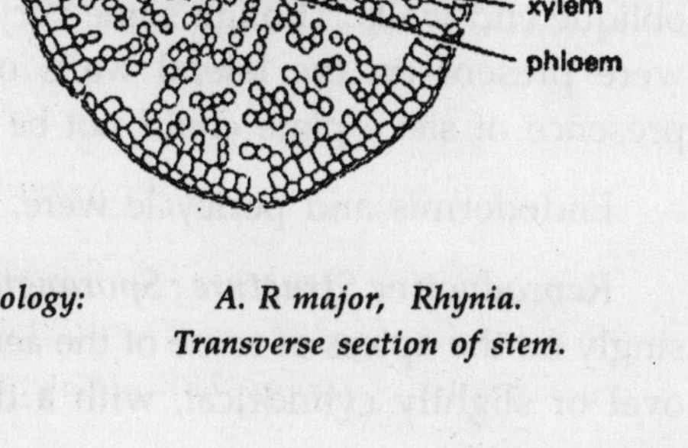

A. R major, Rhynia. Transverse section of stem.

Internal Structure : The internal organization of the rhizome as well as of the aerial shoot was very simple. The following regions are discernible (Fig.).

Epidermis : It was the outermost single-layered envelope covered by a thick layer of cuticle. The epidermis of the aerial shoot was interrupted by stomata. Each stoma consisted of two guard cells, surrounded by many subsidiary cells.

Cortex : The epidermis was followed by a well organised and broad cortex, differentiated into an outer and an inner region. The outer cortex consisted of 1-4 layers of compactly arranged polygonal parenchymatous cells without any intercellular spaces. This region perhaps represents hypodermis. The inner cortex was composed of spherical parenchymatous cells with large intercellular spaces which maintained continuity with the outer atmosphere through the stomata present in the epidermis. This indicates that the inner cortex was the chief photosynthetic region.

In the cortex of some aerial shoots, inter and intracellular fruiting bodies of fungal hyphae were also present. But it is not definitely known whether the fungi were symbiotic or saprophytic.

Central Cylinder : A protostele was present in the central part of the axis as well as rhizome. In the stele the xylem was surrounded by the phloem. The xylem was composed of only tracheids with annular or spiral thickenings. The phloem was represented by 4-5 layers of thin walled elongated cells with oblique end walls. Though some very minute sieve-like areas were present on the lateral walls of the phloem cells, the presence of sieve plate could not be ascertained.

Endodermis and pericycle were, however, absent.

Reproductive Structure : Sporangia : Sporangia were borne singly on the apices of some of the aerial branches. They were oval or slightly cylindrical, with a diameter slightly greater than that of the subtending branch tip. The sporangia of R. major were approximately 12 mm long and up to 4 mm in breadth and those of R. gwynne-vaughani were comparatively small. Each sporangium was surrounded by a multilayered jacket. The outer wall of the jacket was composed of heavily cuticularised thick walled cells. It, was followed by 2-3 layers of thin-walled palisade-like cells. The innermost layer of the jacket was made up of small rounded thin-walled cells, which formed a nutritive layer like tapetum. There were numerous spore tetrads in the sporangial cavity.

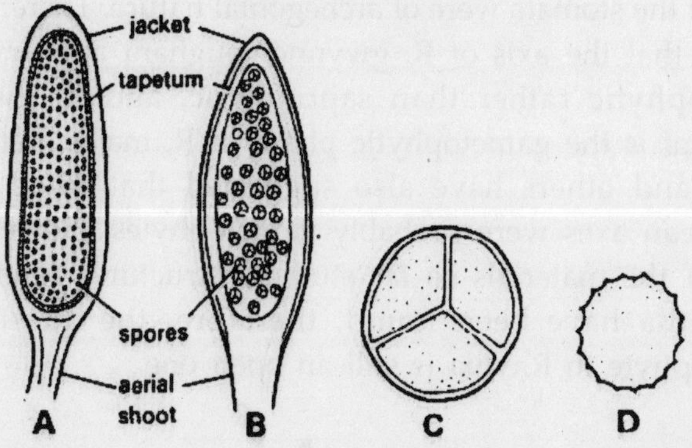

A-D. Rhynia : Sporangium and spores; A-B. Longitudinal section of sporangium of R gwynne-vaughani, C., Spore tetrad, D. Spore.

All spores were of the same type (homosporous). They were heavily cutinised and measured up to 65 microns in diameter. The spores had typical triradiate markings (Fig. B-C). The tetrahedral arrangement of spores reveals that they were formed by reduction division.

The sporangia did not have any specialised mechanism of dehiscence, and liberation of spores perhaps took place by the decay of sporangial jacket.

Gametophyte

We do not have any definite information about the gametophyte of Rhynia. But some palacobotanists suggest that some specimens of Rhynia, which were considered to be saprophytic axes, were probably its gametophytes. But Sporne (1966) did not favour this view as these structures were without archegonia or antheridia. According to Merker (1959), rhizomatous parts of Rhynia represent the gametophyte as some flask-shaped cavities, which probably represent the disintegrated sex organs, have been found in the rhizome. Pant (1962) compared the structure of hemispherical projections (considered to be stomata) in the axis of R. gwynne-vaughani with that of archegonia (Fig.) and he suggested that atleast some of the stomata were of archegonial nature. Therefore, he opined that the axis of R. gwynne-vaughani appears to be gametophytic rather than saprophytic, and R. gwynne-vaughani is the gametophytic phase of R. major. Lemoigne (1968) and others have also suggested that some of the rhynialean axes were probably gametophytes. However, in none of the materials so far studied, structures resembling antheridia have been found, therefore, the question of gametophyte in Rhynia is still an open one.

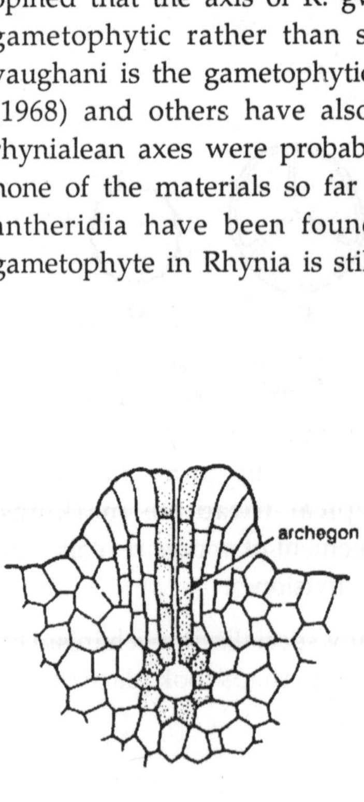

Rhynia gwynne-vaughani : 'Archegonium' (reconstructed; after Lemoigne)

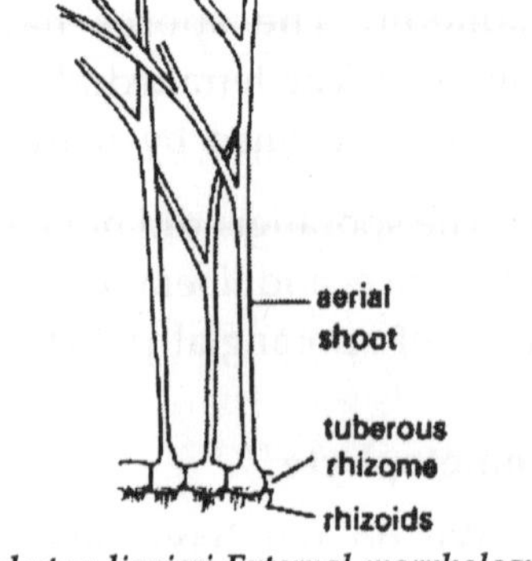

Horneophyton lignieri External morphology.

Horneophyton

Systematic Position

Sub-division	—	Psilophytopsida
Order	—	Psilophytales
Family	—	Rhyniaceae

Homeophyton is a monotypic genus, represented by one species, H. lignieri. It was found along with Rhynia in the chert deposits at Rhynie in Scotland. Earlier, the genus was christened as Homea, but as one member of the family Sapindaceae of the angiosperms is also known by the same name, in 1939 the genus was renamed as Homeophyton.

Structure of Sporophyte

External Structure : The morphology of Horneophyton was similar to that of Rhynia, but it was smaller than Rhynia gwynne-vaughani. The aerial axis was about 13 cm in height and 1 to 2 mm in diameter. It was dichotomously branched and smooth like Rhynia major, adventitious branches were absent.

The structure of the underground rhizome was slightly different from that of Rhynia. It was lobed and tuberous (Fig.), and bore aerial axes on its upper side and numerous unicellular rhizoids on the lower side.

Internal Structure : The internal structure of the aerial axis of Homeophyton was also similar to that of Rhynia. The tissues were differentiated into epidermis, parenchymatous cortex and stele. The vascular tissue was in the form of a cup-like structure in the centre of the stele. The stele of the aerial axis did not continue into the tuber, which was composed of only parenchymatous tissue. In parenchymatous cells of some tubers non-septate fungal hyphae were present in abundance. Since some well preserved tubers were found without fungus, Kidston and Lang (1917) suggested that the fungus was most likely saprophytic in nature which must have invaded the tubers after their death.

Reproductive Structure : Sporangia : The tip of some of the aerial branches had sessile sporangia as in Rhynia. The sporangia varied in size. A feature, peculiar to Homeophyton, was the presence of sterile columella in the sporangium. Some sporangia were bifid and so also their columella. This suggests that the vegetative apex of the axis could be transformed into a sporangium at any time, even during the process of dichotomy.

This fact also rules out the possibility that there were special spore bearing structures, i.e., sporangiophores.

Asteroxylaceae

The sporophytic plant body was differentiated into an underground horizontal rhizome and upright aerial shoots covered with leaves. The internal structure of the aerial axis was much more complex than that of Rhynia. Sporangia were cauline and were definitely not associated with appendages that could be called sporophylls.

The family Asteroxylaceae includes only a single genus Asteroxylon.

Asteroxylon

Systematic Position

Sub-division	—	Psilophytopsida
Order	—	Psilophytales
Family	—	Asteroxylaceae

The genus Asteroxylon is represented by two species, viz., A mackiei and A. elberfeldense. Of these, A. mackiei was found in the Rhynie chert in Scotland along with Rhynia and Homeophyton, whereas A. elberfeldense was reported from middle Devonian rocks near Elberfeld in Germany. The Scottish species (A. mackiei) was preserved only in fragments and was later reconstructed, hence it is not possible to make an estimate of its height. The German species (A. elberfeldense) is, however, better known and it attained a height of about lm.

Structure of Sporophyte : The sporophyte of Asteroxylon was much more complex than that of Rhynia and Hornephyton. It had a naked dichotomously branched underground rhizome which gave rise freely branched upright aerial shoots, clothed with Lycopodium like leaves. The leaves become gradually sparse as they pass downward and the basal part of the aerial shoot was devoid of leaves.

The rhizome did not have rhizoidal hairs, but instead some lateral branches of the rhizome grew downwards. These slender branches probably acted as absorbing organs of the plant.

The leaves were arranged in irregular spirals. They were very small, about 5 mm in length, dorsiventrally flattened, and provided with stomata.

Internal Structure : The internal organisation of Asteroxylon was much more complex than that of Rhynia and Homeophyton. A transverse section of the aerial axis was irregular in outline. The epidermis was indistinct due to the presence of a large number of leaves on the surface of the axis. The cortex had three distinct regions; the outer and the inner cortex were composed of compactly arranged parenchymatous cells while the middle cortex was trabecular. Numerous leaf traces could be seen in the outer and inner cortex. These traces originated from the tips of the protoxylem strands of the central vascular cylinder and faded in the outer cortex without entering the leaves. The leaves were thus veinless. The centre of the axis was occupied by a solid protostele (actinostele). The xylem was solely made up of tracheids with spiral or annular thickenings. It was mesarch with protoxylem present near the tips of the xylem strands. The xylem was surrounded by phloem, composed of thin walled cells.

In rhizomes the stele was haplostele.

Reproductive Structure : Sporangia : Kidston and Lang (1917) found some naked dichotomous branches bearing terminal pear-shaped sporangia, associated but not connected with the preserved fragments of A. mackiei. These sporangia bearing branches, due to their morphological and anatomical similarities with the aerial axis of Asteroxylon, were suspected to be deciduous fertile branches. However, Lyon (1964) discovered some fertile axes of Asteroxylon which bore reniform sporangia among the leaves. Thus the fertile branches with terminal sporangia, described by Kidston and Lang, seems to have been wrongly attributed to Asteroxylon. The presence of reniform sporangia among the leaves suggests a closer affinity of Asteroxylon with Lycopsida than Psilophytopsida.

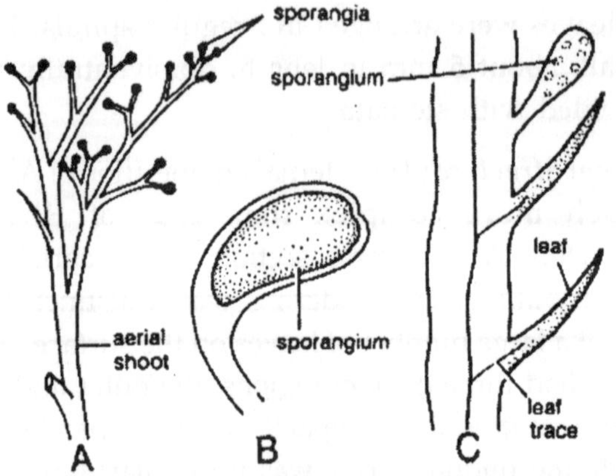

Asteroxylon : Sporangia; A Dichotomously branched axis as originally suggested by Kidston and Lang, B. A single pear shaped sporangium, C. A part of leaf axis showing sporangium amongst leaves (as envisaged by Lyon, 1964).

13

The Fossilisation

Plant life has been in existence on the earth for the past one billion years. The study of the past plant life is known as palaeobotany. It can be made 'through the use of fossils. A fossil, in general, may be defined as a concrete evidence of the existence of a prehistoric organism. Thus, fossils serve as a tool for palaeobotanical studies and the basic material is available in sedimentary rocks.

The process of preservation of living beings or their parts in the form of fossils is known as fossilization. The process of fossilization is a continuing process. Two theories have been proposed to explain the fossilization process in plants.

Replacement Theory : According to this theory, fossilization takes place by the replacement of the molecules of the original substances of the plant one by one by the molecules of minerals in soil solution. The replacement occurs due to hydrolysis or weathering of the organic substances present in the plant body. Several known fossils show such type of replacement.

Infiltration Theory : According to infiltration theory

fossilization takes place as a result of infiltration and precipitation of minerals through the cell wall. After burial, the plant body undergoes partial disintegration, and the free carbon released in this process forms carbonates by reacting with infiltrated calcium, magnesium, etc. Fossilization by infiltration depends upon the disintegration of organic substances present in the cells.

The process of fossilization is affected by a number of factors, such as (i) nature of the plant material (i.e., whether it is thin, delicate or hard), (ii) the rapidity with which the plant material is covered and prevented from organisms of decay, (iii) changes to which rocks are subjected during or after plant parts have been buried, and (iv) extent of mineralization.

Types of Fossils

Fossils occur in a variety of forms. On the basis of the nature of fossilization, fossils may be classified into the following types.

Compression Fossils : These are perhaps the most common type of fossils. As the name indicates, these fossils are formed as a result of burial of plants or its parts in sediments. The buried parts become flat due to compression or overlying pressure of the sediments. Sand stone, diatomaceous earth and

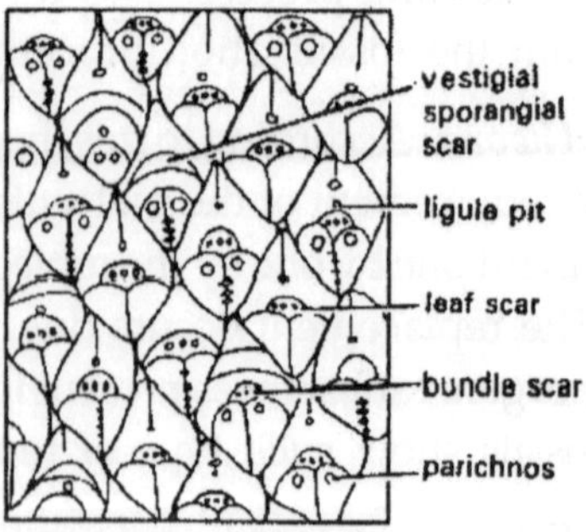

Lepidodendron :Compression fossil (diagrammatic representation).

volcanic ash are some of the materials that cause compression. However, if the compressed plant organ is thin and flat (e.g., leaves), it does not show any appreciable change in its shape. Such fossils occur in the form of a carboniferous layer in between the sediments and they do not show structural details of the cells preserved. Palaeobotanists use compressions to study the external morphology of plants.

Petrifaction Fossils : Those fossils which preserve both external form and internal structure are called petrifactions. In this process, every cell of the plant tissue is preserved and perfectly distended as if it is living. There is molecule-by-molecule replacement of plant parts by some 20 minerals (in the form of carbonates, sulphates, silicates, phosphates, etc.). For petrifaction, the organic part must be protected in some way from rapid decay while mineralization proceeds. In the past this has been accomplished by exclusion of air (as in water saturated sediments), by low pH or by the specific physical and chemical conditions existing under a layer of volcanic dust. Besides, substances like carbon, carboxylic acid, humic acid, hydrogen sulphide, etc., formed by the disintegration of plant tissues are helpful in the precipitation of insoluble petrifying agents such as pyrites. These fossils also show several substances (like cellulose, lignin, etc.) present in living plants. Petrifactions are rare but are most suitable for the study of structural details.

Incrustation Fossils : These are the commonest types of fossils, suitable for the study of morphology of fossil plants. In these fossils, external form of the plant is preserved as a cast, but the internal structure is destroyed. Here, deposition of sediments occurs in the form of a hard coat around the whole plant body or its parts. Incrustations are formed by precipitation of minerals like carbonates on the plant surfaces. Like petrifactions, these fossils do not contain original organic materials and do not show cellular details of the plant.

Impression Fossils : These fossils are just impressions of plants or plant parts on sediments. Hence, they do not contain organic materials as in compressions. Impressions formed in fine and soft materials show better details. They occur abundantly in clay deposits of per year in Western Tennessee. The plant material disorganises and leaves impression, which gives a superficial resemblance of the plant part preserved when clay gets converted into a rock. Impressions are often darker in colour than the rock surface when they contain some organic materials. They show clear details of venation patterns and in some cases even cuticular details can. also be seen. These fossils are extremely useful in the study of external features of various plant parts like the stem, leaf and flower.

Compactation or Mummified Plants : These fossils are plants or plant parts, compressed by the vertical pressure of one against another. In such fossils intervening matrix is not found in between plant parts. Some plant structures like leathery leaves, hard fruits and seeds are preserved in mummified forms. Chemical substances can be infiltrated in such fossils and they may be sectioned for further microscopic investigations.

Coal Balls : Petrified plant organs of roughly spherical shape are known as coal balls. Most of the fossils present in the coal occur in this form. Coal balls are formed by infiltration of calcium carbonate, magnesium carbonate, iron sulphide, etc., in buried plant parts. These substances prevent the conversion of plant organs into coal and result in petrifaction. Hence, these fossils occur in the form of coal balls. They are irregular or sub-spherical masses which show great diversity in their size. Majority of coal balls occur in localised regions and may be congregated in definite bands or in scattered pockets. Coal balls are specifically significant in elucidating palaeobotanical informations regarding the plant communities of geological past.

Ambcr : It is a resinous excretion of certain fossil coniferous trees which flowed due to injuries caused by boring insects or from the decaying branches. The exudate accumulated on the forest floor, gradually hardened to take the shape of amber. As several small plants and animals are found preserved in this resin, amber is considered as a type of fossil.

Leaf Coal or Paper Coal : These fossils consist of thin dead leaves, dispersed in organic matrix. As the inner tissues of leaves are destroyed, there are layer after layer of cuticle.

Actual Remains : Such plant fossils are of relatively young age. Their complete decomposition has been prevented by low temperature and absence of air (i.e., vacuum). Usually materials such as cuticle, spore walls, etc., which show remarkable resistance to natural decomposition, are preserved in this form.

Pseudofossils : Sometimes rocks take the shape of some plant part or animal during their formation and as such give an illusion of fossils. Their detailed studies, however; clearly reveal that they are not plants or animals but are mineral depositions mistaken for plant or animal remains. Such specimens are known as pseudofossils.

Factors Affecting Fossilisation

The utility of fossils depends upon the degree of their preservation. Majority of the known fossils are imperfect, and perfect fossils are very rare. In most cases only external morphological features are preserved and cellular details are hardly visible. The factors affecting the degree of preservation can be classified into the following two groups.

Nature of Tissues Forming the Plant Body : The nature of tissues present in different plants is variable. For instance, most micro-organisms and lower plants like algae and fungi consist of thin-walled soft cells, whereas higher vascular plants have sufficient quantity of thick walled cells, such as sclerenchyma and tracheary elements. The degree of preservation depends upon the mechanical resistance of the cells. Thin walled soft organs like thin leaves, flowers and fleshy fruits are not suitable for preservation, whereas thick walled organs, such as stems and hard seeds are found preserved in the form of petrifactions.

Intensity and Duration of the Destructive Forces : Perfectness of a fossil depends upon the duration and intensity of destructive forces present before or during the fossilization process. The destructive forces are mainly decay, hydrolysis and mechanical pressure of water, sand, soil, sediments and rocks. If destructive forces are more active before fossilization, it results in the formation of imperfect fossils.

For preservation of fossils, the ideal atmosphere is such an enclosed water reservoir in which fine grained sediments accumulate on the surface at a fast rate. In such conditions the plant or plant parts get buried under the sediments easily. In deep waters saprophytic fungi are rare and oxygen content is low, hence, the intensity of disintegration is very slow. Besides, the mechanical pressure of waves is also low in depth. Plants are thus better preserved in deep waters in comparison to shallow waters.

Some useful Techniques for Fossil Study : Several techniques are available for the study of fossil plants. Some important ones are described below.

Ground Thin Sections : This technique is most suitable for the study of petrifactions. The specimen is cut to a convenient size and the surface meant for study is smoothed with # 300 carborundum. The smooth surface is then mounted on a glass slide by a special resin. Thereafter, the fastened specimen is cut as close to the glass as possible to obtain a thin slice. The upper surface of the slice thus obtained is ground further by # 100 carborundum as to make the section so thin and smooth that it appears transparent. The section is then mounted under a coverslip using a suitable mounting resin. Now the slide can be studied under a microscope. The main disadvantage of this technique is that a lot of material is wasted in obtaining thin sections.

Film or Peel Tchnique : This technique is most suitable for the study of well preserved petrifactions. The specimen to be studied is ground and smoothened and then it is etched with hydrofluoric acid (HF) and hydrochloric acid (HCl). The etched surface is gently washed in running water as to remove the acid, air-dried and then covered with a thin film of nitrocellulose. After drying in air for about 6 hours, the film is carefully peeled and mounted on a slide by Canada balsam. The peel is then permanently sealed with a cover glass. The slide can be studied under binocular microscope. The film technique is more advantageous than the ground section because it is less expensive and quicker and the peels obtained are translucent, thinner and durable.

Transfer Technique : This technique is most suitable for the study of compressions. Structural features, such as venation pattern and cuticular structures can be easily observed by this technique. The face of the specimen adjoining the rock surface is coated with two coats of peel solution. When the film is dried the rock matrix is removed by cutting. The specimen is then immersed in 25 per cent hydrofluoric acid to dissolve away the rock matrix. Finally, the specimen is washed, dehydrated and mounted.

Microtomy : The specimen is first softened by treating it with hydrofluoric acid and potassium chlorate ($KClO_3$), followed by a mixture of equal parts of phenol and 95% alcohol. Thereafter, it is embedded in paraffin or celloidin and sectioned with the help of a microtome. This technique is useful for fossil woods which retain much of the cell wall structure intact.

X-rays Radiography : Some fossils are damaged due to acid treatment in the process of peeling or microtomy. Such fossils can be studied by X-rays radiography. This technique is particularly useful for the study of calcified seeds obtained from coal balls. The specimen is photographed by X-rays and difference in density are recorded by noting corresponding differences in shadow effects.

Maceration Technique : This technique is suitably used for peat, lignite and coal. It is very useful for fossil pollen and spore analysis. The specimen is first treated by hot dilute nitric acid and potassium chlorate and then by a dilute solution of sodium hydroxide. In the process, breakdown of matrix takes place and the plant fragments are then washed properly with water. These fragments are mounted on a glass slide.

Excavation Technique : This technique is particularly suitable for the study of minute leaves and sporangiophores. The specimen is removed with the help of a microvibratory machine or a small hammer and steel needles. This process, though slow and tedious, provides highly promising results.

Palaeobotany and Geological Time Scale : The history of plants on the earth is about a billion years old. During this period many forms of plants originated on the earth and disappeared subsequently due to climatic changes or catastrophies like earthquakes, landslides, etc. Although most of the plants of geological periods were decomposed by bacteria after their death, a few plants or their parts were buried deep in the soil and they did not decompose. These plants/parts are found preserved in the sedimentary rocks of different ages. They are called fossils, a word derived from the Latin verb fodere, which means 'to dig'. Arnold (1947) defined fossils as "the relic of some former living things-plants or animals-embedded in or dug out of the superficial deposits in past geological periods". In correct sense the term fossil not only includes preserved plant and animal remains but also everything connected with an organism. For instance, it has been found that many chemical molecules constituting the organic skeleton of plants have undergone no or only minimal change during fossilization due to their resistant nature. Such substances are referred to chemical fossils, and these have proved useful as biological markers.

Nomenclature of Fossils

Rules governing the nomenclature of living plants, also apply to fossils. Fossilization of plants, besides other factors, also depends upon the hardness and softness of different tissues; harder tissues have better chances of fossilization than softer tissues. Consequently, a complete plant is rarely preserved and fossils of only fragments (i.e., stem, root, leaf, etc.) are found. As these fragments are being discovered at different times by different workers, it becomes very difficult to organise scattered fossil fragments and organs into a complete plant and assign a binomial to the plant as a whole. Therefore, at the suggestion of VI International Botanical Congress (1935) each detached plant part or organ is given, a different name which acquires the status of a genus. Unlike living plants, the generic name in fossils is applied only to a plant part, without actually indicating to what plant it belongs. Each part represents a form genus, organ genus or artificial genus. A complete plant is then reconstructed by putting together these form genera on the basis of their similarity and period to which they belong. For example, the fossil plant Calymmatotheca hoeninghausi has been reconstructed by putting together roots called Kaloxylon hookeri, stem known as Lyginopteris oldhamia, leaf as Sphenopteris hoeninghausi, rachis as Rhacopteris aspera and seeds as Lagenostoma lomaxi.

Rules for naming form-genera. In naming a form genus a particular suffix is applied which signifies the part it represents. The suffixes applied to different parts are as follows:

— dendron to a stem; e.g., Lepidodendron, Lyginodendron.

— phyllum to a leaf; e.g., Lepidophyllum, Pttilophyllum, Nipaniophyllum, Stigillariophyllum.

— xylon to a woody part of a plant; e.g., Dadoxylon, Mycloxylon.

— spermum, -carpon, -carpus or -stoma to a seed; e.g., Pterygospertnum, Corystospermum, Lepidocarpon, Mazocarpon, Trigonocarpus, Cordaicarpus, Rotundocarpus, Sterocarpus, Stephanostoma, Lagenostoma.

— theca to a microsporangia, e.g., Polytheca.

— strobus to a cone, e.g., Lepidostrobus, Androstrobus.

Geological Time-table

The history of earth is approximated 350 million years old. Geologists have used two major units-time and strata for sub-dividing the geological history of the earth.

On the basis of time the geological history of the earth has been divided into five Eras—Archaeozoic, Proterozoic, Palaeozoic, Mesozoic and Coenozoic. Each era includes many Periods and each period is further divided into Epoch. Similarly, on the basis of Strata the geological history of the earth is divided into System, Series, Stage and Zone. Various eras and their age, climate and vegetation are listed in Table and a brief description of these eras is given below.

Precambrian Era : Life originated on the earth in Precambrian era. The rocks of this era have only a few fossils because either the environment was not suitable for fosssilisation or the organisms present during that time had tissues not fit for fossilization. In the absence of sufficient fossils, it is difficult to ascertain Precambrian environment.

Palaeozoic Era : The vegetation of this period includes both aquatic and land plants. The aquatic plants like

Nematothallus and Prototaxites showed resemblance with thallophytes.

The first fossils of land plants were found in the Silurian and Devonian period of this era (225-350 million years ago). Manograptus, Sporogonites, Zosterophyllum, Cooksonia, Pseudosporochnus, Rhynia, Homeophyton and Psilophyton were some important genera of Silurian and Devonian period. In the Carboniferous and Permian period, members of Lycopsida, Sphenopsida and Pteropsida formed dense forests. Seed ferns were abundant and as such this period is also known as age of seed ferns. There were perhaps seasonal changes in environment during this period as annual rings were observed in certain fossils like Callixylon erianum and C. newbanyi.

During the late carboniferous period, there were significant changes in the environment which resulted in the development of swampy regions. Such habitat was unsuitable for woody vegetation and in the late Devonian period the vegetation present on the earth was converted into coal due to geological events. The coal , beds of Pennsylvania and West Virginia were formed due to coalification of lycopsids, sphenopsids, seed ferns and gymnosperms of that period. Fortunately, few psilopsids, lycopsids and sphenopsids escaped devastation and eventually gave rise to the present forms.

Mesozoic Era : The environment of the Triassic period of this era was very arid and thus unsuitable for the growth of plants requiring humid climate. Fossil records, however, indicate the presence of many ferns, cycadophytes and conifers in the Triassic period. This period is also known as 'age of cycads'. Although flowering plants also originated in this period, their number was negligible.

Significant changes occurred in the vegetation of the Cretaceous period. Members of Salicaceae, Moraceae, Araliaceae, Lauraceae and Sapinadaceae originated during

Geological time-table

Era	*Period (system)*	*Epoch (Series)*	*Climatic variations*	*Important events in plant-life*	*Dominant plants*
	Recent (30,000 years)	Recent			
	Quaternary	Pleistocene (1 million years)	Development of modern climatic regions; cold climate in polar regions.	Dominance of herbaceous plants.	Herbaceous plants
	Tertiary	**Pliocene** (17 million years)	Development of alpine mountain ranges.	Spread of herbaceous dicoty ledonous plants.	
Coenozoic		**Miocene** (30 million years)		Decrease in number and area of forests.	
		Oligocene (40 million years)		Rise in the number of woody angiosperms .	Angiosperms
		Eocene (60 million years)	Development of Hima layan mountain ranges.	Spread of angiospermic forests.	
		Paleocene (80 million years)			
	Cretaceous (110 million years)	**Upper Cretaceous**		Dominance of angiosperms and decrease in the number of gymnosperms.	
		Middle Cretaceous	Laramide revolution	Rapid increase in the number of angio sperms.	
Mesozoic		**Lower Cretaceous**		Development of cycads, conifers and angiosperms.	

Contd...

Era	*Period (system)*	*Epoch (Series)*	*Climatic variations*	*Important events in plant-life*	*Dominant plants*
	Jurassic (170 million years)		Temperate climate	Rise in the number of leptosporangiate ferns; origin of angiosperms.	Gymnosperms
	Triassic (200 million years)		Warm temperate climate	Seed ferns (Pteridosperms) become extinct; increase in number of cycads, conifers and Ginkgoales.	
	Permian (240 million years)		Glaciation.	Extinction of primitive gymnosperms and arborescent pteridophytes.	
	Carboniferous (280 million years)		Appalachian revolution.	Dominance of lycopods, Calamitales, Lepidodendrales and Cordaitales; Origin of bryophytes.	Pteridosperms
Palaeozoic	**Devonian** (360 million years)		Warm and humid climate.	Origin of land vascular plants (Lycopsida and Psilopsida); algal members in plenty.	Primitive land plants (Nematophytales), then Psilophytales.
	Silurian (375 million years)		Warm climate.	Origin of land plants; dominance of algae.	Algae, mostly calcareous.
	Ordovician (420 million years)		Development of Caledonian ranges.	Dominance.of marine and green algae.	
	Cambrian (550 million years)		Development of Acadian rocky ranges.	Origin of algae.	
Proterozoic (1500 million yrs)			Blue-green algae and bacteria.		
Archaeozoic (3800 million yrs)	**Precambrian**		Warm and aquatic conditions, unicellular organisms.	Organisms were probably very simple and unicellular.	Probably unicellular organisms

this period. Of the 300 Cretaceous species collected from North America, 200 species belong to angiosperms. It shows that angiosperms formed a significant part of the Cretaceous vegetation.

Coenozoic Era : This era is known as 'age of angiosperms'. High mountain ranges like the Himalayas in Asia and the Alps in Europe formed during this period. Consequently, the effect of hot winds was minimised and the ice age commenced. Gradually ice shifted towards the poles (before this the poles were not snow covered). This was accompanied with the development of tropical and temperate regions. These alterations in climate had significant effect on vegetation. The seed plants gradually migrated towards tropical regions and those which could not migrate became extinct. As most plants of the late Coenozoic era were deciduous, it shows that there were seasonal variations too. The vegetation of the temperate and polar regions mostly consisted of annual or biennial herbs which could withstand low temperature.

Objectives of Palaeobotanical Studies : There are mainly two aspects of palaeobotanical studies—botanical and geological. The botanical aspects include (i) study of external and internal structure of fossil plants and their distribution, (ii) to provide evidence for phylogeny and evolution of various plant groups, (iii) to ascertain climate and environment of the geological period, and (iv) to correlate the palaeoclimates with the distribution of plants. The geological aspects include (i) calculation of age of the rocks, (ii) geographical distribution of coal and water, and (iii) to provide useful information in the exploration of fossil fuel like coal and petroleum.

14

The Classification

The vascular plants have long been divided into two groups- Pteridophyta and Spermatophyta. Pteridophyta include vascular cryptogams (e.g., Psilotum, Lycopodium, Equisetum, ferns, etc.), which lack seeds and reproduce by spores. Spermatophyta, on the other hand, are characterised by the presence of seeds and include gymnosperms and angiosperms. However, distinction between these two groups on the basis of the presence and absence of seeds became invalid after the discovery of certain fern like fossil plants which bore seeds (Cycadofilicales). Sinnott (1935), therefore, proposed a new term. Tracheophyta for all those plants which possess sporophyte with well developed vascular system. Thus Tracheophyta include both vascular cryptogams (Pteridophyta) and higher plants (Spermatophyta).

Arthur J. Eames (1936) opined that classification of vascular plants should be based upon several fundamental characters of the plant body, instead of a single character. He classified Tracheophyta into the following four groups on the basis of (i) the nature and relation of leaf and stem, (ii) the vascular anatomy, (iii) the position of sporangia.

TRACHEOPHYTA (Group) → (Order)

- **Psilopsida**
  - Psilophytales (e.g., *Rhynia*)
  - Psilotales (e.g, *Psilotum*)
- **Lycopsida**
  - Lycopodiales (e.g., *Lycopodium*)
  - Selaginellales (e.g., *Selaginella*)
  - Lepidodendrales (e.g., *Stigmaria*)
  - Pleuromeiales (e.g., *Pleuromeia*)
  - Isoetales (e.g., *Isoetes*)
- **Sphenopsida**
  - Hyeniales (e.g., *Hyenia*)
  - Sphenophyllales (e.g., *Sphenophyllum*)
  - Equisetales (e.g., *Equisetum*)
- **Pteropsida**
  - **Filicinae**
    - Coenopteridales (e.g., *Botryopteris*)
    - Ophioglossales (e.g., *Angiopteris*)
    - Filicales (e.g., *Polypodium*)
    - Marattiales (e.g., *Angiopteris*)
  - **Gymnospermae**
  - **Angiospermae**

Tippo (1942) gave Tracheophyta the status of a phylum and divided it into the following four subphyla (equivalent to sub-division).

Phylum-Tracheophyta

Sub-phylum

- **Psilopsida**
  - Class : Psilophytinae
  - Order 1. Psilophytales
    2. Psilotales
- **Lycopsida**
  - Class : Lycopodinae
  - Order 1. Lycopodiales
    2. Selaginellales
    3. Lepidodendrales
    4. Pleuromeiales
    5. Isoetales
- **Sphenopsida**
  - Class : Equisetinae
  - Order 1. Hyeniales
    2. Sphenophyllales
    3. Equisetales
- **Pteropsida**
  - Class : 1.Filicinae
  - Order 1. Coenopteridales
    2. Ophioglossales
    3. Marattiales
    4. Filicales
  - Class 2. Gymnospermae
  - Class 3. Angiospermae

The International Code of Botanical Nomenclature does not recognise the use of terms phylum and sub-phylum in botanical classification. It recommends that all names of divisions end in the suffix - phyta and those of subdivision in the suffix - opsida. Wardlaw (1955), therefore, used division and subdivision for phylum and sub-phylum respectively. An outline of the classification of Tracheophyta proposed by Wardlaw is given below.

Division-Tracheophyta

Subdivision

- **Psilopsida**
  - Class : Psilophytineae
  - Order 1. Psilotales
    2. Psilophytales
- **Lycopsida**
  - Class : Lycopodineae
  - Order 1. Lepidodendrales
    2. Lycopodiales
    3. Pleuromeiales
    4. Isoetales
- **Sphenopsida**
  - Class : Equisetineae
  - Order 1. Equisetales
    2. Sphenophyllales
    3. Calamitales
    4. Pseudoborniales
    5. Hyeniales
- **Pteropsida**
  - Class 1 : Filicinae
  - Order 1. Coenopteridales
    2. Ophioglossales
    3. Marattiales
    4. Filicales
  - Class 2. Gymnospermae
  - Class 3. Angiospermae

Smith (1955) was of the view that it is inappropriate to treat Psilopsida, Lycopsida, Sphenopsida and Pteropsida as subdivisions of a single division Tracheophyta as it minimizes their marked divergence from one another. Therefore, he divided vascular cryptogams (pteridophytes) into the following four divisions.

Bold (1957), Benson (1957) and Zimmermann (1959) also adopted the classification proposed by Smith (1955).

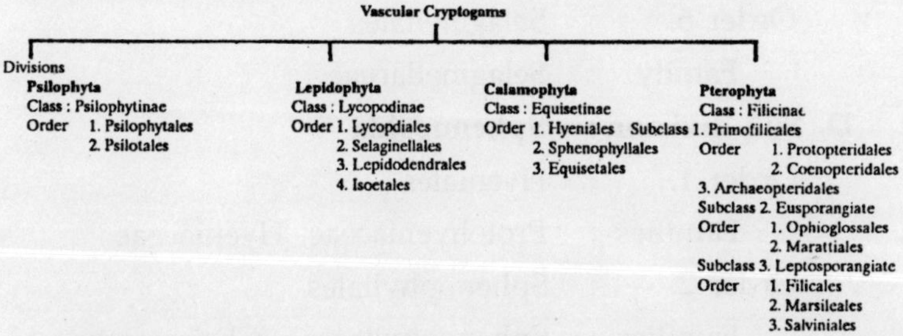

Riemers proposed a classification of Pteridophyta which was published in 1954 edition of Engler's *Syllabus der Pflanzenfamilien.* Sporne (1966) also followed the same classification. Following is an outline of Riemer's classification.

Division	**-**	**Pteridophytes**
A. Sub-division	**:**	**Psilophytopsida**
Order	:	Psilophytales
Families	:	Rhyniaceae, Zosterophyllaceae, Psilophytaceae, Asteroxylaceae
B. Sub-division	**:**	**Psilotopsida**
Order	:	Psilotales
Families	:	Psilotaceae, Tmesipteridaceae
C. Sub-division	**:**	**Lycopsida**
Order 1.	:	Protolepidodendrales
Families	:	Drepanophyceae, Protolepidodendraceae

Order 2.	:	Lycopodiales
Family	:	Lycopodiaceae
Order 3.	:	Lepidodendrales
Families	:	Lepidodendraceae, Bothrodendraceae, Sigillariaceae, Pleuromeiaceae
Order 4.	:	Isoetales
Family	:	Isoetaceae
Order 5.	:	Selaginellales
Family	:	Selaginellaceae
D. Sub-division	:	**Sphenopsida**
Order 1.	:	Hyeniales
Families	:	Protohyeniaceae, Hyeniaceae
Order 2.	:	Sphenophyllales
Families	:	Sphenophyllaceae, Cheirostrobaceae
Order 3.	:	Calamitales
Families	:	Asterocalamitaceae, Calamitaceae
Order 4. Family	:	Equisetales Equisetaceae
E. Sub-division	:	**Pteropsida**
a. Class	:	Primofilices
Order 1.	:	Cladoxylales
Families	:	Cladoxylaceae, Pseudosporochnaceae
Order 2.	:	Coenopteridales
Families	:	Zygopteridaceae, Stauropteridaceae, Botryopteridaceae
b. Class	:	Eusporangiatae
Order 1.	:	Marattiales
Families	:	Asterotheaceae, Angiopteridaceae. Marattiaceae, Danaeaceae, Christenseniaceae

Order 2.	:	Ophioglossales
Family	:	Ophioglossaceae
c. Class	:	Osmundidae
Order	:	Osmundales
Family	:	Osmundaceae
d. Class	:	Leptosporangiatae
Order 1.	:	Filicales
Families	:	Schizaceae, Gleicheniaceae, Hymenophyllaceae, Dicksoniaceae, Matoniaceae, Dipteridaceae, Cyatheaceae, Dennstaedtiaceae, Adiantaceae, Polypodiaceae
Order 2.	:	Marsileales
Families	:	Pilulariaceae, Marsileaceae
Order 3.	:	Salviniales
Families	:	Salviniaceae, Azollaceae

Bierhorst (1971) emphasized to discard categorical differences as new organisms are discovered (especially fossil forms) and new information is coming forward about known ones. He pointed out that categories, such as classes, orders, etc., are recognized because of complete ignorance concerning, or lack of recognition of intermediate forms that we assume to have existed in the past. He dropped the taxon sub-division as a first step away from a pigeonhole system towards a spectral system. An outline of his classification is given below.

Division	:	**Tracheophyta**
Class	:	Rhyniopsida
Order	:	Rhyniales
Class	:	Zosterophyllopsida
Order	:	Zosterophyliales

Class	:	Lycopodiopsida (Lycopsida)
Orders	:	Asteroxylales, Lycopodiales, Protolepidodendrales, Selaginellales, Lepidodendrales, Isoetales
Class	:	Cladoxylopsida
Order	:	Cladoxylales
Class	:	Fquisetopsida (Sphenopsida)
Orders	:	Hyeniales, Pseudoborniales, Sphenophyllales, Equisetales
Class	:	Coenopteridopsida
Order	:	Coenopteridales
Class	:	Fiticopsida
Orders	:	Noeggerathiales. Filicales, Marsileales, Salviniales
Class	:	Ophioglossopsida
Order	:	Ophioglossales
Class	:	Marattiopsida
Order	:	Marattiales
Class	:	Aneurophytopsida
Orders	:	Aneurophytales, Accheopteridales, Protopityales
Class	:	Cycadopsida
Orders	:	Pteridospermales, Cycadales, Cycadeoidales, Caytoniales
Class	:	Coniferopsida
Class	:	Gnetopsida
Class	:	Angiospermopsida
Sub-classes	:	Dicotyledonidae Monocotyledonidac

Riemers (1954) system of classification has been followed here.

Important Characters of Various Sub-Divisions of Pteridophyta

Psilophytopsida : These are the oldest and the simplest vascular plants that originated during Silurian and Devonian period of the Palaeozoic era. Their age and simple form suggest that they represent the earliest vascular land plants. All members of Psilcphytopsida (e.g., Rhynia, Psilophyton, Asteroxylon, Zosterophyllum) are extinct and they are found only in fossil forms.

(1) The plant body was sporophytic and differentiated into a slender horizontal rhizome and an erect cylindrical stem.

(2) The rhizome bore tufts of unicellular rhizoids at intervals. True roots were absent.

(3) The stem was dichotomously or freely branched. It was either naked (e.g., Rhynia, Horneophyton) or was clothed with small leaves (e.g., Asteroxylon).

(4) The axis had a slender central protostele.

(5) Sporangia were borne at the tips of the erect branches either singly (e.g., Rhynia) or in pairs (e.g., Psilophyton). Occasionally they formed synangia (e.g., Yarravia).

(6) Spores were ovoid, obovoid or ellipsoid. All spores were of the same type, i.e., these plants were homosporous.

(7) Their gametophytes are not known.

Psilotopsida : The sub-division Psilotopsida includes only two living members, Psilotum and Tmesipteris. Eames (1936) and Wardlaw (1955) included both these members in Psilopsida, along with fossil members.

(1) The plant body is sporophytic, differentiated into an underground rhizome and erect aerial branches.

(2) The rhizome and aerial branches are dichotomously branched.

(3) True roots are absent, instead absorption is done by rhizoids present on the rhizome.

(4) The aerial branches have some spirally arranged scaly appendages (e.g., Psilotum) or large foliage leaves (e.g., Tmesipteris).

(5) Both the rhizome and aerial branches have a protostele.

(6) The central part of the stele has a solid core of xylem tracheids, usually with annular thickenings. But in the transitional region at the base of the aerial axis, the xylem becomes medullated.

(7) Sporangia are borne in the axils of scaly appendages or foliage leaves. They are bi-or trilocular.

(8) They are homosporous.

(9) The gametophyte is cylindrical or branched, subterranean and colourless.

(10) Sex organs are partially embedded in the prothallus.

Lycopsida : This sub-division includes both living and fossil pteridophytes. These plants developed during the Devonian period of the Palaeozoic era. This sub-division has only five living genera, Lycopodium, Phylloglossum, Selaginella, Isoetes, and Stylites and the rest of the members are known in fossil form.

(1) The sporophytic plant body is differentiated into root, stem and leaves.

(2) Both, the root and stem show dichotomous branching.

(3) The living forms are microphyllous and the leaf has a singie unbranched mid-vein.

(4) Leaves are usually spirally arranged around the stem, but in some members they are opposite or whorled. A ligule is present at the base of the leaf in Selaginella and Isoetes.

(5) The stele is usually protostelic, but some fossil and living forms are siphonostelic or polycyclic.

(6) Secondary growth is absent except for Isoetes.

(7) The leaves bearing sporangia are called sporophylls; sporangia are borne on the adaxial surface.

(8) Sporophylls usually form loose or compact cone (strobilus) at the apex of the branch.

(9) The plants are homosporous (e.g., Lycopodium) or heterosporous (e.g., Selaginella, Isoetes).

(10) The homosporous forms have exosporic gametophyte and the heterosporous forms possess endosporic gametophyte.

Sphenopsida : These pteridophytes evolved during the Carboniferous period of the Palaeozoic era and were represented by large trees like Calamites. This sub-division has only one living genus Equisetum and about 18 extinct forms whose preserved remains are known (e.g., Calamites, Annularia, Sphenophyllum).

(1) The sporophytic plant body is differentiated into root, stem and leaves.

(2) The stem is branched articulated with longitudinal ridges and grooves. It has distinct nodes and internodes.

(3) Leaves are small and scaly, arranged in definite or indefinite whorls at the node.

(4) The stem has a protostele (e.g., Sphenophyllum) or a siphonostele (e.g., Calamites, Calamophyton, Equisetum).

(5) Branches arise from the axils of the scale leaves at the node.

(6) Though living forms of Sphenopsida do not show secondary growth, some extinct forms do possess

secondary thickenings (e.g., Sphenophyllum, Calamites).

(7) Sporangia develop on special stalked structures, known as sporangiophores, which form compact strobilus at the tip of the stem.

(8) The living forms are homosporous, but some extinct forms were heterosporous (e.g., Calamites cashenana).

(9) The gametophyte is green and autotrophic. It is exosporic, i.e., it develops outside the spore wall.

(10) Antherozoids are multiflagellate.

Pteropsida **:** This is the largest and most highly evolved group of Pteridophyta, having nearly 10,000 species, representing about 300 genera. These plants arose during the Devonian period together with other pteridophytes, and were conspicuous element in the flora of the Carboniferous period.

(1) The sporophytic plant body is usually perennial, differentiated into root, stem and leaves.

(2) They grow mostly in moist and shaded terrestrial habitats, but some (e.g., Azolla, Salvinia, Marsilea) are aquatic. Ophioglossum pendulum is an epiphytic species.

(3) Except for some tree ferns (e.g., Angiopteris), most Pteropsida have short and stout rhizome.

(4) Leaves are exstipulate, usually megaphyllous and pinnately compound but the lamina is simple in Ophioglossum.

(5) The rachis is covered with brown hairs, called ramenta.

(6) Young fronds are circinately coiled.

(7) The members of this group have almost all types of

steles, i.e., protostele, siphonostele, solenostele, dictyostele and polycyclic stele.

(8) Vegetative multiplication takes place by fragmentation, adventitious branches, tubers or apogamy.

(9) The plants are homosporous (e.g., Pteris, Pteridium, Dryopteris) or heterosporous (e.g., Marsilea, Regnellidium, Pelularia) and spores are formed in sporangia.

(10) Sporangia develop in groups (called sori) on the ventral surface or on the margins of the leaf (sporophyll).

(11) In some species sporangia are covered by a protective membrane, called indusium.

(12) The sporangium develops either from a single initial cell (leptosporangiate) or from a group of initials (eusporangiate).

(13) The spore, on germination, gives rise to autotrophic prothallus (i.e., gametophyte).

(14) Antherozoids are multiflagellate.

(15) The zygote gives rise to embryo; suspensor is formed only in some primitive forms.

General Characters

Pteridophytes (pteron = feather, phyton = plant, i.e., plants with feather like leaves) are cryptogams with well developed vascular system. They include most primitive living and fossil vascular plants and are also known as vascular cryptogams. They are represented by about 400 living and fossil genera and some 10,500 species. These plants originated about 400 million years ago in the Silurian period of the Palaeozoic era and formed dominant vegetation on the earth during the Devonian period.

On the basis of comparative morphology of living and

fossil forms, two different views have been put forward regarding the origin of pteridophytes. Scott (1900), Bohlin (1901), Church (1919), Fritsch (1916, 1945), Eames (1936), and Arnold (1947) believe in algal origin of pteridophytes. Only Chlorophyceae (monophyletic origin) or Chlorophyceae and Phaeophyceae (polyphyletic origin) have been considered as the progenitor of this group by these workers.

Biochemical and ultrastructural studies also suggest a strong affinity between algae and archegoniate plants (Stebbins and Hill, 1980; Whatley, 1982; Graham, 1982). The supporters of this view consider that the similarity in the sex organs and life-cycles of bryophytes and pteridophytes is due to their parallel evolution. Campbell (1895), Lignier (1903), Zimmermann (1930, 1938), and Bower (1935), on the other hand, postulated that ancestors of pteridophytes were some bryophytes-like plants, especially those resembling Anthocerotales. According to them, both bryophytes and pteridophytes have evolved from a common group of hypothetical archegoniate land plants along different lines.

Characteristic Features of Pteridophytes

(1) The plant body is sporophytic, differentiated into root, stem and leaves. The mature sporophyte is nutritionally independent of the gametophyte.

(2) They show much variation in their form, size and habit. They range from small annuals (e.g., Azolla, Salvinia) to large tree-like perennials (e.g., Angiopteris, Osmunda).

(3) Most of the living pteridophytes are terrestrial, growing in moist and shady places. Some members (e.g., Azolla, Isoetes, Marsilea, Salvinia, are aquatic, usually occurring in permanent ponds. A few species are xerophytic (e.g., Selaginella rupestris, S. lepidophylla, S. densa, Equisetum arvense)

(4) Primary roots are ephemeral and are soon replaced by adventitious roots. The root has a permanent growing apex.

(5) Except for some woody ferns (e.g., Angiopteris. Marattia), most pteridophytes have herbaceous stem. Like root, the stem also has a permanent growing apex.

(6) The leaves are scaly (e.g., Equisetum), small and sessile (e.g., Lycopodium, Selaginella), or large, petiolate and compound (e.g,, ferns). On the basis of leaf structure, pteridophytes are classified into :

(i) *Microphyllous Pteridophytes* : With simple leaves having a single vein, which do not form leaf gap in the stem stele (e.g., Equisetum, Lycopodium, Selaginella), and

(ii) *Megaphyllous Pteridophytes :* With large pinnatified leaves having a complex series of veins; they form prominent leaf gaps in the stem stele (e.g., Reds, Dryopteris, Lastrea).

(7) The stem is usually branched; the branching is monopodial or dichotomous. The branches do not arise from the axils of leaves (axillary) as in spermatophytes.

(8) The leaves and stem possess filiform trichomes. Stomata are present on both the upper and lower surface of the leaf.

(9) The root and stem have well developed vascular system, composed of xylem and phloem. Cambium is, however, absent and hence they donot- show secondary growth. But stems of few living pteridophytes (e.g., Botrychium, Isoetes) do show secondary growth regularly.

(10) The stele is a protostele (e.g., Lycopodium, Selaginella), siphonostele (e.g., Equisetum, Ophioglossum pendulum, Botrychium), dictyostele (e.g., Pteris,

Nephrolepis, Polystichum) or polycyclic (e.g., Angiopteris, Marattia).

(11) The xylem is made up of tracheids and the phloem is composed of sieve cells and phloem parenchyma only. However, vessels are also present in some members (e.g., Pteris vittata, Pteridium aquilinum, Nephrolepis cordifolia, Actinopteris radiata). Presence of sieve tubes have also been reported in the phloem of Equisetum arvense and E. giganteum.

(12) The photosynthetic tissue is restricted mainly to leaves. In megaphyllous leaves, the internal tissue is differentiated into spongy and palisade cells.

(13) They reproduce by spores, produced in sporangia. The sporangia are usually borne on the ventral surface of the leaf, but in some members (e.g., Ophioglossum) a fertile spike, bearing sporangia, arises from the axil of the leaf. In Psilotum and some fossil members (e.g., Rhynia, Homeophyton) sporangia are cauline.

(14) In aquatic forms like Azolla, Isoetes, Salvinia and Marsilea sporangia are present within specialized structures, called sporocarps.

(15) Plants are either homosporous (e.g., Equisetum, Lycopodium, Pteris) or heterosporous (e.g., Selaginella, Marsilea, Salvinia).

(16) The leaves, bearing sporangia are known as sporophylls. Sporophylls with megasporangia are called megasporophylls and with microsporangia as microsporophylls.

(17) The development of sporangia may be :

(i) *Eusporangiate* : The sporangium initiates from a group of superficial cells which divide periclinally, differentiating into an outer layer of jacket initials and an inner layer of primary sporogenous cells;

the jacket initials eventually form the multilayered wall of the sporangium, whereas the primary sporogenous cells, by repeated divisions, give rise to spores; this type of sporangial development is found in Psilotum, Lycopodium, Selaginella and Equisetum, or

(ii) *Leptosporangiate :* The sporangium develops from a single initial cell which divides transversely into an outer and an inner cell; the entire sporangium develops from the outer cell, whereas the inner cell either remains inactive or forms the stalk of the sporangium; this type of sporangial development is seen in Salvinia, Marsilea, Pteris, etc.

(18) Sporophylls are either uniformly distributed (e.g., Pteris, Adiantum, and other ferns) or are aggregated into compact cones (known as strobili) at the and of the stem (e.g., Selaginella, Lycopodium, Equisetum).

(19) In Filicales sporangia are aggregated in clusters, known as sori (singular - sorus). The sori are of the following three types.

(i) *Simple Sorus :* In this type of sorus all sporangia appear, grow and mature at the same time (e.g., Botrychiunt, Ophioglossum).

(ii) *Gradate Sorus :* In this type the central or distal part of the sorus has mature/older sporangia and the peripheral or proximal part has immature/younger sporangia (e.g., Marsilea, Cyathea).

(iii) *Mixed Sorus :* Here mature and immature sporangia are arranged irregularly (e.g., Pteridium, Pteris, Adiantum).

(20) The spore wall is differentiated into an outer exine and an inner intine. The exine consists of an outer layer or ectin, composed of radially placed columnar rods, called

columellae and inner relatively thin and uniform endine. The free and fused ends of columellae form various patterns on the spore wall which may be of taxonomic utility.

(21) The spore, on germination, gives rise to gametophytic generation. The germination of spore may be bipolar (e.g., Lycopodium, Equisetum), tripolar (e.g., Hymenophyllum) or amorphous or irregular (e.g., Angiopteris).

(22) In homosporous species, the development of gametophyte is exosporic (i.e., the gametophyte develops outside the spore wall), whereas in heterosporous species it is endosporic (i.e., the gametophyte develops within the spore wall). In the latter, development of gametophyte is independent of external environment, whereas in the former, it is vulnerable to environment.

(23) There is much variation in the shape of the gametophyte in different genera. In most of the homosporous pteridophytes the prothallus is simple, green and heart-shaped or has aerial branches. In species of Lycopodium, the prothallus is of the following three types.

(i) In L. cernuum it is very small and inconspicuous but autotrophic. It grows on the surface of the soil.

(ii) In species like L. clavatum, L. obscurum and L. selago, the prothallus is tuberous, subterranean and sporophytic.

(iii) In some epiphytic species (e.g., L. phlegmaria) the whole prothallus is heavily associated with endophytic fungal hyphae which show symbiotic association.

(24) Most of the homosporous species are monoecious, i.e.,

antheridia and archegonia are borne on the same prothallus. They are protandrous or protogynous: Some homosporous forms (e.g., Pteris, Pteridium) are also dioecious.

(25) The heterosporous species are always dioecious; the microspore gives rise to the male gametophyte and the megaspore to the female gametophyte.

(26) The sex organs are embedded in the prothallus (e.g., Lycopodium, Selaginella, Equisetum) or are projected (e.g., Psilotum, Pteridium).

(27) The antheridium is surrounded by a sterile antheridial jacket. The jacket is always single layered in species with projected antheridia. The antheridium is sessile or shortly stalked.

(28) The antherozoids are unicellular, uninucleate and spirally coiled with two apical flagella. But they are multiflagellate in the order Psilotales, Equisetales and Pteropsida.

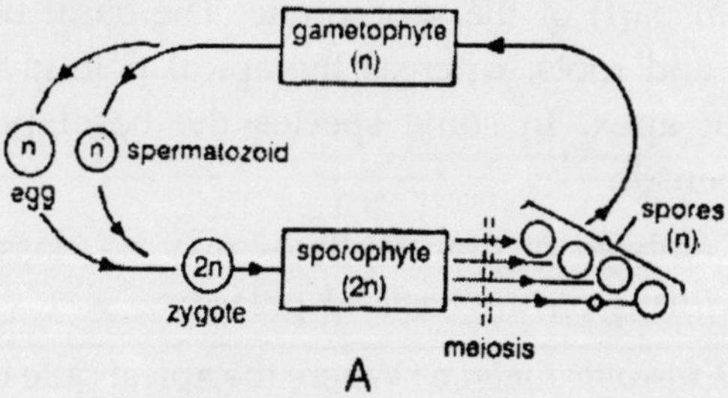

A

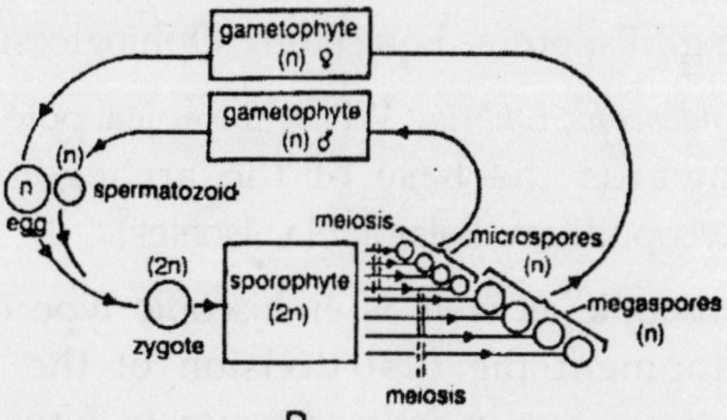

B

(29) The antherozoids are liberated by bursting of the antheridium or separation of the terminal cell, known as opercular cell or cover cell.

(30) The archegonium is differentiated into neck and venter; the neck is projected and the venter is embedded in the prothallus.

(31) The neck is made up of four vertical rows of neck cells and it has usually two neck canal cells, but some species have more than two (e.g., Lycopodium) or only one (e.g., Pteridium). Neck canal cells degenerate before fertilisation and form a mucilagenous substance which attracts antherozoids towards archegonium.

(32) Water is essential for fertilisation. The antherozoid and egg fuse to form a diploid zygote, which functions as the mother cell of the sporophytic generation.

(33) The zygote divides to form embryo. The first division of the zygote determines the polarity (i.e., basal and apical part) of the sporophyte. The basal pole forms foot and roots, whereas the apical pole gives rise to shoot apex. In some species the basal pole forms suspensor.

(34) The embryo of pteridophytes may be classified into two forms on the basis of polarity :

(i) *Exoscopic Embryo* : Where the apical pole is directed outward, i.e., towards the neck of the archegonium (e.g., Psilotum, Equisetum, Ophioglossum); and

(ii) *Endoscopic Embryo* : Where the apical pole is directed towards the base of the archegonium (e.g, Lycopodium, Selaginella, Isoetes).

(35) In exoscopic as well as endoscopic type of embryo development the first division of the zygote is transverse. But in leptosporangiate ferns, the first

division is vertical and the embryo shows lateral polarity.

(36) Pteridophytes show a typical heteromorphic alternation of generations. Morphologically distinct sporophytic and gametophytic generations alternate with each other in regular succession. The main plant body is sporophytic and forms a dominant phase in the life cycle. In homosporous species the haploid spore germinates to form a small and inconspicuous gametophyte, called prothallus. The male (antheridia) and female (archegonia) reproductive structures develop on the gametophytic prothallus. The male and female gametes fuse to form diploid zygote, which is the mother cell of the sporophytic generation. In heterosporous species, the sex of the gametophyte is determined at the spore stage. The microspore always gives rise to male gametophyte and the megaspore to female gametophyte. Thus heterosporous forms are always dioecious.

Affinities of Pteridophytes

Pteridophyta occupy an intermediate position between bryophytes and higher vascular plants (gymnosperms and angiosperms). The simplest living vascular plants (e.g., Psilotales, Lycopodiales) resemble with foliose bryophytes in many of their fundamental characters and may have come from the same ancestral stock but they are nevertheless very diverse and show a great deal of specialization. Pteridophytes resemble with the higher plants in having complex internal organization of the sporophyte, but differ from them in the absence of seed habit. Such specialized features justify independent identity of pteridophytes.

Resemblances of Pteridophytes with Bryophytes

(1) Both show heteromorphic alternation of generation.

(2) The sexual reproduction in both groups is oogamous; the male and female reproductive structures are known as antheridia and archegonia respectively.

(3) In both, sex organs have a protective layer of sterile cells (jacket).

(4) In pteridophytes as well as bryophytes the male gamete (antherozoid) is flagellated and motile and the female gamete (egg) is non-motile.

(5) Water is essential for the opening of mature sex organs and fertilisation in both groups.

(6) Many pteridophytes are homosporous (e.g., Lycopodium, Equisetum, Psilotum, Polypodium, etc.) like bryophytes.

(7) Certain pteridophytes like Rhynia, Psilotum, etc., have rhizoids like bryophytes.

Differences between Bryophytes and Pteridophytes

(1) The most striking characteristic which distinguishes pteridophytes from bryophytes is that the plant body in the former is sporophytic and in the latter gametophytic.

(2) The plant body of pteridophytes is differentiated into root, stem and leaves, whereas in bryophytes it is thalloid or foliose.

(3) The vascular system of pteridophytes is well developed and composed of xylem and phloem, whereas there is no vascular tissue in bryophytes.

(4) Vegetative reproduction is more common in bryophytes than in pteridophytes.

(5) Bryophytes are always homosporous, whereas many pteridophytes show heterospory.

(6) In pteridophytes the sporophyte is large, long-lived

and independent of the gametophyte, but in bryophytes it is completely or partly dependent on the gametophyte.

(7) The sporophyte of some bryophytes has elaters along with spores, but elaters are altogether absent in pteridophytes.

Resemblances of Pteridophytes with Gymnosperms

(1) The plant body in both groups is sporophytic, differentiated into root, stem and leaves.

(2) Young leaves show circinate vernation in both groups.

(3) The vascular system is well developed in both groups, made up of xylem and phloem. The xylem is without vessels (except Gnetales in gymnosperms) and the phloem without companion cells.

(4) Like gymnosperms, many pteridophytes show heterosporous condition.

(5) Antherozods of some gymnosperms (e.g.; Cycas, Ginkgo) are ciliated like those of pteridophytes.

(6) In some pteridophytes (e.g., Selaginella) the megaspore is retained within the megasporangium, a feature normal in gymnosperms.

(7) There is regular alternation of sporophytic and gametophytic generations in both groups.

Differences between Pteridophytes and Gymnosperms

(1) Pteridophytes usually occur in moist and shaded places, whereas gymnosperms grow in xerophytic habitats.

(2) Gymnosperms have tap roots but in pteridophytes the tap root is ephemeral and is soon replaced by adventitious roots.

(3) Pteridophytes lack vascular cambium (hence do not show secondary growth), while cambium is present

in gymnosperms (hence they show secondary growth).

(4) The archegonium in pteridophytes has neck canal cells but they are absent in gymnosperms.

(5) In homosporous pteridophytes, the development of gametophyte is initiated after the dispersal of spores from the sporangium. But in gymnosperms the female gametophyte develops inside the megasporangium (ovule, which is permanently attached to the sporophyte and microspores are liberated at various stages of development of the male gametophyte.

(6) In gymnosperms the pollen grain (microspore) gives rise to a pollen tube on germination, but no such structure is formed in pteridophytes.

(7) The gametophyte of pteridophytes is independent of the sporophyte, whereas it is dependent on the sporophyte in gymnosperms.

PART-II

Gymnosperms

15

The Gnetales

Order Gnetales

The order Gnetales is regarded as the most highly evolved group of Gymnosperms. The order is characterized by :

(i) opposite decussate leaves with unicostate reticulate venation,

(ii) tunics-corpus configuration of the shoot apex,

(iii) presence of vessels in the xylem,

(iv) absence of resin canals,

(v) compound male and female strobili resembling inflorescences,

(vi) a long micropylar tube formed by the prolongation of inner integument,

(vii) tetrasporic development of female gametophyte,

(viii) complete elimination of archegonia,

(ix) absence of free nuclear divisions in embryo development, and

(x) dicotyledonous embryo. The order includes only one family Gnetaceae with a single genus Gnetum.

Gnetum

Systematic Position

Division	—	Chlaitnydospermophyta
Class	—	Gnetopsida
Order	—	Gnetales
Family	—	Gnetacene

Gnetum has about 40 species, confined to the moist tropical forests of the Amazon basin, West Africa, India, South China and Malaysia. Most of these species are endemic within the areas of their distribution. Five species of Gnetum occur in India. G. contractum is found in Conoor in Tamil Nadu, Quilon in Kerala and in Nilgiri hills. G. gnemon var. brunonianum occurs in Golaghat, South Lusai hills and Naga hills. G. gnemon var. grifthii is widely distributed in Kungaba, Naga hills and Sibsagar areas of Assam. G. latifolium is confined to Andaman & Nicobar Islands. G. montanum is prevalent in tropical Himalayas from Sikkim eastwards to Bengal, Assam, Khasi hills and Manipur. G. ula occurs in the moist evergreen forests of Western Ghats, Andhra Pradesh and Orissa.

Young leaves, inflorescences, fruits and seeds of G. gnemon are edible; the bark yields a fibre of high tensile strength, suitable for paper pulp and fishing nets; and the wood is used for junks and rafts. Seed-kernels of G. ula yield an oil used for illumination and cooking and massage in rheumatism. The kernels of G. latifolium are edible and the bark is used for making nets and ropes. G. montanum possesses piscicidal properties.

Sporophyte

External Morphology : Most of the species of Gnetum are trailing or woody climbers, but G. contractum is a scandent shrub and G. gnemon a small tree. G. trinerve is a parasitic species, occurring on Cinchona trees in the forests of Chimborazo. The plant body of Gnetum resembles very much in its general appearance to that of angiosperms. The sporophyte is differentiated into root, stem and leaves.

1. *Root :* It is profusely branched tap root which does not grow very deep.
2. *Stem :* With the exception of G. gnemon, the main stem bears two kinds of branches — dwarf shoots of limited growth, and long shoots of unlimited growth. The stem is articulated and has prominent joints. The joint consists of two portions, one above the node and the other immediately below it, and the two are demarcated by an annular groove. In G. gnemon all branches are of the same type.
3. *Leaves :* The leaves are arranged in a decussate manner and they resemble dicotyledonous leaves in having a broad lamina and reticulate venation. They are exstipulate, shortly petiolate with a simple, entire and coriaceous lamina. Foliage leaves are present only on the dwarf shoots, whereas long shoots bear cup-like pairs of reduced scale leaves.

Internal structure

1. *Root :*

   (a) *Primary Structure :* The root is almost circular in outline in a transection and is differentiated into epiblema, cortex and vascular region. The epiblema (epidermis) represents the outermost layer and is composed of tightly-appressed cells. It gives out unicellular root hairs.

   The cortex is multilayered, consisting of parenchymatous cells, which contain large amount of starch grains. A single-layered endodermis, followed by a multilayered pericycle, separates the central vascular cylinder. The vascular cylinder of the root is di- or triarch, radial and exarch. The primary xylem has only few elements.

(b) *Secondary Growth* : The secondary growth is of normal type. Arcs of cambium are formed internal to the phloem groups and external to the xylem groups and they join to form a complete ring of cambium. The secondary xylem is composed of tracheids, vessels and xylem parenchyma. The tracheids have uniseriate circular-bordered pits on their tangential and radial walls. But in the vessels the pits are multiseriate with smaller diameters. Medullary rays composed of parenchymatous cells impregnated with crystals of calcium oxalate traverse the secondary xylem. The phloem is composed of sieve cells and phloem parenchyma. Thick- walled fibrous elements are sometimes irregularly distributed in the cortex.

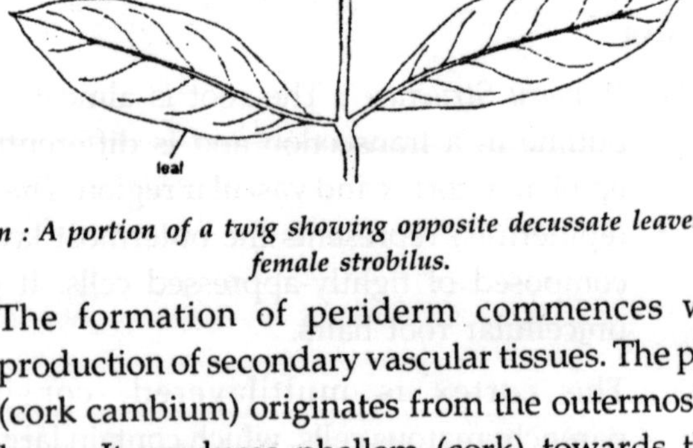

Gnetum : A portion of a twig showing opposite decussate leaves and a female strobilus.

The formation of periderm commences with the production of secondary vascular tissues. The phellogen (cork cambium) originates from the outermost cortical layer. It produces phellem (cork) towards the outer side and phelloderm (secondary cortex) towards the inner side.

2. *Stem :*

(a) *Primary Structure :* The primary structure of Gnetum stem is similar to that of a dicotyledonous stem. A transection of the young stem is more or less circular in outline and shows the following regions.

(i) *Epidermis :* It is the outermost layer composed of compactly arranged and heavily cutinised rectangular papillate cells. The continuity of the epidermis is interrupted by sunken stomata.

(ii) *Cortex :* The cortex is usually differentiated into an outer chlorenchymatous, a middle parenchymatous and an inner sclerenchymatous region. The outer zone consists of 5-7 layers of polygonal cells which contain many chloroplasts. The middle zone is composed of a few layers of thin-walled, compactly arranged cells. Many fibrous cells are also present in this zone. The inner zone has 2-5 layers of thick walled cells. These cells have branched or unbranched pit canals.

(iii) *Endodermis and Pericycle :* The innermost layer of the cortex represents endodermis. However, it is indistinguishable from the cortical layers. The pericycle is also not very conspicuous.

(iv) *Vascular Tissue :* The vascular cylinder is composed of 20-24 primary vascular bundles, arranged in a ring. The vascular bundles are conjoint, collateral, open and endarch. They are separated from one another by broad parenchymatous medullary rays. The xylem consists of both tracheids and vessels. The tracheids have uniseriate bordered pits separated by bars of Sanio. The protoxylem elements are annular or spiral, whereas metaxylem elements have bordered pits. The phloem is made up of sieve tubes and phloem parenchyma.

(v) *Pith :* A parenchymatous pith is present in the centre of the stem. The cells of the pith are usually polygonal in outline.

(b) *Secondary Growth in the Stem :* In some species of Gnetum, like G. gnemon, secondary growth is of normal type. A cambium ring is formed in the usual manner which cuts off secondary phloem towards the outer side and secondary xylem towards the inner side. The amount of xylem formed is much more than the phloem. The primary phloem is crushed due to the pressure of the secondary vascular tissues.

Climbing species (e.g., G. africanum, G. ula) show anomalous secondary growth. Several rings of cambia develop from successively deeper layers in the cortex. These cambial rings are eccentric. They cut off secondary phloem and xylem outward and inward respectively. Due to the formation of medullary rays in the secondary tissue, distinct wedge-shaped vascular bundles appear. When the first ring of cambium becomes non-functional, a second ring appears outside it and produces secondary tissues in similar manner. In this way, several rings of secondary vascular tissue are formed. A phellogen (cork cambium) arises in the epidermis or in the outer layers of the cortex during the third or fourth year of the growth of the stem. It produces periderm on the outside. In G. ula and G. afticanum, due to unequal meristematic activity of the phellogen islands of sclerenchymatous cells and thin walled cells become embedded in the cork layers.

(c) *Secondary Xylem :* In the secondary xylem, there is a considerable variation in the size of vessels. Their diameter is as small as of the tracheids or four or five times of tracheids. There is multiseriate pitting on all walls including the end walls. The end walls soon break down and continuous tubes are formed as in angiosperms. The end walls of the cells, which give rise to vessels, are oblique or perfectly transverse. A single perforation is formed by the enlargement or fusion of bordered pits and the disappearance of the middle lamella. In G. gnemon and G. montanum, scalariform pits also occur. Pit membrane is thin and torus is lacking. Inter-vessel pits have well-developed vestures. In tracheids, vestures may be present or absent.

The medullary rays are usually multiseriate, broad and of considerable heights. Occasionally uniseriate and biseriate rays are also present. The rays are composed of radially arranged thin walled cells with simple pits. Crystals of calcium oxalate occur in ray cells and they appear boat-shaped in a tangential section of the ray.

3. *Leaf :* The leaves of Gnetum resemble closely with those of dicotyledons not only in their general appearance, texture and venation but also in the internal structure. The leaf is differentiated into epidermis, mesophyll and vascular region.

There is an upper and a lower epidermal layer. The epidermal cells have undulate walls. The upper epidermis is covered by a thick layer of cuticle. Stomata are confined to the lower epidermis. The mesophyll is differentiated into palisade and spongy parenchyma.

The palisade is composed of a single layer of elongated cells, rich in chloroplasts. The spongy parenchyma consists of loosely arranged and lobed cells. In the mid-rib region, the parenchyma cells are polygonal and compactly arranged. Latex tubes and groups of fibres are also present in the mesophyll around the mid-rib region. Stellately branched sclereids are fairly common on the abaxial side.

The mid-rib consists of an arc of conjoint, collateral and endarch vascular bundles with an adaxial xylem and abaxial phloem. Each bundle is surrounded by a sheath of thick walled cells. The xylem is composed of tracheids, vessels and xylem parenchyma; and the phloem of sieve cells and phloem parenchyma arranged in distinct files. Transfusion tissue is present in the petiole intermixed with the vascular bundles.

4. *Shoot Apex Organisation :* The shoot apex of Gnetum can be interpreted in terms of the tunica-corpus concept. The shoot apex consists of two regions, the tunica and the corpus. The tunica consists of a single layer of peripheral cells in which divisions are predominantly anticlinal. The corpus consists of a central core of cells in which divisions occur in all planes. The cells at the summit of the apex are large and often more vacuolated than at the flanks.

Reproduction

Gnetum is dioecious, but bisexual strobili also occur rarely as in G. africanum. The reproductive organs of Gnetum, often referred to as flowers, are organised into inflorescence. Both male and female inflorescences are usually axillary in position but terminal inflorescences are also not uncommon. The inflorescence (compound strobilus) consists of a stout long axis, bearing several flowers. The flowers arise in the axils of decussate pairs of bracts. The bracts are connate throughout their length and forming a cup-like collar which surrounds the axis. There are 10-25 collars in each strobilus. They lie very close to each other because of the suppression of internodes. As leaves fall off before the strobili develop, the branches bearing strobili are leafless.

Male Strobilus : The male strobilus has a short stout axis arising in the axil of connate bracts. On the axis 10-25 whorls of bracts are present at short intervals. The bracts at each node fuse to form a cupule-like structure, known as collar. In each collar, there are several male flowers arranged in 3-6 whorls. Above the male flowers, a whorl of abortive female flowers or ovules is present. The strobilus is very compact when young, but at maturity it becomes loose as collars separate from each other due to the elongation of internodes.

1. *Male Flower :* A male flower consists of two coherent bracts (representing the perianth) which enclose two unilocular anthers on short stalks. When the flower matures, the stalks elongate and the anthers are pushed out of the perianth sheath. Sometimes, two or more flowers are fused with each other.
2. *Development of Microsporangium :* The development of microsporangium is of eusporangiate type. There are two or more hypodermal archesporial cells in the young microsporangium. They divide periclinally into outer primary wall cells and inner primary sporogenous cells. The tapetum is derived from the sporogenous tissue and not from the wall layers. The tapetal cells are usually binucleate, but they unite to form polyploid nuclei with many nucleoli. The primary wall cells divide repeatedly to form the epidermis and a middle layer. There is no endothecium.

The sporogenous cells divide in all possible planes and form numerous spore mother cells. Each spore mother cell undergoes meiosis and forms four haploid microspores, arranged in a tetrahedral tetrad. At this stage, tapetal cells degenerate. As the microsporangium matures, all cells between the epidermis and spores break down. The anthers dehisce by longitudinal slits to release pollens.

Female Strobilus : Like male strobilus, the female strobilus also has a stout axis with several collars arranged one above the other at short intervals. A whorl of 4-10 ovules is present above each collar. All the ovules develop at the same rate initially but subsequently the growth of most of the ovules is arrested and only a few are able to mature into seeds.

1. *Female Flower :* The female flower consists of a short stalk and an ovule (megasporangium). The ovule has a prominent nucleus protected by a three-layered envelope. The innermost envelope elongates very much and forms a long micropylar tube. A rudimentary pollen chamber is present at the tip of the nucleus. Nuclear beak is absent.
2. *Vascular Supply :* The main axis of the female strobilus has 12- 24 bundles. Each of these bundles gives off a pair of branches which supply the outermost envelope. A little higher up again some branches arise from these bundles which form a looped complex and terminate below the base of the inner integuments of the ovule.
3. *Morphological Nature of the Envelope :* The morphological nature of the envelope of the ovule is long debated and several views have been put forward from time to time. Some of these views are as follows :

   (1) Van Tieghem (1869) suggested that the outer envelope is equivalent to an ovary or a structure analogous to ovary and the inner two envelopes represent the two integuments of the ovule.

   (2) Strasburger (1872) considered that the three

envelopes are formed by the splitting of a single integument.

(3) Beccari (1877) was of the opinion that the two inner envelopes represent the two integuments as in the angiosperm ovule and the outer most envelope represents the perianth. Coulter (1908) also supported this view.

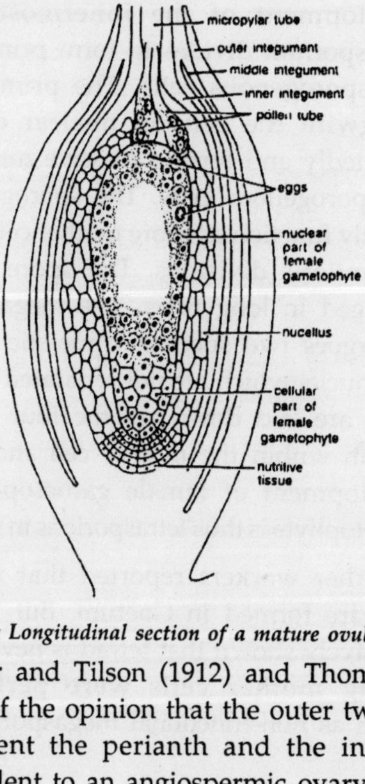

***Gnetum* : Longitudinal section of a mature ovule.**

(4) Lignier and Tilson (1912) and Thompson (1916) were of the opinion that the outer two envelopes represent the perianth and the inner most is equivalent to an angiospermic ovary.

(5) Vasil (1956) observed that in some specimens of G. ula where the innermost envelope is abortive or partially developed, the middle envelope develops into micropylar tube (or 'style') in its place. Sometimes the middle as well as the inner envelopes were observed to grow into micropylar tubes or

'styles'. Following Tilson (1912) and Thompson (1916), she regarded the inner and the middle envelopes as ovaries because of their capacity to form styles.

4. *Development of Megasporangium* : In the hypodermal region of the nucleus, an archesporium of two or four cells is differentiated almost at the time of the development of the innermost envelope. The archesporium divides to form primary parietal cells and sporogenous cells. The primary parietal cells alongwith the cells of nuclear epidermis divide repeatedly and form a massive nuclear tissue above the sporogenous cells. The sporogenous cells either directly function as spore mother cells or may undergo few mitotic divisions. The spore mother cells are arranged in long rows. The megaspore mother cell undergoes two successive meiotic divisions to form four nuclei which are not separated by walls. Since no walls are laid down, all the four megaspore nuclei remain within the mother cell and take part in the development of female gametophyte. The female gametophyte is thus tetrasporic as in some angiosperms.

Some earlier workers reported that normal tetrads of megaspores are formed in Gnetum. But the recent studies have conclusively shown that tetrad is never formed and the degenerating mother cells were perhaps taken into consideration as non-functional megaspores.

Gametophyte

Microspores and tetranucleate megaspores develop into male and female gametophytes respectively.

Development of Male Gametophyte : The microspore (pollen grain) represents the earliest stage of the male gametophyte. It is roughly spherical, uninucleate and is enveloped by a thick and spiny exine and a thin intine. Microspores are shed at 3-nucleate stage. In the development of male gametophyte, the microspore nucleus first divides into two daughter nuclei; one of these nuclei divides again so that three nuclei are formed, which lie free in the cytoplasm. These nuclei are called prothallial nucleus, generative nucleus and tube nucleus. At this stage, pollen grains are released and further development takes place in the pollen chamber. However, some workers believe that there is no prothallial cell. The first division of the microspore nucleus gives rise to a tube nucleus and a generative cell nucleus; and the latter divides further into a body cell and a stalk cell. The stalk cell remains in the grain and only the body cell and the tube nucleus enter the pollen tube. It is believed that in complete elimination of prothallial cell, Gnetum approaches the angiospermic condition, where usually there is no prothallial cell.

The tip of the micropylar tube exudes a drop of sweet fluid to which microspores are carried. As the drop shrinks due to evaporation, microspores are caught in it and ultimately reach the top of the nucleus. Microspores germinate in the shallow pollen chamber and the pollen tube grows into the nucleus. First, the tube nucleus enters the pollen tube but the prothallial and generative cells remain in the microspore. Next, the generative cell moves into the pollen tube. The prothallial cell remains in the microspore. Finally, the generative cell also moves into the pollen tube. The prothallial cell remains in the microspore and eventually degenerates. The nucleus of the generative cell divides to form a pair of sperm nuclei which are surrounded by distinct cytoplasmic sheaths.

The two sperm nuclei differ somewhat in size, and it is believed that the larger one is functional and the other disorganises. The two nuclei of equal size have also been observed, but usually one of them is functional.

Development of Female Gametophyte : The megaspore is the first cell of the female gametophyte. The development of female gametophyte begins with a series of free- nuclear divisions. The number of nuclei formed is 256 in G. gnemon, 512 in G. africanum and 1500 in G. ula. At this stage, the female gametophyte elongates towards the chalazal end and appears almost spathulate in outline. It has a large central vacuole and the free nuclei are arranged in the peripheral layer of the cytoplasm. Wall formation takes place at the chalazal end but nuclei remain free at the micropylar end. Archegonia are absent in Gnetum. Some of the free nuclei of the gametophyte at the micropylar end represent eggs. They can be distinguished from other nuclei by their larger size.

Fertilisation

The microspores (pollen grains) present in the pollen chamber germinate to form pollen tubes which grow through the tissue of the nucleus. The tube nucleus migrates into the pollen tube first and it is followed by generative cell which soon divides in the pollen tube to form two male cells. As soon as the pollen tube comes in contact with the embryo sac, two or three nuclei of the female gametophyte enlarge and become three to four times larger than the normal vegetative nuclei. The pollen tube enters the female gametophyte and ruptures to liberate the male cells. Usually, one of the male cells fuses with an egg and forms zygote. Sometimes, the second male cell may also fuse with the second nucleus of the female gametophyte and in such cases two zygotes are formed. Occasionally, as in G. ula, a few groups of cells develop in the female gametophyte in the vicinity of the pollen tube. Each group consists of 3-8 cells and of these one or two cells become larger and function as egg. The surrounding cells are gradually absorbed as nourishment by the eggs.

Endosperm

The development of endosperm begins even before fertilisation. It develops from the female gametophyte. Before fertilisation only the lower part of the female gametophyte becomes cellular, but after fertilisation wall formation also occurs in the upper part and the whole gametophyte becomes cellular. The wall formation takes place in such a way that the cytoplasm is divided into several compartments, each with one or more nuclei. In the multinucleate cells of endosperm all nuclei fuse and thus polyploid cells of endosperm are formed. In G. gnemon, the female gametophyte is only partly cellular. The development of endosperm in Gnetum is very much delayed and it is formed after fertilisation. But the nature of endosperm is entirely different from angiosperms, where it develops from the endosperm nucleus after the process of triple fusion.

Embryo Development

In G. gnemon the zygote produces a protuberance on one end, which grows down towards the endosperm tissue. Sometimes, more than one protuberances may develop and then only one of the protuberances receives the nucleus and survives and the rest die out. The surviving tubular outgrowth elongates, becomes branched and grows in different directions within the intercellular spaces of the endosperm. All primary suspensor tubes generally stay coiled around each other. At the tip of the primary suspensor tube, a small cell is cut off which soon divides first by a transverse wall and then by a longitudinal wall forming four cells. This is followed by irregular divisions and thus a group of cells is formed. Some of these cells divide further and then elongate, forming secondary suspensor; the rest of the cells form an embryonal mass.

However, in G. ula, a small cell is cut off at the tip of the primary suspensor tube and this cell is known as peculiar cell. The peculiar cell is responsible for the formation of embryo. It divides to form two cells, both of these cells divide again to form a four-celled embryonal mass, and the third division is transverse and thus an eight-celled embryo is formed. Further divisions are irregular. As the embryonal mass increases in size, some cells lying towards the primary suspensor tubes elongate to form the secondary suspensor. The cells lying at the lower end of secondary suspensor eventually form the embryo proper.

Whatever may be the mode of formation, the cells of the embryonal mass are small, and densely cytoplasmic and those of the secondary suspensor are uninucleate, thin-walled and highly vacuolate. The primary and secondary suspensors push the embryo into the endosperm. At the tip of the embryonal mass, a stem tip with two lateral cotyledons emerges. On the other side, the root tip with a root cap is developed. At the same time, a lateral hump appears which forms the so-called feeder. A mature embryo consists of two orange-pink cotyledons, a stem tip, a large feeder and a root clothed by root cap.

Seed : The mature seed is oval or slightly elongated and green to red in colour. The nucleus persists in the seed as a thin strip at the apex. The endosperm, alongwith embryo, is protected by a three- layered envelope. The outer envelope is fleshy, green to red in colour and is free from the other envelopes. It is made up of parenchymatous cells with many sclereids and fibres. The middle layer is stony and is the main protective covering of the seed. Sclereids, fibres and latex tubes are present in this layer. Besides, a layer of palisade like cells is also present below the epidermis. The innermost layer of the envelope consists of thin walled parenchymatous cells and lignified cells. This envelope is coalescent with the nucleus up to two third of its length. Its free part encircles the nuclear cap and projects it beyond the nucleus. A massive endosperm lies within the envelope, which projects upwards in the form of a tent pole. It has thick and lignified walls and it supports the nuclear cap.

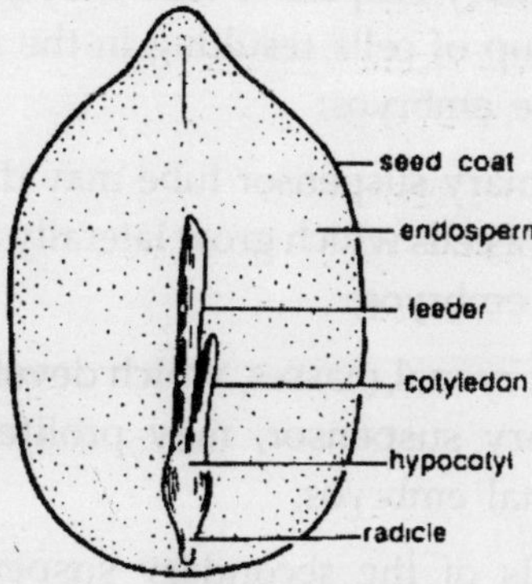

Gnetum* : *Longitudinal section of a seed.

Germination of Seed : Seeds are shed at a stage when secondary suspensors are differentiating. Maturation of the seed takes place after shedding. In G. ula, seeds germinate after 11-12 months of shedding from the mother plant.

The germination of the seed is epigeal. The root emerges first, it bends down, enters into the soil and forms the tap root system. As the hypocotyl elongates, cotyledons come out of the seed and above the soil. The cotyledons are the first green leaves. The plumule produces the first pair of foliage leaves that lie at right angles to the cotyledons. Cotyledons wither at a later stage. The feeder still remains inside the seed and continues to absorb food from the endosperm till the second or third pair of leaves have developed in the seedling.

Polyembryony

Polyembryony is of common occurrence in Gnetum. The extra embryos may originate either by simple polyembryony (where more than one egg gets fertilised and each one develops in an embryo) or by the cleavage of a single fertilized egg. The cleavage polyembryony may occur in the following ways :

(1) Single fertilized egg develops many primary suspensor tubes and each of these may produce an embryo at its tip;

(2) The primary suspensor tube may produce more than one group of cells resulting in the formation of more than one embryos;

(3) The primary suspensor tube may divide and produce a chain of cells which grow laterally and produce many normal embryos;

(4) The embryonal masses, which develop at the tip of the secondary suspensor, may proliferate and produce additional embryos;

(5) The cells of the secondary suspensor may become meristematic at any place and develop embryos.

Relationships

Gnetum resembles with other Gymnosperms on the one hand and with Angiosperms on the other.

Resemblances with Gymnosperms : Gnetum shares the following characters with Gymnosperms.

(1) Ovules are naked and style and stigma are absent.

(2) A prothallial cell is present in the male gametophyte.

(3) A small pollen chamber is present in the nucleus; microspores enter the ovule through the micropyle and get directly lodged on the nuclear surface.

(4) Pollination is anemophilous.

(5) The endosperm is a haploid or polyploid tissue and starts developing before fertilisation.

(6) Presence of simple or cleavage type of polyembryony.

(7) Xylem tracheids are with bordered pits and sieve cells are present in the phloem.

Resemblances with Angiosperms : Gnetum has long held a key position in any discussion relating to the origin of angiosperms. Arber and Parkin (1907) considered Gnetales as the most closely related to angiosperms. Both were derived from a common stock Hemiangiosperms; and had parallel evolution. Thompson (1916) also shared the view that the ancestors of angiosperms were not remote from Gnetum. According to Hagerup (1934), there is a strong affinity between Gnetales and Piperaceae and differences between these two groups are of minor significance. Gnetales actually belong to angiosperms or Piperaceae originated from the gymnospermic branch via Gnetales.

Some important features which Gnetum shares with angiosperms are as follows:

(1) In general habit sporophyte of Gnetum (climber or tree like) resembles angiosperms.

(2) The leaves of Gnetum are broad and green with reticulate venation and opposite decussate arrangement as in angiosperms.

(3) All the main types of sclereids found in angiosperms also occur in Gnetum.

(4) The development of stomata in Gnetum is similar to that of angiosperms.

(5) The shoot apex organization of Gnetum is similar to that of angiosperms.

(6) Vessels similar to those of angiosperms occur in the secondary wood of Gnetum, though their mode of development is different.

(7) The embryo-sac of Gnetum is tetrasporic as in many angiosperms.

(8) There is complete elimination of archegonia in the female gametophyte of Gnetum as in angiosperms.

(9) In the occurrence of free-nuclear divisions in the embryo-sac, Gnetum is on the angiospermic line.

(10) The embryo of Gnetum has two cotyledons like that of a dicotyledonous embryo.

(11) The male and female reproductive organs of Gnetum are more close to angiosperms than Gymnosperms.

Gnetum resembles angiosperms in several respects such as habit, presence of two cotyledons, reticulate venation of leaves, presence of vessels in the secondary wood, catkin like inflorescences as in some amentiferous angiosperms, the reduction and loss of archegonium in the female gametophyte, and the presence of two integuments in the ovule. Thus, at one time Gnetum was considered as closely allied to angiosperms and their possible ancestors.

However, it has been subsequently shown that many of these resemblances are superficial. Angiosperms are considered to have been derived from polycotyledonous rather than dicotyledonous ancestors. There is a general tendency towards reduction in gametophyte not only within Cinetales but among the vascular plants as a whole.

The development of vessels in Gnetum and angiosperms is quite different. In angiosperms perforation of vessels arose through the dissolution of pit membranes in scalariform bordered pitting, whereas in Gnetum they developed through modifications of circular bordered pits.

The resemblances between the inflorescences of Gnetum and amentiferous angiosperms are superficial. It is now generally believed that Amentifera have reduced flowers which are highly evolved. The primitive angiosperm flower was a bisexual strobilus with a well-developed perianth and numerous free stamens and carpels. It is difficult to derive such flowers from unisexual, scaly-bracteate, compound strobili of Gnetum. The resemblances between Gnetum and angiosperms in habit, reticulate venation and the presence of double integument in the ovule are not enough to suggest their relationships.

16

The Ephedrales

Order Ephedrales

The order Ephedrales includes a monotypic family Ephedraceae, represented by Ephedra. It is one of the highly evolved groups of gymnosperms. The plants are mostly small trees or perennial herbs or shrubs. They usually grow in xerophytic habitats. The leaves are minute and scaly. Resin canals are absent and the medullary rays are multiseriate. The plants are dioecious and bear compound strobili (also known as flowers.

At one time, Ephedra was included alongwith Gnetum and Welwitschia in the family Gnetaceae of the order Gnetales. However, due to the heterogeneous nature of the family, the order Gnetales was divided into three families, each comprising one genus. Florin (1934) and Eames (1952) pleaded for the establishment of three distinct orders: Ephedrales, Welwitschiales and Gnetales with a monogeneric family in each. Pant (1957) placed Gnetum and Welwitschia together in Chlamydospermophyta and treated Ephedra under Coniferophyta. The same treatment is followed here.

Ephedra

Systematic Position

Division	—	Coniferophyta
Class	—	Ephedropsida
Order	—	Ephedrales
Family	—	Ephedraceae

There are about 42 species of Ephedra, widely distributed in both Eastern and Western hemispheres. These species are equally divided between the New and Old World, occurring in the temperate and sub-tropical regions. Eight species of Ephedra are known from India; of these, seven (E. pachyclada, E. intermedia, E. saxatilis, E. gerardiana, E. nebrodensis, E. major and E. regeliana) are confined to the north-west Himalayan region and only one (E. foliata) occurs in the plains of Rajasthan and Punjab.

Ephedra is the source of a drug, ephedrine, obtained from species like E. equisetina, E. gerardiana, E. major, E. sinica, E. intermedia and E. nebrodensis. It is used against cold, respiratory disorders, asthma and hay fever. E. antisyphilitica is used in the treatment of syphilis. An aromatic beverage, known as Mormon tea is also brewed from the species of Ephedra in southwestern United States. The rhizomes of E. gerardiana are utilised as fuel in Tibet.

The basic karyotype of the genus (n = 7) consists of five chromosomes with median-submedian primary construction and the remaining two chromosomes with sub-terminal primary construction. Polyploidy has been frequent occurrence in the evolution of the species like E. intermedia and E. saxatilis.

Sporophyte

External Morphology : Most of the species of Ephedra are low-growing woody trailing or climbing shrubs rarely more than 30 cm tall. However, E. trifurca attains a height of 2-3 meters; E,. foliata, a climbing species, may reach to a height of 6 meters or more and E. triandra takes the form of a small tree under favourable growth conditions. The sporophyte is differentiated into root, stem and leaves (Fig.).

1. *Root* : It is a long tap root in xerophytic species, but in larger forms it is replaced by adventitious roots.
2. *Stem* : The stem is green, profusely branched, jointed and is differentiated into distinct nodes and internodes. The branches develop in the leaf axils at the nodes. The number of branches is equal to the number of leaves present at the node (Fig.). The internode grows by a meristematic zone present at its base. The meristem dries up at the end of each growing season causing brittleness and shedding of branches. New axillary branches develop in the next season. The young stem is tender, but its older parts become woody as the result of secondary growth.

   Due to the jointed nature of stems and whorled branches, Ephedra is commonly known as jointed fir. Superficially, it resembles with the aerial shoot of Equisetum.
3. *Leaves* : The leaves are inconspicuous, very minute, scaly and sessile. They are arranged in pairs or whorls at the nodes. The leaves at a node are fused at the base to form a sheath (Fig.). Each leaf has two unbranched parallel veins. Since the leaves do not have the capacity of food synthesis, the function of photosynthesis is carried by the green stem.

Internal Structure

1. *Root* : Internally, the root is differentiated into epiblema, cortex and vascular region (Fig.).

   The epiblema is a single layer of cells. Some cells of

epiblema elongate to form root hairs. The cortex is differentiated into an outer and an inner zone. The outer cortex is composed of collenchymatous cells, whereas the inner cortex is parenchymatous. The innermost layer of the cortex forms the endodermis, which is followed by a layer of thin-walled cells, which forms the pericycle.

The vascular cylinder of the root may be diarch or triarch. Broad strips of phloem are present in between the arms of xylem.

2. *Stem :*

   (a) *Primary Structure :* The stem has ridges and grooves and its tissues are differentiated into epidermis, cortex and stele (Fig.).

The epidermis is the outermost layer of thick-walled cells, covered with a thick layer of cuticle. It has haplochelic stomata, which are sunken in circular pits in the furrows. A sclerenchymatous hypodermis is present just below the epidermis in the ridges, which provides mechanical strength to the plant.

The cortex is differentiated into an outer and an inner zone. The outer cortex consists of 2-3 layers of radially, elongated palisade cells. The inner cortex is represented by 2-3 layers of spongy parenchyma with loosely arranged spherical or oval cells. Many chloroplasts are present both in the palisade and spongy parenchyma. The cortex is the main photosynthetic region of the sporophyte.

The innermost layer of the cortex represents an ill-defined endodermis followed by a layer of pericycle.

The vascular bundles are arranged in a ring as in a typical dicotyledonous stem. There are eight vascular bundles; two pairs of small bundles represent the leaf traces and four larger bundles are axial in nature. They are conjoint, collateral and endarch. A thin strip of cambium is present between the xylem and phloem. The xylem consists of tracheids, vessels and xylem parenchyma. In the presence of vessels in the xylem Ephedra resembles angiosperms. But vessels have originated differently in these two groups.

The phloem is composed of sieve tubes, phloem parenchyma and albuminous cells. Broad medullary rays are present in between the vascular bundles.

A parenchymatous pith is present in the centre of the stem. In the nodal region the pith is composed of strongly lignified cells which form a plate-like nodal diaphragm (Fig.). The stem breaks easily at the nodes due to the presence of nodal diaphragm.

(b) *Secondary Growth* : The secondary growth takes place by the activity of vascular cambium. A thin strip of intrafascicular cambium occurs in between the primary xylem and phloem. Some parenchymatous cells between the vascular bundles become meristematic and form intrafascicular cambium. The inter- and intrafascicular cambium strips join to form a complete cambial ring. This cambial ring cuts off secondary phloem towards the outer side. The secondary xylem is composed of tracheids; vessels and xylem parenchyma; but xylem fibres are absent (Fig.).

There is seasonal variations in the cambial activity and xylem elements of different dimensions are formed in autumn and spring. The tracheids of spring wood are longer and have a large diameter. Thus, distinct layers of spring and autumn wood are formed. Xylem increments of a year (autumn and spring wood) form an annual ring. The tracheids have a single row of bordered pits with crassulae both on the radial and tangential walls. The pits are circular or elliptical and possess ephedroid perforations.

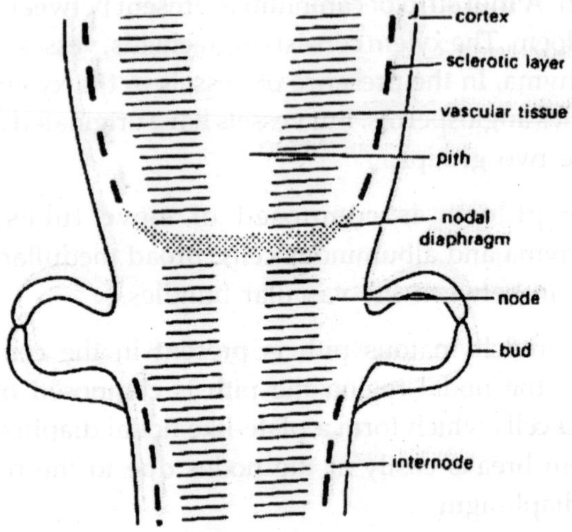

Ephedra : Longitudinal section of node.

In addition to vascular tissues, the cambium also forms secondary vascular rays. These rays become broad and multiseriate due to their vertical growth and fusion of adjacent rays. In radial longitudinal section (R.L.S.), medullary rays are cut lengthwise and their length and height are revealed (Fig.). They are up to 40-50 cells in height and are composed of ray cells and ray tracheids. In tangential longitudinal section (T.L.S.), they are cut transversely and as such their height and breadth can be observed (Fig.).

Phellogen (cork cambium) arises in the outer layers of the cortex during the third or fourth year of growth of the plant. It results in the formation of periderm which displaces all the outer tissues.

3. *Leaf* : The leaf is highly reduced and scale like. Internally, it is differentiated into epidermis, mesophyll and vascular region (Fig.).

The epidermis is a single layer of thick-walled cells, covered by a thick cuticle. It is interrupted by haplochelic stomata.

The mesophyll is undifferentiated and it consists of palisade like elongated cells with large inter-cellular spaces between them. These cells are rich in chloroplasts.

In the mid-rib region of the leaf, two vascular bundles are present in a parenchymatous tissue. These bundles are collateral and closed with adaxial xylem. Transfusion tracheids with bordered pits and thickening bars occur occasionally on the flanks of the xylem.

Reproduction

Most species of Ephedra are strictly dioecious, though occasionally bisporangiate cones may occur in some species, such as E. foliata and E. intermedia. The cones occur in groups at the nodes (Fig.). Unlike other living gymnosperms, in Ephedra both male and female strobili are compound', i.e., the cone axis bears pairs of bracts which subtend either microsporangiate or ovulate shoots.

Male Cones (Strobilus) : Male strobili are borne in whorls in the axils of scale leaves at the nodes of the fertile branch. They represent suppressed modified lateral branches (Fig.).

The male strobilus consists of a short central axis with 2-10 pairs of opposite decussate bracts (Fig.). One or two basal pairs of bracts are sterile, while the others are fertile (Fig.). A single male flower arises in the axil of each fertile bract. Thus a male strobilus with several male flowers can be compared with an inflorescence.

1. *Male Flower :* The male flower has a short stalk known as microsporangiophore or antherophore with 2-8 microsporangia at its tip. At the base of the microsporangiophore two anterioposterior bracteoles are present, which are united at the base (Fig.). These bracteoles are without any vascular supply and have been interpreted as 'perianth'. The microsporangiophore is considered to be the product of phylogenetic fusion of a pair of microsporophylls.

2. *Development of Microsporangium* : Development of microsporangium is of typical eusporangiate type. The microsporangiophore develops as a small protuberance in the axil of a fertile bract. Later, the tip of the microsporangiophore becomes lobed and each lobe represents a microsporangium or anther (Fig.). Some hypodermal cells in each lobe function as archesporial initials. The archesporial cells get divided into two groups by the formation of a band of sterile cells between them. In this way, each sporangium divides into two chambers (Fig.).

The archesporial cells divide periclinally into outer primary wall cells and inner primary sporogenous cells (Fig.). The primary wall cells divide repeatedly to form three layered thick wall. However, Singh and Maheshwari (1962) are of the view that primary wall cells directly form the outermost wall of the sporangium and the primary sporogenous cells further divide by two periclinal divisions to differentiate middle wall layer, inner tapetal layer and sporogenous cells (Fig.).

The sporogenous cells divide in all possible planes and form numerous spore mother cells. At this stage, the middle layer of the sporangial wall is crushed and as a result the sporangium becomes flat. The tapetal cells considerably enlarge and their nuclei divide and thus they become multinucleate in mature sporangium (Fig.). Each spore mother cell undergoes meiosis and forms four haploid microspores, arranged in a tetrahedral tetrad.

After reduction division of spore mother cells, tapetal cells disorganise. A microsporangium has 2-3 loculi and is often called as synangium. Each loculus opens through a terminal slit. The microspores are released through this slit.

Female Cone (Strobilus) **:** The female strobili also arise in the axils of scale leaves at the nodes of the fertile branches. They arise in whorls of 2-4 (Fig. A) and are usually smaller than the male strobili.

The female strobilus consists of a central axis with 2-4 pairs (sometimes more) of decussately arranged bracts. Most of these bracts are sterile, and only two upper bracts have an ovule each in their axils (Fig.). Sometimes, the strobili become uniovulate due to abortion of one of the ovules and the surviving ovule assumes a false terminal position.

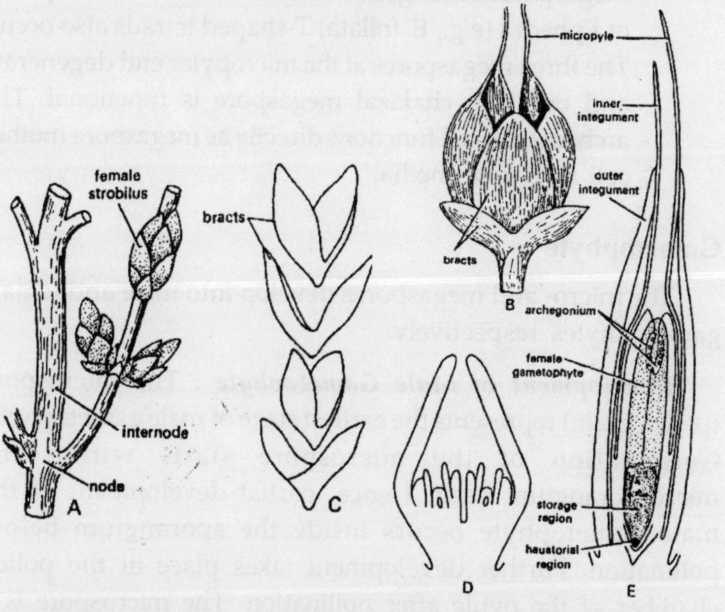

Ephedra* : *Female strobilus; A. A branch with female strobilus, B. Female strobilus, C. Arrangement of bracts in strobilus, D. Female strobilus in longisection, E. Longitudinal section of ovule.

1. *Female Flower* : The female flower consists of a short stalk and an ovule (megasporangium). The ovule has a prominent nucleus protected by a double envelope. The lower half of the inner integument is fused with the nucleus while the upper half is free and is drawn out into an elongated micropylar tube. The outer integument is comparatively short and represents a fused pair of bracts (Fig.).

2. *Development of Megasporangium* : A hypodermal cell of

the nucleus functions as archesporial cell. It divides periclinally into an outer parietal cell and an inner megaspore mother cell. Due to repeated periclinal divisions in parietal cell, the megaspore mother cell becomes deep seated in the nucleus. The megaspore mother cell divides meiotically and forms four haploid megaspores, arranged in a linear tetrad. In some species of Ephedra (e.g., E. foliata) T-shaped tetrads also occur. The three megaspores at the micropylar end degenerate and only the chalazal megaspore is functional. The archesporial cell functions directly as megaspore mother cell in E. intermedia.

Gametophyte

The micro- and megaspores develop into male and female gametophytes respectively.

Development of Male Gametophyte : The microspore (pollen grain) represents the earliest stage of male gametophyte. Germination of the microspore starts within the microsporangium itself. Hence, partial development of the male gametophyte occurs inside the sporangium before pollination. Further development takes place in the pollen chamber of the ovule after pollination. The microspore is a boat-shaped structure, measuring about 75 um in length and 23 um in breadth. The wall of the microspores is two layered—the outer exine and the inner intine. The exine is thick and striated and striations occur parallel to the long axis of spore. The intine is thin and smooth (Fig.). The microspore has a haploid nucleus (Fig.).

The microspore first divides by a transverse wall into a small prothallial cell and a larger outer cell (Fig.). The outer cell divides again by a transverse wall and forms a second prothallial cell and an antheridial cell (Fig.). Both prothallial cells soon disintegrate. The nucleus of the antheridial cell divides to form a small generative cell and a comparatively large tube cell (Fig.). The generative cell does not possess a distinct cell wall. It divides by a periclinal wall into a stalk cell and a body cell. At this stage, the male gametophyte consists of five cells — two prothallial cells, a stalk cell, a body cell and a tube nucleus. The sporangium dehisces at this stage and microspores (male gametophytes) are now ready for pollination (Fig.).

In Ephedra, pollination is anemophilous and the microspores are carried by the wind to the ovulate strobilus. They are deposited on the micropylar canal and then sucked inside by the pollination drop. They come to lie in a deep pollen chamber in direct contact with the neck of the archegonium. Further development of the male gametophyte takes place in the pollen chamber.

In the pollen chamber the microspore germinates, the exine bursts and the intine comes out in the form of a pollen tube. At this stage, the body cell divides to form two male gametes. The gametes are non-motile. The pollen tube grows through the tissue of the long neck carrying the two male nuclei, a tube nucleus and a stalk nucleus (Fig.).

Development of Female Gametophyte : The functional megaspore represents the first stage of the female gametophyte (Fig.). The nucleus of the megaspore divides mitotically and forms two daughter nuclei which occupy the two opposite poles of the megaspore due to the development of a large central vacuole. Both nuclei divide again and the four nuclei thus formed are arranged in equidistant positions on the periphery of the sac (Fig.). These nuclei continue to divide till about 256 free nuclei are formed (Fig.). At this stage, the central vacuole disappears and wall formation starts from the periphery towards the centre. Thus a mass of tissue, known as female gametophyte or endosperm, is formed (Fig.). Later the gametophyte is differentiated into two regions; the lower chalazal region is the food storage tissue and in the upper micropylar region normally 1-3 archegonia develop.

Any superficial cell of the female gametophyte may function as archegonial initial (Fig.). The archegonial initial divides perielinally and forms an outer primary neck cell or neck initial and an inner central cell (Fig.). The primary neck cell divides repeatedly and forms a long neck consisting of 32 or more cells (Fig.). In early stages the neck cells are arranged in definite tiers but later on due to irregular divisions the arrangement is disturbed. The archegonial neck in Ephedra is the largest amongst the gymnosperms. Neck canal cells are not present in the neck. The central cell increases in size and its nucleus divides to form a venter canal nucleus and an egg nucleus. As the archegonium matures, the venter canal nucleus disintegrates and the egg nucleus occupies a central position. The nuclear cells surrounding the central cell divide and form a 2-3 layered archegonial jacket (Fig.).

With the development of the archegonium, the nuclear tissue above the gametophyte disintegrates and a deep pollen chamber is formed.

Fertilisation

In Ephedra fertilisation is siphonogamous. The pollen tube grows through the neck. It reaches the egg, where the tip of the tube ruptures and its contents discharge into the cytoplasm of the central cell. One of the male nucleus fuses with the egg and then an oospore is formed. Later, the remaining nuclei of the pollen tube (one male nucleus, a tube nucleus) and a stalk nucleus disintegrate.

Sometimes, as in E. foliata, the second male nucleus fuses with the venter canal nucleus and a transient tissue is formed which disappears as the embryo develops. It is generally believed that this sort of double fertilisation has no phylogenetic significance and it is simply a natural outcome of a tendency towards the fusion between any two nuclei of opposite sexual potencies those happen to lie free in a common chamber.

Embryo Development

More than one archegonia may be fertilised in an ovule, but only one oospore develops into embryo. Following fertilisation, the oospore nucleus divides and forms eight free nuclei which are evenly distributed in the cytoplasm (Fig.). Later, wall formation takes place and this structure is now known as proembryo (Fig.).. Each cell of the proembryo is capable to develop into an independent embryo. As such polyembryony occurs in Ephedra without any cleavage. Such type of polyembryony is known as embryo-sac polyembryony. All proembryos are not always potential; for example, in E. foliata only six and in E. trifurca only three proembryos are potential.

In its development, the proembryonic cell first produces a tubular outgrowth known as suspensor tube (Fig.). Thereafter, the nucleus of the proembryonic cell divides to form two nuclei which descend into the suspensor tube. At this stage, the suspensor tube grows very fast and reaches the tissue of the female gametophyte after piercing the venter. Later, a wall is laid down between the two nuclei and as such a terminal embryo initial and an elongating suspensor cell are differentiated (Fig.). The embryo initial by further divisions eventually gives rise to a multi-cellular secondary suspensor and an embryo. The embryo consists of two cotyledons and a main axis from which the primary root and shoot are differentiated (Fig.).

Seed : After fertilisation, the ovule is transformed into seed. The mature seed is ovoid to triangular and consists of a large embryo embedded in the female gametophytic tissue and a membranous nuclear layer surrounded by a hard seed coat. As the seed matures, the adjacent bracts of the female strobilus develop into a thick and fleshy layer which forms an additional integument to the seed (Fig.).

Germination of Seed : The seed germinates immediately without undergoing a period of rest. The germination is epigeal. The radicle emerges out through the micropylar end and it develops into a primary root (Fig.). At the same time, the hypocotyl also emerges with two fleshy and linear cotyledons. The cotyledons grow steadily and serve as organs of assimilation for the developing seedling (Fig.). Vivipary is of common occurrence in some species of Ephedra like E. trifurca. The seed germinates within the parent strobilus.

17

The Tales

Order Taxales

The order Taxales incluJes a single family Taxaceae, represented by five genera (Austrotaxus, Amentotaus, Nothotaxus, Taxus and Torreya) and 15 species. The plants are usually small trees or shrubs. The branches bear simple leaves which show spiral arrangement.

The members of the order are characterized by the absence of resin canals in the aerial parts of the sporophyte; they are restricted only to roots. Ray tracheids have tertiary thickenings and the pits are uniseriate. Male cones are borne in the axils of leaves, either singly or in groups.

Ovules are borne terminally on axillary branches; subtended by many scales at the base. At maturity, the ovule is enclosed within a fleshy aril.

Taxus is described here as a representative of the order Taxales.

Taxus

Systematic Position

Division	—	Coniferophyta
Class	—	Taxopsida
Order	—	Taxales
Family	—	Taxaceae

Taxus is represented by nine species which are widely distributed in North America, Europe and Asia extending up to Malaysia. Only one species of Taxus, T. baccata occurs in India, which is distributed all along the Himalayas at an elevation of 1,800-3,300 meters.

Sporophyte

***External Morphology*:** The sporophytic plant body of Taxus is an evergreen, densely branched, medium-sized tree, 9-15 meters in height, but T. canadensis is a low and straggling shrub. It is differentiated into root, stem and leaves.

1. *Root :* The root is much branched deep-feeder tap root. The young root has a distinct root cap and root hairs.
2. *Stem :* The stem is monopodial and profusely branched (Fig.). The main trunk gives off horizontal and spreading branches. The vegetative shoots grow indefinitely and bear green foliage leaves which form a dense canopy. The reproductive shoots are, however, dwarf and they bear scaly leaves only.
3. *Leaves :* The foliage leaves are simple, sub-sessile, persistent, 2-3 cm long and are linear to lanceolate with entire margins, decurrent bases and sharply-pointed apices that contain silica particles (Fig.). They are spirally arranged, but as the leaf bases are twisted they appear to be in two rows. The margins of the leaf are

recurved and there is a single midvein. The upper surface of the leaf is dark green, while the lower surface is pale or rusty red in colour.

The scale leaves are minute and membranous, and are arranged in opposite decussate mariner.

Internal Structure

1. *Root :*

(a) *Primary Root :* The internal structure of the root is similar to that of Pinus.

(i) *Epidermis :* It is the outermost single-layered covering of the root. Some epidermal cells elongate to form long unicellular root hairs.

(ii) *Cortex :* It consists of several layers of parenchymatous cells with narrow intercellular spaces.

(iii) *Endodermis :* The innermost layer of the cortex forms endodermis. The endodermal cells have casparian bands of thickenings on their radial walls.

(iv) *Pericycle :* The endodermis is followed by a multi-layered pericycle, composed of parenchymatous cells. The pericycle contains many resin canals.

(v) *Vascular Tissue :* The stele is diarch and exarch. The xylem consists of only tracheids. Phloem strands alternate with the xylem strands; they consist of sieve cells and phloem parenchyma.

The central part of the stele is occupied by plates of metaxylem tracheids with uniseriate bordered pits in addition to tertiary spirals.

(b) *Secondary Growth in Root :* Secondary growth of Taxus root is exactly similar to that of Pinus root.

Some parenchyma cells along the inner edges of phloem groups become meristematic and form crescent-shaped strips of cambium. Subsequently, the cells of the pericycle opposite the protoxylem strands become meristematic and thus small strips of cambium are also formed outside the xylem strands. The first formed cambial strips below the phloem strands join with the cambial strips outside the xylem strands and thus a wavy ring of vascular cambium is formed. This cambium ring produces secondary phloem on the outer side and secondary xylem on the inner side.

Due to extensive activity of the vascular cambium, enormous amount of secondary tissue is formed. The peripheral tissues of the root are sloughed off as a result of the pressure exerted by the secondary tissues. The peripheral tissues are replaced by a more efficient protective covering, the periderm. The periderm is formed by the activity of the cork cambium or phellogen which originates from the outer layers of pericycle. The cork cambium produces phellem (cork) towards the outer side and phelloderm (secondary cortex) towards the inner side. The first-formed phellogen may persist for a fairly long period and new phellogen layers are formed in the deeper layers after the earlier one becomes non-functional.

2. *Stem :*

(a) *Primary Structure :* A transverse section of the stem is somewhat irregular in outline due to the presence of persistent leaf bases. The internal structure of the primary stem shows the following structures (Fig.). (i) *Epidermis :* It consists of a single layer of parenchymatous cells with a thick cuticle on the outer walls. (ii) *Cortex :* There is a multi-layered parenchymatous cortex. (iii) *Endodermis and*

Pericycle: The innermost layer of the cortex represents endodermis, but it is not differentiated from other cortical layers. The endodermis is followed by a 2-3- layered pericycle composed of sclerenchymatous cells. (iv) *Vascular Tissue :* The vascular bundles are arranged in a ring. They are conjoint, collateral, endarch and open. The protoxylem is composed of tracheids with spiral thickenings, and metaxylem tracheids have uniseriate-bordered pits; xylem parenchyma is, however, absent. The phloem is composed of phloem parenchyma and sieve elements. The medullary rays are narrow. (v) *Pith:* A parenchymatous pith is present in the centre of the stem.

(b) *Secondary Growth in Stem :* A ring of vascular cambium originates in part from the procambium present between the xylem and phloem of the primary vascular bundles (fascicular cambium) and in part from the rejuvenation of interfascicular parenchyma between the bundles (interfascicular cambium). This cambium ring cuts of secondary xylem towards the inner side and secondary phloem towards the outer side. The primary xylem remains intact as the stem increases in girth, but the primary tissues outside the secondary vascular tissues (primary phloem, cortex and epidermis) are pushed outside as they are subjected to pressure from the inside by the enlarging secondary vascular cylinder. With continued secondary growth, the outer tissues are completely crushed (Fig.).

(c) *Secondary Xylem :* The wood of Taxus, mainly composed of secondary xylem, is pycnoxylic. It shows distinct annual rings. The tracheids of secondary xylem have uniseriate bordered pits on their radial walls. They are characterised by the presence of tertiary thickenings in the form of spiral bands on their walls. Bars of Sanio are also present between the bordered pits. The medullary rays are simple and uniseriate.

3. *Leaf* : The leaf is dorsiventral and shows xerophytic features (Fig.).

   (a) *Epidermis* : Both the upper epidermis and lower epidermis are single-layered and composed of compactly arranged parenchymatous cells which appear rectangular in a vertical section. The upper epidermis is smooth and covered with a thick cuticle, whereas the cells of the lower epidermal layer are papillate and covered by a thin cuticle. Stomata are sunken and are confined only to the lower epidermis. They are haplochelic in development.

   (b) *Mesophyll* : The mesophyll is differentiated into palisade tissue and spongy tissue. The palisade is usually 2- layered with compactly arranged, elongated, parenchymatous cells rich in chloroplasts. The spongy parenchyma is composed of loosely-arranged parenchymatous cells with large intercellular spaces between them.

   (c) *Endodermis* : The vascular bundle is surrounded by an endodermal layer of barrel-shaped cells with casparian thickenings on their radial walls.

   (d) *Vascular Bundle* : There is a single vascular bundle in the midrib region with its phloem facing abaxially and xylem adaxially. Transfusion tissue is represented by a group of tracheids and albuminous cells which form wings on either side of the vascular bundle. The parenchyma that occurs on the abaxial and adaxial side of the bundle is also regarded the part of transfusion tissue. Thus the transfusion tissue is present all around the bundle.

Presence of thick cuticle, sunken stomata, transfusion tissue and narrow lamina, and differentiation of mesophyll into palisade and spongy parenchyma indicates xerophytic nature of Taxus.

Shoot Apex Organisation

The shoot apex of Taxus exhibits four distinct cytohistological zonations, i.e., (i) the surface meristem, (ii) the zone of sub-apical initials, (iii) the peripheral or flanking meristem zone, and (iv) the rib meristem zone. The number of cells in the flanking meristem increases by division in the meristem itself as well as by the addition of cells from the surface meristem. The cortex, leaf primordia and procambium develop from this zone. The cells of the rib meristem are comparatively larger in size and are arranged in vertical rows. The cells at the base of these rows divide in all the planes and differentiate into pith cells (Fig.).

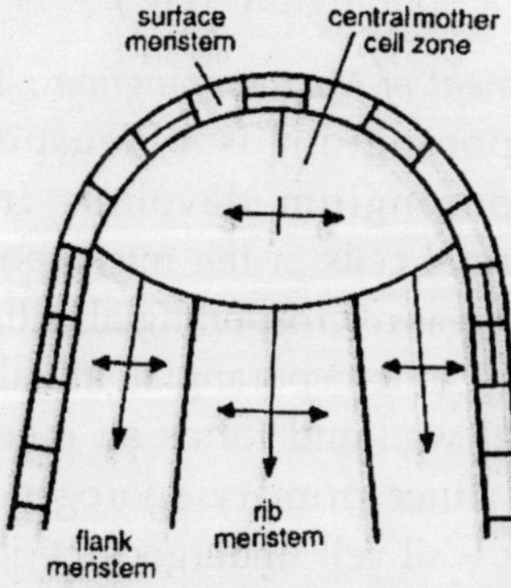

Taxes : Shoot apex organization (diagrammatic representation).

Reproduction

Taxus reproduces only by sexual means. The sporophytic plant body produces two types of spores—microspore and megaspore, which develop into male and female gametophytes respectively. The plants are dioecious, but occasionally male and female shoots develop on the same plant. The flowering shoots are borne on the underside of the shoots of the previous year in the axils of the leaves. The male reproductive organs are organised into definite cones (strobili) but the female reproductive organs are not organised into cones and they are often called female flowers.

Male Cone : The male cone is a stalked, sub-globose, pale yellow structure, developing solitary in the axil of a leaf. Some sterile scales are present at the base of the cone. Young cones are erect, but as they mature they are oriented in such a way as to hang pendent on the lower side of the shoot. There are 6-14 microsporophylls in a cone, crowded at the top of the cone axis. The cone axis has a broad apex which is consumed in the formation of terminal microsporophyll. The microsporophyll is a peltate structure with 6-8 pendent microsporangia. The sessile, elongated, and laterally concrescent sporangia are coherent with the sporophyll stalk. There are numerous yellow microspores in a sporangium (Fig.).

1. *Development of Microsporangium :* The development of microsporangium is of eusporangiate type. A microsporangium develops from one or more hypodermal cells of the microsporophyll. These cells are known as microsporangial initial cells (archesporial initial). A microsporangial initial first divides by a periclinal wall and forms an outer primary wall cell and an inner primary sporogenous cell (Fig.). The primary wall cell undergoes periclinal and anticlinal divisions and forms a 2- layered sporangium wall. The primary sporogenous cell divides repeatedly and forms a mass of sporogenous tissue. The outer sporogenous cells differentiate into a nutritive layer, called tapeturi (Fig.). The remaining sporogenous cells enlarge and function as microspore mother cells. Each microspore

mother cell forms a tetrahedral or isobilateral tetrads of microspores after reduction division (Fig.).

2. *Structure of Microsporangium* : The wall of the microsporangium, besides epidermis, has two wall layers. The epidermis consists of thin-walled cells in the lower part of the microsporangium and thick walled in the upper part. The innermost wall layer functions as nutritive tapetum. Inside the wall layers numerous microspores are present (Fig.). The microspore is an uniaperturate, spheroidal, wingless structure, and it has two wall layers—an outer thick exine and an inner thin intine (Fig.).
3. *Dehiscence of Microsporangium* : The microsporophylls are compactly arranged in a young cone, but as the cone matures they become loose. The sporangium dehisces at this stage by a longitudinal 'slit. The dehiscence starts from the base upwards; the curling of the wall opens the microsporangium. Ultimately, microsporangia separate from the sporophyll stalk. The microspores are dispersed by wind. The thickened epidermal cells perhaps also help in sporangial dehiscence.

Female-flower : In Taxus, female reproductive structures are not organised into cones as in other conifers. The female shoot develops in the axil of a leaf on the previous year's branch as a small bud. It is protected by several closely overlapping scale leaves. It consists of a short primary axis, the growth of which is soon arrested. Then a short secondary axis develops in the axil of one of the distal scales. The secondary axis bears few pairs of scaly leaves and a terminal ovule. The apical bud of the primary axis may sometimes resume growth and then produces another secondary axis which bears a terminal ovule next year. Similarly, the secondary axis may bear another bud in the axil of another scale that produces a tertiary axis. The tertiary axis also bears an ovule. Thus, the secondary axis may have two ovules (Fig.).

1. *Development of Megasporangium or Ovale* : The development of megasporangium (ovule) is eusporangiate. It develops from one or more superficial initial cells which divide repeatedly to form a mass of parenchymatous cells. This mass of cells develops into nucleus. An annular outgrowth is formed at the base of the nucleus which forms a covering around it. This covering forms the integument which completely surrounds the nucleus but for a narrow aperture at its apex the micropyle. The integument is initially free from the nucleus but later its basal part may fuse with the nucleus due to growth of the chalazal region. An another ring, like outgrowth arises from the base of the integument which slowly covers the entire ovule after the embryo has formed. This outgrowth is known as aril (Fig.).

   A hypodermal cell of the nucleus (towards the micropyle) enlarges and functions as archesporial cell. It divides by a periclinal wall into an upper parietal cell and a lower sporogenous cell. The parietal cell divides periclinally and pushes the sporogenous cell deep into the nuclear tissue. The sporogenous cell, which functions as megaspore mother cell, divides meiotically and forms four haploid megaspores, arranged in a linear tetrad. The three megaspores at the micropylar end normally degenerate, and the remaining one is functional. Occasionally, more than one megaspores from the same linear tetrad are functional. In such cases more than one female gametophytes develop but only one of them persists (Fig.).

2. *Structure of Ovule or Megasporangium* : Like other gymnosperms, the ovule of Taxus is also orthotropous. It has the following parts (Fig.).

   (a) *Nucleus* : It is the massive parenchymatous region of the ovule.

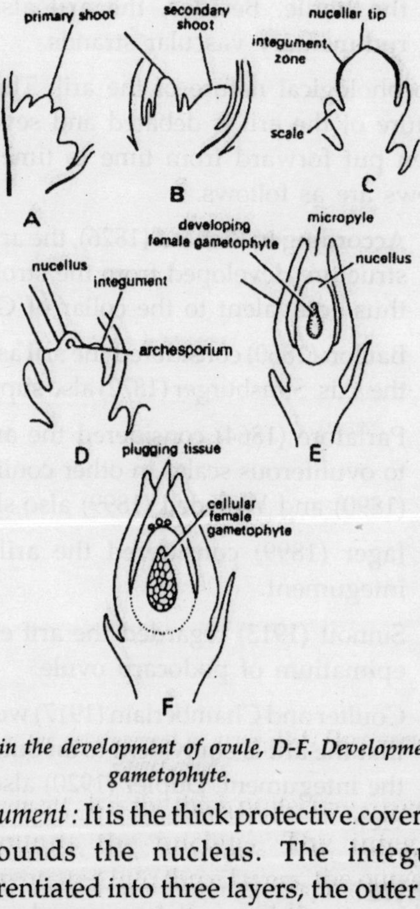

***Taxus* : A-C. Stages in the development of ovule, D-F. Development of female gametophyte.**

(b) *Integument* : It is the thick protective covering which surrounds the nucleus. The integument is differentiated into three layers, the outer and inner fleshy layers and the middle stony layer. The integument remains free from the nucleus throughout.

(c) *Micropyle* : It is a small narrow aperture at the tip of the ovule, formed by the integument.

(d) *Aril* : It is a slow growing ring-like structure which encloses the ovule. The aril is green when young but at maturity it becomes red and juicy.

(e) *Vascular System* : The integument is supplied by two vascular strands that extend up to the base of the ovule. Besides, the aril also receives two rudimentary vascular strands.

3. Morphological nature of the aril. The morphological nature of the aril is debated and several views have been put forward from time to time. Some of these views are as follows.
   (1) According to Richard (1826), the aril is an accessory structure developed from the strobilus axis and is thus equivalent to the collar of Ginkgo ovule.
   (2) Baillon (1860) considered the aril as an expansion of the axis. Strasburger (1872) also supported this view.
   (3) Parlatore (1864) considered the aril as equivalent to ovuliferous scales in other conifers. Celakovsky (1890) and Worsdell (1899) also shared this view.
   (4) Jager (1899) considered the aril as the second integument.
   (5) Sinnott (1913) regarded the aril equivalent to the epimatium of podocarp ovule.
   (6) Coulter and Chamberlain (1917) were of the opinion that the aril is equivalent to an outer fleshy layer of the integument. Dupler (1920) also supported this view.

Gametophyte

Taxus is heterosporous, i.e., it produces two different types of spores — microspores and megaspores. These spores are haploid and represent the first cell of the gametophytic phase. The microspore forms the male gametophyte and the megaspore, the female gametophyte.

Development of Male Gametophyte : The microspore is the first stage in the development of male gametophyte. It is uninucleate, uniaperturate and spheroidal structure with a thick exine and a thin intine (Fig.). The microspores are dispersed at uninucleate stage and carried to the micropyle by the wind. They are caught into the pollination drop at the micropyle. The germination of the pollen thus starts on the nucleus. The nucleus of the microspore divides to form a tube cell and a generative cell (Fig.). Prothallial cells are not formed in Taxus. Then the exine ruptures and the intine protrudes in the form of a pollen tube. The tube nucleus moves into the pollen tube. The generative cell divides to form a stalk cell and a body cell (Fig.). The body cell divides into two unequal male gametes (male nuclei). The pollen tube penetrates the nucleus and reaches up to the archegonial neck. At this stage, the tip of the pollen tube contains a tube nucleus, a stalk nucleus and two male nuclei (Fig.).

Development of Female Gametophyte : The megaspore is the mother cell of the female gametophyte. It is deeply embedded in the parenchymatous tissue of the nucleus. It increases in size considerably and then its haploid nucleus divides repeatedly to form 128 nuclei (256 in T. cuspidata). A central vacuole develops in the megaspore and the multinucleate cytoplasm forms a thin layer at the periphery. Then centripetal wall formation starts. Initially numerous radially-elongated multinucleate tube-like cells called alveoli are formed. Each alveolus later divides to form many uninucleate cells. The uninucleate cells thus formed are known as female prothallus or endosperm. The apex of the female gametophyte forms a flask-shaped structure, called tent-pole, which disappears later (Fig.).

Development of Archegonium : Although all the superficial cells of the female gametophyte at the micropylar end are capable to form archegonia, only a few (usually 2-8, but sometimes up to 25) differentiate into archegonial initials (Fig.). The development of archegonium is similar to Pinus but for minor variations. The archegonial initial is distinguished from the other cells by its large size and dense cytoplasm (Fig.). It divides to form an upper primary neck cell and a lower central cell (Fig.). The primary neck cell divides by intersecting walls to form a 2-4 celled neck (Fig.). The central cell considerably increases in size and its nucleus divides into a venter canal nucleus and an egg nucleus. The former soon disintegrates and thus there is no venter canal cell in Taxus. The central cell increases in size and a nutritive layer is differentiated around it. This layer, known as archegonial jacket, helps in the transfusion of food from the endosperm to the archegonium.

Pollination

The pollination in Taxus is anemophilous. The microspores dispersed by the wind reach the micropyle of the ovule and adhere to the exudates (pollination drop) secreted by it. As the pollination drop sucked in, the microspores are carried to the nucleus, where they germinate immediately.

Fertilisation

Fertilisation usually takes place two months after pollination. The pollen tube is lodged against the female gametophyte even before archegonial initials are formed. As it moves downwards, first it comes in contact with the tent pole. Later, the cells of the tent pole region disorganise and the pollen tube comes in contact with the archegonial neck. The tip of the pollen tube then bursts and all the four nuclei (one tube nucleus, one stalk nucleus, two male nuclei) enter the archegonium (Fig.). The larger male gamete unites with the egg and forms a zygote, whereas the other three nuclei degenerate. Since there are several archegonia in a female gametophyte, many eggs are fertilized, resulting in simple polyembryony. But only a single embryo, however, attains maturity.

Embryo Development

The nucleus of the zygote undergoes free-nuclear divisions and forms about 32 nuclei (Fig.). Now wall formation starts and soon the whole structure becomes cellular and now it is known as proembryo. A few cells at the tip of the proembryo develop into a dicotyledonous embryo, while the cells above it differentiate as suspensor cells.

The suspensor cells elongate and push the embryonal cell deep into the tissue of the food, laden female prothallus. The fully mature embryo is straight and it has a radicle towards the micropylar end and a plumule concealed in between the two cotyledons (Fig.). The hypocotyl is present between radicle and plumule. Only one embryo matures in an ovule. As the embryo consumes the entire endosperm during its development, the seed is non-endospermic.

Seed

With the development of embryo the following changes take place in the ovule during its development into a seed.

(1) The integument of the ovule dries up and forms a hard testa.

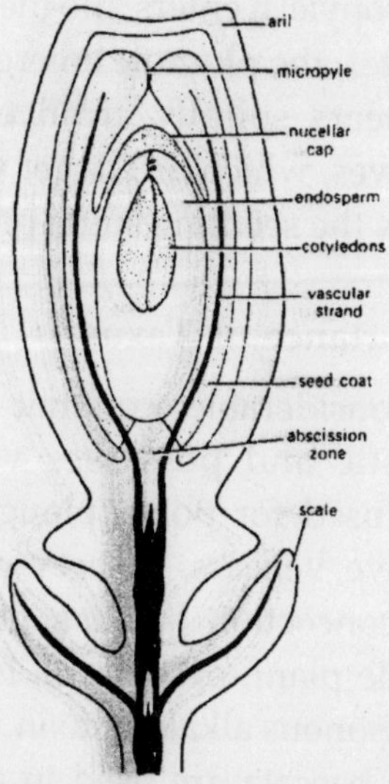

Taxes : Longitudinal section of seed.

(2) The aril grows into a fleshy envelope, giving the seed a berry-like appearance. It becomes red and juicy in the mature seed (Fig.).

(3) The nucleus is completely used by the developing gametophyte.

(4) The endosperm is also consumed by the developing embryo and thus the mature seed is non-endospermic.

(5) The micropyle is retained in the form of a small aperture at the tip of the seed (Fig.). The bright coloured aril helps in attracting birds which help in seed dispersal.

Seed Germination

If suitable conditions are available, seeds of Taxus germinate immediately without undergoing a period of rest. The germination is epigeal, i.e., the cotyledons come above the soil by the elongation of the hypocotyl. During germination, the testa bursts open by absorbing water and the radicle comes out from the micropyle; it enters into the soil and forms primary tap root. Thereafter, the plumule emerges and forms the shoot system which bears spirally arranged foliage leaves. The cotyledonary leaves, which are larger than the foliage leaves, soon fall down as the seedling starts producing foliage leaves.

Economic Importance of Taxus

Taxus is of considerable economic importance. The wood is durable, elastic and possesses very good mechanical properties. It is used for poles, ploughs, cart axles, and for making decorative veneers, turnery and posts. The bark is brewed into a concoction like tea in Laddakh. The aril is edible. The whole plant, except aril, is poisonous due to the presence of a poisonous alkaloid taxin. The seeds are sedative. The leaves of T. baccata are used in asthma, bronchitis and epilepsy. The plants are also grown in gardens as ornamentals.

18

The Coniferales

Order Coniferales

Coniferales is the largest order of gymnosperms with about 52 genera and 550 species. These plants originated in the Carboniferous period. About 30 genera of the order are confined to the Northern and 14 to the Southern hemisphere and rest of the genera cross the equator on either side. They are usually long, branched and evergreen trees. The branches are usually dimorphic and bear needle-like, linear or lanceolate leaves. The wood is pycnoxylic, characterised by the presence of resin canals. The micro- and megasporophylls form compact cones. The male gametes are non-motile and the fertilisation is siphonogamous.

Coulter and Chamberlain (1910) recognised only two families - Pinaceae and Taxaceae - in the order Coniferales, whereas Pilger (1926) divided the order into seven families, Pinaceae, Taxodiaceae, Cupressaceae, Araucariaceae, Podocarpaceae, Cephalotaxaceae and Taxaceae. Chamberlain (1935) recognised only six families as he merged

Cephalotaxaceae with Taxaceae. Pulle (1942) considered Taxales as a separate order and included two families Taxaceae and Cephalotaxaceae in it. Florin (1949), Zimmermann et al. (1959) and Sporne (1965) also recognised Taicales as a distinct order. Pant (1957) elevated the order Taxales to the rank of a class Taxopsida. Sporne (1965) divided Coniferales into nine families Lebachiaceae, Voltziaceae, Palissyaceae, Pinaceae, Taxodiaceae, Cupressaceae, Podocarpaceae, Cephalotaxaceae, Araucariaceae.

Family Pinaceae

Members of this family are monoecious with spirally arranged, needle-like leaves. Each- needle has one or two vascular bundles, enclosed within a common bundle sheath. The pollen grains are mostly winged. The female cone has numerous spirally arranged ovuliferous scales free from the subtending bract scales or adhering only at the base. Each ovuliferous scale bears two ovules on its upper side.

The life history of Pinus is described here as a representative of the order Coniferales.

Pinus

Systematic Position

Division — Coniferophyta

Class — Coniferopsida

Order — Coniferales

Family — Pinaceae

Pinus, one of the most important taxon of the order Coniferales, is represented by about 105 species, distributed throughout the northern hemisphere. The species of Pinus form dense evergreen forests in the north temperate and sub alpine regions.

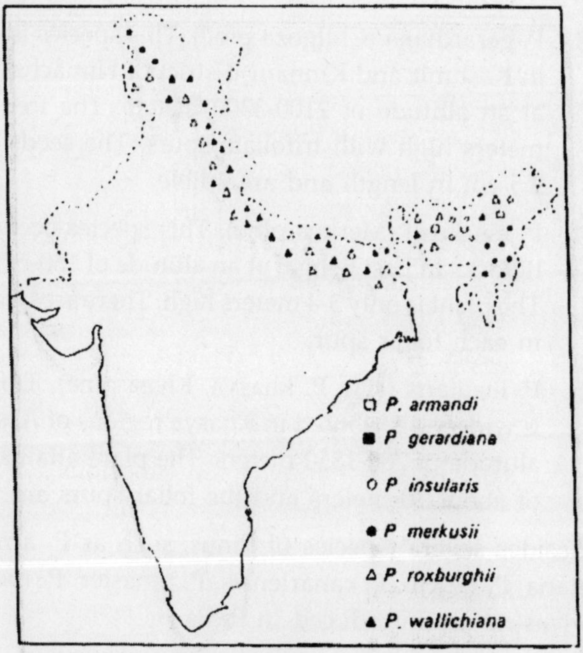

Distribution of Pinus in India.

Five species of Pinus occur in the Indian sub-continent; of these, four species are confined to the north-east and north-west Himalayan regions (Fig.). The distribution of Pinus in India is as follows.

1. P. raburghii (syn. P. longifolia, Chir pine). This species occurs in the hilly regions of Kashmir, Punjab, Himachal Pradesh and Uttar Pradesh at an altitude of 450-2250 meters. The tree attains a height of 35-50 meters. There are three needles in each foliar spur.

2. P. wafchiana (syn. P. ercelsa, Blue pine, Kail). This species grows luxuriantly in the hills of Kashmir, Himachal Pradesh and Punjab at an altitude of 1500-3300 meters. The tree attains a height of about 50 meters. There are five needles in each foliar spur. The female strobili are cylindrical.

3. P. gerardiana (Chilgoza pine). This species is common in Kashmir and Kinnaur district of Himachal Pradesh at an altitude of 2100-3300 meters. The tree is 11-20 meters high with trifoliate spurs. The seeds are 0.5 - 2.5 cm in length and are edible.
4. P. merkusii (Merkus pine). This species occurs on the hillocks in East Bengal at an altitude of 150-600 meters. The plant is only 3-4 meters high. There are two needles in each foliar spur.
5. P. insularis (syn. P. khasya, Khasi pine). This species is widely distributed in Khasya regions of Assam at an altitude of 700-1850 meters. The plant attains a height of about 30 meters and the foliar spurs are trifoliate.

Besides several species of Pinus, such as P. armandi, P. montana, P. laricio, P. canariensis, P. pinaster, P. radiata and P. sylvestris are introduced in India.

Sporophyte

External Morphology : The sporophytic plant body of Pinus is an evergreen tree. It grows luxuriantly in temperate xerophytic habitats. When young, the plant gives a typical pyramidal appearance. The plant body is differentiated into root, stem and leaves.

1. *Root :* The primary root is a typical tap root. It does not penetrate deep into the soil. The lateral roots grow extensively and help the plant to keep firmly in the soil. Root hairs are poorly developed. The roots are covered with fungal hyphae called mycorrhiza. The fungi present in the ectotrophic mycorrhiza are mainly the species of Amantia, Boletus, Clavaria and Seleroderma of the class Basidiomycetes. They remain in symbiotic association with the roots. It has been reported that in the absence of mycorrhiza) association,

the death rate of Pinus seedlings is considerably increased.

2. *Stem :* The stem of Pinus is cylindrical erect, woody and branched. The branches are monopodial and develop spirally in the axils of scaly leaves present on the stem. Such a branching gives a conical appearance to the plant. But in some species, such as P. torreyana, the branches are diffuse or scattered. The stem is dichotomously branched in P. sabiniana.

   In Pinus, the branches are dimorphic. The two types of branches are long shoots or branches of unlimited growth and dwarf shoots or branches of limited growth.

   (a) *Long Shoots or Branches of Unlimited Growth :* These branches are present on the main trunk. The apical buds of these branches grow indefinitely, hence these branches are of unlimited growth. They develop in the axils of scaly leaves and spread horizontally. They shorten gradually towards the tip, thus providing a graceful pyramidal appearance to the tree. These branches bear only scaly leaves.

   (b) *Dwarf Shoots or Branches of Limited Growth :* These branches do not have apical buds and hence show only limited growth. They are also known as brachyblast. They develop in the axils of scaly leaves and bear both scaly and foliage leaves. Dwarf shoots are shed every two or three years, leaving scars on the stem.

3. *Leaves :* The leaves are dimorphic, i.e., there are two types of leaves—scale leaves and foliage leaves.

   (a) *Scale Leaves :* The scale leaves are dark brown, membranous, thin and small, and are present on both long and dwarf shoots. They do not help in photosynthesis. Their main function is protection of the young buds. They fall off as the branches mature. The scale leaves on the dwarf shoots have a distinct midrib and they are called cataphylls.

(b) *Foliage Leaves :* The foliage leaves are green, acicular and needle-like. They are borne only on the dwarf shoot. A dwarf shoot with a group of needle-like foliage leaves is known as foliar spur. The number of needles in a spur varies in different species (Fig.); for instance, P. monophylla is monofoliar with a single needle, P.. sylvestris and P. merkusii are bifoliar with two needles, P. longifolia and P. gerardiana are trifoliar with three needles, P. quadrifolia is quadrifoliar with four needles and P. wallichiana and P. annandi are pentafoliar with five needles per spur.

P. monophylla P. sylvestris P. gerardiana P. excelsa (unifoliar spur) (bifoliar spur) (trifoliar spur) (pentafoliar spur) A B C D Fig. A-D. Pinus : Various kinds of foliar spur.

The foliage leaves are photosynthetic and remain persistent for several years. The needle-like nature of the foliage leaves indicates xerophytic adaptation of these plants.

Internal Structure

1. *Root :*

(a) *Primary Root :* The primary root of Pinus resembles with a typical dicot root. The following regions are discernible in a transection of the young root (Fig.).

(i) *Epidermis :* This is the outermost layer which is composed of tightly appressed cells. It gives out unicellular root hairs.

(ii) *Cortex :* The cortex is composed of 4-5 layers of parenchymatous cells. Some of these cells are filled with resinous substances.

(iii) *Endodermis :* The innermost layer of the cortex

represents endodermis. The cells of this layer have typical thickening bands on their radial walls.

(iv) *Pericycle :* The endodermis is followed by a multi-layered zone of parenchymatous cells which forms pericycle.

(v) *Vascular Tissue :* The roots are di-to hexarch with radial and exarch vascular bundles. The protoxylem bifurcates to form a 'Y' shaped structure and a resin duct lies in between the two arms of protoxylem. The xylem consists of only tracheids; vessels are, however, absent.

Phloem strands are present in alternate positions to the xylem strands. The phloem is composed of sieve cells and phloem parenchyma; companion cells are absent.

(vi) *Pith :* The pith is poorly developed or altogether absent.

(b) *Secondary Growth in Root :* The secondary growth is similar to that of a dicotyledonous root. A few parenchyma cells situated along the inner edges of the phloem groups become meristematic and form crescent-shaped strips of cambium. These strips of cambium produce secondary xylem centripetally and secondary phloem centrifugally (Fig.). Consequently, collateral strands of vascular tissue are formed in between the primary xylem bundles. As the activity of these cambial strips proceeds, the cells of the pericycle lying against the protoxylem also become meristematic and thus small strips of cambium are also formed outside the xylem strands. The first formed cambial strips below the phloem strands join with the newly-formed strips outside the xylem strands. As a result, a complete but wavy ring of cambium is formed which runs outside the xylem and inside the phloem strands. The first formed cambial strips, located along the inner face of phloem, start functioning earlier than the later formed cambium from the pericycle. As a result of the formation of secondary xylem opposite the phloem, the cambium is pushed outwards. Thus the wavy cambium ring becomes circular in outline. The cambium ring produces secondary phloem on the outer side and secondary xylem on the inner side. In secondary xylem the tracheids are comparatively broad and elongated. The primary xylem bundles remain in their original position inside the cylinder of secondary tissue. The primary phloem, endodermis and cortex usually become crushed due to enormous increase in the volume of the secondary tissue.

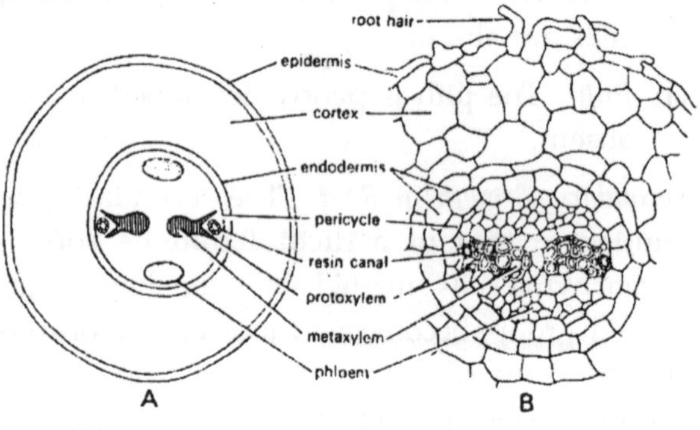

Pinus : Young root; A. Diagrammatic representation of transverse section, B. A part cellular.

The cells of the cambium, which arise in the pericycle opposite the primary xylem poles, do not give rise to secondary vascular tissues. They produce primary vascular rays radiating outwards through the secondary xylem and phloem from the tip of the protoxylem. In addition, short and narrow secondary vascular rays are also formed similarly in the secondary vascular tissue.

The formation of periderm starts soon after the

production of secondary vascular tissue has commenced. The phellogen (cork cambium) originates from the outer layers of pericycle. It produces phellem or cork towards the outer side and phelloderm or secondary cortex towards the inner side. The tissue from endodermis to epidermis is sloughed off and consequently the cork becomes outer protective layer. There are many resin canals in the secondary cortex.

2. *Stem :*

(a) *Primary Structure :* The structure of Pinus stem is similar to that of a dicotyledonous stem. In transection the stem appears wavy or irregular in outline due to the presence of close appressing scale leaves and dwarf shoots. It shows the following regions (Fig.).

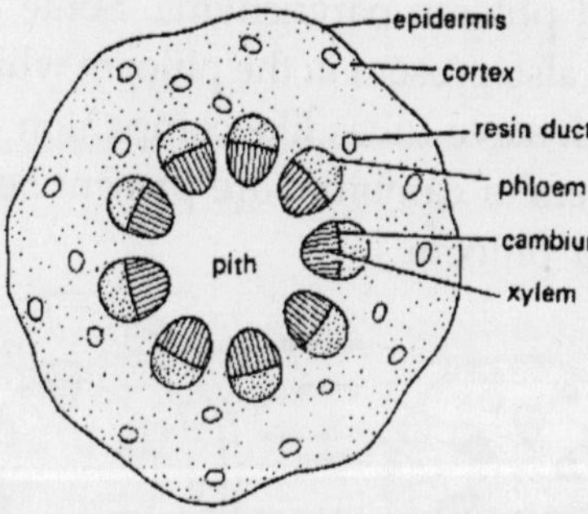

Pinus : Diagrammatic representation of transverse section of young stein.

(i) *Epidermis :* It is the outermost layer composed of compactly arranged and heavily cutinised cells.

(ii) *Cortex :* The outer sclerenchymatous zone of the cortex forms hypodermis. It is followed by several layers of parenchymatous cells. Some cells of this zone are filled with tannins. Several resin canals are also distributed irregularly in the cortex. Each resin canal is bounded by a 2-layered envelope of sclerotic cells, followed by a layer of glandular, resin- secreting epithelial cells.

(iii) *Endodermis and Pericycle* : The innermost layer of the cortex represents endodermis, however, it is indistinguishable from the cortical cells. Next to the endodermis is a 2-3- layered indistinctly pericycle.

(iv) *Vascular Tissue* : The vascular cylinder is composed of 5-9 primary vascular bundles, arranged in a ring. The vascular bundles are conjoint, collateral, open and endarch. As the vascular bundles are closely placed, the medullary rays (parenchymatous zone joining pith and cortex) are narrow. The tracheids are arranged in uniform radial rows. The protoxylem tracheids possess spiral thickenings, whereas those in metaxylem have reticulate thickenings. The phloem consists of sieve tubes and phloem parenchyma. Some albuminous cells are also present in the phloem which are associated with sieve tubes like companion cells. One or two layers of cambium are present between the xylem and phloem.

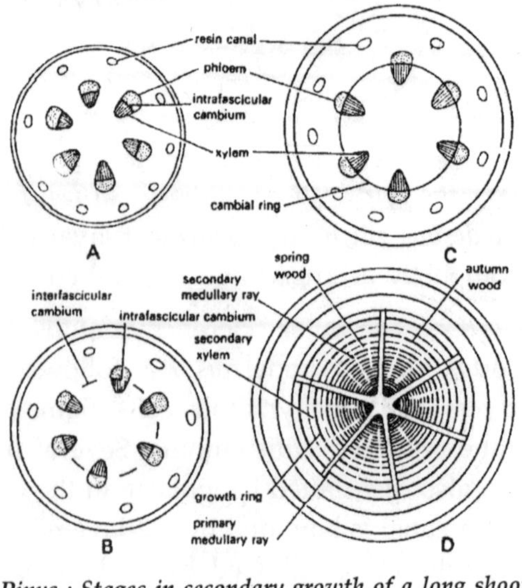

Pinus : Stages in secondary growth of a long shoot.

(v) *Pith* : A parenchymatous pith is present in the centre of the stem. Some pith cells are filled with resinous substances.

(b) *Secondary Growth in Stem* : The stem of Pinus grows in thickness by secondary growth. Interfascicular strips of cambium develop in the primary medullary rays in between the vascular bundles (Fig.). The strips of interfascicular cambium join with the intrafascicular cambium present in the primary vascular bundles and form a complete ring of cambium (Fig.). This cambial ring cuts off secondary xylem towards the inner side and secondary phloem towards the outer side (Fig.). The secondary vascular tissues are thus added continuously by the activity of vascular cambium and the stem increases in girth. The primary phloem is crushed due to the pressure of secondary vascular tissues.

The cambial ring continues to form increments of secondary xylem and secondary phloem every year. The xylem ring formed in a year is differentiated into spring wood and autumn wood. The former is composed of broad tracheids and the latter of narrow tracheids (Fig.). This differentiation is due to seasonal activity of the cambium. In autumn, when there is leaf-fall, active translocation of water and nutrients is not required and hence the tracheids formed in this season are narrow. In spring, new leaves and branches are formed and thus there is active translocation of water and nutrients. Hence the tracheids formed in this season are broad.

The increment of a year consists of a ring of spring wood and a ring of autumn wood; the two together constitute an annual ring (Fig.). The age of a plant can be calculated with the help of these annual rings.

(c) *Secondary Xylem :* The secondary xylem which constitutes the wood of Pinus is pycnoxylic. The tracheids formed in the autumn are narrow and thick-wailed with smaller-bordered pits, whereas those formed in the spring are broad and thin-walled with larger-bordered pits. The pits are present only on the radial walls. An important characteristic feature of the wood of Pinus is the presence of bars of Sanio (Fig.). They occur in the form of crescentic bars in between the pits.

These bars are formed by the deposition of cellulose and pectin on tracheid walls. As the wood matures, the bars are separated from the pits, and the bars of the adjacent pits are fused to form rims of Sanio.

The secondary vascular tissues are traversed by parenchymatous secondary medullary rays which extend from the pith to the cortex. They are formed by the cambial cells simultaneously with the secondary vascular tissues. The medullary rays are uniseriate and 2-12 cells in height. However, the rays which are associated with the resin canals are multiseriate. In radial longitudinal sections, medullary rays are cut lengthwise and thus revealing their length and height (Fig.). In tangential longitudinal sections, they are cut transversely and their height and breadth can thus be noticed (Fig.). The structure of medullary rays is different in secondary xylem and secondary phloem region. In the secondary xylem, the rays consist of thin walled and starch- filled living cells, and one or more layers of ray tracheids occur on both sides of the medullary rays. The lateral and end walls of the tracheids have bordered pits which help in radial diffusion of watery substances. The medullary rays in the secondary phloem region also consist of living starch-filled parenchymatous cells, but in this region tracheidal cells are replaced by albuminous cells on both sides of the medullary rays. The albuminous cells are small, living, thin-walled cells, which are elongated at right angle to the ray axis.

The bordered pits develop on the walls of tracheids by partial spread of the secondary walls over the pits. The pit possesses a well-developed torus (Fig.). The torus is an important structure from functional point of view as it acts as a valve membrane.

(d) *Periderm or Cork :* Concurrently with the secondary growth in the vascular region, a lateral meristem, known as cork cambium (phellogen), develops in the outer region of the cortex. The phellogen divides periclinally to cut cork cells (phellem) towards the outer side and secondary cortex (phelloderm) towards the inner side. As the stem increases in girth, the epidermis ruptures and the cork cells form a protective covering. The cork cells are impervious to water and they check transpiration from the stem surface.

3. *Dwarf Shoot :* The structure of the dwarf shoot is nearly similar to that of the long shoot (Fig.). The internal tissues are differentiated into epidermis, cortex and stele. The epidermis is a single layer of thick-walled cells. It is followed by 4-6-layered parenchymatous cortex. Tannin- filled cells and resin canals are also present in the cortex. The vascular cylinder consists of six conjoint, collateral, endarch and open vascular bundles. The medullary rays, which connect the pith and the cortex, traverse between the vascular bundles. The centre of the dwarf shoot is occupied by a parenchymatous pith.

Dwarf shoots show limited amount of secondary growth. The secondary xylem is small and separated by a thin strip of cambium from the secondary phloem. Secondary medullary rays are also formed as in long shoots.

The tissues of the dwarf shoot at the base of the spur divide into as many parts as the number of leaves in a spur. For instance, in P. longifolia, where spur is trifoliate, the tissues divide into three equal parts.

4. *Foliage Leaf* : In transection the outline of the foliage leaf varies in different species depending on the number of leaves in a spur. For example, it is circular in P. monophylla (with one needle in a spur); semicircular in P. sylvestris (with two needles in a spur) and triangular in P. longifolia (Fig.) and P. gerardiana (with three needles in a spur).

(a) *Epidermis* : The single-layered epidermis consists of heavily cutinised thick-walled cells. It has many deeply-sunken haplochelic stomata. The stomata are arranged in axial rows. Each stoma consists of two guard cells surrounded by 6-9 subsidiary cells. It opens internally into a substomatal cavity.

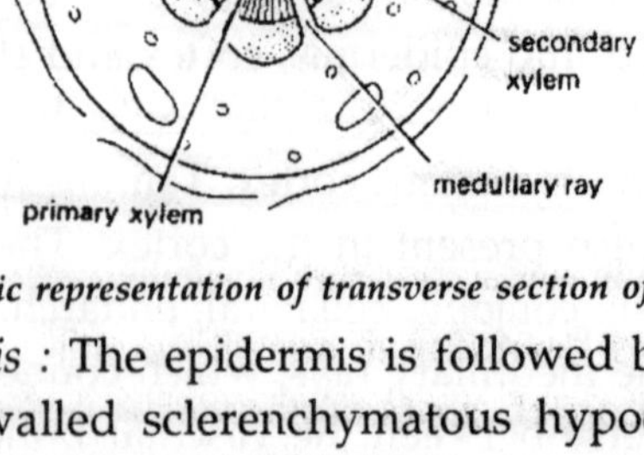

***Pinus* : Diagrammatic representation of transverse section of dwarf shoot.**

(b) *Hypodermis* : The epidermis is followed by 2-3 layers of thick-walled sclerenchymatous hypodermic. The hypodermis is interrupted by the presence of sub-stomal cavities. In P. longifolia, the hypodermic is more developed at the corners.

(c) *Mesophyll* : Next to hypodermis there are 3-5 layers of mesophyll tissue, composed of polygonal parenchymatous cells with dense chloroplasts .and starch grains. The walls of the mesophyll cells give rise to many peg-like infoldings which increase photosynthetic area of these cells. The structural peculiarity of mesophyll cells compensates the reduced area of the needle leaves.

There are usually two resin canals in the mesophyll, just below the hypodermis. Each resin canal has a layer of secretory epithelial cells which is surrounded on outside by a sclerenchymatous sheath.

(d) *Endodermis* : Inner to mesophyll, there is a layer of barrel-shaped cells with prominent casparian thickenings on their radial walls. This layer represents endodermis.

(e) *Pericycle* : A multi-layered pericycle is present next to endodermis. It consists of the following three types of cells.

(i) *Parenchymatous Cells* : Most of the pericycle is composed of parenchymatous cells which are densely filled with starch grains.

(ii) *Albuminous Cells* : These cells occur in close contact with the phloem cells. They are also parenchymatous cells which are packed with proteins and starch grains. They do not possess pits and probably help in translocation of nutrients from mesophyll to phloem cells.

(iii) *Tracheidal Cells* : These are tracheid- like cells which occur close to xylem cells. These cells possess pits and help in conduction of water and minerals to mesophyll. They form transfusion tissue.

Besides the above three types of cells, some sclerenchymatous cells also occur in the pericycle which form a T-shaped girdle above the two vascular bundles in P. roxburghii.

(f) *Vascular Bundles* : The number of vascular bundles

varies in different species of Pinus. There is only a single vascular bundle in P. monophylla and two in P. roxburghii. The vascular bundles are conjoint, collateral and open with abaxial phloem and adaxial xylem.

Narrow acicular form, thickly cuticularised epidermis, sunken stomata, sclerenchymatous hypodermis, peg-like infoldings in the mesophyll cells and the presence of sclerenchymatous tissue over the vascular bundles are some of the features of the leaves which indicate xerophytic nature of Pinus.

Shoot Apex Organization

The shoot apex of Pinus has four distinct cytohistological zones : (i) the surface meristem, (ii) the zone of central mother cells, (iii) the rib meristem, and (iv) the peripheral flank meristem. These zones differ in degree of mitotic activity, directions of growth and in cellular constitution, and show seasonal variations. The surface meristem represents the initiation zone of the entire apex and its cells divide both in anticlinal and periclinal planes. The zone of central mother cells occurs in a median position below the surface layer. Its cells are relatively large, polyhedral, irregularly arranged, and contain numerous vacuoles with large light staining nuclei. Along the sides and the base of the central mother cell zone the other apical regions develop as a result of the diagonal and horizontal divisions of the central mother cells. In this way the peripheral or flank meristem is developed laterally and the rib meristem zone from the base. The cells of the peripheral zone are densely cytoplasmic and from which arise the leaf primordia, cortex and the procambium. Later, the procambium differentiates into the vascular tissues that organise into distinct vascular bundles. The cytoplasm of the rib meristem is less dense and contains numerous small vacuoles. Basally, the rib meristem produces the pith (Fig.).

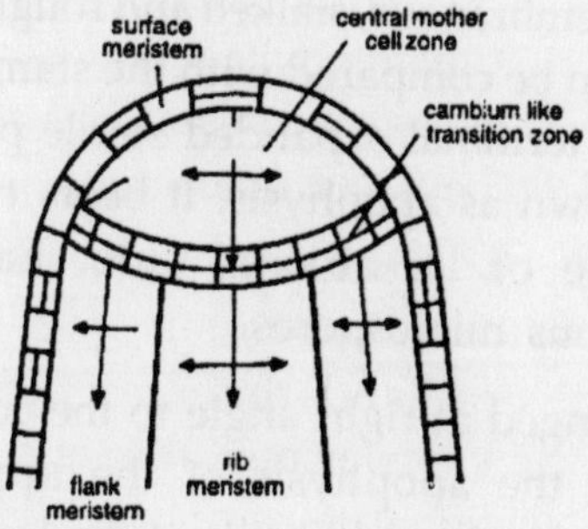

***Pinus* : Shoot apex organisation (diagrammatic representation).**

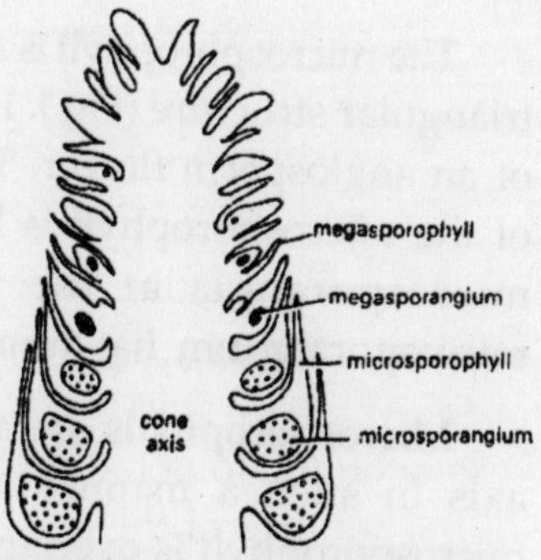

***Pinus maritima* : Longitudinal section of a bisporangiate strobilus.**

Reproduction

The sporophytic plant body forms two types of spores—microspores and megaspores, which develop into male and female gametophytes respectively. The plants are monoecious and male and female strobili are formed on different branches of the same plant. But in some species (e.g., P. montana, P. maritima, P. roxburghii), bisporangiate strobili also occur occasionally (Fig.). Both male and female strobili develop on the branches of the current year.

Male Cone : Male cones develop in the axils of scale leaves on the branches of unlimited growth. As dwarf shoots are normally formed in the axils of scale leaves, the male cones are considered as modified dwarf shoots. The male cones develop in groups just behind the apical bud on the branches of unlimited growth (Fig.). The main shoot thus continues to grow further. The number of cones in a cluster varies from 15 (P. wallichiana) to 140 (P. roxburghii). The male cone is a small oval structure, about 2- 4 cm long and 5-6 mm in diameter (Fig.). The development of male cones starts before the female cones.

The male cone consists of a centrally-located cone axis surrounded by numerous spirally arranged microsporophylls (Fig.). The number of microsporophylls in a cone varies from 60- 150.

The microsporophyll is a membranous, stalked and roughly triangular structure (Fig.). It can be compared with the stamen of an angiosperm flower. The terminal expanded sterile part of the microsporophyll is known as apophysis. It bears two microsporangia at the base of its abaxial side. Each microsporangium has numerous microspores.

Microsporophylls are arranged at right angle to the cone axis in such a manner that the apophysis of the upper microsporophyll is overlapped by the apophysis of the lower microsporophyll (Fig.). Except for a few basal sporophylls, all sporophylls in a male cone are fertile.

1. *Development of Microsporangium* : Development of microsporangium is of eusporangiate type. The microsporangium develops from 3-7 hypodermal cells of the microsporophyll. The microsporangial initial cells divide by a periclinal wall into outer primary wall cells and inner primary sporogenous cells (Fig.). The primary wall cells undergo several periclinal and anticlinal divisions and form a 2-4-layered sporogenous tissue (Fig.). The innermost layer of the sporangium wall differentiates into a nutritive layer, known as tapetum. The cells of the sporogenous tissue separate from each other and function as microspore mother cells. Each microspore mother cell divides meiotically to form four microspores (Fig.).

   The microsporangium is a sac-like sessile structure present on the ventral surface of the microsporophyll. There are a large number of microspores (pollens) in each sporangium. The microspore has an outer thick exine and an inner thin intine. The exine expands on sides in the form of two wings (Fig.).

2. *Dehiscence of Microsporangium* : A longitudinal line of dehiscence is present on the dorsal surface of the microsporangium (Fig.). At maturity, the sporangial

wall bursts along this line and pale yellow microspores are released through the slit. In India, dehiscence of microsporangia takes place in warm and dry environmental conditions. The microspores are released in such a large quantity that pine forests appear yellow at the time of 'dehiscence of microsporangium. This is called shower of sulphur.

Female Cone : Female cones are formed in groups of 1-4 in the axils of the scale leaves towards the tips of long shoots of the current year (Fig.). They replace the branches of indeterminate growth. The female cone takes about three years to mature and the cones of the successive years may be seen below each year's growth of the long shoot on the terminal branches. The first year cones are compact (1-2 cm in length) and red-green with closely arranged sporophylls (Fig.). The second-year cones are large and woody with sporophylls still compactly arranged (Fig.). The third-year cones become loose and the sporophylls separate from one another due to the elongation of the cone axis.

The female cones are much larger than the male cones. They are usually 15-20 cm long, but in P. coultre they are 25-35 cm long and in P. lambertiana they attain a length of about 60 cm. The female cone is much complicated in structure. It has a central axis around which many megasporophylls are arranged spirally (Fig.). Some basal megasporophylls are small and sterile and the rest are fertile. The megasporophyll is a compound structure, consisting of two types of scales : (i) bract scales or cone scales, and (ii) ovuliferous scales or seminiferous scales (Fig.).

Bract scale. It is a small membranous structure which is directly attached to the cone axis just below the ovuliferous scale. At the maturity of the cone, the bract scales curve inwards so that seeds are dispersed easily. Each bract scale has a single vascular bundle with its xylem pointing upwards. The bract scale is sometimes also known as carpellary scale or cover scale.

Ovuliferous Scale : It is a thick, large, woody and brownish structure, attached to the dorsal surface of the bract scale. It is roughly triangular in shape and its upper broad and thick part is known as apophysis (Fig.). In surface view, the apophysis is known as umbo. Two ovules are present at the base of the ovuliferous scale on its dorsal surface. The micropyle of each ovule is directed towards the cone axis.

Morphological Nature of Ovuliferous Scale : There is a great deal of controversy regarding the morphological nature of the ovuliferous scale as it is present in the axil of a bract scale. Several theories have been advanced to explain its morphological nature. Some important ones are briefly discussed below.

(1) According to excrescence or ligular theory of Sachs (1882) and Eichler (1889), the female cone is equivalent to a simple flower. The central cone axis is equivalent to receptacle or thalamus and the bract scales to open carpels. They considered the ovuliferous scale as an outgrowth of the bract scale, similar to the ligule of Selaginella or the placenta of angiosperms.

(2) Robert Brown (1827) considered the ovuliferous scale as an open foliar carpel which is present in the axill of bract scale and bearing naked ovules.

(3) Schleiden (1839) considered ovuliferous scale as an axillary placenta, present in the axil of an axillary leaf (bract scale).

(4) According to Alexander Brown (1842), the ovuliferous scale represents the first two leaves of an axillary shoot which had fused. Growth of the axillary shoot is, however, stopped after the formation of first two leave.

(5) Van Tieghem (1869) considered the bract as a leaf and

the ovuliferous scale as a single leaf branch in the axil of the bract.

(6) Bessey (1902) regarded the ovuliferous scale as a chalazal development of two ovules.

(7) According to the foliar theory of Delpino, (1889) the ovuliferous scale was formed by inward turning and fusion of two lateral lobes of the bract scale. He considered the bract as a tripartite structure with a median sterile and two lateral fertile lobes which were fused together.

(8) Hirmer derived the ovuliferous scale and bract scale from a single structure which forked vertically like the sporangiophore of Sphenophyllales.

(9) Brachyblast theory of Braun suggests that the female-cone is equivalent to an inflorescence rather than a flower. The ovuliferous scale was considered as a determinate axillary shoot bearing 2-3 fertile leaves. Each fertile leaf bears an ovule on the dorsal side.

(10) On the basis of his extensive studies of fossil conifers, Florin (1951) arrived to the conclusion that the female cone of Pinus is comparable to the inflorescence of angiosperms. The cone axis represents peduncle, the bract scale is a true bract and the ovuliferous scale developed in its axil represents a rudimentary female flower (modified reproductive shoot). The female cone of Pinus is thus a compound strobilus. He introduced a combined term seed-scale-complex' for the bract scale and ovuliferous scale. According to Florin, the flattened seed-scale-complex usually has a rudimentary axis, two basal megasporophylls and two or three sterile distal scales, the latter fused to form the so-called ovuliferous scale.

Describing the morphology of the seed-scale-complex in Pinus pinaster, Guedes and Dupuy (1971, 1974) established that only two prophylls make up the scale. On the basis of epidermal features and the developmental order of the various zones of the scale, Lemoine-Sehastian (1973, 1975) visualised that the complex seed scale is composed of an axial portion (which represents the body) possessing the ovules and a foliar part (which corresponds to the apophysis). From the foregoing discussion it appears that morphologically the ovuliferous scale is a reduced shoot and the female cone is an inflorescence.

Development of Megasporangium or Ovule : The development of the megasporangium (ovule) is of eusporangiate type, i.e., the megasporangium develops from more than one superficial initial cells. It appears as a small protuberance on the dorsal surface of the ovuliferous scale. The protuberance develops to form nucleus. The neighbouring cells at the base of the nucleus become meristematic, they grow upwards and form integument. The integument surrounds the nucleus, except at the top, where a small pore, known as micropyle is left. The micropyle is directed towards the base of the ovuliferous scale and opens by a narrow oblique canal at the apex of the nucleus. In early stages the integument remains free from the nucleus but later on due to increased growth of the nucleus towards the chalazal region the basal part of the integument unites with the nucleus. However, the integument remains free from the nucleus at the micropylar end.

At the apex of the nucleus (towards the micropyle), a hypodermal cell is differentiated from the other cells by its larger size. This cell is known as archesporial cell. It divides by a periclinal wall to form an upper parietal cell and a lower megaspore mother cell. The parietal cell divides by periclinal divisions and form a nutritive layer known as tapetum. The megaspore mother cell undergoes reduction division and forms four haploid megaspores, arranged in a linear tetrad. The three megaspores towards the micropylar end degenerate and the remaining one is the functional megaspore.

Structure of Ovule or Megasporangium : The ovule of Pinus is orthotropous as in other gymnosperms. The following parts can be seen in a longitudinal section of the ovule (Fig.).

(a) *Nucleus :* It is the massive parenchymatous region of the ovule.

(b) *Integument :* It is the thick protective covering around the nucleus. It has a very narrow aperture at the apex of the nucleus. The integument is differentiated into three layers; the outer and inner layers are fleshy and the middle layer is stony.

In the upper part of the ovule a space is formed in between the integument and nucleus. This space is known as pollen chamber. After pollination, the pollen grains are collected in the pollen chamber and further development of the male gametophyte takes place inside the pollen chamber.

(c) *Micropyle :* It is a narrow aperture at the apex of the ovule, formed by the integument.

(d) *Vascular System :* The vascular system of the ovule is not very well-developed. It is represented by some tracheidal elements present at the base of the ovule.

Gametophyte

Pinus is heterosporous. The microspores and megaspores are formed in the male and female cones respectively on different branches of the same plant. Both micro- and megaspores are haploid and represent the first cell of the gametophytic phase. The microspore and megaspore develop into male and female gametophytes respectively.

Development of male Gametophyte : The microspore is a uninucleate structure with an outer cuticularised exine and an inner thin intine. The exine expands on the lateral sides to form two balloon-like structures, known as wings. The outer wall of the wing has reticulate sculpturing.

The germination of microspores starts in situ, i.e., they germinate while still inside the microsporangium. The microspores are released from the microsporangium after partial development of the male gametophyte. Further development takes place in the pollen chamber of the ovule after pollination.

1. *Development of Male Gametophyte within the Microsporangium before Pollination* : The microspore divides asymmetrically into a small lenticular cell and a larger cell (Fig.). The former, known as first prothallial cell, does not divide further. The larger cell, known as apical cell, divides asymmetrically and forms a lenticular second prothallial cell and a large antheridial cell (Fig.). The second prothallial cell does not divide further and remains attached to the first prothallial cell. The antheridial cell divides to form a generative cell (adjacent to the second prothallial cell) and a large tube cell (Fig.). At this stage, the male gametophyte has four cells first and second prothallial cells, a generative cell and a tube cell. The microspores are released from the microsporangium at 4-. celled stage. They are disseminated by wind.

2. *Development of Male Gametophyte after Pollination* : After pollination the 4-celled male gametophyte reaches the pollen chamber and there it remains inactive for about 11 months. Further development of the male gametophyte starts in the next spring. The exine bursts and the intine comes out in the form of a pollen tube (Fig.). The pollen tube gradually progresses towards the archegonium, penetrating the nuclear tissue. It is rich in starch grains and it may be branched or unbranched. It acts only as a sperm carrier.

The generative cell of the 4-celled gametophyte divides to form a body cell and a stalk cell. The stalk cell is larger and the body cell remains attached at its free end (Fig.). The nucleus of the body cell divides to form two nuclei just before fertilisation. These two daughter nuclei are known as male nuclei and they swim in the cytoplasm of the body cell (Fig.). The wall of the body cell gradually disappears and both male nuclei pass into the cytoplasm of the pollen tube. At maturity, four nuclei (a tube nucleus, a stalk nucleus and two male nuclei) are present at the tip of the pollen tube (Fig.). The male nuclei function as sperms, which are microscopic, non-flagellate and ephemeral structures.

Development of Female Gametophyte : The megaspore is the mother cell of the female gametophyte. It remains embedded deep inside the parenchymatous tissue of the nucleus. During the development of female gametophyte, the functional megaspore enlarges in size (Fig.) and its haploid nucleus divides repeatedly to form 2,000-2,500 daughter nuclei (Fig.). A vacuole develops in the centre of the megaspore and the multinucleate cytoplasm forms a thin layer near the periphery (Fig.). Thereafter, wall formation starts from the periphery towards the centre. Thus, numerous radially elongated multinucleate tube-like cells are formed, which are known as alveoli (Fig.). Later, each alveolus divides to form many uninucleate cells. They represent endosperm or female prothallus (Fig.). As in other gymnosperms, the development of endosperm in Pinus occurs before fertilisation and hence it is a haploid structure (in angiosperms endosperm is formed by triple fusion and as such it is a triploid (3x) structure. The nuclear cells surrounding the multinucleate endosperm form a 2- 3-layered nutritive layer. This layer is also known as spongy layer or endosperm jacket and is equivalent to the tapetal layer.

1. *Development of Archegonium :* All superficial cells at the micropylar end of the female gametophyte have the capacity to develop into archegonia, but only a few cells differentiate into archegonial initials. The number of archegonial initials varies in different species; for instance, there are 3 archegonial initials in P. strobus,

P. rigida and P. resinosa, 5-6 in P. laricio and up to 9 in P. montana. The archegonial initials remain separated from each other due to the presence of vegetative cells in between.

The archegonial initial can be distinguished from other cells by its larger size (Fig.). The nucleus of the archegonial initial moves towards the periphery and then divides by a transverse wall into an upper smaller primary neck cell and a lower larger central cell (Fig.). The primary neck cell divides by two anticlinal walls at right angle to each other to form four cells. These cells divide transversely to form eight cells which are arranged in two tiers of four each. Thus a 8- celled neck is formed. However, in P. roxburghii and P. wallichiana the neck remains only 4-celled. The central cell increases in size considerably and the cytoplasm becomes vacuolated. It divides asymmetrically into an upper smaller venter canal cell and a larger egg (Fig.). This division takes place just before fertilisation, approximately at the same time when the body cell of the male gametophyte divides to form male nuclei. The nucleus of the venter canal cell degenerates just before fertilisation.

A nutritive layer is differentiated around the central cell and this layer is known as archegonial jacket. The jacket helps in conduction of nutrients from the endosperm to the archegonium.

Pollination : Pinus is wind-pollinated. At the time of pollination a secretion oozes out from the micropyle of the ovule which entangles pollen grains disseminated by wind. This secretion is generally given out at night and remains at the tip of the micropyle till morning. Then it is sucked inside alongwith the pollen grains. The pollen grains are thus collected in the pollen chamber at the tip of the nucleus. Thereafter, the mouth of the micropyle is closed.

At the time of pollination, the female cone assumes vertical position and its bract scales are curved inwards. In this position the pollen grains can easily reach the ovules. In Pinus the pollen grains are formed in enormous quantity but only a few are utilized for fertilisation and the rest go waste.

Fertilisation : Fertilisation takes place after about an year on pollination. The pollen tube, containing four nuclei (tube nucleus, stalk nucleus and two male nuclei), elongates and reaches near the tip of the archegonium, it penetrates neck cells, and when it comes in contact with the egg it releases all its contents in the vicinity of the egg. Out of the two male nuclei only one fuses with the egg and the other male nucleus and the tube and stalk nuclei degenerate. The diploid zygote thus formed represents the first cell of the sporophytic phase.

More than one archegonia in an ovule may be fertilised but the nutrients present in the ovule are sufficient only for the development of one zygote. The other zygotes, if formed, degenerate.

Embryo Development

Development of embryo can be studied in the following two steps.

(1) Development of proembryo from zygote.

(2) Development of embryo from proembryo.

Development of Proembryo from Zygote : Following fertilisation, the diploid nucleus of the zygote migrates towards its base and then divides by two mitotic divisions to form four nuclei (Fig.). All the four nuclei thus formed get themselves arranged in one plane at the base of the zygote. Thus, only two nuclei are visible in lateral view. These nuclei divide further and form eight nuclei which are arranged in two tiers of four each. Thereafter, wall formation takes place in such a way that four complete cells are formed in the lower tier (i.e., these cells possess a wall all around) and four incomplete cells in the upper tier (i.e., these cells have walls only on the lower side; Fig.). The incomplete cells belonging to the upper tier do not take active part in embryo formation.

The cells of the lower tier divide twice to form three tiers of four cells each. At this stage the proembryo consists of 16 cells arranged in four tiers (Fig.). Each tier has a specific function: (i) the lower-most tier is known as the embryonal tier; its cells divide to form embryo, (ii) the cells of the next tier (the suspensor tier) elongate to form suspensor, (iii) the cells of the third tier, known as the rosette tier, conduct nourishment to the embryo, and (iv) the uppermost tier consisting of incomplete cells is known as nutritive tier and it provides nutrition.

Development of Embryo from Proembryo : All the four cells of the suspensor tier elongate considerably and as a result the embryonal cells present at their base are embedded in the endosperm (Fig.). Due to the limitation of space in the endosperm, the primary suspensor cells become coiled to each other (Fig.). At this stage, the cells of the embryonal tier divide by transverse wall to form another tier of cells, known as embryonal tubes or secondary suspensor (Fig.).

The cells of the embryonal tier separate from each other and form four independent embryos. The phenomenon of the formation of more than one embryo from a zygote is known as polyembryony. In Pinus, as the polyembryony occurs by the splitting of a zygote, it is known as cleavage polyembryony. Sometimes cells of the rosette tier may develop into embryo and then it is called rosette polyembryony. But due to limited nutrients available, only a single proembryo develops into embyro and the rest degenerate.

The proembryo first divides by a transverse wall to form two cells. These cells undergo repeated divisions and form embryo. The embryo is differentiated into : (i) cotyledons, usually more than two in number, (ii) plumule or shoot apex, present between the cotyledons in the opposite direction of the micropyle, (iii) radicle, directed towards the micropyle, and (iv) hypocotyl, present in between the plumule and radicle (Fig.). Seed

Concurrently with embryo development, various parts of the ovule, outside the female gametophyte, undergo several changes by which the ovule is transformed into seed.

(1) The outer fleshy layer of the ovule wall disappears and the inner fleshy layer remains in the form of a thin membrane. The middle stony layer, however, develops into seed coat (Fig.).

(2) Due to the growth of endosperm, the nucleus persists in the form of a thin brownish layer, known as perisperm.

(3) The most part of the seed is occupied by endosperm which becomes massive by absorbing nutrients from the nucleus. The embryo remains embedded within the massive endosperm.

(4) The micropyle persists as a small aperture at the apex of the seed.

(5) A thin layer of the ventral surface of the ovuliferous scale is attached to the seed and thus the seed becomes winged (Fig.). This helps in the dispersal of the seed.

(6) At maturity of seeds, cone axis elongates and as such the space between the megasporophylls is increased. Thus seeds are easily dispersed.

Seed Germination

Under favourable conditions seeds of Pinus germinate soon after their dissemination. They do not require any resting period before germination. But they can remain dormant for several years under unfavourable conditions. The germination is epigeal, i.e., cotyledons come outside the soil surface by elongation of the hypocotyl. The seed coat bursts by absorbing water and radicle emerges out through the micropylar aperture and enters the soil. The radicle develops into primary tap root (Fig.). Thereafter plumule emerges, which forms the shoot system (Fig.). The spirally arranged leaves at the apex of the plumule are known as primary or juvenile leaves (Fig.). As the shoot grows, juvenile leaves are reduced to scaly leaves and the branches of limited growth develop in their axils.

Economic Importance of Pinus

The species of Pinus are of considerable economic important as the source of food, timber, oil, resins, etc.

(1) The wood of P. roxburghii (Chir) and P. wallichiana (Kail) is an important timber. It is suitable for making railway sleeper, packing cases, furniture, etc. Besides, it is also used for making match sticks.

(2) Pinus roxburghii, P. wallichiana, P. insularis and P. merkusii are the chief sources of turpentine in India. Rosins, obtained as a residue after the distillation of pine resin, are used in paper sizing, varnish making, enamels and in the preparation of plasters and ointments.

(3) The roasted seeds of P. gerardiana (Chilgoza pine), P. edulis and P. monophylla are edible.

(4) Several species of Pinus are grown as ornamentals in parks, gardens and in landscaping.

(5) Young plants of P. merkush are used in the manufacture of craft paper.

(6) The oleo-resin obtained from P. roxburghii is diuretic.

(7) The fossilized resin, obtained from a fossil species of Pinus (e.g., P. succinifera), is known as amber. It is widely used in ornaments and decoration work.

(8) The wood and female cones of Pinus are used as fuel.

Bibliography

Adil, A. : *Manual of Practical Algae,* Mohit Pub., New Delhi, 2003.

Aggarwal, S.K. : *Vistas in Mycology and Plant Pathology,* APH, New Delhi, 2000.

Ahuja, B. : *Plant Anatomy,* Rawat Pub., New Delhi, 2000.

Antony, T. : *Quarry : Closing in on the Missing Link,* Free Press, New York, 1993.

Arora, J.K.: *Biopesticides and Pest Management,* Ankit Pub., New Delhi, 2003.

Aslam, M.S. : *Genetics-Behavioural and Molecular,* Campus Books, New Delhi, 2003.

Bacon, Mark : *Water use Efficiency in Plant Biology,* Blackwell, London, 2004.

Bahadur, Rai K. Ranga : *A Manual of Botany for India,* Atlantic Publishers, New Delhi, 1994.

Bayleybolfour, I. :*Plant Geography Upon A Physiological Basis,* ABS, London, 2004.

Behnke, H.D. : *Sieve Elements : Comparative Structure, Introduction and Development,* Springer, London, 1990.

Bentley, Robert : *Study of Structural Morphological and Physiological Botany,* Atlantic Publishers, New Delhi, 1994.

Bilgrami, C : *Fundamentals of Botany,* Vikas Publishing House, New Delhi, 2003.

Bilgrami, K.S. : *Physiology of Fungi,* Vikas Publishing House, New Delhi, 2003.

Boodle, L.A. and Sinnott, E.W. : *Student Handbook of Botany and Plant Science,* Atlantic Publishers, New Delhi, 1995.

Bose, J.C. : *Irritabiligy of Plants,* Discovery Publishing House, New Delhi, 1999.

Brooks, F.T., and Smith, M. : *Dictionary of Fungus Diseases of Plants,* Campus Books, New Delhi, 2003.

Busch, L. : *Plants, Power and Profit : Social, Economic, Lawrence Busch, and Ethical Consequences of the New Biotechnologies,* Blackwell, London, 1992.

Chrispeels, M.J. : *Plants, Genes and Agriculture,* Jones & Bart., London, 1994.

Coulter, John Gaylord : *Fundamentals of Plant Science,* Atlantic Publishers, New Delhi, 1994.

Dayal, R. : *Predaceous Fungi,* Campus Books, New Delhi, 2000.

———— : *Predaceous Fungi,* Commonwealth Publishers, New Delhi, 2000.

Dhaliwal, G.S. and Singh, Balwinder : *Pesticides and Environment,* Campus Books, New Delhi, 2000,

Dube, H.C. : *Introduction to Fungi,* Vikas Publishing House, New Delhi, 2003.

Dyson, Freeman : *Origin of Life,* Oxford, New York, 1999.

Frank, S. Smythe : *Flowering Plants of Himalaya,* Commonwealth Publishers, New Delhi, 1994.

Fuller, Harry J. : *University Botany,* Atlantic Publishers, New Delhi, 1994.

Gaston, Kevin J. and spice, John I. : *Biodiversity : An Introduction,* Blackwell, 2004.

Grewal, R.C. : *Medicinal Plants*, Campus Books, New Delhi, 2003.

———— : *Dictionary of Botany*, Arihant Pub., New Delhi, 2003.

Hall, Marion and Halliday, Tim : *Behaviour and Evolution*, Springer, London, 1998.

Hanan, James S. : *The Algorithmic Beauty of Plants*, Springer, London, 1990.

Hardie, D.G. : *Protein Phosphorylation : A Practical Approach*, Oxford, New York, 1999.

Johri, Lata and Tyagi : *A Textbook of Bryophyta*, Dominant Publishers, New Delhi, 2004.

———— : *A Textbook of Fungi*, Dominant Publishers, New Delhi, 2004.

———— : *A Textbook of Gymnosperm*, Dominant Publishers, New Delhi, 2005.

Kains, M.G. and Mcquesten, L.M. : *Fundamentals of Plant Science*, Campus Books, New Delhi, 2002.

———— : *An Introduction to Plant Science*, Atlantic Publishers, New Delhi, 1995.

Kumar, H.D. : *General Ecology*, Vikas Publishing House, New Delhi, 2003.

Kumar, Suresh : *A Textbook of Plant Taxonomy*, Campus Books, New Delhi, 2004.

———— : *Economic Botany*, Campus Books, New Delhi, 2002.

Lang, W.H. : *A Textbook of Botany*, Atlantic Publishers, New Delhi, 1994.

Leung, D.Y.M. : *Superantigens : Molecular Biology, Immunology, and Relevance to Human Disease*, Dekker, London, 1997.

Lozzo, Renato V. : *Proteoglycans : Structure, Biology, and Molecular Interactions*, Dekker, London, 2000.

Madigan, M.T. : *Brock : Biology of Microorganisms,* Prentice Hall, New York, 2000.

Magner, Lois N. : *A History of the Life Sciences,* Dekker, London, 1994.

Magurran, Anne E. : *Measuring Biological Diversity,* Blackwell, London, 2003.

Mishra, S.R. : *Bacterial Plant Diseases,* Discovery Publishing House, New Delhi, 2003.

———— : *Morphology of Plants,* Discovery Publishing House, New Delhi, 2004.

Nairne, Alexander : *Scientific Classification of Flowering Plants,* Atlantic Publishers, New Delhi, 1994.

Narasaiah, M.L. : *Biodiversity and Environment,* Discovery Publishing House, New Delhi, 2004.

Nover, L. : *Plant Promoters and Transcription Factors Results and Problems in Cell Differentiation,* Springer, London, 1994.

Oliver, Daniel : *Study of Indian Botany,* Atlantic Publishers, New Delhi, 1994.

Panday, S.N. : *Economic Botany,* Vikas Publishing House, New Delhi, 2003.

Panday, S.N. and Sinha, B.K. : *Plant Physiology,* Vikas Publishing House, New Delhi, 2003.

Panday, S.N. and Trivedi, P.S. : *A Textbook of Botany,* Vikas Publishing House, New Delhi, 2003.

Pfleiderer, I. : *The Life of Indian Plants,* Atlantic Publishers, New Delhi, 1992.

Prasad, S.N. : *Environmental Biology,* Campus Books, New Delhi, 2003.

———— : *Cell Biology,* Campus Books, New Delhi, 2003.

Pugnaire, Ed. F.I. : *Handbook of functional Plant Ecology,* Dekker, London, 1999.

Rai, Mahendra and Mares, Donatella : *Plant-Derived Antimycotics : Current Trends and Future Prospects*, Blackwell, London, 2003.

Rashid, A. : *An Introduction to Pteridophyta,* Vikas Publishing House, New Delhi, 2003.

Rashid, A., Kumar, Santosh and Tembhre, Manju : *Anatomy and Physiology of Fishes,* Vikas Publishing House, New Delhi, 2003.

Roy, E. Biles, Seymour, E.L.D. : *Encyclopaedia of Horticulture,* Commonwealth Publishers, New Delhi, 1994.

Sandhu, G.S. : *Textbook of Cytology,* Campus Books, New Delhi, 2002.

Sandhu, G.S. and Kumar, Suresh : *Cyto-Genetics,* Campus Books, New Delhi, 2003.

Sarkar, Dilip De : *The Silkworm: Biology, Genetics and Breeding,* Vikas Publishing House, New Delhi, 2003.

Sharma Rajni : *An Introduction to Plant Morphology,* Campus Books, New Delhi, 2004.

———— : *Plant Pathology,* Campus Books, New Delhi, 2004.

———— : *Plant Tissue Culture,* Campus Books, New Delhi, 2003.

Shukla, S.K. and Srivastava, P.R. : *Encyclopaedia of Environmental Pollution,* Commonwealth Publishers, New Delhi, 1994.

Singh, Mahinder : *A Textbook of Biochemistry,* Dominant Publishers, New Delhi, 2005.

Singh, S.K. : *Plant Physiology,* Campus Books, New Delhi, 2000.

Sinha, Asha *Microbes and Plants,* Campus Books, New Delhi, 2001.

Sinha, R.K. :*Biodiversity–Global Concerns,* Commonwealth Publishers, New Delhi, 1996.

Sinnott, E.W. : *Botany and Plant Science,* Commonwealth Publishers, New Delhi, 2000.

Srivastava, M.B.: *Introduction to Forestry,* Vikas Publishing House, New Delhi, 2003.

Stiles, Walter :*Principles of Plant Physiology,* Atlantic Publishers, New Delhi, 1995.

Sundara, Rajan, S. : *Introduction to Cell Biology,* Vikas Publishing House, New Delhi, 2003.

Talbot, Nicholas J. : *Plant-Pathogen Interactions,* Blackwell, London, 2001.

Tassy, Pascal : *The Message of Fossils,* McGraw-Hill, London, 1993.

Transeau, E.N. : *General Botany,* Atlantic Publishers, New Delhi, 1995.

Tripathi, G. and Tripathi, Y.C. : *Biological and Biotechnological Resource,* Campus Books, New Delhi, 2002.

——————— : *Bioresource and Environment,* Campus Books, New Delhi, 2002.

Waisel, Yoav : *Plant Roots : The Hidden Half,* Dekker, London, 1996.

Warren, L. : *Bound Carbohydrates in Nature,* Oxford, New York, 1994.

Wrigglesworth, J. : *Energy and Life : Modules in Life Sciences,* T & F, New York, 1997.

Zaidi, N. : *An Introduction to Algae,* Ankit Pub., New Delhi, 2002.

ENCYCLOPAEDIA OF BOTANY

ENCYCLOPAEDIA OF BOTANY

(Vol. 4)

Prof. Harmeet Kour

ANMOL PUBLICATIONS PVT. LTD.

NEW DELHI - 110 002 (INDIA)

ANMOL PUBLICATIONS PVT. LTD.
H.O.: 4374/4B, Ansari Road, Darya Ganj,
New Delhi-110 002 (India)
Ph.: 23261597, 23278000
B.O.: No. 1015, Ist Main Road, BSK IIIrd Stage
IIIrd Phase, IIIrd Block,
Bangalore - 560 085 (India)
Visit us at: www.anmolpublications.com

Encyclopaedia of Botany

First Published, 2006

ISBN 81-261-2673-6 (Set)

[Responsibility for the facts stated, opinions expressed, conclusions reached and plagiarism, if any, in these volumes is entirely that of the Editor. The Publishers bear no responsibility for them, whatsoever.]

PRINTED IN INDIA

Published by J.L. Kumar for Anmol Publications Pvt. Ltd., New Delhi - 110 002 and Printed at Mehra Offset Press, Delhi.

Contents

PART-III
Microbiology

Preface

Biology comprises mainly two branches of study, namely, Zoology and Botany. Botany pertains to the study of plants, which have always been of great import since the formation of the planet 'Earth'. Plants play a crucial part in the life of living creatures as well as in balancing the environment. In other words, they are as much necessary for life as air or water is.

Apart from numerous benefits, obtained from plants, they have also been a source of medication since days immemorial. Taking all these dimensions of plants into consideration, they have been classified into different groups and every one of them has been given a unique Botanical name. A plant may have different names in different dialects of different parts of the world, but its botanical name remains one and the same the world over.

This multi-volume aims at furnishing a detailed account of plants from every possible dimension. A ready reference book for scholars and students, this encyclopaedia is an authority on the subject without an iota of doubt.

Editor

PART-I

Plant Pathology

1

Controlling Diseases

The prime object of plant disease control is to prevent qualitative and economic losses of crops due to attack by pathogens. In general, the control measures are aimed at saving the population rather than a few individual plants. The use of plant disease control measures is only desirable when the cost incurred on these measures is less than the economic value of the crop. In nature, various types of pathogens are present together and as such their complete eradication is impossible. Hence, disease control programme should be carried in such a way that more than one disease can be controlled in one operation. For the success of disease control programme it is also essential that such programmes should be operated in a large area instead of smaller areas so that the inoculum of the disease from the whole area can be eliminated.

Various methods adopted for control of plant diseases may be classified into the following three categories.

(1) Prophylatic measures

(2) Curative measures

(3) Biological measures

Prophylatic Measures

The prophylatic measures are preventive and include protection of the host from exposure to the pathogen, from infection or from environmental factors favourable to disease development. Some important prophylatic measures in operation are given below.

Quarantine : Quarantine is the legal restriction at the national or international level on the trade of propagating materials like seeds, tubers and roots. This is done to prevent spread of the disease. The first quarantine act was enforced in 1912 in USA. In India, Destructive Insects and Pests Act (DIP Act) was passed by the government in 1914 and it has been amended from time to time. DIP Act allows import of propagating materials (seed, tubers, etc.) from other countries only when it is accompanied by a certificate from the Ministry of Agriculture of the exporting country stating that the propagating material is pathogen free. Moreover, the incoming material is thoroughly checked by the trained staff at entry points. Thus, quarantine regulates entry of the plant material in a very effective way. In India, there are 16 quarantine centres at present, two on land routes, six on air routes and eight on sea routes. However, several diseases have been introduced in India even after the enforcement of DIP Act. These include leaf rust of coffee (caused by Hemileia vestatrix and introduced from Sri Lanka), late blight of potato (caused by Phytophthora infestans and introduced from Europe) and wart disease of potato (caused by Synchytrium endobioticum and introduced from The Netherlands).

Eradication of Alternate Host : The alternate host usually provides shelter to the pathogen during unfavourable periods. Thus eradication of collateral hosts is a good measure for the control of heteroecious pathogen. For example, bearberry eradication in temperate countries helps in the control of black stem rust of wheat caused by Puccinia graminis tritici. Similarly, yellow vein mosaic virus of bhindi persists on several wild hosts in nature and the systematic removal of these hosts helps in control of the disease.

Crop Rotation : Growing a particular crop or similar crops in the same field for several years continuously increase the soil borne pathogens in the field and the disease appears with increasing severity after every year. Ultimately the soil becomes so heavily infested with the pathogen that it can no longer support the growth of the crop. Crop rotation helps in the control of such soil borne diseases. Rotation of various crops reduces survival rate of the pathogen in the absence of the suitable host. Some important diseases that can be controlled by crop rotation are wilt of arhar, wilt of linseed, red rot, wheat bunt, smut of bajra, rice bunt and root rot of vegetable crops.

Sanitation : Sanitation is the simplest method of disease control. It reduces the amount of inoculum to which the host population is exposed. Several types of pathogens are present in underground and decaying parts of plants and in the soil, which act as inoculum for diseases caused by them. Hence, elimination of diseased plants from the fields is necessary. Several infectious diseases like powdery mildew of wheat and peas, downy mildew of maize and peas and wilt of cotton can be controlled successfully by such measure. In green houses, soil borne diseases can be controlled by the use of sterilized soil.

Environmental Manipulation : Environmental manipulation includes such cultural practices which discourage pathogen development. Good soil drainage, proper aeration and irrigation, proper fertilisation or soil amendments, proper spacing of plants, proper method and depth of sowing and weed control are some important factors that discourage growth of the pathogen. Only disease free certified seeds should be sown to achieve the purpose.

Chemical Prophylaxis : Chemical prophylaxis is the

application of chemicals to plants for control of pathogens. Chemicals used in plant protection are known as pesticides and they include fungicides, bactericides, insecticides, nematicides and herbicides. Most of the pesticides are used to control diseases of the foliage and other above ground parts of the plant, but some are used to disinfest and protect seeds, tubers and bulbs from infection.

There are different groups of chemicals which are used as fungicides. A fungicide may be defined as a substance which kills fungi or prevents the development of fungal diseases without causing any damage to the host plant. There are two categories of fungicides.

Contact Fungicides : These fungicides act by direct contact with the pathogenic fungus before or after the infection.

Residual Fungicides : These fungicides are used before the infection of the pathogenic fungus. Hence they form a protective covering over the surface of the plant even before infection.

Fungicides are used in various ways and are mostly applied as sprays or dusts on standing crops. For seed-borne pathogens, fungicides are applied to seeds before they are sown. The powdery fungicide is either thoroughly mixed with seeds before they are bagged and stored or seeds are soaked for a definite time in the fungicide solution. Root pathogens are controlled by soil application of fungicides.

On the basis of their chemical nature, fungicides can be divided into the following two major groups.

Inorganic Compounds

(a) *Bordeaux Mixture :* It is a mixture of copper sulphate, lime and water in the ratio of 10 : 10 : 100 or 10 : 6 : 100. It is prepared by mixing $CuSO_4$ solution with lime stone solution. The mixture should be used immediately after preparation as it deteriorates on storage. Bordeaux mixture is specific against downy mildews, blights, anthracnoses, white rust diseases, etc.

(b) *Burgundy Mixture* : It is a modified form of Bordeaux mixture which consists washing soda (Na_2CO_3) in place of lime. It is comparatively less phytotoxic, hence it can be used safely on tender leaved plants. Burgundy mixture is specifically used for the control of leaf blights and related diseases.

(c) *Chestunt Compound* : It is a mixture of copper sulphate and ammonium carbonate in the ratio of 2 : 1. It is specifically used in the control of damping off disease.

(d) *Mercuric and Mercurous Chloride* : Dilute solutions (1 : 1000) of mercuric chloride ($HgCl_2$) and mercurous chloride (Hg_2Cl_2) are used to control several surface fungal diseases.

(e) *Sulphur Sprays* : Various formulations of elemental sulphur or crude lime sulphur preparations have been developed to control powdery mildews of various crops and leaf curl of peach. Elemental sulphur is used both in dry and wettable form. Sulphur dust is especially used against powdery mildews, Lime sulphur is a mixture of calcium thiosulphate and calcium polysulphide. It is mostly used as spray for powdery mildews. Since lime sulphur is phytotoxic, self-boiled lime sulphur is more commonly used as it is less toxic. In recent times, self-boiled lime sulphur has been replaced by wettable sulphur. The latter is widely used against apple scabs, powdery mildews of grapes and other crops, ornamentals and vegetables.

Organic Compounds : Organic compounds are the most versatile and widely used group of modern fungicides. They are mostly derivatives of dithiocarbamic acid.

(a) *Thiram (Tetramethyl Thiuram Disulphide, TMTD)* : It is

a condensation product of dithiocarbamic acid and is sold in the market under various trade names like Arasan, Tundas, Tulisan, etc. It is commonly used in the control of diseases caused by Pythium, Fusarium and Protomyces.

(b) *Ferbam (Ferric Dimethyl Dithiocarbamate)* : It is widely used as a protectant fungicide against a variety of fungal diseases of fruits and vegetables. It is especially effective in controlling early and late blight of potato. It is sold in the market under the trade name of Fermate or Coromet.

(c) *Maneb (Manganese Ethylene Bisodiumdithiocarbonate)* : It is used as foliar spray against foliage and fruit diseases of several vegetables. It is sold under such trade names as Manzate and Dithane M-22.

(d) *Ziram (Zinc Dimethyl Dithiocarbamate, Cuman)* : It is effective against several foliar diseases of field and orchard crops.

(e) *Nabam (Disodium Ethylene Bisdithiocarbonate, Dithane A-40)* : It is effective against leaf blights and seedling blights.

(f) *Zineb (Zinc Ethylene Bisdithiocarbonate, Dithane Z-78)* : It is a safe and multipurpose foliar and soil fungicide for early and late blight of potato and blast and leaf blight of rice.

Quinones-Chloranil (Spergon, Tetrachloro-p-benzoquinone) : It is an important quinone mainly used for the seed, bulb and flower treatment. Dichlone (Phygon, 2-3 dichloro 0-1, 4 naphthaquinone) is used for the treatment of vegetable seeds and as a protectant spray against some blights, fruit rots and cankers of fruits and vegetables.

Heterocyclic Nitrogen Compounds : They are used as foliage protectants on the fruit and vegetable crops. Captan (n- trichloromethyl thio-4-cyclohexene 1, 2 dicarboximide), a condensation product of perchloromethyl mercaptan with an imido group, is a broad spectrum fungicide which is effective against leaf spots, blights, fruit rots, etc. Folpet is similar to Captan and is also effective against, powdery mildews. Difolatan (n-1, 1, 2, 2-tetrachloro ethyl sulfenyl)-cis-4-cyclohexene 1, 2- dicarboximide) is effective against early and late blights of potato.

Benzene Compounds : Dinitro-o-cresol is used as a spray for some diseases of fruit and ornamental trees. Chloroneb (1,4- dichloro-2, 5 dimethoxy benzene) is effectively used to combat seedling diseases of cotton. Terrachlor (penta-chloro nitrobenzene, PCNB) is effective against various soil-borne diseases of vegetables and is primarily used against Rhizoctonia, Sclerotinia and Plasmodiophora. Dinocap (2-(1-methyl-heptyl), 4, 6 dinitrophenyl carbonate) is specific against powdery mildews. Diazoben (p-dimethyl amino-benzene diazosodium sulfonate) is very effective against several soil-borne phycomycetous fungi causing root rot and damping off disease.

Systemic Fungicides : Systemic fungicides were used for the control of fungal pathogens for the first time in 1966, and since then many major advances have taken place. Systemic fungicides are usually absorbed by roots and are translocated within the plant in the transpiration stream and the latter thus becomes fungitoxic. They can also be injected into the tree trunk and reaches to different plant parts through the xylem stream. They inactivate enzymes and toxins of the pathogen and inhibit growth of the fungus. There are three categories of systemic fungicides-oxanthins, benzimidazoles and pyrimidines. Oxanthiins include primarily carboxin and oxycarboxin. They are effective against some smut and rust fungi, and Rhizoctonia. They adversely affect nitrogen metabolism of the pathogen. Benzimidazoles include several important systemic fungicides, such as benomyl thiabendazole and thiophanate. They are effective against a wide variety of fungal pathogens. Benomyl (methyl-n (butyl carbamyol)-2-benzimidazole carbamate) is a broad spectrum fungicide, highly effective against the infections of Rhizoctonia, Fusarium and Verticillium. Bavistin (2- (methoxy-carbamoyl)benzimidazole) is effective against several ascomycetous and deuteromycetous fungi. Thiabendazole is mainly effective against deuteromycetous fungi causing leaf spots and diseases of bulbs, corms, etc. Pyrimidines include dimethirimol, ethirimol and triarimol, which are mainly effective against powdery mildews.

Antibiotics : Several antibiotics are also useful in the control of plant diseases. They are generally absorbed and translocated systemically, and act directly on the pathogen. They can be applied in the form of spray or dust. Dip application is effective in the treatment of seeds or vegetative propagating parts of fruit trees. Some important antibiotics are as follows.

(a) *Streptomycin :* It is commonly used against several pathogenic bacteria (e.g., Xanthomonas citri, X. juglandis, X. vasicatoria, Pseudomonas phaseoliculosis) and fungi (e.g., Phytophthora parasitica, P. infestans).

(b) *Cycloheximide :* It is effective against the diseases caused by Coccomyces heimalis, Erysiphe polygoni, Ustilago hordei, Sclerotinia fruticola and species of Rhizopus and Botrytis.

(c) *Griseofulvin :* it is effective against Botrytis fabae, Alternaria solani, Uromyces appendiculatus, Sclerotinia fructigena, etc.

(d) *Tetracycline :* It is used in controlling the diseases caused by mycoplasmas.

Curative Measures

Besides prophylatic measures, sometimes infected plants are individually cared by treatment. Such methods are known as curative or therapeutic measures. These measures are used after the appearance of disease symptoms. For the protection of healthy plants, eradication of diseased plants is necessary. The curative measures are, therefore, effective in controlling spread of the disease caused by the inoculum present in seeds or tubers. Besides, these measures are also effective against smut and blight diseases.

The disease can be cured by physical or chemical therapy. The physical therapy includes surgery, temperature modification, radiation and moisture modification. In chemotherapy, plant diseases are controlled by certain chemical compounds which reduce or nullify the effect of pathogen after it has entered the host plant. The chemotherapeutants are used directly on the surface lesions or as solutions to be absorbed through roots. Benomyl, Vitavax, Plantvax and Dithane R-24 are some good chemotherapeutants.

Biological Measures

The biological measures for disease control can be divided into the following three categories.

(1) Breeding resistant varieties

(2) Hyperparasitism

(3) Control through trap crops and antagonistic plants

Breeding Resistant Varieties : Use of varieties which show resistance to infection of pathogenic organisms is an ideal measure for the control of plant diseases. This is also known as painless method of disease control. High yielding resistant varieties have been developed by breeding techniques. Some economically important varieties of wheat which are resistant to diseases are as follows.

1. *Rust Resistant Varieties of Wheat :* Lerma Rojo 64A, Kalyan Sona, HD 1918 (Pratap), HD 2009 (Arjun), UP 262, JNK (4-W-184).
2. *Smut Resistant Varieties of Wheat :* NP 4, NP 710, NP 770.

Hyperparasitism : Hyperparasitism is the control of pathogenic microbes with the help of other microbes which parasitise or antagonise the pathogen. Bacteriophages, mycophages and nematophagous fungi are some common examples of hyperparasitism. Several pathogenic fungi, such as Pythium, Fusarium and Helminthosporium, are infested by other fungi present in the soil. Similarly, Trichothecium roseum attacks uredia of Puccinia graminis tritici. Carnation stem rot, caused by Fusarium roseum, can be controlled by the application of Bacillus subtilis to the stem cuttings or to the soil. Addition of soil amendments favours growth of hyperparasites, mycoparasites or antagonists.

Control Through Trap Crops and Antagonistic Plants : There are several plants which are infested by fungi and nematodes, but after infestation the host secretes certain substances which are inhibitory to the growth of these pathogens. Such hosts are known as trap crops. For example, population of Heterodora nematodes can be reduced in the soil by Solanum nigrum, and Crotalaria spp. trap the larvae of Meloidogyne. The population of several pathogenic microorganisms can be reduced in this way by rotation of trap crops.

2

Plant Diseases

Bacterial Diseases of Plants

Bacteria cause many diseases in plants. Some important plant diseases caused by phytopathogenic bacteria are described in the following pages.

Important Characters of Phytopathogenic Bacteria

(1) They are saprophytic and hence can be cultured in artificial medium.

(2) They do not infect animals.

(3) They are invariably rod-shaped and flagellate.

(4) They are Gram-'ve', except Corynebacterium and Streptomyces which are Gram+'ve'.

Identification of Phytopathogenic Bacteria : The phytopathogenic bacteria can be identified with the help of the following key proposed by Rangaswamy (1962).

Infection of Plants by Pathogenic Bacteria : Pathogenic bacteria enter the host through cuticle free tender organs like root hairs and stigma or through natural pores, such as stomata, lenticels and hydathodes. They may also enter the host through open wounds. They donor have the capability of penetrating the cuticular surface unlike pathogenic fungi.

Mode of Action of Pathogenic Bacteria : Phytopathogenic bacteria affect the host in the following ways.

1. *By Secreting Rot Causing Enzyme :* Several pathogenic bacteria secrete rot causing enzymes which have the capacity to disintegrate pectic compounds of host cell walls. These enzymes are secreted before or after the bacterium penetrates the host. They cause disintegration of host tissues.
2. *By Inducing Hypertrophy :* It has been shown that by stimulation of bacteria, host cells are converted into tumour cells. Both, bacterial pathogen and tumour cells secrete auxins and due to the influence of these auxins tumour cells form galls.
3. *By Secreting Toxic Substances :* Polysaccharides, and cellulolytic and pectinolytic compounds secreted by bacteria stimulate willing symptoms in host plants. These substances are usually not host specific. Bacterial wilt caused by Pseudomonas solanacearum and maize wilt caused by Xanthomonas stewarth are due to secretions of such substances.

Spread of Bacterial Diseases : Bacterial diseases are usually disseminated by seeds, soil, air and water. Some species of Xanthomonas, Corynebacterium and Pseudomonas are seed borne. Soil borne bacteria survive in the soil for a long time and decay plant parts. Pathogenic bacteria are also disseminated by transference of infested soil by air, water, animals or human beings. Besides, insects are also important carriers of bacterial pathogens. Agricultural implements too are responsible for the spread of bacterial diseases.

Important Symptoms of Bacterial Diseases : Bacteria cause various types of disease symptoms. These symptoms depend upon the environmental conditions and host-parasite relationships. Some important symptoms of bacterial diseases are as follows.

1. *Local Lesions :* Some bacterial diseases cause local lesions in the form of spherical to irregular spots on the leaves, stems, fruits, etc. Bacteria enter the host through stomata, establish themselves in the substomatal cavity and disintegrate parenchymatous cells. This causes lesions.
2. *Soft-rot :* Several bacteria secrete enzymes which affect the middle lamella of the cells and as such the cells get separated. A liquid substance also oozes out from the affected parts. These symptoms commonly occur in fleshy fruits.
3. *Canker and Scab :* These are corky or warty outgrowths developed on leaves, branches and other aerial parts of the plants. These symptoms usually appear about 10 days after the primary infection. When infection remains confined to epidermal cells of the host, the outgrowths developed are known as scab, whereas in canker, infection also occurs in sub-epidermal layers, such as cortex and cambium.
4. *Vascular System Diseases :* In some bacterial leaf spot diseases, the pathogen enters the vascular system of the host plant and gradually becomes systemic. Sometimes the bacterial invasion is concentrated in the vascular system and a dense slimy material is secreted by the bacteria. This results in plugging of water conducting vessels and as such typical wilting symptoms appear.
5. *Blights :* In blights, invasion by the bacteria leads to very rapid and extensive necrosis of the affected parts resulting in scorched appearance of the host plant.

Some Important Bacterial Diseases : Some important bacterial diseases of plants are described below.

Citrus Canker : Citrus canker is a widespread disease in all citrus growing areas of the world. The disease perhaps originated in India or Java and then spread to other countries. It is particularly serious in India, China, Japan and Java.

Symptoms : The disease affects all above ground parts of the plant, but leaves, twigs, thorns and fruits are most susceptible. The lesions on the leaf appear as small, round, raised, water soaked brown translucent spots. They are surrounded by a watery yellow halo. In the advanced stages of infection, the lesions rupture in the centre, giving a rough corky and crater-like appearance. The lesions first appear on the lower surface of the leaf, but as the disease progresses the upper surface is also involved. On twigs and fruits, the spots increase in size and coalesce to form elongated lesions. Spots occurring on the petiole and mid-rib cause premature defoliation. On the fruits, lesions have a prominent crater-like appearance and yellow halo is absent.

Causal Organism : The pathogen responsible for citrus canker is a bacterium, Xanthomonas citri. It is a Gram-negative, monotrichous, capsulated bacillus which is strictly aerobic in nature. It is 1.5-3 μm in length and 0.5-1.5 μ m in breadth. The bacterium does not produce endospores and can not reduce nitrates. It shows a yellow, copiously mucoid growth on 2% sucrose peptone agar but on beef agar the colonies are circular and straw yellow to amber yellow.

Disease Cycle : The diseased twigs and leaves are the main source of the inoculum. The pathogen penetrates the host through wounds or natural openings. It multiplies in the intercellular spaces, dissolve the middle lamella and establishes in the cortex, and forms cankerous outgrowths. The exudate released from these outgrowths contain numerous bacteria. The pathogen is chiefly disseminated by wind, rain and insects, but man also plays a major role in its dissemination through infected nursery stock.

Mild temperature and wet weather favour the disease spread. The infection is most severe when the atmospheric temperature is 20-35°C (optimum 30°C).

There are certain disturbances in the physiology of the host due to infection. The respiratory rate increases in the halo region in comparison to the healthy tissues. Total nitrogen, sugar and chlorophyll contents are reduced in the infected regions but total phenolic contents in the halo region increase considerably.

The pathogen can survive up to six months in the infected leaves. It remains viable up to 52 days in the sterilized soil but in the unsterilized soil viability is of only for 9-17 days.

Disease Control : Various methods are adopted for the control of citrus canker. Some important measures are as follows.

(1) Prompt and complete destruction of diseased plants.

(2) Use of disease free nursery stock or disease resistant varieties.

(3) Spraying the host with 1% Bordeaux mixture or other copper fungicides before planting.

(4) Spraying of neem cake at the rate of about 7 kg per acre.

(5) Antibiotic spray with streptomycin (500-1000 ppm) or phytomycin (2500 ppm).

(6) Exclusion by quarantine.

Bacterial Blight of Cotton : This disease, also known as angular leaf spot, vein blight or black arm, is common in all cotton growing regions of the world, especially Egypt, Sudan, U.S.A., Australia, U.S.S.R. and India. In India, the disease was first observed in Tamil Nadu in 1918 and since then it has spread to Maharashtra, M.P., A.P. and U.P.

Symptoms : Pathogen infects the aerial parts of the host. The disease symptoms vary at different stages of plant growth. The earliest symptom of the disease appears on the lower surface of the cotyledons as water-soaked circular lesions. The infection soon spreads and forms brown angular patches. The shape of the cotyledons is distorted and the seedling dries and withers. Finally, the whole seedling may collapse and die.

The symptoms on the leaves appear in the form of angular spots (as such the disease is also known as angular leaf spot of cotton), These spots first appear on the lower surface and then extend to the upper surface of the leaf. The disease advances along the veins, forming finger-like lesions (hence the disease is also called as vein blight). In severe cases, the infection spreads up to the petiole, causing defoliation. From the surface of the infected tissue there is a profuse exudation of slimy material which forms a thin film on drying.

As the host matures, dark brown to black linear lesions appear on the stem and the fruiting branches. If conditions are favourable for the growth of the pathogen, most of the branches of the host are affected. They turn black and finally the plant withers. Due to the presence of these symptoms, the disease is also known as black arm disease.

On fruits (bolls), the symptoms appear in the form of black water-soaked lesions which are invariably sunken on the surface. Infected young bolls may fall down pre-maturely. In severe cases, the infection spreads inside the boll causing considerable reduction in the size of the boll and loss of viability of the seeds.

Causal Organism : The pathogen responsible for bacterial blight of cotton is Xanthomonas malvacearum. It is an aerobic,

Gram '-ve', rod-shaped bacterium measuring 0.3-0.6 μm in breadth and 1.3-2.7, μ m in length. The bacterium is motile with the help of single polar flagellum. It does not form spores or capsules. The colonies on beef extract agar are pale yellow, smooth, round and shining. The bacterium hydrolyses gelatin, casein and soluble starch but does not reduce nitrates.

Disease Cycle : The pathogen is a seed-borne bacterium and hence the primary, infection occurs through seeds. Infected cotton bolls, leaves and twigs which fall on the soil surface are the chief source of infection. The secondary infection occurs through wind, dew or rain.

The bacterium remains viable up to six months on the soil surface but viability decreases when the pathogen is buried in soil. Leaves are infected mainly through stomata. Optimum temperature for the growth of the bacterium is between 25° and 30°C.

Disease Control : The following methods are adopted to control the disease.

(1) Since it is a seed-borne disease, it can be effectively controlled by seed treatment. Delinting seeds before sowing destroys pathogen. Seeds are immersed in sulphuric acid for 10-15 minutes, then rinsed thoroughly with water and finally dried and treated with organomercurial compounds like agrosan GN, ceresan or ethyl mercury phosphate. This treatment, however, does not destroy internal infection.

(2) Treating seeds with antibiotics like streptomycin sulphate eradicates both external and internal infection.

(3) Treating seeds with hot water (56°C) for 10 minutes also eliminates both external and internal infection.

(4) Since pathogen remains viable up to one year in the seed, storing seeds for two years before sowing reduces the incidence of infection.

(5) Growing disease resistant varieties, is the most effective measure. Two species of cotton, Gossypium herbaceum and G. arboreum, are fully immune to the disease.

(6) By removal and destruction of alternate weed hosts and diseased plant debris.

Bacterial Blight of Paddy : Bacterial blight is a very common and widespread disease of paddy in Asia, especially in Sri Lanka, China, India, Japan, Indonesia, Taiwan, Philippines and Thailand. The disease was first reported from Japan in 1881 and it was then presumed that it is confined to that country only. In India, the disease was first reported in 1951 from Bombay. In recent years, it is found to occur in severe condition in all rice-growing areas of the country. The disease causes heavy loss to rice crop.

Symptoms : It is a foliar disease which appears in early or late August and causes maximum damage during the heading stage. The disease occurs in all stages of plant growth, from seedling to grain formation. It first appears as water-soaked, translucent spots on the leaf blade and leaf-sheath. The spots gradually increase to form elongated yellow to straw coloured streaks which may coalesce to form white blotches. The streaks usually develop from the tip towards the base. Sooner or later the tips dry and become twisted and at times the leaf tissues may die. In dry weather, turbid drops of bacterial mass ooze out, which on drying appears as yellowish beads on leaf surface.

In severely infected fields, infection may reach the grains. The glumes become discoloured and water soaked spots develop on them. Finally, the whole plant may wilt and appear completely desiccated (kresek symptoms). The leaves of infected tillers suddenly become dry and roll along the midrib.

Causal Organism : The pathogen responsible for bacterial blight of paddy is Xanthomonas oryzae. It is an aerobic, rod-shaped, noncapsulated, non-spore forming, Gram-negative bacterium, 0.5-0.8 × 1-2 μm in size. It is motile with a single polar flagellum. The bacterium produces waxy yellow coloured colonies on agar medium and is incapable to reduce nitrates. The optimum temperature for its growth is 28-30°C and its thermal death point is 53°C.

Different strains of the pathogen vary in their virulence. Most of the tropical strains are virulent forms, capable of causing considerable damage and 'kresek' symptoms. The strains also differ in their susceptibility to bacteriophages.

Disease Cycle : Seeds and disease stubbles are mainly the primary sources of infection. It is principally a vascular disease. In India, the pathogen also occurs on some grasses which are its collateral hosts. The secondary infection spreads due to wind-borne rain splashes and gains entry into the host through wounds and stomata.

Cloudy weather, high humidity, double cropping and poorly drained conditions are most conducive for disease development. High doses of nitrate, silicate and magnesium in the soil increase disease severity. Relatively high temperatures (above 25°C) during the growth of the crop increase the incidence of disease but severe heat and drought suppress the disease. The rice plant is susceptible to infection at all stages of its growth, but susceptibility to vascular infection decreases with the age.

Disease Control : The following measures can be adopted to control the disease incidence.

(1) One of the good measures to control the disease is to treat seeds with 0.025% water solution of agrimycin plus 0.05% wettable ceresan for 12 hours and then transfer them to hot water at 52-54°C for 3 minutes.

(2) Initial infection can be checked by dipping seeds for 8 hours in 0.1% ceresan wet plus streptomycin (0.3 gm in 2.5 gallons of water).

(3) Spraying agrimycin plus copper oxychloride for five times at 12 day intervals prevent secondary spread.

(4) Chlorination of irrigation water also helps in reducing infection.

(5) Burning of infected straw and stubble to destroy the bacteria present is also a good measure.

(6) By sowing of nursery beds in disease-free, isolated areas to prevent the inflow of primary inoculum into fields. Pruning of top portions of the host should be avoided.

(7) By cultivating resistant varieties such as Sathi, N-22, IR-20, IR-22, W 529, W 348 and Ratna.

Viral Diseases of Plants

Some important viral diseases of plants are described below.

Leaf Roll Disease of Potato : This disease is one of the main causes of degeneration of potato and it occurs commonly in the plains of Northern India.

Symptoms : As the name suggests, one of the important symptoms of the disease is rolling of leaf margins. Due to rolling of margins a trough like structure is formed and the mid-rib lies at the bottom of this trough. In plants with diseased seed tubers, rolling first appears in the basal or lower leaves and then progresses upward. Ultimately the entire plant is affected. In some varieties of potato margins or tips of the leaves become yellow.

The infected leaves become rigid, thick, brittle and leathery, and produce a characteristic rattling sound when brushed with hand. The brittleness and rolling, of leaves is due to the accumulation of large quantities of carbohydrates in the palisade tissue. The plants, which develop from diseased seed tubers, have short internodes and this results in dwarfing of the entire plant. In severe cases the number of tubers and their size is greatly reduced. Sometimes, there is necrosis of primary phloem in the infected plants.

Causal Organism : The pathogen responsible for the disease is leaf roll virus, also known as Potato virus I or Solanum virus 14.

Transmission : In nature, transmission of the virus occurs through infected seed tubers and through the agency of insects. The main insect vector is the aphid Myzus persicae, but other aphids may also carry the disease. The virus persists in the body of the aphid for many days.

There are at least four distinct strains of this virus. Certain other solanaceous plants, such as Datura stramonium, Lycopersicum esculentum, and Solarium dulcamara are also susceptible to this virus.

Damage : In diseased crop the number of tubers per plant and their size is greatly reduced. This considerably affects the commercial value of potato. In diseased tubers germination rate decreases by 6-8%.

Disease Control : Potato seed certification and roguing are recommended for control of the disease. Tubers can be made free from leaf roll viruses by storing them in humid environment at 37.5°C for 20-25 days.

On the basis of the studies carried at Central Potato Research Institute, Shimla, it has been suggested that if the aerial parts of the host are cut in the third or fourth week of December, the crop can be saved from aphid infestation. Though this would reduce the yield, the crop can be saved from virus infection.

Leaf Curl of Papaya : Leaf curl of papaya is widely spread in many states of India, especially Bihar, Delhi, M.P. and U.P. It was first recorded from Madras in 1939.

Symptoms : Crinkling and curling of the leaves accompanied by vein clearing and reduction in the size are the main symptoms of the disease. The infected leaves also become leathery and brittle and on the upper surface the intervenial areas are raised due to hypertrophy. In case of severe infection, the leaf margins turn downwards and form inverted cup like structures. Sometimes all the top leaves of the plant are affected and the petioles are twisted in a zig-zag manner. The infected plants usually have stunted growth and they do not produce flowers and fruits.

Causal Organism : The pathogen responsible for leaf curl is Tobacco virus - 16 or Nicotiana virus - 10.

Transmission : The virus is not transmitted mechanically but its transmission occurs readily by grafting or by sap inoculation. The main insect vector is white fly (Bemisia tabaci). Tomato, tobacco, hollyhock, Zinnia are also susceptible to this virus.

Disease Control : No suitable method is known to control leaf curl of papaya. Once the plant becomes affected it never recovers. The infection spreads even to other healthy plants. Hence, as soon as the disease is noticed the plant must be uprooted and destroyed.

Yellow Vein Mosaic of Bhindi : Yellow vein mosaic is an important disease of bhindi or okra (Abelmoschus esculentus). It is widely spread wherever this crop is grown in India. The disease was first noticed in Bombay in 1940.

Symptoms : Vein clearing, followed by veinal chlorosis of leaves are the main symptoms of the disease. The veins become thick and prominent (Fig.). In case of severe infection, chlorosis may extend to the intervenial areas and this results in complete yellowing of the leaves. The fruits become dwarfed, malformed and yellowish-green in colour and the loss in yield may reach up to 94%. An early infection results into total destruction of the entire crop.

Causal Organism : The pathogen responsible for the disease is Yellow Vein Mosaic Virus (YVMV) or Hibiscus virus I.

Transmission : The virus is transmitted in nature by white flies (Bemisia tabaci) and perhaps also by bhindi leaf hopper (Empoasca devastans). It may be transmitted artificially through grafting.

There are also several weeds, such as Ageratum conyzoides, Croton bonplandianum and Malvastrum tricuspidatum, which are susceptible to the virus.

Disease Control : No suitable method is known to control the disease. Protection of the crop from white flies and other vectors by spraying with 0.3% follidol may reduce the disease incidence. Besides, ekatox (0.02%) or rogor (0.05%) sprays are equally effective to combat the disease. Eradication of weed hosts and growing disease resistant cultivars (e.g., variety Pusa Sawani) are also good measures.

Potato Mosaic : Mosaic diseases cause severe damage to individual plants. Several types of mosaic occur on potato plants. Mild mosaic of potato is caused by Potato virus X (also known as potato mild mosaic virus, potato latent virus, Solanum virus 1, potato mottle virus). The disease, although very widely occurring in most of the potato seed stock, causes slight intervenial mottling or mosaic. The vein-banding severe mosaic of potato is caused by Potato virus Y (also known as potato severe mosaic virus, Solanum virus 2).

The symptoms vary from mild to severe mottle and streak or 'leaf drop streak' with necrosis along the veins of underside of the leaflets. Plants from infected tubers are stunted with brittle and crinkled leaves. The rugose mosaic of potato is caused by a combination of two viruses - Potato virus X (PVX) and Potato virus Y (PVY). The disease causes very severe damage to individual plants. The foliage is not only mottled but is also severely wrinkled, puckered and markedly reduced in size. The lower leaves show black necrotic veins. In severe cases, plants may die before producing any tuber.

Transmission : PVX is transmitted by contact. No vectors are known but grasshoppers can mechanically transmit the virus. On the other hand, PVY is mechanically transmitted by sap inoculation and by grafting. In nature, it is transmitted by infected tubers and by at least 25 species of aphids in the non-persistent manner. Myzus persicae is the most efficient vector.

Disease Control

(1) By systematic roguing of diseased plants and their destruction.

(2) By obtaining certified seed tubers from a reliable source and planting them early in the season.

(3) By raising a mosaic free seed crop by planting large seed tubers during the coldest period of the season.

(4) By detopping (removing foliage) the plants in the 3rd or 4th week of December when aphid infestation starts, and then leave the tubers in soil to mature.

(5) By controlling vectors (particularly Myzus persicae) with systemic insecticides such as aldicarb, phorate, carbofuran, etc.

Mycoplasmal Diseases of Plants

Little Leaf of Brinjal : It is a serious disease of brinjal or egg plant (Solanum melongena) and is found in all brinjal growing areas of India. The disease was first noticed in Coimbatore in 1939.

Symptoms : Bushy appearance of the plant is the main symptom of the disease. The infected plants show extreme reduction in the size of their leaves and nodes. The leaves also become narrow, soft, smooth and yellowish in colour. Severely affected plants develop phyllody and do not bear flowers and fruits.

Causal Organism : The disease is caused by a mycoplasma. The mycoplasma is spherical to ovoid (40-3000 nm) structure occurring in the phloem cells of diseased plants.

Transmission : It is a graft transmissible disease and the disease is transmitted in the field by the vector, Cestius phycitis.

Disease Control : The disease can be controlled by the following methods.

(1) Eradication of weed hosts and affected brinjal plants from the field.

(2) Use of tetracycline therapy.

(3) Spray of metasystox at frequent intervals to control the insect vector population.

Sandal Spike : Spike disease is the most destructive disease of sandal plant (Santalum album). The disease was first reported from Coorg district of Karnataka state in 1899 but now it is also common in parts of Tamil Nadu and Kerala.

Disease Symptoms : Severe reduction in leaf size and shortening of internodes are the main symptoms of the disease. These symptoms are commonly called rosette spike. In advanced stages, leaves become yellowish and finally reddish. Affected parts rarely bear flowers or fruits and phyllody is a common phenomenon. Phloem necrosis is one of the internal symptoms. Ultimately, spiked plants die. In pendulous spike symptoms, continuous apical growth of individual shoots without proportionate thickening of the shoot results in a drooping habit of the shoot.

Causal Organism : The sandal spike disease was earlier thought to be caused by graft transmissible insect-borne virus. But, later electron microscopic studies reveal that the disease is caused by mycoplasma-like bodies (M L B), found in the sieve tubes of leaves and twigs.

Transmission : The pathogen- is transmitted through root contacts and insect vectors (Moonia albimaculata, Nephotettix virescence).

Disease Control : Treatment with antibiotics, like tetracycline, tetramycin, ledermycin, etc. shows remission of symptoms. Besides, use of systemic fungicide - Benlate also brings about temporary remission of symptoms.

Fungal Diseases of Plants

Fungi are a group of such plants which lack root system and chlorophyll. Hence, they lack the capacity of mineral absorption and photosynthesis. They obtain their nutrition from other living or dead plants and animals. Most fungi are saprophytic, i.e., they obtain their nutrition from dead organic matter. There are certain fungi which obtain their nutrition from living cells. These are known as parasitic fungi. The parasitic fungi may be obligate or facultative. Besides, some fungi live symbiotically with other living plants and animals.

Since parasitic fungi obtain their nutrition from the host plant; a close contact between parasite and host is necessary. Infection is successful establishment of pathogen (i.e., parasitic fungus) within the host. Infection by pathogen depends on environmental conditions, besides several other factors.

Parasitic fungi also affect the physiological processes of the host. Some fungi secrete cellulolytic or pectinolytic enzymes, which dissolve the middle lamella of the host cell wall. The toxins produced by certain fungi cause morphological deformity in the host. Such effects of the parasite appear in the host plant in the form of disease symptoms.

Some Important Fungal Diseases

Some important fungal diseases of plants are described below.

Wart Disease of Potato : It is a very destructive disease of potato. It was first discovered in Hungary in 1895 and now the disease occurs in most European countries, North and South America, South Africa, Peru, Bolvia, India, etc.

In India, the wart disease was first noticed in Darjling in 1952 on the 'furor' variety of potato imported from Denmark. The disease is now common in the Nilgiri hills, Shimla, Sikkim and at the border of Sikkim and Darjling. It is most prevalent in Alu bari area of Darjling.

Disease Symptoms : The disease symptoms appear in the underground parts of the plant except roots and the basal part of the stem. They are in the form of dark brown warts or hypertrophy on the infected tubers. The warts may be simple or they are in the form of cauliflower-like excrescences. Sometimes, warts measure more than the tuber. When the infection is severe, galls are produced on aerial shoots.

Causal Organism : Pathogen responsible for the wart disease is an obligate parasitic fungus, Synchytrium endobioticum. The fungus is holocarpic and endobiotic, and it lacks mycelium. The naked thallus of the fungus is uninucleate with amoeboid protoplast. It occurs in the epidermal cells of the host.

Disease Cycle : The tubers are infected by uniflagellate zoospores of the pathogen in the spring season. The naked uninucleate protoplast of the zoospore enters the epidermal cell of the potato tuber in amoeboid way by making a pore. The amoeboid spore increases in size by absorbing nutrients from the host cell and becomes globular. The infected cell also shows an abnormal increase in its size. The zoospore develops a double layered wall; the outer layer is thick and dark golden and the inner is thin and transparent. Now it is called summer spore. The summer spore is located in the lower half of the host cell. At this stage, the adjacent epidermal and cortical cells of the host surrounding the infected cell are stimulated and they divide repeatedly and form a wart or tumour like structure.

The summer spore germinates inside the host cell. A pore is formed in the outer wall of the spore and through this pore the inner thin wall protrudes out to form a vesicle in the upper part of the host cell. The protoplast of the summer spore is transferred to the vesicle. The nucleus of the protoplast divides to form 32 nuclei. This multinucleate structure is known as prosorus. The protoplast of the prosorus is cleaved into 5-9 segments, each surrounded by a thin protoplasmic membrane.

These structures are called zoosporangia or summer sporangia. The nuclei of the zoosporangium undergo repeated mitotic divisions to form about 300 nuclei. Now the protoplast of the zoosporangium divides into uninucleate segments. Each segment metamorphoses into a zoospore.

The mature zoosporangium bursts open by absorbing water, thus releasing zoospores. The zoospores are ovoid, uninucleate, uniflagellate and naked amoeboid structures. They infect the new host.

If conditions are unfavourable, each segment of the prosorus develops into a gametangium instead of zoosporangium. In gametangium motile gametes are formed which are similar to zoospores except for their smaller size. Two gametes from different gametangia fuse to form a diploid biflagellate zygote.

The zygote remains motile for sometime and then the flagella are retracted. At this stage the zygote infects the host cell like the zoospore. It absorbs nutrition from the host cell, increases in size and develops a 2-3-layered wall. It is now known as resting sporangium. Since it remains dormant throughout winter, it is also known as winter sporangium. These sporangia become active in the next spring season. The first division of the nucleus of the sporangium is meiotic, followed by several mitotic divisions. Thus, numerous haploid nuclei are formed in each sporangium. The protoplast of the sporangium is cleaved into many uninucleate segments. Each uninucleate segment metamorphoses into a uniflagellate zoospore. The zoospores are released by rupture of the sporangial wall and they infect the new host cells.

The pathogen may survive in the soil in the form of resting sporangia for several years. Contaminated manure and infected soil are the main sources for the spread of the pathogen. Neutral to slightly acidic soil and 12-24°C temperature range are most favourable for infection.

Disease Control : The following measures may be adopted for the control of wart disease of potato.

(1) Use of resistant varieties of potato.

(2) Strict implementation of quarantine laws at the time of potato import.

(3) Since the disease is prevalent in the north-eastern Himalayan regions, potatoes obtained from this region should not be used as seeds.

(4) Treating the infected soil with soil fungicides.

Green Ear Disease or Downy Mildew of Bajra : This is a very serious disease of bajra (Pennisetum typhoides) occurring in almost all regions where the crop is grown. In India, the disease was first reported by Butler in 1907 in sporadic form. In recent years, it has become a serious disease, causing 27-30% loss in millet production. In 1975, the disease appeared in epidemic form in Karnataka and Maharashtra states, destroying the whole crop. The pathogen also attacks several other species of Setaria, which serve as source of perennation for the pathogen.

Disease Symptoms : There are two stages of the disease, downy mildew stage - dominant on the leaves, and green ear stage - occurs on the ears (inflorescences). The sporangia provide a powdery appearance to the lower surface of the infected leaves. The formation of sporangia is favoured by low temperature and humid atmosphere. The infected leaves become twisted and chlorotic, and sometimes they split longitudinally.

The main symptoms of the disease are produced on the inflorescence. The ear is wholly or partly transformed into loosely arranged, twisted, leaf-like green structures. The glumes become enlarged and green and the bristles of spikelets appear contorted due to hypertrophy. The stamens become leafy or are suppressed to rudimentary peglike stumps. Carpels rarely develop in severely affected plants and they are usually replaced by small, leafy shoots or horny catgrowths.

Causal Organism : The pathogen responsible for this disease is an obligate parasitic fungus, Sclerospora graminicola. It has intercellular, branched, aseptate and coenocytic mycelium. It absorbs food from the host tissues with the help of bulbous—haustoria. The mycelium remains confined to the mesophyll tissue, where it causes hyperplasia.

Disease Cycle : Several erect hyphae, known as sporangiophores, arise from the intercellular mycelium. The sporangiophores develop in the sub-stomatal chambers and emerge through stomata singly or in tufts. The sporangiophore measures about 100 μm in length. Its basal part is stout, aseptate and unbranched, whereas the terminal part is thicker and dichotomously branched. The swollen tips of the branches form sporangia. The sporangia are hyaline, elliptical, papillate and smooth walled. They measure 15-34,μm in length and 12-23 μm in breadth. They are usually formed in the night, and detach from the sporangiophore early in the morning. They soon germinate to produce 3-12 reniform biflagellate zoospores. The zoospore remains active for sometime and then it secretes a wall and becomes dormant. During favourable conditions, the zoospore germinates by producing a germ tube. The germ tube enters through the stoma and forms a new mycelium on reaching the intercellular spaces of the host tissue.

Reproductive structures are usually formed in the leaves and malformed spikelets. Oogonia and antheridia develop in the intercellular spaces. The protoplast of the young oogonium is homogeneous, but later it is differentiated into a peripheral vacuolated periplasm and a central dense ooplasm. As the oogonium matures, only a single nucleus remains in the ooplasm and the rest are transferred to the periplasm. All nuclei present in the periplasm disintegrate before fertilization. The young antheridium is also multinucleate, but at maturity only one nucleus remains functional.

Before fertilization the mature antheridium sticks to the wall of the oogonium. A fertilisation tube emerges from the antheridium, which enters the ooplasm by piercing the oogonial wall and periplasm. The male nucleus, alongwith some protoplast, is transferrred to the oogonium. The male nucleus fuses with the egg and thus a diploid oospore is formed. The oospore is red or brown in colour and measures 26-42, µ m in diameter. The oospores are found scattered in the intercellular spaces of the host tissue. They reach the soil by falling down from the infected plant. They remain viable for more than three years in the soil. The nucleus of the oospore divides meiotically and thus haploid stage is again established. The oospore germinates by producing one or more germ tubes which enter the host through stomata. The optimum temperature for germination is between 20° and 25°C.

Disease Control : Some important methods for the control of the disease are as follows.

(1) Use of resistant varieties is a good measure to control the disease. Some disease-resistant varieties of bajra are HB-5, PHA-10, PHB-14, T-15 and T-65.

(2) Eradication and destruction of infected plants.

(3) Crop rotation at an interval of six years.

(4) Since oospores remain viable for several years in the

soil, disinfection of the soil is the most effective way to combat the disease.

(5) Seed treatment with 0.1% agrosan and 0.4% thiram.

(6) Roguing of diseased plants within a month of planting, followed by diethane-45 spray.

White Rust of Crucifers : White rust is a very destructive disease of several cruciferous plants, such as mustard, radish, cabbage and turnip. The disease occurs throughout India.

Disease Symptoms : The disease shows both localised and systemic symptoms visible on the aerial parts of the host like leaves, sterns, flowers and fruits. It appears in the form of white pustules on the host surface. The pustules usually develop on the lower surface of the leaf, but in severe infections they spread on the upper surface of the leaf and the stem. The infected parts of the host become abnormal and the abnormality is in the form of hypertrophy, fleshyness and distortion. The infected leaves become dwarf, thick and fleshy, while the flowers become sterile and enormously enlarged due to hypertrophy and hyperplasia of the affected cells. The sepals, petals, stamens and carpels become fleshy and malformed. The petals may turn into greenish structures and the stamens into leaf-like or carpelloid structures; the carpels into a sterile carpellary leaf; and the ovules are usually atrophied. The inflorescence axis and stem very often become twisted. A violet pigment appears in the upper epidermis of the host and in severe infection disintegration of soft tissue takes place.

Causal Organism : The disease is caused by Albugo candida (=Cystopus candidus) a fungus which is an obligate parasite and remains within the intercellular spaces of the host tissue. The aseptate coenocytic mycelium is profusely branched, and it has globular or rarely knob-shaped haustoria. Asexual reproduction takes place by multinucleate, nearly spherical and hyaline conidia which occur in long chains on club- shaped conidiophores. The conidia may sometime germinate while still attached to the conidiophore. Germination of the conidium depends upon the prevailing temperature; they do not germinate at temperatures beyond 25°C. Sexual reproduction takes place towards the end of the growing season of the host. The hyphae penetrate deeper into the host tissue and form sex organs. The sexual reproduction is oogamous. Formation of oospore depends on certain environmental conditions.

Disease Cycle : The white rust is a soil-borne disease. The pathogen perpetuates by the oospores lying in the diseased plant debris. The oospores are thus the chief source of primary inoculum. The air borne conidia are responsible for the rapid spread of the infection. The cool and moist weather favours the development of the disease.

Disease Control : The disease may be controlled by the following methods.

(1) By crop rotation at suitable intervals.

(2) By destruction of weeds and infected plants.

(3) By spraying 0.8% Bordeaux mixture (4:4:50).

Powdery Mildews : The diseases caused by the members of the order Erysiphales are commonly known as powdery mildews, as their conidia appear as white powdery mass on the surface of the infected plant. The causal organisms are obligate parasites on flowering plants. They show physiological specialization towards their hosts. Their mycelium is hyaline, septate and uninucleate. The symptoms of the disease usually develop on the leaves, buds, flowers and fruits of the host plant. The cool and shady environment favours the development of the disease. The optimum temperature for the infection of most of the powdery mildews is 11°-28°C. The disease can be controlled by the use, of sulphur dust or chemicals containing sulphur like dinocap.

Some important powdery mildew diseases and their causal organisms are listed below.

Some Common Powdery Mildews and their Causal Organisms

Disease	Causalorganism
1. Powdery mildew of barley	Erysiphe graminis var. hordei
2. Powdery mildew of peas	Erysiphe polygoni
3. Powdery mildew of grape vine	Uncinula necator
4. Powdery mildew of cucurbits	Erysiphe cichoracearum and Sphaerotheca fuliginea
5. Powdery mildew of apple	Podosphaera leucotricha

Powdery mildew of barley is described here in some details.

Powdery Mildew of Barley : It is a very serious disease in the sub-mountain regions of the northern India and in certain parts of Rajasthan, Haryana and Punjab.

Symptoms : The main symptom of the disease is greyish-white powdery growth on the leaf. Normally, the upper surface of the leaf is more severely affected than the lower surface. If conditions for the growth of the disease are favourable, the sheath, stem and even glume may also be affected. On these parts the disease appears in the form of superficial spots with flocculent mycelial growth. These spots are initially white but become greyish after the formation of cleistothecia. The infected plants usually become stunted and the size and the number of leaves is also reduced. The infected leaves become curled and deformed.

Causal Organism : The pathogen responsible for the powdery mildew of barley is Erysiphe graminis var. hordei. The mycelium of the pathogen is superficial, profusely branched and septate with small flat appressoria and haustoria. The primary mycelium gives rise to conidiophores which produce numerous oval and uninucleate conidia in long chains. The secondary mycelium is persistent lanuginose and interlaced.

The cleistothecia, formed rarely are dark black, globose structures, up to 200 μm in diameter. They are provided with simple or branched appendages. There are some 25 asci in a cleistothecium; each ascus has 4-8 elliptical ascospores.

Disease Cycle : It is a soil borne disease. The cleistothecia are able to survive for several years at low temperatures. The optimum temperature for the growth of the fungus is 20°-21°C. Conidia are produced in abundance in a cool and moist environment. The highest germination of conidia is at 12°C and the best growth of the germ tube is at 21°C. The conidia, readily disseminated by wind, are the chief source of secondary infection. Overmanuring increases incidence of the disease. A surplus of carbohydrate supply and sufficient light is necessary for the disease development. Unbalanced application of nitrogen and potassium causes high incidence of disease, whereas overdose of phosphate reduces the disease.

In Indian conditions, complete disease cycle is not known as yet. The disease is introduced in the plains of India by the conidia brought by the wind from hilly areas. The cooler environment of hills is favourable for the survival of the pathogen.

Disease Control : Some important methods adopted for the control of the disease are as follows.

(1) The best way to combat the disease is the use of resistant varieties. NP-710, NP-718, K-53, E-750 and HD-1980 are some important varieties of barley which are resistant to powdery mildew.

(2) Field sanitation by removal and destruction of diseased plant parts.

(3) Spray of sulphur dust and chemicals like potassium sulphide or copper sulphate is very effective in disease control. Spray of 0.1% benlate or 0.1% karathane and calixin on the leaves helps in disease control.

Loose Smut of Wheat : Loose smut, a very serious disease of wheat, is world wide in occurrence and is a serious problem in the humid and semi-humid wheat growing regions. In India, the disease occurs in all wheat growing areas, but its incidence is higher in the cooler and moist northern parts than in the south. According to an estimate, the disease causes about 40% loss in wheat yield every year.

Symptoms : It is a very destructive disease of wheat crop, almost every ear of the affected plants is converted into a black mass of spores and no grains are formed. There are no signs of the disease until the ear formation. The infected plants grow faster than the healthy plants and produce flowers earlier. They do not form healthy grains and instead a black powdery mass (chlamydospores) is filled inside the grain wall. In early stages the chlamydospores remain enveloped by a thin silvery membrane but at the time of flowering the membrane bursts open and the spores are gradually blown away by the wind. In severe infection, only the central axis of the spikelet is left behind on the plant.

Causal Organism : The fungus responsible for the disease is Ustilago nuda var. tritici. The mycelium of this fungus consists of multicellular branched hyphae which are hyaline but turn brown at maturity. The primary mycelium is septate and uninucleate, whereas the secondary mycelium is branched, septate and binucleate. In the vegetative phase the secondary mycelium is profusely branched and spreads in intercellular spaces of the host tissues. The pathogen reproduces by the formation of chlamydospores and basidiospores. The hyphal cells are transformed into olivaceous brown, spherical and echinulate chlamydospores which germinate readily and produce basidia and basidiospores. The haploid basidiospores produce uninucleate primary mycelium on germination.

Disease Cycle : Chlamydospores are carried to healthy plants by wind. They readily germinate and produce a new mycelium. A healthy host plant can be infected in several ways at various stages of its growth :

(i) The host can be infected at the seedling stage from contaminated grains or contaminated soil present in close proximity of the germinating seeds.

(ii) The infection may also occur at blossom time; the spores infect immature ovaries of healthy spikelets, producing therein an internal mycelium.

The mycelium remains dormant in the grain but after seed germination it becomes active and eventually infects the embryonic spikelets on reaching the inflorescence region. Finally, the mycelium forms dense masses of hyphae in the immature ovaries. These hyphae repeatedly divide to form smut spores. A group of such spores is known as smut ball or sorus. The sporogenous hyphae grow in between the anticlinal walls of the epidermis of the infected tissue. The epidermis of the infected parts of the host is thus destroyed and the spores are disseminated by air currents.

The maximum development of disease takes place at 23°C. The disease is less frequent in the areas where humidity is low at the time of flowering.

Disease Control : The disease is internally seed borne and as such spray of fungicides is not effective in controlling the disease. However, the following methods can be adopted for the disease control.

(a) *Hot Water Treatment :* The seeds are first soaked in water for five hours at 20°C, the water is drained off and then they are treated with hot water at 49°C for about a minute and finally with hot water at 52°C for 11 minutes. Immediately after the hot water treatment, the seeds are cooled off by dipping in cold water and dried. The dormant mycelium present inside the seed dies off by this treatment.

(b) *Solar Energy Treatment* : In this method, the seeds are first soaked in water for four hours and then dried in sun for four hours. This duration of sunlight is sufficient for inactivation of the mycelium.

(c) *Use of Systemic Fungicides* : Several fungicides like carboxin, vitavax and benlate are used for seed treatment to reduce the pathogen infectivity. A combination of vitavax with thiram or maneb is very effective for disease control.

(d) *Use of Resistant Varieties* : The most successful method is the use of resistant varieties. NP-710, 718, 761, 770, Bansi 224 and P 9 D are some important varieties of wheat which are resistant to smut disease.

(e) *Crop Rotation* : Crop rotation at suitable intervals is also effective in disease control.

Covered Smut of Barley : The covered smut of barley is common in northern parts of India where it causes considerable damage.

Symptoms : The symptoms of the disease become evident only at flowering. The affected ears are shorter than healthy ears and are usually retained within the sheath for a longer time. They become distinct at maturity of the crop. Chlamydospores are present inside the grain wall or are covered by the glume. The spore masses are held together by the deposition of a fatty substance. The spores are released by breaking off the grain wall at the time of threshing of the oars.

Causal Organism : The causal organism of the disease is Ustilago hordei. The chlamydospores of this fungus are round to elliptical and up to 11 μm in diameter. They are light coloured on one side and possess a smooth epispore. They germinate readily and give rise to a 4-celled promycelium. The promycelium produces basidiospores (sporidia) near the septa and at the apex. The basidiospore forms infection thread on germination, which develops into profusely branched mycelium in the host tissues.

Disease Cycle : Chlamydospores are usually present in seeds. When contaminated seeds are sown in the field, the spores germinate on the seed coat of germinating seeds and the germ tube penetrates the embryonic growing parts of the host through the coleoptile. The host, however, can not be infected by the pathogen after the growth of the primary shoot above the soil surface. The mycelium becomes profusely branched and spreads in the intercellular spaces of the various parts of the host. It derives nutrition from the host cells with the help of haustoria.

At the time of flowering of the host, the mycelium becomes much active and infects the embryonic spikelets on reaching the inflorescence region. Finally, the mycelium forms dense masses of hyphae in the immature ovaries of the flowers. The hyphal cells divide and form a mass of smut spores (chlamydospores or teleutospores). The light brown spores are globose or elliptical with a smooth epispore. Since the sporogenous hyphae do not destroy the host epidermis, the smut ball remains covered by the host epidermis. The spores are thus not disseminated by wind. They are released when the grain wall is broken at the time of threshing of the ears. The seedling infection depends on the soil moisture and soil temperature. The optimum temperature for spore germination is 20°C.

Disease Control : Since the disease is externally seed-borne, fungicidal seed treatment is very effective in disease control. The disease can be completely controlled by the use of ceresan and agrosan 5W. Sulphur dust also helps in disease control. K-12, K-18, K-19, C-50, C-84, CN-294 are some varieties of barley which are moderately resistant to the smut.

Late Blight of Potato : The late blight is one of the

destructive disease of potato. It is the most serious of all the potato diseases when conditions are favourable for its spread. The famous Irish famine of 1845-46 was due largely to the failure of potato crop, due to late blight infection. The disease apparently originated in Mexico from where it spreads along with the potato into the Andes, then into Europe and United States. In India, the disease was first introduced into the Nilgiri hills between 1870 and 1880 and very soon it spreads to Darjling in the himalayan ranges. The first severe outbreaks of the disease were reported between 1912 and 1928 from Assam, Bengal and Bihar. In northern India, the disease was first reported from the plains of western U.P. Since then, the disease appears as a regular feature in the plains causing severe losses to the potato crop.

Disease Symptoms : The symptoms of the disease can be seen on any part of the plant, viz., leaves, petioles, stems and tubers. On the leaves, the symptoms appear in the form of dark brown, oval of irregular water soaked areas. In the early stages, the symptoms develop at the tips or the margins of the old leaves. The infection spreads vigorously when temperature is low and atmosphere is humid, and soon appears in the form of blight. In case of severe infection, practically all the parts of the host become brown and then degenerate. After the tops have been blighted, the infection reaches to the underground tubers. In the infected tubers, the skin becomes slightly sunken and dark in colour. If conditions are humid, the cells of the tuber become soft and dark brown. This symptom is known as wet rot. In dry atmosphere, however, the potato pulp does not rot but its anterior part becomes black. This condition is called dry rot.

Causal Organism : The disease is caused by Phytophthora infestans. The mycelium is endophytic consisting of hyaline, profusely branched, coenocytic hyphae. The hyphae develop intercellularly and form haustoria. Ovoid or lemon-shaped sporangia are produced on sporangiophores. The sporangia are at first terminal but become lateral due to sympodial branching of the sporangiophore.

The sporangium may germinate directly, forming a germ tube at the apex or its protoplasmic contents divide to form a number of biflagellate zoospores which emerge through the papilla. The method of germination is much governed by temperature; low temperature favours zoospore formation, whereas higher temperature, the germ tube development. Sexual reproduction is oogamous type. The antheridium is somewhat elongated and amphigynous, whereas the oogonium is pear-shaped to almost spherical, smooth' and reddish-brown in colour. The fusion of egg and male nucleus results in the formation of a diploid oospore. The oospore germinates by producing a germ tube and the tip of the germ tube ultimately develops into a sporangium.

Disease Cycle : The infected tubers are the main source of infection. The dormant mycelium in the tubers becomes active and grows upward in the stem and sporulates on small shoots. Epiphytotics of the disease is likely to occur when unusually cool weather combined with abundant moisture prevails at the time the sporangia are being produced. The optimum temperature for the sporulation is 21°C. The sporangia germinates best by zoospores at 12°C and by germ tube, 21°C. 100 per cent relative humidity causes abundant production of sporangia. Sporangia are easily detached and disseminated by rain or air. On reaching a suitable host, the sporangia germinate either by germ tubes or by zoospores depending upon the environmental conditions. The spores from the blighted leaves are washed down into the soil where they penetrate to different depths reaching the healthy tubers. Contact of healthy tubers with diseased leaves at the harvesting time is another source of tuber infection.

Several theories have been advanced to explain the yearly occurrence of late blight; but (i) persistence of mycelium in the affected tubers, and (ii) survival of the fungus in fruiting stage or as dormant mycelium in the tubers left in the field from the preceding crop, constituting the primary inoculum are the' most accepted ones. In Indian climatic conditions, the possibility of survival of the pathogen through soil in any form is very remote. Thus, it seems the-infected seed tubers are responsible for the spread of the disease.

Disease Control : The late blight can be effectively controlled by the following methods :

(1) Healthy seed tuber selection.

(2) Field sanitation.

(3) Delayed harvesting and sorting potatoes, from a blighted field.

(4) Tuber treatment before storage.

(5) Storage in a cool, dry and well aerated store house.

(6) Use of disease-resistant varieties, like Kufri red, Kufri neela, Kufri kundan, etc.

(7) Foliar spray of fungicides like Bordeaux mixture, Dithane Z-78, Fytolan, Blitox, etc.

Tikka Disease of Ground Nut : Tikka is one of the most destructive disease of groundnuts (Arachis hypogaea). The disease is common in USA, Phillipines, Africa, Indonesia and Australia. The disease incidence causes defoliation and early maturity, and thereby size and quality of nuts are affected. Losses varying from 15 to 50 per cent have been reported due to disease incidence.

Disease Symptoms : The disease is caused by two fungi, viz. Cercosporidium personatum and Cercospora arachidicola. The symptoms produced by these two pathogens differ in size, shape, and colour of the lesions (spots). All parts of the host plant above soil level are attacked by the fungi. The first signs of the disease are slightly pale areas on the upper surface of leaves. The leaf spots caused by Cercosporidium personatum are circular, 1 to 6 µm in diameter, and the necrotic lesions are dark brown to almost dark colour. There are no halos around the young spots, but bright yellow halos develop around the more mature ones on the upper surface. On the other hand, the spots caused by Cercospora arachidicola are circular to irregular in outline, 4 to 10 µm in diameter and tend to coalesce later. They remain surrounded by a yellow halo from the very beginning. On the upper surface the necrotic areas are reddish brown to black, while on the lower surface they are light brown.

Causal Organism : The tikka disease is caused by two fungi, Cercosporidium personatum (= Cercospora personata) and Cercospora arachidicola both mostly occurring on the same leaves. Mycosphaerella berkeleyii and M. ' arachidicola are their perfect stages respectively. The mycelium of Cercosporidium personatum is entirely internal and ramifies intercellularly by developing haustoria in the mesophyll tissues of the host.

The mycelium develops as dense stroma which produces uniformly olivaceous brown, non-septate (some times 1- to 2- septate), unbranched and geniculate conidiophores. The conidia are obclavate to cylindrical, light coloured, 1 to 7-septate, with bluntly rounded ends. The conidia are restricted to the lower surface and cushions of conidiophores can be seen in concentric circles. The perfect stage of the fungus (M. berkeleyii) is perithecia which are broadly ovate to globose and remain partly embedded in leaf tissues. Asci are cylindrical to clavate and contain eight ascospores which are bi-celled, and constricted at the septum.

The mycelium of Cercospora arachidicola is, however, both intercellular and intracellular and does not produce any haustoria. The conidiophores are usually amphigenous, but on the younger spots they are developed exclusively on the upper surface and do not form concentric rings. The conidia are hyaline or pale yellow, obclavate, 4 to 12-septate, with rounded to distinctly truncate base and subacute tip. The perfect stage (M. arachidicola) is represented by perithecia. The asci are clavate and the ascospores are hyaline, slightly curved and bi-celled.

Disease Cycle : The disease is perpetuated through conidia lying in the soil, on diseased plant debris and conidia carried in shell. Infection of the host is accomplished through both the leaf surfaces by direct penetration or by way of open stomata. The disease is disseminated by wind which blows the inoculum from leaf to leaf. Prolonged low temperature and high relative humidity favour severe infection.

Application of nitrogen and phosphatic fertilizers increases disease incidence while potash decreases it only slightly. Besides, ascorbic acid and riboflavin content of leaves play an important role in susceptibility and resistance to the disease; increase in ascorbic acid is related to susceptibility whereas riboflavin to the resistance.

Disease Control

(1) Eradication of diseased plant debris of previous season.

(2) Crop rotation.

(3) Seed disinfection by 0.5% $CuSO_4$ solution or with agrosan GN.

(4) Foliar spray of fungicides, like Bordeaux mixture, Dithane Z-78, Perenox, etc.

(5) Dusting with sulphur.

(6) Breeding of resistant varieties-G-0120, G-1032, G-0607, Exotic-4.

Whip Smut of Sugarcane : Whip smut is a well known disease of sugarcane. Besides India; the disease has been reported from Java, Taiwan, Philippines, Australia, South

Africa, Italy, etc. It is primarily a disease of wild canes and those sugarcane varieties which approximate the wild ones.

Disease Symptoms : The most conspicuous symptom of the disease is a whip-like black shoot, several feet in length and much curved on itself. In earlier stages, smut spores on the whip remain covered by a silvery, thin membrane. The membrane soon ruptures and flakes off exposing dense black powdery dust consisting of millions of spores. Affected plants usually have slender and thin canes and can be identified easily before the production of smutted whip. The lateral shoots developing from the eyes on the infected canes may also develop whips after infection.

Causal Organism : The disease is caused by Ustilago scitaminea. The mycelium is intercellular and aggregate in dense masses towards the surface of the spore-bearing shoot where the spores are formed. The smut spores (teliospores) are spherical, smooth, light brown and 5 to 10 microns in diameter.

Disease Cycle : The spores from the whip-like structure on affected canes are blown away by wind. Some of these spores fall on the nearby canes and get deposited at the junction of the leaf and the leaf sheath from where they travel down the, sheath and reach the tender nodal region and the young ones. There is sufficient moisture at the base of the leaf sheath facilitating germination of spores.

The spores germinate forming a short promycelium. 100 per cent relative humidity and an optimum temperature of about 25° to 30°C are essential for spore germination. From each of the cells of the promycelium arises a sporidium on shoat stalk. Under favourable conditions, thesporidia may bud off to form more sporidia in short chains. The sporidia germinate to produce the infection thread. The disease is perpetuated by planting setts of smutted canes, by spores borne on buds, by infection of buds on standing canes and by ratooning the smutted canes.

Disease Control : Some control measures to prevent the incidence of whip smut are given below

(1) Removal of smutted whips from the field.

(2) Discourage the practice of ratooning.

(3) Avoidance of planting of setts from smutted canes.

(4) Disinfection of setts by 0.1% mercuric chloride, agallol, aretan, etc. before planting.

(5) Systemic fungicides- like vitavax, benlate, bavistin, etc. can be used from sett treatment to eradicate the internally present dormant mycelium.

(6) Use of resistant varieties-like Co 449, 527, 658, 974, 1148, 6806, 7108, 7319, 62101, BO 22, 24, etc.

Red Rot of Sugarcane : Red rot of sugarcane is a serious disease prevalent wherever sugarcane is grown in the world. The disease was first described in Java in 1893 under the name red smut. In 1906, Butler reported the disease symptoms in several cane varieties in India and renamed it as red rot. Serious epiphytotics of this disease have occurred in Northern India (U.P., and Bihar during 1939-40 and 1946-47). However, localised epidemics occur almost every year.

Disease Symptoms : The disease appears on all aerial parts of the plant. The early symptoms of the disease are yellowing and drooping of leaves. The stems show little indication of the disease in early stages but as severity of the disease increases the cane splits lengthwise and subsequently red blotched appear throughout the length of the cane. These blotches emit a peculiar smell of alcohol fermentation.

The reddening appears mainly in the vascular region. Ultimately, the cane becomes dull in appearance, get rotten and shrinks at the internodes. Besides, the conspicuous symptoms also appear on the leaves. On the mid-ribs of leaves, infection originates as a dark reddish area which elongates rapidly, forming blood-red lesions with dark margins. In later conditions, the centre of the lesions becomes straw-coloured.

Causal Organism : The red rot of sugarcane is caused by Colletotrichum falcatum. The perfect stage of this fungus has been described in Ascomycetes as Glomerella tucumanensis. The mycelium is inter-or intracellular, profusely branched, septate and contains characteristic oil droplets. In later stages the hyphae closely intertwine with one another and form small stromata under the host epidermis. The fungus reproduces asexually by the formation of conidia developed in acervuli. The conidiophores are usually unseptate, unbranched and are arranged compactly like palisade tissue. Usually a single conidium develops at the tip of each conidiophore but occasionally the conidia are produced in acrogenous chains. The conidia are hyaline, necrosis crescent or sickle-shaped. An oil drop is present in the centre of each conidium. The conidia germinate by producing 1 to 4 germ tubes and form the new mycelium.

The perfect stage of the fungus in nature was first reported on sugarcane leaves by Chona and Bajaj (1953). Perithecia are globose and are produced on various parts of the host, and measure 150-300 μm in diameter. Asci are numerous, hyaline, clavate and measure 50-60 × 7-10.5, μm. Each ascus contains eight ascospores arranged biserially. Intermixed with asci are present numerous delicate paraphysis. New physiological races differing in their pathogenicity to sugarcane cultivars have been frequently reported from different parts of the world.

Disease Cycle : Seed sets from diseased canes are the chief means of survival and annual recurrence of the disease. Once, the pathogen establishes, secondary spread occurs by conidia. Ratoon crops also serve as a source of perennation and inoculum multiplication. High humidity, water-logging conditions, lack of proper cultural operations and continuous cultivation of same variety year after year are the main factors responsible for the development of the disease.

Disease Control : The following measures may be adopted for the control of red rot of sugarcane :

(1) Healthy sett selection.

(2) Cultural practices-ratooning results in the multiplication of the disease, hence it must be discouraged.

(3) Field sanitation and crop rotation.

(4) Use of resistant varieties like Co 846, 951, 975, 1007, 1148, 62101, 62399, S109, B022, 24, COLK 7702, 7710, etc.

(5) Treatment of seed setts with organo-mercurials like aretan or agallol (0.25% suspension) helps in eradication of superficial inoculum. Besides, treatment of setts in 0.5% bavistin solution reduces the incidence of red rot from infected setts.

Apple Scab : Scab is one of the most serious diseases of apples, almost wherever in the world they are grown. The, scab disease was first reported in Sweden in 1819, and subsequently it was reported from other countries and mainly from USA in 1834. By now, it occurs in most areas of the world and is most severe in those places which have cool, moist springs and summer. In India, the disease was first recorded in Kashmir Valley in 1930 and by 1973 it had firmly established in most of the apple plantations of the state. During 1977, the disease was detected in Himachal Pradesh and had affected almost the entire apple orchards of the state by 1983. The disease has recently been recorded in Arunachal Pradesh, Sikkim and hilly regions of Uttar Pradesh.

Disease Symptoms : The most striking symptoms appear

on leaves and fruits. In severe attacks, 1-3 years old shoots, petioles, pedicels and blossoms may also be affected. Symptoms first develop on the under surface of the young leaves; later both sides may be infected. Light brown to olive green powdery spots which turn mousy black with age appear on the leaf lamina. Severe spotting results in distortion and dwarfing of leaves and causing premature yellowing and defoliation.

Trees are severely defoliated, weakened and fail to produce fruit buds. On fruits, the scab spots are usually well-defined in shape and appearance against the shining coloured background of the fruit skin. With age, spots become dark, brown and corky. Severe early infection results in mis-shapen, knotty fruits. Cracks often develop in scabbed areas which allow the entry of other organisms causing rot of the fruit. On the twigs, spots develop in the form of small, raised, reddish-brown spots. Besides, in prolonged wet conditions nearing to fruit harvest, mature fruits may even develop scab spots. Fruits may show small, rough, black, circular lesions on their skin when in the field or on being kept in the cold storage after sometime. Affected fruits develop rot due to invasion of scab spots (lesions) by secondary rot pathogens.

Causal Organism : Apple scab is caused by the fungus Venturia inaequalis with the imperfect stage as Spilocaea pomi. The fungus grows sub-cuticularly on host with little or no deleterious effect on the epidermal cells. Eventually epidermal and palisade parenchyma cells die. The fungus invades them after leaf fall. The mycelium is septate, profusely branched with uninucleate cells. In the living tissues the mycelium is generally located between the epidermal cells and the cuticle. With age it becomes compact and several layers in thickness. However in dead leaves, the saprophytic mycelium grows throughout the mesophyll and ultimately gives rise to ascostromata. The conidiophores arise from the hyphal strands or from stroma. Each conidiophore successively cuts off a number of uninucleate conidia at the tip.

The conidia are ovate to lanceolate, generally one-celled (often become two-celled through septation) and smoky brown in colour at maturity. The fungus is heterothallic and requires two different mating lines for further development of a mature ascocarp. Mature ascocarps are spherical, dark brown to black perithecia. They possess a short beak and distinct ostiole around which are present single-celled dark setae. Each ascus is saccade in shape and possesses eight ascospores. The ascospores are oval to boat shaped, hyaline, yellowish or olivaceous in colour and two-celled.

Disease Cycle : The perfect stage of pathogen (Venturia) is saprophytic and grows on dead leaves whereas the conidial stage (Spilocaea) is parasitic on living host tissues. The pathogen remains alive through winter in the dead and fallen leaves of apple on the orchard floor. In late January, the pathogen enters the sexual stage and produces black, pimple like structures (perithecia) embedded in the leaf tissue. Good snow coverage and alternating periods of wetness and dryness in later winter and early spring favour the development of fruiting bodies. Air currents and rain splashes help in the dissemination of ascospores.

The discharge of ascospores starts at a time when the buds on the trees are about to burst open in spring. Ascospores begin to germinate on a wet leaf surface in two to four hours and penetrate the cuticle by means of penetration pegs to form stroma between the cuticle and the epidermal wall. Germination of ascospores and subsequent infection of young leaves and flower buds are optimum at 10° to 22°C. At and above 28°C these processes are inhibited. Usually 9 to 17 days are required for the appearance of scab spots after the spore first infects the leaves or fruits. Within these primary scab spots, large numbers of conidia are produced under favourable conditions (at 8° to 16°C temperature).

The conidia are washed from spots by rain and are spread

to other susceptible tissues (young leaves, fruits, etc.), where they cause secondary infection. High humidity and moderate temperatures favour maximum conidia production in scab spots. Although apple variety influences lesion number and size on the infected portion, lesion (spot) growth and spore production stops generally within 30 to 36 days, and spots become necrotic or bronzed and reddish brown. The spread of the disease continues with each shower of rain throughout the summer and in early autumn. Many cycles of the conidial stage occur during the growing season, and each scab spot produces thousands of conidia.

Disease Control : Several methods are adopted to control the apple scab.

(1) *Sanitation :* By destroying fallen and overwintered leaves, or by making a spray of 5% urea in autumn shortly before general leaf fall to hasten leaf decomposition.

(2) *Fungicide Spray Strategy :* A number of fungicides, like Dikar (combination of mancozeb and karathane), Polyram (dithiocarbamate), Triforin, Difolatan, Captan, Zineb, Benlate, Bavistin, etc. are used to control effectively the apple scab. Systemic fungicides (e.g., carbendazim, benomyl, topsin M) and dodine are used to inhibit established lesions, to suppress spore production and to reduce the viability and germination of existing spores.

(3) By growing resistant varieties.

Club Root of Crucifers : Club root is most dangerous disease of crucifers, especially cabbage and closely related plants. Although the disease is not prevalent in India, it is known to exist in many parts of the world, like south and central part of Sri Lanka.

Disease Symptoms : The symptoms first appear on the

underground parts. The roots become much swollen and the root hairs hypertrophied expanding at their tips to form club-shaped swellings. Seedlings show flagging and symptoms of chronic water deficiency. The leaves appear pale green or yellowish in colour. In advanced stages, the roots die.

Causal Organism : The disease is caused by Plasmodiophora brassicae.

Disease Cycle : The pathogen produces numerous resting spores inside the host cells and when the roots decay, they are released into the soil. The resting spore germinates to produce a single anteriorly biflagellate zoospore. The flagella are of unequal length and of whiplash type. Germination is stimulated by substances diffusing from cabbage roots.

The zoospores infect the root hairs or epidermal cells of the root. Before infecting the host, the zoospores become amoeboid. In the host, the naked protoplast divides to form a small plasmodium with 30 to 100 nuclei. Later the plasmodium divides to form a zoosporangium which produces finally 4-8 anteriorly flagellated zoospores. The zoospores are released into the lumen of the host cells. The released zoospores possibly behave as gametes. They fuse together to form an amoeboid zygote. The zygote develops in large plasmodium that gives rise to resting spores.

The pathogen survives through its resting spores lying free or in crop débris, and multiplies and spreads in the plasmodial stage, causing hypertrophy and hyperplasia of the infected tissue. Hypertrophy of the host cells is brought about by blocking of the mechanism for cell division and is accompanied by enhanced DNA synthesis. Temperature, moisture, aeration, soil reaction, etc. have pronounced effects on the development of the disease. Slightly acidic soil and a temperature range of 17°-25°C are most favourable factors for the incidence of the disease.

Disease Control : The following measures may be adopted for the control of disease

(1) Field sanitation and prolonged crop rotation.

(2) Since the disease is favoured by acidic soils, liming the soil to bring the pH to 7.2 is one of the efficient preventive measure.

(3) Use of sodium methyl dithiocarbamate, aldrin and vapam as soil drenches.

(4) Cultivation of resistant varieties.

Koleroga or Mahli Disease of Areca Palms : Koleroga is a very serious disease of areca palms in Western Peninsular India, Assam and Sri Lanka. The disease is prevalent in the rainy season which lasts from the beginning of June to the end of September.

Disease Symptoms : 'Koleroga' which literally means 'rot', usually makes its appearance two or three weeks after the beginning of the rains. The first sign of the disease is on the nuts on which a water-soaked area usually develops towards the base. In advanced stages, the affected nuts rot and fall off from the bunches. Fallen nuts show a felty, whitish mycelial mass on their surface and very soon the entire nut is enveloped by it.

Causal Organism : The disease is caused by the fungus Phytophthora arecae. The mycelium is hyaline, profusely branched, inter-or intracellular, aseptate (but septate in older parts) and coenocytic. The sporangiophores are entirely aerial and usually emerge through the stomata in groups. The sporangiophore produces ovoid or lemon-shaped sporangia which are at first terminal but become lateral due to sympodial branching of the sporangiophore. Sexual reproduction is oogamous. The antheridia are diclinous, and develop sometime before the oogonia. The oospores are smooth, spherical, and 23 to 26 microns in diameter.

Disease Cycle : The areca palms harbour latent infection in their crowns during the non-monsoon period. Heavy rainfall and constant humid conditions favour the spread of the disease. Besides, the close plantation of trees gives ideal conditions for spread. Over-summering is carried out by mycelium persisting on tree tops and chlamydospores in fallen nuts. Good sunshine is favourable for the formation of sporangia and for the emergence of zoospores.

Disease Control : The disease may be controlled by the following methods

(1) Field sanitation by destruction of diseased trees and fallen infected nuts.

(2) Foliar spray of 0.25% perenox, 1% Bordeaux mixture or fytolan, etc. at suitable intervals.

Early Blight of Potato : The early blight is most destructive disease of potato occurring in nearly every part of the globe. In India, the disease can cause up to 40 per cent loss in yield.

Disease Symptoms : The disease symptoms appear when the plants are 3-4 weeks old. The early symptoms are in the form of small, yellowish brown spots on the leaves, which enlarge to form concentric rings. These are known as target board symptoms. There is usually a narrow chlorotic zone around the spot, which fades into the normal green. Lower leaves are affected first followed by the upper leaves. In dry weather, the spots become hard and the leaves curl. In humid conditions, the disease spreads fast, causing infected tissues die, dry and fall off. The spots formed on the leaf reduce the photosynthetic area. This causes the formation of undersized and reduced number of tubers. In severe infection, entire lamina, petiole and stem is badly damaged. Even tubers are infected, and the potato pulp turns brown and becomes corky in texture.

Causal Organism : The disease is caused by the fungus Alternaria solani. The septate, profusely branched, light to dark brown coloured intercellular mycelium remains localized in lesions caused by the fungus or inhabits saprophytically on decaying plant debris or in the soil. The pathogen reproduces only asexually by conidia formation. The conidia are produced singly or in chains at the tips of the branched or unbranched conidiophores. The conidiophores are relatively short, dark and geniculate and emerge through the stomata from the dead centre of the spot.

The dark coloured conidia are multicellular and obclavate, obovoid or muriform in shape, and are both transversely and longitudinally septate. There are about 5-10 septa in each conidium, and the number of septa is probably dependent on environmental conditions. The conidium measures 30-130 μm in length and 12-30 μm in breadth. The conidia are readily disseminated by wind. On being detached from the conidiophore, they germinate on a suitable substratum by producing germ tubes. Humid conditions favour their germination. The germ tubes enter through the stomata of the host leaf and soon develop into new mycelia within the intercellular spaces, developing symptoms within two or three days of the infection.

Disease Cycle : Mycelium and conidia survive in the soil on diseased plant debris to cause primary infection in the next year's crop. Tubers contaminated with mycelium or conidia are another source of primary inoculum. Secondary spread of disease occurs through conidia developing on primary spots. The optimum temperature for germination of conidia is 28-30°C. Host penetration of the fungus occurs through the stomata but direct penetration through the epidermis is not infrequent. Under favourable conditions (abundant moisture followed by high temperature) the disease becomes serious. However the periods of continued drought checks the spread of the disease. The collateral hosts, such as brinjal, tomato, etc. play an important role in the perpetuation and dissemination of the pathogen.

Disease Control **:** Since the disease is soil borne, following methods are usually used for its control.

(1) Field sanitation by destructing diseased haulms immediately after harvest.

(2) Crop rotation.

(3) Foliar spray of fungicides at regular intervals, like Dithane Z-78, Copper oxychloride, Bordeaux mixture, Blitox-50, Difolatan, Captan, etc.

Blast Disease of Rice (Paddy Blast) **:** Paddy blast is one of the most widely distributed disease of rice, occurring in every region of the world where rice is grown. The disease is more severe in humid areas, with plenty of rainfall. Losses due to blast may be up to 90%, depending upon the part of the plant infected. In India, the disease is most prevalent in southern parts, particularly in the coastal areas.

Disease Symptoms **:** Although chiefly a foliage disease, the symptoms may also appear on leaf sheaths, rachis and the glumes. On the leaves, the symptoms first appear as small bluish flecks about 1 to 3 μm in diameter which may enlarge up to several centimetres in young leaves. In large lesions, the central part becomes pale green or dull greenish green and water soaked where as the outer rim appears dark brown in colour.

In later stages, the centre of the lesion becomes greyish or almost straw coloured. Similar spots are formed on the leaf sheath. Brown to black spots are formed on the rachis of the maturing inflorescence. The most characteristic symptoms appear on the culm. The culms become shrivelled and remain covered with a grey fluffy mycelium. Bluish patches on the culm is the most characteristic feature of the disease. Moreover, due to necrosis, the ears (inflorescences) tend to break and fall off.

Causal Organism **:** The disease is caused by Pyricularia oryzae, a member of class Hyphomycetes of Deuteromycotina. The mycelium is intra or intercellular, consisting of septate, multinucleate branched hyphae. The conidiophores arise singly or in fascicles. They are simple or rarely branched, showing sympodial growth. The conidia are borne in succession, one at a time at the tip of the conidiophores. Each conidium is pyriform, multinucleate and 3-celled structure with pointed or blunt apex. They are hyaline to pale olive and measure 14-40 x 6-13 microns in size.

Disease Cycle : The conidiophores emerge from the host surface, through stomata singly or in groups and produce conidia in succession one at a time. About 4,000 to 6,000 conidia are produced per lesion every day for a fortnight or so. They are released from the conidiophores by dew or rain splashes and disseminated by wind currents. The fungus produces a toxin pyricularin which is phytotoxic in higher concentrations.

Environmental factors, particularly temperature and moisture are important in the spread of the disease. 90% relative humidity is essential for the production of conidia in large numbers. Besides, severity of the infection also depends on the age of the plant and host nutrition. Heavy nitrogenous manuring augments disease proneness. Several weeds, like Digitaria marginata, Brachiaria mutica, Echinochloa crus-Bali, etc., are some important collateral hosts responsible for the survival of the pathogen.

Disease Control : Some important methods for the control of the disease are as follows :

(1) Field sanitation and destruction of alternate hosts.

(2) Seed treatment with organo-mercurials, such as agrosan GN, mixture of aureofungin and copper sulphate or copper oxychloride; fungorene.

(3) Direct foliar spray of fungicides, like 0.5% Copperan,

0.3% Blitox, Bordeaux mixture, Agrosan GN and ceresan lime dust, Hinosan (1.5 ml/1 water), Brestanol (0.44 g/1 water).

(4) Use of antibiotics, such as Blasticidin-S, Kitazin, Blastin, etc.

(5) Use of resistant varieties - Co 4, 25, 26, 29, 30, TKM - 1, A-67, A-90, A-200, A-249, GS-397, GS-480, ADT-20, CH-13, SM-8, T-3, CR-9, etc.

Annual Recurrence of Wheat Rust : Rust epidemic is one of the major causes of instability of wheat production and in India. Wheat crop worth million of rupees is lost every year due to rust disease. In India the wheat crop is chiefly infected by three species of Puccinia and these species are heteroecious requiring two distinct and distantly related hosts (alternate host). The alternate host of Puccinia graminis tritici, causing black rust of wheat, is Berberis vulgaris, that of P. recondita, causing brown rust disease is Thalictrum flavum and that of P. striformis causing yellow rust of wheat is Muehlenbergia huglei.

Five types of spores are formed in the life cycle of P. graminis tritici; of these three types of spores-uredospores, teleutospores and basidiospores develop on wheat plants and pycnidiospores and aeciospores develop on the alternate host, Berberis. Wheat plants may be infected by uredospores and aeciospores, whereas Berberis plants only by basidiospores.

Epidemiology of Stem Rust (P. Graminis Tritici) in India : The stem rust, though occurs all over the country, appears in epidemic form in southern, central and eastern parts of the country, where comparatively high temperature prevails during the crop season. To study the epidemiology of the stem rust of wheat in Indian sub-continent, it is important to consider the following two facts.

(1) In India Berberis spp. occur only in the hills at an altitude of 900-2700 m.

(2) The spores of the rust can not survive in the plains during summer season due to high temperature.

Thus there is no inoculum of rust disease in the plains of India due to high summer temperature. As the disease appears every year between January and March, infection is only possible by the reintroduction of inoculum into plains at a time when temperature is congenial. A systematic study of recurrence of cereal rust in the plains of India was done by Dr. K. C. Mehta during 1922-1950. To ascertain the source of inoculum in the plains he established observatories at Almora, Shimla, Mukteshwar and Narkaunda. His study revealed that over summering of pathogenic spores takes place in sub-Himalayan ranges, particularly central Nepal in the North and Nilgiri and Pulney hills in the South.

Sources of Infection in the Hills : Once it was established that inoculum for the recurrence of rust in plains comes from the hills, it became necessary to ascertain the source of infection in the hills. In the hills of India there may be the following two sources of inoculum, viz.,

(1) aeciospores developing on Berberis may serve as primary inoculum, or

(2) uredospores present on the wheat or collateral host may function as inoculum in the plains.

Role of Berberis in the Recurrence of Disease : Dr. Mehta observed that aecial stages present on Berberis growing in the hills of India are of a different strain that infect the wheat crop in the plains. The strain which infects Berberis belongs to P. graminis var. agropyri and not to P. graminis tritici. It conclusively proved that the aecial stages present on Berberis are of no consequence in the perpetuation of the stem rust in India. Moreover, teleutospores formed on the infected wheat plants remain dormant during summer season and basidiospores are not formed until rainy season. Thus on the hills Berberis is infected not before July while wheat crop is harvested there in May-June and new crop is sown in October. The aeciospores formed on Berberis in July cannot survive till October. This fact further supports Mehta's view.

Importance of Uredospores in the Recurrence of Disease : Dr. Mehta also studied the possible role of grasses in the annual recurrence of rust and recorded the presence of P. graminis tritici on some grasses, such as Brachypodium sylvaticum and Bromus patulus. He observed that uredospores survive on these collateral hosts at a height above 1450 m and the regular wheat crop in hills is infected by these oversummering uredospores in November. During favourable conditions uredospores are also formed on the stem and leaves of wheat plants. Mature spores are dispersed by wind. These spores first reach on the foot hills than in the plains. For instance, in the Tarai region of Uttar Pradesh the disease appears in January, whereas in Agra, Meerut and Bundelkhand regions it appears in March.

Thus Dr. Mehta concluded uredospores are the primary source of infection in the plains and are responsible for the recurrence of disease year after year. He suggested that severity of disease can be checked if the cultivation of wheat in the hills is stopped for sometime.

Present View on Rust Epidemiology : Some important work on the recurrence of wheat rust has also been done by Joshi et al. (1974-1986). Their studies were based on 50 years meteorological data, mobile surveys, trap nurseries and spore dissemination. They concluded that climatic conditions in the Himalayan region and even in parts of North Western plains in winter months are not suitable for the multiplication of stem rust inoculum. In the plains conditions become favourable only 3-4 weeks before harvest and in the hills conditions are congenial much later. Therefore, the stem rust inoculum surviving at different elevations in the Himalayas does not form a source of infection of crops in Indo-Gangetic plains before middle of March. By that time the crop is almost near maturity. Their observations indicate that the stem rust inoculum lies in the Southern hills and North Indian hills contribute very little, if at all, to the spread of the disease.

According to Joshi (1986), the cyclonic currents originating in the Bay of Bengal lift the inoculum and introduce it first to the foot hills and adjacent plains of Karnataka and then gradually to Central and North India.

Control of Wheat Rust : The wheat rust disease can be controlled by the following means.

Eradication of Alternate Hosts : Since the pathogen responsible for wheat rust is a heteroecious fungus, alternate hosts play an important role in the annual recurrence of the disease. These alternate hosts provide shelter to the pathogen during unfavourable conditions. Hence, the eradication of alternate hosts is an important measure for the control of the disease. However, in plains of India, the alternate host (bearberry bushes) does not play any role in the annual recurrence of the rust. Mehta (1952) was of the opinion that unlike western countries, the eradication of bearberry bushes is inadvisable for the control of black rust in India.

Use of Fungicides : Several fungicides are helpful in wheat rust control.

(1) Sulphur dusting at 15 kg per acre at an interval of four days during the period when rust usually appears.

(2) By spraying the wheat crop with Parzat liquid (disodium ethylene Bisdithiocarbonate) plus zinc sulphate at fortnightly intervals.

(3) By spraying zineb, maneb and zinc four times at an interval of 10-14 days.

(4) By using systemic fungicides like vitavax, plantvax, etc.

Use of Resistant Varieties : The rust diseases can be effectively controlled by using resistant varieties. NP-770, 792, 809, 829, 830, 846, NI 146, 284 S, 345, 917 are some important varieties of wheat which are effectively resistant to stem rust. Besides, Mexican S 277, S 307, S 308, S 331, Sonora 63, 64 and Lerma Rojo show resistance towards all the three rust diseases of wheat.

Mixed Cropping : Mixed cropping of wheat and barley is also a good measure to combat the disease. The mixed cropping minimizes the inoculum for secondary infection and spread of the disease.

Besides, reduction in the proportion of N in the NPK fertilizer is also helpful in reducing rust incidence.

3

Roots of Diseases

The disorders produced by the harmful physiological changes in the plant body are known as diseases. These physiological changes produce such structural abnormalities or symptoms which are harmful to the plant or to any of its part or products that reduces the economic value. According to Anon (1950), diseases are harmful changes in the normal biological processes of organisms. Horsfall and Dimond (1959) defined the disease as a malfunctioning process that is caused by continuous irritation. This process results in some suffering and hence the disease is a pathological state.

Thus the disease may be defined broadly as "any departure from health, presenting marked symptoms, malady, illness and disorders." Stakman and Harrar (1957) defined plant disease as "a physiological disorder or structural abnormality that is harmful to the plant or to any of its part or products that reduces the economic value." The branch of science that deals with the study of nature, development and control of plant diseases is known as phytopathology (phyton = plant, pathos = sufferings, logos = knowledge). The plant pathologist is a person who is the master of the science of phytopathology. He is concerned for the learning and understanding of the nature of plant diseases as well as their diagnosis and control.

The phytopathology has four main objectives, viz. (i) etiology, (ii) pathogenesis, (iii) epidemiology, and (iv) control. The etiology concerns with the causal organisms along with the environmental conditions that cause diseases in plants. The pathogenesis is the actual mechanism of disease development and this phage involves complex host-pathogen interactions. The epidemiology is mainly concerned with the epiphytotics, i.e., it involves the interaction of crop, pathogen and environment, and moreover the seasonal carry over and subsequent dispersal of pathogens. The control mainly concerns with the development of suitable methods of controlling the diseases.

The causal organism of a disease is pathogen. It is not necessary that all pathogens should be animate; several abiotic factors, such as nutritional deficiency and environment (light, humidity, aeration, soil, etc.) may also act as pathogens. The animate pathogens can produce disease conditions only when the environmental conditions and the host are suitable for their survival and multiplication. For instance, spores of several fungi are present in the soil, but infection caused by them depends upon the availability of suitable host and favourable conditions of temperature and humidity. Hence, the capacity to produce a disease by a pathogen depends upon the susceptibility of the host and the environmental conditions.

Pathogenicity

Pathogenesis is the process of infection or the actual way in which the disease develops in plant body. Infection is the establishment of a pathogenic micro-organism within the host, following entrance. It signifies the sum of biological processes which take place in the host body after penetration of the pathogen, independent of the fact whether the pathogen causes a disease or not. As the result of infection visible or latent diseases are produced in the host plants. Infact, infection is a process of inter-struggle between the organisms (pathogen and host) living in two different environmental conditions. The potential capacity of infection of any pathogen is called its pathogenicity.

The pathogenicity of every pathogen is its specific feature. This characteristic depends upon the capacity of parasitic adaptation and struggle for existence of the pathogen. In other words, within a pathogenic species different strains have varying degree of disease producing power in one kind of host. The degree of pathogenicity is known as virulence. It is, therefore, an index of the quantitative individual nature of the pathogenic micro-organism. The phenomenon of pathogenesis can be understood easily by studying the three phases of the pathogen, viz., pre-penetration, during penetration and post-penetration phases. These three phases of penetration are briefly discussed below.

Pre-penetration Changes : The pre-penetration phase includes the growth of pathogen before actual entry or penetration into the host. Spores of various pathogenic fungi germinate on the surface of the host. During germination, metabolic activities of spores increase considerably. Spore germination, besides various physical factors (moisture, temperature, light, pH, oxygen, carbon dioxide, etc.), is mainly influenced by non-parasitic micro-organisms present in the rhizosphere and phyllosphere. Certain chemicals secreted in the rhizosphere by roots are known to accelerate or inhibit germination of spores of pathogens. During germination, the spore produces filamentous germ tube from one or more germ pores. When more than one germ tubes develop, one of them grows more rapidly than the others. The formation of germ tube is affected by environmental factors and susceptibility of the plant.

Penetration Process : Spores show various penetration mechanisms. The infection threads of pathogenic fungi may enter the host through natural openings (stomata, lenticels or hydathodes), through wounds or by direct penetration. Although most of the pathogens enter the host only by one method, some pathogens adopt more than one ways.

Most of the rusts and downy mildews enter through stomata. When germ tube reaches near a stoma, its terminal part swells to form a vesicle, known as appressorium. Most of the protoplast of the germ tube accumulates in the appressorium, which is separated from the germ tube by a septum. A blade-like wedge grows from the appressorium through the stomatal slits and it swells to form a substomatal vesicle. The contents of the appressorium pass in the vesicle. One or more penetration hyphae grow from this vesicle and they form inter-or intracellular mycelium.

Stomata, hydathodes and nectaries are the common pathways for the entry of many parasitic bacteria. All powdery mildews and a few downy mildews usually penetrate host tissue through epidermis by mechanical process.

Post-penetration Changes

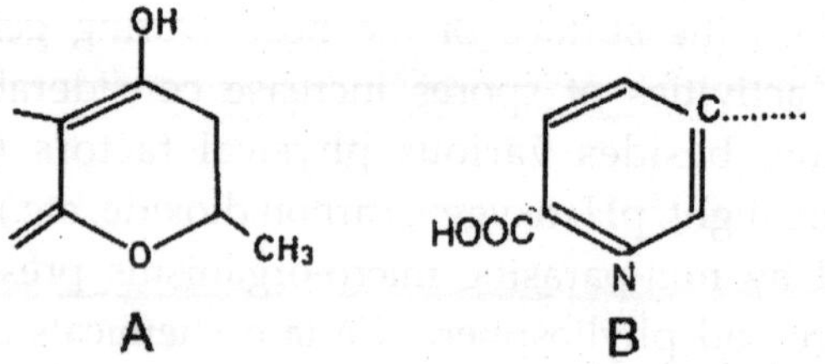

A-B. Toxins : Chemical structure; A. Alternaric acid, B. Fusaric acid.

The post-penetration phase includes the development and growth of the pathogen after penetration. In this phase, colonisation of the pathogen takes place. After successful penetration inside the host, the pathogen secretes several types of substances, such as enzymes, toxins or growth regulators.

These substances bring out physiological, anatomical and morphological disturbances in the host plant.

A cellulosic wall occurs around the plant cell and it is necessary for the pathogen to break this wall to seek entry into the host. Various types of enzymes like cutinase, pectinesterase, polygalacturonase and cellulolytic, secreted by the pathogen carry this work. Some pathogens also secrete certain substances which protect them from the lytic enzymes produced by the host.

Besides enzymes, several pathogens secrete certain- harmful substances, known as toxins. These are poisonous proteins or protein like substances, which are injurious to plants. Toxins differ from enzymes in that they do not attack the structural integrity of tissues. They affect the metabolism in a subtle manner and act directly on the protoplast of the cell. The toxins, may be exotoxins or endotoxins. The former, composed of easily diffusable proteins, are highly toxic, and the latter, composed of firmly bound glucide lipoproteins, are less toxic. Some important toxins are Victorian (secreted by Helminthosponum victoriae), periconin (secreted by Periconia circinata), piricularin (secreted by Pyricularia oryzae), alternaric acid (secreted by Altemaria solani), colletonin (secreted by Colletotrichum falcatum), lycomarasmine (secreted by Fusarium lycopersici) and fusaric acid (secreted by Fusarium oxysporum). Some toxins (e.g., piricularin, alternaric acid, colletonin) are nonspecific, i.e., they affect more than one hosts, whereas others (e.g., Victorian, periconin, photo-alternarin) are host-specific.

Some pathogens produce growth regulators in the host tissues. IAA is produced by several fungi and bacteria. Kinin-like substances increase in the leaves of Phaseolus vulgaris and Vicia faba when attacked by Uromyces phaseoli and U. fabae respectively. Gibberellic acid is produced by Gibberella fujikuroi, causing rot disease of cereals and grasses.

The factor that establishes infection is quantitative. Larger is the dose of the infective pathogen greater are the chances of infection.

Effect of Infection on Host Physiology : As the result of infection, several physiological processes, such as membrane permeability, photosynthesis, respiration and nitrogen metabolism are affected. Disease symptoms appear in the host due to these physiological disturbances.

Effect on Membrane Permeability : The main physiological effect of infection is altered permeability of the cell membrane in diseased plants. Changes in the permeability have been detected before any other symptom of the disease appeared. Permeability changes have been noticed due to infection by all the three types of pathogens-viruses, bacteria and fungi. As the result of infection, membranes become leaky and as such ionic balance of the cell is disturbed. It is believed that toxins combine with some receptors in the susceptible cell and thus bring its disorganization.

Effect on Photosynthesis : The pathogenicity alters the rate of photosynthesis in the host tissues. The rate of photosynthesis is usually lower in the infected tissue in comparison to healthy tissue. There is a decrease of 20-25% in the photosynthetic rate due to viral infection. This is perhaps due to the loss or disintegration of the chloroplast or both. Chlorophyllase enzyme is responsible for the destruction of the chloroplast. This enzyme removes chlorophyll from the lamellar structure of the chloroplast - and crystallizes it into chlorophyllide within the cell.

$$\text{chlorophyll} \underset{\text{enzyme}}{\overset{\text{chlorophyllase}}{\rightleftharpoons}} \text{chlorophyllide+phytol}$$

Similarly, the chlorotic effect produced by the bacterial pathogen is also due to direct action of bacteria on the chloroplast. The chloroplast is disintegrated by toxins produced by bacterial pathogens. Photosynthesis may also be affected indirectly due to occlusion of vascular system by the bacterial pathogen, leading to water stress and wilting.

Effect on Respiration : Respiration is also an index of pathogenesis. Normally, the rate of respiration is higher in infected plants in comparison to healthy ones. Higher respiration rate in infected plants is due to toxins secreted by the pathogen which uncouples phosphorylation from electron transfer in host cells. There is an increase in the respiration rate after inoculation but the increase is very sharp shortly before the symptoms appear.

Effect on Nitrogen Metabolism : Marked increase in nucleic acid and protein synthesis is usually seen in infected plants. The increase in nucleolar volume of hypertrophied host cells is accompanied by an increase in nuclear RNA and non-histone protein, and ultimately the normal transcriptional processes of invaded cells are altered. An increase in the RNA content of the nuclei of diseased cells indicates induced synthesis of proteins.

Host Resistance

Every plant shows some resistance towards pathogens and this resistance reduces the effect of the pathogen or may not allow the pathogen to develop at all. The capacity of resistance of a host towards a specific pathogen depends upon its genetic constitution and environmental conditions. The host resistance is very important as it determines the extent of infection and severity of the disease. The defence mechanisms of the host can be divided into the following two categories.

Morphological or Structural Defence Mechanism : The host develops certain structural features in the epidermal region oi its interior that greatly affect the ability of the pathogen to penetrate. These defence structures are already present or may develop afterwards as the result of host-pathogen interaction. Cutin and wax layers present on the walls of the epidermal cells serve both physical as well as chemical barrier to the entry of the pathogen. Resistance in Berberis to Puccinia graminis and in flax to Melampsora lini is due to the presence of thick cuticle, which acts as a physical barrier. Waxy layer present on the epidermal surface acts as a hydrophobic layer which prevents aggregation of the pathogen on the host surface.

The thick and rigid outer wall of epidermal cells act as barrier for some pathogens. For instance, Pyricularia oryzae enters the leaves of the paddy plant through thin- walled bulliform cells; the walls of the other epidermal cells are resistant to the entrance of hyphae.

Bio-chemical Defence Mechanism : The host also produces some such specific substances whose presence or absence interferes growth and multiplication of the pathogen. These biochemical substances are already present or develop afterwards as the result of the host-pathogen interaction. The pre-infectional biochemical defensive mechanism includes the presence of certain chemicals, such as phenols and chlorogenic acid, which are inhibitory to pathogens. For instance, red variety of onion shows resistance toward Colletotrichum circinans. The scaly leaves of this variety contain certain phenolic compounds like protocatechuic acid and catechol, which inhibit spore germination of the fungus. Similarly, the glandular hairs present on the leaves of Cicer arietinum secrete malic acid which inhibits spore germination and hyphal growth. The resistance of pears to fire blight caused by Erwinia amylovora is due to the presence of a phenolic glucoside, arbutin. The resistance of ripe tomatoes towards Bacterium vesicatorium is due to their acidic nature. Unfavourable pH conditions of host tissues are also inhibitory to the growth of pathogens.

Some hosts produce phytoalexins in response to infection. The rate of formation of phytoalexin is critically important in combination of host variety and parasite strain. Some important phytoalexins produced by hosts are ipomeamarone (induced in Ipomoea batatas by Ceratocystis fimbriata), pisatin (induced in Pisum sativum by Sclerotinia fructicola), phaseollin (induced in Phaseolus vulgaris by Sclerotinia fructicola), isocoumarin (induced in Daucas carota by Ceratocystis fimbriata), rishitin (induced in potato tubers by Phytophthora infestans) and trifolirhizin (induced in Trifolium pratense by Helminthosporium turcicum). In addition to phytoalexins produced in plant tissues in response to infection, induced synthesis of proteins and enzymes play an important role in disease resistance. Some plants produce anti-bacterial substances, such as phytonicides, which provide resistance from the pathogen.

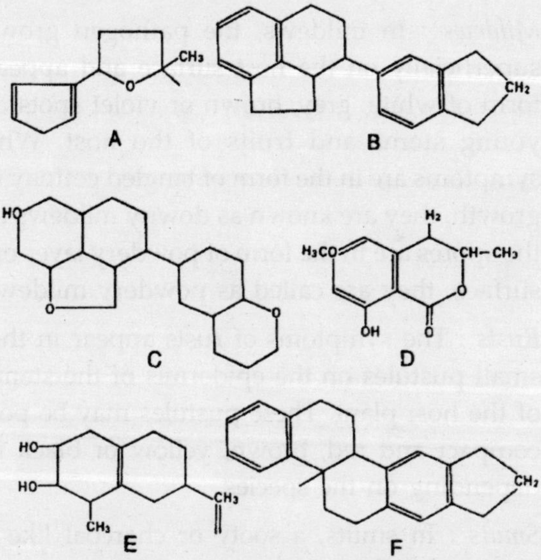

A-F. Phytoalexins : Chemical structure; A. Ipomeamarone, B. Pisatin, C. Phaseollin, B. Isocoumarin., E. Rishitin, E. Trifolirhizin.

General Symptoms of Plant Diseases : As the result of successful penetration of the pathogen, there are several physiological changes in the host, which ultimately alter morphological and anatomical characteristics of the host. Such visible changes in the host are known as disease symptoms.

The symptoms caused by pathogens may be of diverse nature and broadly they fall into the following two categories.

Symptoms Due to the Nature, Structure and Appearance of Visible Pathogen : In most parasitic diseases, vegetative growth of the parasite occurs inside the host tissue and thus the parasite is invisible. However, in some parasites, sex organs or resting structures remain partially or wholly exposed on the host surface. If these structures are of appropriate size, they are visible by naked eye. In some parasites vegetative and reproductive growth occur on the surface of the host and as such the disease symptoms depend on the structure of the pathogen. Some important symptoms are described below.

1. *Mildews :* In mildews, the pathogen grows mostly superficially on the host surface and appears in the form of white, grey, brown or violet spots on leaves, young stems and fruits of the host. When these symptoms are in the form of tangled cottony or downy growth, they are known as downy mildews and when the spores are in the form of powdery layer on the host surface, they are called as powdery mildews.
2. *Rusts :* The symptoms of rusts appear in the form of small pustules on the epidermis of the stem and leaf of the host plant. These pustules may be powdery or compact and red, brown, yellow or black in colour, depending on the species.
3. *Smuts :* In smuts, a sooty or charcoal like powdery mass of spores is produced, usually on the floral organs of the host.
4. *Scabs :* More or less circular, rough or crustaceous lesions developed on the infected parts are known as scabs. They result from the overgrowth of epidermal and cortical tissues, usually modified by rupturing and suberization of cell walls.

5. *Fruiting Bodies* : The fruiting bodies of many fungi occurring on the wood are large and fleshy or woody. These fungi can be easily identified on the basis of the structure of their fruiting bodies.

Anatomical and Morphological Symptoms on Host

Necrosis **:** Death of cells, tissues or organs of the host due to parasitic infection is known as necrosis. It is the most common and destructive symptom of plant diseases. The necrotic symptoms produced by parasites in different hosts are of the following types.

(a) *Spot* : Spots are the most common symptoms produced by fungi. When host cells are killed in a limited area, the necrotic areas appear in the form of spots. The spots may be circular, angular or irregular in shape. Several types of leaf spot disease occur in plants. Tar spots are characteristic leaf spots which appear as raised, black coated bodies. They look like a flat drop of tar on the leaf. In advanced stages of disease, the dead tissue of the necrotic areas shrinks and separates from the healthy tissue. More or less angular and elongated spots appearing on the veins on the lower surface of the leaf 'are known as anthracnose. Leaf spot disease of mango is caused by Pseudomonas mangiferacindicae. Concentric rings of brown and black colour are characteristic of several leaf spot diseases caused by the members of Melaconiales and Moniliales.

(b) *Stripes or Streaks* : When disease symptoms appear in the form of elongated and narrow lesions, they are known as stripes or streaks. These streaks are usually blackish- brown in colour. Stripe disease of barley is caused by Helminthosporium.

(c) *Canker* : The dead areas present in the stem bark or cortex of woody trees are known as cankers. These are often large and rough areas with definite margins which remain embedded in the tissue. The fruiting bodies of cankerous fungi come outside the host by breaking the canker surface. Citrus canker is caused by Xanthomonas citri.

(d) *Blight* : In blights the disease symptoms give a burnt appearance. These symptoms develop due to sudden death of various plant parts, such as leaves, branches or flowers. The affected part usually turns brown or black and soon disintegrates. Late blight of potato is caused by Phytophthora infestans and leaf blight of wheat by Altemaria tenuis.

(e) *Damping Off* : In damping off, the pathogen attacks the base of the stem near the soil surface. The tissues become weak in the infected region and ultimately the plant collapses. Several pathogenic fungi, such as Pythium, Fusarium and Phytophthora cause damping off disease.

Hypertrophy and Hyperplasia : Some diseases cause abnormal growth in a particular part of the plant or the whole plant. This growth may be due to the enlargement of the affected cells (hypertrophy) or due to increased cell divisions resulting in a large number of component cells (hyperplasia). The following are some important forms of hypertrophied symptoms.

(a) *Galls* : These are globose, elongated or irregular malformations or outgrowths developed on the affected parts of the host plant. Small sized galls are known as tubercle or wart, while large sized galls are called knot or excrescence. Wart disease of potato is caused by Synchytrium endobioticum.

(b) *Carl* : Some diseases cause hypertrophy in the cells of some localised areas of the leaf and as such the leaf becomes curled, arched or distorted. These symptoms are known as curl disease. The causal organisms of leaf curl of tomato and leaf curl of papaya are viruses.

(c) *Witches Broom* : Some pathogens incite branching in the host. These branches develop only in certain limited parts and grow vertically instead of horizontally as the normal branches. These branches may become swollen and bear small hypertrophied leaves.

(d) *Floral Abnormalities* : Floral parts of some host plants are enlarged and become fleshy and leaf-like due to the infection of certain fungi, such as Albugo candida.

Classification of Plant Diseases

The plant diseases are usually classified on the basis of the following criteria.

(1) Severity of infection and geographical distribution.

(2) Nature of causal organism.

(3) Mode of primary infection and perpetuation.

(4) Extent of infection.

(5) Nature of the host.

Classification Based on Severity of Infection and Geographical Distribution : On the basis of the severity of infection and geographical distribution, the plant diseases are classified into the following four categories.

Endemic Diseases : Diseases which appear more or less constantly year after year in moderate to severe form in a particular area (district, state or country) are known as endemic. Their pathogens survive in the soil and other media for a long time and the environmental conditions of that particular area do not influence their vitality unfavourably. For example, wart disease of potato, caused by Synchytrium endobioticum, is endemic in Darjling.

Epidemic (Epiphytotic) Diseases : Diseases which occur in severe stage periodically, practically in all regions where a particular crop is grown, are known as epidemic or epiphytotic. Although their pathogens are present in the soil or other similar media like endemic diseases, favourable environmental conditions for their growth are available only for a limited period. Rust diseases of cereal crops are good examples of epidemic diseases.

Sporadic Diseases : Diseases which occur sporadically at irregular intervals and at different locations are known as sporadic diseases, e.g., blotch disease of cucumber.

Pandemic Diseases : Diseases which occur all over the world and cause mass mortality are known as pandemic diseases, e.g., late blight of potato.

Classification Based on Nature of Causal Organism

On the basis of the nature of causal organisms, the following two groups of diseases can be recognised.

Infectious or Parasitic Diseases : Such diseases are incited by living agents which are parasites on the host. These diseases spread by contact of healthy plants with diseased plants. Depending on the causal organisms, these diseases may be mycoplasmal, viral, bacterial, fungal, nematodal or insect diseases, e.g., little leaf of brinjal (mycoplasmal), leaf roll of potato (viral), citrus canker (bacterial), late blight of potato (fungal), ear-cockle of wheat (nematodal), and insect gall disease (insects).

Non-infectious or Non-parasitic Diseases : Such diseases are incited by non-living agents and thus are non-infectious. Unfavourable environmental conditions and soil moisture relations, floods, deficiency of nutrients, injurious atmospheric gases, lightening, injury, etc., are the chief factors that are responsible for such diseases. The common examples of such diseases are tip burn of rice (high soil moisture contents), blossom end rot of tomato (irregular water balance in leaves and flowers), black heart of potato (high temperature and poor ventilation), red leaf of cotton (mineral deficiency) and mango necrosis (black kiln fumes).

Some important parasitic and non-parasitic diseases and their causal agents are listed in Tables.

Some Common Parasitic Diseases of Plants

Name of disease	*Pathogen type*	*Causal organism*
1. Tobacco mosaic	Virus	Tobacco mosaic virus
2. Papaya mosaic	Virus	Papaya mosaic virus
3. Potato mild mosaic	Virus	Potato virus x
4. Potato leaf roll	Virus	Potato leaf roll virus
5. Vein clearing of bhindi	Virus	Hibiscus yellow vein mosaic virus
6. Tristeza (quick decline) of citrus	Virus	Citrus tristeza virus
7. Bunchy disease of banana	Virus	Musa virus I
8. Potato spindle tuber	Viroid	Potato spindle tuber viroid
9. Little leaf of brinjal	Mycoplasma	Mycoplasma like bodies (MLB)
10. Sandal spike	Mycoplasma	Mycoplasma like bodies (MLB)
11. Bacterial blight of paddy	Bacteria	Xanthomonas oryzae
12. Citrus canker	Bacteria	Xanthomonas citri
13. Tundu disease of wheat	Bacteria	Corynebacterium tritici
14. Fire blight of apples and pears	Bacteria	Erwinia amylovora
15. Angular leaf spot of cotton	Bacteria	Xanthomonas malvacearum
16. Black leg of potato	Bacteria	Erwinia carotovora
17. Crown gall of apple	Bacteria	Agrobacterium tumefaciens
18. Red stripe of sugarcane	Bacteria	Pseudomonas rubrilineans
19. Club root of cabbage	Fungi	Plasmodiophora brassicae
20. Wart disease of potato	Fungi	Synchytrium endobioticum
21. Late blight of potato	Fungi	Phytophthora infestans
22. White rust of crucifers	Fungi	Cystopus candidus
23. Downy mildew of grapes	Fungi	Plasmopara viticola
24. Green ear of bajra	Fungi	Sclerospora graminicola
25. Damping off of seedlings	Fungi	Pythium debaryanum

Contd...

Name of disease	Pathogen type	Causal organism
26. Powdery mildew of barley	Fungi	Erysiphe graminis var. hordei
27. Loose smut of wheat	Fungi	Ustilago nuda var. tritici
28. Covered smut of barley	Fungi	Ustilago hordei
29. Black rust of wheat	Fungi	Puccinia graminis
30. Rust of linseed	Fungi	Melampsora lini
31. Tikka disease of groundnut	Fungi	Cercospora personata
32. Early blight of potato	Fungi	Alternaria solani
33. Red rot of sugarcane	Fungi	Colletotrichum falcatum
34. Panama disease of banana	Fungi	Fusarium oxysporum, F. cubense
35. Will of pigeon pea	Fungi	Fusarium oxysporum, F. nudum
36. Bunt of rice	Fungi	Tilletia horrida
37. Peach leaf curl	Fungi	Taphrina deformans
38. Molya disease of barley	Nematodes	Heterodera avenae
39. Ear-cockle of wheat	Nematodes	Anguina tritici
40. Insect gall diseases	Insects	

Some Common Non-parasitic Diseases of Plants

Name of disease	Causal agent
1. Internal bark necrosis of red delicious apple	excess of minerals
2. Red leaf of cotton	mineral deficiency
3. Mottling or foliocellosis	zinc deficiency of citrus
4. Tip burn of paddy	oxygen deficiency
5. Heat canker of linseed	high soil temperature near the soil surface
6. Black heart of potato	high temperature and poor ventilation
7. Black tip or mango necrosis	black kiln fumes; injurious atmosphere
8. Blossom end rot of tomato	nutritional disorders; derangement of water balance

Classification based on Mode of Primary Infection and Perpetuation : The three categories of diseases recognised on the basis of mode of primary infection and perpetuation are :

1. *Soil-borne Diseases :* When the pathogen is present in the soil or infected plant debris, primary infection occurs from the soil under favourable conditions. Such diseases are known as soil- borne diseases, e.g., fruit rot of cucurbits (caused by Pythium butleri) and stem rot of papaya (caused by Pythium aphanidermatum).
2. *Seed-borne Diseases :* When the pathogen is present in seeds or other similar structures and the infection occurs from these structures, the diseases are known as seed-borne diseases, e.g., smut diseases of cereal crops caused by Ustilago.
3. *Air-borne Diseases :* When the pathogen is disseminated by air and the primary and secondary infection also occur from air, they are known as air-borne diseases, e.g., mildews and rusts.

Classification based on Extent of Infection on Host : The following two groups of diseases are recognised on the basis of extent of infection on the host.

1. *Localised Diseases :* Such diseases are confined to a particular region of the plant, i.e., only a part or organ is infected.
2. *Systemic Diseases :* Such diseases spread on the entire plant to varying extents and are associated with nearly every stage of its life cycle.

Classification Based on Nature of the Host Plant : The plant diseases are also classified on the basis of the nature of the host plant, e.g., cereal diseases, forage crop diseases, plantation crop diseases, vegetable crop diseases, fruit diseases and forest diseases.

PART-II

Bryophyta

4

Vegetative Reproduction

Most of the bryophytes are characterised by their ability to propagate by asexual methods. In many dioecious bryophytes sexual reproduction is almost completely absent and they multiply only by asexual methods. Correns (1899) and Cavers (1903) have given a comprehensive account of the vegetative propagation in bryophytes.

The following are some of the common methods of asexual reproduction found amongst bryophytes.

(1) Fragmentation

(2) Formation of tubers

(3) Formation of gemmae

(4) Formation of adventitious branches

(5) Formation of innovations

(6) Formation of primary and secondary protonema

(7) Development of cladia

(8) Persistent apices

Besides, various modifications of above methods have also been reported.

Fragmentation

In bryophytes the growing part is located at the tip of the thallus. The posterior older portion of the thallus undergoes progressive death and decay. As the process of decay reaches up to the point of dichotomy, the young lobes of the thallus get separated (Fig.). These lobes, called fragments, grow into new individual plants.

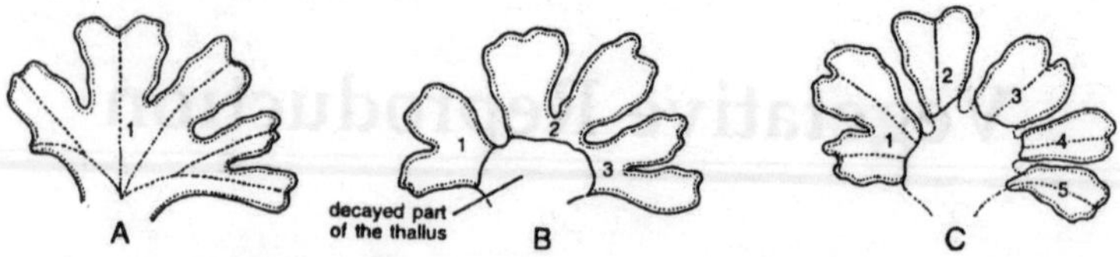

A-C. Vegetative reproduction in bryophytes : Stages of the isolation of younger dichotomies due to progressive decay of the older portions of the thallus.

Fragmentation is the most common method of propagation in the members of Hepaticopsida (e.g., Riccia, Marchantia, Plagiochasma), Anthocerotopsida (e.g., Anthoceros) and some Bryidae (e.g., Sphagnum).

Tubers

Tubers are usually formed in species exposed to desiccation (drought conditions). Such plants develop special subterranean branches towards the end of the growing season which get swollen at their tips to form underground tubers. They are packed with starch grains and are surrounded by 2-3 layers of corky cells, the latter form a waterproof covering. During unfavourable conditions the plant dies but the underground tubers remain unaffected and lie dormant. With the onset of the favourable conditions, the tubers resume growth and develop into new individuals. Thus tubers, besides serving as propagules, are also organs of perennation.

Tubers are frequently found in species of Riccia (e.g., R. discolor, R. vesicata, R. billardieri), Anthoceros (A. himalayensis, A. laevis), Fossombronia (F. himalayensis, F. tuberifera), Sewardiella tuberifera Asterella angusta (Fig.), etc.

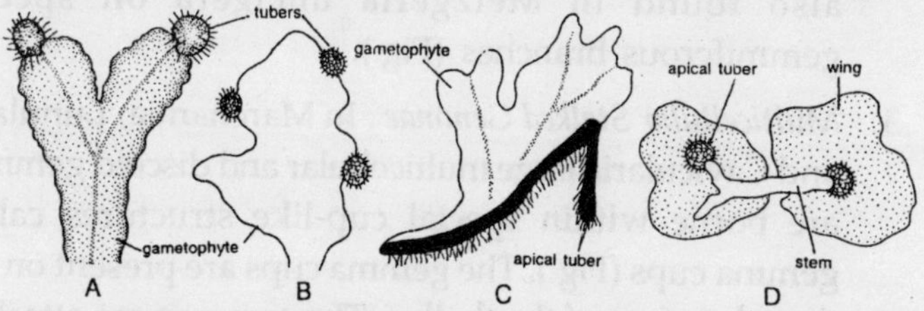

A-D. Vegetative propagation with the help of tubers : A. Thallus of R. billardieri showing tubers, B. Anthoceros laevis thallus bearing tubers, C. Female plant of Fassombronia himalayensis with an apical tuber, D. Forked thallus of Sewardiella tuberifera bearing apical tubers.

Gemmae

These are special propagative organs quite abundant in the members of Hepaticopsida. They are also known to occur in Anthocerotopsida and some true mosses (Bryidae).

Gemmae may occur on stem, leaves, special gemmiferous branches or in gemma cups. Though there is great variation in their form but it is characteristic of a species.

Some common forms of gemmae are described below briefly.

Hepaticopsida

1. *One to Three Celled Gemmae* : In Cephalozia francisci, Lophozia ventricosa, Diplophyllum albicans, Nowellia curvifolia, etc., 1-3 celled gemmae occur on the leaf surface. Similar gemmae are also found on the stem apex of Cephalozia bicuspidata, Lophozia heterophylla, Cephaloziella byssacea, etc. Two celled gemmae have also been reported in some species of Riccardia (e.g., R. multifida, R. sinuta, R. pingus) but they are endogenous, i.e., they occur within the cell (Fig.).

2. *Discoid Multicellular Gemmae* : In some species of Colura, Leptocolea and Radula discoid and multicellular gemmae occur on the leaves (Fig.). Similar gemmae are also found in Metzgeria uncigera on special gemmiferous branches (Fig.).
3. *Multicellular Stalked Gemmae* : In Marchantia, Lunularia and Cavicularia large multicellular and discoid gemmae are borne within special cup-like structures called gemma cups (Fig.). The gemma cups are present on the dorsal surface of the thallus. The gemmae are attached to the floor of the cup by a short stalk.

Anthocerotopsida

In many species of Anthoceros (e.g., A. glandulosus, A. propaguliferous) multicellular gemmae are borne on the dorsal surface of the thallus along the margins.

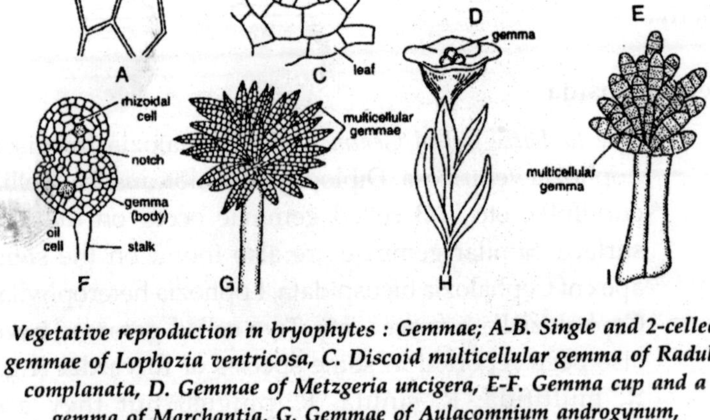

Vegetative reproduction in bryophytes : Gemmae; A-B. Single and 2-celled gemmae of Lophozia ventricosa, C. Discoid multicellular gemma of Radula complanata, D. Gemmae of Metzgeria uncigera, E-F. Gemma cup and a gemma of Marchantia, G. Gemmae of Aulacomnium androgynum, H. A portion of the stem of Tetraphis pellucida bearing gemmae, 1. Leaf tip of Ulota phyllantha bearing gemmae.

Bryopsida

1. *Stalked Multicellular Gemmae :* Such gemmae are found on the stem apices of Tetraphis pellucida and Aulacomnium androgynum (Fig.). In Bryum rubens and B. erythrocarpum multicellular gemmae occur at the base of the stem.

2. *Multicellular Articulate Gemmae :* Multicellular filamentous articulate gemmae are found on the leaves of Ulota phyllantha, Orthotrichum lyllii and Tortula papillosa (Fig.).

Adventitious Branches

In many thalloid bryophytes adventitious branches develop from the midrib on the ventral surface of the thallus. These branches on separation from the parent thallus form a new thallus (e.g., Riccia fluitans, Targionia, Reboulia, Anthoceros, Stephensoniella, Asterella, Sphaerocarpos, etc.). In Pellia the adventitious branches arise from superficial cells on the dorsal surface or the margins of the thallus. In Marchantia and Dumortiera adventitious shoots develop on the archegoniophores.

Innovations : In Sphagnum usually one branch at each node develops more strongly than others. This is called innovation. Unlike other branches (divergent and pendent branches) it forms apical cluster of branches like the main stem. The innovation is separated from the main stem when progressive death of the older part of the stem reaches up to the point of its attachment. It eventually becomes established as independent plant.

Primary and Secondary Protonema : The green filamentous primary protonema of mosses may break into small fragments either accidently or due to the death of intercalary cells. Each fragment grows into a new protonema which bears a fresh crop of leafy gametophores (e.g., Funaria).

Protonema developing from parts other than spores are known as secondary protonema. It may develop from the cells of the rhizoids, leafy gametophores (e.g., Funaria, Sphagnum), injured portion of the leafy gametophore or from the marginal cells of the primary protonema (e.g., Sphagnum). The marginal cells of the secondary protonema form leafy gametophores.

Cladia

These are small detachable branches which help in vegetative propagation. They develop either from the individual cells of the leaf (leaf cladia; e.g., Bryopteris fruticulosa, Plagiochila, Bazzania, Frullania fragilifolia, Mylia cuneifolia) or stem (stem cladia; e.g., Leptolejeunea, Drepanolejeunea and Bryopteris fruticulosa).

Stem cladia develop on the stem at the position of the sexual branches.

Persistent Apices

In many thalloid bryophytes (e.g., Anthoceros, Cyathodium, Athalamia, Cryptomitrium, etc.) at the end of the growing season a portion of the apical part of the thallus becomes thickened while rest of the plant body dies away. The thickened apical part survives during the drought period and as the rainy season sets in, it resumes growth.

Professor Shiv Ram Kashyap (1882-1934) : Professor Shiv Ram Kashyap, the father of Indian Bryology, was born on 6th November, 1882 at Jhelum, Punjab in a meritorious soldier family. He passed matriculation from the Punjab University in 1899. In 1900, he joined the Medical School at Agra and received his medical diploma in 1904, topping the list of successful candidates. He then served for two years in the medical services of U.P. During his student life he was a brilliant scholar. Although, he started his carrier as a medical man, but later turned his attention to Botany. While still a student of Medical school, he passed Intermediate Science Examination of the Punjab University as a private candidate and top the list of successful candidates. In 1906, while serving in the Medical Department of U.P., he passed B.Sc. Examination of Punjab University and again topped the list of successful candidates. In the same year he resigned his post in the medical service and was appointed Assistant Professor of Biology at Government College, Lahore. In 1909, he passed M.Sc. in Botany and stood first in the university. For his distinguished performance in the examination, he was awarded the much-prized Arnold and Maclagan Gold Medals of the University. In 1910, he joined the Cambridge University from where, he took his Honours Degree in the Natural Science Tripos in 1912. On his return home, Professor Kashyap was appointed Professor of Botany at the Government College, Lahore and in 1919, he was appointed Professor of Botany at Punjab University, in which capacity he continued to serve the cause of education until his death.

Professor Kashyap was the first secretary of the Indian Botanical Society, which was founded in 1920. He was its President in 1925 and was the Editor-in-chief of its organ, the Journal of the Indian Botanical Society. He was a member of Asiatic Society of Bengal and an advisory editor of Chronica Botanica, published from Holland. Prof. Kashyap was the first systematic botanist to preside over the annual session of the Indian Science Congress in 1932. He presided over the Botany Section of the Indian Science Congress in 1919 at Bombay. In recognition of his valuable contributions Punjab University conferred on him, Honorary Degree of Doctor of Science in 1933.

Professor Kashyap contributed a lot on various groups of the vegetable kingdom. His contributions to three subjects — the sexual generation of Equisetum, the liverworts of the Western Himalayas and the flora of Tibet, however, stand out above all others. Kashyap's (1914) study of the prothallus of

Equisetum debile aroused much interest both in India and abroad. His contribution to the knowledge of the West-Himalayan liverworts was very great and this may be regarded as his chief subject of study. The frequent tours, he made to the Himalayas, led to the publication of his celebrated work entitled, "Liverworts of the Western Himalayas and the Punjab Plain" in two volumes in 1929 and 1932. Besides establishing four new genera Aitchisonella, Stephensoniella, Sauchia and Sewardiella, he also described about 50 new species from the Himalayan region. He showed that the flora of the Eastern Himalayas is richer than that of the Western Himalayas; and that in the Western Himalayas the most luxuriant liverworts vegetation occurs between the altitudes of 2,000 and 3,000 m, while in the Eastern Himalayas a somewhat lower altitude is more productive. Kashyap's work forms the backbone of the bryophytic literature of our country. From his studies of the Himalayan forms Kashyap (1919) strongly supported Goebel's view of reduction in the Marchantiales and added some new facts in support of it.

He considered Marchantia as the central type and derived other genera from it along various phyletic lines. Regarding the evolution of the bryophytes he contradicted the usual view of a relationship between the algae and the simpler forms of liverworts. According to him the liverworts may have arisen from the pteridophytes by reduction. He was led to this view because of the striking resemblance between the gametophytes, general structure of sex organs and the process of spore formation of some pteridophytes with those of the liverworts. According to him, the first step in simplification was a change from the erect and radial position to a prostrate and dorsiventral habit. Some other outstanding contributions of Kashyap on bryophytes are : (1) Notes on new and little known Himalayan liverworts, (2) Flora of central Punjab, (3) Acrogynous liverworts of Western Himalayas, and (4) Origin of Bryophyta.

The third subject to which Kashyap contributed significantly is the flora of the Western Himalayas and Western and Central Tibet. He was a great traveller and a daring mountaineer. He probably knew more than anybody else about this interesting country. He crossed the Himalayas into Tibet at nine different places throughout its length. The exploration of the Himalayas and the study of its vegetation were his life passion.

With him, his work came first. He died of heart attack on 26th November, 1934.

In some creeping mosses the entire plant dries up during dry season except the growing apex. The cells of the apex secrete a mucilagenous substance which form a protective sheath around the apex, helps the apex to survive during summers. Apices resume active growth in favourable conditions.

Evolution of Sporophyte in Bryophytes : The sporophyte of liverworts and mosses show mostly the same fundamental plan. It is made up of an anchoring and absorbing foot, a stalk-like seta and a capsule, which contains spores and elaters. However, structurally the sporophyte of mosses is much more complex than that of liverworts. A comparative study of the sporophytes of these two groups shows that there has been a progressive reduction in the amount of fertile tissues from Marchantiales to Bryales.

In all bryophytes the oospore, formed as a result of fertilization, is the mother cell of the sporophyte. In the simplest type of sporophyte the entire oospore is utilized in the formation of spores and capsule wall. In certain simple forms, some spore mother cells, which fail to develop into spores, form elaters. In complex forms, the oospore develops into a sporophyte, which is differentiated into foot, seta and capsule. The relative shape and size of these three parts varies in different groups.

The following two contrasting views have been put forward with regard to the evolution of sporophytes in bryophytes.

(1) Evolution of sporophyte by progressive sterilization of potential sporogenous tissue.

(2) Evolution of sporophyte due to progressive reduction or simplification.

These views are discussed here at some length.

Evolution of Sporophyte by Progressive Sterilization of Potential Sporogenous Tissue : This view, also known as 'theory of progressive evolution' or 'theory of sterilization', was put forward by Bower (1908-35) and supported by Cavers (1910) and Campbell (1940). According to this view, the sporophytes of complex forms (e.g., Funaria, Sphagnum, Pogonatum) have evolved due to progressive sterilization of the potential fertile tissue of the simpler forms (e.g., Riccia, Marchantia). In primitive forms, the sporophyte is simple and most of the tissue of the sporophyte is fertile. The sporophyte is devoid of photosynthetic tissue and is completely dependent on the gametophyte for its nutrition.

The progressive sterilization from Riccia to Funaria occurred through the following stages (Fig.).

1. *First Stage :* The simplest known sporophyte among bryophytes is that of Riccia. In all species of Riccia, the sporophyte consists of only capsule, there being no trace of seta and foot. The oospore (the mother cell of the sporophyte) divides first by a transverse and then by a vertical wall to form a four-celled embryo which becomes 20-30-celled by further divisions. Formation of periclinal wall at this stage differentiates the single layered amphithecium from an inner, multicellular mass, the endothecium. The amphithecium forms the capsule wall and the endothecium, the sporogenous tissue. The entire sporogenous tissue is converted into

spores. Occasionally, a few cells fail to develop into spores and these sterile cells (nurse cells) have nutritive function and are supposed to be the ancestors of elaters.

2. *Second Stage :* In Corsinia, the oospore divides by a transverse wall into a hypobasal and an epibasal cell. The former gives rise to a small foot. The derivatives of the epibasal cell form an outer amphithecium and an inner endothecium. The amphithecium gives rise to a single layered capsule wall and the endothecium differentiates into fertile sporocytes and sterile nurse cells. The latter are long elater-like cells but lack characteristic thickening bands. Thus in the sporophyte of Corsinia the sterilization has gone a step further and in addition to nurse cells, a sterile foot is also present.
3. *Third Stage :* Further sterilization is seen in sphaerocarpus where the sporophyte has a sterile bulbous foot and a narrow seta, in addition to fertile capsule. The amphithecium forms a single layered jacket of the capsule and the endothecium, the sporogenous tissue and the sterile nurse cells.
4. *Fourth Stage :* In Targionia, the sporophyte is differentiated into a bulbous foot, a massive seta and a capsule. The amphithecium gives rise to a single layered jacket of the capsule and only about half of the endothecial cells give rise to fertile sporogenous tissue and the remaining half form sterile elaters. Thus in the sporophyte of Targionia there is still further sterilization of sporogenous tissue in comparison to Riccia, Corsinia and Sphaerocarpus.
5. *Fifth Stage :* In Marchantia, the lower half (hypobasal cells) of the octant embryo forms the foot, and seta and the upper half (epibasal cells) the capsule. The amphithecium gives rise to a single layered capsule wall and the endothecium forms sporogenous tissue.

Approximately 50% of the sporogenous cells form spores and the remaining 50% elaters with characteristic thickening bands. Thus a bulbous foot, an elongated seta, a single layered capsule wall and large number of elaters are sterile constituents of the sporophyte of Marchantia.

6. *Sixth Stage* : Further sterilization of the sporophytic tissue can be seen in Pellia. The sporophyte of Pellia is also differentiated into foot, seta and capsule as in Marchantia, but the jacket of the capsule is two or more layered. Furthermore, a fixed sterile elaterophore is present at the proximal end of the capsule with a bunch of elaters.
7. *Seventh Stage* : The sporophyte of Anthoceros is of great interest from the point of view of evolution. The mature sporophyte consists of a bulbous foot and a horn or bristle-like capsule. The capsule wall is made up of 4-6 layers of parenchymatous cells. The outermost layer (epidermis) possesses stomata. The cells of the inner layers of the capsule wall have chloroplasts like those of gametophytic tissue. The sporophyte is thus capable of synthesizing its own organic food and is dependent upon the gametophyte for water and mineral nutrients only.

   Another important feature of the capsule of Anthoceros is the presence of a central column of sterile cells; the columella. It is composed of 16 vertical rows of cells and is surrounded by a dome-shaped archesporium. The tissue of the archesporium differentiates into large fertile spore mother cells and flat and relatively small sterile elater mother cells, alternately. The spore mother cell forms spore tetrad after reduction division, whereas the elater mother cell, by transverse and oblique divisions, gives rise to a 3-4-celled simple or branched elater. The elaters of Anthoceros are known as pseudoelaters as they do not possess thickening bands.

Campbell (1924) observed some specimens of Anthoceros fusiformis which were on the brink of achieving full independence from the ageing gametophyte. In these specimens the basal part of the capsule ceased to produce spores and there was a corresponding increase in the extent of photosynthetic tissue.

Thus, from the point of view of sterility and self-dependence, the sporophyte of Anthoceros is more specialized than those of Hepaticopsida.

8. *Eighth Stage :* The highest degree of sterilization of the sporophyte is found in the class Bryopsida. The regions like foot, seta, capsule wall, columella, apophysis, operculum and peristome in the sporophytes of Funaria, Polytrichum and Pogonatum represent the sterile tissue. The sporogenous tissue is restricted to the spore sacs and constitutes a very negligible part of the sporophyte. The sporophyte in the class Bryopsida shows the culmination of sterilization process.

Thus the sporophyte of Riccia is the simplest amongst the bryophytes, with a very high proportion of fertile tissue and the sterile tissue is very small. On the other hand, the members of the class Bryopsida (e.g., Funaria, Polytrichum, Pogonatum) have the most complex sporophyte with a very high degree of sterility. The evolution of sporophyte in bryophytes as such is considered to have taken place by progressive sterilization of the fertile tissue.

Evolution of Sporophyte by Progressive Reduction or Simplification : Kashyap (1919), Church (1919), Goebel (1930) and Evans (1939) are the supporters of the retrogressive theory of evolution. According to this theory, the apparently simple sporophyte of Riccia is the most advanced and it has evolved by progressive reduction or simplification of the complex forms like Funaria, Polytrichum, etc.

Church (1919) considered that the hypothetical ancestors of bryophytes had foliose sporophytes with complex assimilatory tissue and an epidermis with functional stomata. Such sporophyte had the capability to synthesize its own organic food. Thus the sporophyte was independent of the gametophyte. Church (1919) suggested that progressive simplification, which occurred along the following steps, has led to the evolution of the simplest (but most advanced) sporophyte of Riccia.

(1) During the course of evolution, the independent foliose sporophyte became attached to the gametophyte and gradually lost its leaves.

(2) The photosynthetic tissue gradually reduced and in the sporophytes of Funaria, Polytrichum and Anthoceros chlorophyllous cells are confined to the capsule jacket. In many Bryopsida the capsule wall has stomata in the apophysis region which communicate with the intercellular spaces in the palisade parenchyma-like tissue. In Sphagnum, the photosynthetic tissue is restricted to the expanded apophysis of the capsule.

(3) Stomata, if retained, usually became rudimentary. These vestigial structures (non-functional stomata) in the capsule of Sphagnum proclaim capsule as photosynthetic in the earlier forms.

(4) The multilayered sterile jacket of the capsule (e.g., Funaria, Anthoceros) became single layered (e.g., Riccia, Marchantia) by reduction.

(5) The foot and seta which are well developed in Funaria, Pogonatum, Polytrichum, Pellia, etc., were gradually lost and there is no foot and seta in the simple

sporophyte of Riccia. Moreover, the seta of mosses shows greater tissue differentiation, whereas in liverworts it is composed of mostly thin walled cells.

(6) There has been reduction in the sterile tissue (columella, elaters, etc.) of the capsule and in Riccia the entire endothecium gives rise to sporogenous cells.

Bold (1940) has also opined that autotrophic nature of sporophyte is an indication of its primitiveness. Due to simplification and greater dependence of sporophyte upon gametophyte, the air spaces between the chlorophyllous cells were lost and stomata become non-functional.

5

The Bryidae

The sub-class Bryidae is the most highly evolved group of bryophytes, and its members are known as true mosses. It includes about 14,000 species representing some 675 genera. The following are the characteristic features of this group.

(1) The leaves are more than one cell in thickness; they possess a distinct midrib.

(2) The embryo has two growing points, located in the hypobasal and epibasal halves of the two- celled proembryo. Each growing point of the proembryo has its own apical cell with two cutting faces.

(3) The sporophyte has a distinct foot, a well developed seta and a capsule.

(4) The archesporium develops from the outer layer of the endothecium.

(5) The spore sac is surrounded by septate air chambers.

(6) In most of the genera the capsule has a characteristic operculum and peristomial teeth.

(7) The protonema is usually filamentous and produces leafy gametophores.

The life-histories of Funaria, Polytrichum and Pogonatum are described here as representatives of the sub-class Bryidae.

Funuaria

Systematic Position

Class	—	Bryopsida
Sub-class	—	Bryidae
Order	—	Funariales
Family	—	Funariaceae

Funaria is the most widely distributed and common moss. It includes about 117 species which occur in all parts of the world. The genus is represented in India by about 18 species, which grow in close tufts on damp and shady walls, rocks and tree trunks. They prefer alkaline soil. Funaria hygrometrica is the best known species of the genus.

Mature Gametophyte

External Structure : The mature gametophyte of Funaria consists of an erect leafy axis attached to the substratum with the help of rhizoids (Fig.). The axis is upright, slender, 1-3 cm in height and branched. The branches are lateral as well as extra-axillary; the latter arise below the leaves.

The axis is covered with leaves which are spirally arranged in 3/8 phyllotaxy (i.e., in three complete spirals there are eight leaves). The leaves crowd at the apex of the stem, forming a bud-like head. They are simple, sessile, ovate and have a broad membranous base and a pointed apex (Fig.). The mature leaves present at the base of the stem have a distinct midrib, but the younger leaves at the apex are devoid of midrib. The leaves present on the lower portion of the erect branches and ventral side of the prostrate branches are small, scale-like and without chloroplasts.

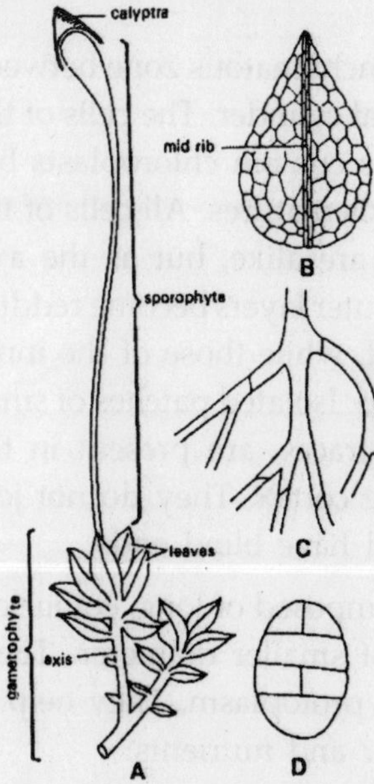

***Funaria* : External structure;**
A. Plant body with sporophyte,
B. Leaf, C. Rhizoids, D. Gemma.

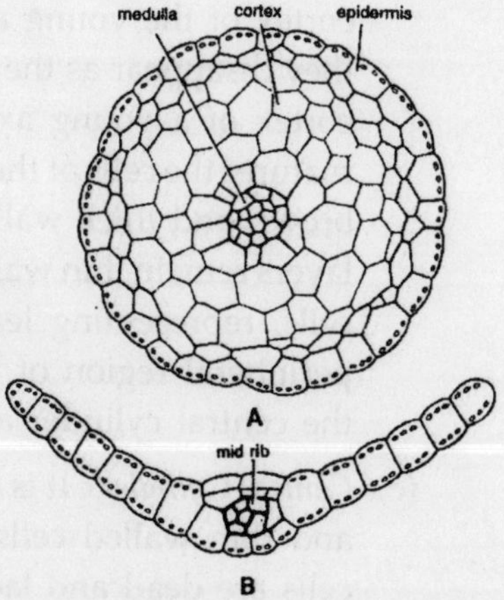

***Funaria* : Internal structure;**
A. Transverse section of axis,
B. Vertical section of leaf.

At the base of the gametophore many branched, slender and multicellular rhizoids are present. The ultimate branches of rhizoids are fine like root hairs. Rhizoids arise from the basal cells of the axis. They are characterized by the presence of oblique septa (Fig.). The young rhizoids are colourless, but at maturity they become strand-like, brown or black, depending upon the colour of the substratum. The branches of rhizoids which are exposed to light develop numerous chloroplasts. The rhizoids function as anchoring and absorbing organs.

Internal Structure

1. *Axis or Stem :* A transverse section of the mature axis shows three distinct regions (i) epidermis, (ii) cortex, and (iii) central cylinder (Fig.).

(a) *Epidermis* : It is the outermost layer, composed of tangentially elongated chlorophyllous cells. It has no pores or stomata.

(b) *Cortex* : It is a large parenchymatous zone between the epidermis and central cylinder. The cells of the cortex of the young axis contain chloroplasts but they disappear as the axis matures. All cells of the cortex of a young axis are alike, but as the axis matures the cells of the outer layers become reddish brown and thick walled, while those of the inner layers remain thin walled. Isolated patches of small cells, representing leaf traces, are present in the peripheral region of the cortex. They do not join the central cylinder and have blind ends.

(c) *Central Cylinder* : It is composed of long, colourless and thin walled cells of smaller diameter. These cells are dead and lack protoplasm. They help in the conduction of water and nutrients.

2. *Leaf* : The leaf has a very simple structure (Fig.). The lamina (also called wing or blade) is composed of a single layer of large thin walled cells which are rich in chloroplasts. These cells form the main photosynthetic tissue of the gametophyte. The midrib, in the central part of the lamina, is composed of a small strand of slightly thickened and narrow cells which help in conduction. The leaf does not have stomata or hairs.

Apical Growth : Apical growth of the axis occurs by a pyramid—shaped apical cell with three cutting faces. The three lateral faces cut off three rows of cells by successive divisions. Each cell thus formed divides periclinally; the inner daughter segment contributes to the inner tissues of the axis and the outer segment to the leaf and the peripheral tissues of the axis. Thus in the young apical part of the axis leaves are arranged in three vertical rows, but this arrangement is disturbed in the lower part of the axis. The branch primordia arise from the superficial initial cells, immediately below the leaves.

Reproduction : Funaria reproduces by vegetative and sexual methods.

Vegetative Reproduction : Vegetative propagation in Funaria takes place by the following methods.

1. *Fragmentation of Primary Protonema :* On germination the spore gives rise to a branched filamentous structure, known as primary protonema. Sometimes it breaks into small fragments either accidentally or due to death of some cells in between. The fragments bear buds and each bud develops into an erect gametophore.
2. *Secondary Protonema :* Under favourable climatic conditions, secondary protonema may be formed on injured rhizoids, stems, leaves or reproductive structures. Like primary protonema, it bears buds which are capable of growing into erect gametophores.
3. *Gemmae :* During unfavourable period, the terminal cells of the protonema form multicellular green bodies, called gemmae. The gemma is both vertically and transversely septate and has 10-30 cells. It remains dormant as long as it is attached to the parent plant. On return of favourable conditions it is detached from the parent plant and germinates to form an erect gametophore. Under certain conditions, gemmae may also develop on the leaf and on the axis of the gametophore.
4. *Bulbils :* These are similar to gemmae in form and behaviour, but they develop on rhizoids and are devoid of chloroplasts. The bulbil grows into a new plant when detached from the parent plant.
5. *Apospory :* Development of the gametophyte from the sporophyte without the formation of spores is known

as apospory. Any unspecialised sterile cell of the sporophyte may form protonema which in turn gives rise to gametophore. Such gametophores are similar to the normal (haploid) gametophores in appearance but have diploid cells (as they develop from the sterile sporophytic cells). The gametes formed on these plants are also diploid, and fusion of these gametes results in the formation of 4n zygotes. Sporophytes developed from such zygotes do not have the capability to form spores and are sterile.

Sexual Reproduction : Funaria is monoecious, i.e., male and female reproductive structures develop on different branches of the same plant (i.e., autoicous). The female branches usually grow more vigorously and become longer than the male branches.

1 *Antheridium:*

(a) *Antheridial Branch :* Antheridia are borne in groups at the expanded convex apex of the main axis. They are surrounded by a rosette of spreading leaves, called perichaetium. These leaves are larger and are arranged more compactly on the axis than the vegetative leaves. A large number of long multicellular capitate hairs, the paraphysis, are intermingled with the antheridia. The paraphysis contain chloroplasts and besides their photosynthetic activity, they protect the antheridial head against loss of water by transpiration. They also hold water between them by capillary action and secrete a mucilagenous substance which helps in the liberation of antherozoids.

(b) *Development of the Antheridium :* The antheridium develops from a superficial cell, the antheridial initial (Fig.). It is located at the apex of the male branch. The antheridial initial becomes papillate and divides transversely to form an outer cell and a basal cell (Fig:). The latter forms the embedded part of the stalk of the antheridium. The outer cell undergoes one or more transverse divisions and forms a 2-4- celled filament (Fig.). The terminal cell of this filament divides by two vertical intersecting walls, thus an apical cell with two cutting faces is differentiated (Fig.). The apical cell cuts off segments in two rows in a regular alternate sequence (Fig.). Now, upper 3-4 cells of this filament divide by diagonal vertical walls, each forming two unequal cells (Fig.). The smaller peripheral daughter cells function as first jacket initials. The larger daughter cells again undergo similar divisions and give rise to second jacket initials on the outer side and primary androgonial cells on the inner side. Thus each cell of the filament gives rise to two jacket initials and one primary androgonial cell (Fig.). The jacket initials divide only anticlinally and form a single layered antheridial jacket (Fig.). The young jacket cells contain chloroplasts which turn into chromoplasts as the cells mature. The apical cell of the filament forms operculum or cap region of the antheridial wall (Fig.). The primary androgonial cells divide repeatedly and the last generation of these cells forms androcyte mother cells.

The androcyte mother cells divide further and each forms two androcytes. The latter differentiate into biflagellate antherozoids (Fig. 4 L, M). All antheridia in an antheridial head do not initiate at the same time and, therefore, different developmental stages of antheridia can be seen in a single head.

(c) *Mature Antheridium :* The mature antheridium has a long multicellular stalk and a club-shaped body of red or dark orange colour. The single layered antheridial jacket encloses a large mass of androcytes. At the tip of the jacket layer there is one (rarely two) relatively large hyaline cell which forms operculum.

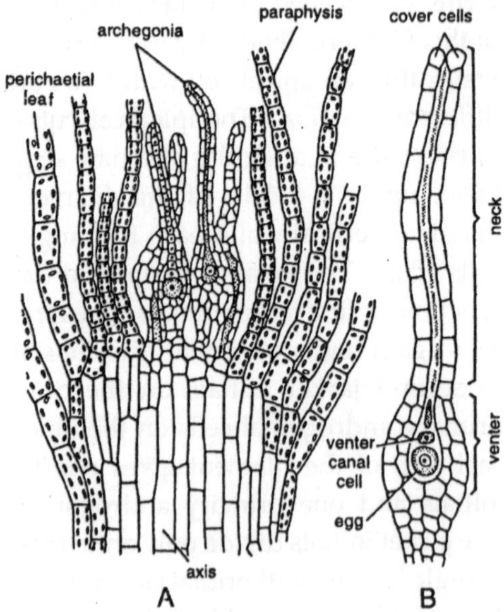

***Funaria* : Female sex organ; A; Archegonial head in longisection, B. An archegonium.**

(d) *Dehiscence of Antheridium* : Dehiscence takes place when the mature antheridium comes in contact with water. The inner and outer wall of the opercular cell absorbs rain or dew water, usually collected in the antheridial head in between the perichaetial leaves. The opercular cell thus becomes mucilagenous and swells up. The pressure within the opercular cell bursts it and consequently a pore is formed at the distal end of the antheridium. A viscous fluid containing androcytes oozes out through this pore. The release of androcytes from the antheridium is also facilitated by the contraction of the antheridial wall and by the hydrostatic pressure of the swollen contents within the antheridium. The androcytes spread out in the form of thin and delicate film. The membranous walls of the androcytes are dissolved in water and spirally coiled biflagellate antherozoids are liberated.

2. *Archegonia:*

   (a) *Archegonial Branch :* The archegonial branches arise laterally at the base of the male branch. Archegonia are borne in groups at the apex of these branches. They are surrounded by a cluster of leaves, known as perichaetial leaves. Paraphyses are also present intermingled with archegonia.

   (b) *Development of Archegonium :* The archegonium develops from a superficial cell, the archegonial initial, which differentiates at the apex of the archegonial branch (Fig.). It divides transversely into a lower cell and an upper cell (Fig.). The former divides repeatedly and forms the basal part of the archegonium which is embedded in the tissues of the archegonial branch. The latter functions as archegonial mother cell and it divides by two intersecting oblique walls, forming an apical cell with two cutting faces (Fig.).

   The apical cell cuts 4-8 segments on both sides alternately. These segments eventually form the stalk of the archegonium (Fig.). The apical cell then divides by three intersecting oblique walls forming three peripheral cells surrounding a central apical cell (Fig.).

   The peripheral cells divide anticlinally into two each. The six jacket cells thus formed give rise to a single layered wall of the venter which becomes double layered by further divisions. In Funaria the neck develops from the derivatives of the axial cell and not by the divisions of jacket cells as in Hepaticopsida.

The axial cell divides transversely into an outer primary cover cell and an inner central cell (Fig.). The latter divides again by a transverse wall into an outer primary neck canal cell and-an inner primary venter (Fig.). The primary neck canal cell undergoes transverse divisions to form a row of neck canal cells which occupy the lower and the median part of the neck (Fig.). At this stage the primary cover cell behaves as an apical cell. It has four cutting faces. The derivatives of the three lateral faces form the jacket of the neck, whereas those of the fourth basal face form a row of neck canal cells. The latter occupy the upper part of the neck. Thus in Funaria the neck canal cells have double origin; those in the lower and the median part of the neck are derived from the primary neck canal cells and those in the upper portion from the basal derivatives of the cover cell. A transverse division of the primary venter cell gives rise to a venter canal cell and an egg (Fig.).

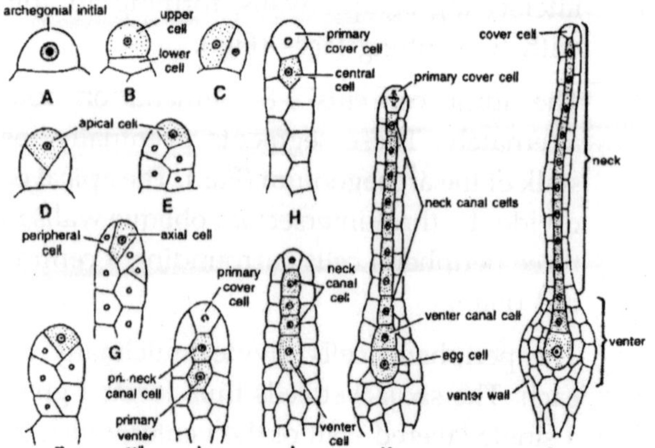

Funaria : Successive stages in the development of archegonium.

(c) *Mature Archegonium* : The mature archegonium has a massive stalk, a swollen venter and an elongated neck. The archegonial jacket is double layered in the venter region and single layered in the neck region. There is a row of six or more neck canal cells in the neck. The venter contains a venter canal cell and an egg.

3. *Fertilization* : Rain or dew water, usually accumulated in antheridial and archegonial clusters, helps in fertilization. The apical cell of the antheridium which forms operculum absorbs water and bursts. The spirally coiled biflagellate antherozoids escape from the antheridium while still enclosed within a vesicle. The vesicle eventually dissolves in water and the antherozoids are set free. At the same time, the neck canal cells and the venter canal cell of the archegonium degenerate and form a mucilagenous mass. The mucilagenous mass absorbs water and as a result the cover cells present at the top of the neck are forced apart. Thus a free passage is formed for the entry of antherozoids.

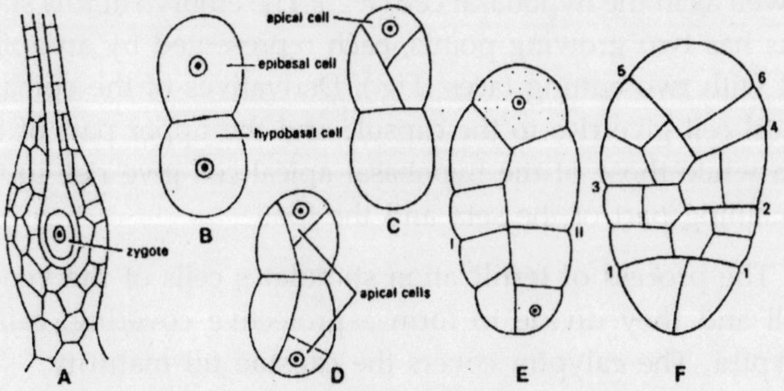

***Funaria* : Development of sporophyte; A. Zygote, B-F. Early stages in the development of sporophyte.**

The transfer of antherozoids from the antheridial head to the archegonial head is brought about by rain drops. The archegonial heads are usually at a lower level than the antheridial heads and as such the water drops containing antherozoids trickle down from the antheridial to the archegonial head. In submerged plants transfer of antherozoids to archegonia is facilitated by water currents. Sometimes transfer of antherozoids is accomplished by microscopic insects.

Entry of antherozoids into the archegonium is due to chemotactic influence of the mucilagenous substances present in the neck. The chemotactic attraction is possibly due to sugars present in the mucilagenous substances. Although a large number of antherozoids enter into the neck of the archegonium, only one of them fuses with the egg nucleus.

Sporophyte : The diploid zygote formed by the fusion of the antherozoid with the egg is the first cell of the sporophytic generation. The zygote soon secretes a wall around itself and increases in size before dividing (Fig.).

Development of Sporophyte : The zygote divides transversely into an upper epibasal and a lower hypobasal cell (Fig.). Both these cells divide by two oblique intersecting walls. As a result a two-sided apical cell is differentiated in the epibasal as well as in the hypobasal cell (Fig.). The embryo at this stage thus has two growing points, each represented by an apical cell with two cutting faces (Fig.). Derivatives of the epibasal apical cell give rise to the capsule and the upper part of the seta while those of the hypobasal apical cell give rise to the remaining part of the seta and the foot.

The process of fertilization stimulates cells of the venter wall and they divide to form a protective covering, called calyptra. The calyptra covers the capsule till maturity.

1. *Development of Capsule :* Development of the capsule can be studied best in serial transverse sections of the young sporophyte. In a transverse section, just below the apical cell, the two derivatives (one each from the

two faces of the epibasal apical cell) form an almost spherical segment (Fig.). Both these derivatives divide by a vertical wall forming a quadrant embryo (Fig.). Each cell of the quadrant divides by a curved anticlinal wall in such a way that a smaller almost triangular cell and a larger more or less rectangular cell is formed (Fig.). Thus at this stage the embryo has eight cells; the four triangular and the four rectangular cells are arranged alternately. Each rectangular cell now divides by a periclinal wall, and at this stage eight peripheral cells, which form the amphithecium, encircle a central core of four cells, the endothecium (Fig.). The amphithecium and the endothecium are the two fundamental embryonic layers of the sporophyte. Cells of both layers divide in a definite sequence and contribute to three regions of the capsule, operculum, theca and apophysis.

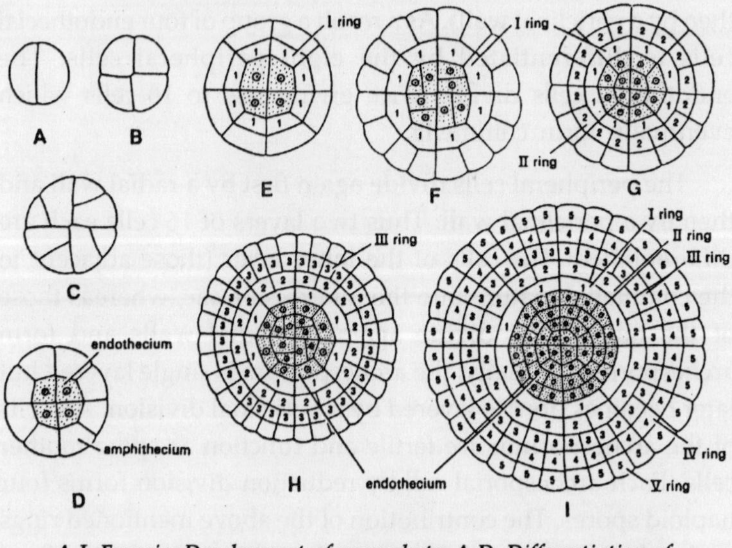

A-I, Funaria: Development of sporophyte; A-D. Differentiation of amphithecium and endothecium, E-I, Formation of five rings of amphithecial cells in the capsular region.

The cells of the amphithecium divide periclinally and form two concentric rings of eight cells each. The inner ring is known as the first ring (Fig.). The cells of the outer ring first divide anticlinally and then periclinally. Thus the outer ring gives rise to two concentric rings of 16 cells each. The inner of these two rings is designated as the second ring. The cells of the outer ring first divide anticlinally forming 32 cells and then periclinally giving rise to two concentric rings of 32 cells each. The inner ring of 32 cells is designated as the third ring and the outer ring again divides periclinally thus forming two rings of 32 cells each. Of these, the inner one is the fourth ring and the outer the fifth ring. In the opercular region, however, an additional ring is formed and thus there are six rings (Fig.).

The four endothecial cells present in the centre repeat the same pattern of divisions that differentiate amphithecium and endothecium (i.e., each cell divides first by a diagonal and then by a periclinal wall). As a result a group of four endothecial cells is differentiated by the eight peripheral cells. The endothecial cells divide twice giving rise to 16 cells which eventually form columella.

The peripheral cells divide again first by a radial wall and then by a periclinal wall. Thus two layers of 16 cells each are differentiated. The cells of the inner layer (those adjacent to the columella) mature into the inner spore sac, whereas those of the outer layer divide again by radial walls and form archesporium. Initially, the archesporium is single layered but later becomes double layered by a periclinal division. All cells of the archesporium are fertile and function as spore mother cells. Each archesporial cell by reduction division forms four haploid spores. The contribution of the above mentioned rings in the development of the opercular region, the fertile theca region and the apophysis region of the capsule is summarised below.

Region	Tissue	Layer	Derivative
Opercular region of young capsule	endothecium —	Central parenchymatous region	
	amphithecium —	1st ring - outer peristomial layer	
		2nd ring - middle peristomial layer	
		3rd ring - inner peristomial layer	
		4th & 5th ring - inner layer of operculum	
		6th ring - epidermis of the operculum and the annulus.	
Theca region of the young capsule	endothecium —	4 central cells - columella	
		inner layer of 16 peripheral arche- sporial cells	inner layer - inner wall of the spore sac.
		peripheral cells outer layer of 16 peripheral arche- sporial cells	outer layer archesporium which forms spore mother cells.
	amphithecium —	1st ring —	outer wall of spore sac
		2nd ring —	trabeculae
		3rd ring —	chlorenchymatous layers
		4th ring —	hypodermis
		5th ring —	epidermis
Apophysis region of the young capsule	endothecium —	central conducting strand	
		4 inner layers —	spongy tissue surrounding the conducting strand
	amphithecium	5th (outermost) layer—	epidermis

Structure of Mature Sporophyte : The mature sporophyte of Funaria is much more complex than that of the Hepaticopsida. It is differentiated into foot, seta and capsule. The foot is a small dagger-shaped conical structure embedded in the archegonial branch. The seta is a long, slender, more or less twisted and stalk-like structure which supports the capsule at its distal end. Internally, the seta is differentiated into a central conducting strand of thin walled elongated cells, surrounded by a cortex of relatively thick walled cells. The conducting strand extends right up to the base of the capsule. It helps in the transport of nutrients and water to the capsule.

The capsule is slightly oblique (asymmetrical). It is erect and green when young, but becomes bright orange and curved at maturity. The complex internal structure of the capsule can be studied in longitudinal section. It has three distinct regions: (i) apophysis, (ii) theca proper, and (iii) apical region.

1. *Apophysis* : It is the basal sterile part of the capsule which connects the capsule with the seta The apophysis gradually narrows down into seta, hence it is also called the neck of the capsule. It consists of a central conducting strand of elongated parenchymatous cells which extend down into the seta. The central strand is surrounded by loosely arranged chlorophyllous cells. The outer most layer of small cells with rather thick walls forms the epidermis. The epidermis is interrupted by stomata through which communication is maintained between the outer atmosphere and the intercellular spaces.

   The presence of chlorophyllous cells in the apophysis shows the photosynthetic capability of the sporophyte. Hence the sporophyte of Funaria is only partially dependent upon the gametophyte.

2. *Theca Proper* : It is the middle part of the capsule which lies between the apophysis and the apical (opercular) region. It is a slightly bent cylindrical structure and constitutes the fertile part of the capsule. It has the following regions.

   (a) *Columella* : It is the central pith-like cylindrical portion of the theca, composed of parenchymatous cells. The distal cone-shaped end of the columella projects into the concavity of the operculum and at the basal constricted end it is connected with the apophysis. The columella provides water and nutrients, required for the development of spores in the spore-sac.

(b) *Spore Sacs* : The columella is surrounded externally by two elongated spore sacs. The outer wall of each spore sac is 3-4-layered, whereas the inner wall is only single layered. In the young sporophyte the spore sacs are packed with many spore mother cells. Each spore mother cell forms four haploid spores. Elaters are absent.

(c) *Air-space* : A large air cavity is present outside the spore sac. It is traversed by delicate filaments of elongated green parenchymatous cells, known as trabeculae. The trabeculae extend from the outer wall of the spore sac to the innermost layer of the capsule wall.

(d) *Capsule Wall* : The capsule wall is composed of many layers of thin-walled parenchymatous cells. The outermost layer forms the epidermis. It is followed by a 2-3-layered hypodermis of well organised colourless parenchymatous cells. Next, there are 2-3 layers of chlorophyllous cells with intercellular spaces, which constitute the photosynthetic tissue of the capsule. The innermost layer of the capsule wall is connected with the outer wall of the spore sac through trabeculae.

3. *The Apical Region* : The apical region of the capsule consists of two important parts, the operculum and peristome. The young sporophyte is covered by calyptra which develops from the cells of the venter wall. As the seta elongates, the calyptra ruptures at the base but its upper part remains at the top of the capsule just like a hood. The hood protects the capsule from drying. It sheds just before the dehiscence of the capsule.

A constriction is present at the junction of the apical region and the theca proper. Immediately below this constriction there is a rim or diaphragm which is composed of 2-3 layers of radially elongated pitted cells. The rim demarcates the upper limit of the theca proper.

The operculum is a conical lid that closes the mouth of the capsule. It is composed of 2-3 layers of thin walled cells. The outermost layer, which forms the epidermis, has cells with thickened outer walls. Few layers of the broader lower (proximal) part of the opercular cap form a ring of large conspicuous cells, the annulus. The annulus is made up of thin walled cells which help in the dehiscence of the capsule.

The peristome lies immediately below the operculum. It is attached beneath the edge of the diaphragm and consists of two rings of long curved triangular teeth. The teeth are simply the strips of cuticle and they are not cellular. In each ring there are 16 teeth, the teeth of the outer ring are thicker, large and red in colour, whereas those of the inner ring are smaller, colourless and delicate. The outer peristomial teeth are also characterised by the presence of transverse thickening bands. The two rings of teeth are arranged in such a way that the outer teeth completely overlap the inner teeth.

Dehiscence of Capsule and Dispersal of Spores : As the sporophyte matures, the water supply of the capsule is cut off. As a result, all tissues of the capsule, except spores, dry up. The thin walled cells of the annulus break and the operculum is thrown away. The spores, however, are not dispersed after shedding of the operculum as the mouth of the theca is covered by peristome. The outer peristomial teeth are hygroscopic and in dry atmosphere they bend outwards with jerky movements, but the inner peristomial teeth remain in their position.

Due to outward movements of the outer peristomial teeth, slits between the inner thin walled peristomial teeth become wider and spores escape through these slits gradually. In a wet atmosphere the outer peristomial teeth absorb moisture and bend inwards and thus closing the slits and prevent the esçape of spores. In Funaria, there is a built-in mechanism which allows the dispersal of spores only when the conditions for germination of spores are favourable. The seta of the mature sporophyte also exhibits hygroscopic movements. In dry weather, by losing water, it twists and bents, thus helping in dispersal of spores.

Young Gametophyte : The spore is the first cell of the gametophytic generation. It is small, more or less round and measuring 12-20μm in diameter. The spore wall consists of two layers, the outer smooth and brown exosporium and the inner colourless endosporium. The cytoplasm of the spore is rich in oil globules and chloroplasts (Fig.).

The spore germinates only under favourable conditions.. The exospore is ruptured and the endospore protrudes as a papilla (Fig.). It soon forms a long filamentous protonema (Fig.). The protonema shows apical growth and forms two types of branches, the chloronemal branches and the rhizoidal branches.

The chloronemal branches are characterised by the presence of chloroplasts in their cells; they are either prostrate growing along the substratum or may become upright. The rhizoidal branches have oblique cross walls and grow down into soil. They have under-developed chloroplasts, but become green when come on the surface of soil. The rhizoidal branches absorb water and minerals from soil (Fig.).

The chloronemal branches give rise to many minute buds. Each bud grows into an erect leafy gametophore (Fig.). Thus many gametophores are formed on each chloronemal branch. The gametophyte consists of leafy stems, rhizoids and protonema. The protonema of moss, which represents the juvenile stage of the gametophyte, is capable of an unusually prolific vegetative propagation.

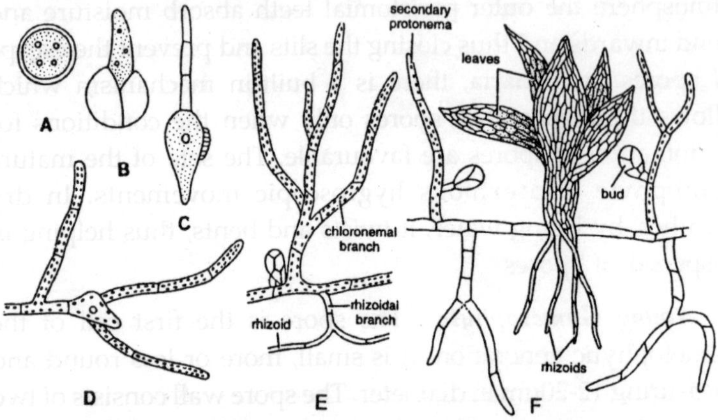

A-F. Funaria : Spore germination; A. Spore, B. Germinated Spore, C-E. Successive stages in the formation of primary protonema. F. Secondary protonema and formation of young gametophyte.

Polytrichum

Systematic Position

Class	—	Bryopsida
Sub-class	—	Bryidae
Order	—	Polytrichales
Family	—	Polytrichaceae

The genus Polytrichum, with nearly 100 species, is widely distributed in temperate and tropical regions of the world. The plants grow on a variety of habitats such as wet sandy banks of rivers and ponds, on branches of trees under shade and on fully exposed rocks. They often form a green carpet on moist and shaded walls. In India, the genus is represented by five species, viz., Polytrichum densifolium, P. juniperinum, P. xanthopilum, P. alpinum and P. formosum. These species are commonly found in the Himalayan ranges.

Gametophyte

External Structure : The mature gametophyte of Polytrichum is differentiated into two parts : a horizontal underground rhizome and an erect leafy shoot (Fig.).

The rhizome is covered by leaves and fluffy rhizoids. The leaves are small, scale-like and usually brown or colourless. They are arranged in three vertical rows (in 1/3 phyllotaxy). The rhizoids are long, thread like, branched and multicellular with oblique septa. Generally a large number of rhizoids coil or twist together and form thick rope-like structures (Fig.). The surface of the rhizoids, as well as the capillaries formed by the twisted rhizoids, absorb water. It is because of this reason that Polytrichum can withstand relatively dry habitats. The rhizoids also provide mechanical support to the gametophores. Buds and gemmae are present on the rhizoids and they help in vegetative propagation.

The erect leafy shoots arise from the underground rhizome. They are well organized and attain a height up to 20 cm (e.g., P. commune). They are branched or unbranched. The branches arise from the primordia present just below the leaves (they do not develop in the axil of leaves like normal branches). Although branch primordia are formed at the base of every 12th leaf on the rhizome and also on the aerial shoots, the primordia on the aerial shoots are usually dormant. The shoot grows by means of an apical cell with three cutting faces.

A short transition zone is present between the rhizome and the aerial shoot. It bears small brown leaves, similar to those present on the rhizome.

The leaves on aerial shoots are different from those present on the rhizome and the transition zone. These are relatively large, and green and are arranged in 3/8 phyllotaxy. The leaf is differentiated into a sheathing leaf-base and an apical limb (Fig.). The former is broad, colourless, unistratose and membranous structure, closely clasped around the stem. The limb is narrow, lanceolate and green. It has a distinct midrib flanked on either side by thin unistratose wings. The margins of the limb are entire (e.g., P. juniperinum) or dentate (e.g., P. commune).

The basal sheathing part of the leaf grows by an intercalary meristem, whereas the growth of the limb is accomplished by a 2- sided apical cell.

Internal Structure

1. *Rhizome* : In most of the species the rhizome is roughly triangular in outline with rounded corners, but sometimes the basal part of the rhizome is more or less circular. A transverse section of the rhizome shows the following regions.

   (a) *Epidermis* : It is the outermost layer of thick walled cells. Several epidermal cells give rise to rhizoids.

   (b) *Cortex* : The epidermis is followed by the cortex consisting of 3-4 layers of thin walled parenchymatous cells. The cortex is interrupted by three hypodermal strands which extend radially from the periphery towards the centre. These strands lie opposite the three ridges of the triangular rhizome. They are composed of prosenchymatous cells with pointed ends, and these cells contain starch grains. The hypodermal strand gradually narrows down towards the centre of the rhizome. If consists of thin walled cells of relatively greater diameter. These thin walled cells, together with the hypodermal strands, are called radial strands.

   Next to the cortex, there is a layer of radially elongated large cells with suberized thickening on their radial and horizontal walls. This layer can be compared with the endodermis of higher plants. Due to the presence of radial strands this layer breaks into three arch-shaped strips.

(c) *Pericycle :* A 2-3- layered parenchymatous pericycle is present in some species between the endodermis and the central conducting strand. The pericycle is considered to be of primitive type and it is also discontinuous like the endodermis.

(d) *Leptoids :* These are thin walled polygonal cells present in furrows opposite to the radial strands. These cells are elongated like sieve-tubes of phloem, hence called leptoids. These are living cells with oblique end walls and are connected with each other through plasmodesmata. The leptoid cells collectively form leptom.

(e) *Amylom :* The innermost layer of leptoids is separated from the central cylinder by a single layer of parenchymatous cells containing starch. This layer is called amylom.

(f) *Central Cylinder :* It is a trilobed structure situated in the centre of the rhizome. It is made up of two types of cells, the steroids and the hydroids. The steroids are thick walled, elongated and living cells, which constitute most part of the central cylinder. The hydroids, on the other hand, are empty dead cells interspersed in between steroids in groups of two or three. The steroids and the hydroids are collectively called hydrom.

2. *Aerial Shoot :* In a transverse section the aerial shoot is irregular or lobed in outline due to the presence of persistent leaf bases. Internally, it is differentiated into the following three regions.

(a) *Epidermis :* It is the superficial uniseriate layer which becomes discontinuous due to the presence of persistent leaf bases. In a mature shoot, epidermis is not distinct.

(b) *Cortex :* The epidermis is followed by a broad cortex. The outer region of the cortex is composed of compactly arranged elongated prosenchymatous cells and the inner cortex is parenchymatous. In young shoots, the cortical cells contain chloroplasts. A large number of leaf traces are also present in the cortex. The leaf traces have a central core of hydroids enveloped by a sheath of parenchyma. They join the central cylinder of the shoot.

(c) *Central Cylinder :* The outermost layer of the central cylinder is discontinuous and is composed of parenchymatous cells. It represents rudimentary pericycle. The cells of this layer have large number of starch grains. It is followed by a 3-4- layered irregular zone of elongated thin walled cells. This zone constitutes leptom mantle. It is regarded as equivalent to the phloem of vascular plants. The leptom mantle is followed by one or two layers of suberized cells which form hydrom sheath. The cells of the hydrom sheath are rich in starch, hence it is also called amylom layer. Next to hydrom sheath is hydrom mantle, which is composed of thin walled empty cells. The center of the shoot is occupied by a hydrom cylinder, made up of thick walled cells. These cells are meant for conduction of water, hence the hydrom cylinder can be compared with the xylem of higher plants.

3. *Leaf :* A transverse section of the limb of the leaf shows a thick midrib, gradually tapering into two rudimentary wings. The ventral (lower) surface is bound by a distinct epidermis. The outer walls of the epidermal cells are relatively thick and are covered with a cuticle. The

epidermis is followed by a sclerenchymatous hypodermis. The ground tissue of the limb is mainly composed of parenchymatous cells. Some small thick walled cells are also found scattered in the ground tissue, either singly or in small groups.

The dorsal (upper) surface of the limb is made up of vertical plates of cells, called lamellae. Each lamella has 4-8 cells; all cells contain chloroplasts except the terminal one. The ultrastructure of these cells is similar to that of a typical chlorenchymatous cell. The terminal cell of the lamella is hyaline and larger than the other cells. Sometime it is bifid (P. commune) or papillose (P. juniperinum).

The arrangement of chlorophyllous cells in lamellae and the presence of air spaces between adjacent lamellae increase the photosynthetic area and thus the efficiency of carbon assimilation. These structural peculiarities compensate for reduced wings. The wing is usually composed of a single layer of hyaline cells. Lamellae are not present in the wing region. The wings become curved when plants grow in dry habitats.

Reproduction : In Polytrichum reproduction takes place by vegetative and sexual methods.

Vegetative Reproduction : Vegetative reproduction takes place by means of bulbils which develop on rhizoids. Besides, fragmentation of underground rhizome, which produces aerial shoots, also helps in propagation.

Sexual Reproduction : Polytrichum is dioecious, the male (antheridia) and the female (archegonial sex organs develop at the apices of main shoots of separate plants.

1. *Antheridium :* Antheridia are borne at the apex of the male gametophore. They are surrounded by a whorl of specialized leaves, called perigonial leaves. The perigonial leaves differ from the vegetative leaves in

their colour and shape. They have broad sheathing bases and short bristle like apices. They are red or brown in colour and form a cup-like structure around the antheridia. Antheridia are produced in groups at the base of the perigonial leaf in the position of lateral bud. The apical cell of the shoot does not contribute to the development of antheridia, therefore, vegetative growth of the male gametophore continues in the following year. Such growth pattern of gametophore beyond the antheridial head is known as proliferation. It is repeated several times.

The mature antheridium is a club-shaped structure, measuring about 1.5 mm in length. It has an elongated body and a short stalk. The body of the antheridium has a single layered jacket of sterile cells, enclosing a mass of androcytes. At maturity, each androcyte metamorphoses into a biflagellate antherozoid. An operculum is present at the tip of the antheridium, consisting of a single large cell or a group of cells. The development of antheridium is similar to that of Funaria.

In the antheridial head, there are also a large number of sterile multicellular hairs, known as paraphysis. The paraphysis are usually simple filamentous structures, but in some species their apical part becomes spathulate (broad spoon-like).

2. *Archegonium :* Archegonia are borne in groups at the apex of the female gametophore. The number of archegonia in a group is 3-6 depending on the species; for example, in P. commune there are three archegonia in a group. The archegonia are also surrounded by an involucre of coloured perichaetial leaves. Some multicellular hair-like structures, known as paraphysis, are also present with archegonia. The apical cell of the

female gametophore acts as archegonal initial thus arresting the further growth of the gametophore. The development of archegonium is similar to that of Funaria.

The mature archegonium is a flask-shaped structure, composed of an elongated neck, a venter and a massive stalk. The neck consists of six vertical rows of neck cells which enclose a narrow cavity, the neck canal. There are about ten neck canal cells. The venter contains an egg and a relatively small venter canal cell.

3. *Fertilization :* Water is essential for fertilization. The jacket cells of the mature antheridium swell by absorbing water. The hydrostatic pressure thus generated bursts the operculum and the antherozoids are liberated. They swim in water held in the antheridial head in between the perigonial leaves and the paraphysis. Almost at the same time the neck canal cells and the venter canal cell of the mature archegonium degenerate forming a mucilagenous mass. The mucilagenous mass swells up by absorbing water and as such the cover cells present at the tip of the archegonium are pushed apart. The antherozoids are transferred to the nearby archegonial head by splashing rain water or small insects. They are attracted towards the archegonia by the mucilagenous substance present in the neck. A number of antherozoids swim down the archegonial neck but only one of them fuses with the egg and thus oospore is formed.

Sporophyte

Development of Sporophyte : The diploid oospore is the mother cell of the sporophytic generation. It soon increases in size and secretes a wall. A transverse wall divides the oospore into an upper epibasal and a lower hypobasal cell. Each of these two daughter cells divide by two intersecting walls and as such an apical cell with two cutting faces is differentiated in both the epibasal and hypobasal cell. Thus the young embryo has two growing points. The derivatives of the upper apical cell give rise to the capsule and the distal part of the seta and those of the lower apical cell form the foot and the proximal part of the seta. The cells destined to form capsule divide periclinally and form an outer amphithecium and an inner endothecium. The amphithecium gives rise to the jacket of the capsule, whereas the endothecium to the columella and the archesporium.

Structure of Mature Sporophyte : The sporophyte of Polytrichum is differentiated into a foot, a long seta and an angular capsule.

1. *Foot :* It is a dagger-shaped structure embedded in the female gametophore. It is composed of thin-walled parenchymatous cells. It acts as an anchoring and absorbing organ.
2. *Seta :* It is a long, slender structure between the foot and the capsule. The length of the seta varies in different species; for example, in P. commune the seta is approximately 5 cm in length and 0.5 mm in diameter. Anatomically, the seta is differentiated into a single layered epidermis, a sclerenchymatous hypodermis, a parenchymatous cortex and a central cylinder. The latter is composed of compactly arranged thin walled hydroid cells. The main function of the seta is to raise the capsule to the required height and conduction of water and nutrients.
3. *Capsule :* The capsule is differentiated into the following three parts.

   (a) *Apophysis :* The distal end of the seta gradually enlarges to form a bulbous apophysis at the base of the capsule. This sterile part is marked off from the fertile theca of the capsule by a distinct constriction. The apophysis has a thick walled epidermis continuous with that of the seta. The epidermis is interrupted by stomata; each stoma has two guard cells. The epidermis is followed by a broad chlorenchymatous cortex with intercellular spaces. It serves as photosynthetic tissue. The central part of the apophysis is occupied by a conducting strand continuation with columella and seta.

(b) *Theca :* It is the middle fertile part of the capsule. It is polygonal in outline due to the presence of many longitudinal grooves. The wall or jacket of the theca is composed of several layers of chlorophyllous cells. The outermost wall layer forms the epidermis which is devoid of stomata. The outer wall of the epidermal cells is relatively thick.

An air space or lacuna (outer air space) is present inner to the wall layers. It is divided into small chambers by transverse filaments of chlorophyll containing cells, the trabeculae. One end of the trabeculae is connected with the inner wall layer and the other with the outer wall of the spore sac. The outer space is followed by spore sac, which is bounded on either side by a layer of thin walled cells. The fertile cells present inside the spore sac constitute archesporial tissue. Initially the archesporium is single layered, but it becomes many layered in later stages of development of the sporophyte. The last generation of archesporial cells differentiates into spore mother cells. Each spore mother cell forms four haploid spores.

The spore sac is also surrounded on the inner side by an inner air space or lacuna. It is also traversed by many transverse trabeculae which are connected with the inner spore sac wall on the outer side and the columella on the inner side.

The central part of the capsule is occupied by a thick column of parenchymatous cells, the columella. This is a sterile tissue which is continuous with the central axis of the apophysis. The upper part of the columella is in contact with the epiphragm.

(c) *Operculum* : It is the apical part of the capsule which forms a cap-like structure at the apex of the theca. It has an extended proximal part, called rostrum and a projected beak-like distal part. The rostrum is attached to the mouth of the theca. The boundaries of the operculum and theca are marked by a distinct constriction, the rim. In Polytrichum, annulus is absent.

The distal end of the columella is expanded into a pale thin walled membranous structure, the epiphragm. It is stretched like a drum-head (tympanum) across the capsule mouth. In the mature capsule the peristome is composed of 32 or 64 peristomial teeth arising from the periphery of the diaphragm. Each tooth is a small, solid and pyramidal structure, made up of several layers of thick walled fibre-like cells. The peristomial teeth are directed upward and are united by their tips with the margins of the epiphragm. They are not hygroscopic like those of Funaria.

Dehiscence of Capsule : At maturity the capsule shrivels as it dries up. The columella disintegrates and the spores come to lie in the cavity thus formed. Further drying and shrivelling of the capsule wall causes the operculum to fall off and thus exposing the peristome. At this stage the capsule is horizontally placed. After exposure of the peristome, the thin walled cells of the epiphragm lying between the peristomial teeth also dry up. Thus several minute holes are formed in the margins of the epiphragm. The spores are dispersed through these holes. Only a few spores are liberated everytime the capsule sways in the wind. This method of spore dispersal is known as censor mechanism.

Young Gametophyte

The Spore : The spores are small rounded structures, measuring 0.005 - 0.01 mm in diameter. They are yellow but turn green immediately after dispersal. The spore wall is differentiated into an outer exospore and an inner endospore. Both wall layers are thin and smooth. In spores the accumulated food is in the form of oil globules. The spores remain viable for a very long period.

Germination of Spores : When spores fall on a suitable substratum, they swell by absorbing water. Many chloroplasts develop in the cytoplasm and the spores become green. The exospore ruptures and the endospore comes out in the form of one or more germ tubes. The germ tube grows rapidly and forms a septate and branched protonema, which grows by an apical cell. The upright branches of the protonema are green and bear small buds at their bases. Each bud grows into an erect leafy gametophore with the help of a pyramidal apical cell. The underground branches or protonema are colourless and form rhizoids.

Pogonatum

Systematic Position

Class	—	Bryopsida
Sub-class	—	Bryidae
Order	—	Polytrichales
Family	—	Polytrichaceae

Pogonatum, the largest genus of the family Polytrichaceae, is distributed in almost all parts of the world. In India, the genus is represented by 34 species, growing abundantly in the Himalayan ranges. Pogonatum microstomum, P. aloides, P. stevensii and P. perichaetiale are some common Indian species.

Gametophyte

External Structure : The gametophyte of Pogonatum is perennial. The adult plant has a basal rhizome-like structure which gives rise to many aerial leafy shoots on the upper side and rhizoids on the lower side. The aerial shoots form thick mat-like cushions on moist rocks, walls and tree trunks.

The rhizoids are multicellular and thick walled, and they have oblique septa. Usually many rhizoids are twisted together to form rope-like thick branches.

The aerial shoot is composed of a central axis, called stem. It may be branched or unbranched and is covered with spirally arranged leaves. The basal part of the leaf is broad, sheathing and somewhat pale in colour and the upper part is extended into a narrow limb of deep red or brown colour with serrate margins. In the basal part of the stem the leaves are small and scaly, whereas in the upper part they are relatively large. The leaf has a thick midrib (which occupies most part of the leaf) and two laterally extended thin and narrow wings or lamina.

Internal Structure

1. *Stem (axis) :* A transverse section of the stem is irregular in outline. It is surrounded by a single layer of epidermis, composed of thick walled cells. The epidermis is discontinuous as it is interrupted by leaf bases. It is followed by a broad cortex, differentiated

into two distinct zones; the outer cortex is dark coloured and is composed of 3-5 layers of collenchymatous cells and the inner cortex is a broad zone of thin-walled parenchymatous cells. The cortical cells are rich in starch grains. The outer region of the inner cortex has many leaf traces, each represented by a small group of thick walled cells. The leaf traces connect the midrib of the leaf to the central hydrom of the stem. Next to cortex is a zone of elongated cells, known as leptoids.

The sheath of leptoid cells forms the leptoid mantle. Although the leptoids are comparable to the sieve cells of the higher plants, they are devoid of sieve plates and sieve areas. The central part of the stem is called hydrom cylinder. It is composed of two types of cells, elongated thin walled hydroids and comparatively thick walled steroids. The former are living, whereas the latter are dead cells.

2. *Leaf* : The midrib forms the major part of the leaf. The upper (dorsal) surface of the midrib is covered by parallel vertical plates of cells, known as lamellae. Each lamella is 4-8 cells in height and all cells, except the terminal one, contain chloroplasts. The terminal cell is hyaline and usually larger than the other cells. In P. microstomum the terminal cell splits into two conical cells. The central part of the midrib is occupied by two types of cells. The thick walled xylem-like cells have narrow lumen and the thin walled phloem-like cells have broader lumen. The lower (ventral) epidermis is distinct and is composed of a single layer of thin walled cells.

Reproduction : In Pogonatum reproduction takes place by vegetative and sexual methods.

Vegetative Reproduction : Vegetative propagation takes place by means of special bud-like gemmae which develop on rhizoids. Besides, secondary protonema may develop from any part of the gametophyte.

Like primary protonema, it also bears multicellular buds which develop into leafy gametophores. In Pogonatum protonema may also develop from any tissue of the sporophyte other than spores. The gametophytes, which develop on such protonema, are diploid. This phenomenon is known as apospory.

Sexual Reproduction : Most of the species of Pogonatum are dioecious but a few (e.g., P. microstomum) are monoecious. The development of sex organs is essentially similar to that of Funaria.

1. *Antheridial Head :* Antheridia develop in groups at the tips of the male branches. They are surrounded by an involucre of red or orange coloured special leaves, the perichaetium. This gives the antheridial head a flower-like appearance. Many sterile hair-like multicellular paraphysis are also present in the antheridial head. Since antheridia develop in the position of the lateral buds, the apical cell of the stem is not utilized and hence the apical growth of the stem is not affected.

   The mature antheridium has a club-shaped body and a multicellular stalk. The body has a single layered jacket of sterile cells enclosing a mass of fertile cells, the androcyte mother cells. Each androcyte mother cell divides to form two androcytes. The androcyte metamorphoses into a curved biflagellate antherozoid.

2. *Archegonial Head* : Archegonia are borne at the tips of the female branches in groups of 3-6. They are surrounded by sterile paraphysis and perichaetial

leaves. In the development of archegonia the apical cell of the stem is utilized and thus the growth of the stem ceases after the formation of archegonia. There is no proliferation of vegetative shoot.

The mature archegonium is a shortly stalked structure, differentiated into a massive venter and an elongated neck. The archegonial jacket is double layered in the venter region and only single layered in the neck region. The venter contains a large egg and a relatively small venter canal cell. The neck is composed of six vertical rows of neck cells. The narrow neck canal contains 6-9 neck canal cells.

3. *Fertilization* : Just before fertilization the rows of neck cells of the archegonium split apart. Almost at the same time the neck canal cells and the venter canal cell disintegrate to form a mucilagenous substance which attracts freely swimming antherozoids towards the neck of the archegonium. One antherozoid fuses with the egg and then oospore is formed. Although more than one archegonia on an archegonial branch may be fertilized, usually only one sporophyte reaches maturity.

Sporophyte **:** The diploid oospore is the mother cell of the sporophytic phase. It gradually increases in size and secretes a wall around itself. It divides by a transverse wall into an upper epibasal and a lower hypobasal cell. Both these cells divide by two successive diagonal walls in such a way that a wedge-shaped apical cell is differentiated in the epibasal as well as in the hypobasal cell. Thus two growing points in opposite directions are established in the young embryo. Although both apical cells form derivatives on two lateral sides, divisions are more regular in the epibasal apical cell. Derivatives of the upper apical cell form the capsule and the upper part of the seta, whereas those of the lower apical cell give rise to the basal part of the seta and the foot.

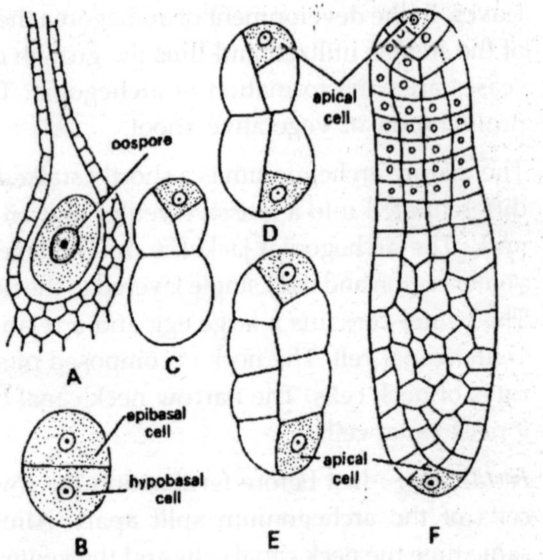

***Pogonatum* : Development of sporophyte; A. Zygote, B.F. Early stages in the development of sporophyte.**

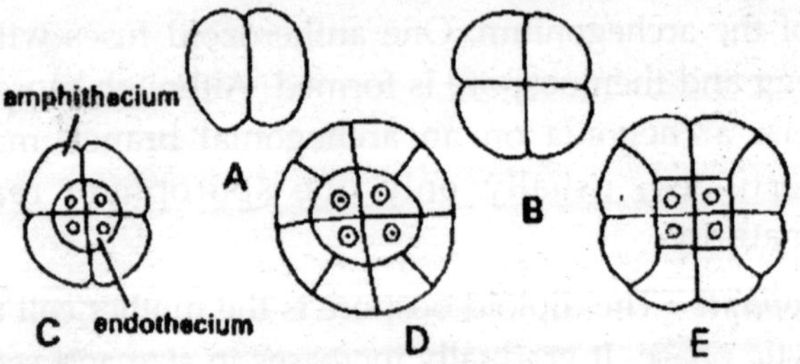

***Pogonatum* : Transection of early stages in the development of sporophyte.**

Development of Capsule

A transverse section of the young embryo, just below the apical cell, shows two cells of almost same shape and size (Fig.). These cells are the derivatives of two lateral faces of the epibasal apical cell. Both these cells divide by a transverse wall, thus forming four-celled (quadrant) embryo (Fig.). The cells of the quadrant divide periclinally, the outer four cells form the amphithecium and the inner four the endothecium

(Fig.). Each cell of the amphithecium divides by an anticlinal wall and thus at this stage the 4-celled endothecium is surrounded by the 8-celled amphithecium (Fig.).

The amphithecium and endothecium are the two fundamental embryonic layers which give rise to different parts of the capsule. Both these layers undergo regular divisions. The amphithecial cells divide anticlinally as well as periclinally, giving rise to seven concentric layers. The cells of the endothecium, on the other hand, divide in all possible planes.

Differentiation of Various Parts of Capsule : The three outer layers of the amphithecium form the capsule wall. As the development of capsule proceeds, these three layers undergo further periclinal and anticlinal divisions, giving rise to a 4-6 layered capsule wall. In the meantime an air space, called outer air space, develops inner to the capsule wall. The cells of the fourth and fifth amphithecial layer extend radially into the outer space, forming trabeculae. The two innermost layers of the amphithecium (sixth and seventh layer) form the outer wall of the spore sac.

The central part of the capsule develops from the endothecium. The peripheral layer of the endothecium gives rise to archesporium, which constitutes the fertile part of the capsule. The inner wall of the spore sac develops from the two layers of the endothecium next to the archesporium. An inner air space, similar to outer air space, develops towards the inner side of the spore sac. This space is also traversed by trabeculae, which develops from the next two layers of the endothecium. The remaining layers of the endothecium form columella. The opercular region of the capsule develops from the derivatives of the amphithecium and endothecium.

Structure of Mature Capsule : The capsule of Pogonatum is differentiated into three parts : (i) the conical apical part, known as operculum or lid, (ii) the middle urn-shaped fertile part, the theca, and (iii) the basal rudimentary sterile part, the apophysis.

1. *Opercular Region* : It is the apical beaked cap-like structure attached to the distal end of theca by a ring-like diaphragm. There is no organised annulus as in Polytrichum. The peristome lies below the operculum and it consists of 32 (e.g., P. microstomum, P. stevensii) or 16 (e.g., P. perichaetiale) peristomial teeth. The teeth are solid pyramidal structures, made up of bundles of fibrous cells. They are attached to the margins of the epiphragm, a thin membrane stretched across the mouth of the capsule just like a drum-head. The epiphragm is the extension of the distal end of columella.

2. *Theca* : It is the urn-shaped fertile region of the capsule. In a transverse section it appears oval or elliptical in outline. The wall of the theca is 4-5 layered. The outermost layer is the epidermis, which has papillate cells. There are no stomata in the epidermis. The wall layers are followed by an outer air space which is traversed by many trabeculae. The trabeculae extend radially from the inner most wall layer to the outer wall of the spore sac. The spore sac is a cylindrical structure bounded by a two cell thick wall. The young spore sac contains archesporium (undifferentiated fertile cells), but the mature spore sac is full of spore tetrads. An air space (inner air space) is also present on the inner side of the spore sac and it is similar to that of the outer air space.

   The central part of theca is composed of sterile tissue and is known as columella. Besides providing mechanical support to the capsule, the columella also helps in the conduction of water and nutrients.

3. *Apophysis :* It is the basal sterile part of the capsule, composed of parenchymatous cells. The proximal end of apophysis is in continuity with seta while the distal end expands to form the theca.

Dehiscence of the Capsule : At maturity of the capsule, the calyptra (forming a protective covering around the capsule) falls off and the operculum is exposed. Some thin walled cells present at the base of the operculum also dry up and consequently the operculum drops off exposing the peristome. The peristomial teeth of Pogonatum are not hygroscopic. Spores are liberated through the minute holes in the epiphragm, which is present in between the peristomial teeth. The discharge of spores from the spore sac is regulated by upward and downward movements of the epiphragm.

Young Gametophyte

Spores : The spores of Pogonatum are small, spherical and smooth walled (Fig.). The spore wall is differentiated into an outer exospore and an inner thin endospore. The cytoplasm of the spore contains many chloroplasts, oil globules and a single nucleus. The spores are haploid and function as mother cells of the gametophytic generation.

Germination of Spores : The spores usually undergo a period of rest prior to germination. Under favourable conditions, after landing on a suitable substratum, they absorb water and increase in size. Then the exospore ruptures and the endospore comes out in the form of one, or more green filamentous germ tubes (Fig.). The germ tube becomes septate and forms filamentous primary protonema. It represents the juvenile stage of the plant. The primary protonema grows by means of an apical cell and forms two types of branches :

(i) chloronemal aerial green branches — these are erect aerial branches, and

(ii) rhizoidal branches — these are non-chlorophyllous branches which grow below the surface of the soil (Fig.). The adult leafy gametophores arise from the lateral buds on the chloronemal aerial branch.

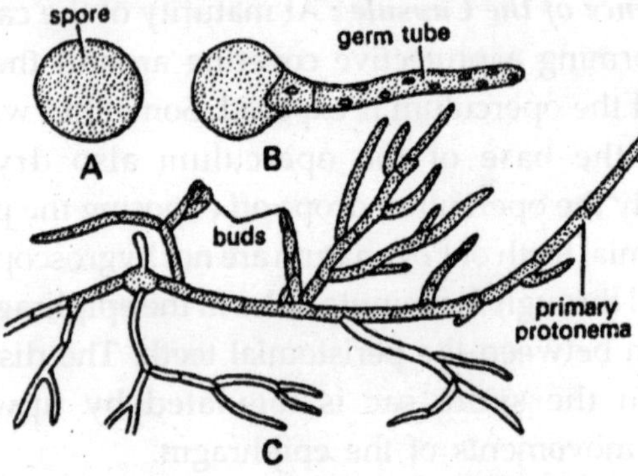

A-C. Pogonatum : Spore germination; A. Spore, B. Germinated spore, C. Protonema.

6

The Bryopsida

The class Bryopsida is the largest group of bryophytes. It includes 660 genera and 14,500 species which occur in almost all parts of the world. They are abundant at altitudes ranging from 1,250 to 2,500 meters. Some species also occur up to an altitude of 6,500 meters. They are better adapted to terrestrial habitats than liverworts, and grow on moist soil, rocks, damp walls of old buildings and on the stem of trees in tropical forests. Tortula desertorum occurs in the deserts of Transcapsia. Some species of Buxbaumia are saprophytes and thrive on decaying wood.

General Characters of Bryopsida

(1) The gametophyte has two growth stages: (i) protonema stage — it is the juvenile stage represented by prostrate, creeping, green and branched filamentous structure; it develops from the spore and is only a transitory vegetative stage; and (ii) leafy stage or gametophore-an erect cylindrical shoot with persistent leaves and sex organs.

(2) The main axis of the gametophore may be branched or unbranched. The branches always arise below the leaves.

(3) The leaves are simple, minute, sessile and usually with a distinct midrib. They are only one cell in thickness.

(4) The rhizoids are multicellular, branched and have oblique septa.

(5) The axis is differentiated into a central conducting tissue and a peripheral cortex.

(6) Vegetative multiplication takes place by fragmentation, stolons, branching of protonema, special leafy shoots, gemmae and persistent apices.

(7) The sex organs are borne in groups at the apices of the branches. The plants are dioecious or monoecious.

(8) The antheridia are club-shaped, narrow and elongated. The antheridial jacket is single layered.

(9) The male gametes (antherozoids) are biflagellate and spirally coiled.

(10) The archegonia are stalked with a much elongated neck and a massive venter.

(11) The sporophyte is usually differentiated into foot, seta and capsule. The elongated seta raises the capsule much above the gametophyte to facilitate the dispersal of spores.

(12) The capsule usually has a peristome which helps in the dispersal of spores.

(13) The spores, on germination, first form protonema, and the leafy gametophore develops from the protonema.

Classification of Bryopsida

Although a few groups of mosses are quite different and stand apart but a great majority are sufficiently alike in structure. This naturally poses a real difficulty in their classification.

Bower (1935), Wettstein (1935) and Campbell (1940) recognised three orders, Sphagnales, Andreaeales and Bryales in the class Bryopsida.

Smith (1938, 1950) divided Bryopsida into three sub-classes, Sphagnobrya, Andreaeobrya and Eubrya.

Reimers recognised the following five sub-classes in the class Musci. His classification was published in Engler's Syllabus der Pflanzenfamilien in 1954.

Class	*Musci*	
Sub-class	Sphagnidae	
Order	Sphagnales	
Sub-class	Andreaeidae	
Order	Andreaeales	
Sub-class	Bryidae	
Order	1. Archidiales	2. Dicranales
	3. Fissidentales	4. Pottiales
	5. Grimmiales	6. Funariales
	7. Schistostegales	8. Tetraphidales
	9. Eubryales	10. Isobryales
	11. Hookeriales	12. Hypnobryales
Sub-class	Buxbaumiidae	
Order	Buxbaumiales	
Sub-class	Polytrichidae	
Order	1. Polytrichales	2. Dawsoniales

The classification of Smith (1950) has been followed here.

Sub-class Sphagnobrya

Sphagnobrya is considered to be the most primitive sub-class of Bryopsida. It is characterized by the following features.

(1) The protonema is flat, thallose and one cell thick. It is attached to the substratum by multicellular rhizoids which have oblique septa. Erect leafy gametophore arises from the protonema.

(2) The leaves are composed of two types of cells : (i) narrow, living and green, and (ii) large, hyaline and dead cells. A distinct midrib is lacking.

(3) The antheridia are borne in the axils of leaves on antheridial branches.

(4) The archegonia are borne at the apices on the archegonial branches.

(5) The sporogenous tissue develops from the amphithecium and the endothecium forms columella only.

(6) The capsule has an operculum but peristomial teeth are absent.

(7) The sporophyte is borne on a slender and leafless pseudopodium. It is analogous to seta present in the advanced members of Bryopsida.

The sub-class Sphagnobrya has only one order, Sphagnales, with a single monotypic family, Sphagnaceae.

***Sphagnum* (bog moss or peat moss)**

Systematic Position

Class	—	Bryopsida
Sub-class	—	Sphagnobrya
Order	—	Sphagnales
Family	—	Sphagnaceae

The genus Sphagnum, commonly known as bog moss or peat moss, includes over 340 species, distributed in almost every part of the world. They are aquatic or semiaquatic plants, growing in swamps, ponds and moist habitats such as along the banks of lakes, wet hill sides, etc. The species of Sphagnum are usually perennial; the upper part of the gametophore grows indefinitely while the lower portion dies progressively. They produce certain acidic substances which render the medium unsuitable for the growth of fungi, bacteria and other pathogens. This slows down the decaying process of the dead parts of the plants. Consequently, a large amount of dead material is added year after year and this partially decomposed material accumulates in the form of a compact dark substance which is rich in carbon. This is known as peat and hence the name peat moss is given to these plants..

The genus is represented in India by 20 species; of these Sphagnum fimbriatum, S. teres, S. khasianum, S. cuspidatum, S. palustre, S. robustum, S. squarrosum and S. ovatum are common.

Gametophyte

The gametophytic phase consists of two distinct stages : (i) juvenile protonema, and (ii) mature leafy or gametophore stage. On germination, the spore gives rise to a thin plate-like irregular juvenile protonema which eventually forms upright leafy axis, the gametophore.

The plants grow in dense clumps. They are always loaded with water and glow with many bright colours (e.g., deep red in S. rubellum and rose-pink in S. plumulosum) due to the presence of anthocyanin pigments.

There are many economic uses of Sphagnum. It has a remarkable water holding capacity, hence it is widely used in packing live plant tubers, bulbs, seedlings, etc., for transportation. It is also used in seedbeds and in moss-sticks.

It has some antiseptic property and was used for filling absorbent bandages in surgical dressings during the World War I.

Sphagnum has a great role in plant succession. Lakes and ponds inhabited by it gradually become unsuitable for the growth of hydrophytes as it multiplies vegetatively and eventually fills the whole area. Such habitats are occupied by mesophytes.

External Structure of Mature Gametophore : The gametophore is differentiated into an upright branched axis and leaves.

1. *Main Axis and Branches :* The main axis has indeterminate growth. It attains a length of several meters in submerged species, but is relatively short in terrestrial forms due to progressive death of the older portions. It is soft and weak when young, but becomes erect and stout (diameter up to 1.2 mm) at maturity. The branches arise single or in tufts of 3-8 from the axil of every fourth leaf on the main axis. Many small branches of limited growth, present at the apex of the main axis, form a compact head, called coma. Infact coma is formed due to condensed growth of the internodes of the main axis near the apex. As the internodes elongate, the branches forming coma resume their normal growth.

   In submerged species (e.g., S. obesum, S. cuspidatum) all branches are alike, but the terrestrial species have two types of branches : (i) Pendent branches — these branches turn downwards and then grow parallel to the main axis. They are also known as flagelliform or decurrent branches, (ii) Divergent branches — these branches are directed outwards and upwards. They are smaller but stouter than the pendent branches. They are also called excurrent branches.

2. *Leaves :* The leaves are borne on the main axis as well as on the lateral branches. They are placed apart on the main axis but are closely set and overlapping on the lateral branches. They are arranged in spiral phyllotaxy. They are small, sessile, entire and without a mid-rib. In submerged species the leaves present on both the main axis and its lateral branches are alike, but in partly submerged species the leaves on the main axis are relatively thin, small and more scaly than those on the lateral branches.

The young gametophore has multicellular rhizoids with oblique septa. But rhizoids disappear as the gametophore matures.

Internal Structure

Main Axis : Internally, the axis is differentiated into three distinct regions.

(a) *Cortex :* The cortex forms the outer part of the axis. In the main axis it is composed of 2-4 layers of thin walled living hyaline cells, but it is always single layered in axillary branches. The mature cortical cells are, however, devoid of protoplast. In S. subsecundum, the cortex of the main axis is also single layered. In lateral branches of S. molluscum and S. tenellum some cortical cells elongate and become bottle or retort-shaped. The neck of these cells is curved and has a pore at the distal end. These cells are called retort cells and they accumulate water. The cortex is surrounded by a single layer of epidermis.

(b) *Prosenchymatous Region :* It lies inner to the cortex and is composed of 4-6 layers of small, thick walled and narrow prosenchymatous cells. The prosenchyma provides mechanical support to the axis. This region is also known as hydrom.

(c) *Central Cylinder :* The innermost region of the axis is known as central cylinder or medulla. Although in most of the species it is composed of vertically elongated thin walled parenchymatous cells, some species (e.g., S. palustre) have a prosenchymatous medulla.

Leaf **:** The leaf is only one cell in thickness. A cross section of a young leaf shows rectangular or square cells of uniform size, but the mature leaf has the following two types of cells.

(a) *Chlorophyllous Cells :* These are small living cells with many discoid chloroplasts. They may be triangular (e.g., S. acutifolium, S. tenellum) or biconvex (e.g., S. squarrosum) in shape. They are also known as photosynthetic cells due to their photosynthetic ability.

(b) *Hyaline Cells :* These are large, polygonal and colourless cells. They are devoid of protoplasts. Their walls usually have spiral or annular thickenings and pores. These cells have a remarkable capacity of absorbing water, hence called capillary cells.

The chlorophyllous and hyaline cells are arranged in an alternate sequence.

Apical Growth

The main axis and lateral branches grow by a tetrahedral apical cell with three cutting faces. Each derivative of the apical cell divides periclinally into an outer and an inner cell. The outer cells, by repeated divisions, form the cortex of the axis and leaves, whereas the inner cells form the central cylinder. The leaves are arranged in three vertical rows on a young stem, but they show phyllotaxy on the older stem.

Reproduction : In Sphagnum reproduction takes place by vegetative and sexual methods.

Vegetative Reproduction : Vegetative propagation takes place by special branches, known as innovations. These branches develop in the axis of the leaves on the main axis just like ordinary branches. The innovations are strong branches resembling the main axis but are directed upwards. Progressive death of the basal part of the main axis results in separation of innovations and then each innovation establishes into a new plant.

In some species the primary protonema also helps in the multiplication of plants. Sometimes, few marginal cells of the thalloid primary protonema become meristematic and form a green multicellular filament. The apical part of this filament forms flat thallus-like secondary protonema. Any cell of the secondary protonema, usually the marginal ones, can grow into leafy gametophore.

Sexual Reproduction : Species of Sphagnum are monoecious or dioecious. In monoecious species antheridia and archegonia are borne on different branches of the same plant. Sex organs are formed in the beginning of the winter season. They develop on special branches which are formed in the axis of the leaves and are much smaller than the vegetative branches.

1. *Antheridium* :

   (a) *Antheridial Branches* : They are small catkin-like branches which are arranged spirally or in straight rows on the main axis. They bear many small, red, yellow or brown leaves; an antheridium is formed in the axil of each leaf. The antheridia develop in acropetal succession, i.e. the youngest antheridium is at the apex and the oldest at the base of the antheridial branch.

   (b) *Development of Antheridium* : The antheridium develops from a superficial cell of the antheridial branch (Fig.). The antheridial initial grows into a papillate outgrowth and then divides transversely to form a, small filamentous structure (Fig.). The terminal cell of this filament undergoes two oblique intersecting divisions so that an apical cell is differentiated (Fig.). The apical cell has two cutting faces and forms derivatives in both planes. Thus a 12-15 celled structure is formed, 2-5 distal cells differentiate into the body of the antheridium and the remaining cells form the stalk (Fig.).

The distal cells, destined to form the body of the antheridium, divide periclinally. Each gives rise to an outer jacket initial and an inner primary androgonial cell (Fig.). At this stage the antheridium consists of 2-5 primary androgonial cells, surrounded by jacket initial. The primary androgonial cells divide in all possible planes and the last generation of these cells forms androcytes (Fig.). The jacket initials also divide and form a single layered jacket of the antheridium (Fig.). A unique feature of Sphagnum is that the apical cell of the antheridium eventually stops dividing and functions as a jacket cell.

(c) *Mature Antheridium* : The mature antheridium has a globular, body borne on a long stalk, composed of 2-4 rows of cells (Fig.). The body is surrounded by a single layered jacket of sterile cells, enclosing a mass of androcyte cells. Each androcyte cell metamorphoses into a spirally coiled biflagellate antherozoid (Fig.).

(d) *Dehiscence of Antheridium* : At maturity the apical cells of the antheridial jacket swell by absorbing water. As a result of the turgor thus generated, the antheridial wall breaks into a number of irregular lobes at the apex. These lobes (or valves) turn backward thus exposing the mass of androcytes. Soon after the liberation of androcytes from the antheridium, antherozoids come out of androcytes and then swim freely in water.

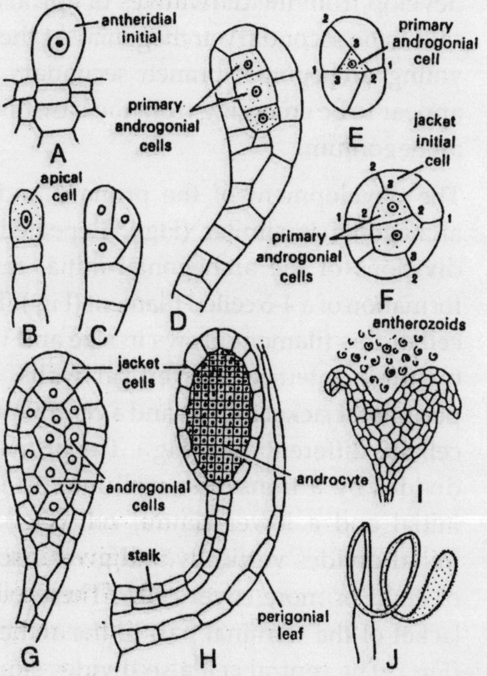

A-J. Sphagnum : Antheridumi, A-G Successive stages in the development of antheridum, H. Mature antheridium, I. Dehisced antheridum, J. Antherozoid.

2. *Archegonium* :

(a) *Archegonial Branch* : Archegonia are borne at the apices of short bud-like archegonial branches. These branches are smaller than the antheridial branches. They bear green leaves, which are larger than those present on the vegetative branches. The upper leaves of these branches, which protect archegonia and young sporophytes, are called perichaetium.

(b) *Development of Archegonium* : Archegonia develop at the tip of the archegonial branches, singly or in groups of 2-5. The central archegonium of a group is formed first and develops from the apical cell of the archegonial branch. It is called primary archegonium. Other archegonia of the group develop from the derivatives of apical cell and are said to be secondary archegonia. At the apex of the young archegonial branch secondary archegonia appear to be emerging from the base of the primary archegonium.

The development of the primary and secondary archegonia is similar (Fig.). Repeated transverse divisions of the archegonial initial results in the formation of a 4-6 celled filament (Fig.). The terminal cell of this filament grows in size and then divides by three intersecting vertical walls. Thus three peripheral jacket initials and a central primary axial cell are differentiated (Fig.). The primary axial cell divides by a transverse wall into an upper cover initial and a lower central cell (Fig.). The cover initial divides vertically and gives rise to a group of eight or more cover cells. 'These cells form the jacket of the terminal part of the archegonial neck (Fig.). The central cell also divides simultaneously by a transverse wall into an upper primary neck canal cell and a lower primary venter cell (Fig.). The primary neck canal cell forms a row of 8-10 neck canal cells by repeated transverse divisions, whereas the primary venter cell undergoes a single transverse division forming a venter canal cell and an egg (Fig.).

At the same time, each of the three jacket initials divides anticlinally as well periclinally to form the middle and basal portion of the archegonial jacket (Fig.)· Repeated transverse divisions of the jacket cells result in the formation of six (or five) vertical rows of neck cells (Fig.). Before fertilization, the jacket cells surrounding the venter and the basal part of the neck divide periclinally and as such the jacket becomes 2-3 layered in these regions (Fig.).

(c) *Mature Archegonium :* The mature archegonium is an elongated stalked structure with a massive venter and a long twisted neck (Fig.). The basal and middle part of the neck jacket is 2-3 cells thick. The apical part of the neck is covered by a group of eight (or sometimes more) cover cells, which are not well demarcated from the neck cells. The narrow cavity of the neck is occupied by 8-10 neck canal cells. The venter has an egg and a venter canal cell, both of almost the same size.

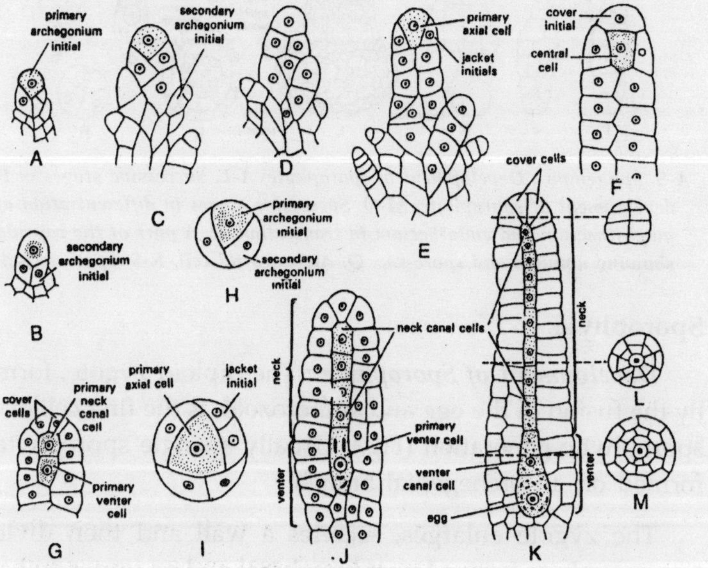

A-M. Sphagnum : Development of archegonium; A-J. Successive stage in the development of archegonium, K. Mature archegonium. L-M. Transverse sections through neck and venture region respectively.

3. *Fertilization :* Fertilization takes place only in the presence of water. The antherozoids, on liberation, swim freely in water and reach near the neck of the archegonium. The neck canal cells and the venter canal cell disorganise and form a free passage for the antherozoids. One of the antherozoids fuses with the egg and thus a zygote is formed.

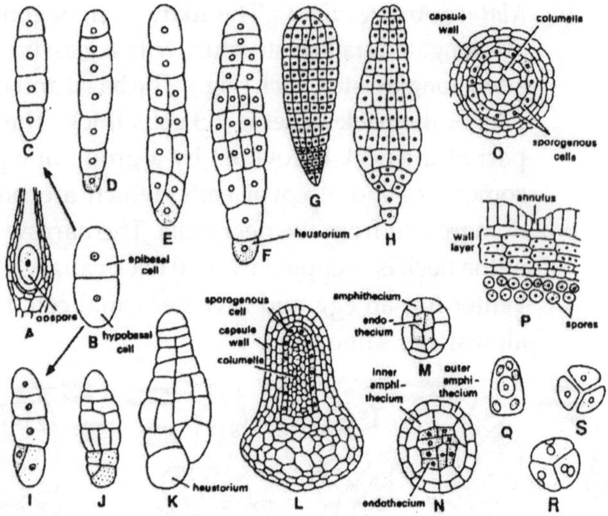

A-S. Sphagnum : Development of sporophyte; A-L. Successive stages in the development of sporophyte, M-O. Successive stages in differentiation of amphithecium and endothecium in transection, P. A part of the capsule showing annulus and spore sac, Q. Archesporial cell, R-S. Spore tetrad.

Sporophyte

Development of Sporophyte : The diploid zygote, formed by the fusion of the egg and antherozoid, is the first cell of the sporophytic generation (Fig.). Usually a single sporophyte is formed on an archegonial branch.

The zygote enlarges, secretes a wall and then divides transversely to form a lower hypobasal and an upper epibasal cell (Fig.). In S. subsecundatum, repeated transverse divisions in both these cells result in the formation of a 6-12 celled filamentous embryo (Fig.). In S. acutifolium, transverse divisions are confined to the epibasal cell, whereas the hypobasal cell undergoes a vertical division, followed by many irregular divisions. The derivatives of the hypobasal cell eventually form a bulbous parenchymatous foot (Fig.). Usually 3-4 upper cells of the filamentous embryo give rise to the capsule and the rest to the foot and the seta (Fig.). The upper cells divide by two vertical division at right angles to each other, thus each cell forms a quadrant. The cells of the quadrant divide periclinally, giving rise to a peripheral amphithecium and a central endothecium (Fig.). The cells of the endothecium undergo repeated divisions and form the central sterile part of the capsule, the columella. The amphithecium divides periclinally and gets differentiated into an outer sterile layer and an inner fertile archesporium (Fig.). The outer sterile layer undergoes further periclinal divisions and forms 3-7 layered capsule wall (Fig.). The fertile layer or the archesporium forms a dome-shaped arch over the columella. The archesporium undergoes further periclinal divisions and gives rise to 2-4 layered sporogenous tissue. All cells of the sporogenous tissue function as spore mother cells and they divide meiotically to form haploid spores (Fig.). The spores are enclosed in a spore sac formed by the surrounding sterile tissue.

Structure of Mature Sporophyte : The mature sporophyte is differentiated into foot, seta and capsule (Fig.). The foot is bulbous or cylindrical and is haustorial in function. It is composed of parenchymatous cells. The seta is an inconspicuous narrow region between the foot and the capsule. The capsule is an almost spherical structure of dark brown or black colour. The wall of the capsule is 3-7 layered. The outermost layer, which has thick walls, forms epidermis. It has many non-functional and rudimentary stomata. The cells of the capsule wall are compactly arranged and they contain chloroplasts. The presence of rudimentary stomata and chloroplasts in the capsule wall show photosynthetic ability of the sporophyte of the ancestors of Sphagnum.

A circular biconvex disc-shaped lid is present at the apex of the capsule. It is called operculum. The operculum is separated from rest of the capsule by a circular groove of thin-walled cells, known as annulus. At maturity, the operculum breaks off from the annulus and allows dispersal of spores.

The central part of the capsule is occupied by a cylinder of sterile cells, called columella. The sporogenous tissue forms a dome-shaped arch over the columella. The young sporophyte is enclosed within calyptra and perichaetium, but at maturity the axis of the archegonial branch elongates, pushing out the capsule from the protective coverings. This elongated and leafless axis of the archegonial branch, present at the base of the sporophyte, is known as pseudopodium. It develops after fertilization. The distal end of the pseudopodium together with the basal part of the calyptta forms a sac-like structure, known as vaginula. The foot of the sporophyte is embedded in vaginula.

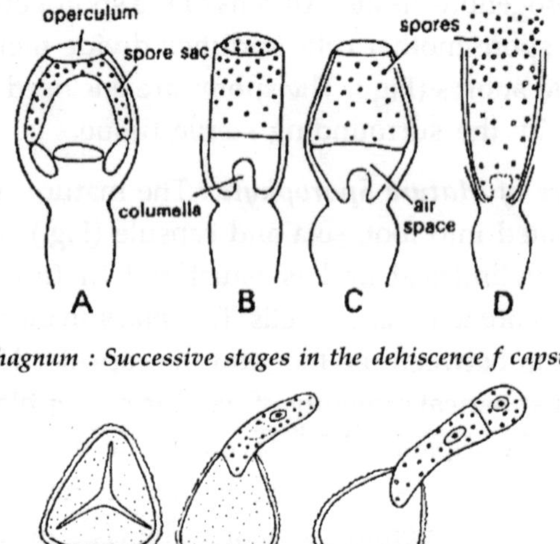

Sphagnum : Successive stages in the dehiscence f capsule.

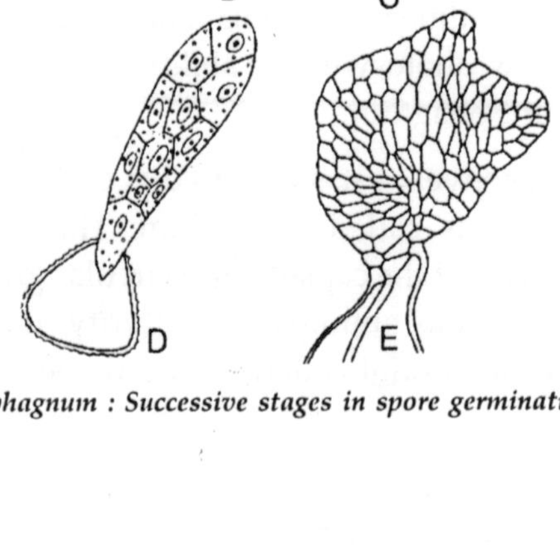

Sphagnum : Successive stages in spore germination.

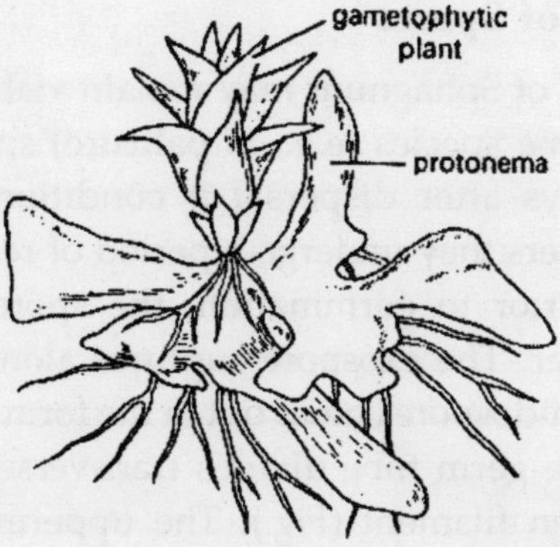

Sphagnum : Mature thalloid protonema with a foliose gametophyte.

Dehiscence of Capsule : The capsule of Sphagnum dehisces usually on a bright sunny day by a special explosive mechanism. The capsule wall and columella dry up and shrivel due to heat. Consequently, an air space develops beneath the spore sac (Fig.) (air probably gets in through the rudimentary stomata present in the capsule wall). At this stage the spherical capsule becomes somewhat cylindrical.

Due to this change in shape, the air within the capsule gets compressed and since there is no outlet for the air, the pressure within causes the operculum to break off at the annulus. This pressure is also responsible for the rupture of spore sac and the yellow or orange coloured mass of dry spores is blown to a height of several centimeters in the atmosphere (Fig.). This explosive mechanism of spore dispersal is also known as air-gun mechanism.

Young Gametophyte

The haploid spore is the first cell of the gametophytic generation. The spores are arranged in tetrahedral tetrads and have a distinct triradiate ridge. The diameter of the spores varies from 18 to 40 µm. The spore wall is differentiated into a smooth, rough or papillose exospore and a relatively thin endospore.

Germination of Spores

The spores of Sphagnum may remain viable for about six months. In some species (e.g., S. palustre) spores germinate within 2-3 days after dispersal if conditions are suitable, whereas in others they undergo a period of rest varying from 2-6 months. Prior to germination, the spore swells up by absorbing water. The exospore ruptures along the triradiate ridge and the endospore comes out in the form of a small germ tube (Fig.). The germ tube divides transversely and forms a 2-4 celled green filament (Fig.). The uppermost cell of this filament divides by two oblique vertical walls forming an apical cell with two cutting faces (Fig.). The derivatives of the apical cell form a flat plate-like structure, called primary protonema (Fig.). The cells of the primary protonema contain chloroplasts. The apical cell loses its activity after sometime and then some marginal cells of the plate-like primary protonema become meristematic and give rise to secondary protonema with rhizoids and leafy buds. These buds eventually develop into gametophores (Fig.).

In some species (e.g., S. girgensohnii), however, the primary protonema instead of giving rise to leafly secondary protonemata, multiplies vegetatively and forms 2-3 or more generations of thallose protonemata before finally bearing leafy buds.

Affinities of Sphagnum

Sphagnum is an interesting genus of Bryopsida. There are certain features in which it shows resemblance with Hepaticopsida and Anthocerotopsida, whereas in others with mosses. In addition, it has some unique features of its own in which it differs from Hepaticopsida, Anthocerotopsida and Bryopsida. The common features which are shared by Sphagnum with these three groups are enumerated below.

Characters Shared with Hepaticopsida

(1) The flat, thallose protonema of Sphagnum resembles with the juvenile stage of some acrogynous Jungermanniales. In both, growth occurs by means of a two-sided apical cell.

(2) The position, structure and dehiscence of antheridium in Sphagnum is similar to that of acrogynous Jungermanniales (e.g., Porella).

(3) The position, structure and development of archegonium in Sphagnum is similar to that of acrogynous Jungermanniales.

Characters Shared with Anthocerotopsida

(1) Young sporogonium does not show apical growth in Sphagnum as in Anthocerotopsida.

(2) The entire endothecium forms columella.

(3) Archesporium develops from the amphithecium and overarches the columella.

(4) Capsule possesses photosynthetic ability.

(5) Sporogonium is differentiated into a large bulbous foot, constriction-like seta and capsule.

Characters Shared with Bryopsida

(1) Gametophore is erect, leafy and radial.

(2) Rhizoids are multicellular and possess oblique septa.

(3) The leaves of Sphagnum with alternate living green cells and hyaline dead cells resemble with those of Leucobrynum.

(4) Development of the leaf, stem and antheridium of Sphagnum is similar to that of other Bryopsida.

(5) Archegonium is stalked and has a massive venter.

(6) Dehiscence of capsule takes place by separation of a lid, known as operculum.

Characters Peculiar to Sphagnum

(1) The adult gametophyte is without rhizoids.

(2) The branches arise in tufts from the axil of every fourth leaf.

(3) The leaves are without midrib.

(4) The stem and leaf show some unique structures. In mature stem the cortex is composed of dead, empty and colourless cells. These cells have large oval pores on their walls in many species and sometimes also develop spiral thickenings. They absorb water like velamen of orchid roots.

(5) Presence of elongated flask-shaped retort cells in the cortex of side branches.

(6) Structurally, Sphagnum shows typical xerophytic characters but physiologically, it is a hydrophyte.

(7) The cell walls of Sphagnum have some organic substance of colloidal nature which absorbs bases and releases acids. It is due to this reason that the water in which Sphagnum grows is highly acidic.

Since Sphagnum shares many features with Hepaticopsida, Anthocerotopsida and Bryopsida, it is generally regarded as a synthetic group.

7

Anthocerotopsida
Anthocerotales

The class Anthocerotopsida is a small group of Bryophyta and is represented by 6 genera and 300 species, widely distributed in both the temperate and tropical regions of the world. The characteristic features of the class are as follows.

(1) The gametophyte is flat, dorsiventral, lobed and thalloid.

(2) The thallus is attached to the substratum with the help of smooth walled rhizoids. Tuberculate rhizoids and scales are absent.

(3) The thallus has a uniform internal organization and is not differentiated into photosynthetic and storage regions. The cells have many chloroplasts with pyrenoids.

(4) Air pores and air chambers are absent but intercellular mucilage cavities are present which open on the ventral surface by slit-like slime pores.

(5) Sex organs are embedded in the thallus. Antheridia develop on the dorsal surface from the hypodermal cells. They occur singly or in groups in antheridial chambers.

(6) Archegonia develop from dorsal epidermal cells.

(7) The sporophyte is differentiated into a long cylindrical capsule, a small meristematic seta and a bulbous foot.

(8) The capsule wall is 4-6 layered and the epidermis has stomata.

(9) Sporogenous tissue develops from the amphithecium.

(10) Elaters do not have thickening bands and as such are called pseudoelaters.

Classification of Anthocerotopsida : The class Anthocerotopsida includes a single order, Anthocerotales. Muller (1940), Proskauer (1951) and Reimers (1954) divided the order Anthocerotales into two families - Anthocerotaceae and Notothylaceae. One representative of each family, i.e., Anthoceros and Notothylus are described here in details.

Family Anthocerotaceae

The family Anthocerotaceae is characterized by the following features.

(1) The capsule is erect and cylindrical.

(2) The capsule wall has stomata.

(3) Elaters are without thickening bands.

(4) Archesporium develops from amphithecium.

The family includes five genera - Anthoceros, Phaeoceros, Aspiromitus, Dendroceros and Megaceros.

Anthoceros

Systematic Position

Class	—	Anthocerotopsida
Order	—	Anthocerotales
Family	—	Anthocerotaceae

Anthoceros, with about 200 species, is the largest and most widely distributed genus of the class Anthocerotopsida. It occurs in both the tropical and temperate regions of the world. The plants grow mostly in moist shady places like river banks and rock crevices. Some species grow on the decaying wood. The species of Anthoceros may be annual (e.g., A. erectus, A. punctatus) or perennial (e.g., A. fusiformis, A. himalayensis).

The genus is represented in India by about 25 species; of these, Anthoceros erectus, A. crispulus, A. longii, A. khandalensis, A. gollani and A. dixitii are common.

Gametophyte

External Structure : The gametophytic plant body is a dark green, dorsiventral and prostrate thallus. The thallus is suborbiculate and lobed (Fig.). It is long and pinnately divided in A. hallii and bilobed in A. himalayensis (Fig.). In A. erectus the thallus is often raised on a thick stalk like structure (Fig.).

The thallus is thick in the middle but it is always without a sharply defined midrib. The dorsal surface of the thallus is smooth (A. laevis), velvety (A. crispulus) or rough (A. fusiformis). The ventral surface bears many unicellular smooth walled rhizoids along the median line (Fig.). They are filled with a granular matter. Tuberculate rhizoids and scales are absent. The mature thalli have erect, elongated and cylindrical sporophytes which usually develop during September-October.

Internal Structure : The internal structure of the thallus is simple without any cellular differentiation (Fig.). All cells of the thallus are similar in structure (Fig.). There is a distinct and continuous epidermis, made up of comparatively small cells. In A. hallii, however, a distinct superficial layer is not present. The thallus is composed of thin walled parenchymatous cells. Each cell has one or more discoid or oval chloroplasts, which show lamellar structure like those of higher plants. The chloroplast contains many pyrenoids, a characteristic feature of the class Anthocerotopsida. The thickness of the middle region of the thallus varies in different species; it is 6-8 cells thick in A. laevis, 8-10 cells in A. punctatus and 30-40 cells in A. crispulus. There are no air pores or air chambers, instead intercellular mucilage cavities are present, which open on the ventral surface through narrow stomata-like slits, called slime pores. These pores are formed by partial separation of two adjacent cells of the epidermis and represent the remnant of a previously existing aerating system. The mucilagenous cavities are invaded by colonies of Nostoc, a blue-green alga. The filaments of this alga enter the thallus through slime pores and then they grow into definite colonies. The presence of Nostoc colonies in the thalli of Anthoceros is not considered a symbiotic association as Anthoceros can grow normally even in sterilized soil, and moreover, there are no mucilagenous cavities in the thalli of some species. Nostoc lives in the thalli of Anthoceros possibly as a space parasite.

Apical Growth : The thallus grows by means of a single apical cell (e.g., A. fusiformis), or by a group of marginal meristematic cells (e.g., A. erectus, A. himalayensis). These cells are located in a shallow depression at the distal end of the thallus and are covered by a mucilagenous substance. The apical cell has four cutting faces — two lateral, a ventral and a dorsal. The derivatives of the dorsal and ventral faces contribute to the thickness of the thallus, whereas those from the lateral faces contribute to its lateral expansion. Sex organs develop from the segments of the dorsal face.

Reproduction : Anthoceros reproduces by vegetative and sexual methods.

Vegetative Reproduction : The common methods of vegetative propagation are as follows.

1. *By Progressive Death and Decay of Thallus :* When gradual death and decay of the posterior part of the thallus reaches to the point of dichotomy, the lobes separate and each grows into an independent plant. In Anthoceros, this process of multiplication is less common than in other liverworts.

2. *By Tubers :* In many species of Anthoceros (e.g., A. hallii, A. laevis, A. pearsoni, A. tuberosus) tubers are formed during unfavourable conditions of prolonged droughts, etc. They store food and also function as perennating organs. Their position varies in different species; for example, they develop behind the growing point in A. laevis and along the margins of the thallus in A. hallii, A. pearsoni and A. tuberosus. In A. himalayensis tubers are stalked and they occur along the margins on the ventral surface of the thallus.

   The tubers have an outer protective corky layer, and hence can resist extreme desiccation. They remain alive even after the death of the thallus. In favourable conditions, when sufficient moisture is available, they germinate and give rise to new plants.

3. *By Gemmae :* Some species like A. glandulosus, A. formosae and A. propaguliferous bear stalked structures on the dorsal surface of the thallus. These are known as gemmae. On separation from the parent thallus, gemmae give rise to new plants.

4. *By Persistent Growing Apices :* Species like A. pearsoni and A. fusiformis grow in areas where summer season is usually dry. Therefore, during summer all parts of

the thallus, except the apical region, become dry. These apical parts remain dormant during dry conditions and on the return of favourable conditions become active and form new thalli.

Sexual Reproduction : Species of Anthoceros are both monoecious or homothallic (e.g., A. fusiformis, A. gollani, A. punctatus) and dioecious or heterothallic (e.g., A. himalayensis, A. hallii, A. erectus, A. pearsoni). The monoecious species are protandrous, i.e., their antheridia mature before archegonia.

Both, male and female sex organs are embedded in the dorsal surface of the thallus. They develop just behind the growing apical region. Their development is probably affected by photoperiod. A diminishing photoperiod generally favours the development of sex organs.

1. *Antheridium :*

(a) *Development :* Antheridia occur singly or in group within closed cavities, known as antheridial chambers. These chambers develop on the dorsal surface of the thallus, but they do not open on the surface of the thallus.

A superficial dorsal cell, lying close to the growing apex, divides by a periclinal wall into an outer roof initial and an inner antheridial initial. The antheridial initial is thus endogenous rather than superficial as in other bryophytes (Fig.). A mucilagenous cavity soon develops between the roof initial and the antheridial initial and this cavity enlarges to form an antheridial chamber (Fig.).

The first division of the roof initial is by a periclinal wall, followed by many anticlinal divisions. This results in the formation of a two layered roof of the antheridial chamber (Fig.). The antheridial initial becomes almost spherical before it divides. In some species of Anthoceros (e.g., A. pearsoni,

A. himalayensis) this cell develops directly into a single antheridium, whereas in others (e.g., A. erectus) it divides vertically into 2-4 or more daughter cells and each of these cells forms an antheridium. The antheridial initial undergoes two vertical divisions at right angles to each other and thus four cells are formed (Fig.). Now, a transverse wall is laid down across all the four cells resulting in eight cells, arranged in two tiers of four each (Fig.). The cells of the lower tier are called stalk cells; they divide by transverse walls and form a multicellular stalk of the antheridium.

The cells of the upper tier form the body of the antheridium. They divide by a transverse wall and form an 8-celled octant (Fig.). Each cell of the octant then divides by a curved periclinal wall so that eight outer primary jacket cells are differentiated from the eight inner primary androgonial cells (Fig.). The primary jacket cells divide only anticlinally and form a single layered wall (jacket) of the antheridium. The jacket cells contain chloroplasts and as such the mature antheridium is green in colour. The primary androgonial cells undergo many regular divisions to form a mass of androcyte mother cells (Fig.). The androcyte mother cell divides to form two androcytes, each of these metamorphoses into a biflagellate antherozoid (Fig.).

(b) *Mature Antheridium* : The mature antheridium has a club-shaped or pouch-like body borne on a slender stalk (Fig.). The wall of the antheridium is composed of a single layer of plastid containing polygonal or elongated cells. These cells enclose a mass of androcytes which eventually metamorphose into antherozoids.

(c) *Dehiscence of Antheridium* : The roof of the antheridial chamber ruptures irregularly as the antheridium matures. At this stage, many antheridia can be seen in a cup-like antheridial chamber. The apical cells of the exposed antheridial wall absorb water and separate from each other, giving rise to an aperture. The mass of antherozoids is liberated to the surrounding film of water (Fig.).

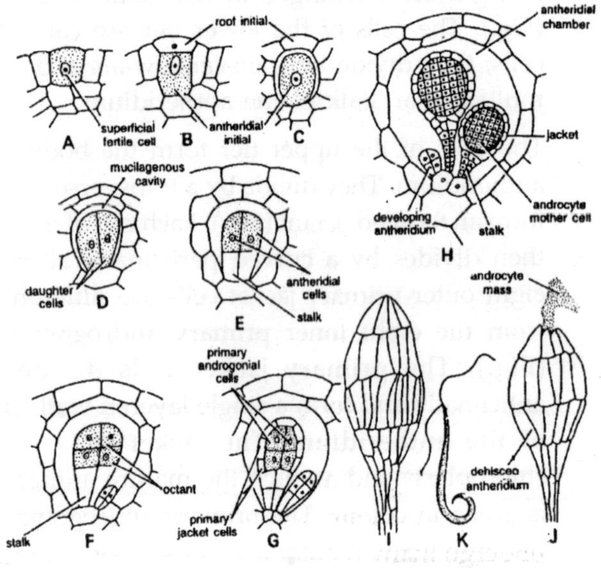

***Anthoceros* : A-H. Successive stages in the development of antheridium, I. Mature antheridium, J. Dehiscence of antheridium, K. Antherozoid.**

2. *Archegonium* : Archegonia arise in acropetal sequence on the dorsal surface of the thallus. In monoecious species, they are formed on the same thallus which had borne antheridia.

(a) *Development* : The archegonium develops from a single superficial dorsal cell that lies immediately behind the growing apex. The archegonial initial divides by a transverse wall to form an outer primary archegonial cell and an inner primary stalk cell. But according to Mehra and Handoo (1953), the archegonial initial directly functions as primary archegonial cell. The primary archegonial cell, by three intersecting periclinal divisions, forms a central cell surrounded by three peripheral jacket initials (Fig.).

The jacket initials divide transversely, forming two tiers of three cells each. The cells of the upper tier divide anticlinally and form six cells. The latter undergo repeated transverse divisions and form the neck of the archegonium which is thus composed of six vertical rows of sterile neck cells. Since the archegonium is embedded in the tissue of the gametophyte, the neck cells are not distinguishable from those of the thallus. The cells of the lower tier, by transverse and vertical divisions, form the venter wall.

Accompanied with the development of the neck and venter wall, the central cell, known as primary axial cell, divides transversely to form two cells of almost equal size. The lower daughter cell functions as primary venter cell (Fig.). The upper daughter cell again divides transversely and differentiates into an upper cover initial and a lower primary neck canal cell (Fig.). The cover initial divides by two vertical walls at right angles to each other and forms a rosette of four cover cells at the tip of the neck. The primary neck canal cell, by repeated transverse divisions, forms 4-6 or rarely more neck canal cells (Fig.). The primary venter cell divides by a transverse wall to form an upper small venter canal cell and a lower large egg cell.

(b) *Mature Archegonium :* The mature archegonium is embedded in the dorsal surface of the gametophyte, but the cover cells protrude slightly above the surface of the thallus. The neck is composed of six vertical rows of neck cells and the narrow cavity of the neck is occupied by an axial row of 4-6 or more neck canal cells. The lower swollen venter contains a large egg and a relatively small venter canal cell (Fig.).

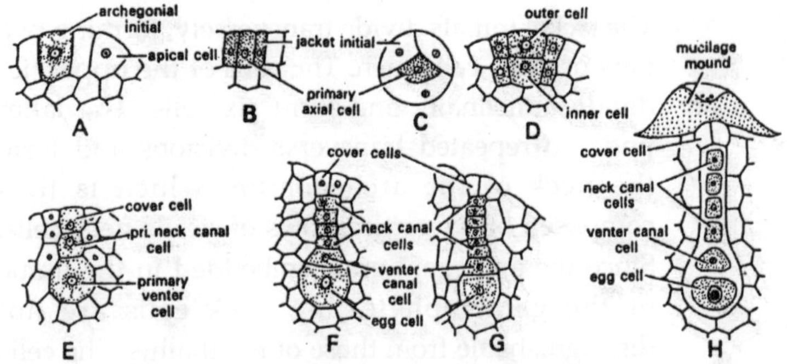

A-H. Anthoceros A-G. Successive stages in the development of archegonium, H. Mature archegonium.

3. *Fertilization* : Prior to fertilization the cover cells present at the tip of the archegonium separate. The neck canal cells and the venter canal cell disintegrate at the same time forming a mucilagenous mass. The mucilage collects in the form of a mound at the tip of the archegonium. Freely swimming antherozoids are attracted towards the neck of the archegonium by the chemicals present in the mucilage. Although many antherozoids enter the neck, only one of them fuses with the egg and thus oospore is formed.

Sporophyte

Development of Sporophyte : The diploid oospore is the mother cell of the sporophytic generation. As the oospore increases in size, it secretes a wall of cellulose around itself (Fig.). The oospore divides by a vertical wall along the long axis of the archegonium, forming two almost equal cells (Fig.). The first division of the oospore is transverse (e.g., A. crispulus) or oblique (e.g., A. laevis). The two daughter cells now divide transversely forming a four-celled embryo (Fig.). These four cells may be equal in size or the upper two are larger. A second vertical division (at right angles to the first one) forms an eight-celled embryo, the octant. In the octant eight cells are arranged in two tiers of four each. Further development of the sporophyte shows variation in different species. For example, in A. erectus the octant does not divide and remains two tiered only, the lower tier forms the foot, and the capsule develops from the upper tier. In A. gemmulosus the cells of one of the two tiers divide by a transverse wall and as such three tiers of four cells each are formed. The lower most tier forms the foot and the two upper tiers the capsule. In A. crispulus, the cells of both the tiers of the octant divide transversely thus forming a four tiered embryo (Fig.). The cells of the upper two tiers give rise to the capsule, while those of the lower two the foot.

The cells of the embryo destined to give rise to foot may divide in regular (e.g., A. erectus) or irregular (e.g., A. himalayensis) fashion. The foot is bulbous and made up of thin walled vacuolated cells. In some species (e.g., A. himalayensis and A. laevis) the cells of the surface layer of the foot grow into short haustoria-like projections. These projections absorb food from the gametophyte for the developing sporophyte. In A. erectus and A. crispulus the superficial cells of the foot are palisade-like (Fig.) and can be easily differentiated from the adjoining gametophytic tissue.

The cells of the upper tier/tiers of the embryo, which are destined to give rise to capsule, divide periclinally. Thus an outer amphithecium is differentiated from an inner endothecium (Fig.). Endothecium forms the central sterile part of the capsule, the columella, composed of 16 vertical rows of sterile cells. The amphithecium, by a second periclinal division, is differentiated into an outer sterile layer of jacket initials and an inner fertile layer of sporogenous cells. The latter forms the archesporium.

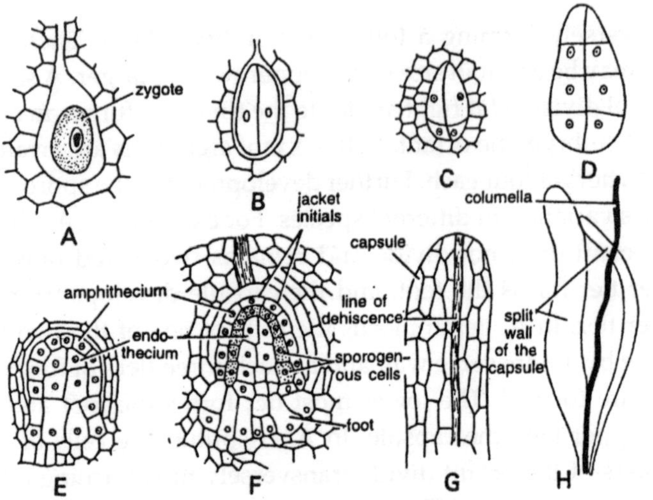

A-H. Anthoceros : A-F. Successive stages in the development of sportophyte, G-Sporophyte showing line of dehiscene, H. Dehisced capsule.

The jacket initials divide both anticlinally and periclinally and form 4-6- layered capsule wall. The outermost layer of the capsule wall, the epidermis, has cells with cutinised outer walls. The epidermis is characterised by the presence of stomata. The cells of the inner layers of the capsule wall have chloroplasts and are photosynthetic.

The archesporial cells divide anticlinally and overarch the rounded apex of the columella. Subsequently, they divide periclinally and thus the mature capsule has two or more layered archesporium. In some species (e.g., A. crispulus, A. erectus), however, the archesporium remains single layered throughout. The young archesporial cells are small, cuboidal and densely cytoplasmic. They mature in basipetal sequence, i.e., the cells of the apical part mature earlier than the basal cells. At maturity the archesporium differentiates into two types of cells : (i) *Spore Mother Cells :* These are large, spherical or oval fertile cells with dense cytoplasm and larger nucleus, and (ii) *Elater Mother Cells :* These are small, sterile cells with comparatively small nucleus. The two types of cells are almost equal in number and are arranged in regular alternate rows in the young capsule (Fig.). But they intermingle as the capsule matures.

The spore mother cells divide meiotically and each forms a spore tetrad (Fig.). The elater mother cells, however, undergo few transverse or oblique divisions and each forms a 3-4-celled simple or branched elater (Fig.). The elaters of Anthoceros do not have thickening bands and as such they are called pseudoelaters. The apical growth of the sporophyte ceases after the differentiation of jacket, archesporium and columella in the capsule. At this stage, an intercalary meristem differentiates at the base of the capsule and this meristem is responsible for further growth of the capsule.

The young sporophyte of Anthoceros is covered by a protective sheath, the calyptra (also known as involucre). It develops mainly from the tissue of the gametophyte surrounding the archegonium. In internal structure, the calyptra resembles the thallus. The cells of the venter wall also contribute to the formation of calyptra. In early stages, growth of the calyptra keeps pace with that of the sporophyte and as such the young sporophyte is completely surrounded by the involucre. Subsequently, the sporophyte grows faster and pushes up through the involucre. In the mature sporophyte, therefore, the calyptra occurs as a tubular sheath or collar at the base.

Structure of Mature Sporophyte

1. *External Structure :* The mature sporophyte is a thin elongated structure, present on the dorsal surface of the thallus. It varies in length from 2-15 cm, depending on the species. A collar like involucre surrounds the base of the sporophyte. The young sporophyte is green in colour, but at maturity it gradually turns dark yellow or black from the apex to the base.

The mature sporophyte consists of a distinct foot and a capsule. The bulbous foot is completely embedded in the tissue of the gametophyte. The capsule appears as a long bristle or horn on the dorsal surface of the thallus and hence the name hornworts has been given to these plants. It is uniformly thick with a tapering tip. There is no distinct seta but an intercalary meristematic zone is present in between the foot and the capsule. This meristem continuously adds new tissues to the capsule from the base.

2. *Internal Structure :*

(a) *Foot :* This basal bulbous part of the sporophyte is composed of vacuolated parenchymatous cells. In some species the superficial cells of the foot become palisade like, whereas in others some of these cells elongate and grow as haustoria in the adjoining tissue of the gametophyte. The foot absorbs water and mineral nutrients from the gametophyte for the developing sporophyte.

(b) *Capsule :* The internal structure of the capsule is a little complex. The various parts are described below.

(i) *Capsule Wall :* It is composed of 4-6 layers of parenchymatous cells. The outermost layer, which forms the epidermis, is composed of vertically elongated cells. It is covered on the outer side by a thick layer of cuticle. The epidermis has stomata which are oriented along the long axis of the capsule. These stomata have guard cells like those of the higher plants. The cells of the capsule wall contain chloroplasts. Thus the sporophyte is capable of synthesizing organic food by itself, but is dependent upon the gametophyte for its water and mineral requirements.

The apical region of the capsule has two or four shallow grooves in the epidermis. The capsule dehisces along these grooves.

(ii) *Sporogenous Tissue :* It extends like a cylinder around the columella from the base to the tip of the capsule. In contrast to other liverworts, the sporogenous tissue in Anthoceros overarches the columella like a dome. The capsule matures in basipetal sequence and as such the differentiation of archesporium into spores and elaters can be studied in serial transverse sections from the base upwards.

The archesporium is single layered and is undifferentiated at the base of the capsule, but in the upper region the spore tetrads and pseudoelaters are arranged in regular alternate blocks. Elaters are simple or branched and are composed of 1-4 cells of irregular shape. In some species, however, elaters are unicellular. The elaters of Anthoceros are characterized by the absence of thickening bands. But they are thick walled in A. physocladus and sometimes the secondary walls even possess helical thickenings.

(iii) *Columella :* It is the central sterile part of the capsule extending almost to entire length. It is composed of 16 vertical rows of thick walled elongated cells. The main function of columella is to provide mechanical support to tender and elongated capsule. It also helps in the dispersal of spores. Under certain conditions the elongated cells of columella also function as water conducting tissue.

Dehiscence of Capsule : In the capsule of Anthoceros spores mature basipetally. At maturity, the apex of the capsule turns yellow, black or brown and loses water. Then narrow slits appear in the capsule wall along the shallow groove, which gradually widen and extend towards the base (Fig.). Thus the capsule wall splits into two or four valves depending on the species. Initially, the valves remain attached at the tip but as the valves dry and twist, they separate from each other, exposing the spores and elaters (Fig.). The dispersal of spores is facilitated by air currents.

Young Gametophyte : The haploid spore is the mother cell of the gametophytic generation. The tetrahedrally arranged spores are yellow (e.g., A. himalayensis), black (e.g., A, erectus) or brown in colour. The spore wall is differentiated into an outer thick, opaque and sculptured exospore and an inner thin and smooth endospore. The spores are uninucleate and contain enough reserve food, oil and colourless plastids.

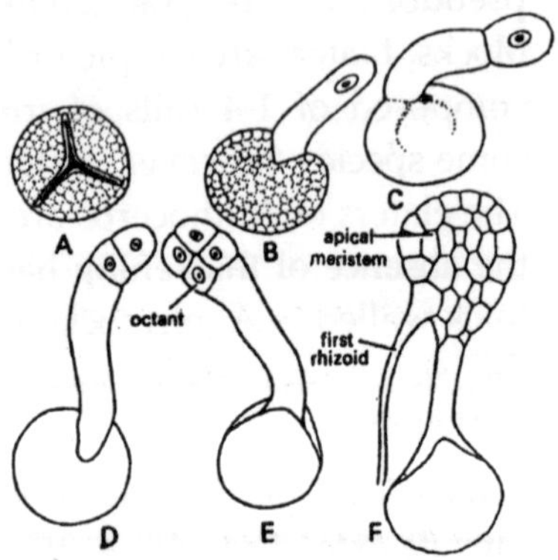

Anthoceros : Successive stages in the germination of spore

Germination of Spores : The spores germinate on suitable substratum either immediately after their liberation (e.g., A. erectus, A. punctatus) or after a rest period extending from a few weeks to few months. Prior to germination, they absorb water and swell. Consequently, the exospore ruptures along the triradiate ridge and the endospore comes out in the form of a long germ tube through the slit thus formed (Fig.). The contents of the spore migrate into the germ tube. The colourless plastids of the spore turn green in the germ tube. Two successive transverse divisions differentiate two cells at the tip of the germ tube (Fig.). Both these cells divide by two vertical walls, at right angles to each other, forming an 8-celled octant at the tip of the germ tube (Fig.). This 8-celled structure is known as sporeling. The four distal cells of the octant function as apical meristem. The activity of these cells results in the formation a new gametophyte (Fig.).

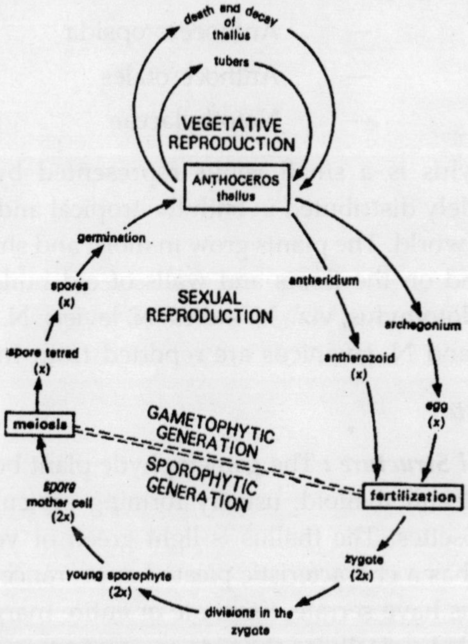

Anthoceros : Diagrammatic representation of life cycle.

Family Notothylaceae

The family Notothylaceae is characterized by the following features.

(1) The capsule is cylindrical and horizontal.

(2) The capsule wall does not contain stomata.

(3) The archesporium develops from the amphithecium in

some species (e.g., N. indica) and from the endothecium in others (e.g., N. chaudhuri).

(4) Elaters have irregular thickening bands.

This monotypic family includes only one genus Notothylus.

Notothylus

Systematic Position

Class	—	Anthocerotopsida
Order	—	Anthocerotales
Family	—	Notothylaceae

Notothylus is a small genus represented by about 13 species, widely distributed in both the tropical and temperate parts of the world. The plants grow in moist and shady places, on rocks and on the floors and walls of old buildings. Five species of Notothylus, viz., N. indica, N. levieri, N. pandei, N. chaudhuri and N. javanicus are reported from India.

Gametophyte

External Structure : The gametophytic plant body is thin, dorsiventral and thalloid, usually forming orbicular or sub-orbicular rosettes. The thallus is light green or yellowish in colour and has a characteristic pleated appearance. The lobes of the thallus have serrate, fimbriate or entire margins. Many smooth-walled unicellular rhizoids are present on the ventral surface of thallus. Tuberculate rhizoids and scales are absent.

Internal Structure : The thallus shows no internal differentiation of tissues. It is 6-8 cells thick in the middle, gradually thinning out towards the margins and is only one cell thick at the extreme margins. The dorsal and ventral epidermis are distinct and are made up of small cells. The centrally situated parenchymatous cells are larger than the epidermal cells. Each cell contains a single large chloroplast with one pyrenoid. All species of Notothylus, with the exception of N. javanicus, have mucilagenous cavities which often contain a blue-green alga, Nostoc. These cavities open to the ventral surface by narrow slits.

Reproduction : Notothylus reproduces by vegetative and sexual methods.

Vegetative Reproduction : The vegetative multiplication takes place by progressive death and decay of the posterior or older parts of the thallus. When the process of decay reaches near the apex of the thallus, the apical lobes separate and each grows into a new individual.

Sexual Reproduction : Species of Notothylus are dioecious as well as monoecious. Of the five Indian species, N. indica, N. chaudhuri and N. levieri are monoecious and protandrous (i.e., antheridia mature before archegonial. However, Kashyap described N. indica and N. levieri as dioecious.

1. *Antheridia :* Antheridia arise endogenously on the dorsal surface, close to the apex of the thallus. They are formed in groups of 2-6 and are enclosed within antheridial chambers (Fig.). The antheridium is large, oval or globose structure, attached to the floor of the antheridial chamber by a short multicellular stalk. The body of the antheridium has a single layered sterile jacket, enclosing a mass of androcytes (Fig.). Each androcyte metamorphoses into a biflagellate antherozoid.
2. *Archegonia :* Archegonia develop from the superficial cells of the thallus close to the growing apex. They are sessile and are embedded in the dorsal surface of the thallus. The mature archegonium consists of a neck and a venter. The neck is short and is composed of six vertical rows of neck cells which enclose 3-5 neck canal cells. A rosette of four cover cells is present at the tip of the neck, which protrudes out on the dorsal surface of the thallus. The venter contains a venter canal cell and an egg (Fig.).

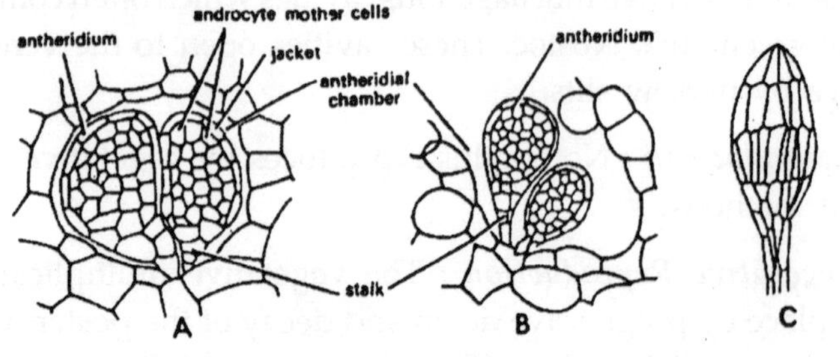

A-C. Notothylus : Antheridium; A. Transection of the thallus showing two antheridia in an antheridial chamber, B. Transection of the thallus showing opening of antheridial chamber, C. A mature antheridium.

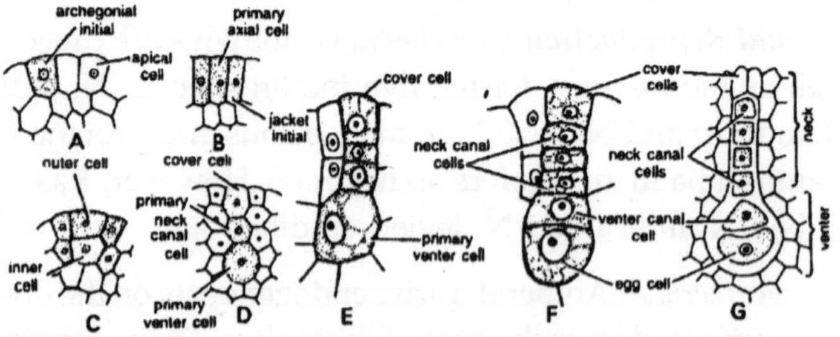

A-G. Notothylus : Successive stages of the development of archegonium.

The development of sex organs and the process of fertilization in Notothylus is similar to that of Anthoceros.

Sporophyte : The zygote, formed after the fusion of the antherozoid with the egg, enlarges in size and occupies almost the entire venter cavity. The development and structure of mature sporophyte varies in different species.

The first division of the zygote may be transverse (e.g., N. indica, N. orbicularis) or vertical (e.g., N. breutellii,. N. javanicus, N. levieri). When the first division of the zygote is transverse, the second division is vertical and when the first division is vertical, the second is transverse. The third division of the zygote is vertical, at right angles to the first vertical division. Thus an eight-celled embryo is formed with two tiers of four cells each. A transverse division in the cells of the upper tier results in a three tiered embryo. The uppermost tier forms the capsule and seta and the two lower tiers, the foot.

The cells of the uppermost tier get differentiated into a peripheral amphithecium and a central endothecium by a periclinal division. Further development of the sporophyte varies and the species of Notothylus can be placed in the following three categories.

1. *Columellate Species :* In species like N. indica and N. orbicularis a central sterile part, the columella, is present. The columella develops from the endothecium, and the archesporium and the capsule wall are derived from the amphithecium.
2. *Non-columellate Species :* In species such as N. chaudhuri, N. flabellata and N. levieri columella is absent. The amphithecium is entirely utilized in the formation of wall layers of the capsule and the endothecium forms the fertile part, the archesporium.
3. *Intermediate Species :* These species form a connecting link between the columellate and non-columellate species (Fig.). For example, in N. javanicus (a columellate species) the endothecium usually develops into a normal columella, but sometimes the columella is much reduced and confined to the basal part of the capsule and in the upper part the endothecium forms fertile sporogenous cells. In N. breutellii (a non-columellate species) though the endothecium develops into archesporium, yet towards the end of the intercalary growth it forms few sterile cells which give rise to a reduced columella.

The Mature Sporophyte

The mature sporophyte is cylindrical, tapering at both the ends and measures 2-3 mm in length and 0.5 mm in diameter. It is placed horizontally along the margins of the thallus between the lobes. In N. levieri, sporophytes arise in pairs. Usually sporophyte is completely enclosed within a membranous involucre, but in N. indica and N. levieri it projects slightly beyond the involucre. The sporophyte is differentiated into foot, seta and capsule. The anchor-shaped foot is composed of parenchymatous cells and is embedded in the tissue of the gametophyte. Some superficial cells of the foot become tubular and absorb nutrition from the gametophytic tissue. The seta is represented by a small narrow intercalary meristematic zone which shows only limited meristematic activity. The capsule wall is made up of four layers. The outermost layer, the epidermis, has relatively thick brown coloured cells. There are no pores or stomata. The three sub-epidermal layers have comparatively thin walled cells. There is little or no chlorophyll in the wall layers and thus the capsule is completely dependent upon the parent thallus for nutrition and water. The capsule wall encloses the sporogenous tissue. In columellate species, the sporogenous tissue surrounds the columella, whereas in non-columellate species, it completely occupies the cavity of the capsule. Sporogenous tissue, at maturity, differentiates into fertile spore mother cells and sterile cells. These two types of cells are arranged in alternate transverse rows. The spore mother cells undergo meiosis to form spore tetrads. The dark brown spores have an opaque spore wall, differentiated into an outer thick exospore and an inner thin endospore. The sterile cells form small unicellular elaters, which are irregular in shape and have curved thickened bands on their walls.

Dehiscence of Capsule : In species like N. indica and N. levieri, the capsule opens like a follicle, usually along one suture only. The mature capsules of N. orbicularis dehisce by 2-4 valves. The dispersal of spores takes place by air currents.

Young Gametophyte : The spores usually germinate immediately after their liberation, if they get a suitable substratum. But sometimes they may undergo a period of rest before germination. Prior to germination, the spore absorbs water, with the result the exospore ruptures and the endospore emerges as a short germ cell. The germ cell divides by repeated transverse and longitudinal divisions, forming a cellular mass which differentiates into a new thallus.

Affinities of Anthocerotopsida : On the one hand, Anthocerotopsida resembles with algae and on the other, they share many features with simple vascular plants—Pteridophyta. The resemblance of Anthocerotopsida with both the immediate lower and higher groups of plants has led bryologists to consider Anthoceros as a synthetic type.

Characters Common with Algae

(1) The gametophytic cell of Anthoceros usually has a single large chloroplast as in algae. However, in Megaceros there are many chloroplasts in each cell and this represents an intermediate condition between Anthoceros and other Embryophyta.

(2) Pyrenoids, which are characteristic of algae, also occur in the chloroplasts of Anthoceros.

(3) The plant body is thalloid and dorsiventrally flattened in both, Anthoceros and algae.

(4) Antherozoids are biflagellate.

Characters Common with Hepaticopsida

(1) The structure of thallus of Anthoceros is very much similar to that of Pellia. Tuberculate rhizoids and scales are absent in both.

(2) The thalli of Anthoceros, as well as of Hepaticopsida, grow by a single pyramidal apical cell with four cutting faces.

(3) Although sex organs differ in structural details, in both they are constructed on the hepatic plan.

(4) In Anthoceros the archesporium differentiates into spores and pseudoelaters, but in the related genera Megaceros and Dendroceros elaters possess spiral thickenings as in Hepaticopsida.

(5) In non-columellate species of Notothylus (e.g., N.

levieri) entire endothecium gives rise to archesporium and the capsule wall is derived from the amphithecium. These species form a connecting link between Anthocerotopsida and Hepaticopsida.

Characters Common with Bryopsida : The sporogonium of Anthocerotopsida shows resemblance with that of Bryopsida in the following characters.

(1) The central solid core of sterile cells of the capsule, the columella, develops from the endothecium.

(2) The capsule wall is photosynthetic and has functional stomata.

(3) The sporogenous tissue is reduced in both, Anthocerotopsida and Bryopsida in comparison to Hepaticopsida.

(4) Archesporium usually differentiates from the inner layer of the amphithecium and overarches the columella.

Characters Common with Pteridophyta

(1) Thalloid plant body of Anthocerotopsida is similar to gametophyte (prothallus) of ferns.

(2) The sex organs in Anthocerotopsida are embedded in the thallus as in Pteridophyta.

(3) The structure of archegonium is similar in both groups.

(4) Many bryologists trace the descent of bryophytes from the simplest known vascular plants — the Psilophytales. The view gets support from the similarity between the semiparasitic sporophyte of Anthoceros, which possesses photosynthetic capsule wall with functional stomata and independent, rootless, leafless, dichotomously branched sporophyte of primitive vascular plants like Horneophyton and Rhynia.

8

Jungermanniales Pellia

Order Jungermanniales

Jungermanniales, with about 244 genera and 9000 species are by far the largest order of the class Hepaticopsida. They are represented in India by genera such as Pellia, Riccardia, Porella, Diplophyllum, Fossombronia, Cephalozia, Sewardiella, Solenostoma, Plagiochila, Radula, Lophozia, Jungermannia, Riella, Aneura, Mastigobryum and Metzgeria.

General Characters of Jungermanniales

(1) The gametophytic plant body is thalloid or foliose. The thalloid forms are simple, dorsiventral and dichotomously branched (e.g., Pellia, Sewardiella, Aneura), whereas the foliose forms (also known as leafy liverworts) are differentiated into axis and leaves (e.g., Porella, Frullania, Radula, Diplophyllum). Besides, some members show a gradual transition from thalloid to leafless habit (e.g., Petalophyllum rolfsii, Fossombronia himalayensis; Fig.).

(2) The thallus does not show differentiation into photosynthetic and storage regions as in Marchantiales. Usually, all the cells of the thallus have chloroplasts. Air pores and air chambers are altogether absent.

(3) The thallus remains attached to the substratum with the help of smooth walled rhizoids. Tuberculate rhizoids and scales are absent.

(4) The thallus grows by a single apical cell.

(5) The vegetative propagation takes place usually by means of gemmae but some members (e.g., Calycularia crispula, C: canarana, Leptolejeunia) multiply with the help of adventitious branches and leafy propagules.

(6) The antheridia are globose and are borne on long stalks.

(7) The neck and venter of the archegonium are almost of the same diameter. The neck is made up of five vertical rows of neck cells.

(8) The sporophyte is differentiated into foot, seta and capsule. The capsule jacket is multilayered; at maturity it splits into four regular valves.

(9) The archesporium develops from the endothecium.

(10) In some members an elaterophore is present in the basal (e.g., Pellia) or apical (e.g., Riccardia) region of the capsule.

(11) Spore mother cells become deeply four lobed prior to sporogenesis.

Classification of Jungermanniales

Verdoorn (1932) divided Jungermanniales into the following two groups which were given the rank of order.

Jungermanniales Anacrogynae or Metzgerineae : The members of this group are usually thalloid. They bear sex organs on the dorsal surface. Archegonia are borne close to the apex, but the apical cell of the thallus is not fully utilized in the formation of archegonia. Hence, the growth of the thallus does not stop after the formation of archegonia. The sporophyte occupies dorsal position, close to the apex. The capsule wall is 2-5 layered.

Metzgerineae are represented by nine families, about 23 genera and more than 550 species. Pellia, Riccardia, Fossombronia, Sewardiella and Metzgeria are some common genera belonging to this group.

Jungermanniales Acrogynae or Jungermannineae : Barring few exceptions, Jungermannineae have foliose gametophyte, differentiated into axis and leaves. Archegonia are borne at the apex of the axis or its lateral branches. The apical cell of the axis (or the branch) is completely utilized in the formation of the last archegonium, terminating the vegetative growth of the axis.

Jungermannineae are a large group that include about 80% of Hepaticopsida. They are represented by 17 families, some 220 genera and 8,500 species.

In this chapter one anacrogynous (Pellia) and one acrogynous (Porella) Jungermanniales are described in details.

Family Pelliaceae

The family Pelliaceae is characterized by the following features.

(1) The plant body is thalloid and irregularly lobed.

(2) Sex organs are borne on the dorsal surface of the thallus.

(3) The capsule is spherical or, oval and is characterized by the presence of a basal elaterophore.

Pellia

Systematic Position

Class	—	Hepaticopsida
Order	—	Jungermanniales
Sub-order	—	Metzgerineae
Family	—	Pelliaceae

Pellia is represented by five species, widely distributed in north temperate regions. The plants commonly occur along the moist banks of rivers and lakes; moist walls of wells and tree trunks. Sometimes they are also found submerged in water but such forms are usually sterile. Four species of Pellia, viz., P. epiphylla, P. endiviaefolia, P. calycina and P. neesiana

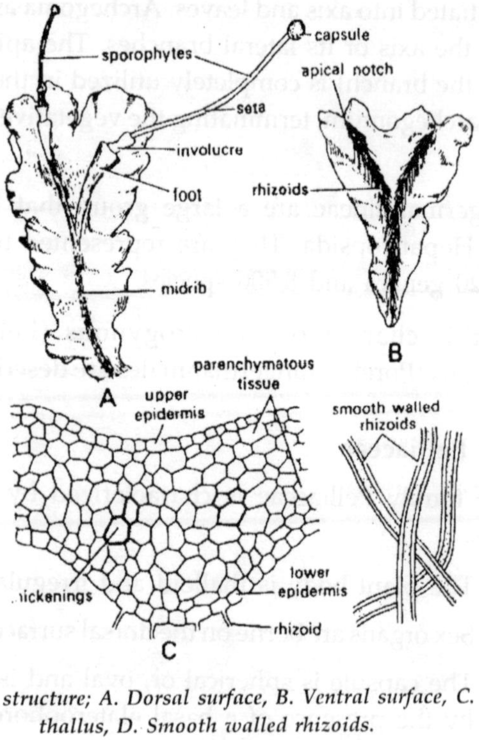

Pellia : Thallus structure; A. Dorsal surface, B. Ventral surface, C. V.T.S. of thallus, D. Smooth walled rhizoids.

have been reported from India. P. endiviaefolia is found throughout the Western and Kumaon Himalayas from 1,700 to 2,500 meters and P. epiphylla and P. neesiana are common in Sikkim and the Eastern Himalayas.

Gametophyte

External Structure : The mature gametophyte of Pellia is simple, prostrate, dorsiventrally flattened, usually thin and dichotomously branched. There is an indistinct midrib on the dorsal surface of the thallus. The thallus is irregularly lobed with sinuous margins (Fig.). The lobes are also dichotomously branched just like the thallus and each lobe of the thallus is supplied with a branch of the midrib. On the ventral surface numerous unicellular smooth walled rhizoids are present (Fig.). They are restricted only to the midrib region. Tuberculate rhizoids and scales are absent. The growing point of the thallus is situated in the apical notch.

The thallus lobes are thin, long, narrow and ribbon like in submerged forms, but are comparatively thick and robust in terrestrial forms.

Internal Structure : The internal structure of the thallus of Pellia is very simple. It is uniformly composed of parenchymatous cells, without any differentiation into photosynthetic and storage regions (Fig.). The dorsal epidermis is continuous; air pores and air chambers are absent. The thallus is 8-16 cells thick in the midrib region and gradually tapers towards margins. The cells of the dorsal epidermal layers and those of the few upper layers of the midrib region and margins have abundant chloroplasts. The cells of the lower layers of the midrib region are devoid of chloroplasts, but they contain starch grains and oil. The lower epidermal cells give rise to unicellular rhizoids in the midrib region.

In P. epiphylla and P. neesiana, some cells of the midrib region become elongated and extend vertically between the upper and the lower epidermis. The walls of these cells are provided with yellow or brown coloured thickening layers.

Reproduction : In Pellia reproduction takes place by vegetative and sexual methods.

Vegetative Reproduction : Vegetative propagation takes place by the following two methods.

1. *By Adventitious Branches :* In some species (e.g., P. endiviaefolia), the dorsal superficial cells of the marginal or apical region form adventitious branches. These branches are in the form of small thalli, and on separation from the parent thallus they develop into independent plants.
2. *By Fragmentation :* The progressive death and decay of the older parts of the thallus results in the separation of lobes. Each lobe develops into a new plant by apical growth.

Sexual Reproduction : Sexual reproduction is of oogamous type. The male and female reproductive structures are known as antheridia and archegonia respectively. Some species of Pellia are monoecious (e.g., P. epiphylla), while others are dioecious (e.g., P. calycina, P. neesiana). The monoecious species are protandrous, i.e., antheridia mature before archegonia.

Antheridium :

(a) *Structure :* Antheridia are present on the dorsal surface of the thallus in irregular rows along the midrib. They appear as wart-like projections in the midrib region. Each projection marks the position of an antheridial cavity in the thallus.

The antheridium is a globular structure attached singly to the floor of the antheridial chamber by a multicellular stalk. The body of the antheridium has a single layered jacket of sterile cells, enclosing a multicellular mass of androcyte mother cells. Each androcyte mother cell eventually forms two antherozoids.

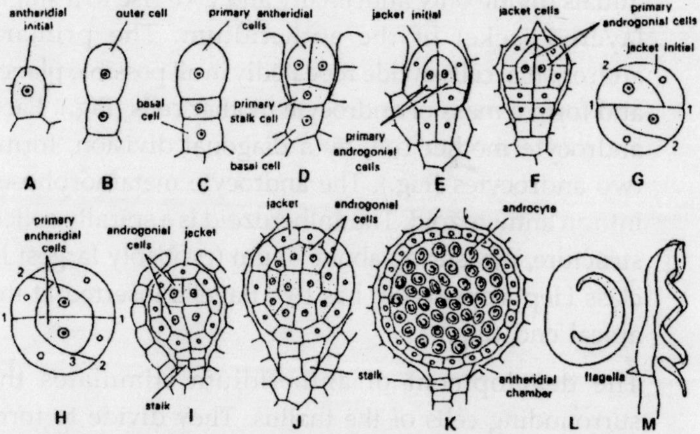

***Pellia* : Development of antheridium; A-J. Successive stages in the development of antheridium, K. Mature antheridium, L. Androcyte, M. Antherozoid.**

(b) *Development* : The antheridium develops from a single superficial dorsal cell of the thallus (Fig.). The antheridial initial is situated 4-5 cells behind the apical cell. A transverse wall soon divides it into an upper outer cell and a lower basal cell (Fig.). The basal cell is embedded in the tissue of the thallus, whereas the outer cell protrudes above the thallus in the form of a papilla. The latter divides again by a transverse wall into a lower primary stalk cell and an upper primary antheridial cell (Fig.). The primary stalk cell undergoes many transverse and vertical divisions and eventually forms a multicellular stalk of the antheridium (Fig.).

The primary antheridial cell, by a median vertical division, forms two daughter cells of almost the same size. Each of these cells divides by a nearly periclinal wall into two unequal cells. The outer or peripheral smaller cell forms the first jacket initial. The inner larger cell again divides by a periclinal wall to form an outer or peripheral second jacket initial and an inner primary androgonial cell (Fig.). The first and the second jacket initials divide only anticlinally and give rise to a single layered jacket of the antheridium. The primary androgonial cells divide repeatedly in all possible planes and form a mass of androcyte mother cells (Fig.). Each androcyte mother cell, by a diagonal division, forms two androcytes (Fig.). The androcyte metamorphoses into an antherozoid. The antherozoid is a spirally coiled structure, measuring about 70 µm (probably largest in class Hepticopsida). It has two flagella inserted at the apical end (Fig.).

The development of antheridium stimulates the surrounding cells of the thallus. They divide to form a ring-shaped involucre around the developing antheridium.

(c) *Dehiscence of Antheridium* : Antheridium dehisces in the presence of water. The cells of the apical part of the antheridial jacket rupture as soon as they come in contact of water. A brown or white mucilagenous mass, containing a large number of androcytes, is violently extruded. It forms a thin film on the surface of water. The androcytes swim freely for sometime and then a single antherozoid is liberated from each androcyte. The entire process of dehiscence takes about 15-20 minutes.

Archegonium :

(a) *Structure* : Archegonia are borne in groups of 4-7, close to the apex of the thallus (Fig.). Each group is, protected by a tubular involucre (Fig.). The archegonia are horizontally placed with their necks directed towards the apex of the thallus. They do not show any definite arrangement, young and mature archegonia are irregularly distributed.

The archegonium is differentiated into a basal venter and a long neck. However, there is little difference in the diameter of the neck and the venter. The neck is composed of five vertical rows of neck cells, which enclose 6-8 neck canal cells. A rosette of four cover cells is present at the apex of the neck. The venter has a venter canal cell and a relatively large egg cell (Fig.). The venter and the basal part of the neck of the archegonium is protected by a two layered sterile jacket.

(b) *Development :* Archegonia are formed in groups on the dorsal surface of the thallus. Each archegonium develops from a single superficial cell, the archegonial initial, which appears as a papillate protrusion on the dorsal surface of the thallus (Fig.). A transverse wall divides the archegonial initial into a basal stalk cell and an upper archegonial mother cell (Fig.). The stalk cell divides repeatedly and eventually forms a multicellular stalk of the archegonium (Fig.). The main body of the archegonium develops from the archegonial mother cell. It divides by three intersecting vertical walls (as in Riccia and Marchantia), forming three peripheral initials, enclosing a primary axial cell (Fig.).

Of the three peripheral initials, two are comparatively large than the third one. Each of the two larger initials undergoes a vertical division but the smaller initial does not divide. Thus a ring of five cells, now called jacket initials, is formed around the primary axial cell (Fig.). The jacket initials divide transversely forming two tiers of five cells each. The upper tier, called neck initials, divides transversely to form the neck of the archegonium. The neck consists of five vertical rows of cells. The cells of the lower tier, called venter initials, form the venter wall (Fig.).

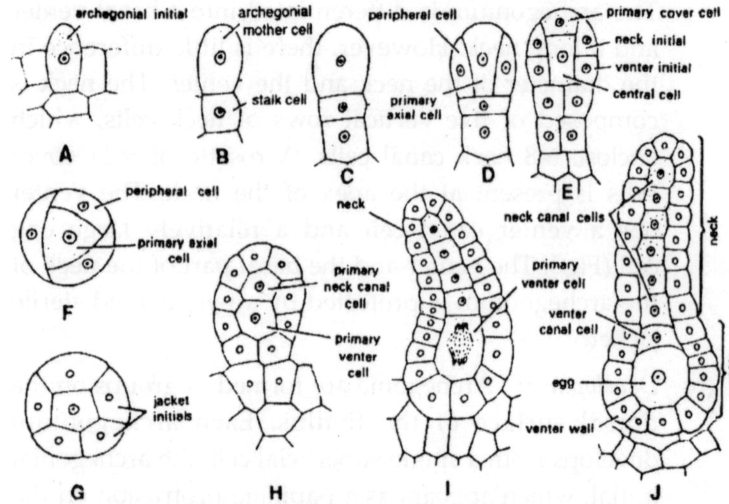

Pellia* : *Development of archegonium; A-E. Successive stages in the development of archegonium, J. Mature archegonium.

The primary axial cell divides transve into an upper primary cover cell and a lower relatively large central cell (Fig.). The primary cover cell undergoes two vertical divisions at right angles to each other, forming a rosette of four cover cells. The central cell, by a transverse wall, differentiates into an upper primary neck canal cell and a lower primary venter cell (Fig.). The primary neck canal cell, by repeated transverse divisions, forms 6-9 or more neck canal cells (Fig.). The primary venter cell also divides transversely into an upper small venter canal cell and a relatively large egg cell (Fig.).

Fertilization : Fertilization takes place only in the presence of water. Prior to fertilization, the neck canal cells and the venter canal cell disintegrate to form a mucilagenous mass. Absorption of water by this mass results in widening of the neck canal and separation of the cover cells from each other. Thus a wide passage is formed for the entry of antherozoids. The antheridia also rupture almost at the same time, liberating antherozoids. When freely swimming antherozoids reach near the archegonium, they are attracted towards its neck. Although a large number of antherozoids enter into the neck of the archegonium, only one of them (usually the one which reaches first) fuses with the egg to form zygote.

Sporophyte : The fertilized egg (Fig.) secretes a wall and functions as the mother cell of the sporophytic generation.

Development of Sporophyte : The zygote increases in size and almost completely fills the venter. The first division of the zygote is transverse and with the result a lower hypobasal cell and an outer epibasal cell are formed (Fig.). The hypobasal cell usually does not divide and it forms unicellular suspensor (or haustorium, Fig.). The epibasal cell divides first by a transverse wall, followed by a vertical division and thus a 4-celled embryo is formed. Each of the four cells of the embryo divides by a transverse wall and eight cells thus formed are arranged in two tiers of four each. The cells of the upper tier form the capsule, whereas those of the lower, the foot and the seta (Fig.).

The cells of the upper tier are differentiated into a peripheral amphithecium and a central endothecium by a periclinal wall. The cells of the amphithecium undergo a periclinal division followed by repeated anticlinal divisions. Eventually a two-layered jacket of the capsule is formed.

The endothecium functions as primary sporogenous tissue. The cells of the endothecium divide repeatedly to form sporogenous tissue. A few sporogenous cells, located at the base, become sterile quite early in the development of the capsule. They elongate and radiate towards the centre of the capsule, and this mass of sterile cells constitutes elaterophore (Fig.).

The fertile sporogenous cells undergo repeated mitotic divisions, forming a large number of spore mother cells, which are diploid in nature. Just before meiosis, each spore mother cell becomes four lobed, the lobes being tetrahedrally arranged.

Now the nucleus of the spore mother cell divides meiotically into four haploid nuclei, one passing into each lobe. This is followed by simultaneous laying down of walls between the lobes and as such a tetrad of four haploid spores is formed.

The cells of the lower tier divide repeatedly to form a parenchymatous foot and a seta. The foot is conical; its upper edges project upwards and surround the basal part of the seta like a collar. The seta is composed of several longitudinal rows of cells.

The young sporophyte is covered by a protective sheath, the calyptra, which develops from the cells of the venter wall.

Structure of Mature Sporophyte : The sporophyte of Pellia is differentiated into foot, seta and capsule (Fig.). The foot is a conical parenchymatous structure. The sporophyte remains attached to the gametophyte with the help of the foot. The distal end of the foot extends like a collar around the base of the seta.

The seta is also composed of nonchlorophyllous parenchymatous cells arranged in definite longitudinal rows. In a young sporophyte, the cells of the seta are relatively small and contain abundant starch, but at maturity the cells elongate rapidly and become devoid of starch.

The mature capsule is an almost spherical structure with a two or more layered jacket. The cells of the outer layer of the jacket are polygonal with radial thickening bands. The cells of the inner layer are relatively thin and have semilunar thickening bands, which help in the dehiscence of the capsule. The jacket encloses a mass of spores and elaters. A characteristic feature of the sporophyte of Pellia is the presence of a basal elaterophore. It consists of 20-100 elongated elaters which radiate from the base of the capsule cavity towards the apex. Each elater has two spiral thickening bands (Fig.).

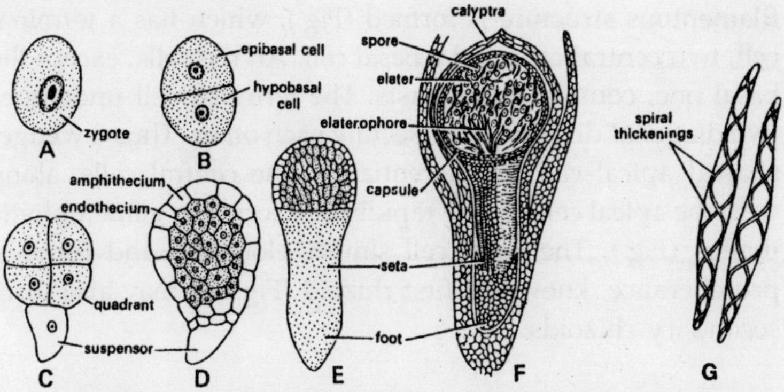

A-G Pellia : Development of sporophyte; A-E. Successive stages in the development of sporophyte, F. Mature sporophyte, G. Elaters.

Dehiscence of Capsule : As the capsule matures, the cells of the seta undergo enormous elongation. Consequently, the capsule is pushed out by rupturing the calyptra. Now, the capsule wall splits into four petal-like valves. These valves become reflexed and hang downwards due to hygroscopic nature of the thickening bands present in the capsule wall. The exposed mass of spores and elaters are set free. In a dehisced capsule, the persistent elaterophore can be seen emerging out from the centre of the four petal-like valves of the capsule wall.

Young Gametophyte

The spore, which is haploid in nature, represents the first cell of the gametophytic generation. It gives rise to a gametophytic thallus on germination.

In Pellia, the germination of spores starts while they are encased within the capsule. It is a characteristic feature of the genus. The spores are liberated at various stages of development in different species of Pellia, e.g., in P. epiphylla they are discharged at 4-6 celled stage and in P. neesiana at 6-9 celled stage.

The first division of the spore is transverse and the two daughter cells again divide transversely. Thus a four-celled filamentous structure is formed (Fig.), which has a terminal cell, two central cells and a basal cell. All the cells, except the basal one, contain chloroplasts. The terminal cell undergoes two diagonal divisions intersecting each other. Thus a wedge-shaped apical cell is differentiated. The central cells, along with the apical cell, divide rapidly and form the gametophytic thallus (Fig.). The basal cell simply elongates and forms a protuberance, known as first rhizoid (Fig.). It may also form secondary rhizoids.

Family Porellaceae

The characteristic features of the family Porellaceae are as follows.

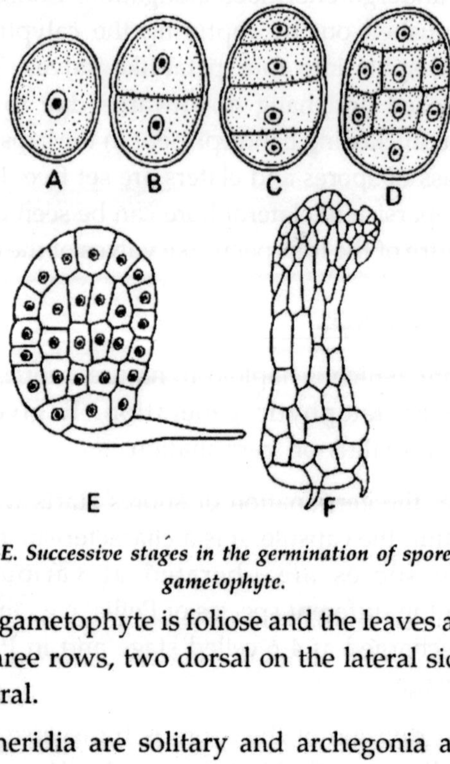

***Pellia* : A-E. Successive stages in the germination of spore, F. Young gametophyte.**

(1) The gametophyte is foliose and the leaves are arranged in three rows, two dorsal on the lateral sides and one ventral.

(2) Antheridia are solitary and archegonia are borne in groups.

(3) The capsule wall splits into four valves which are reflexed only to half way down.

This monotypic family is represented by the genus Porella.

Porella (Madotheca)

Systematic Position

Class	—	Hepaticopsida
Order	—	Jungermanniales
Sub-order	—	Jungermannineae
Family	—	Porellaceae

The genus Porella, with about 180 species, is found in all parts of the world, except arctic and antarctic regions. About 34 species of the genus are recorded from India, mainly from the Kumaon and Garhwal regions of Uttaranchal, Shimla, Chamba and Kullu in Himachal Pradesh and Jammu & Kashmir. Some species have also been recorded from South India.

The gametophytes usually form green patches on moist shaded rocks, tree trunks and on the soil. The aquatic forms of some species, such as P. pinnata, are sterile, but their terrestrial forms are fertile. Some common Indian species of Porella are P. platyphylla, P. plumosa, P. denticulata, P. macroloba, P. acutiphylla, P. variabilis, P. gollani, P. obtusifolia and P. decurrens.

Gametophyte

External Structure : The gametophyte of Porella is dorsiventral and foliose, and grows flat on the substratum (Fig.). The axis or the stem is bi-or tripinnately branched. It bears leaves in three rows, two rows are dorsal (present on lateral sides) and one is ventral. The dorsal leaf is bilobed, the upper anterior lobe, called antical lobe, is larger, while the lower lobe, known as postical lobe or lobule, is much smaller and has an acute apex (Fig.). The dorsal leaves show incubous arrangement, i.e., lower edge of each leaf is covered by the upper edge of the next leaf below. In such an arrangement leaves appear overlapping when viewed from the top.. The leaves of the ventral row, called amphigastria, are small, resembling the postical lobes of the dorsal leaves, but they are broader with decurrent bases. The margins of the leaves may be entire (P. acutiphylla, P. obtusifolia) or dentate (P. denticulata, P. gollani).

A large number of unicellular smooth walled rhizoids arise from the ventral surface of the stem.

Their main function is to anchor the plant to the substratum. In Porella, absorption of water takes place primarily through the leaves and axis.

Internal Structure

1. *Axis* : The internal structure of the axis is simple. It is differentiated into cortex and medulla. The cortex is made up of 2-5 layers of thick walled parenchymatous cells. The cells of the medulla are comparatively thin-walled and elongated (Fig.).
2. *Leaf* : The leaf is composed of a single layer of isodiametric parenchymatous cells with many chloroplasts (Fig.). It is devoid of midrib and vascular tissue. In some species, oil cells are present in the leaf.

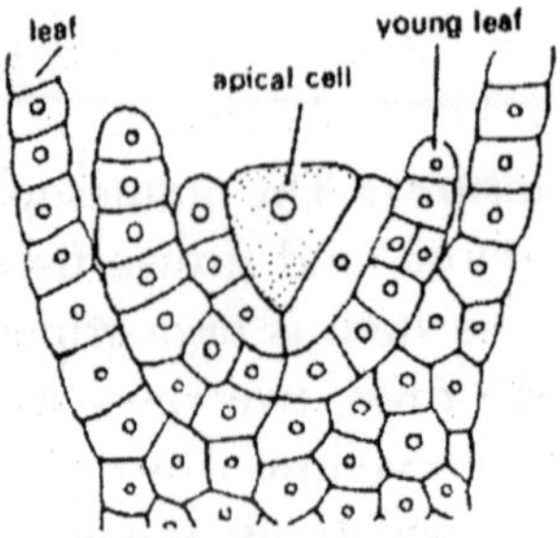

Porella : Longitudinal section of the apex of the axis showing the position of apical cell.

Apical Growth : The growth of the gametophyte takes place by means of a tetrahedral or pyramidal apical cell (Fig.). It has three cutting faces - two dorsal (on lateral sides) and one ventral. They cut off segments in a regular spiral fashion. Derivatives in each plane form a leaf and the subtending axis.

In Porella, the branching is monopodial. The branches arise at the apex of the axis from one of the derivatives of the apical cell which would have otherwise developed into a leaf segment. This type of branching is called terminal.

Reproduction : Porella reproduces by vegetative and sexual methods.

Vegetative Reproduction : Vegetative reproduction takes place by the following two methods.

1. *By Gradual Death and Decay of the Gametophyte :* It is the most common method of multiplication in Porella. The apical growth of the thallus is accompanied by gradual death and decay of the older parts. When the process of decay reaches to the point of dichotomy of the foliose axis, the two branches separate and each grows into a new plant.

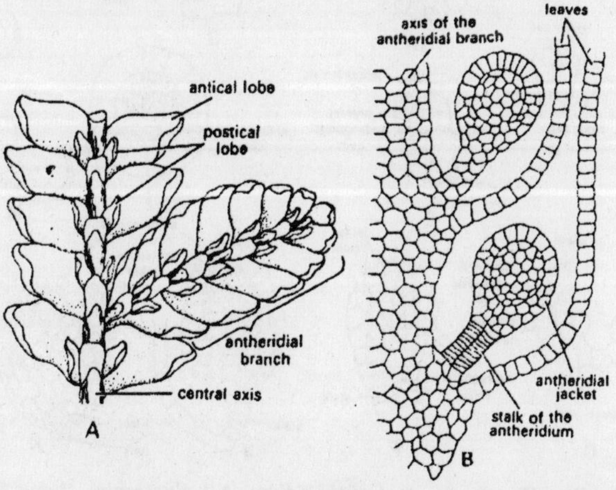

Porella : Male sex organ A. Antheridial branch, B. Longitudinal section of an antheridial branch in part.

2. *By Gemmae* : In some species (e.g., P. rotundifolia), discoid gemmae are present on the lower surface of the leaf, which germinate to give rise to new plants. However, some bryologists are of the view that these spherical thin walled structures, considered to be gemmae, are actually bodies of blue-green algae.

Sexual Reproduction : Porella is dioecious and the male gametophyte is usually smaller than the female gametophyte. Sexual reproduction is of oogamous type. The male and female sex organs are antheridia and archegonia respectively.

Antheridium : Antheridia develop on specialized lateral branches, called antheridial branches (Fig.). These branches project nearly at right angles to the main axis. The dorsal leaves (bracts) are smaller than those on the main branch and are arranged in a close imbricate fashion. The amphigastria (ventral leaves) of the antheridial branches are called bracteoles.

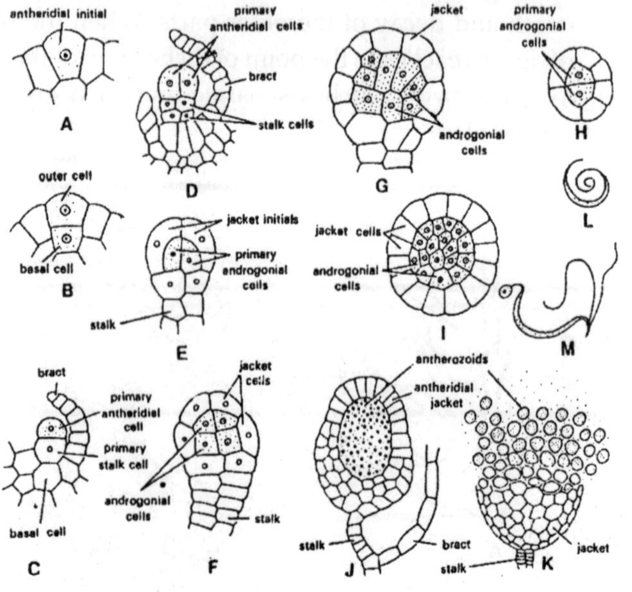

Porella : Development of antheridium; A-I. Successive stages in the development of antheridium, J. Mature antheridium, K. Dehisced antheridium, L. Androcyte, M. Antherozoid.

(a) *Development of Antheridium* : Antheridia arise singly in the axil of the leaf or bract of the antheridial branch (Fig.). An antheridial initial is differentiated at the base of the young bract (Fig.). It divides transversely into an outer cell and a basal cell (Fig.). The basal cell forms the embedded part of the stalk, whereas the outer cell projects above the surrounding tissue and functions as antheridium mother cell. A transverse wall divides the latter into an upper primary antheridial cell and a lower primary stalk cell (Fig.). Repeated transverse and vertical divisions of the primary stalk cell results in the formation of a two-celled thick long stalk of the antheridium (Fig.).

The main body of the antheridium develops from the primary antheridial cell. It divides by a vertical wall, forming two antheridial cells of almost equal size (Fig.). Each of these cells now divides by a periclinal wall into two unequal cells, the outer smaller cell is the first jacket initial and the inner larger the primary androgonial cell (Fig.). The primary androgonial cell divides again to form a second jacket initial. Both, the first and the second jacket initials divide repeatedly by anticlinal walls and form a single layered jacket of the antheridium (Fig.).

Later on, cells of the basal part of the jacket divide periclinally and as such the lower part of the antheridial jacket becomes 2-4-layered (Fig.). The primary androgonial cell divides in all possible planes and forms a large number of rectangular androgonial cells (Fig.). The last generation of these, cells is known as androcyte mother cells. An androcyte mother cell divides diagonally to form two androcytes (Fig.). The androcyte metamorphoses into an antherozoid (Fig.).

(b) *Structure of Mature Antheridium* : The mature

antheridium is differentiated into an elongated stalk and an almost spherical body. The 2-celled thick stalk is attached in the axil of the bract of the antheridial branch. The body of the antheridium is surrounded by a jacket of sterile cells. The jacket is only one-layered in the young antheridium, but later the jacket cells in the proximal part divide periclinally and the jacket becomes 2-3-layered in this region in a mature antheridium. The jacket encloses a mass of fertile cells, called androcytes. The androcytes metamorphose into rod-shaped biflagellate antherozoids (Fig.).

(c) *Dehiscence of Antheridium* : The distal part of the antheridial jacket is single layered and is usually made up of thin walled cells. When this part of the jacket comes in contact with water, it breaks into many irregular lobes which curl back strongly. The whole mass of antherozoids sets free in water (Fig.).

Archegonium : Archegonia are present at the apex of specialized archegonial branches on the female plants. These branches are much smaller than the vegetative branches and have only 2-5 leaves (bracts). The lower bracts form an involucre, whereas the two uppermost bracts coalesce to form a perianth. The latter encloses a group of 10-15 archegonia.

(a) *Development of Archegonium* : Archegonia develop in acropetal succession, i.e., the mature archegonia are at the base and the youngest at the top of the archegonial head. The first 2- 3 segments of the apical cell of the archegonial branch develop into bracts. Thereafter, each derivative of the apical cell functions as archegonial initial. The apical cell of the archegonial branch is eventually used up in the formation of the last archegonium and further growth of the archegonial branch ceases.

The archegonial initial increases in size and forms a papillate outgrowth (Fig.). This outgrowth divides by a transverse wall into a basal or pedicel cell and a distal cell (Fig.). The former undergoes repeated divisions and forms the stalk of archegonium. Distal cell functions as archegonium mother cell (Fig.).

(b) *Structure of Mature Archegonium* : The mature archegonium consists of a neck and a venter. The diameter of venter is not much different from that of the neck. The neck, which is composed of five vertical rows of neck cells, encloses a neck canal with 6-8 neck canal cells. The venter has a small venter canal cell and a relatively large egg. At the top of the neck a rosette of four cover cells is present. The venter and the basal part of the neck has a 2-layered jacket (Fig.).

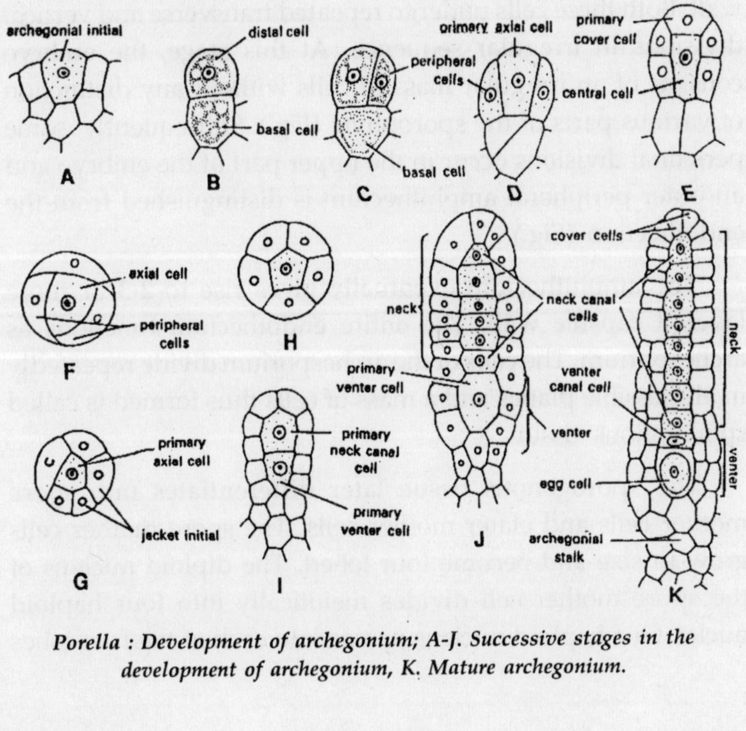

Porella : Development of archegonium; A-J. Successive stages in the development of archegonium, K. Mature archegonium.

Fertilization **:** The male and female plants of Porella usually grow close to each other in dense patches. Water, essential for fertilization, is present in sufficient amounts in between the overlapping leaves. The antherozoids swim freely in water and reach the archegonium in groups. One antherozoid fuses with the egg and thus a diploid oospore is formed.

Sporophyte **:** The oospore is the mother cell of the sporophytic generation. It secretes a wall around itself and enlarges in size (Fig.). The oospore divides by a transverse wall into an epibasal and a hypobasal cell (Fig.). The hypobasal cell does not divide further and simply elongates to form a suspensor which functions as haustorium (Fig.). The various parts of the sporophyte develop from the derivatives of the epibasal cell.

The epibasal cell forms two daughter cells by a transverse wall. Both these cells undergo repeated transverse and vertical divisions in irregular sequence. At this stage, the embryo consists of an irregular mass of cells without any distinction of various parts of the sporophyte (Fig.). Subsequently, some periclinal divisions occur in the upper part of the embryo and an outer peripheral amphithecium is distinguished from the endothecium (Fig.).

The amphithecium eventually gives rise to 2-3 or more layered capsule wall. The entire endothecium functions as archesporium. The cells of the archesporium divide repeatedly in all possible planes and a mass of cells thus formed is called sporogenous tissue.

The sporogenous tissue later differentiates into spore mother cells and elater mother cells. The spore mother cells grow in size and become four lobed. The diploid nucleus of the spore mother cell divides meiotically into four haploid nuclei. One haploid nucleus passes into each of the four lobes and then the lobes are separated into four spores by simultaneous wall formation. The elater mother cell directly develops into an elongated elater with tapering ends. The elater has 2-3 spiral thickening bands (Fig.). The seta and foot develop from the lower portion of the embryo.

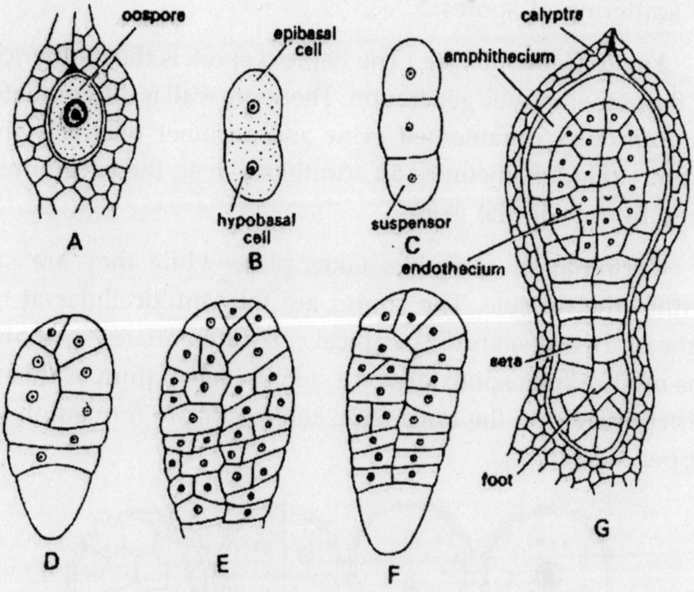

Porella : Development of sporophyte; A. Oospore, B-G. Stages in the development of sporophyte.

Structure of Mature Sporophyte : The mature sporophyte consists of a globose capsule, a short seta and an indistinct foot. The capsule wall is 2-3 or more layered. The cells of the capsule wall are thick walled except for four vertical rows of thin walled cells which demarcate the lines of dehiscence. There are numerous spores and elaters intermingled in the capsule.

The young sporophyte is enclosed by three protective coverings - calyptra, perianth and involucre. The latter is formed by fertile bracts of the archegonial branch.

Dehiscence of Capsule : As the capsule matures, the seta elongates rapidly pushing the capsule out of the protective coverings. The wall of the capsule dries up and splits into four valves from the apex downwards along the lines of dehiscence. The valves, however, remain intact at the base. The jerking movement of elaters (due to their hygroscopic nature) helps in scattering of spores.

Young Gametophyte : The haploid spore is the mother cell of the gametophytic generation. The spore wall is differentiated into an outer ornamented exine and an inner relatively thin intine - (Fig.). Sometimes an additional layer, the perinium; is present outside the exine.

Germination of spores takes place while they are still within the capsule. The spores are thus multicellular at the time of their dispersal. An apical cell differentiates soon after the multicellular spore falls on a suitable substratum. A rhizoid differentiates on the lower side and leaves are formed on the upper side (Fig.).

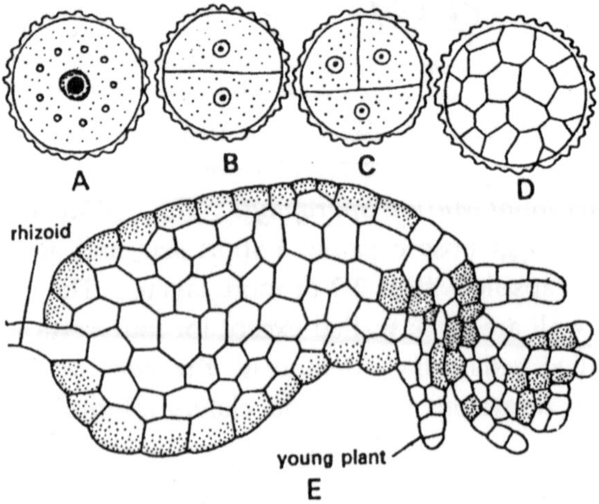

Porella : Germination of Spores; A. Spores, B-D. Stages of spore germination, E. Youg gametophyte.

Frullania

Systematic Position

Class	—	Hepaticopsida
Order	—	Jungermanniales
Sub-order	—	Jungermannineae
Family	—	Frullaniaceae

Frullania is a large genus with about 700 species, mainly distributed in the tropical parts of the world. The genus is represented in India by about 39 species. Most of the species

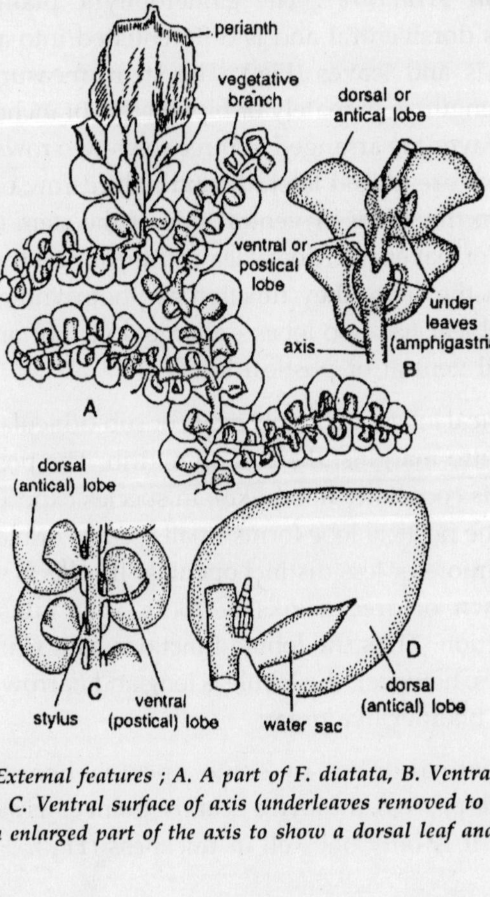

Frullania : External features ; A. A part of F. diatata, B. Ventral (lower) view of the axis. C. Ventral surface of axis (underleaves removed to show stylus). D. An enlarged part of the axis to show a dorsal leaf and stylus.

are terrestrial growing on moist and shady places. They usually form extensive mats of dark green or reddish brown colour. A few species are epiphytic which grow on the trunks of the trees growing in moist woods and swamps.

Frullania squarrosa, F. retusa, F. pyriflora, F. duthiana, F. muscicola, F. gollani, F. himalayensis, and F. gracillima are some Indian species. Of these, F. muscicola, F. gracillima and F. pyriflora are epiphytic while others are found on moist rocks as well as bark of trees.

Gametophyte

External Structure : The gametophytic plant body of Frullania is dorsiventral and is differentiated into a prostrate stem or axis and leaves (Fig.). The stem, measuring about 1-5 cm in length, is pinnately or bipinnately branched. On the stem, the leaves are arranged in three rows-two rows of dorsal leaves which are placed laterally and a third row of leaves is arranged on the lower or ventral side of the stem (Fig.). The lateral row of leaves (i.e., dorsal leaves) are almost transversely inserted on the stem, they function as photosynthetic organ. Each dorsal leaf has two lobes - a large dorsal or antical lobe and a small ventral or postical lobule.

The antical lobe is obliquely ovate or sub-orbicular in shape and has entire margins; there is no midrib. The postical lobe (or lobule) is cocculate (hood-like). In species exposed to long dry spells the postical lobe forms small sac-like or bladder-like organ with more or less distinct opening. It collects water that trickles down on tree trunks or rocks after rains through capillary action. Thus the lobule functions as a water sac. In some species, however, the lobule is long and narrow and does not form a bladder-like organ.

The postical lobe, in some species bears a short membranous process, the stylus near the point of its attachment to the axis. It is only one cell in thickness (Fig.).

The leaves of the ventral row, called amphigastria or underleaves are rounded, cordate or deeply bilobed. The capillary space between stylus and amphigastria retains water.

Internal Structure

1. *Stem :* Internally, the stem is differentiated into outer cortical and inner medullary regions (Fig.). The cortical region is made up of small thick walled cells. In the older part of the stem the cells have pigmented walls. The cells of the medullary region are elongated and have larger diameter; they usually have thin colourless walls. There is no well defined epidermis and conducting tissue.
2. *Leaf :* The leaf consists of a single layer of cells. Each cell of the leaf has several chloroplasts and 3-6 oil bodies (Fig.). The latter are spherical, ellipsoïdal or long fusiform in shape. In young leaves the cells have thin colourless cell walls but in mature leaves the walls are strongly thickened at the corners and pigmented.

In some species (e.g., F. tamarisci and F. asagrayana) a distinct row of large hyaline cells, called ocelli, traverses the antical lobe from its base to near the apex. (Fig.)

Growth : The stem grows by the activity of a pyramidal apical cell with three cutting faces. It cuts off segments in regular spiral succession. Each segment forms a leaf and a portion of the stem. The segments cut off on the dorsal and lateral faces form lateral leaves and those formed on the ventral face form underleaves (amphigastria)

Reproduction : Frullania reproduces by vegetative and sexual methods.

Vegetative Reproduction : Vegetative reproduction takes place by following methods.

1. *Fragmentation :* As in other bryophytes, the posterior portions of the branched gametophyte of Frullania undergo progressive decay and decomposition. This results in the separation of the younger branches. Each branch develops into an independent plant.

2. *Gemmae :* These are 4-6 celled globose structures which develop on the margins and surface of the leaves. The gemmae get detached form the leaf by slightest touch, water drops, insects, etc. Eventually they give rise to new plants when conditions are suitable. Fulford (1959) observed several celled gemmae on the surface of the perianth in F. dilatata.

3. *Adventitious Branches and Caducous Leaves :* In some species (e.g., F. brittoniae, F. reparia) adventitious branches and caducous leaves serve as propagules. In F. fragilifolia the antical lobes of the lateral leaves are caducous and the detached lobes can be induced to give rise to a new plant.

Sexual Reproduction : Species of Frullania are either monoecious or dioecious. In monoecious species though the antheridia and archegonia are borne on the same plant but on different branches. Branches bearing reproductive structures are called fertile branches. The shape and size of the leaves of these branches are different from those of the vegetative branches and they are called bracts. The bracts of the antheridial branches are called perigonial bracts and those of the archegonial branches as perichaetial bracts.

1. *Antheridium :* Antheridia are borne on special short branches called male branches or androecia (Fig.). The male branch bears 2-5 or more pairs of bracts (perigonial bracts). Each perigonial bract is divided into two lobes of almost equal size. There are usually two antheridia in the axil of each bract (Fig.).

The mature antheridium has a stalk and a body (Fig.). The stalk is long, slender and is composed of two rows of cells. The body is nearly spherical; it consists of a mass of spermatocytes enclosed by a single layered sterile jacket. Each spermatocyte metamorphoses into a biflagellate antherozoid.

2. *Archegonium :* Archegonia are borne in groups of 2 or 3 at the apex of the short archegonial branches. Each of these branches has 2-5 pairs of bracts (perichaetial bracts) which are dentate and larger than vegetative leaves. The upper most perichaetial bracts are laterally fused and form an inflated sac-like structure with a short constricted tubular beak at the apex. This structure is known as perianth (Fig.). In a transverse section the perianth is somewhat triangular in outline (Fig.). The group of archegonia is enclosed within the perianth.

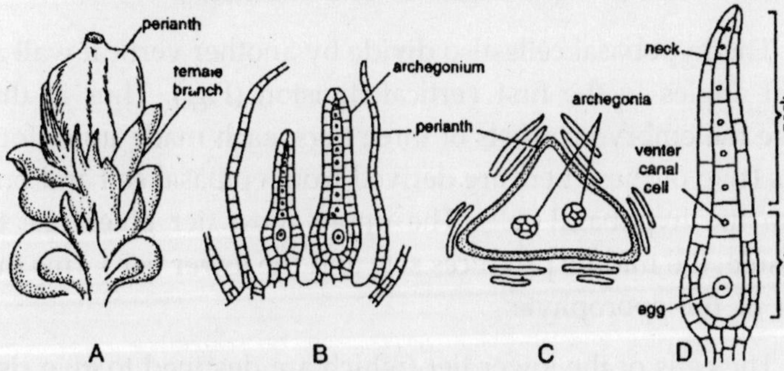

Frullania : Female reproductive structures; A. A female branch, B. Longitudinal section of archegonial head, C. Transverse section of perianth showing two enclosed archegonia, D. An archegonium.

The mature archegonium is a flask-shaped structure differentiated into a constricted neck and swollen venter (Fig.). The neck consists of 5 vertical rows of neck cells and encloses an axial row of 5-8 neck canal cells. The venter wall becomes 2-cell thick towards maturity and has a venter canal cell and an egg.

3. *Fertilization* : As in other bryophytes water is essential for fertilization. The antheridial wall disintegrates in the presence of water and the antherozoids are set free. Approximately at the same time the neck canal cells and the venter canal cell of the archegonium also degenerate forming a free passage for the entry of antherozoids. Fusion of egg with an antherozoid results in formation of zygote.

Sporophyte : The diploid zygote is retained within the venter of the archegonium and serves as the mother cell of the sporophytic generation.

Development of Sporophyte : The first division of zygote is transverse resulting in the formation of an epibasal and a hypobasal cell (Fig.). The epibasal cell divides by a transverse wall while the hypobasal cell divides vertically (Fig.). Subsequently the two upper (epibasal) cells divide by two vertical walls at right angles to one another.

The hypobasal cells also divide by another vertical wall at right angles to the first vertical division (Fig.). Thus at this stage the embryo consists of three tiers, each made up of four cells (two of these tiers are derived from epibasal cell and one from the hypobasal cell). The upper most tier gives rise to capsule, the middle produces seta and the lower tier forms the foot of the sporophyte.

The cells of the lower tier (which are destined to give rise to the foot) divide repeatedly in an irregular manner to form a blunt foot. The surface cells of the foot grow into papillae which absorb nutrition from the adjoining gametophytic tissue (Fig.). Thus foot serves as an absorbing organ. The cells of the middle tier divide by longitudinal walls and give rise to the seta and basal portion of the capsule. The four cells of the upper tier divide periclinally differentiating the outer amphithecium and inner endothecium layers (Fig.).

Cells of the amphithecium undergo repeated anticlinal divisions followed by a periclinal division. As a result a two layered thick capsule wall is formed. The cells of the outer layer of the capsule wall develop rod-like thickenings on the lateral walls and their outer walls remain thin. The cell walls of the inner layer are thickened all over. However, there are four vertical strips of cells in the capsule wall which remain unthickened; these thin walled strips constitute the line of dehiscence. Cells of the endothecium represent archesporium (Fig.). These cells divide repeatedly by longitudinal walls to form a lens-shaped mass of approximately 200 cells. These cells differentiate into two types of cells with great regularity—about 50% of these cells form fertile spore producing cells and remaining 50% cells transform into sterile elater producing cells (Fig.).

The fertile cells undergo repeated transverse divisions to form vertical rows of spore mother cells. The sterile cells do not divide further. They simply elongate to form elaters (Fig.). The upper ends of the elongated elaters remain firmly attached to the top and lower ends to the floor of the capsule. The spore mother cells and elongated elaters are arranged in a regular alternate fashion. Towards maturity of the sporophyte, each spore mother cell undergoes a meiotic division to form four haploid spores. Meanwhile the sterile elater cells also lose their protoplast and a single spiral band of thickening material is deposited on the inner surface of the cell walls (Fig.).

As the zygote develops into the sporophyte, the wall of the venter (where zygote is lodged) is stimulated and it divides to keep pace with the developing embryo. It forms a protective covering called calyptra around the young sporophyte.

Thus the sporophyte is surrounded by two protective coverings - calyptra and perianth. The calyptra lies inner to the perianth (Fig.).

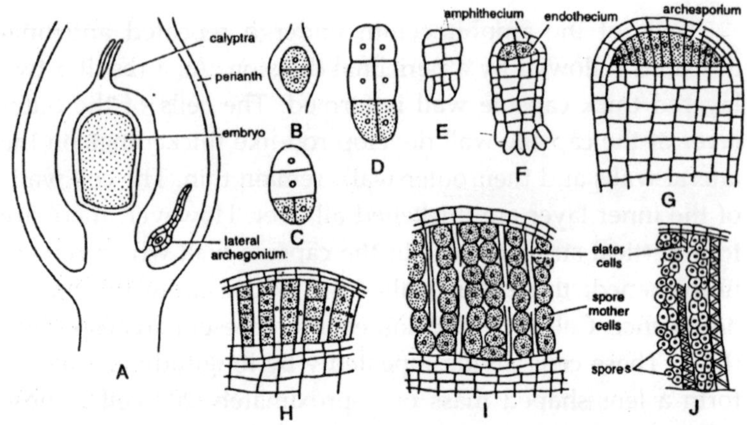

A-J. Frullania : A-I. Stages in the development of sporophyte, J. Longitudinal section of a part of mature capsule showing arrangement of elaters and spores.

Dehiscence of the Capsule : At the maturity of the sporophyte, the seta elongates rapidly and the capsule is thurst out of the protective sheaths - the calyptra and the perianth. Consequently the capsule gets exposed and is now suitably placed at the top of the archegonial branch for dispersal of spores by wind. When the capsule wall dries the thin outer layers of the wall shrink much more rapidly than the relatively thick inner layer. Due to this differential shrinking a tension builds up which causes the capsule wall to split along the lines of dehiscence into four valves (Fig.).

The splits between the valves extend from the top to about two-third of its length. Each valve of the capsule wall has nearly 20 elaters attached firmly near its tip (Fig.). The other ends of these elaters are attached to the floor of the capsule. On further drying the valves bend downwards and as a result of their extensive stretching, the elaters get detached from the floor of the capsule wall. This causes the valves to spring back with a jerk.

Thus at this stage the elaters are attached only to the tips of the valves. Further drying of elaters results in their contraction and their free ends swing into the air. The sudden and rapid movements of the contracting elaters produce a sling-like action. It helps in expelling the spores out of the capsule with a force.

The Young Gametophyte

Spores : The spores of Frullania are large, oblong or spherical measuring 25-55 μ in diameter. The spore wall is differentiated into outer thick and sculptured exospore and inner thin and smooth endospore (Fig.). Each spore is uninucleate and has several chloroplasts.

Germination of Spore : In Frullania the spores begin to germinate while still within the capsule. In early stages, although

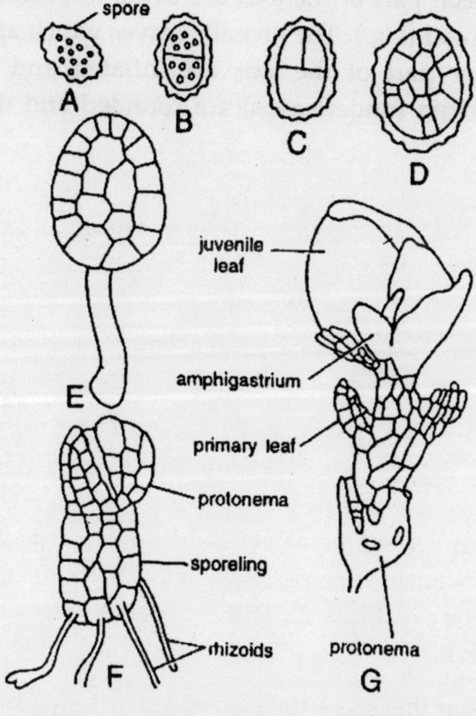

Frullania : A. A spore, B-G. Germination of spores and development of sporeling.

the spore wall is considerably stretched but the exospore does not rupture. The protoplast of the spore divides by a transverse wall and as a result two cells are formed (Fig.). The upper (epibasal) cell now divides by a vertical wall (Fig.). The two daughter cells now divide repeatedly. The divisions occur in all the three planes and as a result a globose mass of about 50 or more cells is formed (Fig.). It is called protonema. Pressure of the expanding protonema causes the spore wall to rupture. At this stage the capsule dehisces and the young sporelings (protonema) are liberated. The thallus- β like sporelings develop rhizoids on reaching a suitable substratum (Fig.). Leafy shoots develop subsequently from the superficial cells of the sporelings. The first formed leaves (primary leaves) present in the basal part of the axis are ovate to oblong and only 2-4 cells broad (Fig.). The juvenile leaves which appear later on the upper part of the axis are inflated and sac-like. The amphigastria (underleaves) are rounded and deeply lobed.

9

Hepaticopsida Marchantiales

The name Hepaticopsida has been derived from the latin word Hepatica, which means liver. In the Medieval period these plants were used in treating liver disorders, hence the members of this class are commonly known as hepatics or liverworts. The class is represented by about 280 genera and 9,500 species, distributed in both tropical and temperate parts of the world.

They require a good deal of moisture during the growing season. In India, liverworts are widely distributed in the outer Himalayas in the range of 1500-2400m. They thrive in heavy rainfall areas. Infact, the number of species and individuals is directly proportional to the amount of rainfall. Their growth slows down during winters due to low temperature and scanty precipitation.

Shiv Ram Kashyap (1929), father of Indian Bryology, has reported some 70 species of liverworts belonging to 33 genera from the Western Himalayas and the Punjab plains. Of these, about 50 species are endemic to India.

General Characters of Hepaticopsida

(1) The gametophyte, which forms the main plant body, is independent, dorsiventral and thalloid or foliose.

(2) In thalloid forms, the plant body is prostrate, lobed and dichotomously branched.

(3) The ventral surface of the thallus bears many unicellular, unbranched, smooth walled and tuberculate rhizoids and scales. Rhizoids absorb water and minerals and also fix the plant to the substratum.

(4) In foliose forms central axis bears leaves in 2 or 3 rows. Leaves are without midrib.

(5) Anatomically, the thallus is either simple or complex and is devoid of mechanical tissue.

(6) Photosynthetic cells contain numerous chloroplasts without pyrenoids.

(7) Vegetative multiplication takes place by death and decay of the older parts of the thallus, and by adventitious branches, gemmae or tubers.

(8) The members of Hepaticopsida are homothallic (monoecious, e.g., Riccia gangetica, R. crystallina, Riccardia multifida, Pellia epiphylla) or heterothallic (dioecious, e.g., Riccia frostii, R. pearsonii, Riccardia indica, Marchantia).

(9) The sex organs occupy dorsal or terminal position on the thallus. They develop from a single superficial initial. In thalloid liverworts, the sex organs are either embedded in the thallus (e.g., Riccia) or are stalked (e.g., Marchantia). In foliose forms, however, they are always stalked.

(10) The sporophyte is simple, represented by capsule only (e.g., Riccia) or differentiated into foot, seta and capsule (e.g., Marchantia).

(11) The sporogenous cells (archesporium) develop from the endothecium. The sporogenous tissue either forms only spore mother cells (e.g., Riccia) or becomes differentiated into fertile spore mother cells and sterile elater mother cells (e.g., Marchantia, Pellia).

(12) The capsule does not have sterile columella. The only sterile cells are elaters.

(13) The dehiscence of capsule is irregular and indefinite.

(14) The spores give rise to the gametophyte on germination.

Classification of Hepaticopsida

Engler (1892) divided the class Hepaticopsida or Hepaticae into the following three orders.

(Orders)

Class Hepaticae —— **Marchantiales**, **Anthocerotales**, **Jungermanniales**

Many bryologists, however, do not include Anthocerotales in Hepaticopsida and Howe (1899) gave it the rank of a class and this treatment was followed by Campbell (1940), Takhtajan (1953), Smith (1955), and others. Thus these workers recognize only two orders - Marchantiales and Jungermanniales in the class Hepaticopsida.

Cavers (1910) removed the family Sphaerocarpaceae from the order Jungermanniales and gave it the rank of an order and as such he recognised the following three orders in Hepaticopsida.

(Orders)

Class Hepaticopsida —— **Marchantiales**, **Sphaerocarpales**, **Jungermanniales**

Verdoorn (1932) divided the class Hepaticae into two sub-classes — Hepaticales and Anthocerotales. He included the following four orders in the sub-class Hepaticales.

Subclass Hepaticales
(Orders)
- **Marchantiales**
- **Sphaerocarpales**
- **Jungermanniales acrogynae**
- **Jungermanniales anacrogynae**

Campbell (1936), placed both acrogynous and anacrogynous Jungermanniales in a single order, and included the order Calobryales in the class Hepaticopsida.

Class Hepaticopsida
(Orders)
- **Marchantiales** (e.g., *Riccia, Marchantia*)
- **Sphaerocarpales** (e.g., *Sphaerocarpos, Geothallus*)
- **Jungermanniales** (e.g., *Pellia, Riccardia, Cryptothallus*)
- **Calobryales** (e.g., *Calobryum, Haplomitrium*)

Schuster (1966) elaborated the classification of liverworts (Class Hepaticae). He recognised following eight orders in the class Hepaticae.

Class 1. Hepaticae	
Sub-class A. Jungermanniae	
Order 1. Takakiales	(includes a single genus Takakia)
2. Calobryales	(includes only Haplomitrium)
3. Jungermanniales	(a very large and diverse order of leafy liverworts)
4. Metzgeriales	(includes thalloid liverworts like Pellia and allied genera)
Sub-class B. Marchantiae	
Order 5. Sphaerocarpales	(includes Sphaerocarpos and Geothallus)
6. Monocleales	(represented by a single genus Monoclea)
7. Marchantiales	(a large group of thalloid liverworts)
Class 2. Anthocerotae	
Order 8. Anthocerotales	(includes Anthoceros and Notothylus)

Order Marchantiales

The order Marchantiales includes prostrate and non leafy liverworts with their thalli differentiated into several morphologically and physiologically distinct tissues. There are about 35 genera and 420 species which are hygrophilous and grow in moist and shady habitats. The common Indian genera of the order are Riccia, Marchantia, Athalamia, Dumortiera, Cyathodium, Conocephalum, Perissia, Plagiochasma, Reboulia and Wiesnerella.

General Characters of Marchantiales

(1) The gametophytic plant body is thalloid, dorsiventrally differentiated and dichotomously branched.

(2) In many forms large polygonal areas, known as areoles, are present on the dorsal surface of the thallus.

(3) The ventral surface of the thallus bears rhizoids (smooth walled and tuberculate) and scales (except for some aquatic species). The arrangement of scales varies in different genera; for example, they occur in a single row on the margins of the thallus (Riccia), in a single row on either side of the midrib (Reboulia and Targionia), in 2-4 rows on either side of the midrib (Marchantia), or are distributed irregularly on the entire ventral surface of the thallus (Corsinia).

(4) Internally, the thallus is differentiated into a dorsal photosynthetic and ventral storage region. In Dumortiera, Monoselenium and Monoclea the differentiation of tissues is poor.

(5) Sex organs are either distributed on the dorsal surface of the thallus (e.g., Riccia) or are borne on distinct receptacles (e.g., Marchantia, Reboulia). The receptacles may be stalked (e.g., Marchantia) or sessile (e.g., Reboulia).

(6) Antheridia and archegonia develop in antheridial and archegonial chambers respectively. Each chamber has only a single sex organ.

(7) The neck of the archegonium consists of six vertical rows of (neck) cells.

(8) The capsule wall is invariably a single layer of sterile cells (unistratose).

Classification of Marchantiales

The order Marchantiales has been variously treated by different workers from time to time. Schiffner (1909) divided Marchantiales into two families, Ricciaceae and Marchantiaceae.

Cavers (1911) included the following five families in the order : (i) Ricciaceae, (ii) Corsiniaceae, (iii) Targioniaceae, (iv) Monocleaceae, and (v) Marchantiaceae.

Verdoorn (1932) divided Marchantiales into the following six families : (i) Marchantiaceae, (ii) Operculatae, (iii) Astroporae, (iv) Targioniaceae, (v) Corsiniaceae, and (vi) Ricciaceae.

Although Evans (1938) also recognised six families in Marchantiales, but these differ from those of Verdoorn. These families are : (i) Marchantiaceae, (ii) Rebouliaceae, (iii) Corsiniaceae, (iv) Targioniaceae, (v) Sauteriaceae, and (vi) Ricciaceae.

Campbell (1940) proposed a revised classification of Marchantiales on the basis of the nature of receptacle bearing reproductive organs and structure of sporophyte. He merged Sauteriaceae and Rebouliaceae (families proposed by Evans) within Marchantiaceae. The five families included under Marchantiales by Campbell are : (i) Ricciaceae, (ii) Corsiniaceae, (iii) Targioniaceae, (iv) Monocleaceae, and (v) Marchantiaceae.

Kashyap, while studying the Himalayan liverworts, came

across many intermediate forms which showed resemblance with Targioniaceae or Corsiniaceae on the one hand and Marchantiaceae on the other. Thus he suggested the inclusion of the members of Targioniaceae and Corsiniaceae in Marchantiaceae. The three families recognised by Kashyap in Marchantiales are : (i) Ricciaceae, (ii) Monocleaceae, and (iii) Marchantiaceae.

To accommodate a newly discovered form, Monocarpus sphaerocarpus (also known as Carpos sphaerocarpos), Carr (1956) and Proskauer (1961) recognised another family, Monocarpaceae or Carpaceae) in Marchantiales.

Most of the modern bryologists include the following six families in Marchantiales (i) Ricciaceae, (ii) Corsiniaceae, (iii) Targioniaceae, (iv) Monocleaceae, (v) Monocarpaceae, and (vi) Marchantiaceae.

The families Ricciaceae and Marchantiaceae are described here.

Family Ricciaceae

The family includes the simplest forms of Marchantiales. It has the following characteristic features.

(1) The gametophyte is prostrate, flat, dorsiventral and dichotomously branched.

(2) The thallus shows differentiation into dorsal photosynthetic and ventral storage regions.

(3) The photosynthetic region consists of vertical filaments of chlorophyll containing cells. They enclose large or narrow air chambers which open on the dorsal surface of the thallus by rudimentary air pores in some species (e.g., Ricciocarpus).

(4) Sex organs are embedded on the dorsal surface of the thallus in the median furrow.

(5) The antheridial and archegonial chambers bear only a single sex organ. The chamber opens on the dorsal surface of the thallus through an ostiole.

(6) The sporophyte of Ricciaceae is the simplest amongst the liverworts. It is represented by capsule only; foot and seta are absent. The capsule is embedded in the archegonial chamber within the thallus.

(7) Almost all cells of the sporogenous tissue form spores. Elaters are absent.

The family Ricciaceae includes three genera (Riccia, Oxymitra and Ricciocarpus) and about 140 species. The life history of Riccia, the most widely distributed genus of the family, is described here.

Riccia

Systematic Position

Class	—	Hepaticopsida
Order	—	Marchantiales
Family	—	Ricciaceae

Occurrence : The genus Riccia has about 130 species, widely distributed in both tropical and temperate regions of the world. The plants grow predominantly on damp soil, moist and shady rocks and other similar terrestrial habitats. Riccia fluitans is, however, an aquatic species floating on stagnant or slowly running water. About 33 species of this genus have been reported from India. They occur mostly in the foot hills and plains. Riccia robusta and R. crystallina occur up to an altitude of 3500 meters in the Western Himalayas. Riccia ciliata, R. pathankotensis, R. robusta, R. gangetica, R. discolor (= R. himalayensis), R. frostii, R. reticulata, R. hirsuta, R. melanospora and R. crystallina are other common Indian species. R. pathankotensis, R. discolor, R. melanospora, R. robusta and R. cruciata are endemic to India.

Gametophyte : The gametophytic phase is predominant in the life cyclè of *Riccia*, hence the main plant body is a gametophyte.

External Structure : The gametophyte is flat, prostrate, dorsiventral, dark green and dichotomously branched. The branches of the thallus are linear to wedgeshaped. In terrestrial forms, the plants usually take a typical rosette form due to the presence of several dichotomies close to each other. These rosettes are up to 15 cm in diameter. The thalli of R. reticulata are relatively thick, sulcate and overlapped. In R. cruciata and R. pandei the thalli show cruciform arrangement. R. melanospora is characterized by the presence of several hairy epidermal outgrowths on the dorsal surface of the thallus. In terrestrial forms the thallus is approximately 5-7 mm in length and 1-3 mm in width.

In most of the spécies each branch of the thallus has a midrib on the dorsal surface. The midrib is represented by a shallow groove, known as dorsal groove or furrow. It usually extends from the base to the tip of the thallus. In the male plants of R. discolor the midrib is restricted to the apical part of the thallus, whereas in the female plants it extends to the entire length of the thallus. In R. frostii, however, midrib is absent and the upper surface of the thallus is flat or convex. At the apical portion of the thallus the midrib ends in a depression, known as apical notch. The growing point of the thallus is located in the notch.

The ventral surface of the thallus bears many rhizoids and scales. The rhizoids, which are simply the outgrowths of lower epidermal cells, are unicellular and unbranched tubular structures. They are of two types : (i) smooth walled rhizoids, their outer as well as inner walls are stretched and smooth, and (ii) tuberculate rhizoids, the inner wall of these rhizoids grows into peg-like or plate-like ingrowths. R. melanospora and R. robusta have mostly smooth walled rhizoids.

The rhizoids attach the thallus to the substratum and they also help in absorption of water and nutrients from the soil.

The scales are multicellular, one cell thick membranous structures. Their pink or violet colour is due to the presence of anthocyanin pigments. They are arranged in a single row along the margins of the thallus. In the apical region, they overlap each other and project forward to protect the growing point. The degree of persistence and size of the scales is dependent upon the habitat. Plants growing in moist terrestrial habitats usually have small, and ephemeral scales, whereas those of dry habitats have large and persistent scales. In some terrestrial species (e.g., R. crystallina, R. cruciata and R. frostii) the scales are either small, thin, delicate and inconspicuous or absent.

Riccia fluitans, an aquatic species, has a long, narrow, flat, ribbon-like, dichotomously branched thallus. It is about 30-50 mm in length and 1 mm in width. It lacks scales and rhizoids, but terrestrial forms of this species develop rhizoids and few small colourless scales at the anterior end of the thallus.

Internal Structure : Anatomically, the thallus is differentiated into a dorsal photosynthetic and a ventral storage region.

1. *Photosynthetic Region :* This region consists of one or more vertical rows of unbranched photosynthetic filaments. All the cells of the photosynthetic filament, except the uppermost one, are similar. These cells contain numerous discoid chloroplasts. The terminal cells are somewhat larger and colourless, and collectively form an ill defined loose and discontinuous upper epidermis. The photosynthetic filaments are separated from each other by narrow (e.g., R. pathankotensis, R. discolor, R. melanospora) or wide (e.g., R. robusta, R. cruciata) air chambers. These air chambers open to the exterior through simple air pores

on the dorsal surface of the thallus. The pores help in gaseous exchange.

In most of the species the air chambers in the photosynthetic region are usually in a single row but in R. robusta, R. cruciata and R. fluitans they occur in many irregular rows. In aquatic forms of R. fluitans, air chambers are almost completely closed by the dorsal epidermis but in terrestrial forms each air chamber opens by a very minute central pore.

2. *Storage Region :* It lies below the photosynthetic region on the ventral side. It is made up of compactly arranged parenchymatous cells. These cells are devoid of chloroplasts and do not have intercellular spaces. They contain starch as reserve food material. The innermost layer of the storage region forms the lower epidermis. Some cells of the lower epidermis give rise to rhizoids and scales.

Apical Growth

The apical growth in Riccia takes place by means of 3-5 or more apical cells situated in the apical notch. These cells appear triangular in vertical longitudinal section and rectangular in horizontal transverse section. The apical cell forms derivatives on dorsal and ventral sides alternately. Most part of the thallus is derived from the cells cut off on the dorsal side. The ventral derivatives of the apical cell form only the lower epidermis, rhizoids and scales.

Reproduction : Reproduction in Riccia takes place by vegetative and sexual methods.

Vegetative Reproduction : Vegetative multiplication of thallus takes place by the following methods.

1. By progressive death and decay of the older parts of the thallus. The thallus of Riccia is dichotomously

branched. The older parts of the thallus gradually die and as the process of death and decay reaches up to the dichotomy, the two branches of the thallus separate and each grows into a now plant.

2. By adventitious branches. Some species of Riccia (e.g., R. fluitans) have adventitious branches on the ventral surface of the gametophyte. These branches, on separation from the parent thallus, grow into new plants.
3. By death of the thallus in drought conditions. During prolonged dry conditions the entire plant, except the apical part dies. The latter, however, grows deep into the soil and becomes thick. It resumes active growth during next season and develops into a new thallus. This is, in fact, a method of perennation rather than multiplication.
4. By tubers. In many species (e.g., R. bulbifera, R. discolor, R. perennis, R. vesicata) the apices of the thallus lobes form perennating structures, the tubers, at the end of the growing season. The tubers remain dormant during unfavourable period and grow into new thalli on the resumption of favourable conditions. Besides multiplication, it is also a method of perennation.
5. By rhizoids. In R. glauca, the apices of rhizoids divide repeatedly and form a cellular mass. These cells contain chloroplasts and have the capability to grow into a new thallus.

Sexual Reproduction : Sexual reproduction takes place by means of well developed sex organs. The male and female reproductive organs are known as antheridia and archegonia respectively. Most of the species of Riccia (e.g., R. melanospora, R. crystallina, R. cruciata, R. pathankotensis, R. glauca and R. robusta) are homothallic or monoecious, i.e., they bear male and female sex organs on the same thallus. However, some species (e.g., R. discolor, R. sanguinea, R. frostii and R. pearsonii) are heterothallic or dioecious and in these species male and female reproductive organs are borne on different thalli. As the thallus matures, reproductive structures develop on the dorsal surface in acropetal succession, i.e., the mature sex organs are present at the posterior end and the younger towards the apex of the thallus. In monoecious species, several alternate groups of antheridia and archegonia occur along the midrib.

In some dioecious species (e.g., R. crustisii), the sex of the gametophyte is genetically determined-two spores of a tetrad give rise to male thalli and the other two to female thalli.

1. *Antheridium :*

   (a) *Structure :* The mature antheridium is a globular, oval or club shaped structure embedded on the dorsal surface of the thallus. It is attached to the bottom of the antheridial chamber by a short multicellular stalk. The antheridial chamber opens on the dorsal surface of the thallus by a narrow pore, called ostiole. The body of the antheridium has a solid central mass of fertile cubical cells, the androcytes, surrounded by a one cell thick sterile jacket layer.

   (b) *Development :* The antheridium develops from a single superficial cell on the dorsal surface of the thallus. This cell, called antheridial initial, usually lies one or two cells behind the apical cell. It is relatively large than other cells and has dense cytoplasm (Fig.). A transverse wall divides the antheridial initial into an outer cell and a lower basal cell (Fig.). The outer cell eventually forms the antheridium and the basal cell forms the embedded part of the stalk of the antheridium.

The outer cell forms a 4-celled filament by two successive transverse divisions. The upper two cells of this filament are called primary antheridial cells and the lower two primary stalk cells (Fig.). The latter form a two celled thick stalk of the antheridium. The primary antheridial cells undergo two vertical divisions at right angles to each other forming two tiers of four cells each (Fig.). The cells of both tiers divide periclinally resulting in eight outer sterile primary jacket cells and eight inner fertile primary androgonial cells (Fig.). The primary jacket cells divide only anticlinally and form one celled thick jacket of the antheridium (Fig.). The primary androgonial cells divide repeatedly in all possible planes and form a mass of androcyte mother cells (Fig.). Each of these cells divides diagonally to form two triangular androcyte cells, enclosed within a common parent wall (Fig.). The androcyte cell metamorphoses into an antherozoid (Fig.).

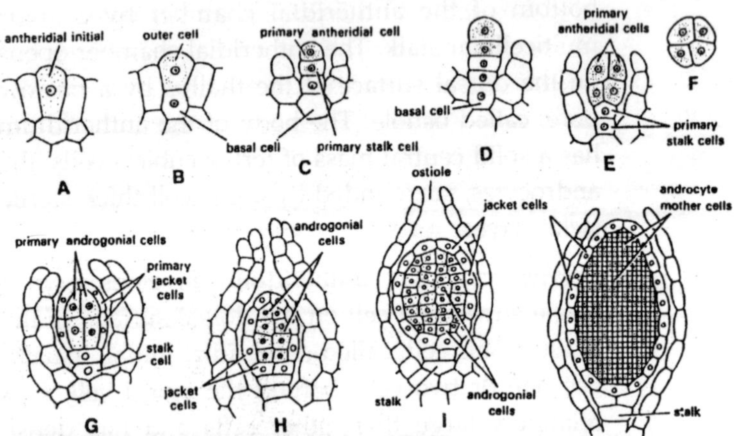

Ricci : Development of antheridium; A-I, successive stages in the development of antheridium, J. Mature antheridium, K-N. Metamorphosis of androcyte into antherozoid, O. An antherozoid

2. *Archegonium :*

(a) *Structure :* Mature archegonium is a flask-shaped structure attached to the bottom of the archegonial chamber by a short multicellular stalk. It is differentiated into a basal swollen venter and an upper elongated and narrow neck. The neck is composed of six vertical rows of cells and each row has six to nine cells. Within the neck lies a uniseriate row of four to six neck canal cells. The tip of the neck is covered by four relatively large cover cells. The venter is surrounded by a one cell thick jacket. It has a large naked egg or oosphere and a small venter canal cell. Although the archegonium is embedded in the archegonial chamber, the upper part of the neck always protrudes above the surface of the thallus.

(b) *Development of Archegonium :* Like antheridium, the archegonium also develops from a single dorsal superficial cell, the archegonial initial which lies close to the apex of the thallus. It is comparatively large, has dense cytoplasm and protrudes on the dorsal surface of the thallus as a small papilla. The archegonial initial is differentiated into an upper outer cell and a lower basal cell by a transverse wall. The basal cell forms the embedded part of the archegonium, whereas the outer cell functions as archegonial mother cell. The archegonial mother cell divides by three vertical eccentric intersecting divisions resulting in three peripheral cells. These cells enclose a central primary axial cell (Fig.). Each peripheral cell divides by a radial wall and as such six jacket initial cells are formed (Fig.). The latter divide transversely forming two superimposed tiers of six cells each. The cells of the upper tier function as neck initials and those of the lower tier as venter initials (Fig.). The neck initials divide transversely to form the neck of the archegonium. The neck consists of six vertical rows of neck cells with six to nine cells in each row (Fig.). The venter initials divide repeatedly and eventually form the wall of the swollen venter.

At the same time, the primary axial cell divides by a transverse wall into an upper primary cover cell and a lower central cell. (Fig.). The primary cover cell forms a rosette of four cover cells by two vertical divisions at right angles to each other. A transverse wall divides the central cell into an upper primary neck canal cell and a lower venter cell (Fig.). The former gives rise to four to six neck canal cells by repeated transverse divisions. The venter cell divides transversely into a small venter canal cell and a large egg cell (Fig.).

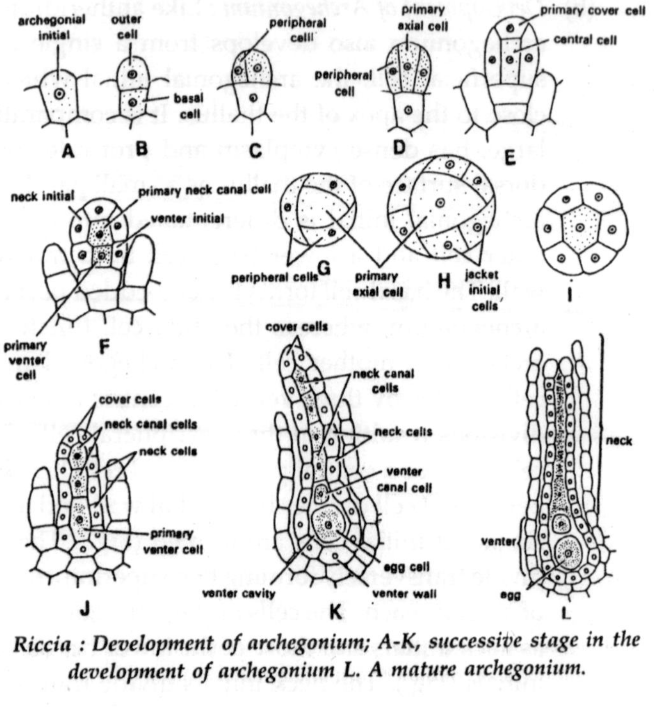

Riccia : Development of archegonium; A-K, successive stage in the development of archegonium L. A mature archegonium.

3. *Fertilization :* Fertilization occurs only in the presence of water as it is essential for (i) the liberation of antherozoids from the antheridium, (ii) the transfer of antherozoids to the archegonium, and (iii) disintegration of neck canal cells and venter canal cell. Usually rain or dew water accumulated in the form of a thin layer in the dorsal furrow helps in fertilization.

As the antheridium matures and the androcytes metamorphose into antherozoids, sterile jacket cells of the apical region of the antheridium are softened by water. These cells eventually disintegrate forming a pore. The antherozoids are thrown out of the antheridium by imbibitional force. They float on the dorsal surface of the thallus. Just before fertilization, the neck canal cells and venter canal cell of the archegonium disintegrate to form a mucilagenous substance which imbibes water. The imbibitional force thus built within the neck results in separation of cover cells from one another. In this way a free passage is formed for the entry of antherozoids in the archegonium. A few drops of mucilage also ooze out of the open neck of the archegonium. It contains certain proteins and potassium salts which attract the freely swimming antherozoids towards the open neck and then direct them towards the egg. Although many antherozoids may enter the neck, only one of them fuses with the egg to form a diploid zygote or oospore.

Sporophyte : The diploid zygote formed by the fusion of male and female gametes is the mother cell of the sporophytic generation. The zygote secretes a wall around itself immediately after fertilization. It increases in size and eventually occupies the entire cavity of the venter (Fig.). The act of fertilization also stimulates the cells of the venter wall. They divide anticlinally as well as periclinally and eventually form a two layered calyptra around the young sporophyte.

Development of Sporophyte : The first division of zygote is usually by a transverse wall, resulting in two cells of almost similar size (Fig.). The second division, at right angles to the first one, results in the formation of a four-celled quadrant (Fig.). Sometimes the second division is parallel to the first one and then a four-celled linear filament is formed. Another vertical division at right angles to the preceding one gives rise to an eight-celled embryo (octant). All cells of the octant are almost of similar size. Subsequent divisions occur in all possible planes until a more or less spherical mass of 20-40 cells is formed. The superficial cells of this mass now undergo periclinal divisions, resulting in an outer amphithecium layer and a central mass of cells, the endothecium. (Fig.). The cells of the amphithecium divide only anticlinally and form a single layered sterile jacket of the sporogonium. The endothecium behaves as archesporium and its cells divide repeatedly to form a mass of sporogenous cells (Fig.). The last generation of sporogenous cells are the potential spore mother cells.

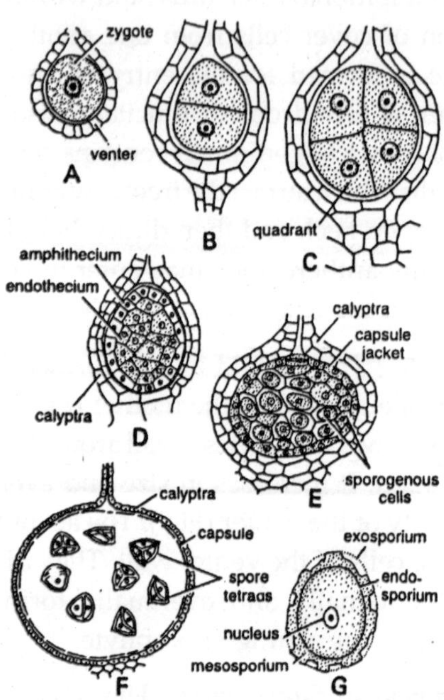

Ricci : Development of sporophyte; A-E. Successive stages in the development of sporophyte, F. A mature sporophyte in longisection, G.Spore.

All spore mother cells are fertile and have the potential to form spore tetrads. But in some species of Riccia there is a gradual sterilization of sporophytic tissue, a characteristic feature of the sporophyte of bryophytes. For example, in R. cerystallina some spore mother cells become abortive and fail to produce spores. The cytoplasm of these cells is watery and vacuolated and provides nutrition to the active sporogenous cells. They are known as nurse cells and are probably forerunners of elaters, present in more advanced forms of Hepaticopsida.

Sporogenesis : The spore mother cell undergoes meiosis to form four haploid spores. This division involves two steps. In the first step, the chromosome number becomes half of the somatic number and in the second step, the two daughter nuclei divide mitotically. After the first step, the wall formation is only partial but towards the end of the second step walls are laid down simultaneously around each haploid nucleus. Consequently, four tetrahedrally arranged haploid spores are formed which remain enclosed within the common wall of the spore mother cell (Fig.). However, in R. pearsonii the four spores form an isobilateral tetrad.

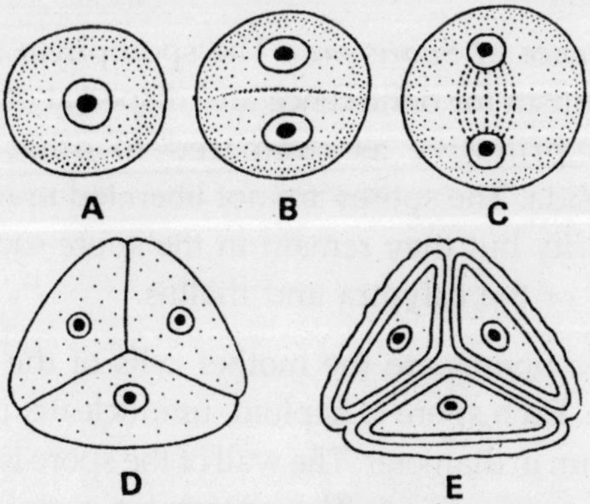

Ricci: Stages in sporogenesis.

In several species of Riccia (e.g., R. crustisii R. pearsonii) the spores are united in permanent tetrads. They are dispersed in groups and all the four spores germinate together to form a group of four plants.

In most species of Riccia, the haploid chromosome number is 8 (e.g., R. crystallina, R. discolor) or 9 (e.g., R. arvensis, R. Californica, R. campbelliana, R. trichocarpa). R. gangetica is, however, a polyploid species (n=24).

Structure of the Mature Sporophyte : The mature sporophyte of Riccia is represented by spore-sac or capsule only, embedded on the dorsal surface of the gametophyte. It lacks both foot and seta (Fig. F). The spore mother cells are enveloped by a single layered sterile jacket, the capsule wall. The capsule has a protective covering of two layered calyptra which, infact, is a part of the gametophyte. The capsule wall degenerates before the division of spore mother cell and after sometime the inner layer of the calyptra also degenerates partly. Thus at this stage the spores of Riccia, which represent the new gametophytic generation, are housed in a cavity of the old gametophyte. This is an anomaly in which the new gametophyte is enclosed within the old gametophyte. Thus in a mature sporophyte spore tetrads lie in a cavity enclosed by a single-layered calyptra.

Dehiscence of Spore-sac : The sporophyte does not have any mechanism for dehiscence and dispersal of spores and as such it is primitive as compared to other members of Hepaticopsida. The spores are not liberated immediately after their maturity but they remain in the spore sac till the death and decay of the calyptra and thallus.

Spores : Spores are the mother cells of the gametophytic generation. Each spore is haploid, uninucleate, pyramidal and 0.05-0.12 mm in diameter. The wall of the spore is differentiated into three layers (Fig.). The outermost cutinised layer, the exosporium is ornamented or sculptured. The middle layer, the mesosporium has three concentric zones, and the innermost pectic layer, the endosporium is homogeneous. These three layers develop in a centripetal order, i.e., the exosporium develops first and the endosporium in the last.

Germination of Spore and Formation of Young Gametophyte : The spores germinate when there is enough water in the soil. The exosporium and the mesosporium rupture at the triradiate ridge and the endosporium comes out in the form of a tubular outgrowth, the germ tube (Fig.). The germ tube is filled with cytoplasm which contains oil globules and chloroplasts. It elongates rapidly and forms a club-shaped structure. A transverse wall at the distal end of the tube forms a small cell (Fig.), which undergoes two vertical divisions at right angles to each other, followed by a transverse division. As a result two tiers of four cells each (i.e., total eight cells) are formed (Fig.). One of the four cells of the upper tier becomes apical cell which regularly cuts off cells on the dorsal and ventral sides (Fig.). These derivatives form a new thallus.

Simultaneously, with the division of the germ tube, a thin filamentous rhizoid develops from its base. This is known as the first rhizoid (Fig.). The young rhizoid has chloroplasts but they disappear subsequently. As the new thallus grows at the tip of the germ tube, a large number of rhizods develop from the ventral epidermis of the thallus (Fig.).

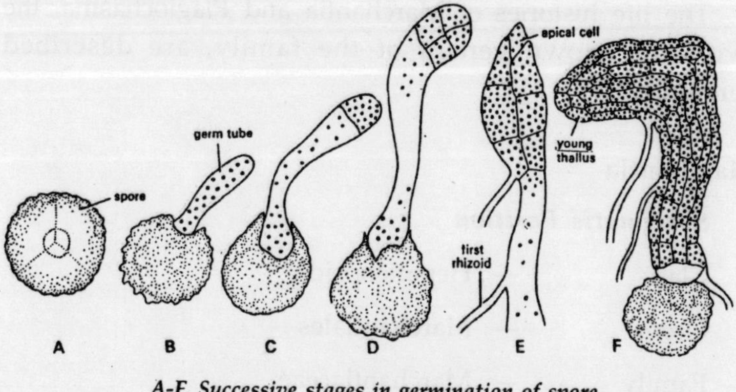

A-F. Successive stages in germination of spore.

Family Marchantiaceae

The family Marchantiaceae is represented by about 23 genera and 250 species. Dumortiera, Marchantia, Plagiochasma, Fimbriaria, Conocephalum, Reboulia, Cryptometrium and Lunularia are some common genera of the family.

The following characteristics of Marchantiaceae differentiate them from Ricciaceae.

(1) The photosynthetic region is divided into many chambers and each chamber has several simple or branched photosynthetic filaments.

(2) The photosynthetic chambers open on the dorsal surface of the thallus by definite pores which may be simple or barrel-shaped.

(3) Sex organs are present on erect and stalked receptacles. The receptacles bearing antheridia and archegonia are known as antheridiophores and archegoniophores respectively.

(4) The sporophyte is differentiated into foot, seta and capsule.

(5) In addition to spores the capsule always has some sterile cells, known as elaters.

The life histories of Marchantia and Plagiochasma, the two well known genera of the family, are described hereunder.

Marchantia

Systematic Position

Class — Hepaticopsida

Order — Marchantiales

Family — Marchantiaceae

The genus Marchantia, with about 65 species, is found in all parts of the world. The species grow best in cool, moist and shady places, Marchantia polymorpha grows as a pioneer in the burnt soil of the forest after fire.

The genus is represented in India by some 11 species growing mainly in the Western Himalayas. Marchantia polymorpha is the most widely distributed species and it occurs at an altitude of 2,500-5,000m on moist river banks and rocks. M. palmata occurs in Kashmir, Kumaon, South India, Bengal and Assam, M. simlana in Himachal Pradesh and M. nepalensis is common in the plains of Punjab, and in Garhwal and Kumaon hills. M. nepalensis and M. simlana are exotic to India.

Thalli of Marchantia with sex organs can be found in the Himalayan ranges during February and March and in the South Indian hills during October and November.

Gametophyte

External Structure : The gametophyte of Marchantia is prostrate, dorsiventral and dichotomously branched like that of Riccia. A distinct midrib is present in each branch on the thallus, which is marked on the dorsal surface by a shallow groove. A close observation of the dorsal surface reveals many polygonal or rhomboidal areas which demarcate the outline of underlying air chambers (Fig.). Each polygonal area has a pore in its centre through which the air chamber is connected with the external environment.

The dorsal surface of the thallus bears gemma cups along the midrib. They are crescent shaped structures with smooth, spiny or fimbriate margins (Fig.). The gemma cups bear gemmae, which help in vegetative propagation.

Each branch of the thallus grows indefinitely by means of a growing point situated at the apex of the thallus in a groove, called apical notch.

The ventral surface of the thallus bears rhizoids and scales on both sides of the midrib. The rhizoids are usually colourless and unicellular. They are of two types : smooth walled and tuberculate. Besides fixing the thallus to the substratum, they also absorb water and nutrients (the function of absorption is mainly performed by tuberculate rhizoids). The scales are multicellular and one cell in thickness. They are usually violet in colour due to the presence of anthocyanin pigments in their cytoplasm. The scales are arranged in two rows on either side of the midrib but in M. polymorpha there are three or four rows on either side of the midrib. Unlike Riccia, the scales in Marchantia are of two types :

(i) *Appendiculate :* These are characterized by the presence of an apical sub-rotund appendage; they usually form the inner row of scales, close to the midrib and

(ii) *Ligulate :* These are relatively small and do not have any appendage; They form the outer (or marginal) row. The primary function of scales is to protect the growing point. Besides, they also retain some water by capillary action.

The sexually mature thalli possess specialized erect branches, the gametophores, which bear sex organs. These branches arise form the growing apex situated in the apical notch. The branches present on the male thalli bear antheridia and are known as antheridiophores, whereas those present on the female thalli bear archegonia and are called archegoniophores.

Internal Structure : Anatomically, the thallus of Marchantia is more complicated than that of Riccia. A cross section of the thallus shows three distinct regions (i) epidermal region, (ii) photosynthetic region, and (iii) storage region.

1. *Epidermal Region :* It includes upper and lower epidermis. The upper or dorsal epidermis forms a

protective covering above the photosynthetic region. It is interrupted by many barrel shaped air pores that open internally into air chambers. Each pore is usually surrounded by 4-8 superimposed tiers of cells and each tier consists of a ring of 4-5 cells. The cells of the lowermost tier project inward giving the pore a star shaped appearance. Half of the tiers project above the level of the dorsal epidermis and the other half protrude into the air chamber below the epidermis. The pores thus have comparatively wide pore passage in the middle than on the margins. The pores are analogous to stomata but the cells of the pore wall do not control the pore size like the guard cells of stomata.

The lowermost layer of the storage region forms the lower epidermis. The cells of the lower epidermis give rise to rhizoids and scales.

2. *Photosynthetic Region :* The upper epidermis is followed by a large number of air chambers. These chambers develop schizogenously and are separated from each other by a single layered partition wall, made up of 4-6 chloroplast containing thin walled cells. The air chambers are usually uniform in shape and are arranged in a single horizontal row. Many simple or branched photosynthetic filaments arise from the floor of each chamber. The filaments are composed of chloroplast containing cells.

3. *Storage Region :* It lies immediately below the photosynthetic region as a compact zone of several layers of polygonal parenchymatous cells. The storage region is thick in the centre and gradually tapers towards the margin. The cells of this region are devoid of chloroplasts and do not have intercellular spaces in between them. Most of the cells of the storage region contain starch or protein granules, and some mucilage

and oil containing cells are also present. The cells of the midrib region are tangentially elongated and have reticulate thickenings.

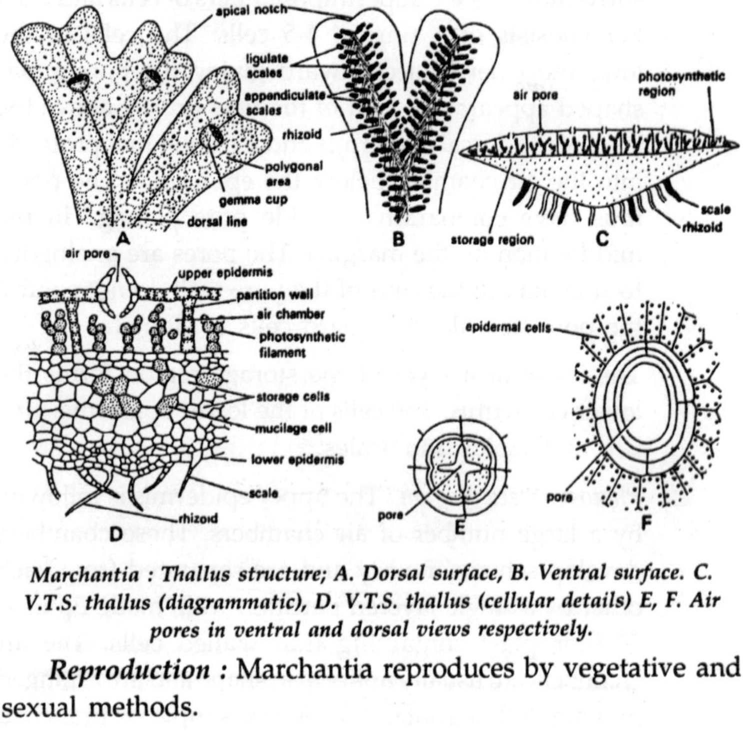

Marchantia : Thallus structure; A. Dorsal surface, B. Ventral surface. C. V.T.S. thallus (diagrammatic), D. V.T.S. thallus (cellular details) E, F. Air pores in ventral and dorsal views respectively.

Reproduction : Marchantia reproduces by vegetative and sexual methods.

Vegetative Reproduction : Vegetative reproduction takes place regularly by the following methods.

1. *By Progressive Death and Decay of Thallus :* The mature cells of the gametophyte are short lived and new cells are regularly formed by the apical cell. The death and decay, therefore naturally starts from the posterior end. As the process of decay reaches the point of dichotomy, two apical parts of the thallus separate and form two independent plants.

2. *By Adventitious Branches :* Some species of Marchantia possess adventitious branches on the ventral surface of

the gametophyte. These branches on separation from parent thallus, grow into new thalli. In species like M. assamica and M. palmata adventitious branches arise from the stalks or discs of archegoniophores. M. nepalensis has apical adventitious shoots.

3. *By Gemmae (singular = gemma)* : These are specialized multicellular asexual reproductive bodies. They are borne in small cupules, known as gemma cups, present on the dorsal surface of the thallus in the midrib region. A gemma cup is approximately 2 mm in diameter and 3 mm in height. Its margins are hyaline, lobed, spinuous or entire. From the floor of gemma cup many small stalked, discoid and biconvex gemmae arise. The gemma is constricted in the middle and the two notches in the constriction possess a row of apical cells. Most of the cells of the gemma contain chloroplasts. But some superficial cells on both sides are colourless and larger than the neighbouring cells. The cytoplasm of these cells is dense and granular. These cells give rise to rhizoids on germination, hence called rhizodal cells.

Some minute club-shaped mucilage hairs also arise from the floor of the gemma cup. The mucilage secreted by these hairs imbibes water and the imbibitional force thus generated helps in dissemination of gemmae from the gemma cup.

The gemmae are bilaterally symmetrical and are not differentiated into dorsal and ventral surfaces. The mature gemmae fall on the ground and if conditions are suitable their germination start immediately. The surface of the gemma which comes in contact of the soil gives out many rhizoids. This surface eventually becomes the lower (ventral) surface of the thallus. Meanwhile the apical cells present in the two lateral notches become active and form two thalli in opposite directions.

Usually the gemmae present on the male thallus form male plants and those on the female thallus give rise to female plants.

Sexual Reproduction : In Marchantia, sex organs are borne on special stalked receptacles, the gametophores. The receptacles bearing male and female sex organs are called antheridiophores and archegoniophores (carpocephalum) respectively. They develop at the distal end of the thallus. In their formation, the growing point of the thallus is fully utilized, therefore, after the formation of gametophores the vegetative growth of the thallus ceases completely.

The gametophore, in fact, is the extension of the vegetative thallus. It shows the characteristics of the thallus, i.e., it has air chambers, air pores rhizoids, etc. The development of sex organs in Marchantia is affected by environmental factors. Usually excess of nitrogenous substances in the soil accelerates their formation. Marchantia is a heterothallic or dioecious plant, i.e., male and female receptacles are present on different thalli. Thus sexual reproduction involves two types of thalli, male (with antheridiophore) and female (with archegoniophore). In some specimens of M. palmata and M. polymorpha both antheridia and archegonia have been found on the same receptacle. Such receptacles are said to be androgynophores.

Antheridiophore : The antheridiophore has a 1-3 cm long stalk that bears a slightly convex usually 8 lobed peltate disc at its apex. However, in M. geminata the peltate disc is 4-lobed and in M. palmata, 3-9 lobed. Each lobe of the peltate disc has a growing point at its tip and this represents the apex of a branch. The internal structure of the peltate disc is similar to that of the thallus.

The upper epidermis is interrupted by a number of barrel-shaped air pores, each opening below into an air chamber. The air chambers have many simple or branched photosynthetic filaments. The antheridial chambers alternate with the air chambers. Each antheridial chamber contains a single antheridium and opens externally by a pore, the ostiole. On each lobe of the antheridial disc antheridia arise in acropetal succession, i.e., the oldest being near the centre and the youngest towards the apex of the lobe. The stalk of the antheridiophore has two longitudinal grooves which contain rhizoids and scales.

(a) *Development of Antheridium* : The development of antheridium in Marchantia is similar to that of Riccia. It develops from a single superficial cell, the antheridial initial, situated on the dorsal surface of the disc of the antheridiophore. The antheridial initial is comparatively large and has dense cytoplasm (Fig.). It differentiates into an outer cell and a basal cell by the formation of a transverse wall (Fig.). The outer cell eventually forms the antheridium, whereas the basal cell forms the embedded part of the stalk of the antheridium.

The outer cell, by two successive transverse divisions, forms a four-celled filament. The upper two cells of the filament function as primary antheridial cells and the lower two as primary stalk cells (Fig.). The primary stalk cells divide to form a 2-celled thick stalk (Fig.). The primary antheridial cells divide by two vertical walls at right angles to each other, thus forming eight cells. These cells are arranged in two tiers of four cells each (Fig.). The cells of both the tiers now undergo a periclinal division forming eight outer sterile primary jacket cells and eight inner fertile primary androgonial cells (Fig.). The primary jacket cells divide only anticlinally and eventually form a single layered jacket of the antheridium. The primary androgonial cells, however, divide in all possible planes and give rise to a mass of androcyte mother cells (Fig.). Each of the latter divides diagonally to form two triangular androcyte cells, which are enclosed within a common parent wall (Fig.). Each androcyte cell metamorphoses into an antherozoid (Fig.).

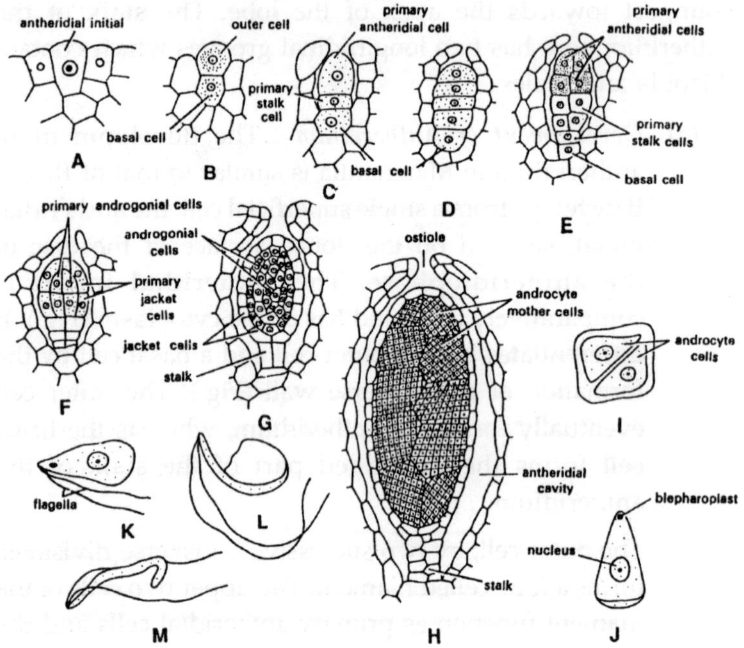

A-m. Marchania : Development of antheridium; A-G. Successive stages in the development of antheridium, H. A mature antheridium, I-L. Metamorphosis of androcyte into antherozoid, M. An antherozoid.

(b) *Mature Antheridium* : The mature antheridium is a globular or more or less pear-shaped structure, attached to the floor of the antheridial chamber by a multicellular stalk. The body of the antheridium has a single layered sterile jacket enclosing a mass of androcytes which eventually metamorphose into antherozoids. The antherozoid is a minute rod like biflagellate structure. When the antherozoid swims one of the flagella beats forward and the other backward.

(c) *Dehiscence of Antheridium :* Water collected on the peltate disc or in the rhizoidal grooves of the antheridiophore enters into the antheridial chambers through the ostiole. Some cells at the distal end of the antheridial jacket disintegrate when they come in contact with water. The antheridium thus ruptures and a large mass of androcytes comes out of the ostiole just like a smoke column. The androcytes spread apart on the surface of the disc and antherozoids come out within minutes.

Archegoniophore or Carpocephalum : The archegoniophore arises at the apical notch of the thallus, just like the antheridiophore. It also consists of a stalk and a terminal disc. The stalk of the archegoniophore is slightly longer (2-5 cm) than that of the antheridiophore. The terminal disc, which is formed by three successive dichotomies of the apex of the young archegoniophore, is usually 8-lobed (7-9 lobed in M. nepalensis and 7-11 in M. palmata). The lobes are dorsiventral and a growing point is located at the tip of each lobe. The entire disc with eight growing points at the margin is equivalent to a branch system as in the rosette-like thalli of Riccia and Marchantia.

(a) *Position of Archegonia on Archegoniophore :* The archegonia are borne on the dorsal surface of the receptacle in acropetal succession, i.e., the mature archegonia are towards the centre and the young ones towards the margin of the disc (Fig.). There are 12-14 archegonia on each lobe of the disc, arranged in a single row. The archegonia are sub-sessile and their necks are directed upwards. After fertilization the stalk of the archegoniophore elongates and the central portion of the dorsal surface of the disc grows more rapidly than the ventral surface. As a result of this differential growth of the two surfaces the marginal region of the disc bearing archegonia is pushed downward and inward (Fig.). Thus, in mature archegoniophore the archegonia lie on the ventral surface of the disc and they are inverted with their necks downwards. In the process of inversion, the arrangement of archegonia is also reversed from centripetal to centrifugal, i.e., now the youngest archegonium lies close to the stalk and the oldest towards the periphery of the disc.

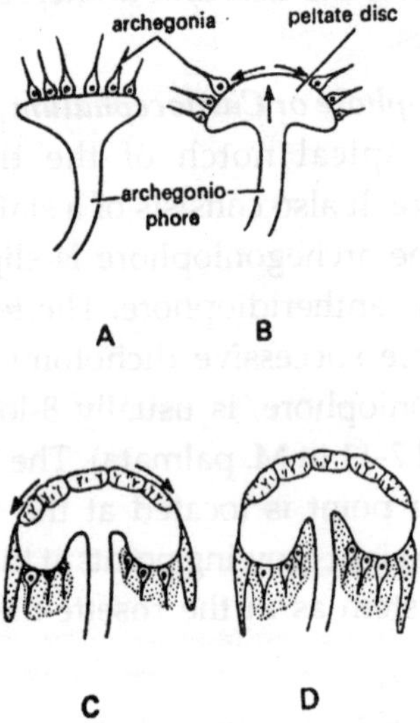

A-D. Marchantia : A. Position of archegonia on archegoniophore before fertilization, B-D. Stages in inversion of archegonia after fertilization.

The process of inversion is accompanied by the development of a one-celled thick plate like tissue, the perichaetium or involucre, on both sides of each archegonial row. Thus in a mature archegoniophore archegonia of a lobe are separated from those of the adjacent one by perichaetium (Fig.). In many species of Marchantia long, stout and green finger like projections develop from the margins of the disc between the groups of archegonia. These appendages, which are usually nine in number, are known as rays.

They give the archegonial disc an umbrella like appearance. The internal structure of the mature peltate disc is similar to that of the thallus. Its dorsal epidermis is interrupted by barrel shaped air-pores which communicate with the underlying air chambers. The air chambers have many simple or branched photosynthetic filaments.

(b) *Development of Archegonium* : The archegonia develop on the dorsal surface of the young receptacle in acropetal sequence. A superficial cell situated near the growing apex functions as archegonial initial. This cell is relatively large than the adjacent cells and has dense cytoplasm (Fig.). It divides by a transverse wall into an upper (or outer) primary archegonial cell and a lower (or inner) primary stalk cell (Fig.). The latter undergoes many irregular divisions and eventually forms a multicellular stalk.

The primary archegonial cell by three successive oblique vertical divisions forms three peripheral cells surrounding a central axial cell (Fig.). The peripheral cells divide anticlinally forming six jacket initials (Fig.). Each jacket initial undergoes a transverse division and thus two superimposed tiers of six cells each are formed. The cells of the upper tier are called neck initials and those of the lower tier as venter initials (Fig.). The neck initials divide transversely to form six vertical rows of neck cells, each with 6-9 cells (Fig.). The venter initials divide irregularly and eventually give rise to the wall of the swollen venter region.

Meanwhile, the axial cell also undergoes division. A transverse wall differentiates it into an upper primary cover cell and a lower central cell (Fig.). The primary cover cell divides by two vertical walls at right angles to each other, forming four cover cells (Fig.). The central cell again divides transversely forming an upper primary neck canal cell and a lower primary venter cell (Fig.). The primary neck canal cell, by two successive transverse divisions, gives rise to a row of four or more neck canal cells, whereas the venter cell undergoes a transverse division forming a venter canal cell and a large egg cell (Fig.).

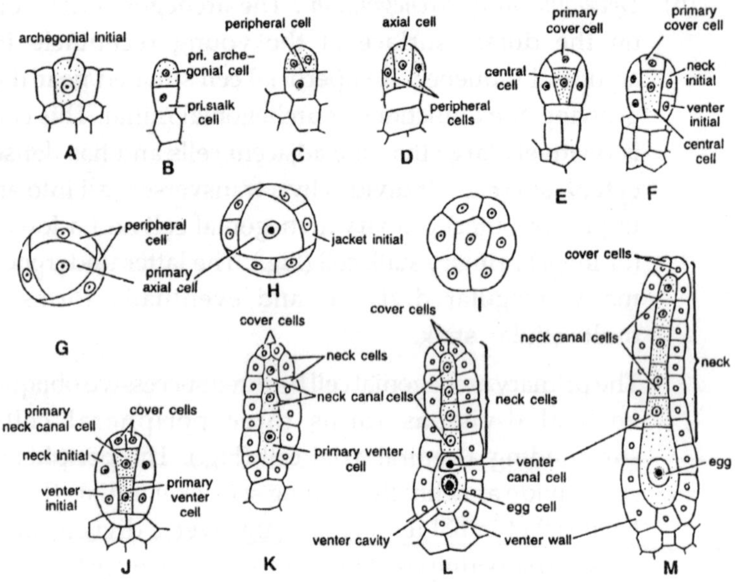

Marchantia: Development of archegonium: A-L. Successive stages in the development of archegonium. M. A mature archegonium

(c) *Structure of Mature Archegonium* : The mature archegonium is pendulous and attached to the ventral surface of the archegonial disc by a short stalk. It is flask-shaped structure with a basal swollen venter and an elongated neck (Fig.) The venter is surrounded by an one-celled thick sterile jacket and contains a large egg and a relatively small venter canal cell. The neck consists of six vertical rows of cells, the neck cells, which enclose a narrow canal with 4-8 neck canal cells. The tip of the neck has a rosette of four cover cells.

Fertilization : Water is essential for fertilization. As most of the species of Marchantia are dioecious (or hetcrothallic), fertilization is possible only when male and female thalli are growing close to each other. Fertilization takes place at a stage when archegonia are placed upright on the dorsal surface of the disc of the archegoniophore. Just before fertilization, the neck canal cells and the venter canal cell of the archegonium degenerate and form a mucilagenous substance. The pressure built by the imbibition of water by this substance results in the separation of cover cells. Thus a free passage is formed for the entry of antherozoids in the archegonium. The antherozoids, which swim freely on the surface of the antheridial disc, are splashed by rain water on the disc of the archegoniophore. They are attracted chemotactically towards the neck of the archegonium. Although many antherozoids may enter in an archegonium, only one fuses with the egg.

Immediately after fertilization, there is elongation of the archegoniophore. It is accompanied by the rapid growth of the central part of the archegonial disc that ultimately leads to inversion of archegonia.

The development of rays and perichaetium also takes place after fertilization.

Sporophyte : The diploid zygote (oospore) is the mother cell of the sporophytic generation.

Development of Sporophyte : The zygote enlarges in size and almost fills the cavity of the venter. The first division of the zygote is by a transverse wall (i.e., at right angles to the long axis to the archegonium), resulting in an outer epibasal cell and an inner hypobasal cell (Fig.). Further development of the embryo show variations in different species. In M. polymorpha and M. domingensis, the second division of the zygote is at right angles to the first division, resulting in a quadrant embryo (Fig.).

In M. chenopoda and some .other species, the second division of the zygote is parallel to the first division (i.e., at right angles to the long axis of the archegonium and it results in the formation of a 3-celled filamentous embryo. In a quadrant embryo, the epibasal cells form capsule and a part of seta, whereas the hypobasal cells form the remaining part of seta and foot. In a filamentous embryo the epibasal cell forms capsule, the middle cell seta and the hypobasal cell develops into foot.

The early development of zygote is accompanied by many important changes in the gametophytic tissue surrounding the embryo. The act of fertilization stimulates the cells of the venter wall which divide periclinally and eventually give rise to a 2-3 layered calyptra. The calyptra forms a protective covering around the young sporophyte. Repeated divisions also occur in the basal cells of the archegonium and as such a one-celled thick collar- like cylindrical outgrowth is formed outside the calyptra. It is called perigynium or pseudoperianth.

The quadrant (as well as the filamentous embryo) undergoes one or two vertical divisions and subsequently the cells destined to form capsule divide periclinally. Thus a peripheral single layered amphithecium becomes distinct from a central many celled endothecium (Fig.). The cells of the amphithecium divide only anticlinally and form a single layered wall of the capsule. As the sporophyte matures, the cells of the capsule wall develop annular thickenings. All cells of the endothecium function as archesporium. Repeated divisions of these cells result in a massive sporogenous tissue. Initially all cells of this tissue are almost of the same shape and size but subsequently about half of the sporogenous cells, by repeated transverse divisions, form vertical rows of more or less cuboidal spore mother cells.

The number of spore mother cells derived from a

sporogenous cell varies in different species. For example, in M. polymorpha each sporogenous cell undergoes five successive divisions giving rise to 32 spore mother cells, whereas in M. domingensis and some other species a sporogenous cell divides 3-4 times forming 8-16 spore mother cells. Each spore mother cell divides meiotically to form a spore tetrad.

The remaining cells of the sporogenous tissue, which do not differentiate into spore mother cells, function as sterile elater mother cells. These cells simply elongate to form long, slender spindle-shaped elaters. Unlike spores, the elaters are diploid cells and they have many spiral thickening bands. They are hygroscopic in nature and their squirming movements help in scattering of the spores.

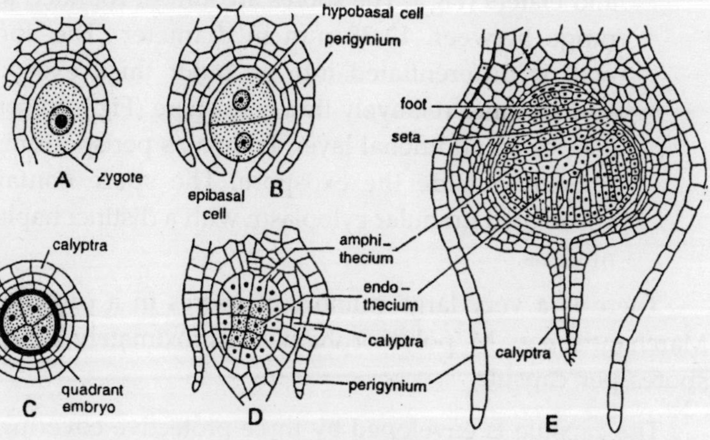

A-E. Marchantia : Successive stages in the development of sporophyte.

Structure of Mature Sporophyte : The sporophyte of Marchantia is differentiated into foot, seta and capsule (Fig.).

1. *Foot :* It is the basal bulbous part which anchors the sporophyte to the disc of the archegoniophore. It is composed of parenchymatous cells. The foot also absorbs nutrition from the gametophyte.

2. *Seta* : It is a short and stout stalk that connects foot to the capsule. It is made up of parenchymatous cells. In a young sporophyte the seta is represented by a small zone of relatively narrow cells. As the capsule matures, these cells undergo repeated transverse divisions and elongation. Thus the seta increases in length; it ruptures the calyptra and pushes the mature capsule out through the perigynium and perichaetium. At maturity, the cells of the seta become highly vacuolated.
3. *Capsule* : The mature capsule is a yellow coloured oval structure. It has a single layered sterile jacket. The cells of the jacket layer have annular thickening bands. The capsule contains numerous spores and elaters (Fig.). The spores are almost rounded and range between 12-30 mm in diameter. The spore wall is differentiated into an outer thick exospore and an inner relatively thin endospore (Fig.). In some species an additional layer, known as perisporium, is present outside the exospore. The spore contains vacuolated granular cytoplasm with a distinct haploid nucleus.

There is a very large number of spores in a capsule of Marchantia; e.g., M. polymorpha has approximately 3,00,000 spores per capsule.

The capsule is enveloped by three protective coverings - calyptra, perigynium and perichaetium.

Dehiscence of Capsule : As mentioned earlier, the seta, as it increases in length, ruptures the calyptra and pushes the capsule out from the protective coverings (Fig.). The single-layered capsule wall then splits into a variable number of longitudinal valves which extend from the apex to the middle of the capsule (Fig.). The annular thickenings in the jacket cells cause the valves to roll backward exposing the spores and elaters. The jerky movements of the elaters due to their hygroscopic nature assist in loosening up of the spore mass and scattering the spores in the air.

Germination of Spores : The spores of Marchantia remain viable for about a year. Under favourable conditions they absorb moisture from substratum and considerably increase in size. Chloroplasts reappear at this stage (Fig.). The first division of the spore is transverse and it results into two unequal cells (Fig.). The smaller cell forms the first rhizoid (Fig.), whereas the larger cell undergoes irregular divisions and forms a 6-8 celled filamentous structure (Fig.). The apex of the filament is usually 2- celled (Fig.), one of these cells functions as apical cell. The apical cell has two cutting faces on the lateral sides. Derivatives from both the faces contribute to the development of a new thallus (Fig.).

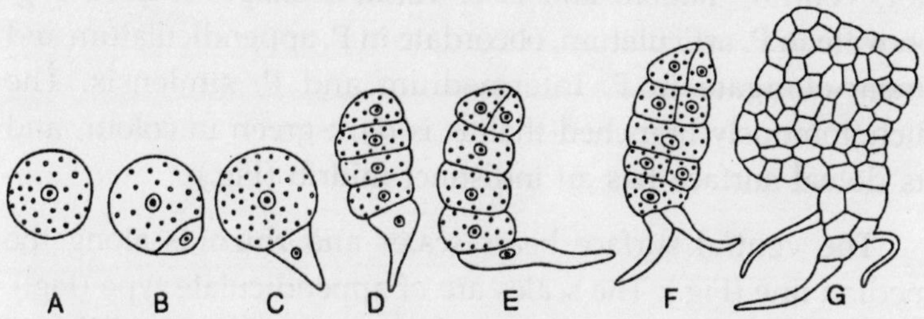

A-G. Marchantia : Successive stages in the germination of spore.

Marchantia is a dioecious plant. Two spores of a tetrad grow into male and the other two into female thalli.

Plagiochasma

Systematic Position

Class — Hepaticopsida

Order — Marchantiales

Family — Marchantiaceae

The genus Plagiochasma is represented by about 20 species,

widely distributed in most parts of the world. Some 10 species of the genus occur in India. Most of the Indian species are xeromorphic and are more adapted to drier habitats. P. appendiculatum occurs in the plains of South India, Gujarat, Rajasthan and Madhya Pradesh. P. simlensis is found near water pools and streams in Shimla hills and P. articulatum inhabits alpine region in the Himalayas at an altitude of about 4500 meters.

Plagiochasma appendiculatum, P. articulatum, P. simlensis, P. intermedium, P. cordatum and P. quadricornutum are some common Indian species. Of these, P. articulatum. P. simlensis, P. cordatum and P. quadricornutum are endemic to India.

Gametophyte

External Structure : The plant body is large, prostrate, dorsiventral, thalloid and is of variable shape. It is oblong-obovate in P. articulatum, obcordate in P. appendiculatum and linear-elongate in P. intermedium and P. simlensis. The dichotomously branched thallus is dark green in colour, and its dorsal surface has an indistinct midrib (Fig.).

The ventral surface bears scales and rhizoids along the median line (Fig.): The scales are of appendiculate type (Fig.). The number and shape of the appendages vary in different species, e.g., there is only one appendage in P. appendiculatum and P. simlensis and two in P. articulatum and P. intermedium. The appendages are rotund (e.g., P. appendiculatum) or acicular (e.g., P. simlensis) in shape. The scales are multicellular and one cell in thickness. They occur in a single overlapping row on either side of the median line. They protect the growing apex of the thallus. The unicellular rhizoids are of two types:

(i) smooth walled, and

(ii) tuberculate.

The inner wall of the smooth walled rhiziods is smooth while those of the tuberculate rhizoids has peg-like ingrowths, which appear as small circular dots in surface view (Fig.). The rhizoids help in anchorage and absorption of water and nutrients from the substratum. The growing point of the thallus is situated in a notch at the apex.

During xeric conditions, the thallus forms characteristic apical innovations (adventitious branches) which help in vegetative propagation. Sex organs are borne on special receptacles present on the dorsal surface of the mature thallus. The male receptacles (antheridiophores) are sessile and have a characteristic horse shoe-shape: The female receptacles (archegoniophores) have a short stalk and are disc like. They initiate in sub-terminal position along the median line but are subsequently pushed to the dorsal surface due to the growth of the thallus.

Internal Structure : In a transverse section the thallus of Plagiochasma shows three distinct regions : (i) epidermal region, (ii) photosynthetic region, (iii) storage region.

1. *Epidermal Region :* This region includes both the dorsal and ventral epidermis. The dorsal (upper) epidermis forms a uniseriate covering over the photosynthetic region. It is interrupted by simple pores that are connected internally with air chambers. The air pores are slightly raised above the surface of the thallus. Each pore is surrounded by 3-8 concentric rings, each with 6-8 cells. The ventral (lower) epidermis follows the storage region. It is made up of thin-walled cells. Many cells of the lower epidermis give rise to rhizoids and scales.
2. *Photosynthetic Region :* It consists of an irregular network of complex air chamber system. The walls of the air chambers are made up of a single layer of chlorophyllous cells. The air chambers are empty, i.e.,

without any photosynthetic filaments. They are interconnected through narrow passages. This region of the thallus is mainly concerned with the manufacture of food and is also called assimilatory region.

3. *Storage Region :* The photosynthetic region is followed by a well-developed storage region made up of parenchymatous cells. These cells are rich in starch grains and reserve food material. Inter- and/or intracellular mycorrhiza is occasionally present in the storage region.

Reproduction : Plagiochasma reproduces both vegetatively and by sexual methods.

Vegetative Reproduction : Vegetative reproduction usually takes place by means of certain adventitious branches, known as innovations. These branches develop at the apex of the thallus. When detached from the parent thallus, they grow into new independent thalli.

Sexual Reproduction : The plants may be homothallic (e.g.' P. appendiculatum) or heterothallic (e.g., P. intermedium, P. articulatum, P. cordatum, P. quadricornutum). Sex organs develop on specialized structures, known as receptacles. Both the male and female receptacles are distinctly dorsal in many species (e.g., P. appendiculatum), hence the vegetative growth of the thallus never stops. But in some species receptacles are terminal and the apical adventitious shoot develops after they have formed (e.g., P. articulatum). In such cases there is always a distinct line of articulation between the old and new adventitious lobe. The onset of the reproductive phase is associated with the change in the metabolic status of the gametophyte.

Male Receptacle : Male receptacles usually appear in early growing season on the dorsal surface of the thallus. They are sessile and typically horse shoe-shaped, a shape attributed to the presence of two growing points in the receptacle. The internal structure of the receptacle is similar to that of the thallus. It contains several antheridial chambers, each with a single antheridium.

Each antheridial chamber opens on the surface of the receptacle by a pore. A number of simple pores are also present on the receptacle and they open internally into air chambers. Antheridia develop acropetally on the surface of the male receptacle.

(a) *Structure of Antheridium :* The mature antheridiun is a club-shaped structure attached to the base of the antheridial cavity by a multicellular stalk. It has one-cell thick jacket layer of elongated or isodiametric cells which encloses a mass of androcyte cells. The androcyte cells metamorphose into slender filiform antherozoids. Each antherozoid has two flagella attached to its anterior end.

The development of the antheridium is similar to that of Marchantia.

(b) *Dehiscence of Antheridium :* Water enters inside the antheridial chamber through the pore, and some jacket cells at the distal part of the antheridium disintegrate as soon as they come in contact with water. This leads to the rupture of the antheridium and a mass of androcytes emerges in a long smoke-like column. On reaching the surface, the androcyte mass breaks up into individual androcytes. The antherozoids are liberated from the androcytes within few minutes.

Female Receptacle : The archegonia are present on special upright branches known as archegoniophores or carpocephalum. They are sessile or stalked. Archegoniophores arise in the midrib region, close to the apex, on the dorsal surface of the thallus. They do not involve the apical cell and as such are sub-terminal in position. The mature archegoniophores are, however, pushed back on the dorsal surface due to subsequent growth of the thallus.

The archegoniophore consists of a short stalk and a terminal disc. There is no rhizoidal furrow on the stalk as in Marchantia. The archegoniophore is encircled by minute, ovate to linear scales organised in a compound imbricate manner. The disc of the receptacle is 2-9 lobed and there is only one archegonium on each lobe. The anatomy of the archegoniophore is similar to that of the thallus, but it has barrel-shaped air pores.

(a) *Position of Archegonia on Archegoniophore* : On the young receptacle the archegonia are borne acropetally on the upper surface of the disc, and the necks of archegonia are directed upward. After fertilization, due to rapid growth of the central part of the disc, the archegonia come to lie on the flanks of the disc with their necks placed horizontally, i.e., parallel to the lobes of the receptacle.

(b) *Structure of Archegonium* : Archegonia lie in grooves formed by the overgrowth of the involucre. A mature archegonium is a flask shaped structure attached to the lobe of the disc by a short stalk. It has a broad venter and a narrow neck. The venter contains a large egg and a venter canal cell. The neck of the archegonium is composed of six vertical rows of cells and it has 8-9 neck canal cells. There are four cover cells at the tip of the neck. Shortly before fertilization, the neck canal cells and the venter canal cell disintegrate to form a mucilagenous mass.

This mucilagenous mass absorbs water and swells up

thus separating the cover cells apart. 3. Fertilization. Antherozoids are carried to the archegonia by splashes of rain water or by water currents. Several antherozoids reach an archegonium but only one unites with the egg to form the oospore. At the time of fertilization there are slow circulatory movements of the neck of the archegonium and this facilitates the process of fertilization.

Sporophyte : The diploid oospore is the mother cell of the sporophytic phase. It secretes a cell wall and increases in size until it almost fills the cavity of the venter (Fig.). It eventually develops into a sporophyte.

Development of Sporophyte : The oospore divides first by a transverse wall, forming an outer epibasal cell and an inner hypobasal cell (Fig.). The epibasal cell divides again by a transverse wall and as such a 3- celled filamentous embryo is formed (Fig.). The upper cell of this filamentous embryo eventually gives rise to the capsule, the middle to the seta, and the lower to the foot. All the three cells of the filamentous embryo divide transversely (Fig.). The derivatives of the middle and the lower cells then divide vertically. Thus at this stage the embryo has only a single tier of cells in the upper region and two or more tiers of cells in the middle and lower regions.

The cells of the upper region of the embryo divide periclinally, differentiating a single-layered outer amphithecium from the inner endothecium (Fig.). The cells of the amphithecium divide only anticlinally and form a single-layered jacket of the capsule. The jacket cells are thin-walled and are almost uniform throughout. The endothecium divides repeatedly in all planes and gives rise to archesporium, which forms the sporogenous tissue (Fig.).

Each undifferentiated cell of the sporogenous tissue divides to form two daughter cells. One of the daughter cells gives rise to one or more spore mother cells and the other acts as elater mother cell (such division of sporogenous cell into fertile and sterile cells is designated as "spore-elater division"). The spore mother cell, by a reduction division, gives rise to a tetrad of haploid spores whereas the elater mother cell elongates and forms the elater. The elater develops characteristic helical thickenings on its walls.

The cells of the middle region divide almost exclusively in one plane and give rise to a short seta, composed of several vertical rows of cells. As the sporophyte matures, there is a sudden increase in the length of seta due to elongation of its cells. As seta grows in length, it ruptures the calyptra and pushes out the capsule.

The cells of the lower region of the embryo divide in all possible planes, forming a thick bulbous foot. The foot serves as an anchorage and absorbing organ.

Accompanied with the development of the sporophyte, some important changes occur in the gametophytic tissue surrounding the archegonium. The cells of the venter wall get stimulated by fertilization and they divide to form a calyptra. The calyptra is composed of two or more cell layers and it surrounds the sporophyte till maturity (Fig.).

Structure of Mature Sporophyte : The sporophyte of Plagiochasma is differentiated into a bulbous foot, a short seta and a spherical capsule (Fig.).

1. *Foot* : It is the basal bulbous part of the sporophyte, composed of thin walled parenchymatous cells. It is an anchoring and absorbing organ of the sporophyte and derives nutrients from the gametophyte.
2. *Seta*. It is present between the foot and the capsule. It is composed of several vertical rows of parenchymatous cells.

3. *Capsule :* The capsule is a nearly spherical structure. It has a single layered jacket of thin walled cells, enclosing spores and elaters. The spore wall is differentiated into three layers, the outermost perinium shows reticulate ornamentation. The elaters are long, narrow, spindle-shaped cells with 2-4 spiral thickening bands (Fig.). In P. intermedium, thickening bands are absent and instead the elaters are uniformly thickened. The elaters are hygroscopic and they coil and uncoil with the changes in atmospheric humidity. Thus they help in separation and scattering of spores.

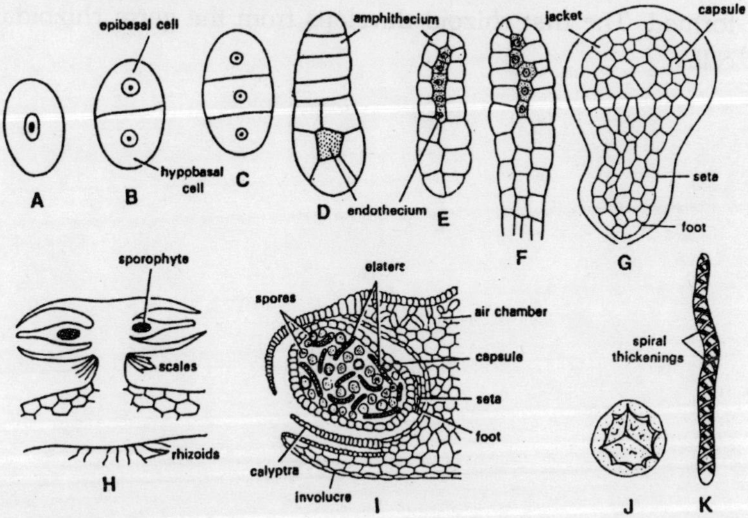

***Plagiochasma:* Development of sporophyte; A-G. Successive stages in the development of sporophyte, H. V.T.S. of thallus passing though sporophytes, I. Mature sporophyte in longisection, J. Spore K. Elater.**

Dehiscence of Capsule : At maturity of the capsule, the upper part of the jacket breaks irregularly but the lower portion remains intact like a cup. The dispersal of spores is assisted by elaters which show hygroscopic movements. These movements are due to unequal expansion of the cellulose wall and the spiral bands of the elaters.

Young Gametophyte

Spores germinate immediately after dispersal, if favourable conditions are available. They are viable for about six months under storage conditions. They absorb water from the substratum and increase in size considerably. Then the spore divides into a large germ tube cell and a small germ rhizoidal cell. As the germ tube elongates, chloroplasts and oil bodies make their appearance.

The germ tube divides repeatedly and forms a germ plate. An apical cell is differentiated at the apex of the germ plate and through the activity of this apical cell a new thallus is formed. The first rhizoid develops from the germ rhizoidal cell.

10

General Characters

Bryophyta (Greek : Bryon = moss, phyton = plants) is a group of the simplest and primitive plants of the class Embryophyta. These plants occupy a position intermediate between the green thallophytes (algae) and pteridophytes. The group is represented by about 960 genera and 24,000 species, occurring on most parts of the earth wherever there is sufficient moisture to sustain plant life. Bryophytes though regarded as terrestrial plants, occur in diverse habitats. On one extreme, they grow in situations where they are almost constantly submerged in water (e.g., Riccia fluitans, Ricciocarpus natans, Riella), on the other, they thrive on dry rocks with scanty soil and moisture (e.g., Porella platyphylla). Most of the species, however, occur in moist situations such as damp rocks, in ravines, moist soil and walls and on decaying logs in forests. They require water for fertilization. This group is, therefore, regarded as plant amphibians. Many mosses and almost all species of Dendroceros grow as epiphytes on the stems of the plants of tropical rain forests. Buxbaumia minakate, B. aphylla and Cryptothallus mirabilis are saprophytic species.

Fossil bryophytes have been recorded from the tertiary beds of the Coenozoic era (e.g., Muscites yallournensis, Jungermannites) and of the Mesozoic era (e.g., Sphagnum sp., Ricciopsis florinii, Marchantiolites porosus, Naiadita lanceolata) and the Palaeozoic era (e.g., Marchantites sp., India vermicularies, I. falciformis, Salaria longtfolia, Hepaticites langi, H. lotatus). In India, most of the species of bryophytes are confined to the Northern and Southern Himalayas and the Nilgiri Hills. But a few species of Riccia, Marchantia and Funaria also occur in the plains of South India, Rajasthan, M.P., U.P. and Gujarat. The plant body (i.e., gametophyte, which forms the dominant phase of life cycle) is generally small, attaining a length of a few centimeters, but some species are up to 70 cm in length (e.g., Dawsonia sp., Polytrichum commune).

General Characters : The general characteristics of bryophytes are as follows.

(1) Gametophytic and sporophytic phases are present in the life- cycle of bryophytes and both these phases are morphologically distinct (heteromorphic).

(2) The gametophytic phase is more conspicuous, long lived, independent, green and freely branched, whereas the sporophytic phase is short lived and completely dependent upon the gametophyte.

(3) In primitive forms (e.g., Riccia and Marchantia) the gametophyte is prostrate and thalloid, but in mosses the plant body is erect, differentiated into the stem (axis) and lateral appendages (leaves).

(4) True roots are absent and the function of anchorage and absorption is performed by filamentous structures, known as rhizoids.

(5) The rhizoids are either unicellular and unbranched (Hepaticopsida and Anthocerotopsida) or multicellular and branched (Bryopsida).

(6) The members of the order Marchantiales also have multicellular scales in addition to rhizoids, which protect the growing region of the thallus and help in the absorption of water.

(7) The plant body is composed of parenchymatous cells only and lacks fully differentiated xylem and phloem, characteristic of true vascular plants.

(8) They reproduce by vegetative and sexual methods.

(9) Vegetative multiplication takes place by decay and death of the older parts of the thallus, by adventitious branches or by special structures like tubers, gemmae, etc.

(10) Sexual reproduction is of oogamous type, the sex organs are multicellular and considerably more complicated than those of the thallophytes.

(11) The male sex organs, known as antheridia, are stalked, globose or somewhat elliptic. They have an outer sterile one cell thick jacket, which surrounds a solid mass of fertile cells, the androcytes. The androcyte eventually metamorphoses into a motile biflagellate antherozoid.

(12) The female sex organ, known as archegonium, is a flask shaped structure, having a basal swollen venter and somewhat more slender elongate upper part, the neck. The venter and neck arc surrounded by a jacket of sterile cells.

(13) Water is necessary for fertilization. Although many antherozoids may enter the neck, only one (usually the one which approaches the egg first) fuses with the egg to form a diploid zygote.

(14) The zygote does not have any resting period and it divides immediately after fertilization.

(15) The first division of the zygote is always transverse

and the outer cell gives rise to embryo. Thus they show exoscopic mode of embryo development. The embryo develops within the venter of the archegonium.

(16) The sporophyte (also known as sporogonium), developing from the embryo, is a simple structure without rhizoids, stem or leaves. It is completely dependent upon the gametophyte. The sporophyte is a projecting structure, differentiated into foot, seta and capsule (Fig.). In some members, however, the sporophyte is represented either only by capsule (e.g., Riccia) or by foot and capsule (e.g., Corsinia) (Fig.).

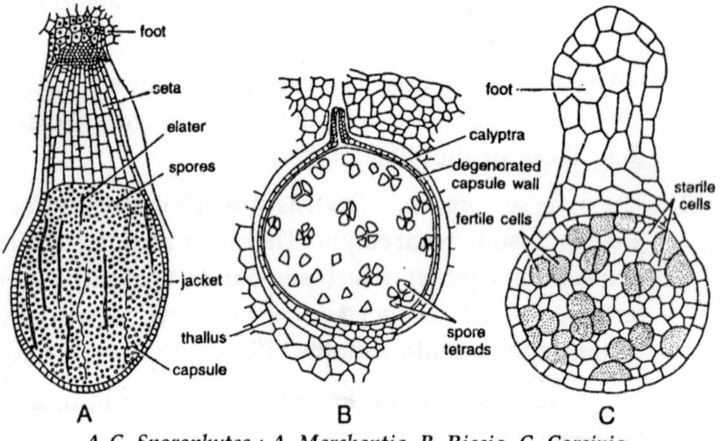

A-C. Sporophytes : A. Marchantia, B. Riccia, C. Corsinia.

(17) The sporogenous cells present in the capsule form haploid spores (meiospores) after reduction division. All spores are similar in shape and size.

(18) Spores are non-motile and they disseminate exclusively by wind.

(19) Under favourable conditions, the spore either forms a filamentous germ tube which divides to give rise to a young gametophyte (e.g., Riccia, Marchantia) or forms a protonema as in mosses. The protonema bears many buds which develop into erect gametophores.

(20) In bryophytes sex is determined either by morphologically distinct sex chromosomes (e.g., Bryum argenteum, Sphaerocarpos donellii) or by environmental factors (e.g., Splachnum luteum and S. ampullaceum).

Alternation of Generations

The life-cycle of bryophytes is characterised by regular alternation of two morphologically distinct phases. One of these is independent, autotrophic, haploid and bears gametes, hence called gametophyte. It develops from the spores produced by the sporophtic, and the gametes represent the last stage of the gametophytic phase. The male and female gametes fuse and give rise to a diploid zygote which functions as the mother cell of the sporophyte phase. The sporogonium is a simple structure, completely dependent upon the gametophyte. The sporogenous tissue present in the capsule forms meiospores after reduction division. These haploid spores on germination give rise to gametophytes.

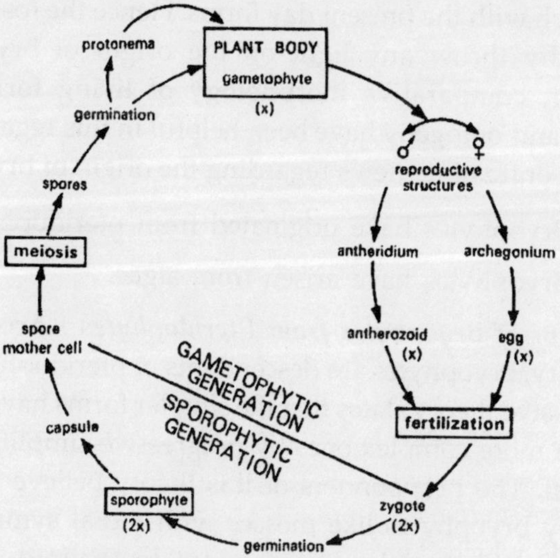

Life-cycle : Diagrammatic representation.

Thus alternation of generation is obligatory in the sense that the gametic union results in the formation of diploid zygote which on germination always develops into a sporogonium. The sporogonium produces haploid spores which function as mother cells of the gametophytic phase.

Bryophytes also show apospory and apogamy. Apospory, in which the gametophyte develops from the vegetative cells of the sporophyte (not from the spores), has been reported in Anthoceros and mosses. Phenotypically, the aposporous gametophyte resembles with the normal gametophyte but possesses diploid number of chromosomes. In apogamy, sporophyte develops from gametophyte without gametic union. It has been reported in Funaria hygrometrica, Physcomitrium coorgense, P. pyriforme and Desmatodon randii.

Origin of Bryophytes

Because of the soft plant body, bryophyte fossils are very meagre and moreover most of the fossils available resemble very much with the present day forms. Hence the fossil record can hardly throw any light on the origin of bryophytes. However, comparative morphology of living forms, their analogy and ontogeny have been helpful in this regard. There are two contrasting views regarding the origin of bryophytes:

(1) Bryophytes have originated from pteridophytes.

(2) Bryophytes have arisen from algae.

Origin of Bryophytes from Pteridophytes : According to this theory, bryophytes are descendants of pteridophytes. This retrogressive theory states that the simpler forms have evolved from the more complex ones by progressive simplification or reduction. The propounders of this theory believe that erect and leafy bryophytes like mosses with radial symmetry are primitive and they have given rise to thalloid forms by progressive simplification. This view was supported by Wettstein (1903-1908), Lang (1917), Kidston and Lang (1917), Kashyap (1919), Scott (1924), Mehra (1953), Christensen (1954), Pande (1960), Zimmermann (1966) and Udar (1970).

This concept is based on the similarity in the sporophyte of some bryophytes like Anthoceros, Sphagnum and Andreaea with certain members of Psilophytales (e.g., Rhynia, Homeophyton, Sporogonites), a primitive order of Pteridophyta. The sporophyte of Psilophytales is simple (without any differentiation into root, stem and leaves), dichotomously branched and bears terminal sporangia. Such a sporophyte resembles with that of Anthocerotales amongst bryophytes. Although the sporophyte of Anthocerotales is not physiologically independent like those of Psilotales, in Anthoceros fusiformis it does show some degree of physiological independence. However, the sporophyte of pteridophytes shows continuous apical growth and has vascular tissue but these features are completely lacking in bryophytes.

According to Christensen (1954), bryophytes have evolved from pteridophytes possibly from one of the following three sources :

(1) From pteridophytes having leaves on the stem of both gametophytes and sporophytes.

(2) From leafless members of the family Rhyniaceae.

(3) Polyphyletically, from both foliose and leafless pteridophytes.

Proskauer (1960) was of the opinion that if bryophytes have a polyphyletic origin, then atleast the order Anthocerotales has evolved from Horneophyton.

Algal Origin of Bryophytes : Although fossil records do not show any relationship between bryophytes and algae, bryophytes do resemble with algae in their amphibious nature, presence of flagellated spermatozoids and necessity of water for fertilization.

According to Lignier (1903), aquatic algal ancestors have given rise to some primitive terrestrial forms, said to be Prohepatics. Bryophytes and pteridophytes have possibly evolved from prohepatics along different lines. Bower (1908) also supported the evolution of bryophytes from algae. Bryophytes show resemblance with green algae (Chlorophyceae) in the presence of : (i) photosynthetic pigments like chlorophyll-*a*, chlorophyll-*b*, carotenes, lutein and violaxanthin, (ii) starch as reserve food material, (iii) cellulose and pectose in the cell wall, and (iv) filamentous protonema in the juvenile stage.

Fritsch (1916, 1945), on the basis of comparative study of various algal groups, concluded that the algae with the following features have given rise to archegoniate plants: (i) heterotrichous habit (as in Chaetophorales), (ii) parenchymatous structure of erect filaments (as in Laminariales and Fucales), (iii) establishment of apical growing region (as in Phaeophyceae and Rhodophyceae), (iv) gradual reduction of prostrate system, and (v) differentiation of conductive tissue in aerial parts (as in Sargassum and Microcystis). According to him, amongst archegoniate group, Bryopsida (mosses) are the only plants which have both prostrate and erect systems as in heterotrichous algal ancestors. The filamentous protonema of mosses is equivalent to the prostrate system and the foliose gametophore is equivalent to the erect system.

Church (1919), in his article "Thallasiophyte and the Subaerial Transmigration" also supported algal origin of land plants, but he considered marine algae as the ancestors of land plants instead of fresh water forms as suggested by Bower (1908).

Smith (1955), who strongly supported algal origin of

bryophytes, was of the opinion that bryophytes have originated from Ulotrichales (Chlorophyceae). The primitive forms are those which became established in areas with plenty of water during change from aquatic to terrestrial environment. Further adaptation led to the establishment of a more massive plant body in which reduced surface, in proportion to volume, tended to prevent undue drying of the thallus. Since these plants grew in moist habitats, there was no need of an elaborate water absorbing system, and it was represented only by rhizoids. The abundance of moisture has also helped in the retention of motile gametes, a characteristic of algae.

It seems probable from the above description that primitive amphibious plants have evolved from algae. Although it is now universally accepted that bryophytes arose from algae, there is controversy about the class of algae which has given rise to these plants. According to Church (1919), marine algae are the ancestors of bryophytes, but Fritsch (1945) and Smith (1955) considered heterotrichous green algae belonging to the order Chaetophorales (such as Fritschiella, Coleochaete and Draparnaldiopsis as their ancestors.

Affinities of Bryophytes

Bryophytes are a group of amphibious plants intermediate between aquatic (algae) and terrestrial (pteridophytes) plants, and they show affinities with both algae and pteridophytes.

Affinities with Algae

1. *Resemblances of Bryophytes with Algae* : The bryophytes resemble with algae in the following characters.

   (1) Presence of thalloid and gametophytic plant body.

   (2) Autotrophic nature of plants.

   (3) In both groups chloroplasts contain chlorophyll-*a*, chlorophyll-*b*, α- and β carotene, lutein, violaxanthin and xeoxanthin.

(4) In Chlorophyceae (green algae) and Anthocerotales plastids are with pyrenoids.

(5) Starch as reserve food material.

(6) Absence of vascular tissue.

(7) Cellulose as chief constituent of cell wall.

(8) Gametophytic phase dominant in the life-cycle.

(9) Presence of motile and flagellate antherozoids and whiplash type of flagella.

(10) Water essential for fertilization.

(11) A filamentous protonema, formed in the juvenile stage of mosses, resembles with algal filaments in structure.

Although bryophytes resemble with algae in many characters, yet there are several differences due to which they have been placed in separate groups.

2. *Differences between Bryophytes and Algae* : Bryophytes differ from algae in the following characters.

Differences between Algae and Bryophytes

Algae	*Bryophytes*
1. Most of the algae are aquatic.	These are terrestrial plants, mostly occur in shady and moist habitats.
2. Plant body is unicellular, multicellular, filamentous or pseudoparen-chymatous.	Plant body is multicellular, parenchymatous and thalloid or differentiated into rhizoids, axis and lateral appendages.
3. Filaments are unbranched or show regular or irregular branching.	Thallus is usually dichotomously branched.
4. Plant body does not show division of labour.	Although plant body does not show distinct division of labour, cells in different parts have specific functions.

Contd...

Algae	*Bryophytes*
5. Structure like pores and stomata are altogether absent.	Pores and stomata are present.
6. Except for some heterotrichous forms, rhizoids are usually absent.	Rhizoids are present in all species. They are smooth walled or tuberculate.
7. In most of the algae all cells of the plant body are capable of divisions.	Divisions are restricted to certain cells only.
8. Asexual reproduction is common and it takes place by zoospores, aplanospores or akinetes.	Asexual reproduction is absent, but some species multiply by vegetative means.
9. Sexual reproduction is isogamous, anisogamous or oogamous.	Sexual reproduction is oogamous.
10. Sex organs are usually unicellular and do not have a covering of sterile cells.	Sex organs are multicellular and have a protective covering (jacket) of sterile cells.
11. Female sex organs are unicellular oogonia.	Female sex organs are multicellular archegonia.
12. Zygote usually undergoes a period of rest after liberation from the parent plant.	Zygote divides immediately after fertilization to form sporophyte.
13. Zygote does not form embryo, instead haploid spores are formed.	Zygote forms embryo.
14. Sporophyte phase (also called diploid phase) is represented mostly by diploid zygote and it is independent.	Sporophyte is well developed and dependent upon the gametophyte.

Affinities with Pteridophytes

1. *Resemblances of Bryophytes with Pteridophytes* : Bryophytes resemble with pteridophytes in the following features.

   (1) Simple leafless and rootless sporophytes of certain

primitive pteridophytes (members of Psilophytales) can be compared with the sporophytes of bryophytes.

(2) In both groups plants are archegoniate, and the structure of archegonium is similar.

(3) Antheridium of bryophytes as well as pteridophytes is surrounded by a jacket of sterile cells.

(4) Antherozoids are flagellated.

(5) Water is necessary for fertilization.

(6) Zygote forms embryo.

(7) The terminal sporangia with columella of Psilophytales are similar to moss capsules.

(8) Both groups are characterized by the presence of heteromorphic alternation of generation.

2. *Differences between Bryophytes and Pteridophytes* : Bryophytes differ from pteridophytes in the following features.

Bryophytes	*Pteridophytes*
1. Gametophyte is the dominant phase in the life-cycle.	Sporophyte is the dominant phase in the life-cycle.
2. Plant body is thalloid or foliose.	Plant body is differentiated into root, stem and leaves.
3. Vascular tissue is absent.	Vascular tissue is present.
4. Sporophyte is completely dependent upon the gametophyte.	Sporophyte is autotrophic and independent.

Classification of Bryophytes

The term Bryophyta was used for the first time by Braun (1864), who included algae, fungi, lichens and mosses in this group. In later systems of classification, algae, fungi and lichens were placed in a separate division Thallophyta, and liverworts and mosses in Bryophyta. Eichler (1813) recognised the following two classes in the division Bryophyta.

Engler (1892) subdivided these classes into three orders each.

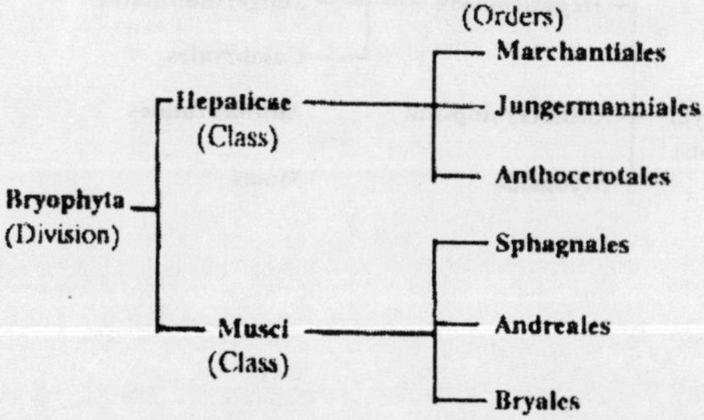

Many bryologists including Bessey (1911), Verdoorn (1932), Bower (1935), Wettstein (1935), Evans (1938), and Engler, Melchior and Werdermann (1954) have followed the basic classification proposed by Eichler and Engler. But Howe (1899) considered order Anthocerotales, containing Anthoceros and other related genera, as quite distinct from other members of the class Hepaticae. Therefore, he suggested that the order Anthocerotales should be given the rank of a class. Howe's view has been supported by Campbell (1940), Takhtajan (1953), Smith (1955), Schuster (1958) and Wardlaw (1962). Thus the division Bryophyta was divided into three classes.

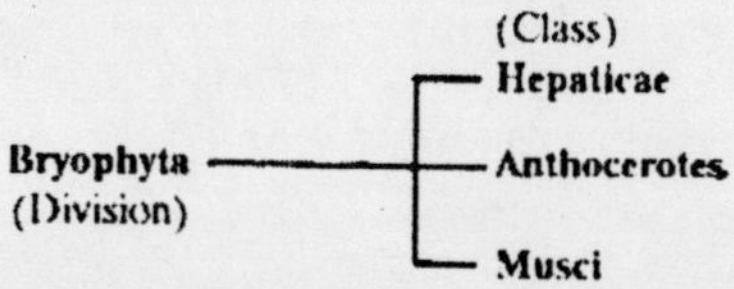

Takhtajan (1953), however, preferred to call Anthocerotes as Anthocerotae. Rothmaler (1951) and Proskauer (1957) although supported the above classification, following the

International Code of Botanical Nomenclature they called the class Hepaticae as Hepaticopsida, Anthocerotae as Anthocerotopsida and Musci as Bryopsida. A brief outline of the classification of Proskauer is given below.

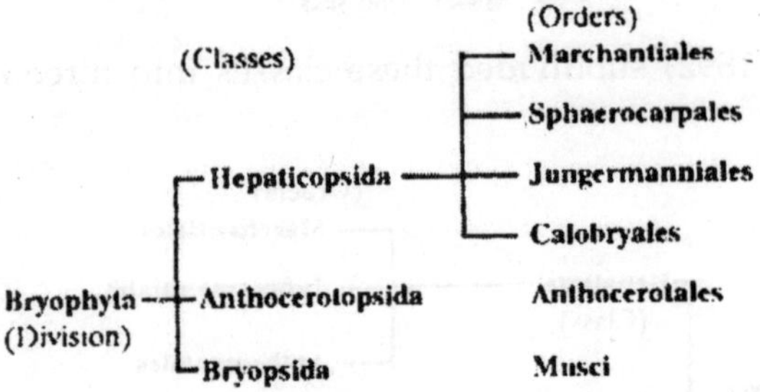

11

The Comparison

Liverworts	Mosses
A. GAMETOPHYTE	
1. The gametophyte is usually a prostrate and dorsiventral thallus, except for leafy Jungermanniales (e.g., Porella, Frullania) where the gametophyte consists of a leafy axis.	The gametophyte phase has a juvenile stage with creeping, green, branched and filamentous protonema and an adult leafy stage represented by upright, slender and leafy axis.
2. Branching is dichotomous.	Branches arise laterally, just below the leaves rather than in their axils.
3. The gametophyte is green and photosynthetic.	The adult plant body is green and better adapted for terrestrial habitats than liverworts. It manufactures organic food by photosynthesis.
4. In thalloid forms the dorsal epidermis of thallus is provided with pores.	Pores are absent in leaves.

Contd...

Liverworts	*Mosses*
5. Thalli in certain forms (Riccia, Marchantia) have air chambers.	Air chambers are absent.
6. Internally, the thallus is either homogeneous (e.g., Pellia, Anthoceros) or differentiated into photosynthetic and storage regions (e.g., Riccia, Marchantia).	The main axis is differentiated internally into cortex and axial cylinder.
7. The ventral surface of the gametophyte bears tuberculate and smooth walled rhizoids (some forms have only smooth walled rhizoids). The rhizoids are always unicellular and unbranched.	Rhizoids usually arise from the base of the stem. They are multicellular and usually have oblique septa; in some forms they are coiled together to form rope-like structure.
8. In some liverworts (e.g., Riccia, Marchantia) multicellular plate like scales are present on the ventral surface of the gametophyte.	Scales are absent.
9. Leaves are without midrib.	Leaves have a distinct midrib.
Reproduction	
10. Gametophytes are dioecious or monoecious.	Gametophytes are always monoecious.
11. In thalloid liverworts sex organs are usually embedded in the gametophyte (e.g., Riccia, Pellia) or in the receptacle (e.g., Marchantia, Plagiochasma). In leafy members they are borne on short lateral branches in the axils of leaves.	Sex organs are either axillary (e.g., Sphagnum) or terminal (e.g., Funaria) in position.

Contd...

Liverworts	Mosses
12. Archegonia are sessile or shortly stalked; the neck of the archegonium is short.	Archegonia usually have a long massive stalk; the neck of the archegonium is long.
13. Neck canal cells are 6-8.	Neck canal cells are 10 or more.
14. Antheridia are borne singly or in groups.	Antheridia arise either singly in the axil of leaves or in groups at the tip of the axis.
15. Antheridial jacket is single layered.	As in liverworts.
16. Water is essential for fertilization.	As in liverworts.
B.SPOROPHYTE	
17. Sporophyte of all liverworts (except for Riccia) is differentiated into foot, seta and capsule.	Sporophyte has well developed foot, seta and capsule.
18. The seta is delicate and an ephemeral structure; it is green when young and colourless at maturity, internally it is undifferentiated.	Seta is stronger and much long-lived structure; it shows differentiation of tissues.
19. Amphithecium forms either (i) only capsule jacket as in Hepaticopsida and few species of Notothylus, or (ii) capsule jacket and archesporium as in Anthocerotopsida.	Amphithecium forms (i) only capsule jacket in Bryidae, and (ii) capsule jacket and archesporium in Sphagnidae.

Contd...

Liverworts	Mosses
20. Endothecium gives rise to (i) archesporium as in Hepaticopsida and in some species of Notothylus, and (ii) columella in most species of Anthoceros.	Endothecium forms (i) only columella in Sphagnidae, and (ii) archesporium and columella in Andreaeidae and Bryidae.
21. Sporogenous tissue forms sporocytes and abortive cells or elaters or pseudoelaters.	Sporogenous tissue forms only sporocytes; elaters are absent.
22. Capsule wall is single layered as in Hepaticopsida and multilayered in Anthocerotopsida.	Capsule wall is multilayered.
23. Capsule does not show external differentiation.	Capsule is differentiated into operculum, theca and apophysis.
24. In simple liverworts there is no special mechanism for the dehiscence of the capsule (e.g., Riccia, Corsinia, Sphaeracarpus) and spores are liberated after the death and decay of the surrounding gametophytic tissue and capsule wall; in others the capsule wall splits into 2 or more valves (e.g., Marchantia, Pellia, Anthoceros).	In cleistocarpous mosses (e.g., Archidium, Ephemerum) operculum is absent and the liberation of spores takes place by decay or irregular rupture of the capsule wall; in other mosses the dehiscence is facilitated by shedding of operculum at maturity of spores (Sphagnum, Polytrichum, Funaria).
25. Sporophytes of liverworts are relatively simple and primitive; amongst liverworts sporophyte of Riccia is the simplest and that of Anthoceros is most complex and advanced.	Sporophytes of mosses show a very complex structure; capsule wall is photosynthetic and it is only partially dependent upon the gametophyte; these are considered to be more advanced than those of liverworts.

Comparative Account of the Gametophytes ;and Sporophytes of Riccia, Marchantia, Pellia, Porella, Anthoceros, Sphagnum, Polytrichum and Funaria.

A. Comparative Study of The Gametophyte

	Character 1	Riccia 2	Manchantia 3	Pellia 4	Porella 5	Anthoceros 6	Anthoceros 7	Polytrichum 8	Funaria 9
1.	**Systematic position**	Hepaticopsida Marchantiales Ricciaceae	Hepaticopsida Marchantiales Marchantiaceae	Hepaticopsida Jungermanniales Metzgerineae Pelliaceae	Hepaticopsida Jungermanniales Jungermannieae Porellceae	Anthocerotopsida Anthocerotales Anthocerotaceae	Bryopsida Sphagnobrya Sphagnales Sphagnaceae	Bryopsida Bryidae Polytrichales Polytrichaceae	Bryopsida Bryidae Funariales Funariaceae
2.	**Number of species recorded from India**	200	65	5	180	200	342	100	117
3.	**Habitat**	Mostly terrestrial, growing on moist soil, Riccia Fluitans is aquatic.	Grow in moist and shady terrestrial habitats.	Grows along the moist river blanks, wet walls of the wells and damp bark of the trees; few species are submerged.	Mostly grows on moist and shady rock surfaces and trees trunks; sometimes found submerged in water but such forms are sterile.	Grows on moist and shady terrestrial habitats; few species grow on decaying logs.	These are aquatic or semiaquatic plants growing in ponds, marshy or moist habitats.	All species are terrestrial grwoing on moist river banks, rocks or on the branches of the shades trees.	All species are terrestrial, forming dense and compact patches on wet wall and tree trunks.
4.	**Plant body**								
i.	Plant form	Plant body is thalloid, prostrate, flat, dorsiventral and has a distinct midrib; each lobe has an apical notch at the distal end in which growing point of the thallus is located.	Thalloid, flat dorsiventral with a distinct midrib; an present at the apex of each lobe as in Riccia.	Thalloid, flat and dorsiventral with sinuous overlapping apical notch is than Riccia and Marchantia; midrib is distinct and the growing point is located in the apical notch.	Plant body is prostrate, flat, dorsiventral, having margins; thinner axis with two dorsal and one ventral row of leaves.	Thalloid, flat, prostrate, dorsiventral and irregularly a central branched indistinct.	Mature gametophyte has an erect and radially lobed; midrib is gametophore bearing green sessile leaves.	Gametophore is differentiated into underground rhizome symmetrical symmetrical, erect aerial and leafy branches.	Mature gametophore is upright, radial and leafy. and radially

Contd...

	Character 1	Riccia 2	Manchantia 3	Pellia 4	Pcrella 5	Anthoceros 6	Anthoceros 7	Polytrichum 8	Funaria 9
ii.	Branching	Dichotomously branched.	Dichotomously branched.	dichotomously branched.	Monopodial.	Dichotomously branched but due to unequal growth of two branches the thallus becomes irregular.	The branches arise from the axil of every fourth leaf; in aquatic species all the branches are alike but semiaquatic species have two typesof branches- (a) long pendent branches, (b) divergent branches.	Branching is lateral; branches arise from the buds situated just below the leaves (they are not axillary).	As in Polytrichum.
iii.	Rhizoids	Unicellular and unbranched; rhizoids are present on the ventral surface in the midrib region; they are of two types (i) smooth walled, and (ii) tuberculate.	Unicellular and unbranched, rhizoids of two types are present as in Riccia	Unicellular, unbranched and only smooth walled rhizoids are present.	As in Pellia.	As in Pellia.	yound plants have soft, colourless and multicellular rhizoids with oblique septa.	Underground rhizome has long, branched, multicellular rhizoids which are usually coiled together to from rope-like sturcture	Many slender muiticellular and strand-like rhizoids arise from the base of the gametophore.
iv.	Scales	Multicellular scales are present in a single row along the margins of the thallus.	Multicellular scales occur in 2-3 rows on either side of the midrib. They are of two types- (i) ligulate and (ii) appendiculate.	Absent	Absent	Absent	Absent	Absent	Absent

Contd...

	Character 1	Riccia 2	Manchantia 3	Pellia 4	Porella 5	Anthoceros 6	Anthoceros 7	Polytrichum 8	Funaria 9
v.	Leaves	Absent	Absent	Absent	Leaves are arranged in 3 rows-two dorsal (lateral) and one ventral; dorsal leaves are unequally lobed, midrib is absent	Absent	Small thin and sessile leaves are present in three vertical rows on the mature stem in 2/5 phyllotaxy; mature leaves are without midrib.	Leaves are present on underground rhizomes as well as on the aerial parts; leaves on the rhizome are colourless or brown in 1/3 phyllotaxy while those on aerial branches are relatively large and green in 3/8 phyllotaxy; midrib present.	Simple sessile leaves are spirally arranged on the stem in 3/8 phyllotaxy; a distinct midrib is present
5.	Apical growth	By a row of 3-5 apical cells present in the apical notch, each cell divides in two planes.	As in Riccia	By a single apical cell which divides in 4 planes.	By a single tetrachedral apical cell which divides by 3 lateral faces.	By one or more apical cells,	By a single pyramid-shaped apical cell which divides in 3 planes.	As in Sphagnum.	As in Sphagnum.
6.	Internal Structure								
i.	Thalloid plants	Thallus differentiated into upper photosynthetic and lower storage regions.	Differentiated into photosynthetic and storage regions like Riccia.	Thallu: undifferentiated (homogeneous).	—	Thallus undifferentiated as in Pellia.	—	—	—

Contd...

	Character 1	Riccia 2	Manchantia 3	Pellia 4	Porella 5	Anthoceros 6	Anthoceros 7	Polytrichum 8	Funaria 9
(a)	Upper epidermis	Indistinct; the terminal cells of the photosynthetic filaments function thick as epidermis.	Well defined; wall of the epidermal cells are relatively thick.	Not differentiated from other cells of the thallus.	—	As in Pellia	—	—	—
(b)	Pores	Rudimentary; air spaces present in between photosoyn-thetic filaments from the pores.	Well developed barrel shaped; composed of 4-8 superimposed tiers of cells; each tier has 4-5 cells; the cells of the pore are projected partly above the surface of the thallus.	Absent	—	Absent	—	—	—
(c)	Photo synthetic region	Composed of vertical columns of chloroplast contai-ning cells—the photosynthetic filaments; all cells of the filaments, except the upper most ones, are alike and contain discoid chloroplasts.	Divided into a number of chambers; each chamer has many small simple or branched photosynthetic filaments.	Dorsal cells of the middle region and those of the wings form the photosynthetic region.	—	All the cells of the thallus contain chloroplasts and are photosynthetic.	—	—	—

Contd...

Character 1	Riccia 2	Manchantia 3	Pellia 4	Porella 5	Anthoceros 6	Anthoceros 7	Polytrichum 8	Funaria 9
(d) Air chambers	Vertical photos-ynthetic filaments are alternately arranged with narrow (e.g.R. himalayensis) or broad (e.g. R.robusta) air chambers.	A large number of air chambers are arranged in a single row just below the dorsal epidermis; they are separated from each other by a single layered partition wall.	Absent	—	Absent	—	—	—
(e) Storage region	Ventral part of the thallus is composed of compactly arranged colourless perenchymatous cells.	As in Riccia.	All the cells of the thallus perform storage function.	—	All the cells of the thallus perform storage function; ventral part of the thallus has mucilagenous cavities.	—	—	—
(ii) Foliose Plants								
(a) Rhizome	—	—	—	—	—	—	Rhizome has three distinct regions-epidermis, cortex and central cylinder. The central cylinder is trilobed and is composed of three types of cells-the leptoid, hydroid and steriod.	Absent

Contd...

Character 1	Riccia 2	Manchantia 3	Pellia 4	Porella 5	Anthoceros 6	Anthoceros 7	Polytrichum 8	Funaria 9
(b) Aerial axis Epidermis	—	—	—	Indistinct due to the presence of large number of leaves.	—	Not distinct.	Single layered, interrupted and indistinct due to the presence of leaves.	Single layered, composed of thick walled cells containing chloroplasts; stomata are absent.
Cortex	—	—	—	composed of 2-5 layers of thick walled cells.	—	Differentiated into an outer perenchymatous and an inner prosenchymatous regions.	Differentiated into an outer prosenchymatous and an inner parenchymatous regions.	Composed of parenchymatous cells; these cells also contain chloroplast in young stem.
Central cylinder	—	—	—	Central part, called medulla, is composed of thin walled cells.	—	Usually composed of thin walled cells but few species have thick walled cells.	Surrounded by a discontinuous pericycle- followed by leptom mantle, amylom layer and hydrom mantle.	Made up of a group of thick walled cells.
(c) Leaves	—	—	—	Composed of a single layer of isodiametric paranchymatous cells; midrib or conduction tissue is absent; oil cells are present.	—	Composed of small and living chlorophyllous cells and relatively large, polygonal dead hyaline cells, arranged alternately; midrib is absent.	Limb mainly consists of midrib which is multilayered, its dorsal surface is occupied by vertical lamellae, each consisting of 4-8 cells; a distinct epidermis is present on the ventral surface.	One cell thick except the midrib which has many cells; the cells of the wing contain chloroplasts.

Contd...

	Character 1	Riccia 2	Manchantia 3	Pellia 4	Porella 5	Anthoceros 6	Anthoceros 7	Polytrichum 8	Funaria 9
7.	Vegetative reproduction	(i) By progressive death and deacy of the thallus (ii) by adventitious branches (iii) by tubers (iv) by gammae-like bodies developing at the tips of the rhizoids.	(i) By Progressive death and decay of the thallus (ii) by adventitious branches (iii) by gammae.	(i) By fragmentation (ii) by adventitious branches.	(i) By progressive death and decay of the thallus (ii) by gammae.	(i) By progressive death and decay of the thallus (ii) by tubers (iii) by gammae (iv) by perennating apices.	(i) By innovations (ii) by secondary protonema.	(i) By bulbils which develop on the rhizoids (ii) fragmentation the rhizome (iii) by secondary protonema.	(i) By secondary protonema which develops on stem, leaves or rhizoids of the gametophyte.
8.	Sexual reproduction								
i.	sex differentiation	Mostly monoecious, a few dioecious.	Mostly dioecious, but a few are monoecious.	Both monoecious and dioecious are found.	Plants are dioecious.	Plants are both monoecious and dioecious.	Both monoecious and dioecious special are present.	All species are dioecious.	Monoecious but autoicous.
ii.	Antheridium								
(a)	Position	Present in the antheridial chambers embedded on the dorsal surface of the thallus.	Borne on the lobes of the terminal peltate disc of the antheridiophores.	Embedded in the antheridial chambers on the dorsal surface of the thallus on either side of the midrib.	Borne on special lateral antheridial branches.	Present in closed antheridial chambers on the dorsal surface of the thallus.	Develop singly in the axil of each leaf of the antheridial branch.	Develop in groups at the apex of the male gametophore.	Occur in group at the apex of the male shoots.
(b)	Number of antheridia in each chamber	One in each antheridial chamber.	As in Riccia.	As in Riccia.	One in the axil of each leaf of the antheridial branch.	In each chamber there are many antheridia.	One in the axil of each of the antheridial branch.	Develop in groups at the apex of the male gametophore.	Occur in group at the apex of gametophore.

Contd...

Character 1	Riccia 2	Manchantia 3	Pellia 4	Porella 5	Anthoceros 6	Anthoceros 7	Polytrichum 8	Funaria 9
(c) Structure	Stalked,globose or club-shaped structure; antheridial jacket single layered.	Stalked, oval or conical structure; jacket single layered.	Globose and its structure is like that of Riccia.	Globose and stalked antheridial jacket single layered in the upper part and 2-3 layered in the basal part.	Club-shaped stalked structure; jacket single layered.	Stalked, globose structure; jacket single layered.	Club-shaped with a multicellular stalk; jacket is single layered an operculum is present at the apex of the antheridium; paraphysis also associated with antheridia.	As in Polytrichum.
(d) Dehiscence Hydrostatic pressure generated in the cells the jacket due to	Antherozoids are As in Polytrichum. liberated by the dis- integration of antheri- dial jacket.	As in Riccia.	By disintegration of the apical part of the	Apical part of anther- antheridial jacket. disintegrate. The	idium is ruptured into irregular valves. exposed cells of the antheridial jacket turn backwark mass of androcytes comes out in the form of a smoke column.	Roof of the antheri- irregularly and the irregular valves which eventually to rupture and the exposing the mass of androcytes.	dial chamber breaks water absorption causes the operculum and- rocytes are liberated.	Cells of the anth- eridial jacket break into many o f
(e) Antherozoids	Antherozoids curved and biciliate.	As in Riccia	Spirally coiled with two cilia inserted at the apex.	Rod-shaped biciliate structure.	Rod-shaped or coiled with two cilia.	Spirally coiled with a terminal flagellophore; the latter bears two terminal cilia.	Spirally coiled and biciliate.	As in Polytrichum.

Contd...

Character 1	Riccia 2	Manchantia 3	Pellia 4	Porella 5	Anthoceros 6	Anthoceros 7	Polytrichum 8	Funaria 9
iii. Archegonium								
(a) Position Develop at the apex res in groups of 3-6.	Present in the arche- Develop in groups gonial chambers embedded on the female shoots. dorsal surface of the thallus.	dorsal surface of the terminal disc of archegoniophore.	Develop on the dorsal surface of the thallus on either side of the midrib in groups of 4-12.	Develop on the up to 15 at the apex of the archego- nial branch.	Develop in groups of dorsal surface of the sequence just behind the growing point.	apex of the in thallus in acropetal branches either single or in groups 2-5.	Embedded on the of female gametopho-	Develop at the at the apex of the archegonial
(b) Structure As in Polytrichum.	neck is composed of 6 vertical rows of neck cells; neck canal cells 4-6; a rosette of 4 cover cells is present at the apex of the neck.	As in Riccia	Neck is made up of 5 vertical rows of neck cells, neck canal cells 6-8; neck is capped by a rosette of 4 cover cells.	Neck is composed of 5 vertical rows of neck cells; 6-8.	As in Riccia	The body of arch- egonium is twist- ed; neck made rows of neck cells, neck canal cells 8-10; tip of the neck is capped by 8 or more cover cells.	vertical rows of cells; neck canal cells 8-10; up of 6 vertical present at the top of the neck.	Neck consists of 6 is a single cover cell is

B. COMPARATIVE STUDY OF THE SPOROPHYTE

Character 1	Riccia 2	Manchantia 3	Pellia 4	Porella 5	Anthoceros 6	Anthoceros 7	Polytrichum 8	Funaria 9
1. Parts of the sporophyte	Represented only by capsule; foot and seta are absent.	Differentiated into foot, seta and	As in Marchantia. capsule.	As in Marchantia.	Differentiated into foot and capsule; meristematic tissue is present between the two parts.	Differentiated into capsule and an intercalary represented by a very small neck- like suppressed tissue.	Differentiated into foot seta and foot; seta is	As in Polytrichum. capsule.

Contd...

Character 1	Riccia 2	Manchantia 3	Pellia 4	Porella 5	Anthoceros 6	Anthoceros 7	Polytrichum 8	Funaria 9
2. Foot	Absent	Well developed and parenchymatous.	Conical and its distal part form a collar-like structure at the base of seta.	Well developed and globose.	Bulbous; its surf-ace cells elongate to form haustoria or palisade like cells which increase the absorbing surface.	Well developed, bulbous and massive.	Relatively small and dagger-shaped.	Poorly developed and concila.
3. Seta	Absent	Small in young sporophyte but elongates rapidly at maturity.	Relatively thin; elongates at the maturity of the capsule.	As in Pellia.	Seta is absent but nerustenatic tissue present between foot and capsule. functions as seta.	Seta is absent but a narrow neck-like part is present between foot and capsule; it does not show meristematic activity like that of Anthoceros.	Elongated, cylindrical and stalk-like.	As in Polytrichum.
4. Capsule i. Shape	Globose	Oval	Globose	As in Pellia.	Bristle or hornlike.	Globose	Polygonal	Pear-shaped, curved and asymmetrical.
ii. Wall	Capsule wall is single layered; it disorganises very early.	Capsule wall single layered; persistent.	Capsule wall 2 or more layered; persistent.	As in Riccia	Capsule wall 4-6 layered; persistent.	As in Anthoceros.	multi-layered. persistent.	Capsule wall As in Polytrichum

Contd...

	Character 1	Riccia 2	Manchantia 3	Pellia 4	Porella 5	Anthoceros 6	Anthoceros 7	Polytrichum 8	Funaria 9
iii.	Arches porium	Develops from endothecium.	As in Riccia.	As in Riccia.	Elaterophore is absent but the elater is similar to that of pellia.	Develops from the inner layer of the amphithecium.	As in Antho-ceros.	Develops from the outer layer of the endothecium.	As in polytrichum.
iv.	Elaters	Absent	Long spindle-shaped; with two spiral thickening bands, hygroscopic.	Present in cluster of 20-100 at the base of the capsule (elaterophore); each elater is spindle-shaped with 2-3 spiral thickenings.		2-3-called, simple or branched pseudo-elaters without thickening bands are present.	Absent	Absent	Absent
v.	Columella	Absent	Absent	Absent	Absent	Present	Present	Present	Present
5.	Protective coverings	Young sporophyte enclosed within a double layered calyptra.	Sporophyte has three protective coverings-calyptra, perigynium and perichaetium.	Young sporophyte is enclosed within a double layered calyptra.	Young Sporophyte is enclosed by calyptra, perianth and involucre (of bracts).	Young sporophyte is covered by a calyptra.	Young sporophyte is enclosed within calyptra and perichaetium.	Calyptra completely covers the young capsule but at maturity it cover only the apex of the capsule.	Completely covered by calyptra which is shed at maturity.
6.	Dehiscence of capsule	No special mechanism of dehiscence; spores are liberated after the death and decay of the thallus.	Capsule wall splits longitudinally into many valves; hygroscopic nature of elaters helps in the dispersal of spores.	Capsule wall splits into 4 longitudinal valves; elaters help in dispersal of spores.	As in Pellia.	Capsule wall splits longitudinally into 1-4 valves; elaters, columella and valves of the capsule wall help in the dispersal of spores.	Dispersal of spores takes place by a special exposive mechanism after the separation of operculum.	Dispersal of spores takes place by censor mechanism with the help of epiphrag and peristomial teeth.	Dehiscence of capsule takes place by hygroscopic movement of peristomial teeth after the separation of operculum.

PART-III

Microbiology

12

The Bacteria

The environment around us has a fascinating world of micro-organisms which affect our life directly or indirectly. The existence of microbial world was unknown until the invention of microscope at the beginning of the seventeenth century, which opened the realm of micro-organisms to systematic scientific exploration.

Antony van Leeuwenhoek (1632-1723), a Dutch cloth merchant, was the first to give an illustrated description of micro-organisms. The thousands of microscopic organisms that he saw in a drop of rain water were collectively named as animalcules. However, it was not until the development of the compound microscope that biologists became aware of the tremendous numbers and diversity of these organisms in nature. As further information accumulated about these microscopic organisms, it became evident that they did not belong to two familiar kingdoms of living world, i.e., plants and animals and thus shared the characters of both plants and animals. Therefore, they could not be placed judiciously in either of these kingdoms. This necessiated the creation of a new group, and Ernst Haeckel, a German biologist, in 1894, proposed a third kingdom Protista. It included all unicellular and microscopic organisms and also many large multicellular forms which do not have any extensive development of tissue.

The protists thus were further divided into two large groups, which were called lower and higher protists. The lower protists are characterized by prokaryotic cellular organization, and they include bacteria and blue-green algae (Cyanobacteria). The higher protists, on the other hand, have eukaryotic cell organization and include protozoa, algae and fungi.

Murray (1968) included all lower prokaryotes (which are characterized by the absence of nuclear membrane and basic proteins in nucleoplasm) in Prokaryotae. Bergey (1974), in the Manual of Determinative Bacteriology, placed bacteria in the kingdom Prokaryotae.

General Characters

Bacteria are microscopic and least differentiated living organisms, believed to be amongst the first primitive organisms on the earth. They show typical prokaryotic characters and have resemblance with both plants and animals. They have been placed in a separate class Schizomycetes under Thallophyta.

Bacteria show considerable variations, but the characters listed below are common to most of the bacteria.

(1) They are omnipresent, found in all possible habitats one can think of.

(2) Most of the bacteria have heterotrophic mode of nutrition, i.e., they obtain their readymade food directly from external agency, and may be parasitic, saprophytic or symbiotic. Some bacteria are autotrophic; they possess bacteriochlorophyll-a photosynthetic pigment.

(3) They are unicellular and morphologically least complex of all the living organisms.

(4) The cell wall of bacteria is rigid and made up of two types of polymers, amino acid subunits and saccharide subunits. Cellulose, characteristic of the cell wall of higher plants, is absent in bacteria.

(5) A well organised nucleus, characteristic of eukaryotes, is lacking and discrete chromosomes are also absent. The nuclear material is not surrounded by nuclear membrane.

(6) Chlorophyll pigments, if present, are located within involuted cytoplasmic membranes; well organised plastids are absent.

(7) Mitochondria are absent and their function is carried out by complex localized infoldings in the cell membrane, known as mesosomes.

(8) The organelles like endoplasmic reticulum and Golgi apparatus are absent.

(9) Ribosomes are abundant in bacterial cell but they sediment somewhat more slowly in a centrifuge than do ribosomes of other organisms.

(10) Binary fission is the most common method of multiplication.

(11) True sexual reproduction is absent. However, recombination of genetic material occurs by conjugation, transformation and transduction.

(12) The motile bacteria may possess one or more flagella, composed of eight parallel chains of flagellin (a protein) molecules.

(13) The Gram negative bacteria possess minute hair-like cytoplasmic appendages, known as pili, extruding through the cell wall. These appendages are composed of a protein, called fimbrillin.

Classification of Bacteria

Bergey's Manual of Determinative Bacteriology is the standard reference book which deals with the classification and identification of bacteria. The classification of bacteria given in the first edition of the manual,, published in 1923, was mainly based on phenotypic characterization. But as new information poured in as the result of researches at biochemical and molecular levels, the composition and arrangement of higher taxonomic groups-orders, families and tribes have changed substantially in successive editions.

In the 8th edition of the Manual, published in 1974, there is a radical departure in the scheme of classification from earlier editions. Bacteria have been distinguished from other micro-organisms on the basis of their typical prokaryotic structure and they have been placed in a separate kingdom Procaryotae. Further treatment of this kingdom was based on cellular characteristics rather than organismal properties. Procaryotae were divided in two divisions-Cyanobacteria and Bacteria. Bacteria were further subdivided into 19 groups, distinguishable by some readily determined criteria. There were certain genera of uncertain affiliation which could not fit perfectly within the set limits of a particular group.

An outline of bacterial classification as adopted in the latest edition of the Bergey's Manual (1974) is given below.

Kingdom Prokaryotae

Division I. Cyanobacteria

They are photosynthetic prokaryotes and their photosynthetic process is similar to that of higher plants (unlike other photosynthetic bacteria where the process is aerobic and electron donor is water). The photosynthetic pigments are chlorophyll and phycobiliproteins. These bacteria usually occur as simple or branched chains of cells. They reproduce by binary fission, spores or fragmentation.

Division II. Bacteria

They are unicellular or occasionally show simple arrangements. They are characterised by the presence of a rigid cell wall of peptidoglycans. Their photosynthetic process is anaerobic, the electron donor is a substance other than water. Their photosynthetic pigments are bacteriochlorophylls.

The division Bacteria has been further divided into 19 groups on the basis of their photosynthetic ability, type of movement, response to Gram stain, etc.

Group 1. Phototrophic Bacteria

These are photosynthetic bacteria, characterized by the presence of a chlorophyll-like pigment, bacteriochlorophyll. They occur mostly in aquatic environments. They are motile or non-motile.

This group includes a single order with 3 families and 18 genera.

Group 2. Gliding Bacteria

These bacteria produce slime and show gliding movements. Some forms produce brightly coloured macroscopic fruiting bodies. They occur abundantly in soil, decomposing plant matter and aquatic environment.

This group includes 2 orders and 8 families, representing 21 genera. In addition, the group also includes 6 genera of uncertain affiliation.

Group 3. Sheathed Bacteria

They are rod-shaped or filamentous bacteria, surrounded by a sheath of insoluble compounds of iron and manganese. They are motile or non-motile, and Gram-negative and occur in aquatic environments, sludge, etc.

This group is represented by 7 genera.

Group 4. Budding and/or Appendage Bacteria

Some bacteria of this group produce a filamentous outgrowth, called prostheca, from the body of the cell, and the others have holdfasts. These appendages help them in attachment to some object. They multiply by budding and fission. They occur in soil or in aquatic environment.

This group includes 17 genera.

Group 5. Spirochaetes

These bacteria are slender, flexuous and helically coiled, varying in length from 3-500 μ m. Some of them are saprophytes and the others are parasites. The parasitic forms are pathogenic to human beings and other animals. They are motile and they multiply by transverse fission.

This group includes 5 genera placed in a single family and a single order.

Group 6. Spiral and Curved Bacteria

These bacteria are also helically coiled like spirochaetes but have rigid cell wall. In some species the spiral is not complete and as such they become 'comma' shaped. They are free living, saprophytes or parasites. These bacteria move with the help of flagella and are Gram-negative.

This group includes 2 genera, placed in a single family. The group also has 4 genera of uncertain affiliation.

Group 7. Gram-negative Aerobic Rods and Cocci

This group includes a large number of Gram-negative aerobic bacteria. Although similar in morphology, they vary in biochemical characteristics, such as metabolism of nitrogen and carbon compounds. They are pathogenic to humans and animals. These bacteria occur in terrestrial as well as aquatic environment. This group includes 5 families represented by 14 genera. There are also 6 genera of uncertain affiliation.

Group 8. Gram-negative Facultative Anaerobic Rods

Most of the common facultative anaerobes belong to this group. They are morphologically similar but differ in biochemical, physiological and serological characteristics. They are motile (peritrichous) or non-motile. These bacteria are pathogenic to human beings, plants and animals. They occur in all types of environments.

This group includes 2 families represented by 17 genera. There are also 9 genera of uncertain affiliation.

Group 9. Gram-negative Anaerobic Bacteria

They are pleomorphic obligate anaerobes which occur in the oral cavity, intestine and faeces of human beings. The motile forms are peritrichous or monotrichous. Some species are pathogenic to human beings and animals. This group includes 3 genera, placed in a single family. There are also 6 genera of uncertain affiliation.

Group 10. Gram-negative Cocci and Coccobacilli

They are Gram-negative, non-motile aerobic Cocci and Coccobacilli which occur in the mucous membranes of humans and other animals. These bacteria have only limited ability to break down carbohydrates and proteins.

This group has only one family represented by 4 genera. There are also 2 genera of uncertain affiliation.

Group 11. Gram-negative Anaerobic Cocci

These are Gram-negative, non-motile, anaerobic bacteria. They are spherical in shape and occur in pairs, masses or in chains. The delimitation of the species within this group is largely on the basis of biochemical characteristics. They occur in the respiratory and intestinal tracts of humans and animals, but are not regarded as pathogenic.

This group has one family, represented by 3 genera.

Group12. Gram-negative Chemolithotrophic Bacteria

These are autotrophic bacteria which derive energy from the oxidation of inorganic compounds. Their shape is rod-like, spherical or spiral. The motile forms of the group possess flagella. These bacteria are non-pathogenic and occur in soil and aquatic environment.

This group includes 2 families, represented by 17 genera.

Group13. Methane-producing (Methanogenic) Bacteria

The bacteria of this group are characterised by their ability to produce methane under anaerobic conditions. They occur in terrestrial as well as in aquatic environment. Most of these bacteria are Gram-positive, but some are Gram-negative.

This group has 8 genera, placed in a single family.

Group 14. Gram-positive Cocci

These are Gram-positive, pathogenic or saprophytic Cocci, which occur singly, in pairs, in clusters or in chains. They commonly inhabit soil, fresh water and the mucous membranes of warm blooded animals.

This group includes 3 families represented by 12 genera.

Group 15. Endospore-forming Rods and Cocci

These bacteria are characterised by their ability to produce endospores. Most of these bacteria are rod-shaped, and are either aerobic or anaerobic. They are pathogenic and cause infections in humans as well as animals.

The group has one family with 5 genera. One genus of uncertain affiliation is also included in this group.

Group 16. Gram-positive Asporogenous Rod-shaped Bacteria

These are non-motile, Gram-positive, rod-shaped (bacilli) bacteria. They are anaerobic and mostly occur in milk and milk products (lactobacilli). They ferment milk sugar (lactose) into lactic acid and other acids. They are non-pathogenic and occur in oral cavity, intestine and vagina of humans and animals.

This group is represented by a single genus, Lactobacillus. There are also 3 genera of uncertain affiliation.

Group 17. Actinomycetes and Related Organisms

These bacteria show pleomorphism and have a tendency to form branched filaments. They are non-motile, Gram-positive and aerobic or anaerobic. Many species are pathogenic to humans and animals.

This group includes one order with 8 families, represented by 31 genera. There are also 4 genera of uncertain affiliation.

Group 18. Rickettsias

These are among the smallest bacteria, ranging from 0.3 to 0.7 μ m in width and 1.0 to 2.0 μ m in length. They are Gram-negative, non-motile, obligate parasites and cause several diseases in humans and animals.

This group is represented by 2 orders, 4 families and 18 genera.

Group 19. Mycoplasmas

The organisms of this group lack a true cell wall and the cellular contents are enclosed by a triple-layered nonrigid cell wall. They are Gram-negative, non-motile and facultative anaerobes. They occur in respiratory tract and lower genital tract of mammals and birds.

This group has one order, 2 families and 2 genera.

Distribution : Bacteria are cosmopolitan in distribution, occurring in all natural habitats. Their wide distribution is primarily due to the fact that they can withstand great extremes of temperature, moisture, acidity and salinity and are adapted to wide variety of energy sources. They occur in the atmosphere to an height of about six kilometers and on the sea floor five kilometers below the mean sea level. Some are adapted to a normal aerobic environment and others thrive well in the absence of oxygen (i.e., anaerobic). They exist in hot springs and can also survive below freezing point in the Atlantic ice. Similarly, their tolerance to hydrogen ion concentration ranges from pH near 0 to pH 11. Although bacteria can survive high temperature or high acidity, they can not survive both simultaneously; in neutral or alkaline hot springs they can occur up to the boiling point of water, but at a lower pH the upper temperature limit is relatively low. Normally bacteria live at a pressure of one atmosphere but they can tolerate a pressure of 3000-6000 atmospheres. The highest pressure in the ocean depth approximates about 1000 atmospheres, thus pressure has no limiting effect on bacteria. Many species of bacteria are free living, and others are symbiotic, parasitic or saprophytic.

The immense biological success of bacteria is undoubtedly due to their small size, rapid reproductive rate, ability to survive under adverse conditions and metabolic adaptations.

Size : Bacteria are very small microscopic organisms. An average bacterial cell ranges from 0.5 (1μm = 0.001mm) to 2.0μm in diameter. A single drop of water may contain as many as 5×10^6 bacteria. In one gram top soil their number may vary from 1×10^3 to 1×10^9. Their size also varies with the shape. For example, the radius of spherical or coccus bacteria ranges from 0.5 to 2.5μm and rod-shaped or bacillus bacteria are larger than cocci and are 0.3 to 15μm in width. Thiophysa volutans, a sulphur bacterium, measuring about 18μm in diameter, is perhaps the largest amongst all bacteria.

Shape : The bacterial cells show considerable variation in their shape, but all individuals of a species have approximately the same shape. Thus from the point of view of their identification, shape of bacteria is an important characteristic. Although conventional light microscope provides little structural details, it is possible to establish the major morphological groups on this basis. Three cell forms, characteristic of true bacteria, are as follows.

Bacillus or Rod-shaped Bacteria : This is probably the most common form of bacteria. These are rod-shaped, cylindrical or elongate and are motile or non-motile cells. The rods may have rounded or blunt ends; some are very short and almost indistinguishable from spherical bacteria, while others are long and narrow. An average bacillus is about 1.5μm in length and 0.5μm in diameter.

Examples : Bacillus anthracis, B. fastidious, B. polymyxa, Lactobacillus.

The bacillus bacteria are classified into two groups on the basis of their arrangement.

Diplobacillus : When bacillus bacteria occur in pairs, they are called diplobacillus; e.g., Corynebacterium diphtheriae.

Streptobacillus : The bacilli bacteria of this group occur in long chains e.g., Bacillus tuberculosis, B. cereus. streptobacillus C.

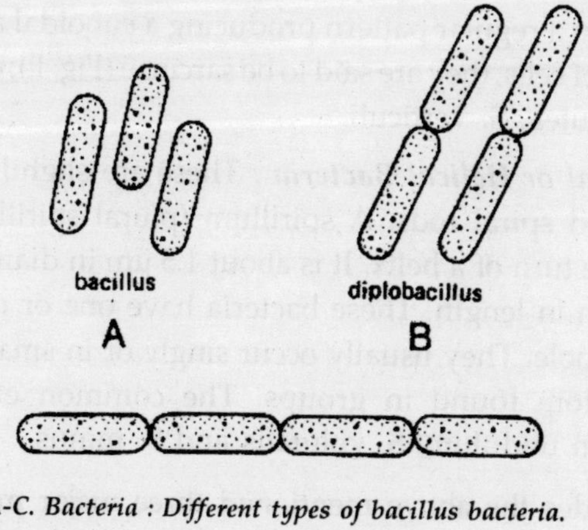

A-C. Bacteria : Different types of bacillus bacteria.

Coccus or Spherical Bacteria : The spherical or ellipsoidal bacteria are called cocci (singular coccus) (Fig. A). They measure 0.5-2.25μm in diameter. They are non-motile, atrichous and often occur in chains or in clusters of variable size and shapes. On the basis of the arrangement and the number of cells in a cluster, the cocci are classified into the following six groups.

1. *Micrococci :* When cocci occur singly, they are known as micrococci; e.g., Micrococcus cerolyticus, M. cryophilus, M. luteus.
2. *Diplococci :* When cocci occur in pairs, they are called diplococci (Fig. B); e.g., Diplococcus pneumoniae.
3. *Streptococci :* Spherical bacteria when occur in long chains, they are called streptococci (Fig. C); e.g., Streptococcus lactis, S. pyogenes.
4. *Tetracocci :* When they form groups of four cells, they are said to be tetracocci (Fig.D); e.g., Pedicoccus cerevisiae, Neisseria.
5. *Staphylococci :* An irregular group of many spherical bacteria is known as staphylococcus (Fig. E); e.g., Staphylococcus albus, S. aureus.
6. *Sarcinae :* When spherical bacteria divide in three planes in a regular pattern producing a cuboidal arrangement of cells, they are said to be sarcinae (Fig. F); e.g., Sarcinae lutea, S. verticuli.

Spiral or Helical Bacteria : These are slightly larger and elongated spiral rods. A spirillum (plural spirilla) has more than one turn of a helix. It is about 1.5 μm in diameter and up to 15 μm in length. These bacteria have one or more flagella at each pole. They usually occur singly or in small chains but are seldom found in groups. The common examples are Spirillum undulum, S. volutans and S. minus.

Besides the above mentioned three major morphological types, bacteria also occur in the following shapes.

1. *Vibrios :* These are slightly curved rods of half turn. They resemble with the sign of comma (,), and are commonly known as comma bacteria'. A vibrio bears a single flagellum at its tip and is about 10mm in length and 1.5-1.7 mm in width. Examples—Vibrio coli, V. cholerae.
2. *Filamentous :* Some bacteria such as Beggiatoa and Thiothrix are filamentous.
3. *Pleomorphic :* Some bacteria are capable of changing their shape and size temporarily in response to changes in the surrounding environment. As such a single bacterium may occur in more than one shape in its life cycle. For example, Acetobacter may occur as bacillus (single rods) or streptobacillus (chain of small rods), depending on the environment.

Structure of Bacterial Cell

The conventional light microscope provides only a vague notion of the internal structure of the bacterial cell. Therefore, the cytological details of these organisms could be possible only after the discovery of electron microscope and development of staining techniques. The bacterial cell shows a typical prokaryotic structure. It is enclosed by three layers, the outermost covering of slime, the middle cell wall and the innermost cytoplasmic membrane.

It is a gelatinous layer present on the outer surface of the cell wall, usually composed of polysaccharides (e.g., dextran, dextrin, lavan) and/or polypeptide chains of amino acids. When its constituents are only polysaccharides, which form a viscous layer, it is said to be slime layer, but when nitrogenous substances (i.e., amino acids) are also present along with polysaccharides, then it is called capsule. Various strains of the same species may have capsules of several different strain-specific polysaccharides.

The slime layer/capsule protects the cell from desiccation and antibodies. The capsulated bacteria also remain unaffected by phagocytosis. They are usually atrichous.

Cell Wall : The cell wall is characteristic of plant kingdom. In higher plants and algae, it is composed of cellulose and shows a fibrillar structure. In contrast, the bacterial cell wall has a granular structure but is tough and rigid. It varies in thickness from 50 to 100Å. The three main constituents of cell wall are as follows : (i) N-acetyl-glucosamine (NAG), (ii) N-acetyl muramic acid (NAM), and (iii) a peptide chain of four or five amino acids. These together form a polymer called peptidoglycan or mucopeptide. The NAG and NAM molecules which are arranged alternately, run in one direction and the peptide chains run crosswise. The latter are also cross-linked to one another. The rigidity of bacterial cell wall is due to the presence of this polymer. Besides above mentioned three constituents, some other chemicals such as teichoic acid, protein polysaccharides and lipoproteins and lipopolysaccharides are also deposited on the cell wall.

The function of bacterial cell wall appears to be wholly mechanical, giving the cell its shape and rigidity. If the cell wall is removed by laboratory procedures, the naked protoplast becomes spherical. The rigidity of cell wall can be judged by the fact that it can withstand an osmotic pressure of about eight atmospheres per square centimeter.

Staining of Bacteria (Gram's Stain) : The staining techniques have proved very useful in taxonomic grouping of bacteria. One of the important cytological features of bacteria is their reaction to a simple staining procedure discovered by a Danish physician, Christian Gram, in 1884. The procedure involves staining the cells first with crystal violet and then with iodine solution. The group of bacteria which retains the stain even after decolourization with alcohol is called Gram-positive, whereas those which lose stain after the treatment with alcohol are called Gram-negative. This differential reaction of two types of bacteria to crystal violet-iodine stain is due to different amount of lipids in their cell wall. The Gram-negative bacteria have relatively high lipid content. Alcohol dissolves the lipids which allows the leakage of stain. The Gram +ve bacteria, on the other hand, have less lipids and hence less susceptible to alcohol action. This differential staining property of the cell correlates surprisingly well with many other characteristics. For example, all spore (endospore) forming bacteria are Gram-positive and all polarly flagellated forms are Gram-negative; most of cocci (spherical bacteria) are Gram-positive and bacilli (rod-shaped bacteria) are Gram-negative. The Gram-negative bacteria are, as a rule, more resistant to many antibiotics than Gram-positive.

The fundamental difference between Gram-positive and Gram-negative bacteria is in the nature of their cell wall. Although both groups have peptidoglycans as the basic constituent of the cell wall, there are certain other polymers which are specific to each group.

The flow-chart of Gram's staining procedure is as follows.

Prepare a smear of bacteria on the slide

↓

Place few drops of crystal-violet stain on the slide
(bacteria take a violet stain)

↓

After 30 seconds rinse the slide in water

↓

Place few drops of iodine on the slide
(bacteria take a deep violet or purple stain)

↓

After 20 seconds rinse the slide in water

↓

Wash the slide with 90% ethyl alcohol

↓

Gram-negative bacteria are decolourized

↓

Gram-positive bacteria retain the violet colour

Acid-fast Stain : Some bacteria have a covering of waxy or fatty material on their surface due to which they cannot be stained by simple staining techniques. At the same time if once they are stained it is difficult to destain them. Their decolourization is not possible even with acid alcohol. Such bacteria are called acid-fast. These bacteria can, however, be stained by Ziehl-Neilson technique. Mycobacterium tuberculosis and M. leprae are examples of acid-fast group.

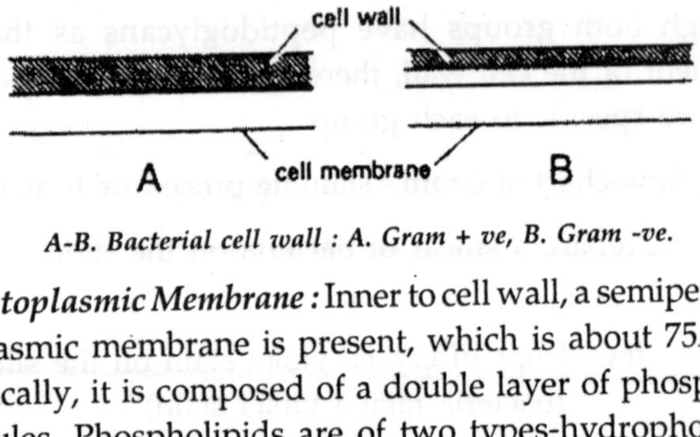

A-B. Bacterial cell wall : A. Gram + ve, B. Gram -ve.

Cytoplasmic Membrane : Inner to cell wall, a semipermeable cytoplasmic membrane is present, which is about 75Å thick. Chemically, it is composed of a double layer of phospholipid molecules. Phospholipids are of two types-hydrophobic and hydrophilic. The hydrophilic phospholipid molecules are present towards the outerside and the hydrophobic molecules towards the inner side.

Proteins are found embedded in the lipid bilayer. Prokaryotic membrane is characterized by the absence of sterols which perhaps account for the enormous resistance of these organisms to antibiotics. The chemical nature of cytoplasmic membrane in Gram-positive and Gram-negative bacteria is basically the same.

Cytoplasmic membrane is the bounding layer of the cytoplasm and it is the centre of multifarious activities of the cell. Enzymes of various metabolic pathways, such as synthesis of lipopolysaccharides, phospholipids, teichoic acid, etc., are located in the cytoplasmic membrane. It also controls the entry of organic and inorganic nutrients in the cell.

Cytoplasm and Cytoplasmic Inclusions : The bacterial cytoplasm is a complex mixture of carbohydrates, proteins, lipids, minerals, nucleic acids and water. It stores organic material in the form of glycogen, volutin and poly-β–hydroxybutyrate. Some photosynthetic and non-photosynthetic bacteria also accumulate sulphur and iron in their cytoplasm.

Besides fluid portion and storage particles, the bacterial cytoplasm also contains a chromatic or nuclear area and some other inclusions. The bacterial cell is devoid of mitochondria, endoplasmic reticulum, centrosome and Golgi bodies. Although a well organised chloroplast is absent in bacteria, the photosynthetic bacteria have chromatophores in their cytoplasm.

Nuclear Material : Basic dyes, which selectively stain chromatin of eukaryotic nucleus, stain most bacterial cells due to the abundance of ribosomes in their cytoplasm. It confers an unusually high nucleic acid stain on the cytoplasmic region. Therefore, to stain the bacterial nucleus selectively, fixed cells are first treated with ribonuclease or HCl_2 which hydrolyses the ribosomal RNA. Subsequent staining with a basic dye reveals the bacterial nucleus as a centrally located body of irregular outline. More than one such chromatin bodies may be present in actively growing cells.

The characteristic feature of the bacterial nucleus is the absence of nuclear membrane, nucleolus, chromonemata and nuclear sap. Such a nucleus is called nucleoid or genophore. Under electron microscope the nucleoid appears to be fibrillar and composed of a double or single stranded DNA. It has approximately 5 x 10^9 base pairs and a molecular weight of about 3 x 10^9. The DNA molecule is approximately 1000μm long, usually forming ring like structure or sometimes remains diffused throughout the cytoplasm of the cell. The nucleoid of Escherichia coli has a central core of RNA, surrounded by 12-82 supercoilings of DNA. A few protein molecules are also associated with DNA.

The bacterial DNA is devoid of histones hence can not be compared with the chromosomes of eukaryotic cells. However, its double stranded DNA is usually referred to as bacterial chromosome.

Plasmids : Bacterial cells also contain some extrachromosomal hereditary determinants which are either independent of bacterial chromosomes or are integrated with them Lederberg (1952) coined the term plasmids for such extrachromosomal hereditary determinants. They are rings of DNA of various sizes, in the range of 20-100 kbp (the whole bacterial cell is 4,000 kbp).

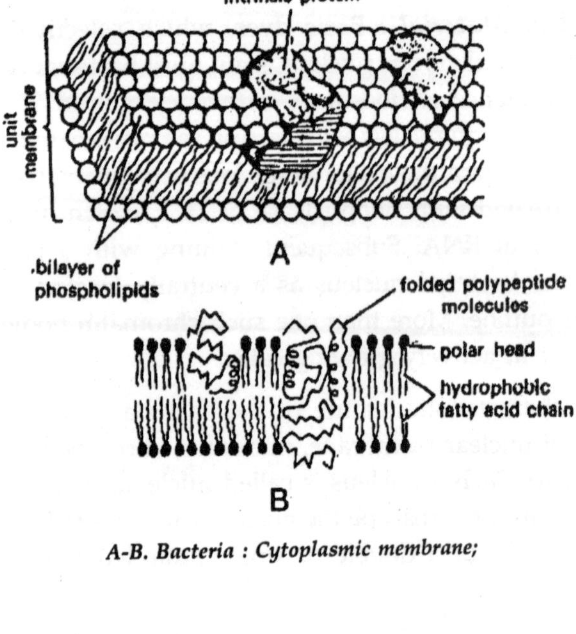

A-B. Bacteria : Cytoplasmic membrane;

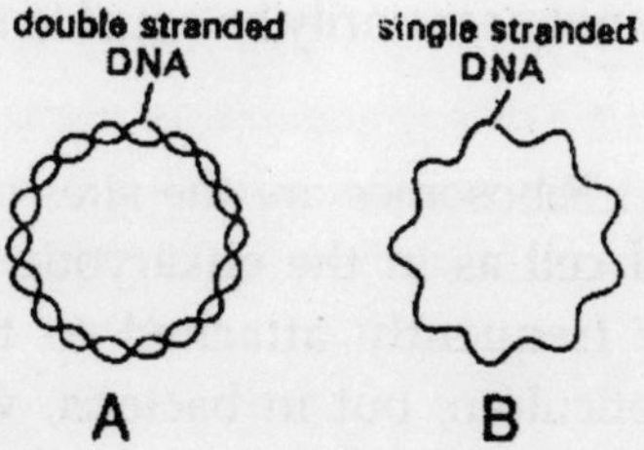

A. Fluid mosaic' model of cytoplasmic
B. Single stranded. membrane, B. Lipid bilayer.

A-B. Bacterial DNA: A. Double stranded,

Each plasmid carries non-esssential genes and has no role in viability and growth of bacteria. Hence they are also said to be dispensable autonomous elements. One good example of plasmids is the F - factor which determines the maleness in bacteria. It is an autonomous element, separate from the bacterial chromosome. It is transmitted from cell to cell by contact or by external agencies. A number of bacterial properties are now known which are determined by plasmid carried genes. Plasmids have been classified on the basis of the host properties which led to their detection. The following nine types of plasmids are known : (i) F-factor (for fertility) plasmids, (ii) R-factor (for resistance) plasmids, (iii) col-factor (for colicinogeny) plasmids of the Gram-negative bacteria, (iv) plasmids conferring pathogenecity to mammals, (v) degradative plasmids of Pseudomonas, (vi) mercury-resistance plasmids, (vii) tumor inducing plasmids of Agrobacterium tumifaciens, (viii) penicillinase plasmids of Staphylococcus aureus, and (ix) cryptic plasmids.

Plasmids are circular double stranded DNA molecules. Their molecular weight ranges between 5×10^7 and 7×10^7 (some cryptic plasmids are even smaller). Each plasmid may contain as many as 100 genes. The general features of plasmid replication are similar to that of chromosome replication, but plasmid seems to operate under its own genetically determined system of replication control. Thus the rate of plasmid replication may not necessarily be equal to that of chromosome replication.

Ribosomes : Ribosomes are the sites of protein synthesis in the bacterial cell as in the eukaryotic cell. In eukaryotes ribosomes are frequently attached to the surface of the endoplasmic reticulum, but in bacteria, which do not have endoplasmic reticulum, ribosomes are free in the cytoplasm. Their number varies from 10,000 to 15,000 in a cell.

Bacterial ribosomes are of 75 S type (eukaryotec ribosomes are of 80 S type) and consist of two subunits. The sedimentation constant of the larger subunit is 50 S and that of the smaller subunit is 30 S. The former has two molecules of RNA and 35 amino acids and the latter has one molecule of RNA and 21 different amino acids.

In a young bacterium ribosomes may occur in groups of 4-6 or more. They are held together by a special RNA molecule, known as messenger RNA. These groups of ribosomes are known as polyribosomes.

Mesosomes : These are complex localized infoldings of the cytoplasmic membrane. There may be 2-4 mesosomes in a cell and the number is usually higher in bacteria which show high respiratory activity, such as nitrifying bacteria. It has been

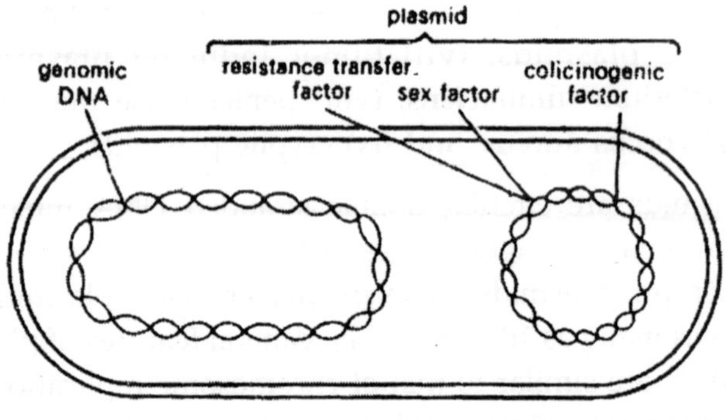

Bacteria : Genome and plasmid (extrachromosomal factors) of a bacterial cell.

suggested that the highly infolded membrane system of mesosomes perhaps serves to accommodate more centres of respiration. But the absence of enzymes like ATPase, dehydrogenase and cytochrome in mesosomes indicates that they are not the sites of respiration. They probably participate in the formation of transverse septum during cell division.

Flagella : The locomotion in bacteria is accomplished by thin hair-like appendages, called flagella. Each flagellum is a whip-like structure of almost uniform thickness, originating from the cytoplasm just beneath the cytoplasmic membrane.

Flagella are characteristic of all spiral bacteria and they also usually occur in bacillus bacteria. Coccus bacteria are, however, devoid of flagella and are non-motile. The following categories of bacteria are recognised on the basis of the presence or absence of flagella.

Atrichous : These bacteria are devoid of flagella, hence they are nonmotile; e.g., Lactobacillus, Pasteurella.

Trichous : These bacteria possess flagella, hence they are motile. The flagella are distributed over the surface of bacteria in a characteristic fashion. Their number, position and arrangement varies in different, species. They may be restricted to the ends of the cell (polar flagella) or uniformly distributed all over the surface of the cell (peritrichous or non-polar flagella). Peritrichous flagellation is of wide occurrence among bacteria, whereas polarly flagellated bacteria form a rather homogeneous assemblage of rods and spirals. Various types of polar and peritrichous flagellation are as follows.

(a) *Polar Flagellation :* It is usually found in Gram- negative bacilli and spirilla. The following four types of polar flagellations are recognised.

(i) *Monotrichous :* If only a single flagellum is present

at one end of the bacterial cell, it is known as monotrichous, e.g., Vibrio cholerae.

(ii) *Amphitrichous* : When the bacterial cell possesses one or more flagella at both the ends. It is said to be amphitrichous. It is a relatively uncommon pattern of flagellation; e.g., Nitrosomonas.

(iii) *Cephalotrichous* : When bacterial cell possesses two or more flagella at one end only, it is called cephalotrichous; e.g., Pseudomonas fluorescens

(iv) *Lophotrichous* : Those bacteria which have a group of flagella at one end only, it is called lophotrichous; e.g., Spirillum volutans.

(b) *Non Polar Flagellation* : In non-polar or peritrichous flagellation the flagella are evenly distributed throughout the surface of the cell; e.g., Proteus vulgaris, Bacillus typhosus and Clostridium.

Structure of the Bacterial Flagellum : The bacterial flagella are quite unlike those of other organisms. The flagella and cilia of higher protists and other organisms have an identical' 9+2 fibrillar structure. The molecular architecture of bacterial flagellum, derived on the basis of chemical analysis, reveals that it is composed of several chains of flagellin (a protein) molecules forming a more or less cylindrical filament, The diameter of a flagellin molecule is about 40Å and that of a flagellum is about 120-150Å. The orientation of flagellin molecules may vary in different species. The length of flagella (4-5 μm long) is usually more than that of bacterial cell.

The swimming ability of bacteria is due to the presence of flagella. If the cell is mechanically deflagellated, it becomes immotile. The locomotion is brought about by clockwise or anti-clockwise spins of flagellum around its own axis. In a liquid medium the speed of movement is more than 20 μm per second.

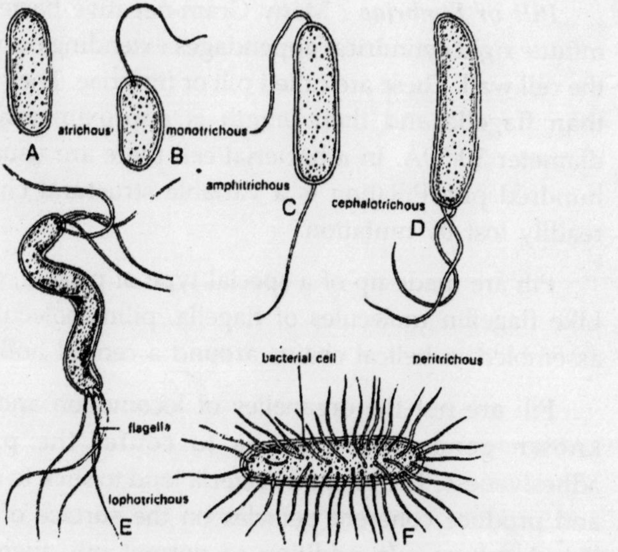

A-F. Bacteria : Different types of flagellation.

A-B. Bacteria flagellum : Models showing probable helical arrangement of protein molecules; A. Flagellum, B. Pilus,

Pili or Fimbriae : Many Gram-negative bacteria possess minute rigid cylindrical appendages extending outwards from the cell wall. These are called pili or fimbriae. They are smaller than flagella and their length is approximately 2μm and diameter 30-50Å. In a bacterial cell there are usually several hundred pili. Piliation is a variable structural characteristic, readily lost by mutation.

Pili are made up of a special type of protein, called pilin. Like flagellin molecules of flagella, pilin molecules are also assembled in helical chains around a central hollow core.

Pili are not the organelles of locomotion and their only known general function is to confer the property of adhesiveness. The piliated bacteria tend to stick to one another and produce coherent pellicles on the surface of unagitated liquid cultures. In addition to normal pili, there are some special type of pili, known as sex pill. There are usually 1-5 sex pili per cell. They are helpful in the transfer of genetic material, probably by acting as genetic tubes.

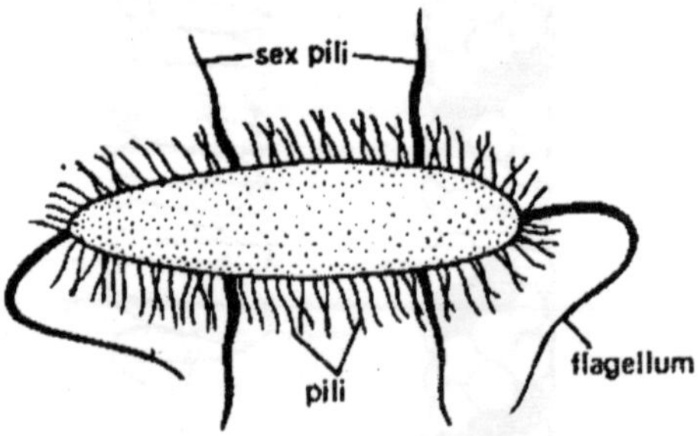

Bacteria : Polar flagella and pili.

Nutrition of Bacteria

On the basis of their mode of nutrition two groups of bacteria recognised are autotrophic bacteria, and heterotrophic bacteria.

Autotrophic Bacteria : Those bacteria which are capable of synthesizing their food by themselves from organic and inorganic substances are called autotrophic. The autotrophs can be divided further into two subgroups : (i) phototrophs, and (ii) chemotrophs, on the basis of energy source they use. The division may also be based on the nature of substances oxidised; lithotrophs utilising inorganic substances and organotrophs oxidising organic matter for their energy.

Photosynthetic Bacteria : These are also known as photoautotrophic or photolithotrophic bacteria. Like higher plants, they are capable of converting radiant energy into chemical energy. The generalized reaction of photosynthesis is

$$CO_2 + 2H_2A \longrightarrow (CH_2O) + H_2O + 2A$$

Where H_2A is an oxidizable compound and A is the corresponding oxidation product. In higher plants oxygen substitutes A, 'thus the hydrogen donor is water and the process produces free oxygen.

$$CO_2 + H_2O \longrightarrow (CH_2O) + O_2$$

In bacterial photosynthesis, however, various substances may substitute A, but never oxygen. Oxygen is not released during photosynthesis and as such these bacteria are anaerobic.

Photoautotrophic bacteria are of the following types.

(a) *Green Sulphur Bacteria* : These are small nonmotile, rod-shaped bacteria which are strictly anaerobic photoautotrophs. Their photosynthetic pigment is chlorobium chlorophyll, located in the invaginations of cytoplasmic membrane. These bacteria use H_2S or other reduced inorganic sulphur compounds (sulphite or sulphide) as electron donor. Elemental sulphur formed as a byproduct in this process is deposited extracellularly. Chlorobium, Prosthecochloris,

Pelodictyon, and Clathrochloris are four well known examples of green sulphur bacteria.

$$6CO_2 + 12H_2S \longrightarrow C_6H_{12}O_6 + 6H_2O + 12S$$

(b) *Purple Sulphur Bacteria* : These are also autotrophs and their photosynthetic pigments are bacteriochlorophyll a and / or b. The reducing power is provided by H_2S which is oxidised anaerobically, via elemental sulphur, to sulphate. The sulphur is deposited in them entracellularly (extracellularly in Ectothiorhodospira). The overall reaction can be represented schematically as:

$$2CO_2+H_2S+2H_2O \xrightarrow[\text{bacteriochlorophyll}]{\text{light}} 2(CH_2O) + H_2SO_4$$

Some purple sulphur bacteria can use other reduced sulphur compounds (thiosulphate, sulphite, etc.) in place of H_2S as exogenous reductants.

$$6CO_2 + 9H_2O + 3Na_2S_2O_3 \xrightarrow[\text{bacteriochlorophyll}]{\text{light}} C_6H_{12}O_6 + 6NaHSO_4$$

Thiospirillum, Chromatium, Thiocystis, Thiocapsa, Lamprocystis and Amoebobacter are some common examples of purple sulphur bacteria.

(c) *Non-sulphur Purple Bacteria* : These are motile bacteria which do not produce gas vacuoles and never accumulate sulphur within the cell. Their photosynthetic pigment is also bacteriochlorophyll. In their metabolism some organic compounds replace sulphur and the extent of CO_2 reduction depends upon the organic substrate. When the substrate is reduced more than the cell material (CH_2O), i.e., the ratio of hydrogen to oxygen is greater than two to one, the excess hydrogen is used to reduce CO_2. On the other hand, if there is greater oxidation of the substrate, CO_2 is given off just as in respiration.

For example, if propionic acid is the substrate, CO_2 is reduced

$$2C_3H_6O_2 + H_2O + CO_2 \longrightarrow 7(CH_2O)$$
(propionic acid)

However, if malic acid is the substrate, CO_2 is released

$$C_4H_6O_5 \longrightarrow 3(CH_2O) + CO_2$$
(malic acid)

Non-sulphur purple bacteria can grow in the presence of oxygen unlike purple sulphur bacteria.

Comparison between Gram-positive and Gram-negative Bacteria

Characteristics	*Gram-positive bacteria*	*Gram-negative bacteria*
1. Cell wall structure.	Cell wall single layered and 150-200Å thick.	Cell wall triple layered and 75-120Å thick.
2. Chemical composition.	Peptidoglycans account for about 80% of the cell wall and the rest are polysaccharides.	Peptidoglycans account for only about 3-12% of the cell wall. It is mainly composed of lipoproteins and lipid polysaccharides.
	Teichoic acid present. Low in lipids (1-4%). Highly responsive to triphenylmethane.	Teichoic acid absent. High in lipids (11-22%). Show little response to triphenyl methane.
	Resistant to alkalies and insoluble in 1% KOH solution.	Show sensitivity to alkalies and soluble in 1% KOH solution.
3. Rigidity of cell wall.	Cell wall is very rigid due to high proportions of peptidoglycans.	Cell wall is elastic due to plastic nature of lipoprotein-polysaccharide mixture.
4. Susceptibility to penicillin.	High susceptibility.	Low susceptibility.
5. Nutritional requirements.	Relatively complex in many species.	Relatively simple.

Chemosynthetic Bacteria : Photosynthesizing bacteria represent only a small fraction of Schizophyta, but chemosynthetic or chemotrophic bacteria are most abundant and are important geochemical agents. These are non-photosynthetic but autotrophic bacteria. They derive energy from ammonia, nitrate, nitrite, ferrous iron, hydrogen sulphide and other inorganic compounds for the synthesis of their food. Thus chemoautotrophs grow in a strictly mineral medium in dark. They are characterized by (i) high specificity with respect to inorganic energy source, and (ii) frequent inability to use organic compounds as energy and carbon sources. Their growth is sometimes adversely affected by organic compounds.

On the basis of substrate specificity, chemoautotrophs can be classified into the following four groups.

(a) *Sulphur Bacteria* : These bacteria occur in sulphur containing terrestrial and aquatic environments. They derive energy by the oxidation of reduced sulphur compounds.

The sulphur bacterium, Beggiatoa, oxidises hydrogen sulphide to elemental sulphur.

$$H_2S + 1/2O_2 \longrightarrow S + H_2O$$

The elemental sulphur is deposited in the filaments as minute granules.

Another sulphur bacterium, Thiobacillus thiooxidans, utilizes free sulphur and produces sulphuric acid.

$$S + H_2O + 1.5\ O_2 \longrightarrow H_2SO_4 + \text{energy}$$

(b) *Iron Bacteria* : These bacteria (e.g., Sphaerotilus, Gallionella, Ferrobacillus and Leptothrix) form natural colonies in fresh water ponds and springs with high contents of reduced sulphur salts. They oxidise ferrous compounds into ferric forms and the energy released in this process is utilized in the synthesis of organic compounds.

$$Fe^{2+} \longrightarrow Fe^{3+} + e^-$$

The ferric iron is deposited as insoluble ferric hydroxide. The overall reaction is as follows;

$$2Fe(HCO_3)_2 + O + H_2O \longrightarrow 2Fe\ (OH)_3 + 4CO_2 + energy$$

(c) *Hydrogen Bacteria* : Many species of chemoautotrophic bacteria have the ability to grow with molecular hydrogen. They oxidise molecular hydrogen and in this process water and energy are obtained.

$$H_2 + 1/2O_2 \longrightarrow H_2O$$

The common H_2-oxidising bacteria are Pseudomonas saccharophila, P. facilis, Alcaligenes eutrophs, A. paradoxus, Nocardia opaca and Paracoccus denitrificans (previously all these bacteria were placed in a separate genus Hydrogenomonas, but now they are included in a series of genera).

(d) *Nitrifying Bacteria* : These are soil-borne obligate autotrophs incapable of growing in the absence of specific inorganic energy source. They oxidize ammonia to nitrate and help greatly in the economy of nitrogen in nature. The nitrification process is carried out in two steps, each step by a very specialized groups of bacteria. The first step involves oxidation of ammonia to nitrate with the help of Nitrosomonas.

$$2NH_3 + 3O_2 \longrightarrow 2HNO_2 + 2H_2O + 78K\ Cal.\ energy$$

This oxidation is the sole source of energy for Nitrosomonas.

The second step of nitrification involves oxidation of nitrite to nitrate.

$$2HNO_2 + O_2 \longrightarrow 2HNO_3 + 21.6K\ Cal.\ energy.$$

Here the oxidation is carried out by Nitrobacter and oxidation of nitrite is the only source of energy for the bacterium.

Heterotrophic Bacteria : These bacteria obtain their readymade food from any organic source. They can be distinguished into three major nutritional categories.

Parasitic Bacteria : Those bacteria which feed on living organisms are known as parasitic bacteria. The organism from which parasitic bacteria obtain their food is known as host. Some parasitic bacteria are known to cause diseases in plants (e.g., citrus canker) and animals (e.g., pneumonia, typhoid), and such bacteria are called pathogenic bacteria.

Saprophytic Bacteria : These bacteria grow on dead and decaying organic matter. They obtain their food by decomposing the complex organic molecules into simple inorganic constituents. The decomposition of carbohydrates and proteins by saprophytic bacteria is technically known as fermentation and putrifaction respectively.

Breakdown of sugars by yeast is a common example of fermentation. In this process CO_2 is released. Similarly, lactic acid fermentation of milk is carried out by lactic acid bacteria (Lactobacillus). In putrifaction reduction of protein to peptone, polypeptide, peptides and amino acids takes place by enzymes secreted by anaerobic bacteria. This process does not require oxygen, but many saprophytic bacteria which breakdown amino acids into methane, CO_2, NH_3, H_2 and N_2 do require oxygen.

Symbiotic Bacteria : Those bacteria, which grow in close (beneficial) association with other living organisms, are known as symbiotic bacteria. In terms of their association with the host, symbiotic bacteria may be ectosymbionts (when bacteria live on the surface of the host) or endosymbionts (when bacteria live inside the host tissue). The bacteria inhabiting intestine of man and animals are good examples of symbiotic bacteria. The enzymes secreted by these bacteria are helpful in the digestion of cellulose and in return they obtain their food from the host. Similarly, root nodule bacteria (Rhizobium), present in the roots of leguminous plants, fix atmospheric nitrogen to augment their nitrogen supply and in return the plants provide them shelter and carbohydrates.

Reproduction : Bacteria reproduce mainly by asexual methods and therefore they have a dominant haploid phase in their life-cycle. They do not have sex organs or gametes but they definitely show genetic recombination, i.e., exchange of genetic material.

Asexual Reproduction : Asexual reproduction in bacteria takes place by the following methods.

Binary Fission : It is the simplest and most common method of multiplication in bacteria. Under favourable conditions the cell divides into two daughter cells by a transverse wall. During fission the cell elongates and there is division of the nuclear material. In Gram-negative bacteria it is followed by the

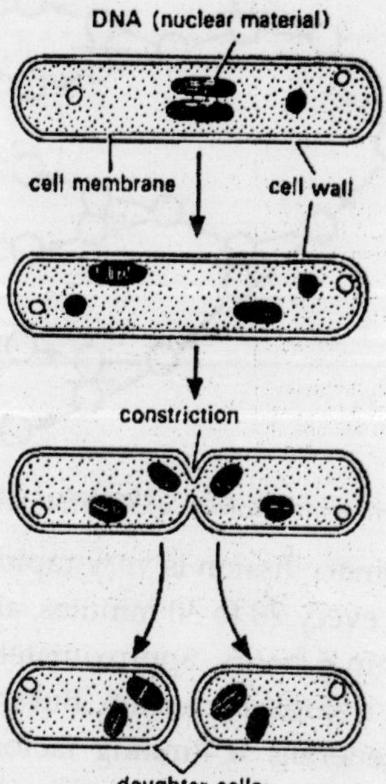

Bacteria : Binary fission.

development of a simple median constriction that finally results in complete separation of two daughter cells (without cross wall formation). In Gram-negative bacteria, however, first a transverse cell membrane is laid down between the two nucleoids of a dividing cell and then the centripetal growth of the cell wall forms complete cross wall. The transverse wall between the daughter cells is very often incomplete and then the daughter cells remain connected at the transverse membrane. When the planes of the successive divisions are parallel, long chains may be formed and if perpendicular to one another, sheets of packets of cells are formed.

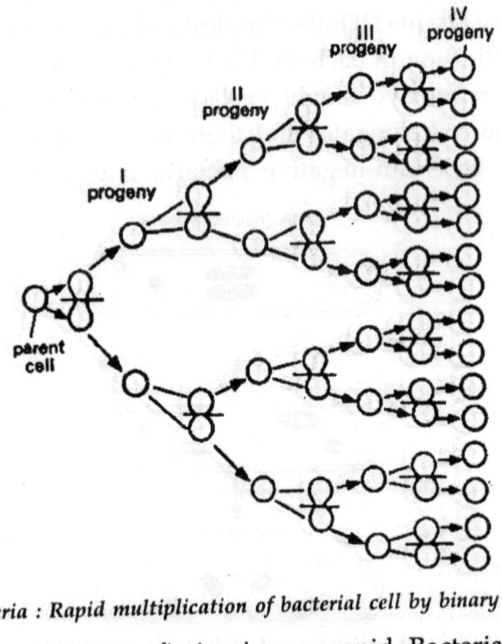

Bacteria : Rapid multiplication of bacterial cell by binary tission.

Process of binary fission is very rapid. Bacterial cells may undergo fission every 20 to 30 minutes, and under favourable conditions, within 6 hours, approximately 2,50,000 cells may be formed from a single bacterium. But in nature the essential nutrients are generally a limiting factor and moreover the accumulated waste products of bacteria also restrict their rapid growth. Normally, growth and multiplication is very fast in the early stages, reaching at the optimum and then declining rapidly.

Fission is different from normal mitotic division as it does not involve spindle formation during the division of nuclear material. It is not definitely known whether in fission the genetic material is equally distributed between the daughter cells. Fission is characteristic of all types of bacteria and that is why they have been placed in the class Schizomycetes (Schiz = split or cut, mycetes = fungi; split, cut or fission fungi).

Budding : In this type of multiplication the bacterial cell gives out many outgrowths. These outgrowths are filled with cytoplasm and other cell organelles also move into them. These outgrowths, called buds, eventually get detached from the parent cell by a constriction and each grows into a new bacterial cell (Fig. A). Hyphomicrobium vulgare and Rhodomicrobium vannielia are common examples of budding bacteria.

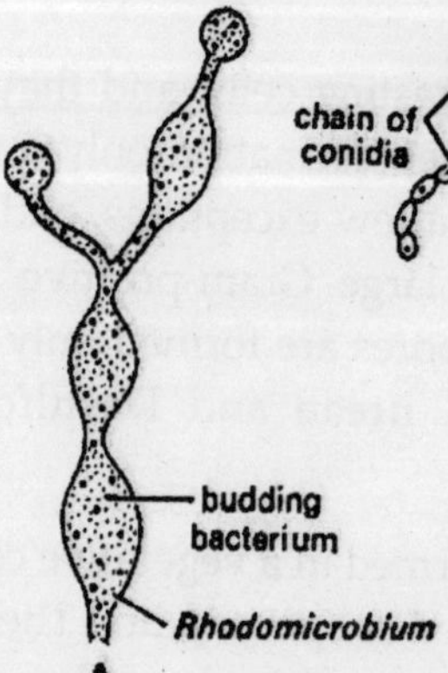

A

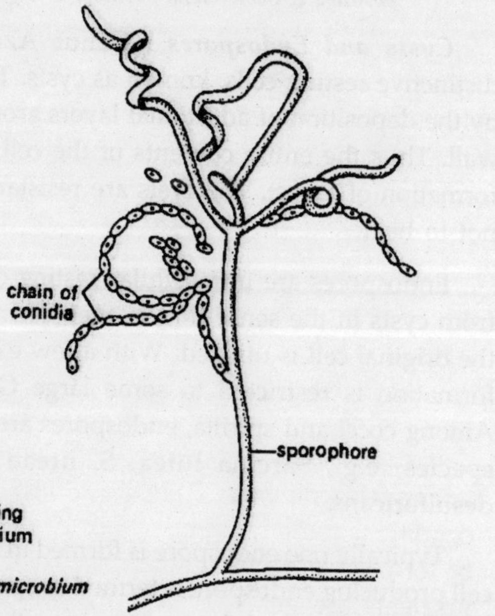

A-B. Bacteria : Asexual reproduction; *A.D. Bacteria : Asexual reproduction;*

Conidia : Many filamentous bacteria (e.g., Streptomyces) form chains of small, spherical spore-like conidia at the tips of the filaments. A conidium develops by the formation of a transverse wall at the tip of the filament. The filaments bearing conidia are known as conidiophores. After liberation each conidium gives rise to a new filamentous bacterium, provided conditions for germination are favourable.

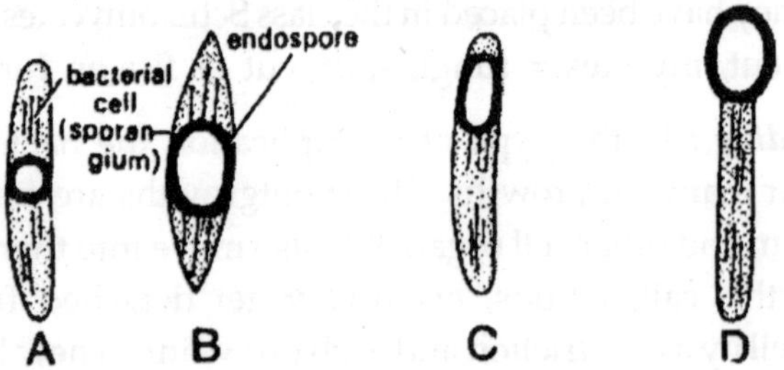

A. Budding in Rhodomicrobium, Various positions of endospores in bacterial cells. B. Conidia formation in Streptomyces. A-B. Central position, C. Subterminal position, D. Terminal position.

Cysts and Endospores : Genus Azotobacter produces distinctive resting cells, known as cysts. The cysts are formed by the deposition of additional layers around the existing cell wall. Thus the entire contents of the cell are involved in the formation of a cyst. The cysts are resistant to desiccation but not to heat.

Endospores are intracellular resting cells, and they differ from cysts in the sense that in their formation only a part of the original cell is utilized. With a few exceptions, endospore formation is restricted to some large Gram-positive bacilli. Among cocci and spirilla, endospores are formed only in few species, e.g., Sarcina lutea, S. ureae and Desulfovibrio desulfuricans.

Typically one endospore is formed in a vegetative cell. The cell producing endospore is termed as sporangium. The shape, size and position of the endospore within the cell may vary in different species. Endospores are oval or spherical and their diameter may be larger or smaller than that of the vegetative cell. Their position within the sporangium may be central, subterminal or terminal.

Most of the endospore forming bacteria are soil borne, and few are pathogenic to insects and vertebrates. Pathogenic spore forming bacteria cause disease by toxin production.

(a) *Formation of Endospores* : The ability to produce endospores is normally never expressed during the active phase of growth. It is expressed only when the vegetative growth reaches the stationary phase due to nutrient limitations. It is believed that sporulation is controlled by protein metabolism of the cell and a spore is formed only when the vegetative proteins of the cell are converted into spore proteins or when such new proteins are formed.

At the onset, of the sporulation, the two nucleoids present in a cell coalesce to form an axial chromatin thread. Then a transverse septum is formed near one pole of the cell, which separates the cytoplasm and DNA of the smaller daughter cell from that of the larger cell. The smaller cell, which is destined to form spore, is known as spore primordium. The septum formation is not followed by the development of transverse wall (as in normal cell division), instead the membrane of the larger cell rapidly grows around the spore primordium. The primordium gradually increases in size and within few hours of sporulation there is enormous growth in its size and it becomes opaque and highly refractive. The entire process of sporulation is completed in 18-20 hours. After the maturation of endospore, the sporangium (i.e., the bacterial cell in which endospore is formed) remains alive only for a short period. The endospore is liberated due to autolysis of the sporangium.

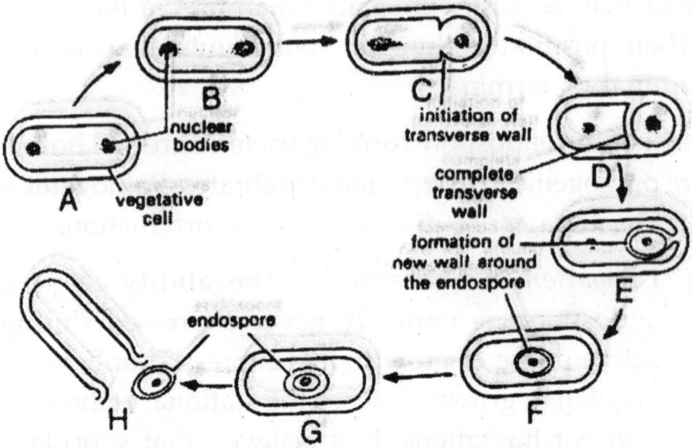

A-H. Bacteria : Asexual reproduction; Diagrammatic representation of enfospore formation.

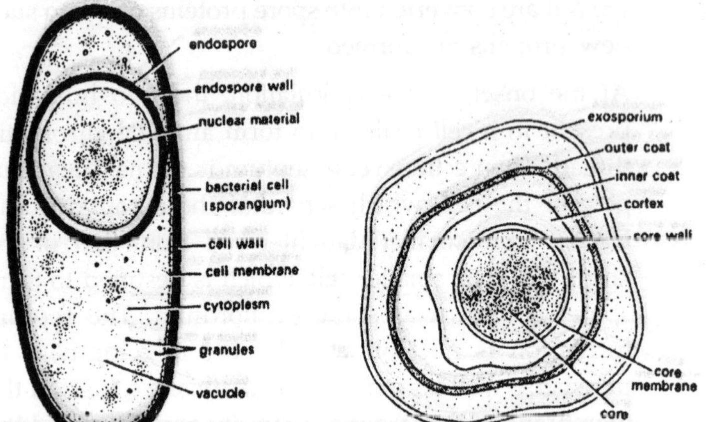

A-B. Bacteria : Asexual reproduction; A. A bacterial cell with endospore, B. Transverse section of an endospore of Bacillus (diagrammatic representation).

(b) *Structure of Endospore* : The endospore consists of a central protoplast, the core, mainly composed of DNA, tRNAs, ribosomes, enzymes and accessory factors. It is enclosed by a thin membrane, variously called - core membrane, germ cell membrane or inner membrane. The multilayered region around the core is known as cortex, and it is largely composed of calcium—dipicolinic acid-peptidoglycan complex. Outer to the cortex there is an electron dense layer, called spore coat, which is differentiated into an inner coat and an outer coat. The spore coat is mainly composed of proteins and accounts for 30-60 per cent of the dry weight of the spore. It also contains about 80% of the total spore proteins. Endospores of certain bacteria (e.g., Bacillus cereus) have an additional outer covering, the exosporium, made up of polysaccharides, proteins and some lipids.

Endospores are tough enough to last for years, enduring many stresses. Some of the spore producing bacteria are serious pathogens. For example, anthrax is caused by Bacillus anthracis, and its tough spores remain viable for years on the wool from an infected animal. The disease gets passed to weavers and other people through cuts on their hands.

(c) *Germination of Endospore* : Free endospores do not show any detectable metabolism, but they retain the potential capacity to germinate and develop into vegetative cells. They are highly resistant to heat, ultraviolet and ionizing radiations and many toxic chemicals. Endospores of some species of Bacillus can tolerate temperatures as high as 150-170°C. The extreme resistance of spores to physical and chemical agents is attributed to dipicolinic acid—calcium complex associated with the spore wall.

Endospores germinate under favourable conditions. Sufficient nutrition, humidity and suitable temperature are essential for their germination. These conditions trigger a series of metabolic reactions. Spores imbibe water and swell, losing their refractivity. The spore coat ruptures and new vegetative cell grows out; the spore wall forms the wall of the vegetative cell.

Each bacterial cell forms only one endospore, which in turn gives rise to a single bacterium. Therefore, sporulation can not be considered as a process of multiplication. It is, in fact, a method of perennation.

Genetic Recombination (Sexual Reproduction) **:** Sexuality in bacteria was first demonstrated by Tatum and Lederberg (1947) in Escherichia coli. In bacteria sexual reproduction does not involve fusion of gametes as in other organisms but each heritable character is determined by a portion of its genetic material (DNA) as in other organisms. The most important

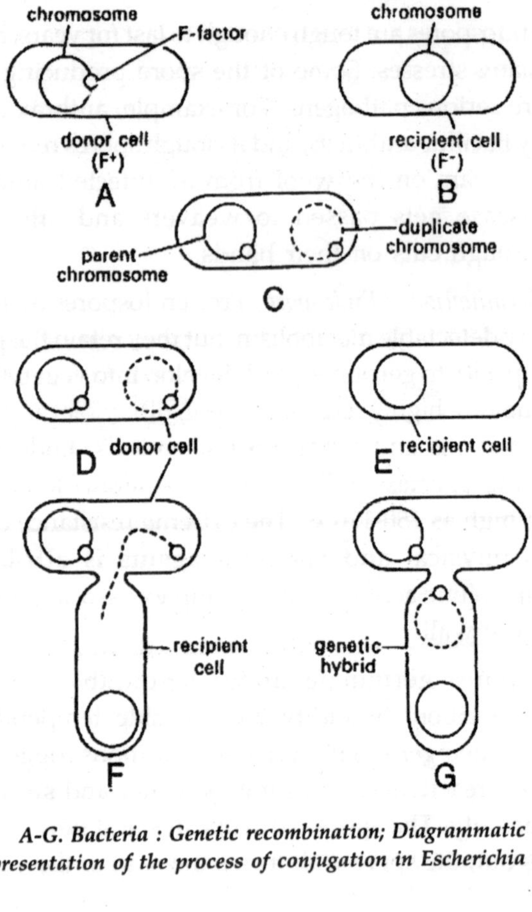

A-G. Bacteria : Genetic recombination; Diagrammatic representation of the process of conjugation in Escherichia coli.

characteristic of sexual reproduction in bacteria is interchange of genetic material. This is accomplished either by conjugation or by exotic methods. In conjugation, exchange of genetic material between two cells takes place by a conjugation tube, but in exotic method the two bacterial cells never come in contact and exchange of genetic material takes place by transformation and transduction.

Conjugation : It is known to occur in several bacteria but has been studied in detail in Escherichia coli 'by Lederberg (in Wisconsin), Hayes (in London) and Woolman (in Paris). They demonstrated that in K_{12} strain of E. coli exchange of genetic material is brought about by physical contact between two cells. This strain has two mating types and both of these are haploid. But they (two mating types) do not have the same role in the process of recombination.

One of the mating types is donor cell or fertile cell (represented by F^+) and the other is recipient cell (represented as F^-). The donor cell; which is also known as male type, is characterized by the presence of fertility ' factor (called F-Factor). When the fertility factor is transferred to the recipient cell (also known as female type), the latter is transformed into fertile cell (F^+, i.e., male cell).

F-factor, like other genetic elements, is also composed of DNA. However, the amount of DNA associated with the F-factor is only a small fraction of the DNA of the bacterial cell. It exists in the cytoplasm as an autonomous element (i.e., plasmid) or may be integrated into the chromosome of the cell. When F-factor exists in the cytoplasm, it replicates independently of the chromosome but when integrated into the bacterial chromosome, it replicates along with it like any other chromosomal element.

In the process of conjugation two cells of the opposite strains (i.e., F^+ and F^-) become temporarily attached to each other with the help of sex pili. A conjugation tube is formed at the point of their contact and through this tube the duplicate chromosome (replicate DNA) of the donor cell (F^+) is transferred to the recipient cell (F^-). If the donor (male) cell has F-Factor in its cytoplasm, it is transferred to the recipient (female) cell and as such the latter becomes male. The chromosome of the male cell, however, is not transferred to the female cell and, therefore, recombinants are not produced. Due to this reason, the males with F-factor in their cytoplasm are called infertile males.

The ringed chromosome (DNA strand) of those males where F-factor is incorporated in the chromosome, replicates before conjugation. The duplicate chromosome breaks near the point of F-factor, but its double helical structure is not disturbed. During the conjugation duplicate chromosome from the male cell is transferred to the female cell.

The passage of chromosome into the female cell is slow and takes about 110 minutes. The F-factor is always at the rear end of the chromosome. Usually the process of conjugation continues for a short period only and hence the transfer of entire chromosome is not possible. Consequently, only those male characters that are transferred would appear among the recombinants.

Thus the newly formed recombinant cell or genetic hybrid is only partially diploid. As the conjugation normally terminates before the transfer of F-factor present at the rear end of the duplicate chromosome, hence the donor cell continues to function as a male cell. Moreover, the recombinant cell leaves the chromosome obtained from the donor cell after some time, hence the diploid nature of the recipient cell (i.e., recombinant cell) is only short-lived.

Transformation : Transformation is unusual form of recombination in which the molecules of DNA that carry the genetic information of the cell move from the donor to the recipient cell through the liquid, medium in which they are growing. It was discovered by Frederick Griffith (1928) to a mixture of two strains of Diplococcus pneumoniae while experimenting with laboratory mice. Later, in 1944, O. Avery, C.M. Macleod and M. McCarty demonstrated it in vitro. Since then it has been shown in several other bacteria, such as Bacillus, Hemophilus and Neisseria. Those cells in which transformation can be affected are known as competent cells, and that part of DNA strand in which quality of transformation lies is called transfer factor.

The first step in transformation is the entry of exogenous DNA (which is used as transfer factor) in a living bacterial cell. The absorption of exogenous DNA takes place through minute pores or certain absorption sites present on the bacterial cell wall. The DNA, thus absorbed, subsequently gets incorporated in the genetic material of the recipient cell. Thus a new type of chromosome (i.e., transformed chromosome) is formed. The transformed cell is diploid at this stage and when it divides, transformed haploid daughter cells are formed.

Transformation takes place only under certain special condition of the bacterial cell and this condition is known as state of competance. The transforming ability of the cell is, however, lost in the presence of high temperature, ultraviolet radiations, chemical mutagens, etc. On the contrary, increase in the concentration of inorganic phosphates in the culture medium increases the frequency of transformation. The optimum temperature range for transformation is 29-32°C. The effectiveness of the process is determined by many factors such as temperature, composition of the nutrient medium and physiological state of the recipient cell. In nature, the frequency of transformation in different bacteria varies from 0.047 to 0.0004%.

The basic differences between transformation and conjugation are as follows.

(1) In transformation exchange of genetic material takes place between the members of the same species, whereas conjugation occurs between the members of two different genera; e.g., Escherichia and Shigella, Salmonella and Serratia, Salmonella and Escherichia.

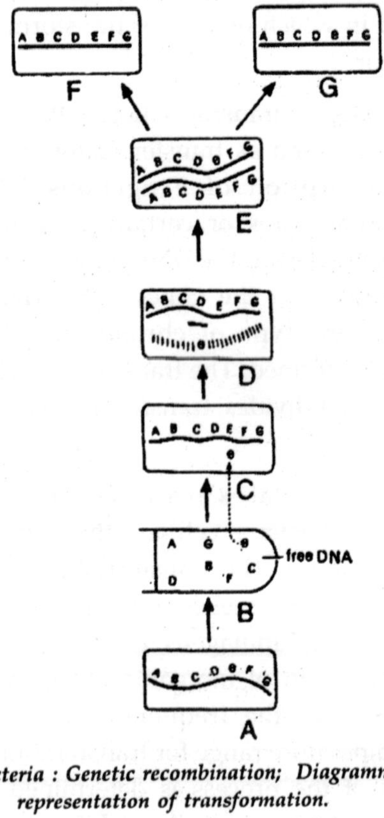

A-G. Bacteria : Genetic recombination; Diagrammatic representation of transformation.

(2) Genetic recombinants obtained through conjugation are temporary, whereas genetic characters incorporated in a transformed cell are permanent.

(3) *Transduction* : Transduction is a special method of genetic recombination in which genetic material is

transferred from the donor to the recipient cell by means of a temperate bacteriophage. It was discovered by Zinder and Lederberg (1952) in Salmonella typhimurium. In transduction, a small piece of bacterial chromosome (carrying specific character x) is incorporated into an attacking phage particle, and when this particle infects a new host cell, it injects the genetic material (with new piece of chromosome) into it.

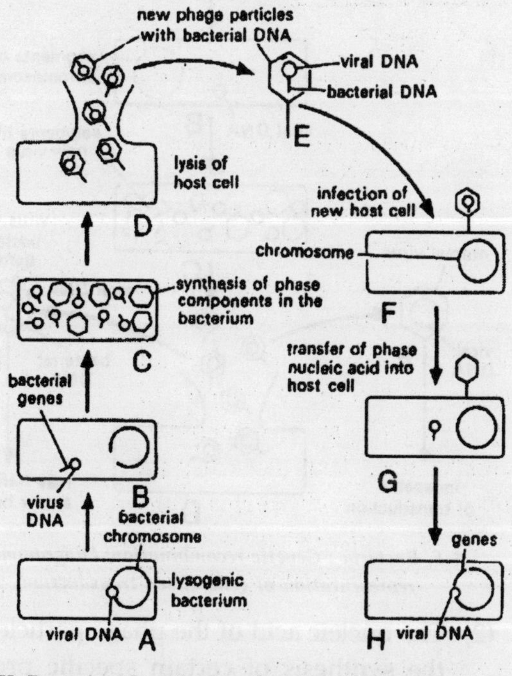

A-H. Bacteria : Genetic recombination; Diagrammatic representation of specialized transduction.

Transduction may be of two types- Specialized transduction and Generalized transduction.

(a) *Specialized Transduction :* In this type of transduction transfer of only few specific genes of a bacterial chromosome takes place. The main steps of specialized transduction are given below.

(1) The bacteriophage gets attached to the bacterial cell on the receptor site and the nucleic acid of phage particle is transferred to the cytoplasm of the bacterial cell.

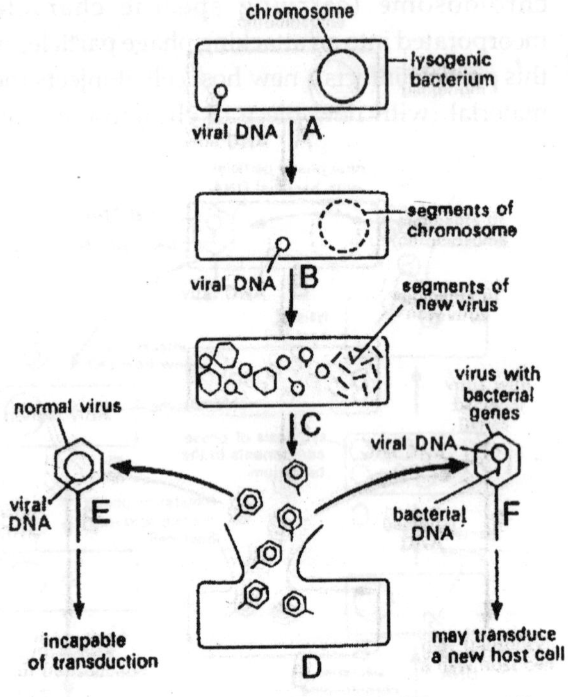

A-F. Bacteria : Genetic recombination; Diagrammatic representation of generalised transduction.

(2) The nucleic acid of the phage particle is coded for the synthesis of certain specific proteins in the bacterial cell. These proteins are known as repressor proteins, and their function is to check the synthesis of phage particles in the bacterial cell. The phage DNA occurs in the bacterial cell in the form of small fragments, known as prophage. These fragments are either free in the cytoplasm or are attached to the chromosome. A bacterial cell with prophage is lysogenic, and it may remain lysogenic

for several generations. During this period the phage DNA keeps on dividing along with bacterial chromosome. A stage comes when synthesis of the repressor proteins stops in the bacterial cell and synthesis of phage components starts.

(3) Under such conditions the phage DNA, which was so far attached to the bacterial chromosome, separates and starts synthesizing phage proteins.

(4) When phage DNA breaks off the bacterial chromosome, few genes of the bacterium remain attached to it. These genes keep on replicating along with the phage DNA and become a part and parcel of the phage particle. Such phage particles when infect any other bacterial cell, then the bacterial genes present in phage particles are incorporated in the chromosome of the new bacterial cell (i.e., recombinant cell). Thus the recombinant cell, besides its own chromosome, also contains few genes of the parent bacterial cell.

(b) *Generalized Transduction* : It is a more common process than the specialized transduction. It involves only those prophage particles which are present in the cytoplasm of the infected cell (and not in its chromosome). The main steps of the generalized transduction are as follows.

(1) The phage DNA, present in the lysogenic bacterial cell, starts synthesizing new phage components. In this process the chromosomes of some bacterial cells get fragmented. Eventually these segments are incorporated in some new phage particles. Thus some of the phage particles present in the lysogenic cell have segments of bacterial chromosome incorporated in them, while others have only phage DNA.

(2) If a phage particle with segment of bacterial chromosome in its DNA attacks a bacterium of any other strain the genes of the parent bacterium are transferred to the new cell. Such phage particles (with genes of bacterial cell) are thus capable of transduction. On the contrary, the particles with only phage DNA are incapable of transduction.

Transduction differs from transformation in the following two features.

(1) In transduction transfer of genetic material from donor to recipient cell is through a temperate phage, whereas in transformation genetic material is transferred through a solution.

(2) Deoxyribonuclease (an enzyme) completely checks the process of transformation but there is no effect of this enzyme on transduction.

Economic Importance

Bacteria are both our foes as well as our friends. On the one hand, they cause a number of serious diseases of plants and animals, including man. They may threaten the very existence of human race, if there is a bacterial warfare. On the other hand, they run the carbon cycle and the nitrogen cycle thus maintaining the balance of carbon and nitrogen and a number of industries such as dairy, vinegar, tea, tobacco and leather are bound on them. Thus they play an important role in our daily life. Some important harmful and useful activities of bacteria are described below.

Harmful Activities of Bacteria

Pathogenic Activities : Pathogenic ability of bacteria was first recognised by Burrill in as early as 1878. He discovered that fire blight of apple is caused by a bacterium. Today we know that more than 180 diseases of plants are caused by bacteria.

The infection of bacteria to plants gives rise specific symptoms, such as galls, imperfect flowers and fruits, blight, brooming, canker, leaf distortion, leaf spot, dwarfing, rot and wilting. Some important diseases of plants caused by bacteria are listed in the Table.

There are also many pathogenic bacteria which occur in animals and human beings. They have a characteristic portal of entry and exit into the organism and mode of transmission. They infect almost all systems of the human body and are easily transported to various parts through circulatory system. Some important diseases of human beings caused by bacteria are listed in the Table.

Some Important Bacterial Diseases of Plants

Diseases	*Pathogens*
Leaf spot of cherry	Coccomyces hiemates
Ring spot of potato	Xanthomonas solanacearum
Blight of walnut	Xanthomonas juglandes
Blight of paddy	Xanthomonas oryzae
Bacterial spot of peach	Xanthomonasprund
Citrus canker	Xanthomonas citri
Angular leaf spot of cotton	Xanthomonas malvacearum
Blight of bean	Pseudomonas phaseolicola
Crown gall of sugar beet	Agrobacterium tumcfaciens
Wilt of tobacco	Phytobacterium solanacearum
Soft rot of mango	Bacterium cartovorus
Tundu of wheat	Corynebacterium tritici

Some pathogenic bacteria destroy the cells of the host, but most cause diseases because they produce poisonous substances that damage the host's metabolism. These substances are called toxins. Bacterial toxins may be divided into two classes : endotoxins and exotoxins. Endotoxins (inside toxins) are lipopolysaccharides in the cell walls of Gram -ve bacteria; they produce fever and damage to the circulatory system. Exotoxins (outside toxins) are much less common than endotoxins; they are proteins and are secreted from the bacterium into the surrounding medium. They may be carried around the host's body in the blood stream. Bacteria causing diphtheria, tetanus, cholera, dysentery, etc. produce exotoxins.

Some Important Bacterial Disease of Human beings

Diseases	*Pathogens*
Cholera	Vibrio cholerae
Bacterial dysentery	Shiegella dysenteriae
Plague	Yersinia pestis
Rheumatic fever	Streptomyces sp.
Tuberculosis	Mycobacterium tuberculosis
Pneumonia	Streptococcus pneumoniae
Typhoid	Salmonella typhi Jaundice
	Leptospiraictero
	haemorrhagiae
Diphtheria	Corynebacterium diptheriae
Gastro-enteritis	Escherichia coli
Syphilis	Treponema pallidum
Gonorrhoea	Neisseria gonorrhoeae
Bacterial conjunctivitis	Haemophilus influenzae
Tetanus	Clostridium tetani
Diarrhoea	Bacillus coli

Food Spoilage : Most of the food materials are excellent media for the growth of bacteria. They grow luxuriantly when temperature and humidity are favourable. Bacteria change the flavour, smell and appearance of food material thus affecting its quality. Carbohydrate fermenting bacteria convert carbohydrate foods into acids, alcohols and gases; lipolytic bacteria convert fatty foods into fatty acids and glycerol and proteolytic bacteria convert protein foods into amino acids, amines, ammonia, etc.

Milk is sterile when it leaves a healthy cow, but it contains several types of bacteria by the time it reaches the table.

Species of Lactobacillus, Streptococcus, Micrococcus and Proteus are responsible for spoilage of milk and milk products. The exotoxins produced by these bacteria are the cause of food poisoning. Besides, bacteria causing diseases like tuberculosis (Mycobacterium bovis) and brucellosis (Brucella abortus) are transmitted through cow's milk. The botulism disease is caused by the exotoxins produced by Clostridium botulinum. Swelling of tongue, double vision and respiratory paralysis are main symptoms of this disease.

Water Pollution : Many bacteria multiply in water and makes it unpotable. Bacteria causing typhoid fever (Salmonella typhi), bacterial dysentery (Shiegella dysenteriae) and cholera (Vibrio cholerae) are generally transmitted through water.

Denitrification : In this process, decomposition of nitrates and nitrites into ammonia and free nitrogen takes place under the influence of denitrifying bacteria, such as Bacillus licheniformis, Pseudomonas aeruginosa and Escherichia coli. These bacteria use nitrate as oxidant in respiration. In this process the combined nitrogen of the soil is removed in the form of gas. Thus the soil is depleted of an essential nutrient for plants and thereby decreasing agriculture productivity.

Useful Activities of Bacteria

Increase Soil Fertility : Although there is 78% nitrogen in the atmosphere, plants in general do not have the capability of utilizing it directly. However, certain free-living as well as symbiotic bacteria can fix molecular nitrogen into nitrogenous compounds. The nitrogen fixing bacteria can be grouped into the following three categories.

Ammonifying Bacteria : Many saprophytic bacteria hydrolyse the proteins of dead plant and animal matter present in the soil into amino acids and organic nitrogenous bases. The ammonifying bacteria convert amino acids into ammonia (the process is known as ammonification). The free ammonia combines with water and carbon dioxide (present in the soil) to form ammonium carbonate. These ammonium compounds are used by many crops as a source of nitrogen. Bacillus mycoides, B. ramosus and B. vulgaris are some important ammonifying bacteria.

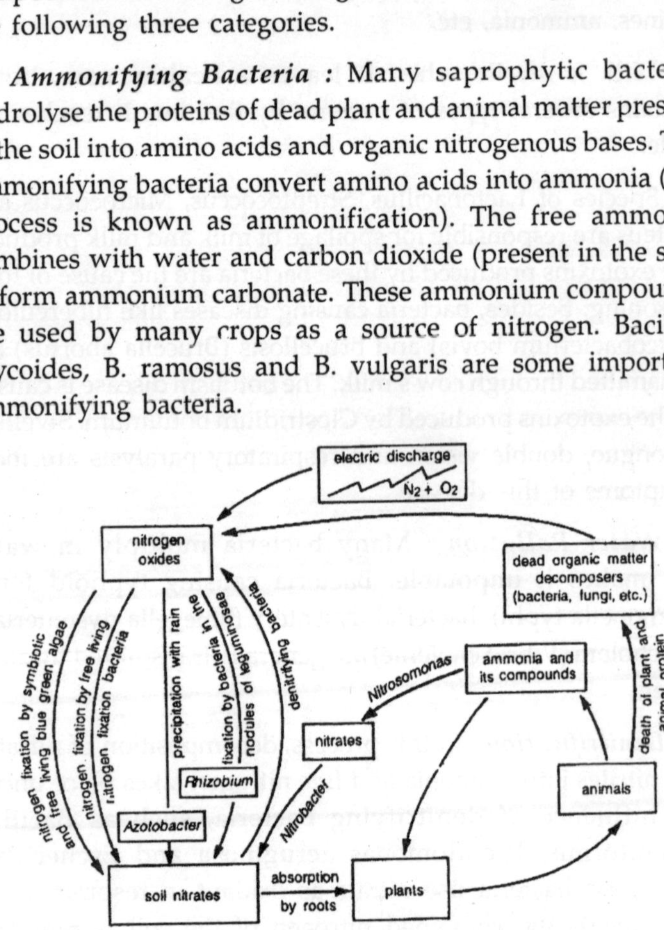

Nitrogen cycle.

Another important source of ammonia in the soil is urea. Animal population of the world excretes more than 2,00,000 tons of urea [CO $(NH_2)_2$] per day which is decomposed by Bacillus pasteuri and Sarcina ureae.

$$CO(NH_2)_2 + 2H_2O \longrightarrow (NH_4)_2CO_3$$

urea (carbamide)

$$(NH_4)_2CO_3 \longrightarrow 2NH_3 + CO_2 + H_2O$$

Nitrifying Bacteria : Oxidation of ammonia (or ammonium salts) to nitrate is known as nitrification. In nature it is brought about by two highly specialized groups of obligately aerobic chemoautotrophic bacteria. It occurs in two steps.

In the first step, ammonia is oxidised to nitrite; the oxidation is brought about by Nitrosomonas, Nitrococcus or Nitrosospira.

$$NH_4 + 2O_2 \xrightarrow{\text{Nitrosomonas}} NO_2 + 2H_2O + 158 \text{ Cal.}$$

In the second step, nitrite is oxidised to nitrate by Nitrobacter.

$$2NO_2 + O_2 \xrightarrow{\text{Nitrobacter}} 2NO_3 + 43.2 \text{ Cal.}$$

Nitrogen Fixing Bacteria : These bacteria fix the free nitrogen and are of the following two types.

(a) *Symbiotic Nitrogen Fixers :* These bacteria belong to the genus Rhizobium. They occur in the root nodules of leguminous plants. Carbon compounds (i.e., carbohydrate) are supplied to these bacteria by the plant and the bacteria fix nitrogen, producing ammonia and various amino acids. In a field containing legume plants 100-400 kg of nitrogen per hectare can be fixed in a year. Thus leguminous plants increase the utilizable nitrogen in the soil and are therefore, used in crop rotation.

(b) *Free-living Forms :* It is a large and heterogeneous group of nitrogen fixing bacteria which occur free in the soil. Species of Azotobacter and Clostridium belong to this category. These forms are probably not as important as the symbiotic nitrogen fixers. These are aerobic forms and can also use ammonia as a nitrogen source. Hence they do not fix nitrogen if ammonia is present in the soil. In each hectare of ordinary soil 25-50 kg nitrogen is fixed by these bacteria in a year.

Importance in industry

Microorganisms have been used by man in the preparation of food, drink and textiles since prehistoric times. Several traditional microbial processes are still used in many industries with astonishing degree of perfection. Advancement in microbiology during the last few decades has led to the development of entirely new industries based on the use of microorganisms.

Dairy Industry : Manufacture of milk products like butter, cheese, butter milk, and yoghurt involves the use of microorganisms, among which lactic acid bacteria are particularly important. These bacteria convert the lactose sugar of milk into lactic acid and the resulting decrease in pH coagulates milk protein rendering the medium unsuitable for the growth of most other microorganisms. The growth of lactic acid bacteria is, therefore, a mean of preserving milk, a highly unstable foodstuff. Lactobacillus bulgaricus, L. acidophilus, L. helviticus, L. plantarium, Streptococcus cremoris, S. laths and S. thermophilus are some common bacteria used in dairy industry.

Dextran Production : Dextran is a polysaccharide of high molecular weight, varying from 15,000 to 20,000,000. It is produced extracellularly by some lactic acid bacteria belonging to the genus Leuconostoc. Some water insoluble derivatives of dextran can act as molecular sieves by retarding the passage of certain molecules. These are used for physical fractionation of solutes which differ in molecular weight.

Production of Acetone and Butanol : Species of Clostridium are used in the commercial production of industrial solvents like acetone and butanol. Many clostridia ferment sugars into butyric acid and some carry out further reactions converting butyric acid to butanol and acetic acid to ethanol and acetone. Application of this technology requires careful microbiological control because the medium used for the culture of Clostridium acetobutylicum is also favourable for the growth of lactic acid bacteria. The growth of latter makes the medium acidic and thus rendering it unsuitable for the growth of Clostridium.

Vinegar (Acetic Acid) Industry : If wine and beer are exposed to air, they turn sour. It is due to the conversion of alcohol to acetic acid, mediated by strictly aerobic acetic acid bacteria, such as Acetobaeter aced and Clostridium acetobutylicum. In industries this process is carried out in wooden vats. The vats are partially filled with wine and the acetic acid bacteria develop as a gelatinous pellicle on the surface of the liquid. The process takes several weeks for completion.

Curing of Tobacco and Tea : Curing of tobacco and tea leaves is done under controlled activity of microorganisms. The colour, flavour and aroma of these products depend on curing and ripening.

Decomposition of Dead Organisms : Bacteria are natural scavangers. They break down dead organic matter into inorganic constituents. The inorganic substances are used by plants as nutrients. Thus bacteria increase the fertility of soil by decomposition of dead organic matter.

Microbial Control of Insects : Many bacteria secrete crystalline proteins which are highly toxic to the larvae of caterpillars and insects belonging to Lepidoptera. These proteins are, however, nontoxic for other animals and plants. Thus they provide an ideal agent for the control of many serious insect pests which damage plant crops. In practice, instead of isolating proteins, crystal-producing bacilli are grown on large scale and their spores are mixed with dusting powder for spraying in fields.

Retting Process : Retting is a controlled microbial decomposition of pectin (the cementing substance which keeps plant tissues together) without simultaneous decomposition of fibres. Thus it results in the separation of fibres. The stems of flax, hemp, etc. are immersed in water for long periods where decomposition sets in. This process is facilitated by anaerobic butyric acid bacteria; such as Clostridium botulinum, C. tetani and C. perfringens. These bacteria primarily decompose the plant pectin, thus freeing fibres. If the process of retting continues for a long period, then cellulose fermenting bacteria develop and destroy the fibres.

Degradation of Petroleum Hydrocarbons : Many bacteria, such as Pseudomonas, Achromobacter, Micrococcus and Candida are capable of degrading petroleum hydrocarbons both in marine and fresh water habitats. Amongst these, Pseudomonas is the most important genus that can utilize petroleum hydrocarbons. Thus these bacteria can be used to check pollution after oil spillage.

Sewage Disposal : Sewage is the domestic and/or industrial waterborne waste. The domestic sewage includes human excreta and wash water that goes into sewage system. The industrial sewage includes acids, oils, gases and animal and vegetable matter discharged by factories. Sewage contains 99-99.5 per cent water and 0.5-1.2 per cent solids in the form of microscopic particles, colloids, organic compounds and microorganisms. Besides, raw sewage may contain millions of bacteria per milliliter including coliforms, streptococci, anaerobic spore forming bacilli and also types originating from the intestinal tract of human beings.

The discharge of untreated sewage into rivers and lakes causes :

(i) greater dissemination of pathogenic microorganisms,

(ii) pollution of drinking water, grasslands and agriculture fields,

(iii) increased danger to aquatic life, and

(iv) offensive smell. Therefore, it is essential to treat sewage prior to its discharge as this results in the mineralization of the organic components in sewage. A variety of microorganisms (including bacteria, fungi and protozoa), collectively known as sewage fungi, help in the biological degradation of organic matter.

Use of Bacteria in Medicines

Bacteria are also the source of many antibiotics, serums and vaccines. Antibiotics are the chemical substances secreted by certain microorganisms which inhibit the growth and development of other microbes.

The antibacterial property of culture filtrates of Penicillium notatum was first observed by Alexander Fleming in 1928 and the active principle of the filtrate, penicillin, was isolated by H.W. Florey in 1944. Since then a number of antibiotics have been isolated from bacteria mainly from the species of Streptomyces and Bacillus.

Antibiotics show the following characteristics:

(1) They have inhibitory effects only on parasites but do not affect the growth of the host.

(2) They are effective only when a certain minimum dose is provided.

(3) They do not interfere with the natural defence mechanism of the host.

(4) They do not produce undesirable effects (such as allergy, desiccation, etc.) on the host.

(5) They do not affect the natural microflora of the host.

Some important antibiotics and their source, range of action and mode of action are given in the Table.

Some Important Antibiotics and their Source, Range of Action and Mode of Action.

Antibiotics	*Source*	*Range of action*	*Mode of action*
Bacitracin	Bacillus subtilis	Gram +ve bacteria	Inhibits synthesis of bacterial cell walls
Polymyxin	G Bacillus polymyxa	Gram +ve bacteria	Destroys cyto plasmic membranes
Colistin (Polymyxin E)	Bacillus colstins	Gram -ve bacteria Gram -ve and cocci	Destroys cell membrane Inhibits formation of bacterial
Tyrothrycin	Bacillus brevis		cell wall

Bibliography

Adil, A. : *Manual of Practical Algae,* Mohit Pub., New Delhi, 2003.

Aggarwal, S.K. : *Vistas in Mycology and Plant Pathology,* APH, New Delhi, 2000.

Ahuja, B. : *Plant Anatomy,* Rawat Pub., New Delhi, 2000.

Antony, T. : *Quarry : Closing in on the Missing Link,* Free Press, New York, 1993.

Arora, J.K.: *Biopesticides and Pest Management,* Ankit Pub., New Delhi, 2003.

Aslam, M.S. : *Genetics-Behavioural and Molecular,* Campus Books, New Delhi, 2003.

Bacon, Mark : *Water use Efficiency in Plant Biology,* Blackwell, London, 2004.

Bahadur, Rai K. Ranga : *A Manual of Botany for India,* Atlantic Publishers, New Delhi, 1994.

Bayleybolfour, I. :*Plant Geography Upon A Physiological Basis,* ABS, London, 2004.

Behnke, H.D. : *Sieve Elements : Comparative Structure, Introduction and Development,* Springer, London, 1990.

Bentley, Robert : *Study of Structural Morphological and Physiological Botany,* Atlantic Publishers, New Delhi, 1994.

Bilgrami, C : *Fundamentals of Botany,* Vikas Publishing House, New Delhi, 2003.

Bilgrami, K.S. : *Physiology of Fungi,* Vikas Publishing House, New Delhi, 2003.

Boodle, L.A. and Sinnott, E.W. : *Student Handbook of Botany and Plant Science,* Atlantic Publishers, New Delhi, 1995.

Bose, J.C. : *Irritabiligy of Plants,* Discovery Publishing House, New Delhi, 1999.

Brooks, F.T., and Smith, M. : *Dictionary of Fungus Diseases of Plants,* Campus Books, New Delhi, 2003.

Busch, L. : *Plants, Power and Profit : Social, Economic, Lawrence Busch, and Ethical Consequences of the New Biotechnologies,* Blackwell, London, 1992.

Chrispeels, M.J. : *Plants, Genes and Agriculture,* Jones & Bart., London, 1994.

Coulter, John Gaylord : *Fundamentals of Plant Science,* Atlantic Publishers, New Delhi, 1994.

Dayal, R. : *Predaceous Fungi,* Campus Books, New Delhi, 2000.

———— : *Predaceous Fungi,* Commonwealth Publishers, New Delhi, 2000.

Dhaliwal, G.S. and Singh, Balwinder : *Pesticides and Environment,* Campus Books, New Delhi, 2000,

Dube, H.C. : *Introduction to Fungi,* Vikas Publishing House, New Delhi, 2003.

Dyson, Freeman : *Origin of Life,* Oxford, New York, 1999.

Frank, S. Smythe : *Flowering Plants of Himalaya,* Commonwealth Publishers, New Delhi, 1994.

Fuller, Harry J. : *University Botany,* Atlantic Publishers, New Delhi, 1994.

Gaston, Kevin J. and spice, John I. : *Biodiversity : An Introduction,* Blackwell, 2004.

Grewal, R.C. : *Medicinal Plants*, Campus Books, New Delhi, 2003.

——— : *Dictionary of Botany*, Arihant Pub., New Delhi, 2003.

Hall, Marion and Halliday, Tim : *Behaviour and Evolution*, Springer, London, 1998.

Hanan, James S. : *The Algorithmic Beauty of Plants*, Springer, London, 1990.

Hardie, D.G. : *Protein Phosphorylation : A Practical Approach*, Oxford, New York, 1999.

Johri, Lata and Tyagi : *A Textbook of Bryophyta*, Dominant Publishers, New Delhi, 2004.

——— : *A Textbook of Fungi*, Dominant Publishers, New Delhi, 2004.

——— : *A Textbook of Gymnosperm*, Dominant Publishers, New Delhi, 2005.

Kains, M.G. and Mcquesten, L.M. : *Fundamentals of Plant Science*, Campus Books, New Delhi, 2002.

——— : *An Introduction to Plant Science*, Atlantic Publishers, New Delhi, 1995.

Kumar, H.D. : *General Ecology*, Vikas Publishing House, New Delhi, 2003.

Kumar, Suresh : *A Textbook of Plant Taxonomy*, Campus Books, New Delhi, 2004.

——— : *Economic Botany*, Campus Books, New Delhi, 2002.

Lang, W.H. : *A Textbook of Botany*, Atlantic Publishers, New Delhi, 1994.

Leung, D.Y.M. : *Superantigens : Molecular Biology, Immunology, and Relevance to Human Disease*, Dekker, London, 1997.

Lozzo, Renato V. : *Proteoglycans : Structure, Biology, and Molecular Interactions*, Dekker, London, 2000.

Madigan, M.T. : *Brock : Biology of Microorganisms,* Prentice Hall, New York, 2000.

Magner, Lois N. : *A History of the Life Sciences,* Dekker, London, 1994.

Magurran, Anne E. : *Measuring Biological Diversity,* Blackwell, London, 2003.

Mishra, S.R. : *Bacterial Plant Diseases,* Discovery Publishing House, New Delhi, 2003.

———— : *Morphology of Plants,* Discovery Publishing House, New Delhi, 2004.

Nairne, Alexander : *Scientific Classification of Flowering Plants,* Atlantic Publishers, New Delhi, 1994.

Narasaiah, M.L. : *Biodiversity and Environment,* Discovery Publishing House, New Delhi, 2004.

Nover, L. : *Plant Promoters and Transcription Factors Results and Problems in Cell Differentiation,* Springer, London, 1994.

Oliver, Daniel : *Study of Indian Botany,* Atlantic Publishers, New Delhi, 1994.

Panday, S.N. : *Economic Botany,* Vikas Publishing House, New Delhi, 2003.

Panday, S.N. and Sinha, B.K. : *Plant Physiology,* Vikas Publishing House, New Delhi, 2003.

Panday, S.N. and Trivedi, P.S. : *A Textbook of Botany,* Vikas Publishing House, New Delhi, 2003.

Pfleiderer, I. : *The Life of Indian Plants,* Atlantic Publishers, New Delhi, 1992.

Prasad, S.N. : *Environmental Biology,* Campus Books, New Delhi, 2003.

———— : *Cell Biology,* Campus Books, New Delhi, 2003.

Pugnaire, Ed. F.I. : *Handbook of functional Plant Ecology,* Dekker, London, 1999.

Rai, Mahendra and Mares, Donatella : *Plant-Derived Antimycotics : Current Trends and Future Prospects*, Blackwell, London, 2003.

Rashid, A. : *An Introduction to Pteridophyta*, Vikas Publishing House, New Delhi, 2003.

Rashid, A., Kumar, Santosh and Tembhre, Manju : *Anatomy and Physiology of Fishes*, Vikas Publishing House, New Delhi, 2003.

Roy, E. Biles, Seymour, E.L.D. : *Encyclopaedia of Horticulture*, Commonwealth Publishers, New Delhi, 1994.

Sandhu, G.S. : *Textbook of Cytology*, Campus Books, New Delhi, 2002.

Sandhu, G.S. and Kumar, Suresh : *Cyto-Genetics*, Campus Books, New Delhi, 2003.

Sarkar, Dilip De : *The Silkworm: Biology, Genetics and Breeding*, Vikas Publishing House, New Delhi, 2003.

Sharma Rajni : *An Introduction to Plant Morphology*, Campus Books, New Delhi, 2004.

——— : *Plant Pathology*, Campus Books, New Delhi, 2004.

——— : *Plant Tissue Culture*, Campus Books, New Delhi, 2003.

Shukla, S.K. and Srivastava, P.R. : *Encyclopaedia of Environmental Pollution*, Commonwealth Publishers, New Delhi, 1994.

Singh, Mahinder : *A Textbook of Biochemistry*, Dominant Publishers, New Delhi, 2005.

Singh, S.K. : *Plant Physiology*, Campus Books, New Delhi, 2000.

Sinha, Asha *Microbes and Plants*, Campus Books, New Delhi, 2001.

Sinha, R.K. :*Biodiversity–Global Concerns*, Commonwealth Publishers, New Delhi, 1996.

Sinnott, E.W. : *Botany and Plant Science,* Commonwealth Publishers, New Delhi, 2000.

Srivastava, M.B.: *Introduction to Forestry,* Vikas Publishing House, New Delhi, 2003.

Stiles, Walter :*Principles of Plant Physiology,* Atlantic Publishers, New Delhi, 1995.

Sundara, Rajan, S. : *Introduction to Cell Biology,* Vikas Publishing House, New Delhi, 2003.

Talbot, Nicholas J. : *Plant-Pathogen Interactions,* Blackwell, London, 2001.

Tassy, Pascal : *The Message of Fossils,* McGraw-Hill, London, 1993.

Transeau, E.N. : *General Botany,* Atlantic Publishers, New Delhi, 1995.

Tripathi, G. and Tripathi, Y.C. : *Biological and Biotechnological Resource,* Campus Books, New Delhi, 2002.

———— : *Bioresource and Environment,* Campus Books, New Delhi, 2002.

Waisel, Yoav : *Plant Roots : The Hidden Half,* Dekker, London, 1996.

Warren, L. : *Bound Carbohydrates in Nature,* Oxford, New York, 1994.

Wrigglesworth, J. : *Energy and Life : Modules in Life Sciences,* T & F, New York, 1997.

Zaidi, N. : *An Introduction to Algae,* Ankit Pub., New Delhi, 2002.